Decoherence, Entanglement and Information Protection in Complex Quantum Systems

NATO Science Series

A Series presenting the results of scientific meetings supported under the NATO Science Programme.

The Series is published by IOS Press, Amsterdam, and Springer (formerly Kluwer Academic Publishers) in conjunction with the NATO Public Diplomacy Division.

Sub-Series

I. Life and Behavioural Sciences	IOS Press
II. Mathematics, Physics and Chemistry	Springer (formerly Kluwer Academic Publishers)
III. Computer and Systems Science	IOS Press
IV. Earth and Environmental Sciences	Springer (formerly Kluwer Academic Publishers)

The NATO Science Series continues the series of books published formerly as the NATO ASI Series.

The NATO Science Programme offers support for collaboration in civil science between scientists of countries of the Euro-Atlantic Partnership Council. The types of scientific meeting generally supported are "Advanced Study Institutes" and "Advanced Research Workshops", and the NATO Science Series collects together the results of these meetings. The meetings are co-organized by scientists from NATO countries and scientists from NATO's Partner countries — countries of the CIS and Central and Eastern Europe.

Advanced Study Institutes are high-level tutorial courses offering in-depth study of latest advances in a field.
Advanced Research Workshops are expert meetings aimed at critical assessment of a field, and identification of directions for future action.

As a consequence of the restructuring of the NATO Science Programme in 1999, the NATO Science Series was re-organized to the four sub-series noted above. Please consult the following web sites for information on previous volumes published in the Series.

http://www.nato.int/science
http://www.springeronline.com
http://www.iospress.nl

Series II: Mathematics, Physics and Chemistry –Vol. 189

Decoherence, Entanglement and Information Protection in Complex Quantum Systems

edited by

V. M. Akulin
Laboratoire Aimé Cotton,
Campus d'Orsay,
Orsay Cedex, France

A. Sarfati
Laboratoire Aimé Cotton,
Campus d'Orsay,
Orsay Cedex, France

G. Kurizki
Chemical Physics Department,
Weizmann Institute of Science,
Rehovot, Israel

and

S. Pellegrin
Chemical Physics Department,
Weizmann Institute of Science,
Rehovot, Israel

Springer

Published in cooperation with NATO Public Diplomacy Division

Proceedings of the NATO Advanced Research Workshop on
Decoherence, Entanglement and Information Protection in Complex Quantum Systems
Les Houches, France
26–30 April 2004

A C.I.P. Catalogue record for this book is available from the Library of Congress.

ISBN 10 1-4020-3282-X (PB)
ISBN 13 978-1-4020-3282-0 (PB)
ISBN 10 1-4020-3281-1 (HB)
ISBN 13 978-1-4020-3281-3 (HB)
ISBN 10 1-4020-3283-8 (e-book)
ISBN 13 978-1-4020-3283-7 (e-book)

Published by Springer,
P.O. Box 17, 3300 AA Dordrecht, The Netherlands.

www.springeronline.com

Printed on acid-free paper

Printed in the Netherlands.

Contents

Acknowledgments

We gratefully acknowledge the joint effort of our Topical Editors: Miriam Blaauboer, Maciej Krzystyniak, Tomáš Opatrný, David Petrosyan and Jane Salo, the financial support of our sponsors: NATO and EC (through the Research Training Network on Quantum Complex Systems, "QUACS") and the technical support of Laboratoire Aimé Cotton, Orsay, France.

The Editors

EDITORIAL ARTICLE

How to control decoherence and entanglement in quantum complex systems?

V. M. Akulin,[1] G. Kurizki,[2] S. Pellegrin[2] and A. Sarfati[1]

[1]*Laboratoire Aimé Cotton, Bât. 505, Campus d'Orsay, 91405 Orsay Cedex, France*
vladimir.akulin@lac.u-psud.fr

[2]*Chemical Physics Department, Weizmann Institute of Science, Rehovot 76100, Israel*
gershon.kurizki@weizmann.ac.il

1. Complex quantum ensembles: a new paradigm?

Theory and experiment have not fully resolved the *apparent dichotomy*, which has agonized physics for the past eighty years: on the one hand, the description of microsystems by quantum mechanics and, on the other, the description of macrosystems by classical dynamics or statistical mechanics. Von Neumann's axiomatic formulation of quantum mechanics underscored this dichtonomy, by contrasting the reversible unitary evolution of a free system with the projection postulate describing the irreversible outcome of its measurement [von Neuman 1955]. Later derivations of the time-irreversible Liouville equation for an open quantum system, based on projecting out its environment, have narrowed the gap between the quantum and classical statistical descriptions [Kubo 1963; Agarwal 1974]. Yet our "classical" intuition continues to be confronted by quantum-mechanical results like the Einstein-Podolsky-Rosen paradox [Einstein 1935; Mann 1995] that challenges the classical notion of locality, or the quantum Zeno effect [Misra 1977; Kofman 2000], which suggests that the isolation of a system is not the only way to preserve its quantum state.

There are two key concepts in any discussion of such issues. The first, which is responsible for the most salient nonclassical properties, is *entanglement*, that is partial or complete correlation or, more generally, inseparability, of the elements comprising a *quantum ensemble*. Even after their interaction has ceased, this inseparability, originating from their past interaction, can af-

V.M. Akulin et al. (eds.), *Decoherence, Entanglement and Information Protection in Complex Quantum Systems*, 1–12.

fect the state of one element when another element is subject to a nonunitary action, such as its measurement, tracing-out, or thermalization. The second key concept is *decoherence* of open quantum systems, which is the consequence of their entanglement with their environment, a "meter" or a thermal "reservoir", followed by the tracing-out of the latter. Despite new insights into entanglement [Bouwmeester 2000] and decoherence [Zurek 2000; Blanchard 2000], such as the emergence of a "pointer" basis in the entanglement of a system with a "meter", there are still no complete, unequivocal answers to the fundamental questions of the transition from quantal to classical behavior: How do irreversibility and classicality emerge from unitarity as systems and their environments become increasingly *complex*? At what stage does system-meter entanglement give rise to a classical readout of the meter? Is there an upper limit on the size or complexity of systems displaying entanglement?

The recent upsurge of interest in such fundamental questions can be attributed to spectacular progress in two areas: (a) *manipulations* of quantum states of matter, namely atoms [Cohen-Tannoudji 1998], bosonic or fermionic gases [Inouye 1998; de Marco 1999] and molecules [Zewail 2000; Shapiro 2003] and their promising applications in metrology, interferometry and chemical reaction control and (b) the encoding, transmission and processing of *quantum information*, whose potentially revolutionary implementations include quantum teleportation, quantum cryptography, and the coveted quantum computing [Nielsen 2000]. These major developments have opened new vistas into the basic quantum phenomena of interference, which is the key to quantum-state manipulation, and *controlled entanglement*, which is the resource of quantum information processing. However, these developments have mainly focussed on ensembles of simple two- or three-level systems that are either thoroughly isolated from their environment, such as atoms in high-Q cavities and optical lattices or trapped ions [Special Issue on Quantum Computation 2000]. By contrast, controlled entanglement has not been attempted in the more complex molecular or condensed-matter systems. Treatments of decoherence in quantum computing have mostly assumed that only a single or a few elements of the quantum ensemble may simultaneously undergo an uncontrolled intervention - a quantum error. These assumptions define a rather limited subgroup of the group of all possible uncontrolled interventions in the entire Hilbert space of the quantum ensemble, thus considerably restricting the admissible variety of quantum errors. Decoherence-control protocols for more general types of errors are still lacking.

In order to resolve the outstanding issues of the quantum-classical transition, and study the control of entanglement and decoherence without the foregoing restrictions, we must venture into the domain of *Quantum Complex Systems* (QUACS), either consisting of a large number of inseparable elements or having many coupled degrees of freedom. Modern statistical physics copes with quantum complexity by resorting to the Wigner-Dyson concept of ensemble averaging and the ensuing *random matrix theory* [Akulin 2004], which describe *universal* quantum phenomena in almost all large systems, irrespective of their Hamiltonians and external perturbations. This approach has singled out the class of *mesoscopic* systems where interference phenomena are crucial, in spite of their large size and high Hilbert-space dimensionality [Altshuller 1991]. However, the basic questions of entanglement and decoherence are not dealt with by this prevailing approach of mesoscopic physics.

Our conviction is that fundamental understanding and manipulation of entanglement within QUACS or their entanglement with the environment or a meter, call for the creation of a new conceptual framework or paradigm, that would encompass phenomena common to cold atoms in laser fields, large molecules, Josephson junctions, quantum gases and solids, with the view of employing these systems for quantum information processing and computing. Progress within this paradigm should allow us to answer the questions: Does entanglement play an essential role in the evolution of large collections of complex systems? What are the size or complexity limits of systems and ensembles still controllable by an external intervention? What are the most appropriate decoherence protection schemes and control algorithms? A related question of great phenomenological importance is: Does inter-particle entanglement occur in condensed media, in the presence of the fast decoherence prevailing under ambient conditions? The standard answer is negative, as indicated by the complete absence of entanglement from the lore of material science or chemistry. However, as discussed below, evidence is mounting that entanglement may be manifest even in condensed media at ambient temperatures.

Clearly, the development of the envisioned conceptual framework that would provide answers to these questions *necessitates a merger of usually orthogonal disciplines*, aimed at describing and controlling entanglement and decoherence in QUACS and ensembles thereof by unconventional methods, different in many respects both from those currently in use for simpler systems and from the methods of mesoscopic physics. Rather than look for large-scale universal behavior, our aim should be to reveal and steer, or "engineer" the intricate correlations within QUACS and the entanglement with their *specific* environment.

2. Synopsis

This book addresses the challenge of understanding in depth and manipulating the basic quantum properties of optical, atomic, molecular and condensed-matter QUACS, and large ensembles thereof. Many of the presented results have been obtained in the framework of the QUACS Research Training Network sponsored by the EU. They have been supplemented by results of leading experts with pioneering accomplishments in several disciplines, presented at the Workshop on Decoherence, Entanglement and Information in Complex Systems (DEICS) cosponsored by NATO, which was held in Les Houches in April 2004.

The articles in this book have been grouped in seven parts that are briefly summarized in what follows. Each part is preceded by a short introduction by its contributors, for the purpose of reviewing the topic and provinding an extended list of references. The following overview will prove that the various parts address related questions, albeit in different domains of quantum physics.

Part I: Role of entanglement in quantum computing and information processing

This part shows that though entanglement is a major resource for the speedup of quantum information processing, its role in quantum computing is not completely understood as yet. Quantum information processing using subgroups of the complete group of operators that can act in the entire Hilbert space are also addressed:
(a) S. L. Braunstein introduces the main concepts and theorems of quantum information theory showing the difference between quantum entanglement and classical correlations in quantum systems. He describes pseudo-pure states for NMR quantum computers and discusses the role of entanglement in quantum computations based on open systems.
(b) Y. Ozhigov discusses the influence of fermionic statistics in a large ensemble on quantum computations.

Part II: Multiatom and multiphoton entanglement

Efficient theoretical and experimental tools are presented for the control of entanglement in multiatom and multiphoton systems. The problems of storage of quantum information encoded in multiphoton beams and its processing via measurements or controlled interactions with two- and multi-level atomic species are discussed:
(a) A. I. Lvovsky proposes a qualitative description of the preparation of a single photon via conditional measurements.
(b) Ph. Walther and A. Zeilinger discuss the fundamentals of entanglement examplified by photons, and explain how can entanglement and its implementation for quantum logic gates be experimentally realized in the framework of linear optics, by means of nonunitary operations, *i.e.*, quantum measurements.
(c) L. I. Childress, M. D. Eisaman, A. Andre, F. Massou, A. S. Zibrov, and M. D. Lukin describe schemes of pulse shaping and control of the quantum states of light, based on the entanglement of the electromagnetic field with an atomic ensemble displaying strong memory effects.
(d) D. Petrosyan describes the enhancement of controlled photon entanglement in coherently driven atomic systems.

(e) Yu. P. Malakyan, D. Budker, S. M. Rochester, D. F. Kimbal, V. V. Yashchuk, and W. Gawlik consider a practical way of performing quantum computations with the help of polarized photons by amplification of their controlled entanglement via driven non-linear Kerr interactions in multilevel atomic systems.
(f) G. Yu. Kruchkyan and H. H. Adamyan consider possibilities of control and quantum entanglement of lasing field modes with large numbers of photons, in the presence of decoherence resulting from dissipation.

Part III: Controlled entanglement and protection from decoherence

New powerful approaches are suggested for the control of decoherence and quantum state protection:
(a) E. Brion, V. M. Akulin, D. Comparat, I. Dumer, G. Harel, N. Kébaïli, G. Kurizki, I. E. Mazets, and P. Pillet discuss the principles of the protection of an entangled quantum state via tailored external interventions and repeated measurements in the Zeno regime and its possible experimental realization in atomic systems.
(b) L. B. Ioffe demonstrates that by sharing the quantum information among a large number of individual qubits, realized on Josephson junctions, and by imposing topological constrains via aggregation of these qubits in a network, one can achieve an inhibition of the decoherence rate, which scales exponentially with the number of qubits.
(c) C. Mewes, S. Pellegrin, M. Fleischhauer and G. Kurizki suggest a scheme for the inhibition of decoherence by controlled interventions in a level-band system, particulary the inhibition of radiative decay via changes of the atom-photon entanglement in photonic crystals. In a second part they propose to inhibit the decoherence rate by distributing the quantum information over a large ensemble of atoms, and show that the decoherence rate then scales as the inverse square root of the number of atoms in the ensemble.
(c) G. Kurizki, A. G. Kofman, V. M. Akulin, E. Brion, and J. Clausen show that it is possible not only to inhibit but also to accelerate the process of decoherence in the anti-Zeno regime of frequent interventions or measurements.

Part IV: Non-markovian decay and decoherence in open quantum systems

This part is dedicated to the interaction of a quantum system with a slowly relaxing environment and the influence of temporal correlations in the reservoir on the system dynamics:
(a) J. Salo, S. Stenholm, G. Kurizki, and A. G. Kofman discuss the treatment of memory effects in the interaction of a quantum system with a reservoir and their entanglement, and show how these can be incorporated in master equations describing the time-dependent decoherence of the system.
(b) J. Clausen, V. M. Akulin, J. Salo, and S. Stenholm consider the role of re-

peated measurements of one part of a quantum system on the dynamics of the other part, and single out an important case of a nondemolishing interaction between the parts, which results in a master equation of a higher rank.
(c) I. E. Mazets and G. Kurizki show that the interaction of a particle with a large, distinctly quantal ensemble, such as a Bose condensate, may result in appreciable modification of the coherence loss and its time-dependence.

Part V: Coherence and entanglement in multidimensional systems

This part is concerned with the quantum dynamics of molecules and ensembles of trapped cold atoms and the effect of internal-translational entanglement on interference and diffraction:
(a) S. Haroche, M. Brune, and J. M. Raimond review experiments centred around the interaction of atoms with mesoscopic quantized fields.
(b) M. Arndt, L. Hackermüller, K. Hornberger, and A. Zeilinger experimentally show that the quantum phenomena of two-beam interference and diffraction persist even for complex quantum objects in a strongly mixed internal quantum state, such as large molecules at high-temperatures, provided their translational degrees of freedom are isolated from the internal ones.
(c) J. Sherson, B. Julsgaard and E. S. Polzik experimentally demonstrate distant entanglement of macroscopic gas samples, via the interaction of these samples with a pulse of light, and discuss the implementation of various quantum information protocols that draw upon this resource.
(d) T. Opatrny, M. Kolar, and G. Kurizki discuss the concept of translational EPR entanglement pertaining to positions and momenta and illustrate it for atomic pairs that are trapped in an optical lattice and entangled by dipole-dipole interactions.
(e) R. Garcia-Fernandes, A. Ekers, B. W. Shore, J. Klavins, L. P. Yatsenko, and K. Bergmann demonstrate that the entanglement between the vibrational and electronic degrees of freedom of atomic dimers can result in considerable decrease of their spontaneous relaxation rate, when the dimers are subject to an appropriate external intervention via interaction with a laser field.
(f) M. Leibscher and S. Stenholm suggest that the spontaneous transition rates in molecules can be controlled via laser-induced shaping of the coordinate-dependence of molecular-term couplings.
(g) P. Pillet, D. Comparat, M. Muldrich, T. Vogt, N. Zahzam, V. M. Akulin, T. F. Gallagher, W. Li, P. Tanner, M. W. Noel, and I. Mourachko present the experimental signature of the dipole-dipole entanglement accompanying the coherent dynamics of excitations in a large disordered ensemble of frozen Rydberg atoms, which manifests itself in non-Lorentzian shapes of atomic resonances.

Part VI: Entanglement and decoherence of proton scatterers in solids

This part is focussed on the intriguing question of quantum entanglement in large statistical condensed-matter ensembles and its possible influence on the cross-section of neutron scattering from solids:

(a) J. Mayers and T. Abdul-Redah present experimental results showing an appreciable reduction of the observed differential cross-section of hyperthermal (sub-keV) neutrons scattered off hydrogenated condensed media compared to that given by the Rutherford formula.

(b) T. Abdul-Redah, M. Krzystyniak, and C. A. Chatzidimitriou-Dreismann suggest that the subfemtosecond entanglement between protons in such media and their electronic environment may modify the standard cross-section, pertaining to statistically independent proton scatterers.

(c) C. A. Chatzidimitriou-Dreismann, T. Abdul-Redah, M. Krzystyniak, and M. Vos extend this consideration to measurements of scattering of 20 keV electrons off the same media, stressing that these electrons have approximately the same de Broglie wavelength as 1eV neutrons.

(d) F. Fillaux reports on similar measurements of hyperthermal neutron scattering off crystals of potassium hydrocarbonate.

(e) H. Naumann, T. Abdul-Redah, and C. A. Chatzidimitriou-Dreismann present experimental results that demonstrate a similar anomaly for X-ray scattering in such media.

(f) E. B. Karlsson suggests an explanation of the foregoing scattering anomaly based on pairwise entanglement of proton spin states in the ensemble of scatterers.

(g) I. E. Mazets, C. A. Chatzidimitriou-Dreismann, and G. Kurizki remark that, in principle, a scattering anomaly may originate from memory effects in the interaction of the scatterers with their environment.

(h) S. Stenholm and C. A. Chatzidimitriou-Dreismann argue that decoherence-induced irreversible dynamics may reduce the scattering cross-section.

Part VII: Coherence and entanglement in mesoscopic systems

This part addresses mesoscopic systems as diverse as SQUIDS, Bose condensates, trapped atomic ensembles, electronic inerferometers and micromasers. It shows that coherence and entanglement can persist in such systems:

(a) M. Blaauboer, G. Kurizki, and V. M. Akulin predict unusual interference effects in probe scattering off fluctuating mesoscopic ensembles of atoms trapped in optical lattices.

(b) D. O'Dell, S. Giovanazzi, and G. Kurizki consider the laser-induced dipole-dipole interaction in the quantum ensemble of Bose-condensed atoms and demonstrate that this controlled intervention results in long-range spatial entanglement and quantum correlations of the atoms.

(c) N. Katz, E. Rowen, R. Ozeri, J. Steinhauer, E. Gershnabel, and N. Davidson survey the experimental state-of-the-art in the manipulation of external potentials trapping mesoscopic atomic ensembles and Bose-condensates.
(d) Y. Ji, Y. Chung, D. Sprinzak, F. Portier, and M. Heiblum describe experiments with a mesoscopic electronic interferometer.
(e) M. Blaauboer considers the possibility of decoherence-free transport of quantum information through a mesoscopic quantum dot sructure.
(f) A. Barone, A. G. Kofman, and G. Kurizki suggest a scheme for decoherence suppression in mesoscopic Josephson junctions by bias-current modulation acting as a controlled intervention.
(g) E. Sarnelli, G. Testa, A. Monaco, M. Adamo, and D. Perez de Lara discuss the possibility of employing the mesoscopic grains of high-T_c superconductors as qubits.
(h) A. M. Dykhne and A. Rudavets suggest a scheme for the invesigation of the conductance properties of Bose-condensates that has analogies with mesoscopic transistors.

3. Where Do We Stand?

In view of the interdisciplinary character of this book, we find it is expedient to look at the presented articles not only according to the physical objects they describe, but also from a unifying standpoint. Several *unifying themes* may be discerned:
(a) memory effects in dephasing and relaxation;
(b) protection of quantum information by its distribution throughout the ensemble;
(c) control of decoherence and entanglement by time-dependent interventions or measurements;
(d) manifestations of entanglement in scattering off statistical ensembles;
(e) measurement-induced dynamics;
(f) coherent dynamics and interference in mesoscopic or macroscopic ensembles.

In what follows, we wish to dwell on these unifying themes and their impact upon the experimental and theoretical state-of-the-art and its foreseen progress, in the domain of dynamics and control of QUACS, as they transpire from this book.

3.1 Experimental state-of-the-art and trends

We first comment ont the present status of experiments pertaining to QUACS and its conspicuous trends:

3.1.1 Particle interferometers. Particle interferometers, which have been among the main experimental tools of quantum mechanics since its inception, are developing in two principal directions: (a) increase in resolution and in the ability to study more and more massive objects such as large molecules (see M. Arndt *et al.*), and (b) reduction in the size of the devices, which now reaches the mesoscopic scale (see Y. Ji *et al.*). A laser-based interferometer for molecular dimers is within the present state-of-the-art (see R. Garcia Fernandez *et al.*).

3.1.2 Probes with high spatiotemporal resolution. Traditionally, neutrons have either been used as probes of nuclear reactions in the high-energy domain or as thermal scatterers at meV energies, while the hyperthermal sub-keV-energy region has been considered as exotic. Yet in the latter regime, the nuclear collision time is of the order of the photon time-of-flight over atomic distances. Hence, scattering of such beams (see J. Mayers *et al.* and F. Fillaux) may act as a *sub-femtosecond* probe of molecular and other condensed media under ambient conditions in a wide variety of organic and inorganic substances. These experiments, complemented by experiments with electrons (see C. A. Chatzidimitriou-Dreismann *et al.*) and X-rays (see H. Naumann *et al.*) of the same short-wavelength scale, have revealed scattering anomalies, not yet understood completely, that may have their roots in quantum entanglement (see S. Stenholm *et al.*, E. B. Karlsson, and I. E. Mazets *et al.*), particularly, in the entanglement of protons with their environment. These results, supplemented by studies of probe scattering off fluctuations correlated ensembles (see M. Blaauboer *et al.* and D. O'Dell *et al.*) suggest that high-resolution *probe-scattering and their statistical correlations can reveal multiparticle entanglement* in QUACS which may render invalid concept of pairwise entanglement of the probe and scatterer.

3.1.3 Nonlinear and linear quantum optics. Current progress in optics offers unprecedented possibilities of controlling quantum states of atoms, molecules, and electromagnetic fields. In particular, field-atom interactions can considerably slow-down and even stop (see L. I. Childress *et al.*) light pulses propagating in coherent media, thereby allowing highly efficient nonlinear optical interactions. Yet, in the limit of a single pair of photons, required for performing deterministic quantum gate operations, the nonlinear optics approach still remains a desirable perspective (see D. Petrosyan, Yu. P. Malakyan *et al.*). Therefore, alternative, probabilistic approaches, have emerged. These aproaches are based on linear optics and conditional measurements, requiring high-performance photon detectors (see Ph. Walther *et al.* and A. I. Lvovsky *et al.*).

3.1.4 Manipulation of quantized electromagnetic fields. Control of electromagnetic fields containing a large number of photons under mesoscopic conditions, where essentially quantum phenomena are still observable, has been realized experimentaly (see S. Haroche *et al.*) in high-Q atomic micromasers. This motivates theoretical suggestions of schemes for the control of entanglement and decoherence in laser fields (see G. Yu. Kryuchkyan *et al.*).

3.1.5 Bose condensates, trapped cold atoms, and cold Rydberg gases. Optical and magneto-optical traps, that allow one to confine up to 10^6 atoms and cool them down towards Bose condensation, have become routine experimental tools during the last decade. Experiments with mesoscopic quantum billiards controllable by laser confinement (see N. Katz *et al.*) and with ensembles of cold Rydberg atoms (see P. Pillet *et al.*) give direct access to verification of basic theoretical models of complex quantum ensembles. Suggested new experiments concerning conductance (see A. M. Dykhne *et al.*) and dipole-dipole induced entanglement (see D. O'Dell *et al.*) in Bose condensates should offer a deeper insight into the role of correlations in quantum ensembles.

3.1.6 Microlithography, material science and low temperatures. Considerable technological progress in lithography, low temperature techniques, and material science have allowed the creation of various quantum mesoscopic devices. This technology is potentially capable of creating compact networks of high-T_c superconducting devices (see E. Sarnelli *et al.*) suitable for topological quantum error protection (see L. Ioffe).

3.2 Theoretical methods and models

Here we focus on emerging theoretical concepts related to entanglement, decoherence and their control in QUACS:

3.2.1 Models of entanglement. In order to implement entanglement-based quantum algorithms in large ensembles, it sufices to demonstrate control over many qubits via a *limited* number of standard operations, forming a group of transformations in the exponentially large Hilbert space of such ensembles. The manipulation of ensembles in mixed quantum states (see S. L. Braunstein) constitutes a new step in this direction.

3.2.2 Topological protection. Recent developments of coherence protection algorithms concentrate on errors restricted to a subgroup (see Sec. 1. One of the most promising extensions relies on the topological structuring of Josephson-circuit ensembles, resulting in their reducted susceptibility to quantum errors (see L. B. Ioffe). This approach invokes a direct analogy to topologically stabilized excitations in quantum field theory.

3.2.3 Control algorithms and coding theory methods extended to the quantum domain. An alternative to topological stabilization is the idea of non-holonom control, which relies on unitary transformations of the entire exponentially-large Hilbert space of the ensemble. This approach generalizes the basic ideas of classical coding theory to the quantum domain. Its two main ingredients are the procedure of quantum coding (see E. Brion *et al.*), supplemented by algorithms for the construction of the required external interventions. These interventions act in the augmented Hilbert space spanned by the entanglement of the protected system with an auxiliary system. The interventions amount to repeated measurements of the auxiliary system in the Zeno regime, which disentangle it from the subspace of the protected system and leave the latter intact. Another type of Hilbert space augmentation is achieved via mapping the system onto a collective state (see M. Fleischhauer *et al.*) with an auxillary system.

3.2.4 Role of quantum statistics. When considering complex systems by methods of statistical physics, one operates with their time-dependent distributions. In fermionic systems (see Yu. Ozhigov), statistical requirements imply that we must replace the independent-particle description by a quasiparticle formalism for quantum information processing. Effects of statistical fluctuations on coherent scattering processes (see M. Blaauboer *et al.*) suggest the need for furher exploration of the role of statistics on the dynamics of entangled systems.

3.2.5 Description of decoherence. A novel approach to decoherence consists in dynamically analyzing the evolution of *information flow* between the system and its reservoir, allowing for the memory or spectral response of the reservoir (see J. Salo *et al.*, A. G. Kofman *et al.*, J. Clausen *et al.*, I. E. Mazets *et al.*). The main aim is to determine the space-time scales of the transition from unitarity to irreversibility and classicality for each experimentally pertinent observable

3.2.6 Decoherence control schemes. A new universal control strategy for the inhibition of decoherence relies on appropriately designed time-dependent interventions into open quantum ensembles, which affects the interaction among the elements of the ensemble and their coupling to reservoirs (see G. Kurizki *et al.*, A. G. Kofman *et al.*). Such interventions include either frequent measurements or temporal modulation of the system-reservoir coupling. This strategy can be useful for controlling chemical reactions or the transition to chaos [Prezhdo 2000; Kaulakys 1997], nonadiabatic protection from decoherence near the edge of a continuum (see S. Pellegrin *et al.*) and control of decoherence in Josephson junctions (see A. Barone *et al.*). Intramolecular decoherence may be affected via control over the electronic-vibrational entanglement by laser fields (see M. Leibscher *et al.*).

4. Conclusions

The concerted discussion of the topics outlined above should help us advance the new paradigm that addresses our abilities to diagnose and manipulate the entangled states of complex quantum objects and their robustness against decoherence. These abilities are required for quantum information (QI) applications or matter-wave interferometry in molecular, semiconducting or superconducting systems. On the fundamental level, this book may help establish the notion of dynamical *information exchange* between quantum systems and chart in detail the route from unitarity to classicality.

Further developments within the outlined paradigm should yield novel, advantageous QI processing schemes and high-sensitivity interferometers, owing to decoherence suppression and effective control of many degrees of freedom. In the long run, these strategies may prove to be the first step towards the (hitherto unattempted) use of entanglement in nanotechnology, metrology and chemistry, with potentially remarkable novel applications.

I

ROLE OF ENTANGLEMENT IN QUANTUM COMPUTING AND INFORMATION PROCESSING

QUANTUM ENSEMBLES AND QUANTUM INFORMATICS

V. M. Akulin,[1] G. Kurizki,[2] S. L. Braunstein[3] and Y. Ozhigov[4]

[1] *Laboratoire Aimé Cotton, Bât. 505, Campus d'Orsay, 91405 Orsay Cedex, France*

[2] *Chemical Physics Department, Weizmann Institute of Science, Rehovot 76100, Israel*

[3] *Department of Computer Science, The University of York, York YO10 5DD, UK*

[4] *Institute of Physics and Technology, Russian Academy of Sciences, Nakhimovsky pr. 34, Moscow, 117218, Russia*

Keywords: Quantum ensembles, group theory

Traditionally, theoretical physics employs two main approaches to the many-body problem - the concept of quasiparticles, and the statistical description. In an ensemble of interacting elements, the first approach consists in identifying new elementary collective excitations as nearly independent quasiparticles that live much longer than this inverse interaction energy. The statistical approach is just the opposite, namely, the assumption that the motion of individual particles comprising the ensemble is completely chaotic or even ergodic. None of these extremes applies to systems designated for information processing, where both interactions and single-particle dynamics must be known in detail.

In quantum ensembles of qubits used for information processing, each qubit serves as a register for the encoded information, thus selecting in the ensemble a preferable Hilbert space, a so-called computational basis composed by the direct products of the qubit eigenstates. Information processing is carried out as a result of application of a certain sequence of unitary transformations either to a single qubit or to pairs of qubits at once. The latter introduce entanglement into the ensemble. The total number of such transformations is limited to a few standard operations that form a group.

Group theory aspects of information processing in quantum ensembles is discussed in the first paper of this chapter. It presents the basic concepts of the information processing and teleportation, based on the group theory approach. It also addresses the question whether the information processing can be carried out when the ensemble of qubits is not in a pure quantum state, being subject to thermalization prior to the quantum computation.

V.M. Akulin et al. (eds.), Decoherence, Entanglement and Information Protection in Complex Quantum Systems, 15–16.

While considering ensembles of identical particles, one has to take into account their quantum statistical properties. This imposes additional symmetry restrictions on the ensemble dynamics and requires the inclusion of additional symmetry operations into the group of operations. This situation is discussed in the second paper of the chapter for Fermi systems, where the additional operation is antisymmetrization over the wave functions of individual fermionic qubits.

ENTANGLEMENT IN QUANTUM INFORMATION PROCESSING

S. L. Braunstein
Department of Computer Science, The University of York,
York YO10 5DD, UK
schmuel@cs.york.ac.uk

Abstract I briefly review the status of entanglement in quantum information processing. Its role is mostly well understood for pure-state processing, but less so for mixed-state processing. I consider a case where we can answer whether entanglement is necessary in mixed-state quantum computation.

Introduction

In this article I will discuss entanglement and its role in quantum information processing - especially, but not exclusively in the context of quantum computation.

Before I get to anything as fancy as entanglement or computation, let me start with some basics: let us start with representations of information. On conventional computers, we have bits that take on one of two values. These bits are representations of physical systems sitting in a computer, but ultimately, at the microscopic level these systems and therefore information obey the laws of quantum mechanics - or so we would like to think.

Moving from bits to qubits, a qubit can take one of two orthogonal states, $\alpha\,|0\rangle + \beta\,|1\rangle$, just like bits or any superposition of these states. Similarly, if we have two qubits, they span a four-dimensional space. If the qubits are correlated we say they are entangled (I will define this more carefully later).

The key point is that as the number of qubits increases, the dimensionality of the Hilbert space grows exponentially. In some sense, we can store an exponential number of classical configurations, which - if we can manipulate them - can give us access to vast computational power.

This realization is at the heart of quantum computation and led, in 1993, to Peter Shor's astounding result of a polynomial-time algorithm for factoring on a quantum computer [Shor 1994]. In contrast, the best known classical algorithms for factoring are virtually exponential in run time.

V.M. Akulin et al. (eds.), Decoherence, Entanglement and Information Protection in Complex Quantum Systems, 17–26.

1. Computational complexity

Formally, the language we use for assessing the performance of algorithms and comparing their performance on different platforms is the language of computational complexity. Roughly speaking, computational complexity asks how the 'time' to complete an algorithm scales with the size of the input it is given. The reason that computational complexity only measures the scaling of the time to complete an algorithm is so that it can give a robust measure independent of the underlying instruction set used by the computer performing the task. More precisely, it is a robust measure of complexity for machines that can simulate each other in polynomial time. This measure has been used to characterize algorithms into a polynomial hierarchy.

Within this hierarchy, factoring is non polynomial (NP) - although we do not know of any classical algorithm that runs in polynomial time, it is easy to test the solution in polynomial time. Curiously, until about a year ago, it had been believed that another problem, primality testing, was not polynomial. But then a polynomial time algorithm was discovered. Thus, the structure of the polynomial hierarchy is not yet set in stone.

Of interest to us is that quantum computers add a completely new class to this hierarchy: the class of non-deterministic algorithms that can be solved in polynomial time on a quantum computer, so-called BQP [Bernstein 1997]. The key to these fantastic quantum algorithms, or so everybody believed until recently, was the presence of in these systems.

2. Picturing entanglement

So, what is entanglement? It is actually very simple for pure states: if a pure state cannot be written as a tensor product $|\psi\rangle \neq |A\rangle_{\text{Alice}} \otimes |B\rangle_{\text{Bob}} = |AB\rangle$, then it is entangled. Otherwise it is unentangled, or so-called separable. There is the famous example of the Bell state, $|00\rangle + |11\rangle$, where each qubit can be either zero or one, just so long as they are in the same state (either both zero or both one). Because of the correlations between the two qubits it is easy to see that there is no possible way to write this state as a single tensor product of two independent qubit states.

3. Computation as unitary evolution

Moving one step ahead to actual computation, within the quantum world everything is unitary, so the dynamics of computation can be described as $U : |\psi\rangle \longrightarrow U|\psi\rangle$. An important result is that any unitary operator U may be simulated by a set of one- and two-qubit operations called gates [Barenco 1995 (b)]. Such a decomposition makes implementation feasible, which is good news. To see why this is so powerful, consider an arbitrary (entangled)

n-qubit state

$$|\psi\rangle = \sum_{j_1, \ldots, j_n = 0, 1} \alpha_{j_1 \ldots j_n} |j_1\rangle \otimes \cdots \otimes |j_n\rangle.$$

To write down the action of even a one-qubit gate would require our updating an exponential number of coefficients, *e.g.*,

$$U_1^{(i)} : \alpha_{j_1 \ldots j_i \ldots j_n} \longrightarrow \sum_{k_i = 0, 1} M_{j_i k_i} \alpha_{j_1 \ldots k_i \ldots j_n},$$

yet for a quantum computer this would be as easy as flipping a qubit.

4. Entanglement as a resource

It is easy to see that this immense power comes from the entanglement of the qubits, because if they were not entangled then the state could be written down as a tensor product of independent qubits, in which case the coefficients would have the form $\alpha_{j_1 j_2 \ldots j_n} = a_{j_1} b_{j_2} \ldots d_{j_n}$, and an operation on one or two of them would not affect the other qubits - not much different from classical dynamics.

This is exactly what Richard Feynman thought (and all those who followed) when he asked *"Can a quantum system be probabilistically simulated by a classical universal computer? ... the answer is certainly, No!"* Put differently by Carlton Caves as *"Hilbert space is a big place."* And echos of these ideas exist anonymously in contemporary culture as *"Size matters."*

But only recently has a formal result been proved by Richard Jozsa and Noah Linden [Jozsa 2003; Vidal 2003], which may be roughly stated as

Theorem: Pure-state quantum algorithms may be efficiently simulated classically, provided there is a bounded amount of global entanglement.

I am not going to define precisely what they mean by a "bounded amount of entanglement", but I do want to note that efficient simulation means within polynomial time, which means that if a classical algorithm can efficiently simulate a quantum algorithm then the algorithms have the same computational complexity. This is because when we talk of one machine simulating another, we ignore polynomial-time differences since this potentially amounts to merely changing the instruction set.

5. Caveat emptor ...

Coming back to the Jozsa-Linden theorem, we might naively interpret the theorem as saying that to obtain an exponential speed-up, the entanglement must grow with the size of the input. This is a very powerful result, but there are some caveats:

i) The converse is not true, there is a theorem that even with entanglement we can sometimes efficiently simulate the quantum evolution with a classical computer.

ii) Everything I have talked about so far applies only to pure states, there are more complicated states called mixed states to which none of this applies, this is important because to-date the single experimental system that has produced the most results uses NMR and is based on mixed states.

iii) As I mentioned above, any result in computational complexity is based on scaling behavior and only applies asymptotically - so it does not apply to small systems and also does not apply to more modest speed-ups (such as found in Grover's search algorithm).

iv) Finally, all this talk of speed-up clearly only applies to computation and not to other forms of information processing for instance, communication.

After this lengthy introduction, I will say something about each of these caveats.

6. Gottesman-Knill theorem

Let us start by considering the Pauli group P_n which is generated by the n-fold tensor product of the Pauli matrices σ_i, the identity matrix $\mathbb{1}$ and the factors ± 1 and i. Subgroups of P_n have compact descriptions, since they consist of up to n generators each of which is such an n-fold tensor product. Indeed, since each generator may be described by $2n + 2$ bits of information any such subgroups may be described by no more than $2n(n + 1)$ bits. The compact description of these subgroups is important because the subgroups themselves may be used to represent certain Hilbert space subspaces. One such representation is based on those states which are so-called *stabilized* by the elements of the subgroup. An operator stabilizes a state if that state is the eigenstate with a plus-one eigenvalue. Each independent generator will reduce the dimensionality of the subspace that is stabilized by a factor of two. Thus, a subgroup with n generators stabilizes a unique Hilbert space state.

For example, the subgroup $\left\langle \sigma_z^{(1)}, \ldots, \sigma_z^{(n)} \right\rangle \subset P_n$ uniquely stabilizes $|0\rangle \otimes \ldots \otimes |0\rangle$. Similarly, the two-qubit stabilizer $\langle \sigma_x \otimes \sigma_x, \sigma_z \otimes \sigma_z \rangle \subset P_2$ uniquely stabilizes the state $|0\rangle \otimes |0\rangle + |1\rangle \otimes |1\rangle$.

Thus, at least some states with entanglement across all n qubits will have a compact representation in terms of its stabilizers. The description of these compact forms have a size that grows as n^2, instead of 2^n for an arbitrary pure state as would be required by the conventional expansion in the computational basis. Further, we note that the set of operators consisting of the Pauli matrices σ_i, the Hadamard gate H, the phase gate $\sqrt{\sigma_z}$ and the controlled-NOT gate, map products of Pauli matrices to products of Pauli matrices. This set of gates forms the so-called Clifford group. Thus, under conjugation the gates from the Clifford group map subgroups of the Pauli group to subgroups of the

Pauli group. That is, they map states with compact descriptions to states with compact descriptions. All this may be summarized by the Gottesman-Knill theorem [Gottesman 1997], which may be roughly stated as
Theorem: Any computation restricted to these gates may be simulated efficiently within the stabilizer formalism.

Thus, despite these gates being sufficient to create unbounded amounts of global entanglement between the qubits of a quantum computer they may still be efficiently classical simulated. Unfortunately, this set of gates is not computationally universal. So even though it does not seem like we can do any interesting computation with such a limited set of gates, this caveat to the Jozsa-Linden theorem still gives us an important insight and in particular that entanglement, although necessary, is not sufficient. This takes us to the next caveat, that the theorem does not apply to mixed states.

7. Mixed state entanglement

Let us start by defining what they are. A pure state may be written as a projection operator $\rho_j = |\psi_j\rangle\langle\psi_j|$. A mixed state, or so-called density matrix, is defined to be any convex combination of such projectors

$$\rho = \sum_j p_j \, \rho_j, \; p_j \geq 0.$$

For a density matrix ρ_{AB} defined on a tensor producted Hilbert space $\mathcal{H}_A \otimes \mathcal{H}_B$ this state is unentangled if it may be written as a convex sum of unentangled states, namely

$$\rho_{AB} = \sum_j p_j; \; \rho_A^j \otimes \rho_B^j, \; p_j \geq 0,$$

otherwise the state is entangled. Because there can be classical correlations in such states, there is no general test for entanglement - *i.e.*, we do not know in general how to distinguish classical from quantum correlations in mixed states.

8. Liquid-state NMR quantum computation

So we have defined mixed states. How are they used in NMR quantum computation in the liquid state?

The basic idea is that these are experiments on molecular spins in a fluid at room temperature. At this temperature, the states tend to be completely mixed (a statistical mixture of all states) but adding an external field introduces some correlations that can be written as small perturbations

$$\rho_\epsilon = \frac{1-\epsilon}{2^n}\,\mathbb{1} \; + \epsilon \, |\psi\rangle\langle\psi|.$$

For ϵ sufficiently small we may treat this perturbation as a pure state, and the whole state is called a pseudo-pure state and obeys the following very nice properties: for any unitary transformation U the state evolves to $U\ \rho_\epsilon\ U^\dagger$, which is the pseudo-pure state with $|\psi\rangle$ replaced by $U|\psi\rangle$. Thus, the algorithm unfolds as usual as on the pure-state perturbation. Further, for traceless observables A, their expectation value is just $\langle A\rangle = \epsilon\ \langle\psi|A|\psi\rangle$. Finally, because there are roughly 10^{10} or even many more such molecules in the fluid we get a strong classical signal from such measurements.

This is fantastic: it is at room temperature, we have effectively pure-state evolution, we measure observables up to some attenuation and amplification is easy. It looks like we have got a quantum computer.

So indeed, ever since the scheme was invented in 1997 there have been dozens of papers with a large impact (here is just a small selection) with algorithms performed including Grover [Jones 1998], Deutsch-Jozsa [Chuang 1998], Shor [Vandersypen 2001] recently and even teleportation [Nielsen 1998].

The only problem is that we do not really understand what is going on because these are mixed states! Soon enough this started a community-wide debate [Fitzgerald 2000]!

9. Does NMR computation involve entanglement?

However, first-things-first, we have to get back to the question of entanglement testing and find out if there is any entanglement in these systems or if all the correlations are, in some sense, classical. Here is how we do it.

To start with, any n-qubit mixed state may be written

$$\rho = \int d^n\Omega\ w(\vec{r}_1,\ \ldots,\ \vec{r}_n)\ P_{\vec{r}_1} \otimes \ldots \otimes P_{\vec{r}_n},$$

where the single-qubit projectors may be written

$$P_{\vec{r}} \equiv \frac{1}{2}(\mathbb{1}_2 + \vec{r}.\vec{\sigma}) = \frac{1}{4\pi}\int d\Omega_j\ (\mathbb{1}_2 + 3\,\vec{r}\ \ldots\ \vec{r}_j)P_{\vec{r}_j}.$$

This expression for the kernel function $w(\vec{r}_1,\ \ldots,\ \vec{r}_n)$ may be inverted via

$$w(\vec{r}_1,\ \ldots,\ \vec{r}_n) = \frac{1}{(4\pi)^n}\ \mathrm{Tr}\ \{\rho\ [(\mathbb{1}_2 + 3\,\vec{r}_1.\vec{\sigma}) \otimes \ldots \otimes (\mathbb{1}_2 + 3\,\vec{r}_n.\vec{\sigma})]\}\,.$$

Note that the most negative eigenvalue of the tensor-producted operator within the square brackets is simply $4^{n-1}(-2) = -2^{2n-1}$. Because of this we may immediately place a lower bound that any kernel function may take as

$$w(\vec{r}_1,\ \ldots,\ \vec{r}_n) \geq -\frac{2^{2n-1}}{(4\pi)^n},$$

whereas when $w(\vec{r}_1, \ldots, \vec{r}_n) \geq 0$, ρ is guaranteed to be unentangled.

Applying this to pseudo-pure states

$$\rho_\epsilon = \frac{1-\epsilon}{2^n}\, \mathbb{1} + \epsilon\, |\psi\rangle\langle\psi|,$$

gives the pseudo-pure state kernel as

$$\begin{aligned} w_\epsilon(\vec{r}_1, \ldots, \vec{r}_n) &= \frac{1-\epsilon}{(4\pi)^n} + \epsilon\, w_1(\vec{r}_1, \ldots, \vec{r}_n) \\ &\geq \frac{1-\epsilon\,(1+2^{2n-1})}{(4\pi)^n}, \end{aligned}$$

in terms of the kernel for the pure component. If $w_\epsilon(\vec{r}_1, \ldots, \vec{r}_n) \geq 0$ then ρ is unentangled. Therefore provided that the perturbation parameter is bounded by $\epsilon \leq (1+2^{2n-1})^{-1}$, then the pseudo-pure states being used are unentangled.

Indeed, in current liquid-state NMR experiments $\epsilon \simeq 3 \times 10^{-5}$ on $n < 10$ qubits. So we may conclude that in all liquid-state NMR experiments to-date no entangled state has been accessed [Braunstein 1999]. So that resolves one issue. Our first intuition was that entanglement is everything in quantum computation, but recall that the Jozsa-Linden theorem does not actually say that for mixed states. Maybe one can still obtain a speed-up without entanglement? It turns out that this was no less controversial than the first question.

10. Can there be speed-up in NMR quantum computation?

So now we are asking, can there be speed-up in NMR quantum computation given that there is no entanglement? For Shor's algorithm, Linden and Popescu showed that in the absence of entanglement, no speed-up is possible with pseudo-pure states [Linden 2001]. However, this result (like the Jozsa-Linden theorem) relies on the computational complexity measure of speed-up and so only applies asymptotically as the number of qubits tends to infinity. Unfortunately, current experiments involve less than 10 qubits. To obtain a non-asymptotic answer to this question, one that can be directly relevant to today's experiments, we need to move away from computational complexity to other measures of complexity and to other algorithms. This brings us to the next caveat of the Jozsa-Linden theorem where I want to talk about an algorithm that only gives a modest polynomial speed-up and I want to be able to measure speed-up for even just a few qubits. But this will still be in the context of NMR quantum computation and mixed states. For the algorithm we shall choose Grover's search algorithm [Grover 1997]: suppose we seek a 'marked' number x_0 from $x \in \{0, \ldots, N-1\}$ satisfying

$$f(x) = \begin{cases} 1, & \text{if } x = x_0 \\ 0, & \text{otherwise.} \end{cases}$$

Classically, finding x_0 would take $O(N)$ queries of the function $f(x)$. Thus, the (classical) query complexity for this problem would be $O(N)$. In fact, the best classical algorithm requires $\mathcal{N}_{\text{class}} = (N+2)(N-1)/(2N)$ queries.

Because we can count every use of the 'oracle' that evaluates the function $f(x)$, we have a measure of speed-up which applies even in small systems. We shall now apply this measure to NMR.

We shall ask whether Grover's quantum algorithm for solving this problem on an entanglement-free pseudo-pure state based machine can offer any speed-up over classical performance. The key feature of Grover's algorithm for a pure-state implementation is that at step k the quantum computer is in state

$$|\psi_k\rangle = \frac{\cos\theta_k}{\sqrt{N-1}} \sum_{x \neq x_0} |x\rangle + \sin\theta_k \, |x_0\rangle,$$

where the virtual database is of size $N = 2^n$ for n qubits and $\theta_k = (2k+1)\theta_0$ with $\sin\theta_0 = 1/\sqrt{N}$. By selecting out any of the n qubits (it does not matter which) we can partition the state $|\psi_k\rangle$ into a bipartite system. In the Schmidt basis this state takes the form

$$|\psi_k\rangle = \sqrt{\lambda_1(k)} \, |g_k\rangle \, |e'_k\rangle + \sqrt{\lambda_2(k)} \, |e_k\rangle \, |g'_k\rangle,$$

where $\sqrt{\lambda_j(k)}$ are the Schmidt coefficients at step k in the computation.

In an NMR implementation of this algorithm the system will be in the pseudo-pure state

$$\rho_k = \frac{1-\epsilon}{N} \, \mathbb{1}_N + \epsilon \, |\psi_k\rangle\langle\psi_k|,$$

at step k. Since all the action of Grover's algorithm is happening within the four-dimensional subspace spanned by $\{|g_k\rangle|g'_k\rangle, |g_k\rangle|e'_k\rangle, |e_k\rangle|g'_k\rangle, |e_k\rangle|e'_k\rangle\}$ we may project the pseudo-pure state into this subspace to obtain

$$\rho_k^{(4)} = \frac{N}{4+\epsilon(N-4)} \left(\frac{1-\epsilon}{N} \, \mathbb{1}_4 + \epsilon \, |\psi_k\rangle\langle\psi_k| \right).$$

Since projection into a subspace cannot create entanglement we know that whenever $\rho_k^{(4)}$ is entangled so is ρ_k. However, because $\rho_k^{(4)}$ lives in a four-dimensional Hilbert space we can completely characterize its entanglement. In fact, it is entangled whenever $\epsilon > \epsilon_k \equiv [1 + N\sqrt{\lambda_1(k)\,\lambda_2(k)}]^{-1}$. So if the state is entangled, the perturbation parameter must be bounded. Thus, if ρ_k is unentangled then it must be the case that $\epsilon \leq \epsilon_k$. Since we know it is unentangled, let us plug this bound in and compute the query complexity.

Before we precede, however, there is just one minor point: when we count the query complexity, we have to count the amplification factor (since each molecule does separate queries).

The amplification factor is determined as follows: for a single molecule, the probability of success at step k is $p(k) = \langle x_0|\rho_k|x_0\rangle$ which is less than 1 and will tell us the minimum number of repetitions $1/p(k)$ required. Now, if the algorithm terminates early at step k each repetition involves $k+1$ function evaluations. Thus, the unentangled query complexity will be

$$\mathcal{N}_{\text{pseudo}}^{(\text{min})} \equiv \min_k \frac{k+1}{p(k)},$$

computed for $\epsilon \leq \epsilon_k$. The minimum over k allows for the possibility that an early termination may lead to a more efficient result.

Table 1. Query complexity of Grover's algorithm for an unentangled pseudo-pure state implementation compared to the classical query complexity for up to 8 qubits.

n	N	k_{opt}	$\mathcal{N}_{pseudo}^{(min)}$	$\mathcal{N}_{class}$
1	2	0	2.00	1.00
2	4	1	2.00	2.25
3	8	1	5.48	4.38
4	16	2	12.89	8.44
5	32	0	32.00	16.47
6	64	0	64.00	32.48
7	128	0	128.00	64.49
8	256	0	256.00	128.50

What do we find? We computed the unentangled query complexity for up to 8 qubits and compared it to the best classical query complexity. The results are found in Tab. 1.

Except for the exceptional case for $n = 2$ qubits, there is basically a factor of two difference! (The exceptional case of $n = 2$ does not apply to NMR since the perturbation parameter must be so large to obtain a speed-up that the assumptions behind the pseudo-pure state description break down.) Not only is the classical algorithm more efficient, but the NMR scheme reduces to random guessing - which is just what one might expect from an essentially random statistical mixture. Further, the trend we see extends beyond $n = 8$ to all higher values. Thus, we may conclude that we find that entanglement is necessary for obtaining speed-up for Grover's algorithm in liquid-state NMR quantum computation [Pati 2002].

The good news is that these results reinforce our intuition that entanglement really is important for speed-up. The bad news is that we still do not have a general understanding and in this sense, even without speed-up NMR continues to be a fantastic test bed for this sort of question.

11. Entanglement in quantum communication

Finally, I would like to say a few words about entanglement in communication, as the last caveat to the Jozsa-Linden theorem. I will just use the example of quantum teleportation.

In quantum teleportation, Alice is given a state $\rho_{\text{in}} = |\psi\rangle\langle\psi|$ whose identity is unknown to her. She may do anything she wishes to this state and then she communicates with Bob via only a classical communication channel. Bob's aim is to create a state ρ_{out} which best resembles the original state.

In the absence of shared entanglement between Alice and Bob the mean fidelity of the output state $F = \langle\psi|\rho_{\text{out}}|\psi\rangle$ is bounded. For example, for teleporting qubits, $F \leq 2/3$, whereas for the teleportation of coherent states in an infinite-dimensional Hilbert space it is bounded by $F \leq 1/2$ [Braunstein 2000 (b)].

Fidelities beating these bounds have been achieved experimentally when the sender and receiver share entanglement. So teleportation is a success and entanglement really matters.

To close the loop, teleportation and indeed many other quantum communication protocols do not need a universal set of gates and do quite well with just those gates covered by the Gottesman-Knill theorem [Bartlett 2002]. Although these protocols may be easily simulated, they have capabilities beyond classical communication channels - clearly simulation is not everything.

Summary

To conclude, we may say that the role of entanglement in quantum information processing is not yet well understood.

For pure states, unbounded amounts of entanglement are a rough measure of the complexity of the underlying quantum state. However, there are exceptions as we saw for states encompassed by the Gottesman-Knill theorem.

For mixed states, even the unentangled state description is already very complex. Nonetheless, entanglement seems to play the same role (for speed-up) in all examples examined to-date, an intuition which extends down to few-qubit systems.

However, in quantum communication, entanglement is much better understood and clearly sets apart quantum protocols as a distinct class.

I hope I have given you some glimpse into one of the more subtle areas in quantum information processing. Both in terms of some of the progress that we have made in recent years towards understanding when and how entanglement really matters, and in terms of the still open questions.

EASY CONTROL OVER FERMIONIC COMPUTATIONS

Y. Ozhigov
Institute of Physics and Technology, Russian Academy of Sciences,
Nakhimovsky pr. 34, Moscow, 117218, Russia
ozhigov@ftian.oivta.ru

Abstract Quantum fermionic computations on occupation numbers proposed in [Bravyi 2000] are studied. It is shown that a control over external field and tunneling would suffice to fulfill all quantum computations without valuable slowdown in the framework of such model when an interaction of diagonal type is fixed and permanent. Substantiation is given through a reduction of some subset of this model to the conventional language of quantum computing and application of the construction from [Ozhigov 2002 (b)].

Introduction and background

Quantum computing is an unprecedented examination of the modern physics because it requires a level of control over microscopic sized objects which has never been reached artificially. While a mathematical theory of quantum computations is well developed, its physical implementation represents a serious challenge to our understanding of nature. This is why it is important to look for its simplest possible realization which would depend only on fundamental principles of quantum mechanics and contain minimal technological difficulties. Two requirements can be formulated for such scheme: adequate description of states forming a basis of computational Hilbert space, and realistic method of control. Conventionally, a computational element - qubit is represented as some characteristic like spin, charge or position of some elementary particle. This approach works well for one isolated qubit. For a system of many qubits it meets a serious difficulty. It comes from a fundamental physical principle of identity of elementary particles of the same type[1]. To control a computation, we must be able to address a separated qubit whereas different particles are indistinguishable according to the principle of identity. Of course, we can distinguish particles by their spatial positions, placing them fairly far one from another, but in this case it will be difficult to keep them in an entangled state that is necessary for quantum computations. A solution to this dilemma was

V.M. Akulin et al. (eds.), Decoherence, Entanglement and Information Protection in Complex Quantum Systems, 27–32.

found by Bravyi and Kitaev, who proposed to use a Fock space of occupation numbers for the description of quantum computations (see [Bravyi 2000]). It uses in fact a natural identification of qubits with energy levels in Fock space where one is treated as occupied level and zero as free level. This approach gives a universal quantum computing for a high price: it requires a control over external field, tunneling, diagonal interaction and changes of numbers of particles (contact with a superconductor) that is to control over coefficients α, β, γ in (7) and over additional summand $\delta\, a_k^\dagger\, a_j^\dagger + \delta^\star\, a_k\, a_j$.

We shall see how to reduce this price using an idea of fixed and permanent interaction. To do this we need two things: assumption that the corresponding Hamiltonian consists only of diagonal and tunneling summands and modified correspondence between states in occupation number representation and Hilbert spaces. Then to fulfill any quantum computation it is required to control over only an external field and tunneling. Such kind of control is in principle realizable by lasers. The main scheme is presented in Sec. 2 and is in fact based on the idea of computational model with fixed permanent interaction proposed in [Ozhigov 2002 (a); Ozhigov 2002 (b)] adapted to the language of Fock spaces. Sec.1 contains a short description of the occupation numbers formalism.

1. Formalism of occupation numbers

In this paper we shall consider a system of n identical fermions. At first let us a make a nonphysical assumption that they can be reliably distinguished. Then its state belongs to the Hilbert space of all states with the basis $\psi(r_1,\, r_2,\, \dots,\, r_n) = \psi_{j_1}(r_1)\, \psi_{j_2}(r_2)\, \dots\, \psi_{j_n}(r_n)$, where $\{\psi_j\}$ are some basis for one particle states, where j_s belongs to the general set of indexes $1,\, 2,\, \dots,\, J$, r_j includes spatial and spin coordinates. A choice of basis means that the system after measurement can be found in one of the basis states. In a real system of identical particles, they cannot be distinguished. Hence any basis state must contain all summands of the form $\psi_{j_1}(r_1)\, \psi_{j_2}(r_2)\, \dots\, \psi_{j_n}(r_n)$ with some factors. Such state must change the sign after permutation of every two fermions and it is convenient to assume that a basis state for n fermions system is given by

$$\Psi = \frac{1}{\sqrt{N!}} \begin{vmatrix} \psi_{j_1}(r_1) & \psi_{j_1}(r_2) & \dots & \psi_{j_1}(r_n) \\ \vdots & \vdots & \vdots & \vdots \\ \psi_{j_n}(r_1) & \psi_{j_n}(r_2) & \dots & \psi_{j_n}(r_n) \end{vmatrix}. \qquad (1)$$

This state may be considered as a situation when only states ψ_{j_s} for $s = 1,\, 2,\, \dots,\, n$ are occupied by particles from our system and all other ψ_k for $k \in \{1,\, 2,\, \dots,\, J\}$ which have not form j_s are free. If ψ with index denotes the eigenvectors of one particle Hamiltonian, we speak about occupied or free energy levels, but generally speaking ψ_k may form arbitrary basis in the space of states for one particle. A state of the form (1) may be represented as a symbol $|\bar{n}_\Psi\rangle = |n_1,\, n_2,\, \ldots,\, n_J\rangle$, where n_k is one if k^{th} energy level is occupied and zero if it is free. This is a representation of states of fermionic ensemble in the form of occupation numbers. Such vectors $\bar{n}$ form a basis of Fock space and the general form of state of our system will be $\sum\limits_{\bar{n}} \lambda_{\bar{n}}|\bar{n}\rangle$ with amplitudes λ.

An operator of annihilation a_j of a particle on j^{th} level and its conjugated $a_j^\dagger$ (creation) are defined by

$$a_j\,|n_1,\,\ldots,\,n_J\rangle = \delta_{1n_j}\,(-1)^{\sigma_j}|n_1,\,\ldots,\,n_{j-1},\,n_j-1,\,n_{j+1},\,\ldots,\,n_J\rangle, \quad (2)$$

where $\sigma_j = n_1 + \ldots + n_j$. They possess the well-known commutative relations:

$$\begin{aligned} a_j^\dagger\, a_k + a_k\, a_j^\dagger &= \delta_{jk}, \\ a_j\, a_k + a_k\, a_j = a_j^\dagger\, a_k^\dagger &+ a_k^\dagger\, a_j^\dagger = 0. \end{aligned} \quad (3)$$

Assume that any interaction in nature goes between no more than two particles. Hence any interaction in many-particle system may be expanded into the sum of one and two particles interactions of the form $H = H_{one} + H_{two}$ with the corresponding potentials $V_1(r)$ and $V_2(r,\, r')$. Each of them can be represented by the operators of creations and annihilations as

$$H_{one} = \sum_{k,\, l} H_{kl}\, a_k^\dagger\, a_l, \;\; \text{and} \;\; H_{two} = \sum_{k,\, l,\, m,\, n} H_{klmn}\, a_l^\dagger\, a_k^\dagger\, a_m\, a_n \quad (4)$$

where

$$\begin{aligned} H_{kl} &= \langle\psi_k|\, H_{one}\, |\psi_l\rangle = \int \psi_k^\star(r)\, V_1(r)\, \psi_l(r)\, dr, \\ H_{klmn} &= \langle\psi_l,\, \psi_k|\, H_{two}\, |\psi_m,\, \psi_n\rangle \\ &= \int \psi_k^\star(r)\, \psi_l^\star(r')\, V_2(r,\, r')\, \psi_m(r)\, \psi_n(r')\, dr\, dr'. \end{aligned} \quad (5)$$

Hence, given potentials of all interactions and all basis states ψ_i, we can, in principle, obtain its representation in terms of creations and annihilations, *e.g.*, in the language of occupation numbers.

Consider an ensemble with Hamiltonian of the form

$$H = \sum_i H^i_{ext.\ f.} + \sum_{i,\, j} (H^{ij}_{diag.} + H^{ij}_{tun.}) \quad (6)$$

where Hamiltonians of external field, diagonal interaction and tunneling are represented by means of the operators of creations and annihilations by

$$H^{i}_{ext.\ f.} = \alpha_i\, a_i^\dagger\, a_i,\ \alpha_i \in \mathbb{R},$$

$$H^{ij}_{diag.} = \beta_{ij}\, a_i^\dagger\, a_i\, a_j^\dagger\, a_j,\ \beta_{ij} \in \mathbb{R}, \tag{7}$$

$$H^{ij}_{tun.} = \gamma_{ij}\, a_i^\dagger\, a_j + \gamma_{ij}^\star\, a_j^\dagger\, a_i.$$

Note that it would not be easy to implement a control over diagonal part of Hamiltonian. Assume that the diagonal interaction is fixed and acts permanently whereas external field and tunneling are subjects of control. Then it is possible to fulfill every quantum computation. This type of control seems to be fairly realistic because the tunneling may be controlled by laser impulses.

2. Computation controlled by tunneling

Instead of simple correspondence between occupation numbers and Hilbert spaces described above, we now establish another correspondence that makes possible to transfer a universal computing with fixed permanent interaction [Ozhigov 2002 (a); Ozhigov 2002 (b)] to the language of fermionic computing in Fock space of occupation numbers.

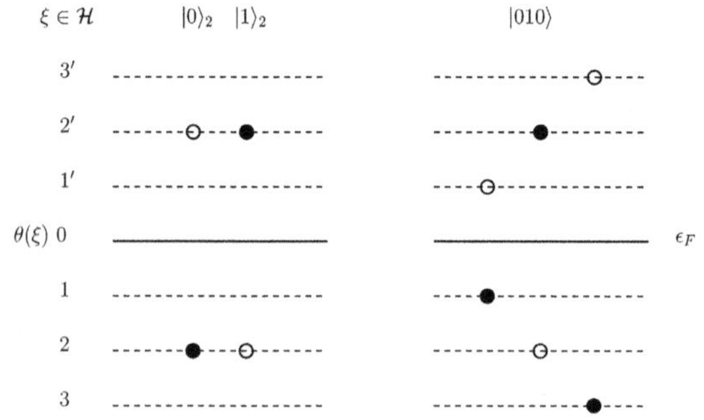

Figure 1. Correspondence between Fock and Hilbert spaces

Let us fix some partitioning of all energy levels to two equal parts and choose some one-to-one correspondence between them. Say we can consider k^{th} level down from Fermi bound ϵ_F and agree that it corresponds to k^{th} level up from ϵ_F. We shall denote j^{th} level down from Fermi bound by ordinary letter and j^{th} level up from Fermi bound by j'. Call the first level j^{th} lower level and the second one j^{th} upper level. Fock space $\mathcal{F}$ can be represented as $\mathcal{F} = \mathcal{F}_1 \otimes \mathcal{F}_2 \otimes \ldots \otimes \mathcal{F}_k$ where each $\mathcal{F}_j$ corresponds to j^{th} pair of the corresponding energy levels. Consider a subspace F_j in $\mathcal{F}_j$ which is spanned by two following states. The first one is: "j'^{th} level is occupied and j^{th} level is free", the second is "j^{th} level is occupied and j'^{th} is free". Denote them by $|1\rangle_j$ and $|0\rangle_j$, correspondingly. We shall deal with subspace $F = F_1 \otimes F_2 \otimes \ldots \otimes F_k$ in Fock space $\mathcal{F}$. Now determine a function θ that maps our Hilbert space $\mathcal{H}$ to F by the following definition on basis states: $\theta(|\xi_1, \xi_2, \ldots, \xi_n\rangle) = |\xi_1\rangle_1 \otimes |\xi_2\rangle_2 \otimes \ldots \otimes |\xi_n\rangle_n$ where all ξ_j are ones and zeroes. Thus θ establishes unconventional correspondence between Hilbert and Fock spaces (see Fig. 1).

One qubit state in Hilbert space corresponds to the two qubits state in conventional assignment of qubits for Fock space – each occupation number to each qubit. But we shall see that this assignment corresponds to the task of control over computation better than conventional assignment.

Now the door is open for the representation of unitary transformations in Hilbert space required for quantum computing by transformations in Fock space. Consider Hermitian operator H in one-dimensional Hilbert space $\mathcal{H}$. It has the form $H_0 + H_1$ where

$$H_0 = \begin{pmatrix} d_1 & 0 \\ 0 & d_2 \end{pmatrix}, \; H_1 = \begin{pmatrix} 0 & d \\ \bar{d} & 0 \end{pmatrix}. \tag{8}$$

It can be straightforwardly verified that for the operators $\tilde{H}_0 = d_1\, a_k^\dagger\, a_k + d_2\, a_{k'}^\dagger\, a_{k'}$ and $\tilde{H}_1 = d\, a_k^\dagger\, a_{k'} + \bar{d}\, a_{k'}^\dagger\, a_k$ (that is external field and tunneling) we have the equalities $\tilde{H}_i\, \theta = \theta\, H_i$ for $i = 0,\ 1$. Using linearity of θ, we obtain $(\tilde{H}_0 + \tilde{H}_1)\theta = \theta\, H$. Now consider one qubit unitary transformations U in Hilbert space. It has the form e^{-iH} for Hamiltonian H (we choose appropriate time scale to get rid of Plank constant and time). Using linearity of θ and the equality $\theta^{-1}\, H^s\, \theta = (\theta^{-1}\, H\, \theta)^s$ for natural s, we obtain that for every one qubit unitary transformation U we can effectively find the corresponding Hamiltonian in Fock space containing only external field and tunneling that makes diagram A from the Fig. 2 closed.

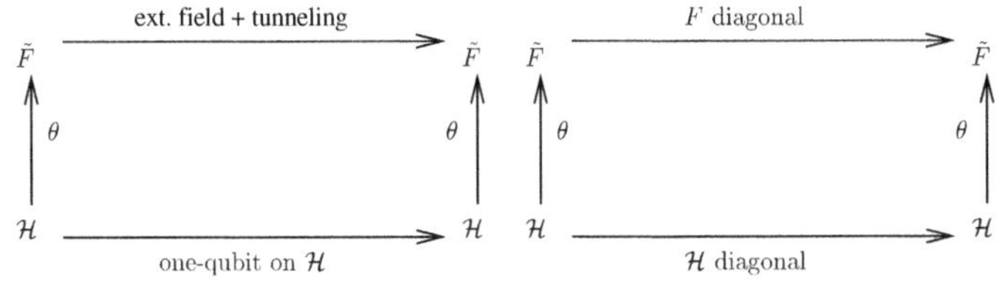

Figure 2. Correspondence of transformations in Fock and Hilbert subspaces. $\tilde{F} = F_j \bigotimes F_k$.

Consider now two qubits transformations in Hilbert space. Since all diagonal matrices commute, for all diagonal transformation in space $\mathcal{F}_k \otimes \mathcal{F}_j$ we can effectively find the corresponding diagonal transformation in the corresponding Hilbert space that makes diagram B from Fig. 2 closed. Note that for entangling $\mathcal{V}$ the transformation V will be entangled as well. Now all is ready to transfer a trick from [Ozhigov 2002 (b)] with one qubit controlled universal computations to Fock space. Combination of diagrams from Fig. 2 gives the diagram of Fig. 3.

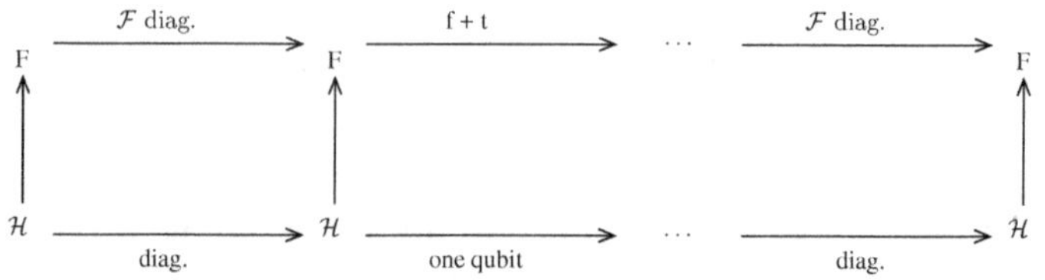

Figure 3. Correspondence of computations in Fock and Hilbert spaces.

Let the diagonal part of interaction in Fock space be fixed and act permanently. Then we can find the corresponding diagonal interaction on Hilbert space making all diagonal parts of diagram in Fig. 3 closed. According to the result of [Ozhigov 2002 (b)], we can choose one qubit transformations implementing arbitrary quantum computations in Hilbert space in the form represented by the lower sequence of the diagram. At last we can find field + tunneling control over states in Fock space making all diagram closed. Note that all operators of creations and annihilations considered in the whole Fock space are not local due to the factor $(-1)^{\sigma_j}$ depending on a given state. For the diagonal operators $a_j^\dagger\, a_j\, a_k^\dagger\, a_k$ and external fields these factors are compensated. The tunneling operator $a_j^\dagger\, a_{j'}$ in space F brings factor $(-1)^{\sigma'}$, where

$$\sigma' = \sum_{s+j}^{j'-1} n_s = j' - j \tag{9}$$

that is independent of a given state $|\bar{n}\rangle \in F$ because for such state exactly half of levels between j and j' are occupied. Hence a sign can be factored out of the complete state and ignored.

Thus we obtain a universal quantum computer on states in occupation numbers space controlled only by external field and tunneling.

Notes

1. I express my thanks to Sergei Molotkov who attracted my attention to this problem and referred me to the paper [Bravyi 2000]. I am also grateful to Alexander Tsukanov and Alexander Kokin for useful comments to the preliminary version of this paper.

II

MULTIATOM AND MULTIPHOTON ENTANGLEMENT

MULTIATOM AND MULTIPHOTON ENTANGLEMENT

D. Petrosyan,[1] L. I. Childress,[2] G. Yu. Kryuchkyan,[3] A. I. Lvovsky,[4] Yu. P. Malakyan,[5] and Ph. Walther[6]

[1] *Institute of Electronic Structure & Laser, Foundation for Research & Technology - Hellas, Heraklion 71110, Crete, Greece*

[2] *Physics Department, Harvard University, Cambridge, MA 02138, USA*

[3] *Yerevan State University, A. Manookyan 1, 375049, Yerevan, Armenia*
Institute for Physical Research, National Academy of Sciences, Ashtarak-2, 378410, Armenia

[4] *Department of Physics & Astronomy, University of Calgary, Alberta T2N 1N4, Canada and Fachbereich der Physik, Universität Konstanz, D-78457 Konstanz, Germany*

[5] *Institute for Physical Research, National Academy of Sciences, Ashtarak-2, 378410, Armenia*

[6] *Institut für Experimentalphysik, Universität Wien, Vienna, Austria and Institut für Quantenoptik & Quanteninformation, Österreichische Akademie der Wissenschaften, Vienna, Austria*

Keywords: Photon entanglement, all-optical quantum computation, nonlinear optical media, atomic ensembles, atomic memory, electromagnetically induced transparency, spontaneous parametric down conversion.

Entanglement is a vital information resource employed in quantum teleportation, dense coding and quantum computation [Nielsen 2000]. The fundamental role played by the entanglement in quantum information science was discussed in part I; this part of the book is devoted to the generation and characterization of the entanglement of photons and their usage in quantum communication and computation protocols.

Light is very robust and versatile carrier of information, as attested by an ever growing number of its applications in information processing, telecommunication and defense industries. The usage of photons for quantum information applications has not escaped attention as well. Typically, the two orthogonal polarization states of single photons in a well defined spatial mode serve as qubit basis, although there are alternative but essentially equivalent representations as well, *e.g.*, [Chuang 1995]. Successful experimental demonstrations of quantum teleportation and cryptography protocols employing photons were reported in, *e.g.*, Refs. [Bennett 1992; Muller 1996; Bouwmeester 1997; Furusawa 1998; Hughes 2000]. The realization of quantum computation with

V.M. Akulin et al. (eds.), Decoherence, Entanglement and Information Protection in Complex Quantum Systems, 35–39.

photon-polarization qubits is, however, in a less developed stage, the main restraining factor being associated with the difficulties of experimentally implementing two qubit quantum logic gates. This is because strong nonlinearities capable of entangling very weak, ideally single-photon, pulses can not be attained in conventional nonlinear media. A possible avenue to circumvent these difficulties has been opened up by a proposal to use linear optical elements, such as beam-splitters and phase-shifters, in combination with single-photon sources and detectors, to achieve the probabilistic photon-photon entanglement conditioned on the outcome of successful measurement performed on auxiliary photons [Knill 2001]. Generation of single-photons in a well defined spatiotemporal mode is another problem too. Typically, one employs spontaneous parametric down-conversion to generate a pair of polarization- and momentum-correlated photons. Conditional upon the outcome of measurement on one of the photons, the other photon is projected onto a well-defined polarization and momentum state, as discussed in the first article of this part by A. I. Lvovsky. In the following article by Ph. Walther and A. Zeilinger, explicit schemes to implement the photonic controlled-NOT (CNOT) and nonlinear sign-shift (NS) (known also as controlled-phase - CPHASE) gates, using linear optical elements and correlated photon pairs produced by the parametric down conversion, are studied both theoretically and experimentally. Although these gates work essentially probabilistically, the mere fact of their successful operation undoubtedly brightens the prospects of realizing a scalable, all-optical quantum computation and communication.

An essential ingredient of quantum cryptography, distributed quantum computation or teleportation is a reliable, long distance quantum communication. Classical telecommunication with light uses fiber optical cables connecting the end users separated from each other by hundreds and even thousands kilometers. The inevitable attenuation of light signals is compensated by the use of repeaters whose main task is to amplify the signal at every hundred or so kilometers. One can then envisage long distance quantum communication with photons. Straightforward usage of repeaters to recover degraded by the noise and attenuation quantum states, as it is done in classical communication channels, is, however, problematic. The main difficulties are associated with the fact that in order to reproduce an arbitrary quantum state, one should either know it beforehand or be able to produce its copies - clones. Both however, are forbidden by the laws of quantum mechanics [Nielsen 2000]. In particular, an arbitrary quantum state cannot reliably be inferred from a single quantum measurement; also, such state cannot be cloned. Usage of entanglement purification offers a possible solution to the problem of decoherence, provided the two ends of the quantum communication channel are not too far apart from each other, because otherwise the existing purification protocols would require an exponentially large number of entangles states. An alternative, more realistic protocol would divide a long communication channel into a number of closely spaced quantum nodes and combine the entanglement generation between the nodes with the entanglement purification. To implement the complete protocol, one thus would need to find an efficient scheme for the entanglement generation and purification between the nodes, that would also allow for a reliable storage of that entanglement, to be later used for the communication. Such scheme has been recently proposed in Ref. [Duan 2001]. It employs atomic ensembles connected to each other only via linear optical elements. The crux of the scheme is that a pair of atomic ensembles simultaneously irradiated by weak classical fields, emit with a small probability a single photon that is detected after it passes through a beam splitter. In analogy with the probabilistic entanglement protocol of Ref. [Knill 2001], the passage through the beam splitter effectively erases the information on where the photon came from. The photon is strongly correlated with the atomic spin excitation. As a result, the atomic ensembles become entangled. This entanglement is coherently stored in the ensembles and can be retrieved on demand at a later time. In order to do that, one employs the recently developed technique based on the electromagnetically-induced transparency (EIT) in atomic media [Harris 1997], which allows one to coherently convert a photonic excitation into the spin excitation, store it for a controllable time, and then retrieve on demand [Fleischhauer 2000; Liu 2001; Phillips 2001]. In the third article of this part by L. I. Childress *et al.*, the authors demonstrate, both theoretically and experimentally, the main building block of the described protocol, namely, the probabilistic generation of the spin excitation in an atomic ensemble via Raman scattering of a weak light pulse, and the subsequent retrieval of the atomic spin excitation in the form of photon wavepackets, using the EIT. Their study also extends this technique to the generation of multiphoton wavepackets with controllable propagation direction, timing, and pulse shapes.

Electromagnetically-induced transparency in optically dense atomic media is a quantum interference effect that results in a dramatic reduction of the group velocity (down to complete stop) of a weak probe field in the absence of absorption [Harris 1997; Hau 1999; Fleischhauer 2000]. The essence of EIT is a phenomenon of coherent population trapping consisting of that the application of two laser fields to a three-level atomic system creates a specific coherent superposition of the atomic states -"dark state" - which is stable against the absorption of the fields. Recently, EIT has attracted much interest as a basic mechanism for a number of novel nonlinear optical phenomena, *e.g.*, Refs. [Fleischhauer 2000; Matsko 2000; Andre 2002]. In particular, it was suggested to use coherently driven EIT atoms with N-configuration of levels to achieve an appreciable nonlinear phase shift of extremely weak optical fields [Schmidt 1996; Harris 1999] or a two-photon switch [Harris 1998], which could in principle be used for implementing the deterministic quantum logic gates, *e.g.*, CPHASE or CNOT. The main hindrance of such schemes is the mismatch between the group velocities of the field that is subject to EIT and its nearly-free propagating partner, which severely limits their effective interaction length and the resulting nonlinear coupling [Harris 1999]. This mismatch can be compensated by employing a second species of three-level atoms that slow down the otherwise free-propagating field via EIT [Lukin 2000 (a)]. The experimental realization of the proposed schemes [Schmidt 1996; Harris 1999; Lukin 2000 (a)] is, however, demanding, as achieving purely dispersive cross-phase modulation between the two interacting fields requires extremely cold atomic vapors. Relaxing this constrain, and designing an explicit experimentally realizable scheme for archiving a π cross-phase shift, and thereby the CPHASE gate between two single-photon pulses are thus the important practical issues addressed in the following article by D. Petrosyan. The strong nonlinearities attainable by such systems are shown to be capable of implementing the two-photon quantum logic gates, which may pave the way to deterministic all-optical quantum computation and communication.

Coherent population trapping and the associated with it EIT can be observed as well in other atomic systems having larger number $n_{\text{gr}} > 2$ of ground states. While in the case of N-systems with $n_{\text{gr}} = 2$ and the corresponding number of interacting quantum fields, the attainable $\chi^{(3)}$ Kerr nonlinearity can serve to implement the two qubit logic gate as outlined above, atomic systems with $n_{\text{gr}} > 2$ can support larger order nonlinearities $\chi^{(2n_{\text{gr}}-1)}$ [Zubairy 2002], allowing for the implementation of n_{gr}-qubit logic gates. For the coherently driven atoms, the order of these nonlinearities is associated with the order of ground state coherence of the atomic Zeeman sublevels. Thus, the detection and characterization of the decoherence rate of high order atomic coherences is of vital practical importance. The article by Yu. P. Malakyan *et al.* is devoted to the issue of production and detection of the high order atomic coherences.

While the quantum information applications described above employ the manipulation of quantum systems with discrete spectrum representing the qubits, many quantum variables, such as, *e.g.*, the position and momentum of a particle or the quadrature amplitudes of electromagnetic fields, are essentially continuous. The original quantum state introduced by Einstein, Podolsky and Rosen [Einstein 1935] that unveiled the nonlocality of quantum mechanics, is in fact a position-momentum entangled state of two particles. An alternative avenue for the quantum information processing and communication has been opened up recently by the studies of continuous variables [Braunstein 2003]. Quantum teleportation [Vaidman 1994; Furusawa 1998], dense coding [Braunstein 2000 (a)], quantum key distribution [Grosshaus 2003] and quantum computation [Lloyd 1999] with continuous variables, are some of the most representative examples. Other applications of the entangled continuous variables include the fundamental tests of quantum mechanics in the spirit of Einstein, Podolsky and Rosen, and ultrasensitive interferometric measurements using squeezed entangled states. In the laboratory, such states are usually realized using coherent light whose two quadrature amplitudes constitute the two conjugate continuous variables [Walls 1994]. Typically, bright entangled light beams are achieved using the optical parametric oscillator - a nonlinear crystal possessing $\chi^{(2)}$ nonlinearity pumped by a strong coherent light. Various modes of operation of such parametric oscillators placed in a dissipative environment, *i.e.*, low-Q optical cavity, and the resulting entanglement of the output light fields is studied in the last article of this part by G. Yu. Kryuchkyan and H. H. Adamyan. The results obtained by these authors demonstrate an essential improvement of the degree of entanglement in nonlinear optical parametric oscillators driven by time-modulated pump field, which paves the way towards the generation of bright, continuous variable entangled light beams with well-localized phases, which can be used in precise interferometric measurements as well as in quantum information applications.

ON THE ADVANCED WAVE MODEL OF PARAMETRIC DOWN-CONVERSION

A. I. Lvovsky,[1] and T. Aichele[2]

[1] *Department of Physics & Astronomy, University of Calgary, Alberta T2N 1N4, Canada and Fachbereich Physik, Universität Konstanz, D-78457 Konstanz, Germany*

[2] *Nano-Optics, Physics Department, Humboldt University, D-10117 Berlin, Germany*

lvov@ucalgary.ca, http://qis.ucalgary.ca/quantech/

Abstract The spatiotemporal optical mode of the single-photon Fock state prepared by conditional measurements on a biphoton is investigated and found to be identical to that of a classical wave due to a nonlinear interaction of the pump wave and Klyshko's advanced wave. We discuss the applicability of this identity in various experimental settings.

Keywords: Parametric down-conversion, conditional preparation, quantum imaging

Introduction

Preparation of single-photon states (SPSs) by means of conditional measurements on a biphoton generated through parametric down-conversion (PDC) has been proposed and tested experimentally in 1986 by Hong and Mandel [Hong 1986] as well as Grangier, Roger and Aspect [Grangier 1986] and has since become the workhorse for many quantum optics experiments. The idea of the method is based on the fact that in PDC, photons are necessarily born in pairs. Two generated photons can be separated into two emission channels according to their propagation direction, wavelength and / or polarization. Detection of a photon in one of the emission channels (labeled $trigger$) indicates that the quantum ensemble in the remaining ($signal$) channel is also a SPS.

This method has recently been utilized by the Konstanz quantum optics group, who has prepared the signal photon in a well-defined, highly pure spatiotemporal mode which can be matched with, coupled into, or caused to interfere with, a classical laser mode. Based on this achievement, we have demonstrated homodyne tomography of the single-photon Fock state [Lvovsky 2001]

V.M. Akulin et al. (eds.), Decoherence, Entanglement and Information Protection in Complex Quantum Systems, 41–48.

and applied the developed know-how to implementing new tools of quantum-optical information technology [Babichev 2004].

Conditional preparation of a photon from a down-converted pair is a highly complex quantum-mechanical process which involves a collapse of a hyper-entangled biphoton state onto a particular signal mode upon the measurement in the trigger channel. When trying to develop insight into this process and relevant experimental techniques [Aichele 2002], we have found the visualization tool proposed by D. N. Klyshko [Klyshko 1988 (a); Klyshko 1988 (b); Klyshko 1988 (c)] to be of great help. This is the concept of *advanced waves* which assumes that "one of the two detectors, say number 2, at the moment t_2 of registration of a photon emits back in time and space a short δ-like pulse. [...] This pulse interacts within the excited (in a coherent state) atom[1] and the latter emits a growing wave [...] with a converted carrier frequency" [Klyshko 1988 (a)]. According to Klyshko, one can sometimes think to the conditionally prepared photon (CPP) as a wave resulting from the nonlinear interaction between the pump and the fictitious advanced wave.

In the author's opinion, it is unfortunate that Klyshko's ingenious idea has not been developed further in subsequent years. Indeed, many experimental schemes that explore nonclassical correlations in a down-converted pair can be reformulated in a way compatible with the advanced wave model. Within this scope are experiments on entanglement in polarization [Kwiat 1995], frequency-time [Franson 1989], angular-momentum [Mair 2001] and arrival time [Brendel 1999] domains, four-photon interference [Pan 1998], Einstein-Podolsky-Rosen-type nonlocality [Howell 2004], quantum imaging [Pittman 1995; Bennik 2002] and many others. Our understanding of entanglement could be greatly enhanced if the visual power of the Klyshko model were employed to its full extent.

Furthermore, as demonstrated in our earlier theoretical work, the optical mode of the conditionally prepared photon is in fact *completely identical* to that of the difference-frequency field generated by a properly defined advanced wave [Aichele 2002]. In other words, the advanced wave concept is not merely an informal visual tool, but a rigorous mathematical model which possesses analytic capability that approaches that of the canonical quantum theory. In the present paper, we review the theory associated with the advanced wave concept and discuss its applicability in various experimental situations.

1. The conditionally prepared single photon

We start with a general calculation of the quantum state of a single photon state prepared by a conditional measurement on a PDC biphoton state (Fig. 1). We restrict our consideration to the pulsed regime. In all calculations in this pa-

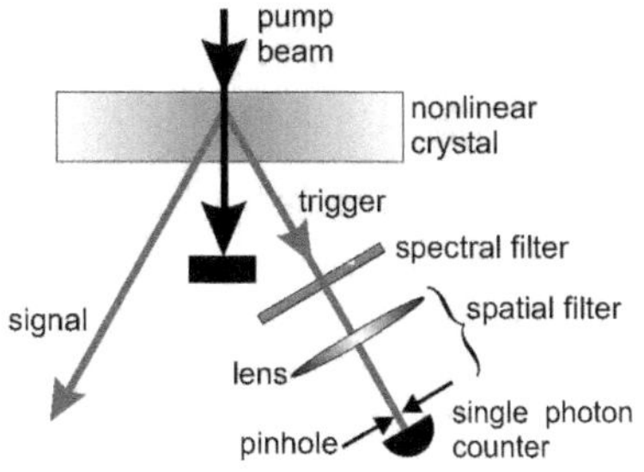

Figure 1. Preparation of single photons by conditional measurements on a biphoton state.

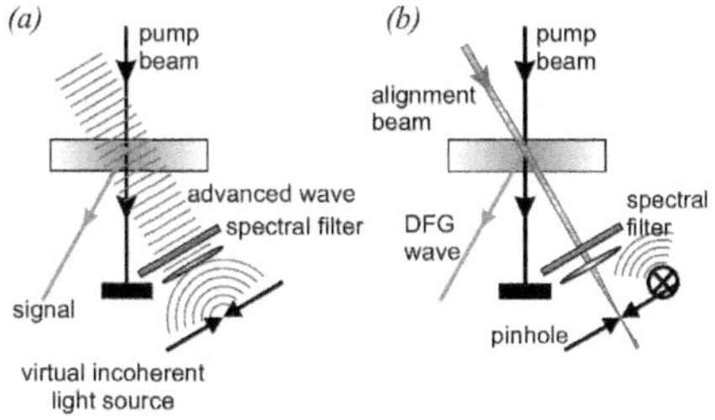

Figure 2. (a) Interaction of Klyshko advanced wave with the pump generates a DFG mode that mimics that of the CPP. (b) In experiments, a laser beam aligned so as to obtain maximum transmission can play the role of the advanced wave.

per, we neglect polarization entanglement (polarization is assumed to be well-defined in both PDC channels) and refraction inside the crystal.

The interaction Hamiltonian of parametric down-conversion is given by [Ou 1989]

$$\hat{V}(t) = \alpha \int \tilde{K}(\vec{r})\ \hat{\tilde{E}}_t^{(-)}(\vec{r},\ t)\ \hat{\tilde{E}}_s^{(-)}(\vec{r},\ t)\ \tilde{E}_p^{(+)}(\vec{r},\ t)\ d^3r + h.c., \quad (1)$$

where α is proportional to the second order nonlinear susceptibility and is assumed frequency independent, $\tilde{K}(\vec{r})$ describes the nonlinear crystal volume and is one inside and zero outside the crystal. We treat the fields in the signal (s) and trigger (t) channels as quantum operators, with their positive-frequency components given by

$$\hat{\tilde{E}}_{s(t)}^{(+)}(\vec{r},\ t) = \int e^{-i\left(\vec{k}_{s(t)}.\vec{r} - \omega_{s(t)}t\right)}\ \hat{a}_{\vec{k}_{s(t)},\ \omega_{s(t)}}\ d^3k_{s(t)}\ d\omega_{s(t)}; \quad (2)$$

the coherent pump field is treated classically:

$$\tilde{E}_p^{(+)}(\vec{r},\ t) = \int E_p^{(+)}(\vec{k}_p,\ \omega_p)\ e^{i\left(\vec{k}_p.\vec{r} - \omega_p t\right)}\ d^3k_p\ d\omega_p. \quad (3)$$

For all fields, quantum or classical, the Hermitian electric field observable is written as $\tilde{E}_p(\vec{r},\ t) = \tilde{E}_p^{(+)}(\vec{r},\ t) + \tilde{E}_p^{(-)}(\vec{r},\ t)$, with $\tilde{E}_p^{(-)}(\vec{r},\ t) = [\tilde{E}_p^{(+)}(\vec{r},\ t)]^\dagger$.

Assuming the signal and trigger modes to be initially in the vacuum state and restricting the consideration to the first order perturbation theory, we write the resulting biphoton state as

$$|B\rangle = |0\rangle_s \, |0\rangle_t - i \int_{-\infty}^{\infty} \hat{V}(t)\, dt. \tag{4}$$

Performing the integration we obtain:

$$|B\rangle = |0\rangle_s \, |0\rangle_t - i \int d^3k_s\, d\omega_s\, d^3k_t\, d\omega_t \\ \Psi\left(\vec{k}_s,\, \omega_s,\, \vec{k}_t,\, \omega_t\right) \left|1_{\vec{k}_s,\, \omega_s}\right\rangle_s \left|1_{\vec{k}_t,\, \omega_t}\right\rangle_t, \tag{5}$$

with

$$\Psi\left(\vec{k}_s,\, \omega_s,\, \vec{k}_t,\, \omega_t\right) = \alpha \int E_p^{(+)}\left(\vec{k}_p,\, \omega_s + \omega_t\right)\, K(\Delta\vec{k})\, d^3k_p. \tag{6}$$

Here $K(\vec{k})$ is the Fourier transform of $\tilde{K}(\vec{r})$ and the $\vec{k}$-vector mismatch is $\Delta\vec{k} = \vec{k}_p - \vec{k}_s - \vec{k}_t$.

The trigger photon is then selected by spatial and frequency filters and is detected by a single-photon counter. Conditioned on the detection event the non-local biphoton state collapses into a single photon state in the signal mode. The properties of this mode are determined by the optical mode of the pump photon and the spatial and spectral filtering in the trigger channel:

$$\hat{\rho}_s = \mathrm{Tr_t}\left(\hat{\rho}_\mathrm{t}\, |\mathrm{B}\rangle \langle \mathrm{B}|\right), \tag{7}$$

where the trace is taken over the trigger states and $\hat{\rho}_t$ denotes the state ensemble selected by the filters:

$$\hat{\rho}_t = \int T\left(\vec{k}_t,\, \omega_t\right) \left|1_{\vec{k}_t,\, \omega_t}\right\rangle_t \left\langle 1_{\vec{k}_t,\, \omega_t}\right|_t d^3k_t\, d\omega_t \tag{8}$$

with $T(\vec{k},\, \omega)$ being the spatiotemporal transmission function of the filters.

An explicit calculation of the quantum state (7) of the photon in the signal channel yields

$$\hat{\rho}_s = \int d^3k_s\, d\omega_s\, d^3k'_s\, d\omega'_s\, \Phi\left(\vec{k}_s,\, \omega_s,\, \vec{k'}_s,\, \omega'_s\right) \left|1_{\vec{k'}_s,\, \omega'_s}\right\rangle_s \left\langle 1_{\vec{k}_s,\, \omega_s}\right|_s, \tag{9}$$

where

$$\Phi(\vec{k}_s,\, \omega_s,\, \vec{k'}_s,\, \omega'_s) = |\alpha|^2 \int d^3k_t\, d\omega_t\, d^3k_p\, d^3k'_p\, E_p^{(-)}\left(\vec{k}_p,\, \omega_s + \omega_t\right) \\ \times E_p^{(+)}\left(\vec{k'}_p,\, \omega'_s + \omega_t\right)\, T\left(\vec{k}_t,\, \omega_t\right) K^\star\left(\Delta\vec{k}\right)\, K\left(\Delta\vec{k'}\right), \tag{10}$$

with $\Delta\vec{k}$ as above and $\Delta\vec{k'} = \vec{k'}_p - \vec{k'}_s - \vec{k}_t$.

2. Modeling the single photon mode with a classical wave

We now calculate the field correlation function

$$\Gamma(\vec{k},\ \omega,\ \vec{k'},\ \omega') = \langle E^{(-)}(\vec{k},\ \omega)\ E^{(+)}(\vec{k'},\ \omega')\rangle \tag{11}$$

of the difference frequency (DFG) wave generated through nonlinear interaction of the advanced wave and the pump. For the nonlinear polarization inside the crystal, we write

$$\tilde{P}_{\rm DFG}\left(\vec{r},\ t\right) \propto \tilde{E}_A\left(\vec{r},\ t\right)\ \tilde{E}_p\left(\vec{r},\ t\right). \tag{12}$$

Here $\tilde{E}_p(\vec{r},\ t)$ and $\tilde{E}_A(\vec{r},\ t)$ are the electric fields of the pump and advanced waves, respectively. The mode of the DFG field is obtained from Eq. (12) via a Fourier transform which is restricted to the crystal volume:

$$\begin{aligned} E^{(+)}_{\rm DFG}\left(\vec{k}_s,\ \omega_s\right) = \ & \beta\,\delta\left(k_s - \frac{\omega_s}{c}\right)\int d^3k_A\,d\omega_A\,d^3k_p\,E^{(-)}_A\left(\vec{k}_A,\ \omega_A\right) \\ & \times E^{(+)}_p\left(\vec{k}_p,\ \omega_s + \omega_A\right)\ K\left(\Delta\vec{k}\right). \end{aligned} \tag{13}$$

The proportionality coefficient β represents the nonlinearity of the medium, $\Delta\vec{k} = \vec{k}_p - \vec{k}_s - \vec{k}_A$. If the advanced wave field is partially incoherent and is characterized by a correlation function $\Gamma_A(\vec{k}_A,\ \omega_A,\ \vec{k'}_A,\ \omega'_A)$, the above equation generalizes to

$$\begin{aligned} \Gamma_{\rm DFG}\left(\vec{k}_s,\ \omega_s,\ \vec{k'}_s,\ \omega'_s\right) = |\beta|^2\,\delta\left(k_s - \frac{\omega_s}{c}\right)\ \delta\left(k'_s - \frac{\omega'_s}{c}\right) \\ \int d^3k_A\,d\omega_A\,d^3k'_A\,d\omega'_A\,d^3k_p\,d^3k'_p\,E^{(-)}_p\left(\vec{k}_p,\ \omega_s + \omega_A\right) \\ E^{(+)}_p\left(\vec{k'}_p,\ \omega'_s + \omega'_A\right)\ \Gamma^{\star}_A\left(\vec{k}_A,\ \omega_A,\ \vec{k'}_A,\ \omega'_A\right)\ K^{\star}\left(\Delta\vec{k}\right)\ K\left(\Delta\vec{k'}\right). \end{aligned} \tag{14}$$

We define the correlation function of the advanced wave in the following fashion. Suppose the single photon detector is replaced by an incoherent source continuously emitting omnidirectional incoherent light into a wide spectral range backwards in time. This completely incoherent light is characterized by the correlation function $\Gamma_0(\vec{k'},\ \omega',\ \vec{k},\ \omega) \propto \delta^{(3)}(\vec{k'} - \vec{k})\ \delta(\omega' - \omega)$ which, upon passing through the spatial and spectral filters, transforms into

$$\Gamma_A\left(\vec{k'},\,\omega',\,\vec{k},\,\omega\right) = T\left(\vec{k},\,\omega\right)\,\delta^{(3)}\left(\vec{k'}-\vec{k}\right)\,\delta\left(\omega'-\omega\right). \tag{15}$$

The advanced wave then enters the nonlinear crystal and interacts with the pump wave whenever and wherever it is present in the crystal. The nonlinear interaction of Klyshko's advanced wave with the pump pulse produces a pulse of DFG emission into the signal channel (Fig. 2 (a)). Substituting the correlation function (15) of the advanced wave into Eq. (14) as Γ_A we find that

$$\Gamma_{\mathrm{DFG}}\left(\vec{k}_s,\,\omega_s,\,\vec{k'}_s,\,\omega'_s\right) \equiv \Phi\left(\vec{k}_s,\,\omega_s,\,\vec{k'}_s,\,\omega'_s\right). \tag{16}$$

The correlation function of the DFG pulse generated through the nonlinear interaction of the advanced wave and the pump pulse is identical to the density matrix of the single photon prepared by conditional measurements on a biphoton performed in the same optical arrangement.[2]

3. Discussion

Unlike Klyshko, who said that the advanced wave is a δ-function pulse, we consider it to be a continuous, partially incoherent wave. The duration of the advanced wave is in fact determined by the uncertainty of the photon arrival time measurement. With modern detectors, it amounts to at least tens of picoseconds. If the down-conversion experiment is performed in an ultrashort pulsed setting, this uncertainty substantially exceeds the pump pulse width, so the advanced wave can be considered continuous. On the other hand, if the pump laser is continuous, the situation is more complicated and the timing uncertainty must be taken into account more rigorously in order to determine the correct correlation function of the DFG wave and the density matrix of the conditional single photon.

Can a completely incoherent advanced wave give rise to a highly coherent, transform-limited difference frequency pulse? The answer is positive provided that the filtering in the trigger channel is sufficiently narrow [Aichele 2002]. Indeed, when propagating through the spectral filter, the initially incoherent advanced wave acquires some finite degree of temporal coherence, quantified by the coherence time equal to the inverse filter width $\tau_c = \sigma_t^{-1}$. Further, we notice that the nonlinear interaction between the advanced wave and the pump is restricted by the time window determined by the duration τ_p of the pump pulse. If the latter is much shorter than τ_c, the advanced wave can be considered almost coherent within this window, and the DFG pulse is almost transform limited. Reformulating this in the language of single photons, we find that in order to obtain a conditional photon in a pure temporal mode, the filter in the trigger channel must be much narrower than the inverse pulse width: $\sigma_t \ll \tau_p^{-1}$. This result confirms the conclusion of Ou [Ou 1997].

Same considerations are valid for spatial coherence. As the advanced wave passes through a narrow aperture, it gains some degree of transverse coherence according to the van Cittert-Zernike theorem. Because the nonlinear interaction is restricted to the area where the pump field is present, the resulting signal (DFG) field is also partially coherent provided the pump beam diameter is smaller than the coherence width of the advanced wave in the plane of the crystal [Aichele 2002].

Identity (16) can be easily generalized to optical filters of random configuration, more complex than a combination of spatial and spectral filters described by Eqs. (8) and (15). Its applicability is also independent from other features of the experimental setup, such as the configuration of PDC, properties of the pump beam, geometry of the crystal, walk-off and group velocity dispersion effects, *etc.* and appears to be very general. The only restriction that has to be taken into account is the first order perturbation theory, which implies that the probability of generating two or more biphotons at a time is negligible.

By varying the configuration of the filter in the trigger channel one has some freedom in forming the CPP mode with the required spatiotemporal properties. This possibility can be considered as an example of *remote state preparation* in the sense discussed by Bennett *et al.* [Bennett 2001]. The original biphoton state is highly entangled in the frequency-momentum space and this entanglement plays an essential role in generating the Fock state. The signal mode *does not exist* unless and until the trigger photon passes through the filters and is registered. A detection event results in a non-local preparation of a single photon in an optical mode whose characteristics are determined by the way in which the measurement in the trigger channel is performed.

Apart from its theoretical implications, the result (16) finds its use in experimental practice, namely when a need arises to model a CPP with a classical wave. This is necessary, for example, when the mode of the photon needs to be matched to a classical mode [Lvovsky 2001], or to that of another CPP from a different source [Pan 1998]. The traditional procedure of matching two classical modes with each other - by observing interference fringes and optimizing their visibility - is not applicable to the situation when one of the modes is a single photon. There is no laser beam to mode match to. The only information available to the experimentalist is the remote location and the width of the trigger filter and the parameters of the pump. Although the spatial location of the CPP can be approximately determined by detecting coincidences between the photon count events in the signal and trigger [Pittman 1996], optimizing the mode matching requires adjustment of a much larger set of degrees of freedom, such as the beam direction, divergence, spatial and temporal width, optical delay, *etc.* Reliable adjustment of these parameters cannot be achieved through sole optimization of the coincidence rate.

This is where the Klyshko model comes into play. Although the advanced wave propagates backwards in space and time and is thus a purely imaginary object, it can be modeled by an *alignment beam* inserted into the trigger channel so that it overlaps spatially and temporally with the pump beam inside the crystal and passes through the optical filters (Fig. 2 (b)). Nonlinear interaction of such an alignment beam with the pump wave will produce difference frequency generation into a spatiotemporal mode similar (albeit no longer completely identical) to that of the CPP.

As mentioned in the Introduction, the universality of the result (16) permits its application in the analysis of many experimental settings involving parametric down-conversion. One obvious application area is quantum coincidence imaging [Pittman 1995; Bennik 2002]. Employing the advanced wave model allows one to analyze the quantum image formation in a completely classical view frame of geometrical optics, and helps one solve frequently debated issues such as the role of entanglement [Abouraddy 2001] or the configuration of the photon detector in the trigger channel, as well as evaluate the resolution and coherence properties of the image.

A perhaps less straightforward application of the advanced wave model is multimode entanglement. In the case of polarization entanglement, for example, detection of a polarized photon in the trigger channel remotely prepared a photon of the same polarization in the signal channel, leading to a violation of the Bell inequality [Kwiat 1995]. In the framework of the advanced wave model, detection of a photon with a certain polarization is equivalent to emission of an advanced wave with this polarization. If the spatial mode of the advanced wave lies within one of the emission cone intersection areas [Kwiat 1995], both its polarization components will interact with the pump wave, giving rise to a DFG wave also containing two polarization components of the same relative intensity. In a similar manner, one can understand other multimode entanglement experiments, such as angular-momentum [Mair 2001] or time-of-arrival [Brendel 1999] entanglement.

In summary, we have investigated the spatiotemporal optical mode of the single-photon Fock state prepared by conditional measurements on a biphoton and found it to be identical to that of a classical wave generated due to a nonlinear interaction of the pump wave and Klyshko's advanced wave. We discussed the applicability of this identity in various experimental settings.

Acknowledgments

This work has been sponsored by the Deutsche Forschungsgemeinschaft.

QUANTUM LOGICS BASED ON FOUR-PHOTON ENTANGLEMENT

Ph. Walther and A. Zeilinger
Institut für Experimentalphysik, Universität Wien, Vienna, Austria
Institut für Quantenoptik & Quanteninformation, Österreichische Akademie der Wissenschaften, Vienna, Austria

1. Entanglement

Strikingly, quantum information processing has its origins in the purely philosophically motivated questions concerning the nonlocality and completeness of quantum mechanics sparked by the work of Einstein, Podolsky and Rosen in 1935 [Einstein 1935]. In experiments using entanglement, the system of spin-$\frac{1}{2}$ particles is realized by the usage of single photons, whose properties are defined by their polarization. Considering the H / V basis, a logical $|0\rangle$ corresponds to a horizontally polarized photon $|H\rangle$, and a logical $|1\rangle$ corresponds to a vertically polarized photon $|V\rangle$, respectively. A single qubit can be written as a coherent superposition of the form $|\psi\rangle = \alpha\,|H\rangle + \beta\,|V\rangle$, where the the probabilities α^2 and β^2 sum up to $\alpha^2 + \beta^2 = 1$. For the two qubit case, the four different maximally entangled Bell states are defined as:

$$
\begin{aligned}
|\Phi^{\pm}\rangle_{12} &= \frac{1}{\sqrt{2}}\left(|H\rangle_1\,|H\rangle_2 \pm |V\rangle_1\,|V\rangle_2\right), \\
|\Psi^{\pm}\rangle_{12} &= \frac{1}{\sqrt{2}}\left(|H\rangle_1\,|V\rangle_2 \pm |V\rangle_1\,|H\rangle_2\right).
\end{aligned}
\tag{1}
$$

The Bell states have a unique feature that all the information on polarization properties is completely contained in the (polarization-) correlations between the separate photons, while the individual particle does not have any polarization prior to measurement. In other words, all the information is distributed among two particles, and none of the individual systems carries any information. This is the essence of entanglement. At the same time, these (polarization-) correlations are stronger than allowed classically, since they violate bounds imposed by local realistic theories via the Bell-inequality [Bell 1964] or they lead to a maximal contradiction between such theories and quantum mechanics as signified by the Greenberger-Horne-Zeilinger theorem

V.M. Akulin et al. (eds.), Decoherence, Entanglement and Information Protection in Complex Quantum Systems, 49–62.

[Greenberger 1989; Greenberger 1990]. Distributed entanglement thus allows to establish non-classical correlations between distant parties and can therefore be considered as a quantum analogue of the classical communication channel, a quantum communication channel.

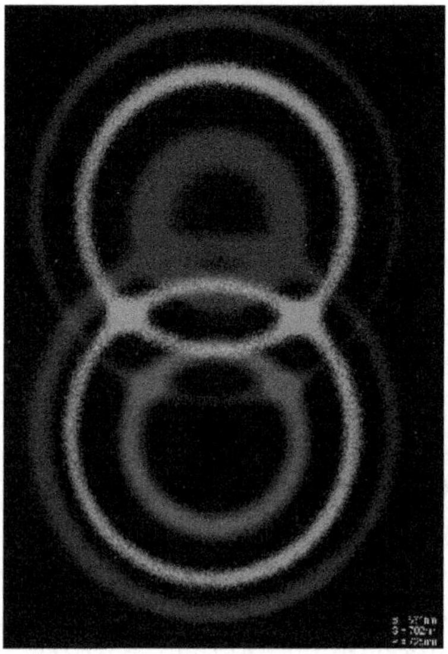

Figure 1. Photograph of the light emitted in type-II parametric down-conversion (false colours). The polarization-entangled photons emerge along the directions of the intersection between the green rings and are selected by placing small holes there

The most widely used source for the polarization-entangled photons today utilizes the process of spontaneous parametric down-conversion in nonlinear optical crystals [Kwiat 1995]. A typical picture of the emerging radiation is shown in Fig. 1.

2. Quantum gates

A promising system for quantum computation uses single photons to encode the quantum information [Bouwmeester 2000]. This is due to the robustness of photons against decoherence and the availability of single-qubit operations. Polarization-encoded qubits are well suited for information transmission in quantum information processing. In recent years, the polarization state of single photons has been used to experimentally demonstrate quantum dense coding [Mattle 1996], quantum teleportation [Bouwmeester 1997] and quantum cryptography [Jennewein 2000; Naik 2000; Gisin 2001]. However, due to the difficulty of achieving quantum logic operations between independent photons, the application of photon states has been limited primarily to the field of quantum communication. More precisely, the two-qubit gates suitable for quantum computation generically require strong interactions between individual photons, implying the need for massive, reversible non-linearities well beyond those presently available for photons, as opposed to other physical systems [Schmidt-Kaler 2003].

Remarkably, Knill, Laflamme and Milburn (KLM) [Knill 2001] found a way to circumvent this problem and implement efficient quantum computation using only linear optics, photo-detectors and single-photon sources. Indeed, they showed that measurement induced nonlinearity was sufficient to obtain efficient quantum computation.

A crucial requirement of both KLM's and Nielsen's constructions is *classical feed-forwardability*. Specifically, it must be in principle possible to detect when the gate has succeeded by measuring ancilla photons in some appropriate state. This information can then be fed-forward in such a way as to condition future operations on the photon modes.

3. The photonic controlled-NOT gate

In former experiments [Sanaka 2002; O'Brien 2003; Pittman 2003] destructive linear optical gate operations have been realized. As they necessarily destroy the output state, such schemes are not classically feed-forwardable. The first realization of a CNOT gate, which operates on two polarization qubits carried by independent photons and that satisfies the feed-forwardability criterion, was done by this experiment.

A CNOT gate flips the second (target) bit if and only if the first one (control) has the logical value 1 and the control bit remains unaffected. The scheme we use to achieve the CNOT gate was first proposed in Ref. [Pittman 2001] and is shown in figure 2.

This scheme performs the CNOT operation on the input photons in spatial modes a_1 and a_2; the output qubits are contained in spatial modes b_1 and b_2. The ancilla photons in the spatial modes a_3 and a_4 are in the maximally entangled Bell state

$$|\psi_{a_3\,a_4}\rangle = \frac{1}{\sqrt{2}}\left(|H\rangle_{a_3}|H\rangle_{a_4} + |V\rangle_{a_3}|V\rangle_{a_4}\right). \qquad (2)$$

In the following, H (horizontally polarized photon) and V (vertically polarized one) will denote our logical 0 and 1. The CNOT operation for the polarization encoded qubits can be written as

$$\begin{gathered}|H\rangle_c|H\rangle_t \to |H\rangle_c|H\rangle_t, \quad |H\rangle_c|V\rangle_t \to |H\rangle_c|V\rangle_t, \\ |V\rangle_c|H\rangle_t \to |V\rangle_c|V\rangle_t \text{ and } |V\rangle_c|V\rangle_t \to |V\rangle_c|H\rangle_t,\end{gathered} \qquad (3)$$

where the indices c and t denote the control and target bits.

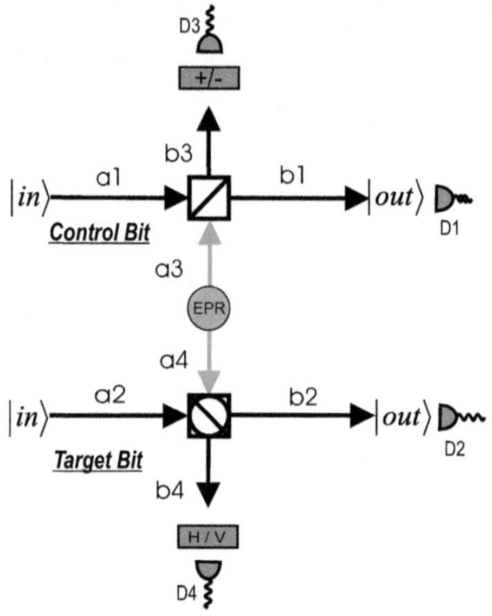

Figure 2 A scheme to obtain a photonic realization of CNOT gate with two independent qubits. The qubits are encoded in the polarization of the photons. The scheme makes use of linear optical components, polarization entanglement and postselection. When one and only one photon is detected at the polarization sensitive detectors in the spatial modes b_3 and b_4 and in the polarization H, the scheme works as a CNOT gate.

The scheme works in those cases where one and only one photon is found in each of the modes b_3, b_4; when both photons are H-polarized no further transformation is necessary on the output state (usually this is referred to as passive operation). The scheme combines two simple gates, namely, the destructive CNOT and the quantum encoder. The first gate can be seen in the lower part of Fig. 2 and is constituted by a polarizing beam splitter (PBS2) rotated by $45°$ (the rotation is represented by the circle drawn inside the symbol of the PBS), which works as a destructive CNOT gate on the polarization qubits, as was experimentally demonstrated in Ref. [Pittman 2001]. The upper part, comprising the entangled state and the PBS1, is meant to encode the control bit in the two channels a_4 and b_1. The photons in the spatial modes a_3 and a_4 are in the maximally entangled Bell state (2).

Thanks to the behavior of our polarizing beam splitter (PBS), that transmits the horizontally polarized photons and reflects the vertically polarized ones, the successful detection at the port b_3 of state $|+\rangle$ (the symbols $+, -$ stand for $H + V$ and $H - V$) post-selects the following transformation of an arbitrarily input state in a_1

$$\alpha\, |H\rangle_{a_1} + \beta\, |V\rangle_{a_1} \rightarrow \alpha\, |HH\rangle_{a_4\, b_1} + \beta\, |VV\rangle_{a_4\, b_1}. \tag{4}$$

We thus have the control bit encoded in a_4 and in b_1, the photon in a_4 will be the control input to the destructive CNOT gate, and will thus be destroyed, while the second photon in b1 will be the output control qubit.

For the gate to work properly, we want the most general input state

$$|\Psi_{a_1\,a_2}\rangle = |H\rangle_{a_1}(\alpha_1\,|H\rangle_{a_2} + \alpha_2\,|V\rangle_{a_2}) + |V\rangle_{a_1}(\alpha_3\,|H\rangle_{a_2} + \alpha_4\,|V\rangle_{a_2}) \tag{5}$$

to be converted to the output state

$$|\Psi_{a_1\,a_2}\rangle = |H\rangle_{a_1}(\alpha_1\,|H\rangle_{a_2} + \alpha_2\,|V\rangle_{a_2}) + |V\rangle_{a_1}(\alpha_3\,|V\rangle_{a_2} + \alpha_4\,|H\rangle_{a_2})\,. \tag{6}$$

Let us consider first the case where the control photon is in the logical zero (H polarization state). The control photon will then travel undisturbed through the PBS, arriving in the spatial mode b_1. As required, the output photon is H polarized. In order for the scheme to work, a photon has to arrive also at the detector D_3 in b_3: given the input photon already in the mode b_1, this additional photon comes necessarily from the EPR pair, and is H polarized as it is transmitted by the PBS1. We know that the photons in a_3 and a_4 are correlated (2), so the photon in a_4 is also in the horizontal polarization. Taking into account the – 45° rotation of the polarization on the paths a_2, a_4 operated by the half-wave plates, the input state in the PBS2 will then be state $|--\rangle_{a_2\,a_4}(|-+\rangle_{a_2\,a_4})$ for a target photon H (V) polarized. This state will give rise, with a probability of 50%, to the state in which two photons go through the PBS2 $(|HH\rangle \pm |VV\rangle)_{b_2\,b_4}$ which, after the additional rotation of the polarization, and the subsequent change to the H / V basis (where the measurement will be performed) acquires the form $(|HH\rangle + |VV\rangle)_{b_2b_4}((|HV\rangle + |VH\rangle)_{b_2b_4})$. The expected result in the mode b_2 H (V) for the case when the photon in b_4 is horizontally polarized. We can see in a similar way that the gate works also for the cases when the control photon is vertically polarized, or is polarized at 45°. Our experimental setup is shown in Fig. 3.

An ultraviolet pulsed laser, centered at wavelength 398 nm, with pulse duration 200 fs and a repetition rate 76 MHz, impinges on a BBO crystal [Kwiat 1995] producing probabilistically the first pair in the spatial modes a_1 and a_2: these two photons are fed into the gate as the input qubits. The UV laser is then reflected back by the mirror M1 and, passing through the crystal for the second time, produces the entangled ancilla pair in spatial modes a_3 and a_4. Half-wave plates and non-linear crystals in the paths provide the necessary birefringence compensation, and the same half-wave plates are used to adjust the phases between the down converted photons (*i.e.*, to produce the state ϕ^+) and to implement the CNOT gate.

We then superpose the two photons at Alice's (Bob's) side in the modes a_1, a_3 (a_2, a_4) at a polarizing beam splitter PBS1 (PBS2). Moving the mirror M1, mounted on a motorized translation stage, allows us to change the arrival time to make the photons as indistinguishable as possible. A further

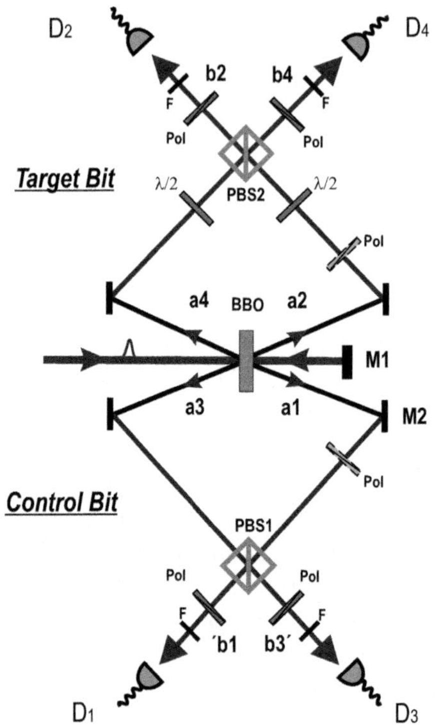

Figure 3 The experimental setup. A type II Spontaneous parametric down-conversion is used both to produce the ancilla pair (in the spatial modes a_3 and a_4) and to produce the two input qubits (in the spatial modes a_1 and a_2). In this case initial entanglement polarization is not desired, and it is destroyed by making the photons go through polarization filters which prepare the required input state. Half-wave plates have been placed in the photon paths in order to rotate the polarization; compensators are able to nullify the birefringence effects of the non-linear crystal and of the polarizing beam splitters. Overlap of the wavepackets at the PBSs is assured through spatial and spectral filtering.

degree of freedom is afforded by the mirror M2, whose movement on a micrometrical translation stage corrects slight asymmetries in the arms of the setup. The indistinguishability between the overlapping photons is improved by introducing narrow bandwidth (3 nm) spectral filters at the outputs of the PBSs and monitoring the outgoing photons by fiber-coupled detectors. The single-mode fiber couplers guarantee good spatial overlap of the detected photons; the narrow bandwidths filters stretch the coherence time to about 700 fs - substantially larger than the pump pulse duration [Zukowski 1995]. The temporal and spatial filtering process effectively erases any possibility of distinguishing the photon-pairs and therefore leads to interference.

The scheme we have described allows the output photons to travel freely in space, so that they may further be used in quantum communication protocols, and this is achieved by detecting one and only one photon in modes b_3 and b_4. The fact that we do not yet have single-photon detectors for this wavelength at our disposal actually forces us to implement a four-fold coincidence detection to confirm that photons actually arrive in the output modes b_1 and b_2.

So far, we have analyzed only the ideal case where exactly one pair is produced at each passage of the UV light beam through the BBO crystal. It is easy to see that our gate cuts out those unwanted cases when two pairs are produced in the spatial modes a_3, a_4: if the two photons in the spatial mode a_3 are both horizontally or vertically polarized, either two or no photons will arrive at the detector D3. The alternative case is that the pairs in each spatial mode (a_3, a_4) have two photons that are in the opposite polarization states; this could produce either two or no photons in paths b_3 and b_4. It is thus proved that no four-fold event may occur from an event of this kind.

Unluckily, this kind of noise is present for superposed qubits; however, it may easily be measured and eliminated covering the spatial modes a_3 and a_4 and measuring the four-fold coincidences. Anyway, we note that the noise is not intrinsic in the setup and is only due to practical drawbacks. Indeed, an unbalancing method like the one used in Ref. [Pan 2003] would allow one to increase the signal to noise ratio to any desired value.

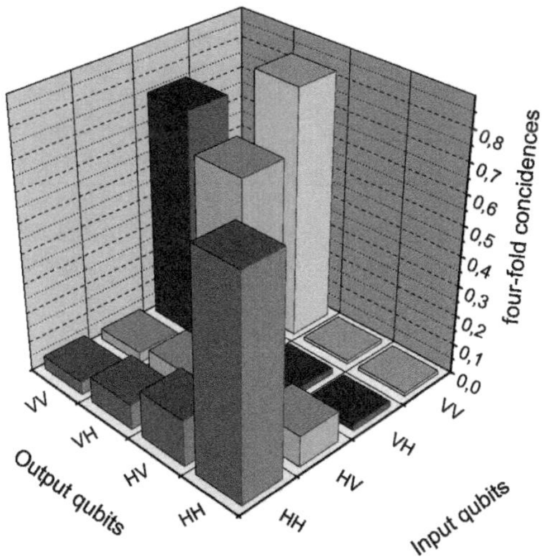

Figure 4 This graph shows that the scheme indeed works for the linear polarizations H, V. Four-fold coincidences for all the 16 possible combinations of inputs and outputs are shown. When the control qubit has the logical value 0 (HH or HV), the gate works as the identity gate. In contrast, when the control qubit has the logical value 1 (VH or VV), the gate works as a NOT gate, flipping the second input bit. Noise is due to the non ideal nature of the PBSs.

To experimentally demonstrate that the gate works, we first verify that we obtain the desired CNOT (appropriately conditioned) for the input qubits in states HH, HV, VH and VV. In Fig. 4 we compare the count rates of all 16 possible combinations. Then, it was proven that the gate also works for a superposition of states. The special case where the control input is a 45° polarized photon and the target qubit is a H photon is very interesting: we expect that the state $|H+V\rangle_{a_1}\ |H\rangle_{a_2}$ evolves into the maximally entangled state $(|HH\rangle_{b_1\ b_2} + |VV\rangle_{b_1\ b_2})$. We input the state $|+\rangle_{a_1}|H\rangle_{a_2}$; first we measure the count rates of the 4 combinations of the output polarization (HH, ..., VV) and then after going to the $|+\rangle$, $|-\rangle$ linear polarization basis a Ou-Hong-Mandel interference measurement is possible; this is shown in Fig. 5.

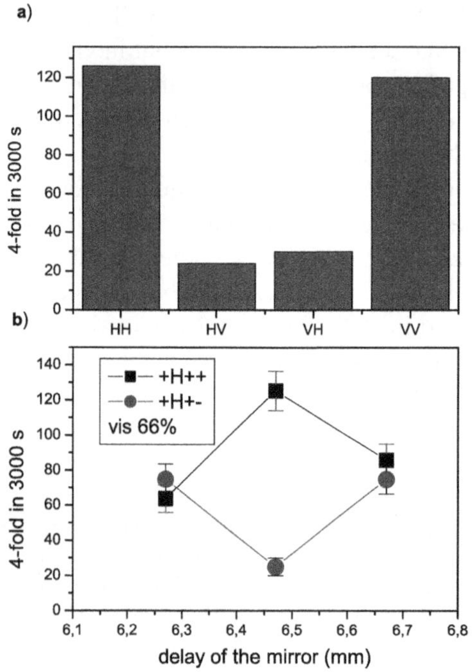

Figure 5 Demonstration of the ability of the CNOT gate to transform a separable state into an entangled state. In a) the coincidence ratio between the different terms HH, ..., VV is measured, proving the birefringence of the PBS has been sufficiently compensated; in b) the superposition between HH and VV is proved to be coherent, by showing via Ou-Hong Mandel dip at 45° that the desired $(H + V)$ state of the target bit emerges much more often than the spurious state $(H - V)$. The fidelity is $81\% \pm 2\%$ in the first case and $77\% \pm 3\%$ for the second.

4. The nonlinear sign shift gate

In the original KLM scheme, the fundamental element is not the CNOT operation, but the more or less equivalent nonlinear sign-shift (NS) operation from which the two-qubit conditional sign flip gate can be constructed. Similar to the accessibility of the CNOT operation, universal quantum computation becomes possible with such a two-qubit gate [Barenco 1995 (a); Sleator 1995]. Recently, our group has experimentally demonstrated the NS operation using photons produced via parametric down-conversion. In contrast to the KLM scheme, our method to observe the NS operates in the polarization basis and therefore does not require interferometric phase stability.

A simplified version of NS operation is shown in Fig. 6 a. An input state, $|\Psi_{IN}\rangle = |n\rangle$, impinges on a beam-splitter (BS) with reflection probability R; a single ancilla photon, $|1\rangle$, impinges from the other side of the beam splitter. The two input modes, 1 and 2, undergo a unitary transformation into two output modes, 3 and 4, described by

$$\begin{aligned} \widehat{a}_1 &\rightarrow \widehat{a}_3\,\sqrt{R} + \widehat{a}_4\,\sqrt{1-R} \\ \widehat{a}_2 &\rightarrow -\,\widehat{a}_3\,\sqrt{1-R} + \widehat{a}_4\,\sqrt{R}. \end{aligned} \qquad (7)$$

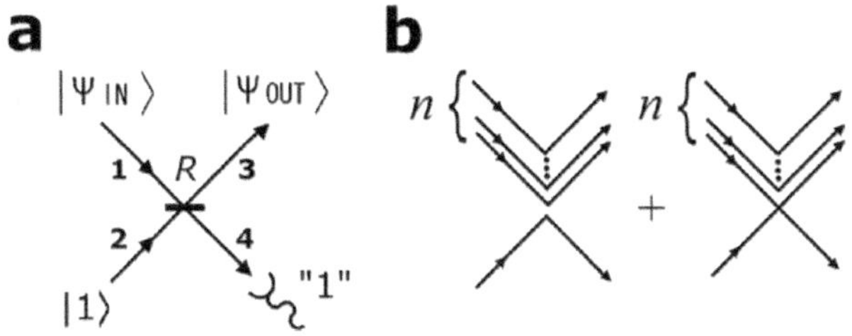

Figure 6. a - Schematic of a simplified version of nonlinear sign-shift (NS) operation constructed by a non-polarizing beam splitter of reflectivity R. $|\Psi_{IN}\rangle$ and $|\Psi_{OUT}\rangle$ are the quantum states of input and output photons. The operation is successful when the single-photon detector in ancilla mode 4 counts a single photon. b - The two paths that lead to the detection of exactly one photon in output mode 4. As long as all of the photons are indistinguishable, these two paths can interfere.

The NS operation is successful when one and only one photon reaches the detector in mode 4. Provided the photons are indistinguishable, the two paths leading to exactly one photon in mode 4 will interfere. The two interfering processes are depicted in Fig. 6 b for n input photons. Either all $n+1$ photons are reflected, or $n-1$ photons in mode 1 are reflected and 1 photon in each mode 1 and 2 is transmitted. When a single photon ends up in mode 4, the photon number state undergoes the following transformation:

$$|\Psi_{IN}\rangle = |n\rangle \rightarrow |\Psi_{OUT}\rangle = (\sqrt{R})^{n-1}[R - n(1-R)]\,|n\rangle, \qquad (8)$$

where the unusual normalization of the output state reflects the probability amplitude of success. The sign of the phase shift depends on the number of incident photons and the reflection probability of the BS. For $n < R/(1-R)$, the sign of the amplitude is unchanged and for $n > R/(1-R)$ it picks up a negative sign. For the critical case, where $n = R/(1-R)$, the output probability amplitude becomes zero [Hong 1987].

In the original KLM proposal, the NS gate is achieved using a phase sensitive interferometer. In our experiment, we induce the phase shift between two polarizations in the same spatial mode and therefore have much less stringent stability requirements. The extension of the NS operation to include second polarization mode is straightforward. We inject a horizontally-polarized ancilla photon into the BS in Fig. 6 a and consider only the cases when the single photon detected in mode 4 is horizontally polarized. The transformation for the horizontal polarization is the same as in Eq. (8). There is only one possible path which leads to no vertically-polarized photons in mode 4, it is for all vertically-polarized photons to be reflected. This operation for the input state with m vertically-polarized photons and n horizontally-polarized photons is given by:

$$|\Psi_{IN}\rangle = |m_V, n_H\rangle \xrightarrow{\text{NS}}$$
$$|\Psi_{OUT}\rangle = \left(\sqrt{R_V}\right)^m \left(\sqrt{R_H}\right)^{n-1} [R_H - n_H(1 - R_H)] \, |m_V, n_H\rangle \tag{9}$$

where R_V and R_H are the reflection probabilities for the vertical and horizontal polarizations, respectively. As expected, the vertical photon number m, does not appear in the square bracket nonlinear-sign term. The only change the vertically-polarized photons contribute to is the reflection amplitude raised to the power of m.

A quantum phase gate for the KLM scheme can be implemented using two such NS gates when the BS has reflection probabilities $R_V = 5 - 3\sqrt{2} \approx 0.76$ and $R_H = (3 - \sqrt{2})/7 \approx 0.23$ [Sanaka 2004]. For the experiment, we use input states with $m + n = 2$ and a typical 50 / 50 BS, where $R_V = R_H = 1/2$. The three possible input states are transformed by the NS operation according to

$$|2_V, 0_H\rangle \rightarrow \frac{1}{2\sqrt{2}} |2_V, 0_H\rangle, \tag{10a}$$

$$|1_V, 1_H\rangle \rightarrow 0, \tag{10b}$$

$$|0_V, 2_H\rangle \rightarrow -\frac{1}{2\sqrt{2}} |0_V, 2_H\rangle. \tag{10c}$$

The operation with this set of input parameters serves to change only the phase of the input state $|0_V; 2_H\rangle$. The input state $|1_V; 1_H\rangle$ is "annihilated" by this operation [Hong 1987]. This means that for that input state the condition of having exactly one horizontally polarized photon in mode 4 never occurs. The NS operation using this particular BS reflectivity is important for a related "quantum filter" protocol [Hofmann 2002].

In the experiment of Fig. 7, a frequency-doubled pulses from a mode-locked Ti:Sapphire laser (center wavelength 394.5 nm, 200 fs pulse duration, 76 MHz repetition rate) makes two passes through a type-II phase-matched 2 mm BBO crystal (BBO1). Through the spontaneous parametric down-conversion, there is some probability for one pair of entangled photons to be created on the first pass and another pair on the second. Additional 1 mm crystals (BBO2, BBO3) and a half-wave plate (HWP1) are used for the compensation of the birefringence effect inside BBO1 and also for the selection of appropriate

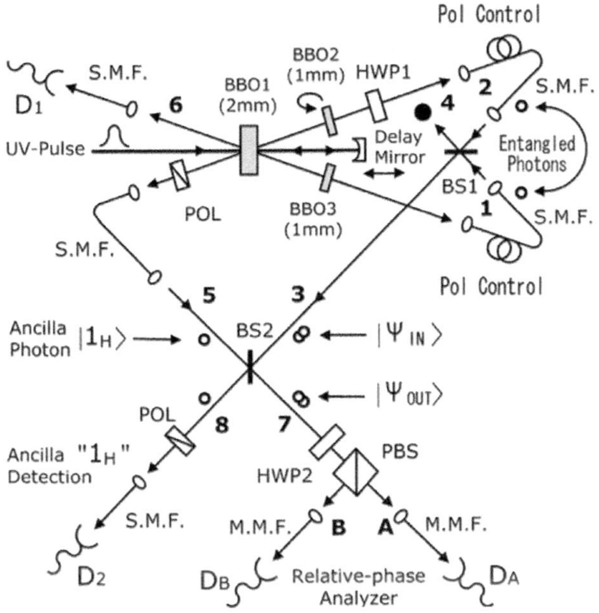

Figure 7. Experimental setup for the demonstration of nonlinear sign-shift (NS) operation using double-pass parametric down-conversion. Photon pairs created from first pass are used for the input of NS operation and pairs from the second pass are used for the triggered single photon source as ancilla. Successful operation is identified through four-fold coincidence counts between all four detectors.

polarization-entangled photons. The first pair (right going modes 1 and 2 in Fig. 7) serves as the input to the NS operation in mode 3. The second pair (left going modes 5 and 6) is used to produce the ancilla photon. Upon detection of a "trigger" photon in mode 6, a single photon state will be present in mode 5 with high probability. BS2 is a normal 50 / 50 BS and its reflectivity determines the NS operation. Four-photon events are post-selected by single-photon counting detectors D_1, D_2, D_A, and D_B. We first verified the operations (10a) and (10c). The state of the polarization-entangled photons in modes 1 and 2 was prepared as $|\Phi_\theta\rangle = (|1_V\rangle_1 \, |1_V\rangle_2 + e^{i\theta} \, |1_H\rangle_1 \, |1_H\rangle_2)/\sqrt{2}$, where the relative phase θ was controlled by tilting the compensation crystal BBO2. BS1 (also 50 / 50) transforms this polarization-entangled state into a photon-number entangled state. The photons in mode 3 are in state $|\Psi_{IN}\rangle = (|2_V, \, 0_H\rangle_3 + e^{i\theta} \, |0_V, \, 2_H\rangle_3)/\sqrt{2}$. The pairs created from the second pass are used as a source of triggered single photons. A translatable mirror on the pump allows for the relative creation time of the two pairs to be varied. Down-converted photons are coupled into single-mode fibers (S.M.F.) for mode filtering. The photons come out of the fibers to free space again for interference in BS2 (also 50 / 50 BS). The polarization of the ancilla photon $|1_H\rangle$ in mode 5 is set to horizontal using a polarizer (Pol). The entangled photons are sent to BS2 and combined with the horizontally polarized ancilla photon in mode 5. A successful operation occurs when the single photon detector D_2 in output mode 8 counts a single photon in horizontal polarization state. The output state in mode 7 is analyzed using HWP2, a polarizing beam splitter (PBS), multi-mode fibers (M.M.F.) and single-photon counters placed in modes A and B ($D_A$and D_B) - together these form a relative-phase analyzer. Successful operation is identified through the four-fold coincidence counts between all four detectors (D_1, D_2, D_A and D_B).

For our first measurement we act on the photon-number entangled states in mode 3. Since the NS operation is an interference effect, it only proceeds when the entangled photons and the ancilla photon arrive at BS2 within their coherence time τ_{coh}. In this case, the operation performs the following:

$$
\begin{aligned}
|\Psi_{IN}\rangle &= \frac{1}{\sqrt{2}}\left(|2_V,\ 0_H\rangle_3 + e^{i\theta}\,|0_V,\ 2_H\rangle_3\right) \\
\longrightarrow |\Psi_{OUT}\rangle &= \frac{1}{4}\left(|2_V,\ 0_H\rangle_7 - e^{i\theta}\,|0_V,\ 2_H\rangle_7\right).
\end{aligned} \tag{11}
$$

To analyze this effect, we use HWP2 to rotate the polarization states by $45°$. Under this operation,

$$
\frac{|2_V,\ 0_H\rangle + |0_V,\ 2_H\rangle}{\sqrt{2}} \rightarrow \frac{|2_V,\ 0_H\rangle + |0_V,\ 2_H\rangle}{\sqrt{2}} \tag{12}
$$

(*i.e.*, it is invariant), however

$$
\frac{|2_V,\ 0_H\rangle - |0_V,\ 2_H\rangle}{\sqrt{2}} \rightarrow |1_V,\ 1_H\rangle \tag{13}
$$

(*i.e.*, it can produce coincidences between D_A and D_B). We can verify the transformation (11) by measuring one vertically- and one horizontally-polarized photon using the PBS and photodetectors D_A and D_B. Fig. 8 a shows the observed variations of the count rate as functions of the pump mirror position, which varies the relative arrival times of the entangled photons and the ancilla photons at BS2. Solid squares (circles) show the four-fold coincidence counts for $\theta = 0$ (π) in (11) as a function of the pump delay. At zero delay, the coincidences for D_1 D_2 D_A D_B are enhanced (suppressed) by the NS operation (11). When the arrival time difference is larger than τ_{coh}, we obtain coincidence counts from the state $|\Psi_{IN}\rangle$ in (11) and also accidental coincidence counts between the ancilla photon and one entangled photon. If one could resolve the sub-picosecond level time differences, the size of the dip of one curve would be equal to the size of the peak in the other. The fidelity of the sign-shifted entangled photons can be estimated from the data to be 77 ± 6 %. When taking the fidelity 92.9 ± 0.5 % of the initial state into account, this confirms the high quality of our NS operation.

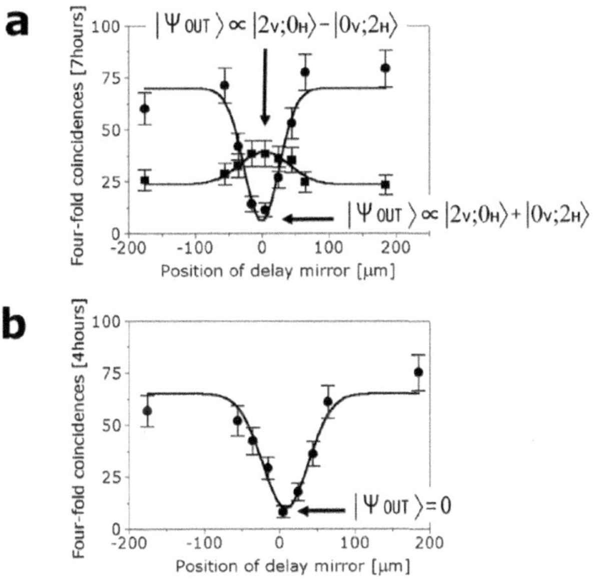

Figure 8. a - The four-fold coincidences as a function of the pump delay mirror position for the photon-number entangled photons. HWP2 is set to rotate the polarization state by 45°. Solid squares (circles) show the four-fold coincidence counts for the input state where $\theta = 0$ (π). The peak and dip of the two curves are different in size only because of accidental coincidences occurring when the delay is much larger than the photons' coherence length. b - The four-fold coincidences as functions of the pump delay mirror position for the input photons in the state $|1_V\ 1_H\rangle_3$. The HWP2 is set to leave the input polarization states unchanged. At zero delay, the coincidences are suppressed nearly to zero by the HOM effect.

We then set the pump delay to zero so that the NS operation can proceed with maximum efficiency. We analyze the output mode 7 by tilting the compensation crystal BBO2 - this allows for the variation of the correlations with θ to be directly observed. For state $|2_V,\ 0_H\rangle + e^{i\theta}\ |0_V,\ 2_H\rangle$ in mode 7, the coincidence rate between detectors D_A and D_B is proportional to $\sin^2(\theta/2)$. We input state $|2_V,\ 0_H\rangle + e^{i\theta}\ |0_V,\ 2_H\rangle$ into BS2 and record both the two-fold D_A D_B and the four-fold D_1 D_2 D_A D_B coincidences with the ancilla path open. The two-fold coincidence rate reflects the initial correlations for the input entangled-photon pair, whereas the four-fold coincidence rate shows the correlations after a successful NS operation. Fig. 9 shows the observed coincidence rates (two-fold are solid triangles and four-fold are solid diamonds) at

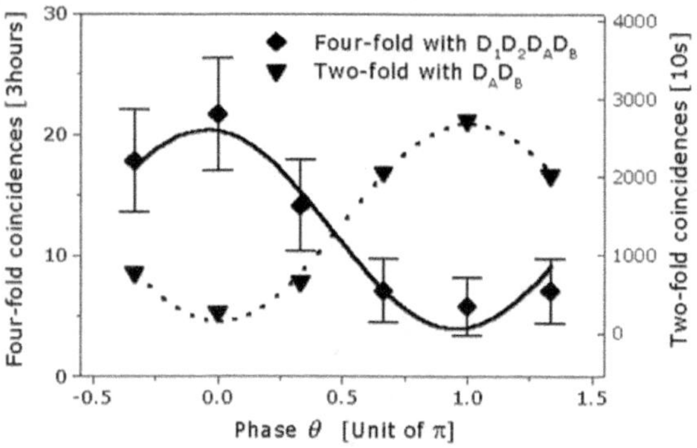

Figure 9. The observed variation of the coincidence count rates as functions of the phase of entangled photons θ at zero delay for the photon-number entangled photons. Solid triangles represent the two-fold coincidences with D_A D_B and show the phase of input photons. Solid diamonds represent the four-fold coincidences with D_1 D_2 D_A D_B which show the phase of the output photons demonstrating a successful NS operation. Error bars are based on the usual Poisson fluctuation in the number of counts on the uncorrected data (error bars of two-fold coincidences are too small to display). The phase of four-fold coincidence is shifted (1.05 ± 0.06) π against two-fold coincidence in agreement with the expected π phase shift.

zero delay for different phase angles set by BBO2. Clearly, the phase of the correlations has been changed by the NS operation; the relative phase between the two curves is (1.05 ± 0.06) π in excellent agreement with the expected shift of π.

To complete the experimental confirmation of the NS operation, we also verified its action on the input state $|1_V, 1_H\rangle$ (Eq. 10b). We prepare the input state $|\Psi_{IN}\rangle = |1_V, 1_H\rangle_3$ from the polarization-entangled photons that was prepared as $|\Psi^+\rangle = (|1_V\rangle_1 |1_H\rangle_2 + |1_H\rangle_1 |1_V\rangle_2)/\sqrt{2}$. The HWP2 is set such that it does not rotate the polarization. Fig. 8 b shows the four-fold coincidences as a function of pump delay. At zero delay, the four-fold coincidences D_1 D_2 D_A D_B are suppressed nearly to zero because of the effect at BS2 [Hong 1987]. The visibility of the fringe is about 89 ± 4 %.

TOWARDS QUANTUM CONTROL OF LIGHT: SHAPING QUANTUM PULSES OF LIGHT VIA COHERENT ATOMIC MEMORY

L. I. Childress,[1] M. D. Eisaman,[1] A. Andre,[1] F. Massou,[1] A. S. Zibrov,[1,2,3] M. D. Lukin[1,2]

[1] *Physics Department, Harvard University, Cambridge, MA 02138, USA*

[2] *Harvard-Smithsonian Center for Astrophysics, Cambridge, MA 02138, USA*

[3] *P. N. Lebedev Institute of Physics, Moscow, 117924, Russia*

Abstract We describe a technique for generating pulses of light with controllable, well-defined photon numbers, propagation direction, timing, and pulse shapes. The technique is based on preparation of an atomic ensemble in a state with a desired number of atomic spin excitations, which is later converted into a photon pulse. Spatio-temporal control over the pulses is obtained by exploiting long-lived coherent memory for photon states and electromagnetically induced transparency in an optically dense atomic medium. Using photon counting experiments we observe collective behavior of atomic spins in a room-temperature ensemble, enabling controlled generation and shaping of few-photon sub-Poissonian light pulses. We discuss the prospects for controlled generation of n-photon Fock states.

Keywords: Electromagnetically induced transparency, Raman scattering, atomic ensembles, nonclassical photon source, atomic memory, Fock states, collective excitations

Introduction

In recent years much effort has been directed toward generating quantum-mechanical states of the electromagnetic field with a well-defined number of light quanta (*i.e.*, photon-number or Fock states). In addition to being of fundamental interest, these states represent an essential resource for the practical implementation of many ideas from quantum information science such as quantum communication [Briegel 2000]. Over the past decade, tremendous progress has been made in generating single-photon states by using photon pairs in parametric down-converters [Hong 1986; Lvovsky 2001], quantum dots [Michler 2000; Santori 2002], color centers [Kurtsiefer 2000; Beveratos 2001], and single atoms in high-finesse cavities [Kuhn 2002; McKeever 2004].

V.M. Akulin et al. (eds.), Decoherence, Entanglement and Information Protection in Complex Quantum Systems, 63–76.

While parametric down-conversion techniques have recently been used to generate multi-photon states [Waks 2004], it remains experimentally challenging to implement schemes that allow for simultaneous control over both photon number and spatio-temporal properties of the pulse.

In this article we describe a novel technique for generating pulses of light with controllable, well-defined photon numbers, propagation direction, timing, and pulse shapes by exploiting long-lived coherent memory for photon states in an optically dense atomic medium [Lukin 2003]. We experimentally demonstrate key elements of this technique, including simultaneous control over the timing, shape, and quantum states of few-photon pulses. Our experiment combines different aspects of earlier studies on "light storage" [Liu 2001; Phillips 2001] and Raman preparation and retrieval of atomic excitations [Kuzmich 2003; van der Wal 2003]. This approach is particularly important in the context of long-distance quantum communication [Duan 2001], as well as in the context of electromagnetically-induced transparency (EIT) based on quantum nonlinear optics [Schmidt 1996; Lukin 1999; Braje 2003].

1. Overview: preparation and retrieval of an atomic spin wave

In our approach we first prepare a large ensemble of N atoms with a three-state Λ configuration of atomic states (see Fig. 1 A) in the ground state $|g\rangle$ via optical pumping. Spontaneous Raman scattering [Raymer 1996] is induced by a weak, off-resonant laser beam with Rabi frequency Ω_W and detuning Δ_W, referred to as the write laser. This two-photon process flips an atomic spin into the metastable state $|s\rangle$ while producing a correlated frequency-shifted photon (a so-called Stokes photon). Energy and momentum conservation ensure that for each Stokes photon emitted in certain direction there exists exactly one flipped spin quantum in a well-defined spin-wave mode. The number of spin wave quanta and the number of photons in the Stokes field thus exhibit strong correlations, analogous to the correlations between photons emitted in parametric down-conversion [Heidmann 1987; Aytur 1990; Smithey 1992]. As a result of these atom-photon correlations, the measurement of the Stokes photon number n_S ideally projects the spin-wave into a nonclassical collective state with n_S spin quanta [Duan 2001]. It is important to emphasize that the resulting atomic state is collective in character - it can not be represented as a product of the individual atomic states. After a controllable delay time τ_d (see Fig. 1 B), the stored spin-wave can be coherently converted into a light pulse by applying a second near-resonant laser beam Ω_R (retrieve laser), see Fig. 1 A. The physical mechanism for this retrieval process involves EIT [Harris 1997; Scully 1997; Boller 1991; Lukin 1999; Fleischhauer 2000] and is identical to that employed in previous experiments [Liu 2001; Phillips 2001].

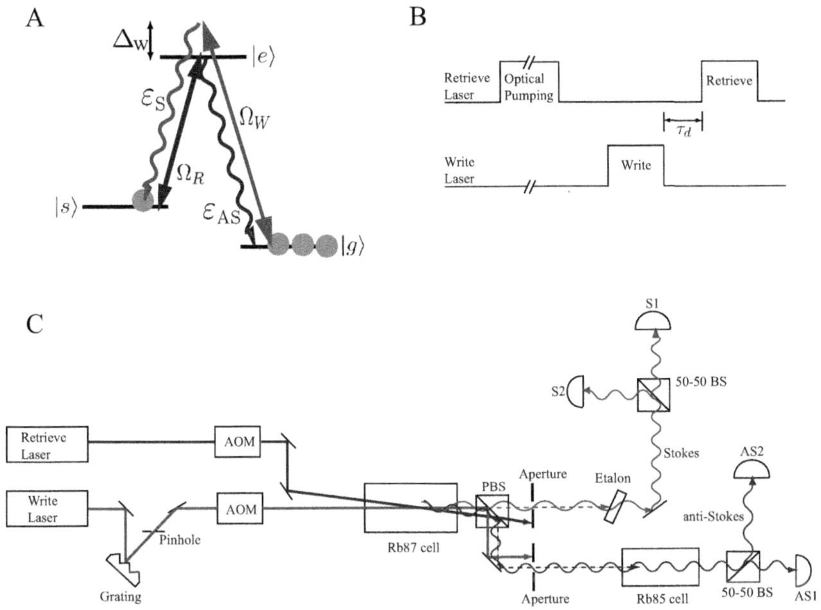

Figure 1. A - ^{87}Rb levels used in the experiments (D_1 line; Zeeman sublevels are not shown in this simplified scheme; the ground-state hyperfine splitting is 6.835 GHz). The write (retrieve) laser and Stokes (anti-Stokes) beam are illustrated in red (blue). B - After the optical pumping pulse (provided by the retrieve laser), the $1.6\,\mu s$-long write pulse is followed by the retrieve pulse after a controllable delay τ_d. C - Schematic of the experimental setup. S1, S2 (AS1, AS2) denote the avalanche photodiodes for the Stokes (anti-Stokes) light.

The direction, delay time τ_d, and rate of retrieval are determined by the direction, timing, and intensity of the retrieve laser, allowing control over the spatio-temporal properties of the retrieved pulse (referred to as the anti-Stokes pulse). Since the storage and retrieval processes ideally result in identical photon numbers in the Stokes and anti-Stokes pulses [Lukin 1999], this technique should allow preparation of an n-photon Fock state in the anti-Stokes pulse conditioned on detection of n Stokes photons.

2. Experimental setup

Following earlier work [van der Wal 2003], we use long-lived hyperfine sublevels of the ^{87}Rb electronic ground state as the storage states $|g\rangle = |5^2S_{1/2}, F = 1\rangle$ and $|s\rangle = |5^2S_{1/2}, F = 2\rangle$ (see Fig. 1 A). The write laser couples $|g\rangle$ to the excited state $|e\rangle$ (two nearly degenerate hyperfine levels $|5^2P_{1/2}, F' = 1\rangle$ and $|5^2P_{1/2}, F' = 2\rangle$) with a detuning $\Delta_W \approx 1$ GHz; after passing through a grating to eliminate undesired frequency components, it is focused to a waist of $100\,\mu$m. The near-resonant retrieve laser beam couples

$|s\rangle$ to $|e\rangle$ with Rabi frequency Ω_R, and has a waist of 1 mm. Both lasers pass through acousto-optical modulators (AOMs) to allow pulsing, and then overlap at a small angle ($\sim$ 10 mrad) inside a magnetically-shielded ^{87}Rb vapor cell held at $\approx 75^o$C.

The primary experimental challenge lies in transmitting the few-photon Stokes and anti-Stokes pulses while simultaneously blocking the write and retrieve laser beams. A polarizing beam splitter (PBS) separates the write (retrieve) laser from the Stokes (anti-Stokes) Raman light, and further filtering is provided by an etalon or an optically-pumped ^{87}Rb cell in the write channel, and a ^{85}Rb cell in the read channel; this combined filtering separates the write (retrieve) laser from the Stokes (anti-Stokes) Raman light to one part in 10^9 (10^{12}). The filtered Stokes and anti-Stokes beams are then split by a 50-50 beam splitter, and the photon number in each arm is detected with an avalanche photodiode (APD); this setup reduces the influence of dead-time effects and enables us to measure the photon shot noise (PSN) level and second-order correlation function $g^{(2)}$ for the Stokes and anti-Stokes channels. Experimentally, we take advantage of the long coherence time of the atomic memory ($\sim 3\mu$s in the present experiment, see Fig. 1 B) to create few-photon pulses with long coherence lengths ($\sim$ few μs) that significantly exceed the time-resolution of the APDs ($\sim$ 50 ns). This allows us to directly count the photon number in each of the pulses and to directly measure the pulse shapes by averaging the time-resolved APD output over many experimental runs.

3. Pulse shapes: theory and experiment

Fig. 2 A shows the average number of detected Stokes photons per unit time (photon flux) in the write channel as a function of time during the 1.6 μs-long write pulse. The magnitude of the photon flux (and hence the total number of photons in the pulse) is controlled by varying the excitation intensity. After a time delay τ_d, we apply the retrieve beam to convert the stored spin wave into anti-Stokes photons, as shown in Fig. 2 B. The duration and peak flux of the anti-Stokes pulse can be controlled by the intensity of the retrieve laser.

These pulse shapes can be quantitatively understood by examining the propagation of the Stokes and anti-Stokes fields through an optically dense atomic ensemble. As shown in Fig. 2 B, the results of this analysis are in an excellent agreement with experimental observations. We theoretically model the medium as N three-level Λ-type atoms as shown in Fig. 1 A. The two laser fields are treated classically, whereas the generation and propagation of one transverse mode of the Stokes (anti-Stokes) quantum field is described by an effective one-dimensional slowly varying operator $\mathcal{E}_S(z,\, t)$ ($\mathcal{E}_{AS}(z,\, t)$), which corresponds to the vacuum field at the entrance of the cell $z = 0$. Our model allows us to treat a small number of transverse spatial modes which evolve

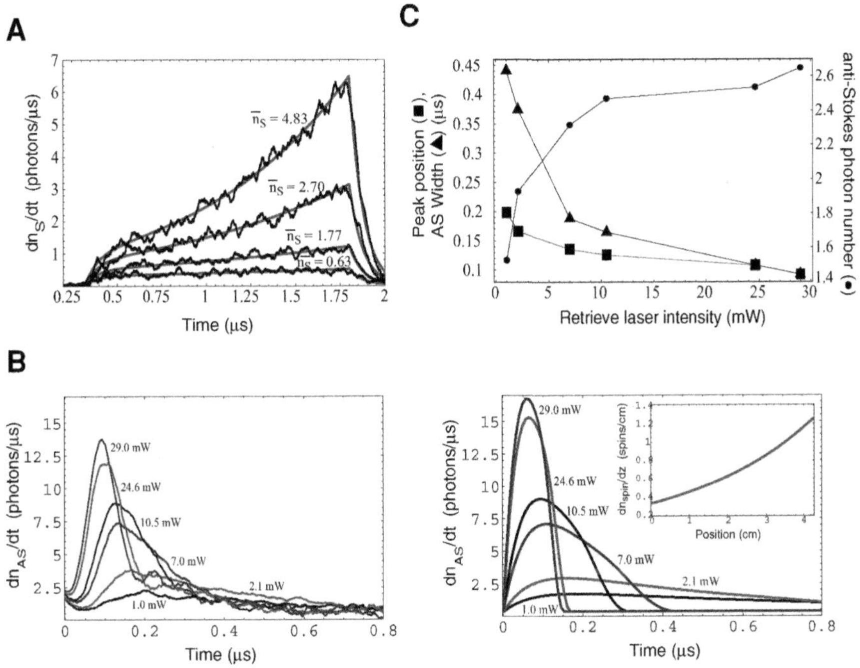

Figure 2. A - Experimentally measured and theoretically calculated values of dn_S/dt, the number of Stokes photons per unit time emitted from the atomic vapor cell. For each plot, $n_S = \int dt\ dn_S/dt$ represents the total number of photons emitted from the cell. The write laser power is varied from 25 mW to 100 mW. B - Experimentally measured and theoretically calculated values of dn_{AS}/dt, the number of anti-Stokes photons per unit time emitted from the atomic vapor cell. The experimental pulse shapes correspond to a Stokes pulse with $n_S \approx 3$ photons, and the theoretical curves assume an initial spin wave with $n_{spin} = 3$ excitations and an optical depth of $\simeq 20$. Each curve is labeled with the power of the retrieve laser. Inset: theoretical calculation of the number of flipped spins per unit length dn_{spin}/dt as a function of position in the atomic cell, for $n_{spin} = 3$. C - Measured anti-Stokes pulse width (full-width at half-maximum) and total photon number as a function of the retrieve laser intensity. Lines are intended only to guide the eye.

both in time and the longitudinal spatial coordinate. The atomic properties are described by collective, slowly varying operators [Drummond 1991]; in particular $\mathcal{S}(z,\ t)$ is the spin-flip operator corresponding to the atomic coherence $|g\rangle\langle s|$ at position z (which is also in the vacuum state at $t = 0$).

The theoretical curves in Fig. 2 A show the number of Stokes photons emitted per unit time at the end of the cell $dn_S(t)/dt = c\ \langle \mathcal{E}_S^\dagger(L,\ t)\ \mathcal{E}_S(L,\ t)\rangle\ /L$; for each curve the total number of emitted Stokes photons $n_S(t) = \int_0^t dt'\ dn_S(t')/dt'$ is matched to the corresponding experimental value. In both theoretical and experimental results, the shape of the Stokes pulse depends on the total number of photons it contains: for pulses containing on average one photon or less, the flux is constant in time (more generally it follows the shape of the write laser), as expected in the spontaneous regime of Raman scattering [Raymer 1985]. For pulses containing more than one photon, however, the flux increases with time due to a Bose-stimulation effect.

The observed evolution of the Stokes pulses can be understood qualitatively by considering the mutual growth of the photon field and spin excitation: the first flipped spin *stimulates* subsequent spin excitations which are accompanied by increased probability of Stokes photon emission. This process is governed by the collective Raman scattering rate $\xi = \eta\, |\Omega_W|^2\, \gamma/\Delta_W^2$, which is equal to the product of the optical depth η and the single atom scattering rate $|\Omega_W|^2\gamma/\Delta_W^2$, where γ is the decay rate of $|e\rangle$. To accurately describe our experimental conditions, our model accounts for multiple transverse modes by adding the intensities of several independent Stokes (and anti-Stokes) modes. Taking a basis of Hermite-Gaussian modes, we can choose the waist of the mode expansion independently of the excitation pulse waist. The number of Stokes photons detected in the fundamental mode is then $dn_S/dt = \bar{N}\,\xi(1+\xi\,t+\ldots)$, where $\bar{N}$ is the effective number of transverse modes. In the present case, we infer $\bar{N} \sim 4$. The predicted pulse profile agrees with our observation that the flux will increase with time for $n_s \sim \xi\, t \geq 1$. The observed dynamics result from growth in the coherent spin-wave mode rather than from individual incoherent atomic transitions, thus providing evidence for the collective nature of the atomic spin excitations.

This transition from a spontaneous to stimulated nature of the Stokes process also affects the spatial distribution of the atomic spin wave. Using our theoretical model, we calculate the number of flipped spins per unit length at position z and time t, $dn_{spin}(z)/dz = \langle \mathcal{S}^\dagger(z,\, t)\; \mathcal{S}(z,\, t)\rangle/L$. For $n_s \leq 1$ the excitation is calculated to be uniformly distributed in the cell, while for $n_s > 1$ the spin-wave amplitude grows toward the end of the cell (see Fig. 2 B, inset). Again, this qualitative change is due to the Bose-stimulation effect. While the first spin-flip is equally likely to occur anywhere in the cell, the probability of Stokes emission (and an accompanying spin-flip) at position z is enhanced by the number of spin flips in $[0,\, z]$, as "seen" by the propagating Stokes field, resulting in an increasing emission probability with increasing position z

The shape of this spin excitation is directly mirrored in the shape of the retrieved anti-Stokes pulse. The resonant classical read laser beats with the ground-state coherence $\mathcal{S}(z,\, t)$ and converts the atomic spin wave into an anti-Stokes field $\mathcal{E}_{AS}(s,\, t)$; the number of anti-Stokes photons emitted per unit time from the end of the cell is $dn_{AS}(t)/dt = c\, \langle \mathcal{E}_{AS}^\dagger(L,\, t)\; \mathcal{E}_{AS}(L,\, t)\rangle/L$. Note that the read laser establishes an EIT configuration for the generated anti-Stokes field, so that the anti-Stokes light propagates unabsorbed through the cell. In the ideal limit of perfect EIT and large optical depth, the temporal shape of the anti-Stokes pulse is equivalent to the spatial shape of the atomic spin coherence, delayed by the time required to propagate out of the atomic cell at the group velocity v_g

$$\frac{dn_{AS}(t)}{dt} = \frac{v_g(t)}{L}\, n_{spin}\left(z - \int_0^t dt'\, v_g(t'),\; t = 0\right),$$
$$\text{with } v_g(t) = c\,\frac{|\Omega_R(t)|^2}{g^2 N} \tag{1}$$

where, g is the coupling constant between the anti-Stokes field and the $|g\rangle \rightarrow |e\rangle$ transition. For larger (smaller) retrieve laser intensity $\sim |\Omega_R(t)|^2$, the excitation is released faster (slower), while the amplitude changes in such a way that the total number of anti-Stokes photons is always equal to the number of spin-wave excitations. We can thus use the intensity of the retrieve laser $\propto |\Omega_R|^2$ to control the bandwidth of the anti-Stokes pulse.

In practice, decay of the spin coherence during the delay time τ_d and finite optical depth flatten and broaden the anti-Stokes pulse, reducing the total number of anti-Stokes photons which can be retrieved within the coherence time of the atomic memory. For weak retrieve laser intensities, the total photon number per pulse increases with increasing laser power because the time required to read out the spin wave is longer than the characteristic decoherence time of our atomic memory ($\sim 3\mu s$, see Fig. 3 b). After accounting for dead-time effects, we find that once the retrieve laser power increases to $\approx 25\,\text{mW}$, all of the spin wave is retrieved in a time shorter than the decoherence time, resulting in a constant anti-Stokes number versus retrieve power.

Including these effects of spin decay and finite optical depth in our theoretical model, we find

$$\mathcal{E}_{AS}(L,\,t) = \sqrt{\frac{v_g(t)}{c}}\; e^{-\gamma_0 t} \int_0^L dz\, \frac{\exp\left[-\left(\frac{L-z}{2\,\Delta z(t)}\right)^2\right]}{\sqrt{\pi}\,\Delta z(t)} \tag{2}$$
$$\times \mathcal{S}\left(z - \textstyle\int_0^t dt'\, v_g(t'),\; t = 0\right)$$

where γ_0 is the ground-state coherence dephasing rate,

$$\Delta z(t) = \sqrt{\frac{L}{\eta}\int_0^t dt'\, v_g(t')}, \tag{3}$$

where $\eta = 3N\,\lambda^2 L/4\pi\,V)$ is the on-resonance optical depth. From this we can calculate the anti-Stokes pulse shape $dn_{AS}(t)/dt = c\langle\mathcal{E}^\dagger_{AS}(L,\,t)\mathcal{E}_{AS}(L,\,t)\rangle/L$ which is shown in Fig. 2 B.

Finally, we should note that for some of the larger retrieve powers used in Fig. 2 B, the photons are retrieved fast enough that dead-time effects become noticeable (for correlation measurements, much longer pulses were used to avoid these effects). Therefore, in generating the theoretical curves for Fig. 2 B, dead-time effects were included by using the model $n_d = n_i(1 - \zeta\, n_i)$, where n_d is the number of detected photons, n_i is the number of incident photons, and ζ is a parameter determined by the specifics of our experiment. The theoretical curves in Fig. 2 B represent the detected photons as calculated from this model. The resulting detailed comparison between theory and experiment in Figs. 2 A and 2 B suggests that the bandwidth of the generated anti-Stokes pulse is close to being Fourier-transform limited, while the transverse profile effectively corresponds to only a few spatial modes.

4. Correlations in the Stokes and anti-Stokes pulses

At fixed laser intensities and durations, the number of Stokes and anti-Stokes photons fluctuates from event to event in a highly correlated manner [van der Wal 2003]. In order to quantify these correlations, we directly compare the number of Stokes and anti-Stokes photons for a large number of pulsed events (each with identical delay times τ_d and laser parameters). The variance of the resulting distributions (Fig. 3 a) is then compared to the photon shot noise level $\text{PSN}_{th} = \bar{n}_S + \bar{n}_{AS}$, which represents the maximum degree of correlations possible for classical states [Mandel 1995]. In the experiment, great care is taken to eliminate systematic sources of errors, in particular APD dead-time effects. To this end we experimentally determine photon shot noise for each channel by using a 50-50 beam-splitter and two APDs per detection channel (see Fig. 1 C) which allows us to accurately determine the measured $\text{PSN}_{meas} = \text{var}(AS1 - AS2) + \text{var}(S1 - S2)$ value for each experiment. For correlation measurements we typically choose the excitation intensities such that the average number of photons in each channel is on the order of or smaller than unity. Under such conditions the measured PSN approaches the expected, theoretical value of PSN. To quantify the correlations, we consider the normalized variance $\text{V} = \text{var}(\{n_{AS} - n_S\})/\text{PSN}_{meas}$, which is one for classically correlated pulses and zero for pulses exhibiting perfect number correlations. Using this method, we measure $V = 0.942 \pm 0.006$ for the data shown in Fig. 3 a at delay time $\tau_d = 0$.

Fig. 3 b shows the normalized variance V as a function of storage time τ_d. Non-classical correlations ($V < 1$) between Stokes and anti-Stokes pulses are clearly observed for storage times up to a few microseconds. The time scale

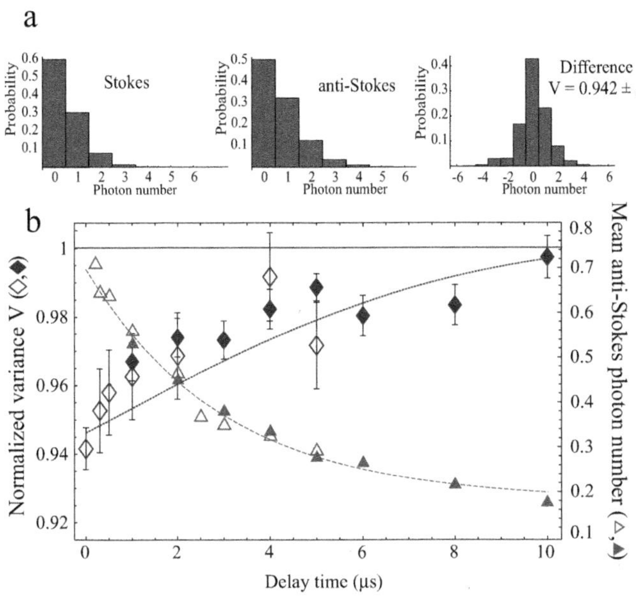

Figure 3. a - Observation of nonclassical correlations. Stokes, anti-Stokes and difference (anti-Stokes minus Stokes) photon-number histograms for pulsed experiments in which a write pulse is followed by a retrieve pulse after a controlled time delay $\tau_d = 0$ ns. The Stokes and anti-Stokes histograms have means of 0.51 and 0.72, and variances of 0.52 and 0.81 respectively. Due to the several transverse spatial modes excited in these experiments and the presence of background photons, these values correspond more closely to Poissonian statistics rather than the expected Bose-Einstein statistics. b - Normalized variance V (blue) and mean number of anti-Stokes photons (red) versus delay time τ_d, shown with statistical error bars of one standard deviation. The open and closed symbols represent two experimental runs with similar experimental parameters, for which the focus was the behavior of V for short and long delay times, respectively. To calculate the variance at each delay time τ_d, the Stokes and Anti-Stokes channels were weighted to balance the signals on the two channels. The dotted line is a fit to the exponentially decreasing mean anti-Stokes number, and yields a characteristic decay time of ~ 3 μs. The solid line is the result of a theoretical model including the effects of loss, background, and more than one spatial mode on the Stokes and anti-Stokes channels. Based on experimental measurements, the overall detection efficiency (α) and number of background photons (n_{BG}) used in the model are $\alpha_S = 0.07$, $n_S^{BG} = 0.3$ ($\alpha_{AS} = 0.21$, $n_{AS}^{BG} = 0.12$) on the Stokes (anti-Stokes) channel, with a retrieval efficiency characteristic decay time of 3 μs and 4 transverse spatial modes assumed. For these measurements, an etalon was used to filter the Stokes light from the write laser beam. For these experimental conditions we estimate that dead-time effects reduce the measured $\bar{n}_{AS}$ by 2.5% from its actual value and increase V by less than 0.2% from its actual value (well within the $\sim 0.6\%$ error bars).

over which the correlations decay is determined by the coherence properties of the atomic spin-wave: nonclassical correlations are obtained only as long as the coherence of the stored excitation is preserved. The solid line is the result of a theoretical model including the effects of loss, background, and more than one spatial mode on the Stokes and anti-Stokes channels. Fig. 3 b also shows that the retrieval efficiency (the ratio of the average number of anti-Stokes photons to the average number of Stokes photons) decreases in a similar manner as τ_d is increased. A fit to this time dependence (dotted line) yields a characteristic decoherence time $1/\gamma_c$ of about $3\ \mu s$, consistent with the timescale for atomic diffusion from the laser beam. These results demonstrate that within the spin coherence time, it is possible to control the timing between preparation and retrieval, while preserving nonclassical correlations.

It is important to point out that even at $\tau_d = 0$ the observed value $V = 0.942 \pm 0.006$ is far from its ideal value of $V = 0$. One important source of error is the finite retrieval efficiency, which is limited by two factors. Due to the atomic memory decoherence rate γ_c, the finite retrieval time τ_r always results in a finite loss probability $p \approx \gamma_c\, \tau_r$. For the correlation measurements we use a relatively weak retrieve laser (~ 2 mW) to reduce the number of background photons and to avoid APD dead-time effects. The resulting anti-Stokes pulse width is on the order of the measured decoherence time, so the atomic excitation decays before it is fully retrieved. Moreover, even as $\gamma_c \rightarrow 0$ the retrieval efficiency is limited by the finite optical depth η of the ensemble, which yields an error scaling as $p \sim 1/\sqrt{\eta}$. The measured maximum retrieval efficiency at $\tau_d = 0$ corresponds to about 0.3. In addition to finite retrieval efficiency, many other factors reduce correlations, including losses in the detection system, background photons, APD afterpulsing effects, and imperfect spatial mode-matching.

5. Conditional generation of nonclassical few-photon states

These correlations between Stokes and anti-Stokes pulses allow for the conditional preparation of the anti-Stokes pulse with intensity fluctuations that are suppressed compared with classical light. In order to quantify the performance of this technique, we measured the second-order intensity correlation function $g^{(2)}_{n_S}(AS)$ and mean number of photons $\bar{n}^{AS}_{n_S}$ for the anti-Stokes pulse conditioned on the detection of n_S photons in the Stokes channel (see Fig. 4). (For classical states of light, $g^{(2)} \geq 1$, whereas an ideal Fock state with n photons has $g^{(2)} = 1 - 1/n$.) Note that the mean number of anti-Stokes photons grows linearly with n_S, while $g^{(2)}_{n_S}(AS)$ drops below unity, indicating the nonclassical character of the anti-Stokes photon states. In the presence of background counts, $g^{(2)}_{n_S}(AS)$ does not increase monotonically with n_S, but instead exhibits a minimum at $n_S = 2$. The Mandel Q parameter [Mandel 1995] can be calculated using $Q^{AS}_{n_S} = \bar{n}^{AS}_{n_S}(g^{(2)}_{n_S}(AS) - 1)$; from these measurements we

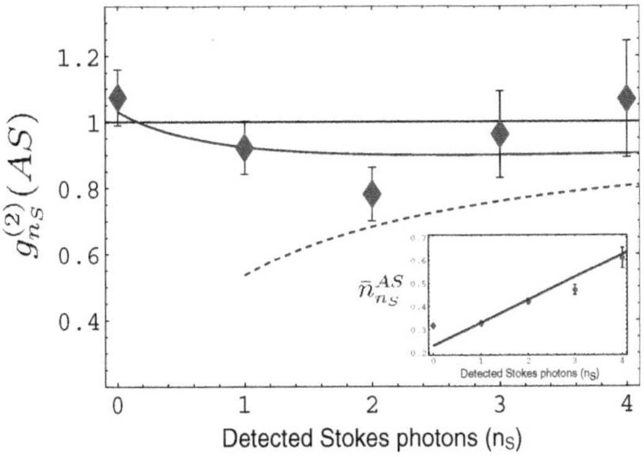

Figure 4. Conditional nonclassical state generation. Conditional $g^{(2)}_{nS}(AS)$ as a function of the number of detected Stokes photons. Diamonds show experimentally measured values, which are calculated from the two arms of the anti-Stokes beam-splitter via $g^{(2)}(AS) = \langle AS_1 \cdot AS_2\rangle / \langle AS_1\rangle\langle AS_2\rangle$ (see Fig. 1 C). The measured mean photons number on the Stokes and anti-Stokes channels were $\bar{n}_S = 1.06$ and $\bar{n}_{AS} = 0.36$ respectively. The solid line shows the result of a theoretical model including background and loss on both the Stokes and anti-Stokes channels. The overall detection efficiency (α) and number of background photons (n_{BG}) used in the model were $\alpha_S = 0.35$, $n^{BG}_S = 0.27$ ($\alpha_{AS} = 0.1$, $n^{BG}_{AS} = 0.12$) on the Stokes (anti-Stokes) channel, and were estimated from experimental measurements. For these measurements an optically-pumped ^{87}Rb cell was used to filter the Stokes photons from the write laser. The dotted line represents $g^{(2)}_{nS}(AS)$ corrected for loss and background on the anti-Stokes channel, obtained by setting the anti-Stokes channel loss and background to zero in this model. Inset: measured mean anti-Stokes number $\bar{n}^{AS}_{nS}$ conditioned on the Stokes photon number n_S. The solid line represents $\bar{n}^{AS}_{nS}$ as predicted by the model.

determine $Q^{AS}_{nS=2} = -0.09 \pm 0.03$ for conditionally generated states with $n_S = 2$ ($Q \geq 0$ for classical states and $Q = -1$ for Fock states).

The observed reduction in the intensity fluctuations is imperfect due to background and losses in both the preparation and the retrieval detection channels. These can be accounted for in a theoretical model that yields reasonable agreement with experimental observations (solid curve in Fig. 4). The dotted line represents the same model as the solid line, but corrected for loss and background on the retrieval detection channel. Using this model, we estimate $(\bar{n}^{AS}_{nS=2}, Q^{AS}_{nS=2})$ corrected for the retrieval detection efficiency and background to be approximately $(2.5, -0.85)$.

Although the corrected parameters are closer to the ideal limit, they still do not correspond to a perfect Fock state. This is due to loss and background in the preparation channel, which prevent measurement of the exact number of created spin excitations. In principle, the conditional state preparation can be made insensitive to overall Stokes detection efficiency α_S by working in the regime of a very weak excitation [Duan 2001]; however, Stokes channel background counts n_S^{BG} prevent one from reaching this regime in practice. Nevertheless, minor improvements in the Stokes detection efficiency and background could enable high quality Fock state generation. In the following section, we will show that a qualitative condition for high quality Fock state generation is that $\zeta = n_S^{BG}(1-\alpha_S)/n_S\,\alpha_S \ll 1$; in our experiments, $\zeta \sim 0.3$ is already significantly below one.

6. Toward Fock state generation

We now turn to a quantitative examination of the feasibility of conditional Fock state generation using our preparation and retrieval technique. For applications in long-distance quantum communication, the quality of the atomic state preparation is the most important quantity. Assuming perfect atom-photon correlations in the write Raman processes, we can find the density matrix $\hat{\rho}$ for the number of atomic spin-wave excitations conditioned on the detection of n_S Stokes photons. Here we consider only the spin-wave modes correlated with our detection mode. For example, in the absence of losses and background, the conditional atomic density matrix is simply $\hat{\rho}(n_S) = |n_S\rangle\langle n_S|$. Loss on the Stokes channel (characterized by transmission coefficient α_S) leads to a statistical mixture of spin-wave excitations,

$$\hat{\rho}_{\alpha_S}(n_S) = \frac{1}{P_{\alpha_S}(n_S)} \sum_{n=n_S}^{\infty} P^{SW}(n) \left(\begin{array}{c} n \\ n_S \end{array} \right) \alpha_S^{n_S}(1-\alpha_S)^{n-n_S} |n\rangle\langle n| \quad (4)$$

where $P^{SW}(n)$ is the unconditional probability to obtain n spin-wave excitations, $P_{\alpha_S}(n_S)$ is the probability to detect n_S Stokes photons (ensuring normalization), and $\left(\begin{array}{c} n \\ n_S \end{array} \right)$ represents the binomial coefficient. Background on the Stokes channel further complicates the picture, as one cannot distinguish between Stokes signal photons and background photons. Including the probability for m background counts $P_B(m)$ in our model, we find

$$\hat{\rho}_{\alpha_S}^{(B)}(n_S) = \frac{1}{\tilde{P}_{\alpha_S}(n_S)} \sum_{i=0}^{n_S} \hat{\rho}_{\alpha_S}(n_S - i)\, P_{\alpha_S}(n_S - i)\, P_B(i) \quad (5)$$

where the normalization factor $\tilde{P}_{\alpha_S}(n_S)$ is the overall probability to have n_S counts of either Stokes signal or background photons.

Since we are primarily interested in the conditional quantum state of the atomic ensemble, we can use the density-matrix formalism to calculate the fidelity with which we create the desired atomic state, $|n_S\rangle$:

$$F = \langle n_S | \hat{\rho}_{\alpha_S}^{(B)}(n_S) | n_S \rangle. \quad (6)$$

We can furthermore calculate the Mandel Q-factor associated with this atomic state

$$Q = \frac{\langle n_{SW}^2 \rangle - \langle n_{SW} \rangle^2}{\langle n_{SW} \rangle} - 1, \tag{7}$$

where $\hat{n}_{SW}$ is the number operator for spin-wave excitations. This quantity can also be interpreted as the Mandel Q-factor associated with a perfectly retrieved anti-Stokes pulse.

To determine the unconditional probability distribution for the spin-wave excitations $P^{SW}(n)$, we must find the effective number of transverse modes which contribute to the Raman processes. We identify two extreme regimes which permit analytic treatment: a single mode regime where the number of excitations in the ^{87}Rb cell follows Bose-Einstein (thermal) statistics and a multimode regime where it follows Poisson statistics. We find in both cases that the quantities F and Q depend on two experimental parameters θ ($\sim$ number of lost Stokes photons) and υ ($\sim$ noise to signal ratio), which are defined in Tab. 1.

Table 1. Scaling for the anti-Stokes pulse Q-paramcter and Fock state fidelity F. $\bar{n}$ refers to the mean number of excitations in the rubidium cell, $\bar{n}_S^{BG}$ is the mean photon number of background counts in the write channel (we assume they are mainly due to leak of the write drive and so follow Poisson statistics), α_S is the Stokes detection efficiency and n_S is the number of Stokes photons on which we condition. The mean number of atomic excitations is calculated via $\langle n_{SW} \rangle = \text{Tr}\,(\hat{\rho}_{\alpha_S}\, \hat{n}_{SW})$; similarly $\langle n_{SW}^2 = \text{Tr}\,(\hat{\rho}_{\alpha_S}\, \hat{n}_{SW}^2)$. The subscript T (P) refers to thermal (Poisson) photon statistics of the unconditional Stokes light.

Regime	*Single mode*	*Multimode*
Parameters	$\theta_T = \frac{\bar{n}}{1+\bar{n}}\,(1-\alpha_S)$	$\theta_P = \bar{n}(1-\alpha_S)$
	$\upsilon_T = \frac{\bar{n}_S^{BG}\,(1+\alpha_S\,\bar{n})}{\alpha_S\,\bar{n}}$	$\upsilon_P = \frac{\bar{n}_S^{BG}}{\alpha_S\,\bar{n}}$
Q (no background)	$\frac{\theta_T}{1-\theta_T} - \frac{n_S}{n_S+\theta_T}$	$-\frac{n_S}{n_S+\theta_P}$
Q with background	$\sim Q[\upsilon_T = 0] + \upsilon_T\,(1+\theta_T)$	$\frac{-\,n_S\,(1+\upsilon_P)^{-1}}{[n_S+\theta_P(1+\upsilon_P)]}$
F (no background)	$(1-\theta_T)^{n_S+1}$	$e^{-\theta_P}$
F (small background) ($\upsilon \ll 1$)	$\sim (1-\theta_T)^{n_S+1}(1-\upsilon_T)$	$\sim \frac{e^{-\theta_P}\,[1+O(\upsilon_P\theta_P)]}{(1+\upsilon_P)^{n_S}}$

We now turn to the conditions under which high-quality Fock states can be generated in the presence of loss and background. In principle, one can compensate for loss in the Stokes channel by decreasing the probability for excitations in the atomic ensemble, so that the probability for more than one spin excitation is negligible. In this situation, a lossy Stokes channel will rarely register a photon, but when a photon is detected it almost certainly came from a single spin-wave excitation. Conversely, one can compensate for background in the absence of loss by increasing the number of spin-wave excitations so that $\bar{n} \gg \bar{n}_S^{BG}$. When both losses and background blur the signal, there exists an optimum value for $\bar{n}$ that maximizes fidelity and minimizes the Mandel Q-parameter. Tuning the mean number of excitations to this optimum gives us the smallest possible Q. However, this minimum is not always negative, which means that for certain conditions we cannot conditionally produce nonclassical states. For example, in the multimode case, the border between classical and nonclassical state generation is determined by the parameter

$$\zeta = \frac{v_P\,\theta_P}{n_s} = \frac{\bar{n}_S^{BG}}{n_S}\left(\frac{1-\alpha_S}{\alpha_S}\right). \tag{8}$$

Provided that $\zeta \ll 1$, one can always find a specific number of excitations in the cell that allows conditional production of number-squeezed atomic states. For instance, if $\zeta \sim 0.1$, the optimal Q is found to be ~ -0.5 in both multimode and monomode cases. This indicates that highly nonclassical states $(Q < 0)$ can be produced with parameters in the range of our current experimental.

High fidelities require very high Stokes efficiencies α_S or significantly reduced background. We expect that refinements in the Stokes detection system and better transverse mode selection should permit far better fidelity for conditional Fock-state generation, thereby providing a basic building block for long-distance quantum communication [Duan 2001] as well for other applications in quantum information science [Briegel 2000; Duan 2002 (a); Blinov 2004; Scully 2004].

Acknowledgments

We gratefully acknowledge T. P. Zibrova, J. MacArthur, D. F. Phillips, R. Walsworth, P. R. Hemmer, A. Trifonov, and C. H. van der Wal for useful discussions and experimental help. This work was supported by the NSF, DARPA, the David and Lucille Packard Foundation, the Alfred Sloan Foundation, the Fannie and John Hertz Foundation (LC) and Ecole des Mines de Paris (FM).

DETERMINISTIC ENTANGLEMENT OF SINGLE PHOTONS VIA COHERENTLY DRIVEN ATOMS

D. Petrosyan
Institute of Electronic Structure & Laser,
Foundation for Research & Technology - Hellas,
Heraklion 71110, Crete, Greece
dap@iesl.forth.gr

Abstract A simple, experimentally feasible scheme for achieving large cross-Kerr nonlinear coupling of weak optical fields is proposed and studied in detail. The scheme is based on the attainment of electromagnetically induced transparency in a medium of coherently driven four-level atoms in tripod configuration. The resulting giant cross-phase modulation accompanied by negligible absorption can be used to implement the controlled-phase (CPHASE) universal logic gate between two single-photon pulses.

Keywords: All-optical quantum computation, electromagnetically induced transparency

Introduction

Electromagnetically induced transparency (EIT) in atomic media is a quantum interference effect that results in a dramatic reduction of the group velocity of propagating weak probe field accompanied by vanishing absorption [Harris 1997; Scully 1997; Hau 1999]. As the quantum interference is usually very sensitive to the system parameters, various schemes exhibiting EIT are attracting growing attention due to their potential for significantly enhancing nonlinear optical effects. Some of the most representative examples include slow-light enhancement of acusto-optical interactions in doped fibers [Matsko 2000], trapping light in optically dense atomic media by coherently converting photonic excitation into spin excitation [Fleischhauer 2000; Liu 2001; Phillips 2001] or by creating photonic bang gap via periodic modulation of the EIT resonance [Andre 2002], and nonlinear photon-photon coupling using N configuration of atomic levels [Schmidt 1996; Harris 1998; Harris 1999].

EIT is based on the phenomenon of coherent population trapping [Harris 1997; Scully 1997; Liu 2001], in which the application of two laser fields to a three-level Λ system creates the so-called "dark state", which is stable against absorption of both fields. Dark states are also found in several other

V.M. Akulin et al. (eds.), Decoherence, Entanglement and Information Protection in Complex Quantum Systems, 77–90.

multilevel systems, one of them being four-level atoms interacting with three laser fields in tripod configuration. Tripod atoms proved to be robust systems for "engineering" arbitrary coherent superpositions of atomic states [Vewinger 2003] using an extension of the well-known technique of stimulated Raman adiabatic passage (STIRAP) [Bergmann 1998]. Parametric generation of light in a medium of tripod atoms, prepared in a certain coherent superposition of ground states, has been recently discussed in Ref. [Paspalakis 2002]. Here show that two weak (quantum) fields propagating through a medium of tripod atoms under the EIT conditions, can impress very large nonlinear phase-shift upon each other [Petrosyan 2004].

The main motivation for the present work is its relevance to the field of quantum information (QI), which is attracting broad interest in view of its fundamental nature and its potentially revolutionary applications to cryptography, teleportation and computing [Nielsen 2000]. Among the various QI processing schemes of current interest [Cirac 1995; Turchette 1995; Imamoğlu 1997; Kane 1998; Loss 1998; Knill 2001], those based on photons [Turchette 1995; Imamoğlu 1997; Knill 2001] have the advantage of using very robust and versatile carriers of QI. Yet the main impediment towards their operation at the few-photon level is the weakness of optical nonlinearities in conventional media [Boyd 1992]. As mentioned above, EIT schemes with atoms having N-configuration of levels have opened up a possibility of achieving enhanced nonlinear coupling of weak quantum fields at the single-photon level [Schmidt 1996; Harris 1998; Harris 1999]. The main hindrance of such schemes is the mismatch between the group velocities of the pulse subject to EIT and its nearly-free propagating partner, which severely limits their effective interaction length [Harris 1999]. This drawback may be remedied by using an equal mixture of two isotopic species, interacting with two driving fields and an appropriate magnetic field, which would render the group velocities of the two pulses equal [Lukin 2000 (a)]. In contrast, the present scheme relies solely on an intra-atomic process, without resorting to two isotopic species and using just one driving field [Petrosyan 2004]. Under the appropriate conditions specified below, two orthogonally-polarized quantum fields, acting on adjacent transitions of tripod atoms, propagate with the same group velocity and impress large conditional phase shift upon each other, thereby enabling full entanglement of the single-photon pulses.

1. Formulation

We consider near-resonant interaction of two weak, orthogonally (circularly) polarized optical fields $\mathcal{E}_1$ and $\mathcal{E}_2$ and a strong driving field E_d with a medium of atoms with tripod configuration of levels (Fig. 1 (a)). The medium is subject to a weak longitudinal magnetic field B that removes the degener-

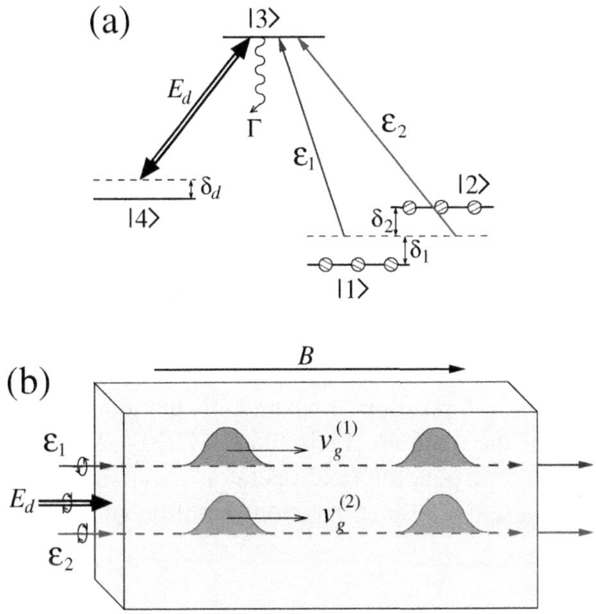

Figure 1. (a) Level scheme of tripod atoms interacting with two weak fields $\mathcal{E}_1$ and $\mathcal{E}_2$, strong driving field E_d and weak magnetic field B that removes the degeneracy of Zeeman sublevels $|1\rangle$ and $|2\rangle$. (b) Copropagating circularly left- and right-polarized weak field $\mathcal{E}_{1(2)}$ pass through the atomic medium that is subject to the circularly polarized driving field E_d and longitudinal magnetic field B.

acy of the ground state sublevels $|1\rangle$ and $|2\rangle$ whose Zeeman shift is given by $\hbar\,\Delta = \mu_{\rm B}\, M_F\, g_F\, B$, where $\mu_{\rm B}$ is the Bohr magneton, g_F is the gyromagnetic factor and $M_F = \pm 1$ is the magnetic quantum number of the corresponding state. All the atoms are assumed to be optically pumped to the states $|1\rangle$ and $|2\rangle$ which thus have the same incoherent populations equal to $1/2$. The weak fields $\mathcal{E}_1$ and $\mathcal{E}_2$, having the same carrier frequency ω_p and wavevector k_p parallel to the magnetic field direction, act on the atomic transitions $|1\rangle \to |3\rangle$ and $|2\rangle \to |3\rangle$, with the detunings $\delta_{1(2)} = \omega_p - \omega_{31}^0 - k_p\, v \mp \Delta$, where $\omega_{31}^0 = \omega_{32}^0$ is the frequency of the unshifted atomic resonance and $k_p\, v$ is the Doppler shift for the atoms having velocity v along the propagation direction. The strong classical cw field E_d, having frequency ω_d and wavevector $k_d \simeq k_p$, is driving the atomic transition $|3\rangle \leftrightarrow |4\rangle$ with the Rabi frequency $\Omega_d = \wp_{34}\, E_d/\hbar$, where $\wp_{\mu\nu}$ is the dipole matrix element on the transition $|\mu\rangle \to |\nu\rangle$. In the collinear Doppler-free geometry shown in Fig. 1 (b), the driving field should be circularly polarized so as to couple to a single magnetic sublevel $|4\rangle$. Its Zeeman shift $\hbar\,\Delta' = \mu_{\rm B}\, M_{F'}\, g_{F'}\, B$ is incorporated in the detuning of the driving field via $\delta_d = \omega_d - \omega_{34}^0 - k_d\, v + \Delta'$, where ω_{34}^0 is the atomic resonance frequency for zero magnetic field.

We describe the medium using collective slowly varying atomic operators

$$\hat{\sigma}_{\mu\nu}(z,\, t) = \frac{1}{N_z} \sum_{j=1}^{N_z} |\mu_j\rangle\langle\nu_j|, \tag{1}$$

averaged over small but macroscopic volume containing many atoms $N_z = (N/L)\, dz \gg 1$ around position z, where N is the total number of atoms and L is the length of the medium [Fleischhauer 2000]. The quantum fields are described by the corresponding field operators $\hat{\mathcal{E}}_{1(2)}$. In a frame rotating with the frequencies ω_p and ω_d, the interaction Hamiltonian has the following form

$$\begin{aligned} H = \;& \hbar\, \frac{N}{L} \int_0^L dz\, \Big[\, \delta_1\, \hat{\sigma}_{11} + \delta_2\, \hat{\sigma}_{22} + \delta_d\, \hat{\sigma}_{44} \\ & - g\, \Big(\hat{\mathcal{E}}_1\, \hat{\sigma}_{31} + \hat{\mathcal{E}}_2\, \hat{\sigma}_{32}\Big) - \Omega_d\, \hat{\sigma}_{34} + \text{h. c.}\Big] . \end{aligned} \tag{2}$$

Here $g = \wp_{31} \sqrt{\omega_p/(2\hbar\, \epsilon_0\, A\, L)}$, with A being the cross-sectional area of the quantum fields, is the atom-field coupling constant, which is the same for both fields $\hat{\mathcal{E}}_{1(2)}$ due to the symmetry of the system ($|\wp_{31}| = |\wp_{32}|$ while the opposite signs of the Clebsch-Gordan coefficients on the transitions $|1\rangle \to |3\rangle$ and $|2\rangle \to |3\rangle$ can be incorporated into the atomic eigenstate via the transformation $|1\rangle \to e^{i\pi}|1\rangle$). Using the slowly varying envelope approximation, we obtain the following propagation equations for the quantum field operators

$$\left(\frac{\partial}{\partial t} + c\, \frac{\partial}{\partial z}\right) \hat{\mathcal{E}}_1(z,\, t) \;=\; ig\, N\, \hat{\sigma}_{13}, \tag{3a}$$

$$\left(\frac{\partial}{\partial t} + c\, \frac{\partial}{\partial z}\right) \hat{\mathcal{E}}_2(z,\, t) \;=\; ig\, N\, \hat{\sigma}_{23}. \tag{3b}$$

The equations for the atomic coherences are given by

$$\frac{\partial}{\partial t}\,\hat{\sigma}_{12} = [i\,(\delta_1-\delta_2)-\gamma_c]\,\hat{\sigma}_{12} - ig\,\hat{\mathcal{E}}_1\,\hat{\sigma}_{32} + ig\,\hat{\mathcal{E}}_2^{\dagger}\,\hat{\sigma}_{13} + \hat{F}_{12}, \quad \text{(4a)}$$

$$\frac{\partial}{\partial t}\,\hat{\sigma}_{13} = \left(i\,\delta_1 - \frac{\Gamma}{2}\right)\hat{\sigma}_{13} + ig\,\hat{\mathcal{E}}_1\,(\hat{\sigma}_{11}-\hat{\sigma}_{33}) + ig\,\hat{\mathcal{E}}_2\,\hat{\sigma}_{12} + i\,\Omega_d\,\hat{\sigma}_{14} + \hat{F}_{13}, \quad \text{(4b)}$$

$$\frac{\partial}{\partial t}\,\hat{\sigma}_{14} = [i\,(\delta_1-\delta_d)-\gamma_c]\,\hat{\sigma}_{14} - ig\,\hat{\mathcal{E}}_1\,\hat{\sigma}_{34} + i\,\Omega_d^{\star}\,\hat{\sigma}_{13} + \hat{F}_{14}, \quad \text{(4c)}$$

$$\frac{\partial}{\partial t}\,\hat{\sigma}_{23} = \left(i\,\delta_2 - \frac{\Gamma}{2}\right)\hat{\sigma}_{23} + ig\,\hat{\mathcal{E}}_2\,(\hat{\sigma}_{22}-\hat{\sigma}_{33}) + ig\,\hat{\mathcal{E}}_1\,\hat{\sigma}_{21} + i\,\Omega_d\,\hat{\sigma}_{24} + \hat{F}_{23}, \quad \text{(4d)}$$

$$\frac{\partial}{\partial t}\,\hat{\sigma}_{24} = [i\,(\delta_2-\delta_d)-\gamma_c]\,\hat{\sigma}_{24} - ig\,\hat{\mathcal{E}}_2\,\hat{\sigma}_{34} + i\,\Omega_d^{\star}\,\hat{\sigma}_{23} + \hat{F}_{24}, \quad \text{(4e)}$$

$$\frac{\partial}{\partial t}\,\hat{\sigma}_{34} = -\left(i\,\delta_d + \frac{\Gamma}{2}\right)\hat{\sigma}_{34} - ig\,\hat{\mathcal{E}}_1^{\dagger}\,\hat{\sigma}_{14} - ig\,\hat{\mathcal{E}}_2^{\dagger}\,\hat{\sigma}_{24} + i\,\Omega_d^{\star}\,(\hat{\sigma}_{33}-\hat{\sigma}_{44}) + \hat{F}_{34} \quad \text{(4f)}$$

where γ_c is the ground state coherence (spin) relaxation rate, Γ is the decay rate of the excited state $|3\rangle$ and $\hat{F}_{\mu\nu}$ are δ-correlated noise operators associated with the relaxation.

We now outline the solution of Eqs. (4a) to (4f) in the weak field limit. To this end, we assume that the Rabi frequencies $g\,\mathcal{E}_{1(2)}$ of the quantum fields are much smaller than Ω_d and the number of photons in $\hat{\mathcal{E}}_{1(2)}$ is much less than the number of atoms, therefore $\hat{\sigma}_{11} = \hat{\sigma}_{22} \simeq \mathbb{1}/2$ while $\hat{\sigma}_{33} = \hat{\sigma}_{44} = \hat{\sigma}_{34} \simeq 0$. We may thus treat the atomic equations perturbatively in the small parameters $g\,\hat{\mathcal{E}}_{1(2)}/\Omega_d$, obtaining in the first order

$$\hat{\sigma}_{13}^{(1)} = \frac{g\,\hat{\mathcal{E}}_1(\delta_1-\delta_d)}{2\,|\Omega_d|^2}, \quad \hat{\sigma}_{23}^{(1)} = \frac{g\,\hat{\mathcal{E}}_2(\delta_2-\delta_d)}{2\,|\Omega_d|^2}. \quad \text{(5)}$$

Assuming that the temporal width T_p of the weak pulses is large enough so that the atoms follow the fields adiabatically, from Eq. (4a) we have

$$\hat{\sigma}_{12} = \frac{g\,\hat{\mathcal{E}}_1\,\hat{\sigma}_{32}^{(1)} - g\,\hat{\mathcal{E}}_2^{\dagger}\,\hat{\sigma}_{13}^{(1)}}{i\,\gamma_c - 2\,\Delta}. \quad \text{(6)}$$

Substituting the above expressions into

$$\hat{\sigma}_{14} = -\frac{g\,\hat{\mathcal{E}}_1}{2\Omega_d} - \frac{g\,\hat{\mathcal{E}}_2}{\Omega_d}\,\hat{\sigma}_{12} - \frac{i}{\Omega_d}\left[\left(\frac{\partial}{\partial t} - i\,\delta_1 + \frac{\Gamma}{2}\right)\hat{\sigma}_{13}^{(1)} - \hat{F}_{13}\right],$$
$$\hat{\sigma}_{24} = -\frac{g\,\hat{\mathcal{E}}_2}{2\Omega_d} - \frac{g\,\hat{\mathcal{E}}_1}{\Omega_d}\,\hat{\sigma}_{21} - \frac{i}{\Omega_d}\left[\left(\frac{\partial}{\partial t} - i\,\delta_2 + \frac{\Gamma}{2}\right)\hat{\sigma}_{23}^{(1)} - \hat{F}_{23}\right], \tag{7}$$

we arrive to the following set of equations

$$\hat{\sigma}_{13} = -\frac{i}{\Omega_d^\star}\left[\frac{\partial}{\partial t} - i(\delta_1 - \delta_d) + \gamma_c\right]\hat{\sigma}_{14} + \frac{i}{\Omega_d^\star}\hat{F}_{14} \tag{8a}$$

$$\hat{\sigma}_{14} = -\frac{g\,\hat{\mathcal{E}}_1}{2\Omega_d}\left[1 + \frac{(\delta_1 + i\,\frac{\Gamma}{2})(\delta_1 - \delta_d)}{|\Omega_d|^2}\right.$$
$$\left. + \frac{g^2\,\hat{I}_2\,2\Delta}{|\Omega_d|^2(i\gamma_c - 2\Delta)}\right] + \frac{i}{\Omega_d}\,\hat{F}_{13} \tag{8b}$$

$$\hat{\sigma}_{23} = -\frac{i}{\Omega_d^\star}\left[\frac{\partial}{\partial t} - i(\delta_2 - \delta_d) + \gamma_c\right]\hat{\sigma}_{24} + \frac{i}{\Omega_d^\star}\hat{F}_{24} \tag{8c}$$

$$\hat{\sigma}_{24} = -\frac{g\,\hat{\mathcal{E}}_2}{2\Omega_d}\left[1 + \frac{(\delta_2 + i\,\frac{\Gamma}{2})(\delta_2 - \delta_d)}{|\Omega_d|^2}\right.$$
$$\left. - \frac{g^2\,\hat{I}_1\,2\Delta}{|\Omega_d|^2(i\gamma_c + 2\Delta)}\right] + \frac{i}{\Omega_d}\,\hat{F}_{23} \tag{8d}$$

where $\hat{I}_j \equiv \hat{\mathcal{E}}_j^\dagger\,\hat{\mathcal{E}}_j$ is the dimensionless intensity (photon-number) operator for the jth field.

From now on we focus on the case of $\omega_p = \omega_{31}^0 = \omega_{32}^0$. Substituting Eqs. (8a) to (8d) into (3a) and (3b), the equations of motion for quantum fields are obtained as

$$\left[\frac{\partial}{\partial z} + \frac{1}{v_g^{(1)}}\frac{\partial}{\partial t}\right]\hat{\mathcal{E}}_1 = -\,\kappa_1\,\hat{\mathcal{E}}_1 - i(\Delta + \Delta_d)(s_1 - \eta_1\,\hat{I}_2)\,\hat{\mathcal{E}}_1 + \hat{\mathcal{F}}_1 \tag{9a}$$

$$\left[\frac{\partial}{\partial z} + \frac{1}{v_g^{(2)}}\frac{\partial}{\partial t}\right]\hat{\mathcal{E}}_2 = -\,\kappa_2\,\hat{\mathcal{E}}_2 + i(\Delta - \Delta_d)(s_2 - \eta_2\hat{I}_1)\,\hat{\mathcal{E}}_2 + \hat{\mathcal{F}}_2 \tag{9b}$$

where $\Delta_d = \omega_d - \omega_{34}^0 + \Delta'$ is the driving field detuning,

$$\kappa_{1(2)} = \frac{N\,g^2}{2c\,|\Omega_d|^2}\left[\gamma_c + \frac{\Gamma\,(\Delta \pm \Delta_d)^2}{|\Omega_d|^2}\right]$$
$$s_{1(2)} = \frac{N\,g^2}{2c\,|\Omega_d|^2}\left[1 + \frac{\Delta\,(\Delta \pm \Delta_d)}{|\Omega_d|^2}\right] \tag{10}$$

are, respectively, the linear absorption and phase modulation coefficients,

$$\eta_{1(2)} = \frac{N\,g^4\,2\Delta}{2c\,|\Omega_d|^4(2\Delta \mp i\gamma_c)} \tag{11}$$

are the cross-coupling coefficients, $v_g^{(1(2))} = (1/c + s_{1(2)})^{-1}$ are the group velocities of the corresponding fields, and $\hat{\mathcal{F}}_{1(2)}$ are the noise operators having the properties [Scully 1997]

$$\left\langle \hat{\mathcal{F}}_i(z) \right\rangle = \left\langle \hat{\mathcal{F}}_i(z)\,\hat{\mathcal{F}}_i\,(z') \right\rangle = \left\langle \hat{\mathcal{F}}_i^\dagger(z)\,\hat{\mathcal{F}}_i^\dagger\,(z') \right\rangle = 0$$
$$\left\langle \hat{\mathcal{F}}_i(z)\,\hat{\mathcal{F}}_j^\dagger\,(z') \right\rangle = 2\kappa_i\,\delta_{ij}\,\delta\,(z - z') \tag{12}$$

In deriving Eqs. (9a) and (9b), we have assumed that the usual EIT conditions $|\Omega_d|^2 \gg (\Delta \pm \Delta_d)\,k_{p,\,d}\,\bar{v}, \gamma_c\,(\Gamma + k_{p,\,d}\,\bar{v})$, where $\bar{v}$ is the mean thermal atomic velocity, are satisfied, allowing us to neglect the Doppler induced absorption. On the other hand, since the terms containing $k_p\,v$ enter Eqs. (8a) to (8d) linearly, the net phase-shift of the quantum fields, due to the Doppler shifts of the atomic resonance frequencies, averages to zero. Note also that if states $|1\rangle$, $|2\rangle$ and $|4\rangle$ belong to different hyperfine components of a common ground state, the frequencies ω_p and ω_d of the optical fields differ from each other by at most a few GHz, $\omega_p - \omega_d \simeq \omega_{41}^0 \ll \omega_{p,\,d}$. Then, as seen from Eqs. (8a) to (8d), the difference $(k_p - k_d)\,v$ in the Doppler shifts of the atomic resonances $|1\rangle, |2\rangle \rightarrow |3\rangle$ and $|4\rangle \rightarrow |3\rangle$ is negligible.

When $\Delta(\Delta \pm \Delta_d) \ll |\Omega_d|^2$, the group velocities of $\hat{\mathcal{E}}_1$ are $\hat{\mathcal{E}}_2$ are practically the same, $v_g^{(1(2))} \simeq v_g$. Expressing the atom-field coupling constant g through the linear resonant absorption coefficient $a_0 = \wp_{13}^2\,\omega_p\,\rho/(\hbar c\,\epsilon_0\,\Gamma)$ for the transitions $|1\rangle, |2\rangle \rightarrow |3\rangle$ as $N\,g^2 = a_0\,c\,\Gamma/2$ and assuming that the density of atoms $\rho = N/(AL)$ is large enough so that $a_0\,c\,\Gamma \gg 4|\Omega_d|^2$, we have $v_g \simeq 4|\Omega_d|^2/(a_0\,\Gamma) \ll c$. Then the solution of Eqs. (9a) and (9b) can be expressed in terms of the retarded time $\tau = t - z/v_g$ as

$$\hat{\mathcal{E}}_1(z,\, t) \;=\; \hat{\mathcal{E}}_1(0,\, \tau)\, \exp\Big[-\, \kappa_1\, z + i\, \hat{\phi}_1(z,\, 0,\, t)\Big] \qquad \text{(13a)}$$
$$+ \int_0^z dz'\, \hat{\mathcal{F}}_1(z')\, \exp\Big[-\, \kappa_1\, (z - z') + i\, \hat{\phi}_1(z,\, z',\, t)\Big]$$

$$\hat{\mathcal{E}}_2(z,\, t) \;=\; \hat{\mathcal{E}}_2(0,\, \tau)\, \exp\Big[-\, \kappa_2\, z + i\, \hat{\phi}_2(z,\, 0,\, t)\Big] \qquad \text{(13b)}$$
$$+ \int_0^z dz'\, \hat{\mathcal{F}}_2(z')\, \exp\Big[-\, \kappa_2\, (z - z') + i\, \hat{\phi}_2(z,\, z',\, t)\Big]$$

where the phase operators are given by

$$\hat{\phi}_1(z,\, z',\, t) = \; -\, s_1(\Delta + \Delta_d)(z - z') + \eta_1(\Delta + \Delta_d) \int_{z'}^{z} dz''\, \hat{I}_2\left(z'',\, \tau + \frac{z''}{v_g}\right) \qquad \text{(14)}$$
$$\hat{\phi}_2(z,\, z',\, t) = \; s_2(\Delta - \Delta_d)(z - z') - \eta_2(\Delta - \Delta_d) \int_{z'}^{z} dz''\, \hat{I}_1\left(z'',\, \tau + \frac{z''}{v_g}\right).$$

These are the central equations of this paper. The first terms in Eqs. (13a) and (13b) describe the linear attenuation and the phase shift of the corresponding quantum field $\hat{\mathcal{E}}_{1(2)}$ upon propagating through the medium, while the second terms account for the noise contribution.

2. Cross-phase modulation and entanglement

From now on we consider the relatively simple case of small magnetic field, such that $\gamma_c \ll \Delta,\, \Delta' \ll \Delta_d$ and $\Delta_d = \omega_d - \omega_{34}^0$. When the absorption is small, $\kappa_{1(2)} z \ll 1,\; z \in \{0,\; L\}$, which requires that $v_g/\gamma_c \gg L$ and $\Delta_d^2 < \gamma_c\, |\Omega_d|^2/\Gamma$, the quantum Eqs. (13a) and (13b) can be cast as

$$\hat{\mathcal{E}}_1(z,\, t) \;=\; \hat{\mathcal{E}}_1(0,\, \tau)\, \exp\Big[i\eta\, \Delta_d\, \hat{\mathcal{E}}_2^\dagger(0,\, \tau)\, \hat{\mathcal{E}}_2(0,\, \tau)\, z\Big] \qquad \text{(15a)}$$

$$\hat{\mathcal{E}}_2(z,\, t) \;=\; \hat{\mathcal{E}}_1(0,\, \tau)\, \exp\Big[i\eta\, \Delta_d\, \hat{\mathcal{E}}_1^\dagger(0,\, \tau)\, \hat{\mathcal{E}}_1(0,\, \tau)\, z\Big] \qquad \text{(15b)}$$

where the cross-phase modulation coefficient is given by $\eta = g^2/(v_g\, |\Omega_d|^2)$, while the linear phase-modulation is incorporated into the field operators via the unitary transformations $\hat{\mathcal{E}}_{1(2)}(z,\, t) \to \hat{\mathcal{E}}_{1(2)}(z,\, t)\, e^{i\, \Delta_d\, z/v_g}$.

Consider first the classical limit of Eqs. (15a) and (15b), where the operators $\hat{\mathcal{E}}_{1(2)}$ are replaced by the corresponding c-numbers $\mathcal{E}_{1(2)}$. The equations for the two weak fields have the form

$$\mathcal{E}_{1(2)}(z) = \mathcal{E}_{1(2)}(0) \, \exp\left[i\eta \, \Delta_d \, |\mathcal{E}_{2(1)}(0)|^2 \, z\right] \tag{16}$$

which indicates that the cross-phase shift grows linearly with the propagation distance z.

In order to rigorously describe the nonlinear interaction between the weak pulsed fields, we now turn to the fully quantum treatment of the system. The traveling-wave electric fields can be expressed through single mode operators as $\hat{\mathcal{E}}_j(z,\, t) = \sum_q \, a_j^q(t) \, e^{i\, qz}$ $(j = 1,\, 2)$, where a_j^q is the annihilation operator for the field mode with the wavevector $k_p + q$. The single-mode operators a_j^q and $a_j^{q\dagger}$ possess the standard bosonic commutation relations $\left[a_i^q,\, a_j^{q'\dagger}\right] = \delta_{ij} \, \delta_{qq'}$. The continuum of modes scanned by $q \in \{-\delta q/2,\, \delta q/2\}$, where $\delta q \leq \delta\omega/c$, is bounded by the EIT window [Lukin 1997]

$$\delta\omega \leq \frac{|\Omega_d|^2}{\Gamma} \, \frac{k_p}{\sqrt{3\pi \, \rho \, L}} \tag{17}$$

The finite quantization bandwidth δq for the field operators leads to the equal-time commutation relations

$$[\hat{\mathcal{E}}_i(z),\, \hat{\mathcal{E}}_j^\dagger(z')] = \delta_{ij} \, \frac{L \, \delta q}{2\pi} \, \mathrm{sinc} \frac{\delta q \, (z - z')}{2} \tag{18}$$

where $\mathrm{sinc} x = \sin x/x$.

Before proceeding, we note that Eqs. (15a) and (15b) are similar to the corresponding equations of Ref. [Lukin 2000 (a)], where the cross-phase modulation between two quantum fields was mediated by atoms with N-configuration of levels [Schmidt 1996], while the group velocity mismatch between the fields was compensated by using a second kind of Λ-atoms controlled by an additional driving field. In contrast, our scheme relies solely on an intra-atomic process employing only one driving field that causes simultaneous EIT for both fields and their cross-coupling. It is therefore deprived of complications associated with using mixtures of two isotopic species of atoms [Lukin 2000 (a)] or invoking cavity QED techniques [Turchette 1995; Imamoğlu 1997].

The most classical of all the quantum states is the coherent state. To compare the classical and quantum pictures, we therefore consider first the evolution of input wavepacket $|\psi_{\mathrm{in}}\rangle = |\alpha_1\rangle \otimes |\alpha_2\rangle$ composed of the multimode coherent states $|\alpha_j\rangle \equiv \Pi_q |\alpha_j^q\rangle$ $(j = 1,\, 2)$. The states $|\alpha_j\rangle$ are the eigenstates of the input operators $\hat{\mathcal{E}}_j(0,\, t)$ at $z = 0$ with the eigenvalues $\alpha_j(t) = \sum_q \, \alpha_j^q \, e^{-i\, q\, c\, t}$: $\hat{\mathcal{E}}_j(0,\, t) \, |\alpha_j\rangle = \alpha_j(t) \, |\alpha_j\rangle$. Upon propagating through the medium, each pulse experiences a nonlinear cross-phase modulation. The expectation values for the fields are then obtained as

$$\left\langle \hat{\mathcal{E}}_{1(2)}(z,\, t) \right\rangle = \alpha_{1(2)}(\tau)\, \exp\left\{ \left[e^{i\,\theta(z)} - 1 \right] \frac{2\pi\, \left|\alpha_{2(1)}(\tau)\right|^2}{L\, \delta q} \right\} \tag{19}$$

where $\theta(z) = \eta\, \Delta_d\, L\, \delta q\, z/(2\pi)$. This equation is similar to the corresponding equations obtained for single-mode [Sanders 1992] and multimode copropagating fields [Lukin 2000 (a)]. It indicates that when the cross-phase modulation is large, upon propagating through the medium, the phases

$$2\pi\, \sin\theta(z)\, \frac{\left|\alpha_{2(1)}(\tau)\right|^2}{L\, \delta q} \tag{20}$$

and amplitudes

$$\alpha_{1(2)}(\tau)\, \exp\left\{ -\, 4\pi\, \sin^2 \frac{\theta(z)}{2}\, \frac{\left|\alpha_{2(1)}(\tau)\right|^2}{L\, \delta q} \right\} \tag{21}$$

of the quantum fields exhibit periodic collapses and revivals as $\theta(z)$ changes from 0 to 2π. In particular, when the phase-shift is maximal, $\theta = \pi/2$, the amplitude of the corresponding field is reduced by a factor of $r_{1(2)} = \exp[-\, 2\pi\, |\alpha_{2(1)}|^2/(L\, \delta q)]$. On the other hand, the maximal dephasing of the multimode coherent field, $r_{1(2)} = \exp[-\, 4\pi\, |\alpha_{2(1)}|^2/(L\, \delta q)]$, is attained for $\theta(z) = (2n+1)\pi$ $(n = 0,\, 1,\, 2,\, \ldots)$, where the phase shift is zero.

We have thus seen that the behavior of weak quantum fields is remarkably different from that of classical fields, as in the quantum regime the nonlinear phase shift is bounded between $\pm\, 2\pi\, |\alpha_{2(1)}|^2/(L\, \delta q)$. This fact severely limits the usefulness of weak coherent states for QI applications. Only in the limit of weak cross-phase modulation $\theta \ll 1$, the quantum Eq. (19) reproduces the classical result

$$\left\langle \hat{\mathcal{E}}_{1(2)}(z,\, t) \right\rangle = \alpha_{1(2)}(\tau)\, \exp\left[i\eta\, \Delta_d\, \left|\alpha_{2(1)}(\tau)\right|^2\, z \right] \tag{22}$$

whereby the cross-phase shift grows linearly with the propagation distance and can attain large values when the field amplitudes are sufficiently high.

Let us now consider the input state $|\psi_{\rm in}\rangle = |1_1\rangle \otimes |1_2\rangle$, consisting of two single photon wavepackets $|1_j\rangle = \sum_q\, \xi_j^q\, a_j^{q\dagger}\, |0\rangle$ $(j = 1,\, 2)$. The Fourier amplitudes ξ_j^q, normalized as $\sum_q |\xi_j^q|^2 = 1$, define the spatial envelopes $f_j(z)$ of the two pulses that initially (at $t = 0$) are localized around $z = 0$,

$$\left\langle 0 | \hat{\mathcal{E}}_j(z,\, 0) | 1_j \right\rangle = \sum_q\, \xi_j^q\, e^{i\, qz} = f_j(z). \tag{23}$$

In free space, $\hat{\mathcal{E}}_j(z,\, t) = \hat{\mathcal{E}}_j(0,\, \tau)$ with $\tau = t - z/c$, and we have $\left\langle 0 \right| \hat{\mathcal{E}}_j(z,\, t) \left| 1_j \right\rangle = f_j(z - ct)$. The state of the system at any time can be represented as

$$|\psi(t)\rangle = \sum_{q,\, q'} \xi_{12}^{qq'}(t) \left|1_1^q\right\rangle \left|1_2^{q'}\right\rangle \tag{24}$$

from where it is apparent that $\xi_{12}^{qq'}(0) = \xi_1^q\, \xi_2^{q'}$.

Since for the photon-number states the expectation values of the field operators vanish, all the information about the state of the system is contained in the intensities of the corresponding fields

$$\left\langle \hat{I}_j(z,\, t) \right\rangle = \left\langle \psi_{\text{in}} \left| \hat{\mathcal{E}}_j^{\dagger}(z,\, t)\, \hat{\mathcal{E}}_j(z,\, t) \right| \psi_{\text{in}} \right\rangle \tag{25}$$

and their "two-photon wavefunction" [Scully 1997; Lukin 2000 (a)]

$$\Psi_{ij}(z,\, t;\, z',\, t') = \left\langle 0 \left| \hat{\mathcal{E}}_j(z',\, t')\, \hat{\mathcal{E}}_i(z,\, t) \right| \psi_{\text{in}} \right\rangle . \tag{26}$$

The physical meaning of Ψ_{ij} is a two-photon detection amplitude, through which one can express the second-order correlation function $G_{ij}^{(2)} = \Psi_{ij}^{\star} \Psi_{ij}$ [Scully 1997]. The knowledge of the two-photon wavefunction allows one to calculate the amplitudes $\xi_{12}^{qq'}$ of state vector (24) via the two dimensional Fourier transform of Ψ_{ij} at $t = t'$:

$$\xi_{ij}^{qq'}(t) = \frac{1}{L^2} \int dz \int dz' \Psi_{ij}(z,\, z',\, t)\, e^{-i\, qz}\, e^{-i\, q'z'} . \tag{27}$$

We first calculate the expectation values of the intensities $\langle \hat{I}_j(z,\, t) \rangle$ by substituting the operator solution (15a) and (15b) into (25),

$$\left\langle \hat{I}_j(z,\, t) \right\rangle = |f_j(-c\,\tau)|^2 = \left| f_j\left(\frac{z\, c}{v_g} - c\, t \right) \right|^2 \tag{28}$$

where $\tau = t - z/v_g$ for $0 \leq z < L$. This equation indicates that upon entering the medium, as the group velocities of the pulses are slowed down to $v_g \ll c$, their spatial envelopes are compressed by a factor of c/v_g [Fleischhauer 2000]. Outside the medium, at $z \geq L$ and accordingly $\tau = t - L/v_g - (z - L)/c$, we have $\langle \hat{I}_j(z,\, t) \rangle = |f_j(z + L\, (c/v_g - 1) - c\, t)|^2$, which shows that the propagation velocity and the pulse envelopes are restored to their free-space values.

Consider next the two photon wavefunction Ψ_{ij}. After the interaction, at $z,\, z' \geq L$, we have the general expression

$$\Psi_{ij}(z,\, t;\, z',\, t') = \; f_i(-\,c\,\tau)\, f_j(-\,c\,\tau')\left\{1 + \frac{f_j(-\,c\,\tau)}{f_j(-\,c\,\tau')}\right.$$
$$\left. \times\, \mathrm{sinc}\left[\frac{\delta\omega}{2}(\tau - \tau')\right]\left(e^{i\,\theta(L)} - 1\right)\right\} \tag{29}$$

where, as before, $\tau = t - L/v_g - (z - L)/c$ and similarly for τ'. The equal-time ($t = t'$) two-photon wavefunction reads

$$\Psi_{ij}(z,\, z',\, t) = f_i\left[z + L\left(\frac{c}{v_g} - 1\right) - c\,t\right] f_j\left[z' + L\left(\frac{c}{v_g} - 1\right) - c\,t\right]$$
$$\times\left\{1 + \frac{f_j\left[z + L\left(\frac{c}{v_g} - 1\right) - c\,t\right]}{f_j\left[z' + L\left(\frac{c}{v_g} - 1\right) - c\,t\right]}\,\mathrm{sinc}\left[\frac{\delta q}{2}\,(z' - z)\right]\left(e^{i\theta} - 1\right)\right\} \tag{30}$$

where $\theta = \eta\, \Delta_d\, L^2\, \delta q/(2\pi)$. For large enough spatial separation between the two photons, such that $|z' - z| > \delta q^{-1}$ and therefore $\mathrm{sinc}[\delta q\,(z' - z)/2] \simeq 0$, Eq. (30) yields

$$\Psi_{ij}(z,\, z',\, t) \simeq f_i\left[z + L\left(\frac{c}{v_g} - 1\right) - c\,t\right] f_j\left[z' + L\left(\frac{c}{v_g} - 1\right) - c\,t\right] \tag{31}$$

which indicates that no nonlinear interaction takes place between the photons, which emerge from the medium unchanged. This is due to the *local* character of the interaction described by the sinc function.

Consider now the opposite limit of $|z' - z| \ll \delta q^{-1}$ and therefore $\mathrm{sinc}[\delta q\,(z' - z)/2] \simeq 1$. Then for two narrow-band (Fourier limited) pulses with the duration $T_p \gg |z' - z|/c$, one has $f_j(z)/f_j(z') \simeq 1$, and Eq. (30) results in

$$\Psi_{ij}(z,\, z',\, t) \simeq e^{i\theta} f_i\left[z + L\left(\frac{c}{v_g} - 1\right) - c\,t\right] f_j\left[z' + L\left(\frac{c}{v_g} - 1\right) - c\,t\right]. \tag{32}$$

Thus, after the interaction, a pair of single photons acquires conditional phase shift θ, which can exceed π when

$$\left(\frac{\delta q\, L}{2\pi}\right)^2 > \frac{v_g\, |\Omega_d|^2}{c\, g^2} \tag{33}$$

To see this more clearly, we use Eq. (27) to calculate the amplitudes of the state vector $|\psi(t)\rangle$:

$$\xi_{ij}^{qq'}(t) = e^{i\theta}\, \xi_{ij}^{qq'}(0)\, \exp\left\{ i(q+q')\left[L\left(\frac{c}{v_g} - 1\right) - c\,t \right] \right\}. \tag{34}$$

At the exit from the medium, at time $t \simeq L/v_g$, the second exponent in Eq. (34) can be neglected for all $q,\ q'$ and the state of the system is given by

$$\left|\psi\left(\frac{L}{v_g}\right)\right\rangle = e^{i\theta}\, |\psi_{\rm in}\rangle. \tag{35}$$

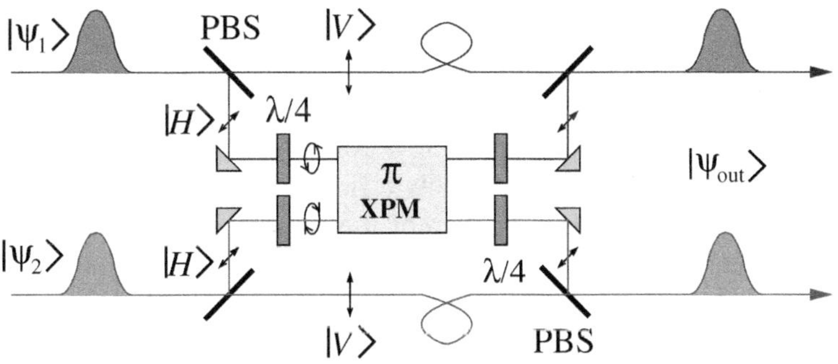

Figure 2. Realization of optical CPHASE logic gate between two single-photon qubits, using polarizing beam-splitters (PBS), $\lambda/4$ plates and π cross-phase modulation (XPM) studied here.

When $\theta = \pi$, the output state of the two photons is given by

$$|\psi_{\rm out}\rangle = -|\psi_{\rm in}\rangle. \tag{36}$$

Utilizing the scheme of Fig. 2, one can realize a transformation corresponding to the *controlled-phase* (CPHASE) logic gate between two photons representing qubits. Namely, let us assume that the qubit basis states are given by the vertical $|V\rangle \equiv |0\rangle$ and horizontal $|H\rangle \equiv |1\rangle$ polarization states of the photon. After being deflected by a polarizing beam-splitter (PBS), each horizontally polarized photon is converted to a circularly polarized one by an appropriately oriented $\lambda/4$ plate. The two circularly polarized photons, upon passing through the active medium, acquire the condition phase-shift π, as per Eq. (36). Finally, each photon is converted back to the horizontal polarization and is recombined with its vertically polarized component on another PBS. The resulting transformation corresponds to the truth table of the CPHASE gate,

$$
\begin{aligned}
|V_1\, V_2\rangle &\to |V_1\, V_2\rangle \\
|V_1\, H_2\rangle &\to |V_1\, H_2\rangle \\
|H_1\, V_2\rangle &\to |H_1\, V_2\rangle \\
|H_1\, H_2\rangle &\to -\ |H_1\, H_2\rangle
\end{aligned}
\tag{37}
$$

Together with the Faraday rotation of photon polarization (implementing single-qubit rotations) and linear phase-shift, the CPHASE gate is *universal* as it can realize arbitrary unitary transformation [Nielsen 2000].

Conclusions

To summarize, we have studied the interaction of two weak quantum fields with an optically dense medium of coherently driven four-level atoms in tripod configuration. We have presented a detailed semiclassical as well as quantum analysis of the system. The main conclusion that has emerged from this study is that optically dense vapors of tripod atoms are capable of realizing a novel regime of symmetric, extremely efficient nonlinear interaction of two multimode single-photon pulses, whereby the combined state of the system acquires a large conditional phase shift that can easily exceed π. Thus our scheme may pave the way to photon-based quantum information applications, such as *deterministic* all-optical quantum computation, dense coding and teleportation [Nielsen 2000]. We have also analyzed the behavior of the multimode coherent state and shown that the restriction on the classical correspondence of the coherent states severely limits their usefulness for QI applications.

SELECTIVE CONTROL OF HIGH-ORDER ATOMIC COHERENCES

Yu. P. Malakyan,[1] D. Budker,[2, 3] S. M. Rochester,[2] D. F. Kimball,[2] V. V. Yashchuk[4] and W. Gawlik[5]

[1]*Institute for Physical Research, National Academy of Sciences, Ashtarak-2, 378410, Armenia*
yumal@ipr.sci.am

[2]*Department of Physics, University of California at Berkeley, Berkeley, California 94720-7300, USA*

[3]*Nuclear Science Division, Lowerence Berkeley National Laboratory, Berkeley, California 94720, USA*

[4]*Advanced Light Source Division, Lawrence Berkeley National Laboratory, Berkeley, California 94720, USA*

[5]*M. Smoluchowski Physical Institute, Jagiellonian University, Reymonta 4, 30-059 Krakow, Poland*

Abstract Resonant enhancement of optical Kerr and higher order nonlinearities in multilevel systems under the conditions of electromagnetically-induced transparency might be useful for quantum signal processing. The main problem with these schemes is related to the selective measurement of high-order atomic coherences that requires multiphoton interactions for the production and detection. In this paper we present a method in which a single laser beam is used for the creation of required nonlinear interactions. It is based on the nonlinear magneto-optical rotation with frequency-modulated light. Using this technique, we measure the nonlinear Faraday rotation caused by four-photon atomic coherence in M-type system. The method is also applicable to the selective control of higher order atomic coherences.

Keywords: atomic coherence, Faraday rotation

Introduction

The fundamental challenge for the implementation of quantum information processing is the realization of large and lossless photon-photon interaction at the level of a few photons [Steane 1998]. The strength of this interaction in conventional nonlinear optics is typically extremely weak. However, recent

V.M. Akulin et al. (eds.), Decoherence, Entanglement and Information Protection in Complex Quantum Systems, 91–104.

studies have shown that the atomic coherence created in multilevel systems via electromagnetically-induced transparency (EIT) [Boller 1991; Feld 1991; Li 1995; Harris 1997] can significantly enhance the resonant nonlinear interaction of a propagating probe field with the atomic medium. Furthermore, the steep dispersion in the EIT media results in a strong reduction of the group velocity of the probe pulse accompanied by vanishing absorption [Budker 1999; Hau 1999; Kash 1999] and, as a result, greatly increases the interaction time of the pulse with the atomic sample. This makes it possible to produce a large Kerr nonlinearity, which is many orders of magnitude greater than that obtained in conventional media, and thus to perform nonlinear optical processes with very low intensities [Schmidt 1996; Schmidt 1997; Harris 1998; Harris 1999; Lukin 1999; Kang 2003].

EIT schemes, capable of producing enhanced Kerr nonlinearity, have been suggested for generation of quantum entanglement. When in a coherently driven medium the EIT conditions are established for two weak light pulses in such a way that they propagate with slow but equal group velocities, the Kerr nonlinearity gives rise to an effective photon-photon interaction. As a result, the combined state of two single-photon pulses acquires a large conditional phase shift, which can easily exceed π, sufficient for the construction of universal two-bit quantum gates. Different schemes have been proposed to realize these expectations [Lukin 2000 (a); Zubairy 2002; Greentree 2002; Matsko 2003; Ottaviani 2003]. The scheme of four-level atoms in $N-$configuration considered by Lukin and Imamoglu [Lukin 2000 (a)] was the first one, where dissipation-free nonlinear interaction of an extraordinary strength at the single photon level has been shown. Zubairy *et al.* [Zubairy 2002] suggested an extension of the N-configuration to a system with an arbitrary even number of states, where the resonant enhancement of not only $\chi^{(3)}$ but also higher-order nonlinearities could be attained. The main problem with these schemes is related to the phase mismatch between different photons, which arises because the group velocities are different for different pulses causing problems in the traveling wave configuration. Recently, two other schemes, free from these shortcomings and without significant optical losses, have been proposed. The five-level M-configuration and generalized multistate systems with high-order atomic coherences created by two or more driving fields have been discussed by different authors [Greentree 2002; Matsko 2003; Ottaviani 2003]. In [Petrosyan 2004] a simple and robust approach has been developed for fully entangling two single-photon pulses in an optically dense medium of four-level atoms with tripod configuration, using only one driving field. Another important advantage of the tripod and M-type level configurations is that in these systems the quantum phase gates can be created on the Zeeman sublevels of alkali atoms, resolved in the magnetic field. Moreover, due to high-order atomic coherences in the M- and generalized multistate-configurations, the nonlinear dispersion of these systems can be even higher than in the Λ-system and this nonlinearity increases as the number of atomic level increases [Greentree 2002]. This can be used for constructing multiphoton phase gates and for creating quantum memory elements, which is essential for coherent processing of quantum information and for distant quantum communication. However, an important question that arises here is how the decoherence rate can be measured in these systems. While in the Λ-scheme the atomic memory can be generated and detected even via linear optics, which is already realized in recent experiments [Kuzmich 2003; van der Wal 2003], as well as the coherence between magnetic sublevels in tripod configuration can be accurately measured by the well-known technique of STIRAP [Vewinger 2003], the situation is much more complicated in the case of the M-configuration and generalized systems, where the high-order coherences require multiphoton interactions for the production and detection. In this paper we present a method in which a single laser beam of sub-mW power can be used for the nonlinear interactions required to produce and probe the high-order atomic coherences. It is based on nonlinear magneto-optical (Faraday) rotation (NMOR) with frequency-modulated light [Budker 2002 (b)]. Using this technique, we have measured the nonlinear Faraday signal (NFS) corresponding to four-photon atomic coherence in the M-system. The method is also usable for selective control of higher order atomic coherences.

In the next section we describe our approach to distinguish the detection of NFS associated with a given order atomic coherence. We then discuss the experimental results obtained for the four-photon atomic coherence in a ^{87}Rb vapor-cell with antirelaxation coating. In particular, from the measured dependence of the widths of Faraday resonances on light power, we were able to find the damping rate of four-photon atomic coherence. We present a low-light-power theory of NMOR with frequency modulated light, the results of which are in a good qualitative agreement with the experimental data.

1. The approach

We consider the NMOR in coherent atomic media, where the basic mechanism of NMOR is the laser-induced coherence between the Zeeman sublevels of atomic ground state and, hence, the detected NFS is sensitive to the damping rate of atomic coherence. An atomic transition is chosen such that both Λ- and M-systems are created. Under usual conditions, the contributions of these systems to the Faraday signal cannot be separated, because their manifestations are similar. On the other hand, it is well known that for a given state the highest order atomic coherence is uniquely associated with the atomic polarization moment (PM) of the same order. This means that if we are able to detect the NMOR signal separately from different PM, the corresponding atomic coherence can selectively be detected. We have shown that this method is realizable with frequency-modulated light.

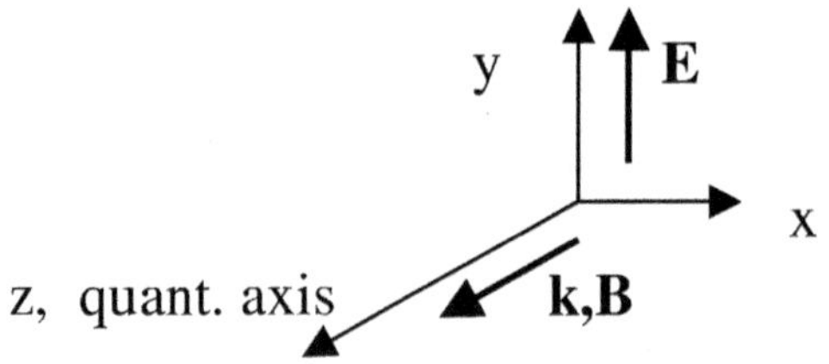

Figure 1 In the atomic coherence picture, the quantization axis **z** is parallel to the light propagation direction and coincides with the orientation of magnetic field: $\mathbf{z} \parallel \mathbf{k}$, $\mathbf{B}$ and $\mathbf{z} \perp \mathbf{E}$.

A convenient experimental scheme for the selective observations of different-order atomic coherences is the $F = 2 \rightarrow F' = 1$ transition in ^{87}Rb atoms with Faraday geometry setup (Fig. 1) that uses a single linearly polarized laser beam. In order to better understand the connection between the light-induced atomic coherences and atomic PM, consider first the atomic coherence picture, where the quantization axis is chosen parallel to the light propagation direction and coincides with the orientation of magnetic field $\vec{B}$. In this case, the linearly polarized field $\vec{E}$ can be represented as a superposition of circularly polarized components $\vec{E}_+$ and $\vec{E}_-$, so that on the transition $F = 2 \rightarrow F' = 1$ both Λ-(dashed) and M-systems (solid) are formed (Fig. 2).

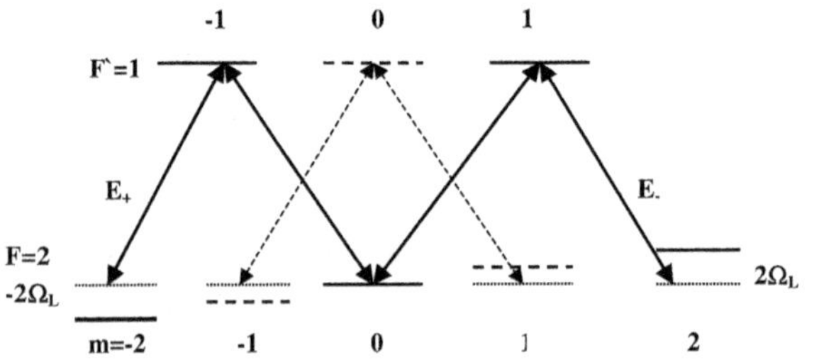

Figure 2 Energy level configuration for $F = 2 \rightarrow F' = 1$ transition in atomic coherence picture. The gyromagnetic factor of the upper level is assumed zero. Ω_L is the Larmor frequency.

The Λ-configuration consists of two ground magnetic sublevels $m = \pm 1$ and of the upper level $F' = 1$, $m = 0$ (Fig. 3). Under the EIT conditions $|\Omega_\pm|^2 \gg \gamma_0\,\gamma$, and at zero magnetic field, an antisymmetric superposition of two ground sublevels $|\mathrm{DS}\rangle = (|-1\rangle - |+1\rangle)/\sqrt{2}$, called the dark state (DS), is created with maximal coherence $\rho^{\mathrm{max}}_{-1,\,1} = -1/2$ between the two sublevels. Here $\Omega_\pm$ are the Rabi frequencies of the circular components of the field on the transitions $|F = 2,\ m\rangle = \mp 1 \rightarrow |F' = 0,\ m = 0\rangle$, respectively, γ_0 is the spontaneous decay rate of the upper level and γ is the ground state atomic coherence relaxation rate. If the atoms are initially in an arbitrary state, after some finite interaction time they are optically pumped into the DS where the linear absorption of resonant atomic medium vanishes, keeping nonlinear dispersion at a very high level, which is governed by the coherence $\rho_{-1,\,1}$.

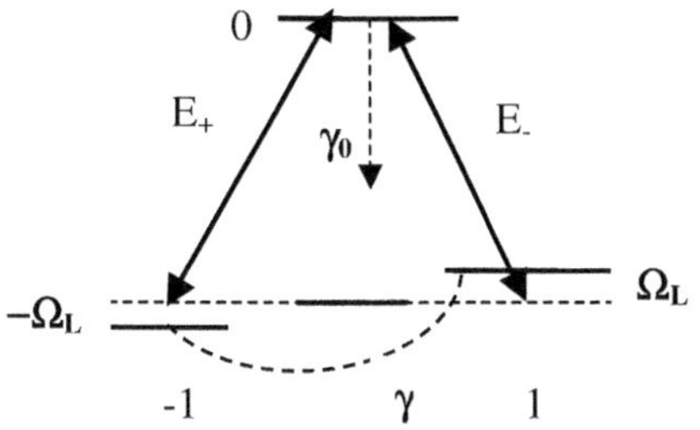

Figure 3. Three-level Λ-configuration in the presence of a rate.

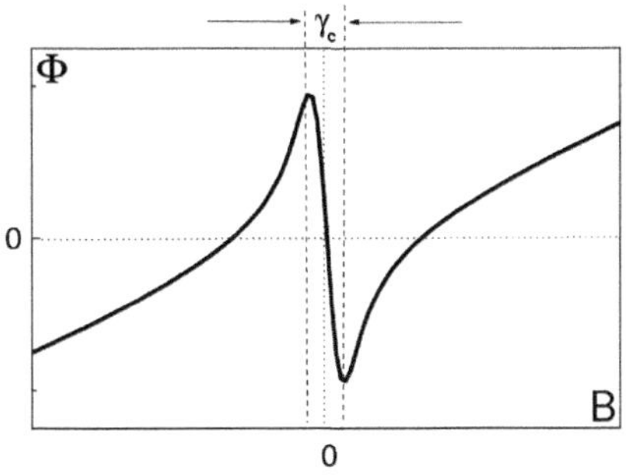

Figure 4. Dispersive-Lorentzian dependence of NMOR on the longitudinal magnetic-field.

The magnetic field destroys the DS and shifts the Zeeman sublevels by an amount proportional to the Larmor frequency $\Omega_L = \mu_B\, g\, B/\hbar$, where μ_B is Bohr magneton and g is the Lande factor for the ground state $F = 2$. We assume that $|\Omega_L| \ll \gamma_0, \Omega_\pm$. This produces a strong nonlinear dispersion for the two circular components of the field, while maintaining their absorption very small. As a result, both components acquire intensity dependent phase shifts of different signs $\varphi_+ = -\,\varphi_-$ which leads to the nonlinear Faraday rotation with the rotation angle $\Phi = (\varphi_+ - \varphi_-)/2$ showing the dispersive dependence on magnetic field $\vec{B}$ (Fig. 4). In the low-power limit, the width of this resonance is given by the decoherence rate γ in the Λ-system. Similar considerations for the M-system lead to the same dispersive behavior of Φ on the magnetic field. The distinctive feature of nonlinear dispersion in the M-scheme is that the four-photon atomic coherence $\rho_{-2,\,2}$ between the states with $m = \pm 2$ (Fig. 2) becomes important. Nevertheless, if the decoherence origin in both schemes is the same, the contributions from Λ- and M-systems are identical provided that the ellipticity of the input field is negligibly small [Matsko 2003] and, therefore, these contributions cannot be distinguished under usual experimental conditions.

To describe the same physical situation in terms of atomic PM, we represent the atomic density matrix ρ through the state multipoles $\rho_q^{(k)}$, which are the PM components with $q = -k, \ldots, k$ and $k = 0, \ldots, 2F$ and, for a state with total angular momentum F, are related to the atomic coherences $\rho_{m,\,m'}$ by

$$\rho_q^{(k)} = \sum_{m,\,m'=\,-F}^{F} (-1)^{F-m'} \left\langle F,\, m,\, F,\, -m'|k,\, q\right\rangle \rho_{m,\,m'}. \qquad (1)$$

The important result following from Eq. (1) is that the highest atomic coherence $\rho_{m,\, m'}$ with $|\Delta m| = 2F$ is uniquely associated with the highest PM ($k = 2F$) for a given F. In particular, the atomic coherence $\rho_{-1,\, 1}$ is simply related to the quadrupole moment ($k = 2$) for $F = 1$, while $\rho_{-2,\, 2}$ is related to the hexadecapole moment ($k = 4$) for $F = 2$. At the same time, for an atomic system with $F = 2$, the laser field induces both the quadrupole and hexadecapole moments. However, taking into account that in the picture with the quantization axis taken along the direction of the linear polarization of the light (y axis) the PM transform under spatial rotations independently from each other, the different axial symmetries of the PM allow one to selectively detect the corresponding atomic coherences. In this picture, for $F = 1 \rightarrow F' = 0$ transition, the laser field optically pumps the atoms into the $|-1\rangle$ and $|+1\rangle$ sublevels (Fig. 5), and the system evolves into a state with quadrupole moment, which for zero magnetic field is aligned along the linear polarization of the light. This situation corresponds to the maximal second-order atomic coher-

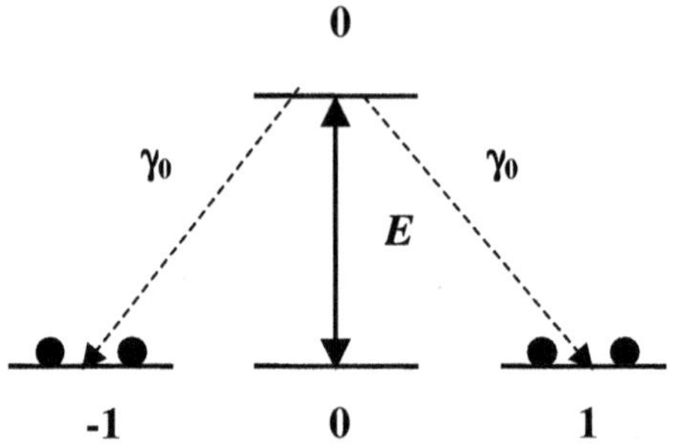

Figure 5 Energy level configuration for the $F = 1 \rightarrow F' = 0$ transition with a linearly polarized optical field, when the quantization axis is parallel to the optical polarization.

ence $\rho^{\max}_{-1,1}$ in the Λ-system. In a magnetic field, the polarized atoms undergo the Larmor precession, and eventually, as a result of the competition between the two processes of optical pumping and Larmor precession, their polarization direction tips from initial y direction towards x by an angle φ depending on the laser intensity and magnetic field. Thus, the propagating linear polarized field experiences an optical rotation by the angle $\Phi \sim \varphi\, L\, N$ (Fig. 6), where L is the length of the atomic sample and N is the atomic number density. Similarly, in the case of $F = 2 \rightarrow F' = 1$ transition, the state with hexadecapole moment is also created, which under optical pumping acquires only two-fold symmetry with respect to the rotations around z and in the absence of magnetic field is oriented with its symmetry axis along the linear polarization of the field. It is precisely this fact that explains why at nonzero magnetic field the quadrupole and hexadecapole moments give inseparable contributions into NFS, though, as has been mentioned earlier, they are decoupled under spatial rotations.

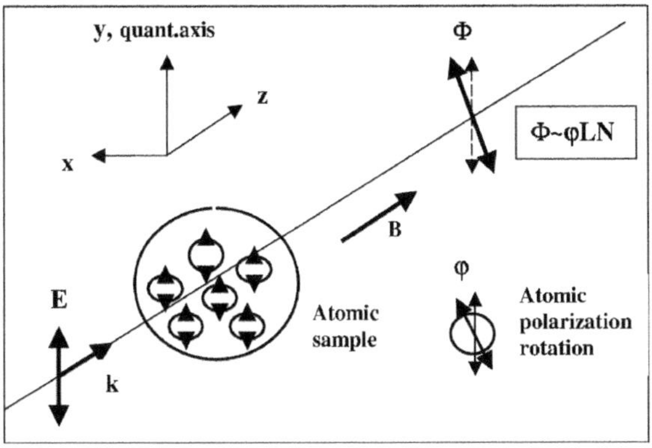

Figure 6 Schematic diagram of NMOR with the quantization axis taken along the linear polarization of the light.

The idea of selective detection of the nonlinear Faraday signal induced by the different PM is based on the use of frequency modulated (FM) light. Suppose that the light frequency varies in time as

$$\omega_L = \omega_0 - \frac{\Delta_0}{2}\,(1 - \cos\Omega_m t) \tag{2}$$

where ω_0 is the frequency of atomic transition, Δ_0 and Ω_m are the amplitude and frequency of the modulation. As the laser frequency is modulated, the optical pumping rate changes depending on the instantaneous detuning of the laser from the atomic transition (Fig. 7 (b)). At the same time, in the presence of magnetic field, the polarized atoms undergo Larmor precession with the frequency Ω_L. Therefore, when the optical pump rate is synchronized with the precession of atomic polarization, a resonance occurs and the atomic medium is pumped into a polarized state whose axis rotates at Larmor frequency. As a result, the optical properties of the medium are modulated at the frequencies $\Omega_m = k\,\Omega_L$, due to the symmetry of atomic PM with rank k. In particular, the resonances occur at frequency $2\Omega_L$ for the quadrupole moment and at $4\Omega_L$ in the case of hexadecapole moment.

It is worth noting that the four-fold symmetry of hexadecapole moment is revealed only at the synchronous pumping and, what is important, at time moments when the hexadecapole moment is precisely aligned with one of its symmetry axis along the linear polarization of the field and the atomic coherence $\rho_{-2,\,2}$ in the M-system has its maximal value. The periodic change of the optical properties of atomic medium modulates the angle of light polarization that leads to the FM NMOR resonances. If the time-dependent optical rotation is measured at the first harmonic of Ω_m, a resonance is seen when $\Omega_m = k\,\Omega_L$ which allows one to separate the NFS produced by different atomic PM. Indeed, in the experiment the in-phase and quadrature amplitudes of optical rotation,

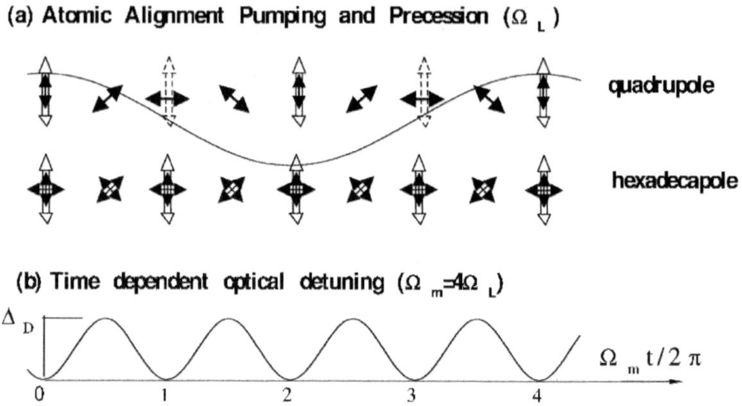

Figure 7. (a) Origin of the NMOR resonances at $\Omega_m = 4\Omega_L$. Atomic alignment precesses with Larmor frequency Ω_L (upper curve). Linearly polarized light periodically interacts with the atoms at frequency $2\Omega_L$ for the quadrupole moment and at $4\Omega_L$ in the case of hexadecapole moment. (b) Time dependence of optical detuning $\Delta_l = \omega_0 - \omega_l$ from the atomic resonance.

$$\varphi_j^{(in)}(L) = \int_0^T dt\ \Phi(L,\ t)\ \cos(j\ \Omega_m\ t) \tag{3a}$$

$$\varphi_j^{(out)}(L) = \int_0^T dt\ \Phi(L,\ t)\ \sin(j\ \Omega_m\ t) \tag{3b}$$

of the NFS at different j harmonics of Ω_m are detected as a function of longitudinal magnetic field, where $T \gg \Omega_m^{-1}, \gamma^{-1}$ is the measurement time. Then, as follows from Fig. 7 (a), the first harmonic signal ($j = 1$ in Eqs. (3a) and (3b)) contains only the contribution of hexadecapole moment at $\Omega_L = \Omega_m/4$ and, hence, it can selectively be measured.

2. Experiment

We have applied the method outlined above to the selective generation and study of polarization moments of ^{87}Rb atoms contained in a vapor cell with antirelaxation coating. Our experimental setup is shown in Fig. 8 [Yashchuk 2002]. The cell is placed between a polarizer and an analyzer oriented at $\approx 45°$ with respect to each other and contains an isotopically enriched sample of ^{87}Rb atoms with a density $\approx 7.10^9$ cm^{-3} at 20°C. The central laser frequency is tuned near various hyperfine structure components of the D_1 line. The typical light power is a few hundred μW and the laser beam diameter is ~ 3 mm. The laser frequency is modulated at $\Omega_m/(2\pi)$ from 50 Hz to 1 kHz, and the frequency modulation amplitude is ~ 40 MHz (peak to peak). The vapor cell is

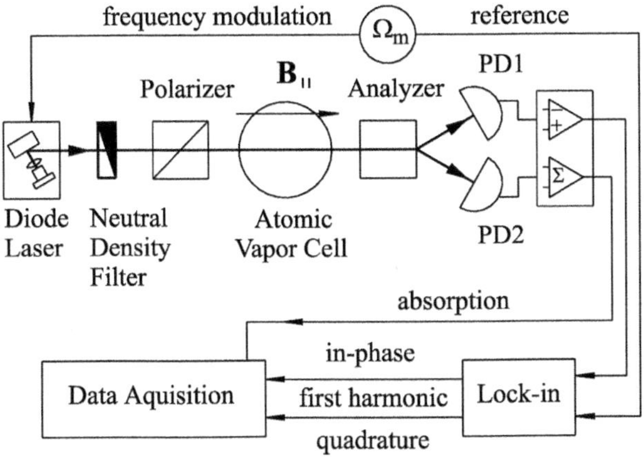

Figure 8 Simplified diagram of the FM NMOR setup. The balanced polarimeter incorporating the polarizer, analyzer, and photodiodes PD1 and PD2 detects signals due to the time-dependent optical rotation of linearly polarized frequency-modulated light.

10 cm in diameter and has an antirelaxation coating and no buffer gas [Alexandrov 2000]. The cell is surrounded with four layers of magnetic shielding. A system of coils inside the innermost shield is used to compensate for the residual fields (at a level $\leq 0.1\ \mu$G) and first-order gradients and to apply a well-controlled, arbitrarily directed magnetic field to the atoms. This allows the observation of FM NMOR resonances with magnetic-field widths of about 1 μG in the low-light-intensity limit.

Fig. 9 shows the magnetic-field dependence of the observed NFS. The central laser frequency was tuned to the low-frequency slope of the $F = 2 \rightarrow F' = 1$ absorption line. At relatively low light power, Fig. 9 (a), there are three prominent resonances: one at $B = 0$, and two corresponding to $2\Omega_L = \Omega_m$. Much smaller signals, whose relative amplitudes rapidly grow with light power, Fig. 9 (b) and 9 (c), are seen at $\Omega_L = \Omega_m$, the expected position of the hexadecapole resonances.

Light-intensity dependences on the detected signals are shown in Fig. 10. The observed low-intensity asymptotic of these curves, which show saturation at higher intensities, scale approximately as I^2 and I^4. These power dependences and many other salient features of the observed resonances can be understood with the help of a simple model of the $F = 2 \rightarrow F' = 1$ transition with independent pumping, evolution in a magnetic field, and probing, in which, to a lowest order, the quadrupole and hexadecapole moments correspond to the first ($\rho^{(2)} \propto I_{pump}$) and second ($\rho^{(4)} \propto I_{pump}^2$) order effects in the light intensity, respectively. The component of the signal due to the induced optical rotation in the probe beam proportional to the quadrupole moment goes as $I_{probe}\, \rho^{(2)}$, while that proportional to the hexadecapole moment goes as $I_{probe}^2\, \rho^{(4)}$. These results coincide with the theoretical predictions for the intensity dependence of NFS amplitudes in the low-power limit (see below).

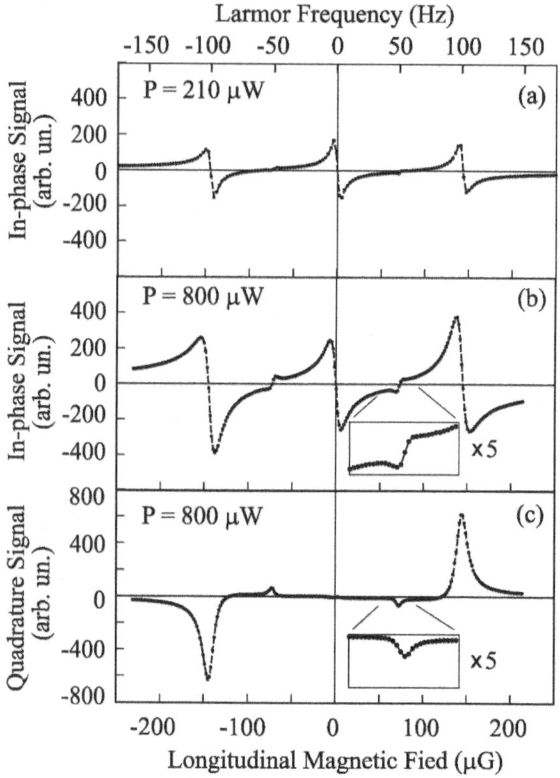

Figure 9 Magnetic-field dependence of the FM NMOR signals, showing quadrupole resonances at $B = \pm\ 143.0\ \mu$G, and the hexadecapole resonances at $\pm\ 71.5\ \mu$G. Laser modulation frequency is 200 Hz, and modulation amplitudes is 40 MHz peak to peak; the central frequency is tuned to the low-frequency slope of the $F = 2 \rightarrow F' = 1$ absorption line. Plots (a) and (b) show the in-phase component of the signal at two different light powers; plot (c) shows the quadrature component. Note the increase in the relative size of the hexadecapole signals at the higher power. The insets show zooms on hexadecapole resonances.

Fig. 11 shows the dependence of the resonance widths on the power at a fixed magnetic field. While both the quadrupole and hexadecapole resonances exhibit power broadening, it is much less pronounced in the latter case. This is important for different applications such as magnetometry [Budker 2000] and construction of non-classical states of light, because it allows for operation at higher light powers with better statistical sensitivity. In the zero-power limit, we find that the resonance widths for the two types of resonances tend to the values near 1 μG with the ratio $\Delta B_q/\Delta B_h = 0.94(4)$. The ratio of these widths should be twice the ratio of the light-independent relaxation rates for the PM with $k = 2$ and 4. It can be shown that the width of the nonlinear Faraday resonance corresponding to a PM of rank k goes as $\gamma_k/(kg\ \mu_B)$, γ_k being the rate of light-independent relaxation for a PM of rank k. Thus, we find that the light-independent relaxation rate for the quadrupole is by a factor of $0.47(2)$ smaller than that of the hexadecapole. Relaxation of the PM in our experiment is dominated by the residual relaxation on the paraffin-coated cell walls and spin-exchange collisions between Rb atoms. The electron-randomization collision model (see, for example, [Knize 1998]), predicts that the quadrupole moment relaxes at a rate $3/8$ of that of the hexadecapole moment, which relaxes at the electron-randomization rate. Thus, the observed ratio of the widths is close to the expected.

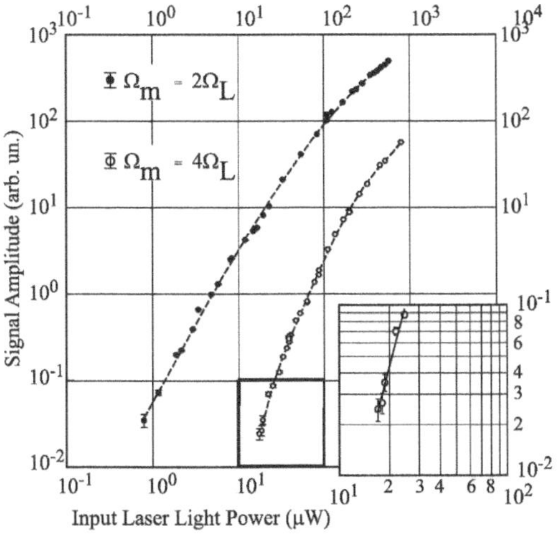

Figure 10 Signal amplitudes vs the input laser power. Filled circles: the quadrupole resonance; open circles: hexadecapole resonance. The inset shows the expanded low-power region for the latter. From these data we determine the initial exponent in the power dependence of the signal to be 1.96(6) and 3.75(37) for the quadrupole and hexadecapole cases, respectively. The expected exponents are 2 and 4.

3. Theory and comparison with experiment

A quantitative theory including complete energy state description of the multilevel structure of Rb atoms and taking into account the saturation effects should be used in the study of the observed effects. As a first step towards a complete theory, we present here the results of our calculations [Malakyan 2004] of quadrupole resonances in a simplified model of three-level Λ-atoms (Fig. 3) in the low-light-power regime that takes into account the Doppler broadening and velocity mixing due to collisions. We find that the results of the theory are in qualitative agreement with the experimental data taken for the ^{87}Rb D_1 line.

We proceed from the master equation for the atomic density matrix including the relaxation of optical and ground state low-frequency coherences. We assume that the upper state spontaneously decays with a rate γ_0, and that the ground state relaxes with a rate γ, due to the escape of the atoms from the light beam, in the case of "transit" effect, or collisions with other atoms or the cell wall, in the case of "wall-induced Ramsey effect" [Budker 2002 (a)]. In the low-light-power limit, the atomic populations are essentially unperturbed by the light. Under these assumptions, the temporal equation for the ground-state coherence is solved, taking into account the periodic time-dependence of the light Rabi frequency. An important feature of the vapor cell with antirelaxation coating is that the ground-state coherence of each velocity group becomes the velocity-averaged quantity due to velocity mixing induced by the collisions of the atoms with the cell walls. We then calculate the medium dispersion for both circular components of the linear polarized light and find the final results for the in-phase and quadrature components (see Eqs. (3a) and (3b)) of the optical rotation in the form

$$\varphi_j^{(in)}(L) = \eta \sum_{n=-\infty}^{\infty} \frac{\gamma\,(2\Omega_L + n\,\Omega_m)\,\overline{a}_n\,(\overline{a}_{n+j} + \overline{a}_{n-j})}{\gamma^2 + (2\Omega_L + n\,\Omega_m)^2} \tag{4a}$$

$$\varphi_j^{(out)}(L) = \eta \sum_{n=-\infty}^{\infty} \frac{\gamma^2\,\overline{a}_n\,(\overline{a}_{n+j} - \overline{a}_{n-j})}{\gamma^2 + (2\Omega_L + n\,\Omega_m)^2} \tag{4b}$$

where $\overline{a}_n$ are the velocity-averaged coefficients which depend on frequency modulation amplitude Δ_0, and the amplitude factor η is defined by

$$\eta = -\frac{1}{6\pi}\frac{\Omega^2\,\gamma_0}{\Gamma_D^2\,\gamma}\,\lambda^2\,NL \tag{5}$$

with $\Omega = |\Omega_\pm|$, Γ_D being the Doppler width and λ the wavelength of the light.

Each term in the sums in Eqs. (4a) and (4b) corresponds to a resonance at $\Omega_L/\Omega_m = -\,n/2$. Near each resonance the in-phase signal is dispersive in shape, whereas the quadrature signal is a Lorentzian with the same width γ. At the same time, in the considered limit of low-light power, the amplitudes of both components appear in the first order in light intensity I and, hence, the observed signals must scale as I^2. The relative amplitudes of the resonances are determined by the ratio of the modulation depth to the Doppler width, Δ_0/Γ_D, and the normalized average detuning $(\omega_0 - \omega_l)/\Delta_0$.

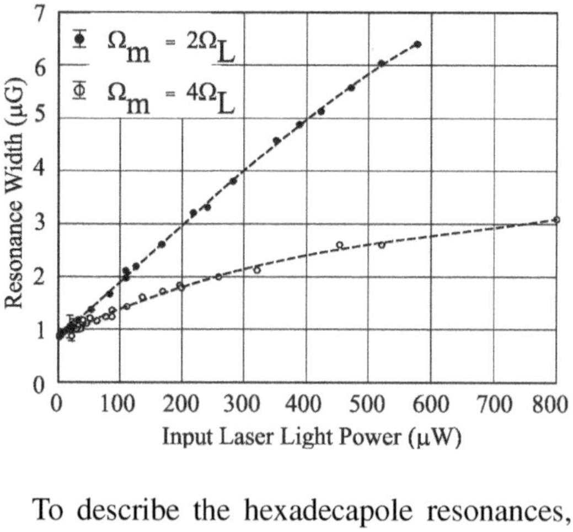

Figure 11 Resonance width vs the input laser power. Filled circles: the quadrupole resonance; open circles: hexadecapole resonance. The widths extrapolated to zero light power are found to be $\Delta B_q = 0.848(4)$ μG and $\Delta B_h = 0.904(33)$ μG for the quadrupole and hexadecapole cases, respectively.

To describe the hexadecapole resonances, a more complicated theory is needed, because the presence of the hexadecapole moment requires ground-state angular momentum ≥ 2 and second-order interactions with the light.

Nevertheless, some properties of these resonances can be deduced from the general considerations. In particular, in the low light-power limit, the optical rotation amplitudes of the hexadecapole resonances are obviously proportional to the square of the light intensity I^2, as they are induced by the four-photon atomic coherence. The corresponding signals scale as I^4, as observed experimentally (Fig. 11). Moreover, the light-independent relaxation rate for hexadecapole resonances must be large as compared to γ, if the relaxation of atomic polarization is conditioned by the processes which are different for the PM of different rank k. As we have seen above, this expectation is confirmed experimentally.

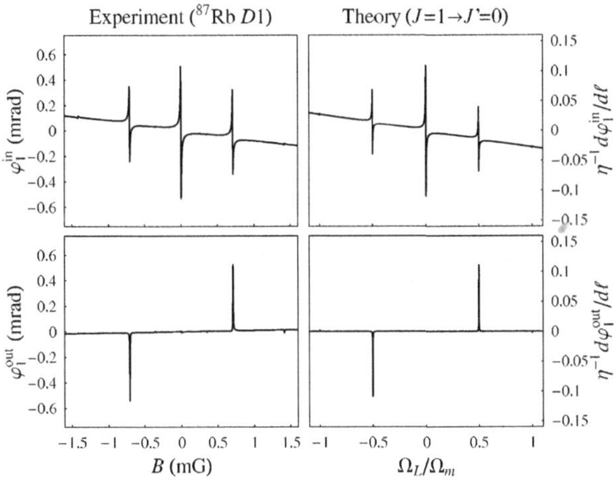

Figure 12. Measured (left column) [Budker 2002 (b)] and calculated (right column) [Malakyan 2004] in-phase (top row) and quadrature (bottom row) first-harmonic amplitudes of quadrupole resonances. The experimental signals, plotted as functions of the magnetic field B applied along the light propagation direction, are obtained with light tuned to the wing of the $F = 2 \to F' = 1$ absorption line of the ^{87}Rb D_1 spectrum. The laser power is 15 μW, beam diameter is $\sim$ 2 mm, $\Omega_m = 2\pi \times 1$ kHz, and modulation amplitude is $2\pi \times 220$ MHz. All resonances have widths $\sim 1\ \mu$G corresponding to the rate of atomic polarization relaxation in the paraffin-coated cell. The normalized calculated signals $\varphi_j^{(in)}/L$ and $\varphi_j^{(out)}/L$ (Eqs. (4a) and (4b)) are plotted as functions of normalized Larmor frequency Ω_L/Ω_m. For these plots, the parameters $\Delta_0/\Gamma_D = 0.7$, $\Omega_m/\gamma = 500$, $(\omega_0 - \omega_l)/\Delta_0 = 1$ are chosen to match the experimental parameters given above.

In Figs. 12 and 13 we compare the results of the present theory for quadrupole moments with the experimental data obtained for the D_1 line of rubidium [Budker 2002 (b)]. The calculations for our simpler system reproduce many of the qualitative aspects of the experimental data for Rb. As seen, at the center of the in-phase plots of Figs. 12 and 13, zero-field resonances arise, while resonances centered at the magnetic-field values for which $|\Omega_L/\Omega_m| = 1/2$ and

1 correspond to the first and second harmonic quadrupole signals. In the latter case, both dispersively shaped in-phase signals and out-of-phase (quadrature) Lorentzian components are peaked at the centers of these resonances. The resonances also exhibit linear light-power dependence on the optical rotation amplitudes, as predicted by the theory and observed in experiments at low power.

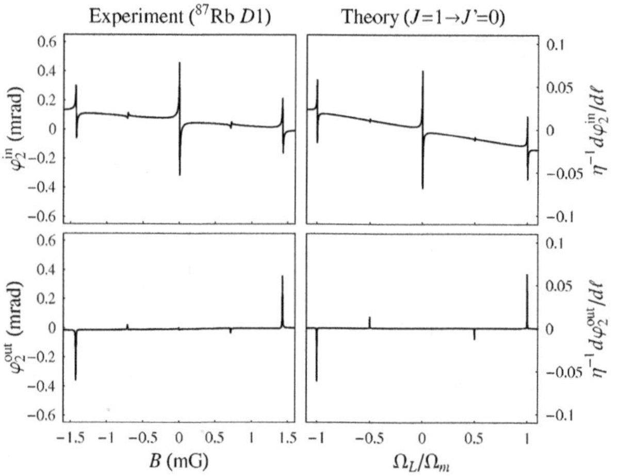

Figure 13. Measured [Budker 2002 (b)] and calculated [Malakyan 2004] second-harmonic amplitudes of the quadrupole resonances (see caption of Fig. 12). For the experimental signals, light is tuned to the wing of the $F = 2 \rightarrow F' = 1$ absorption line of the ^{87}Rb D_1 spectrum. The laser power is 15 μW, beam diameter is $\sim$ 2.5 mm, $\Omega_m = 2\pi \times 1$ kHz, and modulation amplitude is $2\pi \times 440$ MHz. All resonances have widths $\sim$ 1 μG corresponding to the rate of atomic polarization relaxation in the paraffin-coated cell. The parameters for the theoretical signals are $\Delta_0/\Gamma_D = 1.4$, $\Omega_m/\gamma = 500$, $(\omega_0 - \omega_l)/\Delta_0 = 0.2$.

Conclusions

In conclusion, we have presented a reliable method for selective production and detection of high-order atomic polarization moments based on the nonlinear magneto-optical effects with frequency modulated light. This method can be used for the selective control of higher order atomic coherences in multilevel systems exploiting large Kerr nonlinearities for the construction of all-optical quantum phase gates.

Acknowledgments

This work has been supported by the U. S. Office of Naval Research, NSF, U. S. - Armenian Bilateral Grant No. CRDF AP2-3213 / NFSAT PH 071 - 02, U. S. - Poland NRC Twinning grant and NATO.

STRONG ENTANGLEMENT OF BRIGHT LIGHT BEAMS IN CONTROLLED QUANTUM SYSTEMS

G. Yu. Kryuchkyan and H. H. Adamyan
Yerevan State University, A. Manookyan 1, 375049, Yerevan, Armenia
Institute for Physical Research, National Academy of Sciences, Ashtarak-2, 378410, Armenia
gkryuchk@server.physdep.r.am, adam@unicad.am

Abstract We investigate some aspects of continuous-variable (CV) entanglement of light-ields in controlled quantum dissipative systems. We discuss entangling resources of a nondegenerate optical parametric oscillator (NOPO) in transition through the generation threshold as well as in the regime of lasing. We review our proposal for the creation of CV entanglement in the presence of phase-localizing processes and discuss properties of entangled light-fields with well-defined distinct phases in application to quantum communications. We propose a new approach for the generation of strongly entangled Einstein-Podolsky-Rosen states of intense light beams based on the time-modulation of quantum dissipative dynamics. We demonstrate an essential improvement of the degree of entanglement in time-modulated NOPO in comparison with the ordinary one, if the frequency of modulation is close to the decay rate of the dissipative processes.

Introduction

Continuous-variable (CV) entangled states of light beams provide excellent tools for testing the foundations of quantum physics as well as are of increasing interest for quantum information technologies [Braunstein 2003] and ultra-precise physical measurements such as gravitational-wave detectors. CVs are fascinating alternatives to discrete ones and lend themselves well to quantum optical implementations. Among the quantum information applications are the realizations of quantum teleportation, dense coding, quantum key distribution, and the proposals for CV quantum cryptography [Furusawa 1998; Braunstein 2000 (a); Li 2002 (a); Ralph 2002; Bowen 2003 (b); Zhang 2003]. The efficiency of quantum information schemes and quantum measurements significantly depends on the degree of entanglement of the employed states. On the other hand, in the majority of real applications, bright stable light beams are required. Important examples are the entanglement-assisted interferometric high-precision measurements, where the phase-stability of light fields as well

V.M. Akulin et al. (eds.), Decoherence, Entanglement and Information Protection in Complex Quantum Systems, 105–126.

as the stabilization of the optical paths and indistinguishably, *i.e.*, frequency and polarization degeneracy, are the main requirements for their implementations. It is therefore highly desirable to elaborate on the reliable sources of light beams possessing the mentioned properties.

The recent development of CV quantum information is stipulated mainly by the preparation of two-mode squeezed states, which approximate the EPR (Einstein-Podolsky-Rosen) entangled states. The generation of these states requires simultaneous squeezing of two physical variables corresponding to canonically conjugate operators, such as photon number sum (difference) and phase difference (sum), or amplitude-quadrature sum (difference) and phase-quadrature difference (sum). An important example in the entangled EPR state of orthogonally polarized but frequency degenerate beams, efficiently generated via nonlinear optical process of parametric down-conversion. This has been pointed out by Drummond and Reid in [Reid 1989; Reid 1998], and has been demonstrated experimentally by Kimble *et al.* [Ou 1992; Pereira 2002] for CV, employing sub-threshold nondegenerate optical parametric oscillator (NOPO). Then a CV entanglement source was built from two independent quadrature squeezed beams combined on a beam-splitter [Furusawa 1998; Bowen 2003 (b); Zhang 2003]. It should be noted that so far most experimental realizations of the CV entanglement on NOPO's have only been operated below threshold. A natural next step is the extension of these investigations to laser-like systems generating entangled bright-light states. Unfortunately, both theoretical and experimental studies of these problems are very complicated and only a few examples are known up to now. Much effort has been devoted to the study of intensity correlated twin beams from NOPO above threshold [Reynaud 1987; Mertz 1991; Rarity 1992; Peng 1998]. Further experimental studies of a bright two-mode entangled state from cw nondegenerate optical parametric amplifier have been made in Refs. [Zhang 2000; Li 2002 (b)]. An ultrastable NOPO, that emits twin beams classically phase-locked at exact frequency degeneracy, has been presented in [Feng 2003; Feng 2004]. In spite of these important developments in the investigation of entangled light, the generation of stable, bright light beams with high degree of CV entanglement up to now meets serious problems. One of these is the degradation of entanglement due to uncontrolled dissipation and decoherence processes which, in particular, takes place during the intracavity generation of light beams. The other problem is the phase-instability of light fields in the regime of above-threshold generation.

Recently several approaches have been proposed to control the quantum dissipative dynamics. It is natural to ask whether the analogous approaches can be used for the improvement of the degree of entanglement of the above mentioned schemes. This paper will give a positive answer. Our analysis also offers new insights into crossovers between two phenomena, the generation of CV entangled light beams and the control of quantum dissipative dynamics. The present paper is a three-pronged contribution to the theory of CV entangled bright light.

Entangling resources of NOPO

The first goal of the paper, explored in Sec. 2, is to discuss physical properties and the presence of CV entanglement for NOPO in the transition through the generation threshold as well as in the regime of lasing. There are various questions that emerge in the study of these problems. Because NOPO displays both monostable and bistable regimes, it is important to understand how the properties of the entanglement depend on the operational regimes. What are the peculiarities of entanglement in the vicinity of threshold? Will entanglement take place in the regime of lasing or how far it can be extended into the high intensity domain? Answering these questions is extremely important for a deeper understanding of quantum entanglement, and also for the prospects of creating an entangled light laser.

Control over continuous-variable entanglement by time-modulation

The second goal of this paper, explored in detail in Sec. 3, is to develop approaches for the creation of non-limited entanglement. Note that because the number of experimentally accessible and feasible multi-wave nonlinear interactions leading to formation of entanglement is rather limited, the class of schemes that can be elaborated in practice is restricted. In this paper we point out that the class of currently proposed schemes for the generation of intense light beams with high degree of entanglement may significantly be extended, if instead of the monochromatic pumping we consider periodically modulated pump fields. As a realization of this idea, we propose in this paper a novel scheme of NOPO in a cavity, driven by an amplitude-modulated electromagnetic field and stress that this scheme provides highly efficient mechanism for the improvement of the degree of CV entanglement, even in the presence of dissipation and cavity induced feedback.

CV entangling resources are usually analyzed in terms of two-mode squeezing through the variances of the quadrature amplitudes of two generated modes. In NOPO, the two-mode integral squeezing, which characterizes the CV entanglement, reaches only 50% relative to the level of vacuum fluctuations, if the pump field intensity is close to the generation threshold [Dechoum 2003; Kryuchkyan 2004]. We will demonstrate in Sec. 1, that the level of two-mode squeezing in the proposed scheme is not limited, which indicates a high degree of quadrature entanglement obeying the condition of EPR-like paradox criterion as quantified by Reid and Drummond [Reid 1989; Reid 1998].

It seems intuitively clear that such an achievement is due to the control of quantum dissipative dynamics through the application of a suitably tailored, time-modulated driving field. Indeed, some interesting examples of suppression of quantum decoherence by the modulation of system parameters have been considered in [Viola 1998; Vitali 1999]. An improvement of sub-Poissonian statistics of an anharmonic oscillator by the application of amplitude-modulated pump field have been demonstrated in [Kryuchkyan 2002; Kryuchkyan 2003].

Marriage of continuous-variable entanglement and self-phase locking

The other problem of our interest is the phase diffusion leading to instability of generated modes in NOPO. It is well known that each of the orthogonally polarized and frequency degenerate fields generated by NOPO is a field of zero-mean values. The phase sum of generated modes is fixed by the phase of the pump laser, while their phase difference undergoes a phase diffusion process [Reid 1989; Reid 1998] stipulated by vacuum fluctuations. As a rule, the NOPO phase diffusion noise is substantially greater than the shot noise, which limits the usage of NOPO in precision phase-sensitive measurements. Various methods based on phase locking mechanisms [Graham 1968; Lane 1988; Nabors 1990; Kryuchkyan 1993; Kryuchkyan 1994; Mason 1998; Fabre 1999; Kryuchkyan 2001; Zondy 2001] have been proposed for reducing such phase diffusion. It was shown in [Graham 1968] that the phase difference noise should become dominated by the Schawlow-Townes drift of the OPO phase difference. The self-phase locking process was observed in type-I OPO [Nabors 1990]. In the comparatively simple scheme realized in the experiment [Mason 1998], self-phase locking was achieved in NOPO by adding an intracavity quarter-wave plate to provide polarization mixing between two orthogonally polarized modes of the subharmonics. The evidence of self-phase locking was provided there by the high level of phase coherence between the signal and idler fields. Following this experiment, the semiclassical theory of such NOPO was developed in [Fabre 1999] and its further more detailed consideration has been presented in [Longchambon 2003 (a)]. Recently, the schemes of multiphoton parametric oscillators based on cascaded down-conversion processes in $\chi^{(2)}$ media placed inside the same cavity and showing self-phase locking have been proposed [Kryuchkyan 2001]. As was demonstrated in [Zondy 2001], the system based on combination of OPO and second harmonic generation also displays self-phase locking. The formation of self-phase locking and its connection with squeezing in the parametric four-wave mixing under two laser fields has been demonstrated in [Kryuchkyan 1993; Kryuchkyan 1994]. An important characteristic of self-phase locked devices concerns the phase structure of generated subharmonics. Indeed, the formation of the variety of distinct phase states under self-phase locked conditions has been obtained in the mentioned Refs. [Kryuchkyan 1993; Kryuchkyan 1994; Mason 1998; Fabre 1999; Kryuchkyan 2001; Zondy 2001]. It was recently noted that the schemes involving phase locking are potentially useful for precise interferometric measurements and optical frequency division because they combine fine tuning capability and stability of type-II phase matching with effective suppression of phase noise. That is why we believe that it will be interesting to consider phase locked dynamics also from the perspective of quantum optics and, in particular, from the standpoint of production of CV entanglement. In this paper we discuss self-phase locked CV entangled states which have been recently investigated by two groups [Adamyan 2004 (a); Longchambon 2003 (b)]. We develop the quantum theory of self-phase locked NOPO, with decoherence included, in application to the generation of such entangled states. This scheme is based on the combination of two processes, namely, type-II parametric down-conversion and linear polarization mixing with cavity-induced feedback.

Further motivation for such a task is connected to the problem of experimental generation of bright entangled light. So far, to the best of our knowledge, there is no experimental demonstration of CV entanglement above the threshold of NOPO. One of the principal experimental difficulties in the advance toward a high-intensity level is the impossibility to control the frequency degeneration of modes above the threshold. We hope that the usage of phase locked NOPO may open a new interesting possibility to avoid this difficulty.

1. Model of NOPO and basic equations

In this section we give a brief description of the generalized scheme of NOPO. As an entangler we consider the combination of two processes in a triply resonant cavity, namely, the type-II parametric down-conversion in $\chi^{(2)}$-medium and polarization mixing between subharmonics in lossless symmetric quarter-wave plate. The Hamiltonian describing intracavity interactions is

$$H = H_{sys} + \sum_{i=1}^{3}\left(a_i\,\Gamma_i^{\dagger} + a_i^{\dagger}\,\Gamma_i\right) \tag{1a}$$

$$\begin{aligned} H_{sys} &= \hbar\Big\{i\Big[E\left(e^{i(\Phi_L-\omega t)}\,a_3^{\dagger} - e^{-i(\Phi_L-\omega t)}\,a_3\right) + k\left(e^{i\Phi_k}\,a_3\,a_1^{\dagger}\,a_2^{\dagger}\right.\\ &- \left. e^{-i\Phi_k}\,a_3^{\dagger}\,a_1\,a_2\right)\Big] + \chi\left(e^{i\Phi_\chi}\,a_1^{\dagger}\,a_2 + e^{-i\Phi_\chi}\,a_1\,a_2^{\dagger}\right)\Big\} \end{aligned} \tag{1b}$$

where a_i are the boson operators for the cavity modes ω_i. The mode a_3 at frequency ω is driven by an external field with amplitude E and phase Φ_L, while a_1 and a_2 describe subharmonics of two orthogonal polarizations at degenerate frequencies $\omega/2$. The constant $k\, e^{i\Phi_k}$ determines the efficiency of the down-conversion process. The last term in (1a) describes the mode damping in the cavity in terms of the reservoir operators Γ_i and Γ_i^+, which determine the damping factors γ_i. So, we take into account the detunings of subharmonics Δ_i and the cavity damping rates γ_i and consider the case of high cavity losses for the pump mode ($\gamma_3 \gg \gamma_1,\ \gamma_2$), when this mode can be adiabatically eliminated. However, in our analysis we take into account the pump depletion effects.

1.1 Measure of continuous-variable entanglement

Before we turn to the main results of this paper, let us briefly explain how we characterize CV entanglement. We note that the state generated in NOPO is non-Gaussian, *i.e.*, its Wigner function is non-Gaussian [Kheruntsyan 2000]. So far, the inseparability problem for bipartite non-Gaussian states is far from being understood. On the theoretical side, the necessary and sufficient conditions for the separability of bipartite CV systems have been fully developed only for Gaussian states [Duan 2000 (a); Simon 2000; Giedke 2001; Werner 2001]. To characterize the CV entanglement, we address to both the inseparability criterion proposed by Duan *et al.* and Simon [Duan 2000 (a); Simon 2000] and the EPR paradox criterion [Reid 1989; Reid 1998]. These criteria could be quantified by analyzing the variances of a relative distance $V_- = V(X_1 - X_2)$ and total momentum $V_+ = V\left(Y_1 + Y_2\right)$ in terms of the quadrature amplitudes of two modes

$$X_k = X_k\left(\Theta_k\right) = \frac{a_k^\dagger\, e^{-i\Theta_k} + a_k\, e^{i\Theta_k}}{\sqrt{2}},\ Y_k = X_k\left(\Theta_k - \frac{\pi}{2}\right),\ (k = 1,\ 2) \tag{2}$$

where $V(x) = \left\langle x^2 \right\rangle - \left\langle x \right\rangle^2$ is a denotation of the variance. The inseparability criterion, or weak entanglement criterion, for the sum of variances quite generally reads as

$$V_+ + V_- < 2. \tag{3}$$

In the case of the perfect symmetry between subharmonic modes ($\gamma_1 = \gamma_2 = \gamma$, $\delta_1 = \delta_2 = \delta$) this criterion is reduced to the following form $V = V_+ = V_- < 1$, while for the product of variances this criterion has the form $V_+\, V_- = V^2 < 1$. The strong CV entanglement criterion shows that when the inequality

$$V_+\, V_- < \frac{1}{4} \tag{4}$$

is satisfied, there arises an EPR-like paradox. Both criteria have been used to characterize the entanglement mainly in spectral measurements [Bowen 2004]. Contrary to that, we confine ourselves to analyzing only the total intracavity variances.

At this point we must note the difference between the focus of our paper and most of the preceding work devoted to the study of two-mode squeezing. It is an established standard to describe squeezing with the spectra of quantum fluctuations of the considered variables. A recent experiment on spectral investigation of criteria for CV entanglement was presented in [Bowen 2003 (a)]. Unlike that, we aim at analyzing the separability properties of the system solely with the help of criteria (3) and (4) which involve integral characteristics of the NOPO quantum state.

1.2 Fokker-Planck equation for an ordinary NOPO

We perform concrete calculations in the complex P-representation [Drummond 1980; McNeil 1983] in the frame of both probability distribution functions and stochastic equations for the complex c-number variables. We follow the standard procedures of quantum optics to eliminate the reservoir operators and to obtain a master equation for the density operator of the modes. The master equation is then transformed into a Fokker-Planck equation for the P-quasiprobability distribution function. In particular, for an ordinary NOPO and in the case of high cavity losses for the pump mode ($\gamma_3 \gg \gamma$), if in the operational regime the pump depletion effects are involved, this approach yields

$$\frac{\partial P(\overline{\alpha},\, t)}{\partial t} = \Bigg[- \frac{\partial \left(-\overline{\gamma_1}\, \alpha_1 + \kappa\, G\, \alpha_2^{\dagger} \right)}{\partial \alpha_1} - \frac{\partial \left(-\overline{\gamma_1^{\star}}\, \alpha_1^{\dagger} + \kappa\, G^{\dagger}\, \alpha_2 \right)}{\partial \alpha_1^{\dagger}}$$
$$- \frac{\partial \left(-\overline{\gamma_2}\, \alpha_2 + \kappa\, G\, \alpha_1^{\dagger} \right)}{\partial \alpha_2} - \frac{\partial \left(-\overline{\gamma_2^{\star}}\, \alpha_2^{\dagger} + \kappa\, G^{\dagger}\, \alpha_1 \right)}{\partial \alpha_2^{\dagger}}$$
$$+ \kappa\, \frac{\partial^2 G}{\partial \alpha_1\, \partial \alpha_2} + \kappa\, \frac{\partial^2 G^{\dagger}}{\partial \alpha_1^{\dagger}\, \partial \alpha_2^{\dagger}} \Bigg] P(\overline{\alpha},\, t). \tag{5}$$

Here $\overline{\alpha} = (\alpha_1,\, \alpha_1^{\dagger},\, \alpha_2,\, \alpha_2^{\dagger})$, α_i, $\alpha_i^{\dagger}$ $(i = 1,\, 2)$ are the independent complex variables corresponding to the operators a_i, $a_i^{\dagger}$, $\overline{\gamma_j} = \gamma_j - i\,\Delta_j$, $(\overline{\gamma_1} = \overline{\gamma_2} = \overline{\gamma})$, $G = (E - \kappa\, \alpha_1\, \alpha_2)/\overline{\gamma_3}$ and $G^{\dagger} = (E^{\star} - \kappa\, \alpha_1^{\dagger}\, \alpha_2^{\dagger})/\overline{\gamma_3^{\star}}$.

The normally-ordered moments of time-dependent operators are calculated through the P-quasiprobability distribution function as

$$\left\langle a_1^\dagger(t)^k \, a_1(t)^l \, a_2^\dagger(t)^m \, a_2(t)^n \right\rangle = \int d\alpha_1^\dagger \, d\alpha_1 \, d\alpha_2^\dagger \, d\alpha_2 \, P(\overline{\alpha}, \, t) \, \alpha_1^{\dagger k} \, \alpha_1^l \, \alpha_2^{\dagger m} \, \alpha_2^n . \tag{6}$$

1.3 Stochastic equations for NOPO with polarization mixer

In Secs. 3 and 4 we will use the stochastic equations in the positive P-representation for the complex c-number variables α_i and β_i corresponding to the operators a_i and $a_i^\dagger$. For the generalized model of NOPO they read as:

$$\frac{\partial \alpha_1}{\partial t} = -(\gamma_1 + i\,\Delta_1)\,\alpha_1 + k\,\alpha_3\,\beta_2 - i\,\chi\,\alpha_2 + R_1 \tag{7a}$$

$$\frac{\partial \beta_1}{\partial t} = -(\gamma_1 - i\,\Delta_1)\,\beta_1 + k\,\beta_3\,\alpha_2 + i\,\chi\,\beta_2 + R_1^\dagger \tag{7b}$$

$$\frac{\partial \alpha_3}{\partial t} = -\gamma_3\,\alpha_3 + E - k\,\alpha_1\,\alpha_2 \tag{7c}$$

$$\frac{\partial \beta_3}{\partial t} = -\gamma_3\,\beta_3 + E - k\,\beta_1\,\beta_2 \tag{7d}$$

Equations for α_2, β_2 are obtained from (7a) and (7b) by exchanging the subscripts 1 and 2. Our derivation is based on the Ito stochastic calculus and $R_{1,2}$ are Gaussian noise terms with zero means and the following nonzero correlators:

$$\left\langle R_1(t)\,R_2(t')\right\rangle = k\,\alpha_3\,\delta\left(t - t'\right), \tag{8a}$$

$$\left\langle R_1^\dagger(t)\,R_2^\dagger(t')\right\rangle = k\,\beta_3\,\delta\left(t - t'\right). \tag{8b}$$

Recall, that we consider the regime of adiabatic elimination of the pump mode. In this approach the stochastic amplitudes α_3 and β_3 are given by

$$\alpha_3 = \frac{E - k\,\alpha_1\,\alpha_2}{\gamma_3} \quad \text{and} \quad \beta_3 = \frac{E - k\,\beta_1\,\beta_2}{\gamma_3}. \tag{9}$$

Substituting amplitudes α_3, β_3 into Eqs. (7a) and (7b), (8a) and (8b) we arrive at

$$\frac{\partial \alpha_1}{\partial t} = -\left(\gamma_1 + i\,\Delta_1\right)\alpha_1 + \left(\varepsilon - \lambda\,\alpha_1\,\alpha_2\right)\beta_2 - i\,\chi\,\alpha_2 + R_1, \quad (10a)$$

$$\frac{\partial \beta_1}{\partial t} = -\left(\gamma_1 - i\,\Delta_1\right)\beta_1 + \left(\varepsilon - \lambda\,\beta_1\,\beta_2\right)\alpha_2 + i\,\chi\,\beta_2 + R_1^{\dagger}, \quad (10b)$$

$$\langle R_1(t)\,R_2(t')\rangle = \left(\varepsilon - \lambda\,\alpha_1\,\alpha_2\right)\delta\,(t - t') \quad (10c)$$

$$\left\langle R_1^{\dagger}(t)\,R_2^{\dagger}(t')\right\rangle = \left(\varepsilon - \lambda\,\beta_1\,\beta_2\right)\delta\,(t - t') \quad (10d)$$

Here $\varepsilon = k\,E/\gamma_3$, $\lambda = k^2/\gamma_3$. So Eqs. (10a), (10b) and the corresponding equations for α_2, β_2 involve the depletion effect of the pump mode, which leads to the appearance of the above-threshold operational regime.

2. Entanglement in the range of critical quantum fluctuations

In this section we will reproduce the results of the paper [Kryuchkyan 2004], where the CV entangling resources for all operational regimes of nondegenerate OPO were investigated in the most basic and explicit way through the P-complex probability distribution. One of the principal problems in this study is the description of quantum fluctuations. In most theoretical works, nonclassical effects and entanglement resources of nonlinear quantum systems are usually described within linear treatment of quantum noise. It is obvious that such approach does not describe the critical ranges (threshold, point of multistability, *etc.*) where the level of quantum noise increases substantially. More adequate approach has been developed within the framework of exact nonlinear treatment of quantum fluctuations via the solution of the Fokker-Planck equation for the quasiprobability distribution function. Deriving the quasiprobability functions so far has been performed only for a few simple models (see [Kryuchkyan 1996 (a); Kheruntsyan 1997] and references therein). For the NOPO, the exact steady-state solution of the Fokker-Planck equation in the complex P-representation was first obtained in [McNeil 1983] and then for a more general case involving also detunings of the modes and hence the regime of bistability in [Kryuchkyan 1996 (b)].

Let us turn to the integral variances. With an appropriate selection of the phases θ_k of the quadrature amplitudes, namely $\theta_1 + \theta_2 = \arg\langle a_1\,a_2\rangle$ we arrive at the following minimum value of $V(\theta_1,\,\theta_2)$

$$V_{\min} = 1 + 2\left(n - |\langle a_1\,a_2\rangle|\right) \quad (11)$$

where $n = \langle a_1^{\dagger}\,a_1\rangle = \langle a_2^{\dagger}\,a_2\rangle$ is the mean photon number of the modes. The squeezed variance in the steady-state regime can be calculated through the formula (11) and the P-function derived in [Kryuchkyan 1996 (b)]. The results are depicted in Fig. 1, which illustrates the dependence of $V_{\min}$ on the pump amplitude for three values of detunings corresponding to monostable, bistable and interjacent regimes. One can see that for monostable dynamics (curve 1) entanglement is realized in the entire range of pump intensities. In all cases maximal degree of two-mode squeezing $V_{\min} = 0.5$ is achieved within the critical region. Above the critical point, mean photon numbers of the modes increase considerably and variance $V_{\min}$ starts to increase too. For the bistable case (curve 3) the growth of $V_{\min}$ is much faster and larger so that the sufficient criterion for inseparability (3) is not fulfilled. Note that $V_{\min}$ has a sharp peak in the critical range of bistability. In any case, in the far-above threshold region, the CV entanglement of the system is guaranteed. More exactly, the asymptotic value of $V_{\min}$ for all detunings is equal to 0.75.

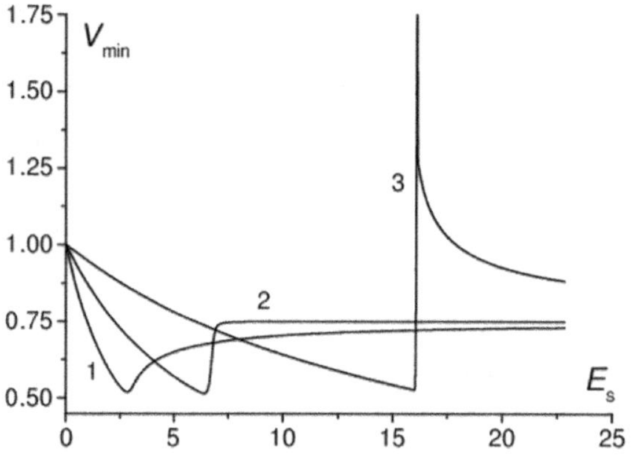

Figure 1 Minimized variance versus dimensionless amplitude of the pump field $E_s = 2k\ E/\gamma\ \gamma_3$ for the parameters: (1) monostable ($\Delta/\gamma = 1$), (2) interjacent ($\Delta/\gamma = 3$) and (3) bistable ($\Delta/\gamma = 7$) regimes ($k/\gamma = 0.5$, $\gamma_3/\gamma = 18$).

Note, that for smaller, experimentally available values of the parameter $\kappa/\gamma \sim 10^{-6} - 10^{-8}$, the behavior of $V_{\min}$ is not changed qualitatively in comparison with the results of Fig. 1.

It is remarkable that the numerical results of Fig. 1 can also be explained analytically. In the far above-threshold range, ($E \gg E_{th}$, where the threshold value of the pump field is $E_{th} = \left|(\gamma - i\Delta)\,(\gamma_3 - i\Delta_3)\right|/\kappa$), a simple analytical result for the squeezed variance was derived

$$\lim_{E\to\infty} V_{\min} = 0.5 - 2\,\delta n. \tag{12}$$

Here the mean photons number of subharmonic modes are represented as the sum $n = n_{cl} + \delta n$ of the semiclassical and quantum parts. Straightforward, but complicated analytical calculations (see details in [Kryuchkyan 2004]) show that $\delta n \to -0.125$ in the limit $E \to \infty$, which leads to the asymptotic value $V_{\min} = 0.75 < 1$. Therefore, as the analysis shows, allowing for quantum fluctuations of arbitrary level, CV entanglement is always achieved in the NOPO above threshold. It is remarkable that a very small quantum correction δn to the semiclassical intracavity photon number plays an essential role in the production of entanglement at the high-intensity level. The obtained result seems interesting as it provides an example of preserving CV entanglement in the high-intensity domain. The numerical and analytical analysis for the quantum critical range of NOPO has provided a clear distinction between the degree of two-mode squeezing for monostable and bistable regimes. Most favorable is the monostable regime, where the CV entanglement can be maintained for all values of the pump intensity and for experimentally available parameters.

3. Non-limited EPR entanglement in time-modulated quantum dissipative dynamics

We have seen that the degree of total two-mode squeezing and hence the CV entanglement are limited in the ordinary NOPO. This is an essential restriction which seems to be carried out for the most devices based on intracavity multimode interactions. Nevertheless, we will show that a different situation is realized for time-modulated systems. In this section we investigate the creation of entangled states of bright light beams obeying the condition of strong Einstein-Podolsky-Rosen-like paradox criterion in the time-modulated quantum dissipative dynamics [Adamyan 2004 (c)]. We consider an ordinary NOPO without polarization mixer for the case where the pump mode a_3 is driven by an amplitude-modulated external field at the frequency $\omega_L = \omega_3$ with amplitude $f(t)$ that is a periodic function with modulation frequency $\delta \ll \omega_L$. The corresponding Hamiltonian (1b) in this case is

$$\begin{aligned} H_{sys} = \; & i\hbar \left(f(t)\, e^{i(\Phi_L - \omega t)}\, a_3^\dagger - f^\star(t)\, e^{-i(\Phi_L - \omega t)}\, a_3 \right) \\ & + i\hbar\, k \left(e^{i\Phi_k}\, a_3\, a_1^\dagger\, a_2^\dagger - e^{-i\Phi_k}\, a_3^\dagger\, a_1\, a_2 \right). \end{aligned} \tag{13}$$

The stochastic equations for two groups of independent complex c-number variables $\alpha_{1(2)}$ and $\beta_{1(2)}$ corresponding to operators $a_{1(2)}$ and $a_{1(2)}^\dagger$ for the case of zero detunings have the form:

$$\frac{d\alpha_1}{dt} = -(\gamma + \lambda\, \alpha_2\, \beta_2)\, \alpha_1 + \varepsilon(t)\, \beta_2 + W_{\alpha_1}(t), \tag{14a}$$

$$\frac{d\beta_1}{dt} = -(\gamma + \lambda\, \alpha_2\, \beta_2)\, \beta_1 + \varepsilon^\star(t)\, \alpha_2 + W_{\beta_1}(t). \tag{14b}$$

Here $\varepsilon(t) = f(t)\, k/\gamma_3$, $\lambda = k^2/\gamma_3$ and equations for α_2, β_2 are obtained from (14a) and (14b) by exchanging the subscripts 1 and 2. The nonzero stochastic correlations are:

$$\langle W_{\alpha_1}(t)\, W_{\alpha_2}(t')\rangle = (\varepsilon(t) - \lambda\, \alpha_1\, \alpha_2)\; \delta\left(t - t'\right), \tag{15a}$$

$$\langle W_{\beta_1}(t)\, W_{\beta_2}(t')\rangle = (\varepsilon^{\star}(t) - \lambda\, \beta_1\, \beta_2)\; \delta\left(t - t'\right). \tag{15b}$$

Note, that in obtaining these equations we used the transformed boson operators $a_i \to a_i\, \exp\left(-i\, \Phi_i\right)$ with Φ_i being $\Phi_3 = \Phi_L$, $\Phi_1 = \Phi_2 = \left(\Phi_L + \Phi_k\right)/2$. This leads to the cancellation of phases at intermediate stages of calculation.

3.1 Modulated semiclassical dynamics

The equations of motion (14a) and (14b) contain time-dependent coefficients. Nevertheless, surprisingly, it is possible to find their analytical solution in the semiclassical approach for an arbitrary but real modulation amplitude $f(t)$.

We shall study the solution of stochastic equations in the semiclassical treatment, neglecting the noise terms, for mean photon numbers n_j and phases φ_j of the modes ($n_j = \alpha_j\, \beta_j$, $\varphi_j = \ln(\alpha_j/\beta_j)/2i$) for time-intervals exceeding the transient time, $t \gg \gamma^{-1}$. The analysis shows (see [Adamyan 2004 (b)] for the details) that similarly to the standard NOPO, the considered system also exhibits threshold behavior, which is easily described through the period-averaged pump field amplitude $\overline{f(t)} = T^{-1}\int_0^T f(t)\, dt$, where $T = 2\pi/\delta$. The below-threshold regime with a stable trivial zero-amplitude solution is realized for $\overline{f} < f_{th}$, where $f_{th} = \gamma\, \gamma_3/k$ is the threshold value. When $\overline{f} > f_{th}$, the stable nontrivial solution exists with the following properties. First, as for usual NOPO, the phase difference is undefined due to the phase diffusion, while the sum of phases is equal to $\varphi_1 + \varphi_2 = 2\pi\, m$. The mean photon numbers for subharmonic modes $n_{oi} = \left\langle a_i^{\dagger}\, a_i\right\rangle = |\alpha_i|^2$ are equal to each other ($n_{01} = n_{02} = n_0$) due to the symmetry of the system, $\gamma_1 = \gamma_2 = \gamma$.

We present the final results for the case of harmonic modulation with modulation amplitude $f(t) = \overline{f} + f_1\, \cos(\delta t + \Phi)$, assuming, without loss of generality, $\overline{f} > 0$, $f_1 > 0$ and $\Phi = 0$. In this case the mean photon number reads

$$\begin{aligned} n_0^{-1}(t) = \; & 2\lambda \int_{-\infty}^{0} \exp\left[2\gamma\, \tau\left(\frac{\overline{f}}{f_{th}} - 1\right)\right] \times \\ & \exp\left(\frac{2\gamma\, f_1}{\delta\, f_{th}}\left[\sin\left(\delta\left(t + \tau\right)\right) - \sin\left(\delta t\right)\right]\right) d\tau. \end{aligned} \tag{16}$$

This result is illustrated in Fig. 2 for the different levels of modulation.

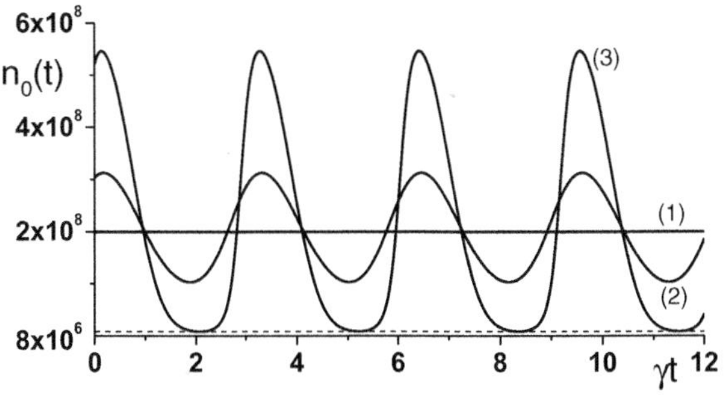

Figure 2 The semiclassical mean photon number versus dimensionless time for the parameters $k/\gamma = 5 \cdot 10^{-4}$, $\gamma_3/\gamma = 25$, $\delta/\gamma = 2$, $\overline{f} = 3f_{th}$; (1) $f_1 = 0$, (2) $f_1 = 0.4\overline{f}$ and (2) $f_1 = 1.2\overline{f}$.

3.2 Improvement of continuous-variable entanglement by time-modulation

The total intracavity variances are expressed through the stochastic variables using the relationships between normally-ordered operator averages and stochastic moments with respect to the P-function. Restoring the previous phase structure of intracavity interaction, we obtain that $V_+ = V_- = V$ and

$$V = 1 + \langle \alpha_1 \beta_1 \rangle + \langle \alpha_2 \beta_2 \rangle - \langle \alpha_1 \alpha_2 \rangle e^{i\Theta} - \langle \beta_1 \beta_2 \rangle e^{-i\Theta} \tag{17}$$

where $\Theta = \Theta_1 + \Theta_2 + \Phi_L + \Phi_k$.

Further, we will calculate the variance in the standard linear treatment of quantum fluctuations. To this end, it is convenient to use the following moments of stochastic variables

$$\begin{gathered} \langle n_+ \rangle = \langle \alpha_1 \beta_1 \rangle + \langle \alpha_2 \beta_2 \rangle, \quad \langle R \rangle = \langle (\alpha_1 - \beta_2)(\beta_1 - \alpha_2) \rangle \\ \text{and} \quad \langle Z \rangle = \left\langle (\alpha_1 \beta_1 - \alpha_2 \beta_2)^2 \right\rangle + \langle \alpha_1 \beta_1 \rangle + \langle \alpha_2 \beta_2 \rangle . \end{gathered} \tag{18}$$

As can be seen, the possible minimal level of variance, realized under appropriate selection of phases $\Theta_1 + \Theta_2 = -\Phi_L - \Phi_k$ in Eq. (17), is expressed through the moment $\langle R \rangle$ as $V(t) = 1 + \langle R(t) \rangle$. Using Itô rules for changing the stochastic variables, we obtain from (14a) and (14b) the following equations:

$$\frac{d \langle n_+ \rangle}{dt} = (2\,\varepsilon(t) - 2\gamma - \lambda) \langle n_+ \rangle - \lambda \left\langle n_+^2 \right\rangle - 2\,\varepsilon(t) \langle R \rangle + \lambda \langle Z \rangle \tag{19a}$$

$$\frac{d \langle R \rangle}{dt} = -(2\,\varepsilon(t) + 2\gamma + \lambda) \langle R \rangle - \lambda \langle n_+ R \rangle - 2\,\varepsilon(t) + \lambda \langle Z \rangle \tag{19b}$$

$$\frac{d \langle Z \rangle}{dt} = -4\gamma \langle Z \rangle + 2\gamma \langle n_+ \rangle . \tag{19c}$$

From Eq. (19c) $\langle Z\rangle$ can be expressed as a function of $\langle n_+\rangle$. Substituting this expression into (19a)and (19b) we get the following equations which are convenient for the perturbative analysis of quantum fluctuations

$$\frac{d}{dt}\langle n_+\rangle = (2\,\varepsilon(t) - 2\gamma - \lambda)\langle n_+\rangle - \lambda\left\langle n_+^2\right\rangle - 2\,\varepsilon(t)\langle R\rangle + 2\gamma\,\lambda\int_{-\infty}^{t} e^{4\gamma(\tau-t)}\langle n_+(\tau)\rangle\, d\tau \tag{20a}$$

$$\frac{d}{dt}\langle R\rangle = -\left(2\,\varepsilon(t) + 2\gamma + \lambda\right)\langle R\rangle - \lambda\langle n_+\, R\rangle - 2\,\varepsilon(t) + 2\gamma\,\lambda\int_{-\infty}^{t} e^{4\gamma(\tau-t)}\langle n_+(\tau)\rangle\, d\tau. \tag{20b}$$

First, we consider the above-threshold regime linearizing quantum fluctuations around the stable semiclassical solutions. In the linear treatment of quantum fluctuations we have the expansions $\langle n_+\rangle = n_{10}+n_{20}+\langle\delta n_+\rangle = 2n_0+\langle\delta n_+\rangle$, $\langle R\rangle = R^0 + \langle\delta R\rangle = \langle\delta R\rangle$, $\langle n_+\, R\rangle = 2n_0\,\langle\delta R\rangle$, $\left\langle n_+^2\right\rangle = 4n_0\,\langle\delta n_+\rangle$, where it was assumed that $n_{10} = n_{20} = n_0(t)$, $\varphi_1+\varphi_2 = 2\pi k$, and hence $R^0 = 0$. Note, that in the current experiments the ratio of nonlinearity to dumping is small, $k/\gamma \ll 1$ (typically 10^{-4} or less), and hence $\lambda/\gamma = k^2/\left(\gamma\,\gamma_3\right) \ll 1$ is a small parameter of the theory. Therefore, the zero order terms in the above expansion correspond to a large classical field of the order γ/λ in accordance with Eq. (16), while the following terms describing the quantum fluctuations are of the order of 1. On the whole, combining the procedure of linearization with the $\lambda/\gamma \ll 1$ approximation we get a linear equation for the variance $V(t) = 1 + \langle\delta R\rangle$

$$\frac{d}{dt}V(t) = -2\left[\gamma + \varepsilon(t) + \lambda\, n_0(t)\right]V(t) + 2\lambda\, n_0(t) + 2\gamma\left(1 + 2\lambda\int_{-\infty}^{t} e^{4\gamma(\tau-t)} n_0(\tau)\, d\tau\right) \tag{21}$$

with the following periodic asymptotic solution

$$V(t) = 2\int_{-\infty}^{t} \exp\left[-2\int_{\tau}^{t}\left(\gamma + \varepsilon\left(t'\right) + \lambda\, n_0\left(t'\right)\right)dt'\right]\times \left[\gamma + \lambda\, n_0(\tau) + 2\gamma\,\lambda\int_{-\infty}^{\tau} e^{4\gamma(\tau'-\tau)}\, n_0\left(\tau'\right)\, d\tau'\right] d\tau. \tag{22}$$

It should be noted that the result (22) is also obtained from initial Eqs. (14a) and (14b) in the photon number and phase variables using, however, more complicated calculations. In the absence of modulation, Eq. (22) coincides with an analogous one for ordinary NOPO. The analysis of the below-threshold regime is more simple and leads to Eq. (22) with $n_0 = 0$.

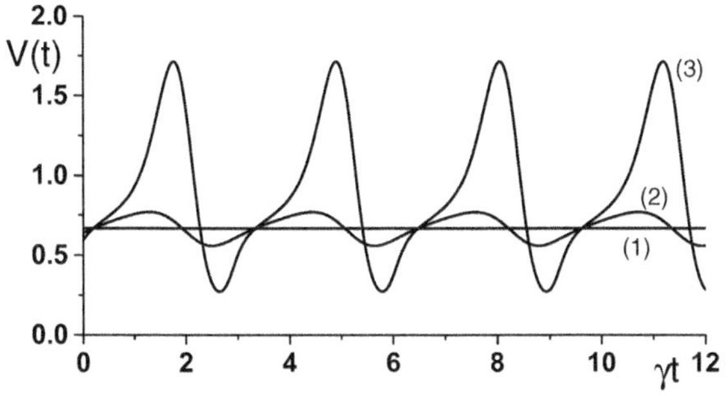

Figure 3 Variance $V(t)$ given by the linear theory, versus the dimensionless time for the same parameters as in Fig. 2.

Let us outline some conclusions from this result for the case of harmonic modulation, using that $\varepsilon(t) = k\left(\overline{f} + f_1 \cos(\delta t)\right)/\gamma_3$ and expression (16). Typical results are presented in Fig. 3 for the above-threshold regime, $\overline{f}/f_{th} = 3$. The variance is seen to show a time-dependent modulation with a period $2\pi/\delta$. The drastic difference between the degree of two-mode squeezing / entanglement for modulated and stationary dynamics is also clearly seen in Fig. 3. The stationary variance (curve (1)) near the threshold having a limiting squeezing of 0.5 (see also Fig. 4, curve (1)) is bounded by quantum inseparability criterion $V < 1$, while the variance for the case of modulated dynamics obeys the EPR criterion $V^2 < 1/4$ of strong CV entanglement for definite time intervals. In particular, the minimum values of the variance and corresponding photon numbers of Fig. 2 at fixed time intervals $t_m = t_0 + 2\pi\, m/\delta$, $(m = 0,\ 1,\ 2,\ \ldots)$ are: $n_0 \simeq 6.16 \times 10^7$, $V_{min} \simeq 0.27$, $t_0 = 2.64\ \gamma^{-1}$ (curve (3)) and $n_0 \simeq 1.71 \times 10^8$, $V_{min} \simeq 0.56$, $t_0 = 2.51\ \gamma^{-1}$ (curve (2)). The dependence of V_{min} on the period-averaged pump field amplitude is shown in Fig. 4 for different levels of modulation. As expected, the degree of EPR entanglement increases with the ratio $f_1/\overline{f}$. Another peculiarity here is that the stationary variance (curve (1)) has a characteristic threshold behavior that disappears in the case of strong modulation (curve (3)). The production of strong entanglement occurs for the period of modulation comparable to the characteristic time of dissipation, $\delta \approx \gamma$. For both asymptotic cases of slow ($\delta \ll \gamma$) and fast ($\delta \gg \gamma$) modulations this effect disappears. It confirms our supposition that such an improvement of the CV entanglement is due to the control of quantum dissipative dynamics through the application of suitably tailored, time-modulated driving field.

The linearized theory is applicable only outside the critical region. As a rule, the quantum corrections diverge at the classical threshold, although,

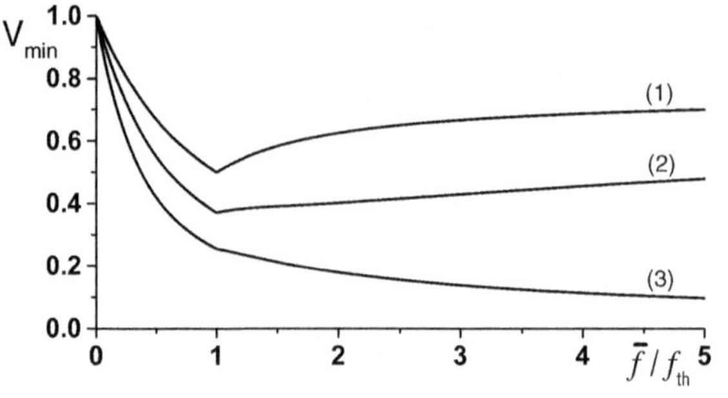

Figure 4 The minimum level of the variance versus $\overline{f}/f_{th}$ for three levels of modulation: (1) $f_1 = 0$, (2) $f_1 = 0.75\,\overline{f}$ and (3) $f_1 = 2\,\overline{f}$. The parameters are: $k/\gamma = 5 \cdot 10^{-4}$, $\gamma_3/\gamma = 25$, $\delta/\gamma = 2$.

surprisingly, the variance (22) is well defined also at the threshold. As our analysis shows, the condition of the validity of linear results for the near-threshold regimes reads $\left|\overline{f}/f_{th} - 1\right| \gg (\lambda/\gamma)\ \exp\left[2(f_1/f_{th})(\gamma/\delta)\right]$. For typical $\lambda/\gamma \ll 1$, this condition is fairly easy to satisfy even for narrow critical ranges, provided that $\delta \approx \gamma$.

3.3 Conclusions

To conclude this part of the paper, we have proposed a new approach for generation of strongly EPR entangled states of intense light beams based on the time-modulation of quantum dissipative dynamics. The obtained results demonstrate an essential improvement of the degree of entanglement in the modulated NOPO in comparison with the ordinary one, if the frequency of modulation is close to the decay rate of dissipative processes. It was shown that, surprisingly, the CV entanglement becomes approximately perfect for high-level of modulation, but is only realized for a periodic sequence of time intervals synchronized with the modulation period. We believe that the obtained results are applicable to a general class of quantum dissipative systems and can serve as a guide for further theoretical and experimental studies of intense light beams with high degree of entanglement.

As it has been mentioned above, the degree of CV entanglement is usually studied through the homodyne detection of quadrature amplitudes of light fields in a spectral domain that leads to the experimental investigations of quantum information protocols in this domain too. Recently, the new method of homodyne measurement in time-domain has been developed [Grangier 2003; Wenger 2004]. It opens a possibility for homodyne measurement of time-dependent quadrature amplitudes and hence for elaboration of time resolved quantum information protocols in addition to the ordinary ones elaborated in spectral domains. We expect that an experimental evidence of modulated entanglement proposed here may be given in this way.

4. Phase-locked entangled states

Our aim in this section is to study the interplay between the phase locking phenomena and CV entanglement for the self-phase locked NOPO. This scheme is based on the combination of two processes, namely, type-II parametric down-conversion and linear polarization mixing with cavity-induced feedback. The Hamiltonian describing the intracavity interaction is given by Eqs. (1a) and (1b). The parametric down-conversion is a standard technique used to produce an entangled photon pairs as well as CV two-mode squeezed states [Schumacher 1985]. The beam splitter including polarization mixer is also considered as experimentally accessible device for the production of entangled light-fields [Tan 1990; Tan 1991; Sanders 1995; van Loock 2000]. Besides these, there have been some studies of a beam splitter for various non-classical input states, including two-mode squeezing states [Kim 2002; Wang 2002 (a)]. It is obvious, and also follows from the results of the mentioned papers [Kryuchkyan 1993; Kryuchkyan 1994; Kryuchkyan 2001; Zondy 2001], that the operational regimes of the combined system with cavity-induced feedback and dissipation drastically differ from those for pure processes. We show below that analogous situation takes place in the investigation of quantum-statistical properties of a combined system such as the self-phase locked NOPO.

4.1 Phase structure of subharmonics

First, we shall study the steady-state solution of the stochastic Eqs. (10a) to (10d) in the semiclassical treatment, ignoring the noise terms for the mean photon numbers n_{j0} and the phases φ_{j0} of the modes ($n_j = \alpha_j \beta_j$, $\varphi_j = \ln(\alpha_j/\beta_j)/2i$). For simplicity we assume the perfect symmetry between the modes, provided that the cavity decay rates and the detunings do not depend on the polarization ($\gamma_1 = \gamma_2 = \gamma$, $\Delta_1 = \Delta_2 = \Delta$).

We list below the semiclassical results for the stable values of photons number and phases of the modes. In the above-threshold operational regime at $\varepsilon \geqslant \varepsilon_{th}$, where

$$\varepsilon_{th} = \sqrt{(\chi - |\Delta|)^2 + \gamma^2} \tag{23}$$

the results take the form

$$n_0 = n_{10} = n_{20} = \frac{1}{\lambda}\left[\sqrt{\varepsilon^2 - (\chi - |\Delta|)^2} - \gamma\right] \tag{24}$$

The phases are found to be

$$\varphi_{10} = \varphi_{20} = -\frac{1}{2}\arcsin\frac{1}{\varepsilon}(\chi + |\Delta|) + \pi k \tag{25}$$

for $\Delta > 0$. For the opposite sign of the detuning, $\Delta < 0$, the mean photon numbers are given by the same Eq. (24), while the phases read

$$\varphi_{10} = \frac{1}{2} \arcsin \frac{1}{\varepsilon} (\chi + |\Delta|) + \pi \left(k + \frac{1}{2} \right) \quad (26a)$$

$$\varphi_{20} = \frac{1}{2} \arcsin \frac{1}{\varepsilon} (\chi + |\Delta|) + \pi \left(k - \frac{1}{2} \right) \quad (26b)$$

($k = 0, 1, 2, \ldots$). In the region $\varepsilon \leqslant \varepsilon_{th}$, the stability condition is fulfilled only for the zero amplitude steady-state solution $\alpha_1 = \alpha_2 = \beta_1 = \beta_2 = 0$.

So, the set of the above-threshold stable solutions for both modes have two-fold symmetry in the phase-space. There are two stable states of each of the subharmonic modes with equal photon numbers, but with two different phases.

4.2 Squeezed variances

Let us now switch our attention to the quantum statistical effects and entanglement production for the case of perfect symmetry between the modes ($\gamma_1 = \gamma_2 = \gamma$, $\Delta_1 = \Delta_2 = \Delta$). In order to obtain general expressions for the variances, we first write them in terms of the boson operators corresponding to the Hamiltonian (1b). We perform the transformations $a_i \rightarrow a_i \exp{(i\,\Phi_i)}$, which restore the previous phase structure of the intracavity interaction and find, quite generally, the variances at some arbitrary quadrature phase angles θ_1, θ_2 as

$$V_{\pm} = V \pm R \cos(\Delta\theta) \quad (27)$$

where

$$V = \frac{V_+ + V_-}{2} = 1 + 2n - 2\,|\langle a_1\, a_2\rangle|\, \cos{(\Sigma\theta + \Phi_{\text{arg}})} \quad (28a)$$

$$R = 2\,\text{Re}\left(\langle a_1^2\rangle\, e^{i\,\Sigma\theta}\right) - 2\,\left|\left\langle a_1^\dagger\, a_2\right\rangle\right| \quad (28b)$$

$$\Delta\theta = \theta_2 - \theta_1 - \Phi_\chi, \quad (28c)$$

$$\Sigma\theta = \theta_1 + \theta_2 + \Phi_l + \Phi_k \quad (28d)$$

and $n = \left\langle a_1^\dagger\, a_1\right\rangle = \left\langle a_2^\dagger\, a_2\right\rangle$ is the mean photon number of the modes, $\Phi_{\text{arg}} = \arg\langle a_1\, a_2\rangle$. So, in accordance with (27), the relative phase Φ_χ between the transformed modes gives the effect of the rotation of the quadrature amplitudes angle $\theta_2 - \theta_1$.

Obviously, the variances $V_\pm$ and hence the level of CV entanglement depend on all the parameters of the system including the phases. The minimal possible level of V is realized for an appropriate selection of phases θ_i, namely for $\theta_1 + \theta_2 = -\arg \langle a_1\, a_2 \rangle - \Phi_l - \Phi_k$. Further, in most cases we assume that this phase relationship takes place, but do not introduce new notations for V and $V_\pm$ for the sake of simplicity. In this case, ($\Sigma\theta = -\Phi_{\text{arg}}$), in correspondence with (27), the variances $V_\pm$ depend only on the difference between phases $\Delta\theta$ and we arrive at

$$V = 1 + 2(n - |\langle a_1\, a_2 \rangle|) \tag{29a}$$

$$R = 2\,\text{Re}\left(\langle a_1^2 \rangle\, e^{-i\,\Phi_{\text{arg}}}\right) - 2\left|\left\langle a_1^\dagger\, a_2 \right\rangle\right| \tag{29b}$$

For the NOPO without additional polarization mixing, $\left\langle a_1^\dagger\, a_2 \right\rangle = \langle a_1^2 \rangle = \langle a_2^2 \rangle = 0$ and, hence, the case of symmetric variances $V_+ = V_-$ is realized. The phase-locked NOPO generally has non symmetric uncertainty region. However, the variances V_+ and V_- become equal for the special case of $\theta_2 - \theta_1 - \Phi_\chi = \pi/2$, when the inseparability condition reads $V < 1$, ($V = V_- = V_+$). We note that the relative phase Φ_χ plays an important role in the specification of entanglement.

4.3 Entanglement in the self-phase locked NOPO: sub-threshold regime

The quantum theory of phase locked NOPO has been developed in the linear treatment of quantum fluctuations in Ref. [Adamyan 2004 (a)]. Using these results, after some algebra, we obtain the minimal variance for the steady-state regime in the following form

$$V = 1 + \frac{\varepsilon\left(\varepsilon\, S^2 - \sqrt{\gamma^2\, S^4 + \Delta^2\,(S^2 - 2\chi^2)^2}\right)}{S^4 - 4\Delta^2\,\chi^2} \tag{30}$$

where S^2 is introduced as $S^2 = \gamma^2 + \chi^2 + \Delta^2 - \varepsilon^2$. Note that $S^2 > 0$ in the below-threshold regime.

It is easy to check that in the limit $\chi \longrightarrow 0$, the variance coincides with the analogous one for the ordinary NOPO, $V = V_- = V_+ = 1 - \varepsilon/(\varepsilon + \sqrt{\gamma^2 + \Delta^2})$. We see that the minimal variance remains less than unity for all values of pump intensity and is a monotonically decreasing function of ε^2. For all parameters, the maximal degree of two-mode squeezing $V \simeq 0.5$ is achieved within the threshold range. It is also easy to check that this expression is well-defined for all values of ε^2, including the vicinity of the threshold. This is not surprising, since the cancellation of infinities in the quadrature amplitude variances at the threshold takes place for the ordinary NOPO also.

One of the differences between the squeezing effects of the ordinary and self-phase locked NOPO is that variances V_- and V_+ for the ordinary NOPO are equal to each other, while for the self-phase locked NOPO they are in general different.

4.4 Entanglement in the self-phase locked NOPO: above-threshold regime

Performing calculations for each of the cases $\Delta > 0$ and $\Delta < 0$, we find the variance V in the above-threshold regime in the following form

$$V = \frac{3}{4} - \left\{ 4\sqrt{1 + \left(\frac{\varepsilon_{th}}{\gamma}\right)^2 \left[\left(\frac{\varepsilon}{\varepsilon_{th}}\right)^2 - 1\right]} \right\}^{-1} + \frac{\chi}{4\,|\Delta|}. \tag{31}$$

To rewrite this expression through the original parameters, we should take into account that $E/E_{th} = \varepsilon/\varepsilon_{th}$ and $\varepsilon_{th}^2/\gamma^2 = (\chi/\gamma - |\Delta|/\gamma)^2 + 1$.

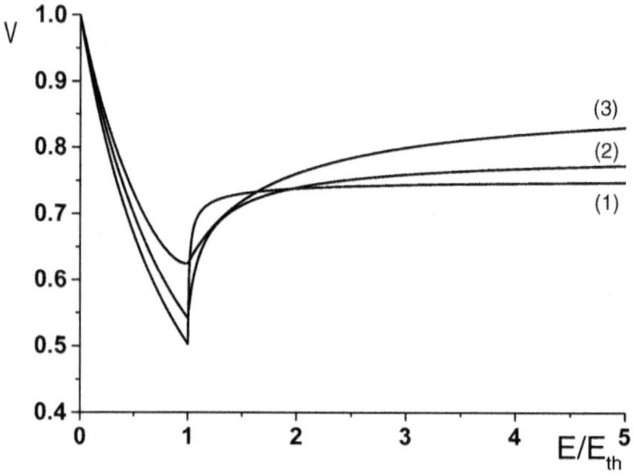

Figure 5 Minimized variance V versus the dimensionless amplitude of the pump field $E/E_{th} = \varepsilon/\varepsilon_{th}$ for both operational regimes. ε_{th} is given by Eq. (23). The parameters are: (1) $\chi/\gamma = 0.1$ (weak coupling), $\Delta/\gamma = 10$; (2) $\chi/\gamma = 0.5$ (moderate coupling), $\Delta/\gamma = 3$ and (3) $\chi/\gamma = 0.5$, $\Delta/\gamma = 1$.

The results (30) and (31) for both operational regimes are summarized in Fig. 5, where variance V is plotted as a function of the amplitude of the pump field. One can immediately grasp from the figure that the sum of variances $V = (V_+ + V_-)/2$ remains less than unity for all nonzero values of the pump field, provided that $\chi < |\Delta|$. This demonstrates the nonseparability of the generated state. The maximal degree of entanglement is achieved in the vicinity of threshold, $V \simeq 0.5$, if $\chi/|\Delta| \ll 1$. Far above the threshold, $E \gg E_{th}$, V increases with mean photon numbers of the modes and reaches the asymptotic value $V = 3/4 + \chi/(4\,|\Delta|)$. It should also be mentioned that the result (31) is expressed through the scaled pump field amplitude $\varepsilon = kE/\gamma_3$ and therefore depends on coupling constants k and χ.

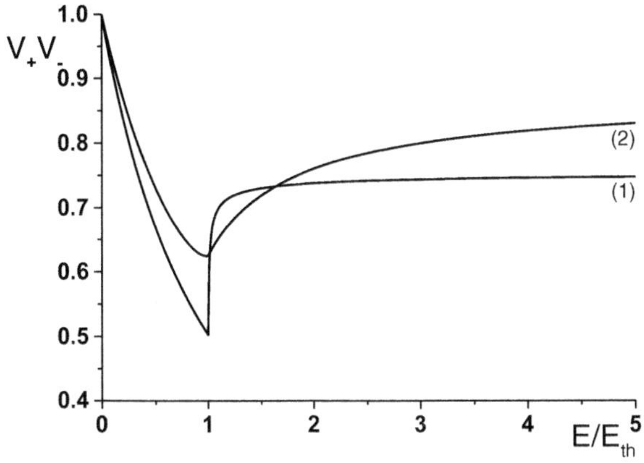

Figure 6 Product of V_+ and V_- versus the dimensionless amplitude of the pump field $E/E_{th} = \varepsilon/\varepsilon_{th}$ for parameters (1) $\chi/\gamma = 0.1$, $\Delta/\gamma = 10$ and (2) $\chi/\gamma = 0.5$, $\Delta/\gamma = 1$.

We stress, that the level of two-mode squeezing in the proposed scheme is limited due to the dissipation and pump depletion effects and reaches only 50% relative to the level of vacuum fluctuations when the pump intensity is near the generation threshold. Such limitation on the level of intracavity two-mode squeezing takes place also for the ordinary NOPO as was demonstrated above by the analysis of quantum fluctuations in the near-threshold range. Analogous conclusion has arrived at for single-mode squeezed light generated in the OPO [Chaturvedi 2002; Drummond 2002]. Recall, however, that these results pertain to the full squeezing of the intracavity variances and not to the spectra of squeezing of the output fields.

In terms of demonstrating the CV strong EPR entanglement, one has to apply another criterion, $V_+V_- < 1/4$. For the case of symmetric uncertainties ($V_+ = V_- = V$), the product of variances $V^2 \geq 1/4$ and hence the strong EPR entanglement can not be realized. In the general case, the product of variances reads $V_+ \, V_- = V^2 - R^2 \, \cos^2(\Delta\theta)$. It seems that $V_+ \, V_-$ lies below $1/4$ at least for the relative phase $\Delta\theta = \pm \, \pi m$, $(m = 1, \, 2, \, \ldots)$, and in the vicinity of the threshold, where $V \simeq 0.5$. However, for such a selection of phases we arrive at

$$V_+ \, V_- = \frac{|\Delta| + \chi}{4\,|\Delta|} \left\{ (2 - \left[1 + \left(\frac{\varepsilon_{th}}{\gamma} \right)^2 \left(\left(\frac{\varepsilon}{\varepsilon_{th}} \right)^2 - 1 \right) \right]^{-1/2} \right\}. \tag{32}$$

It is easy to check that the product of variances exceeds $1/4$ even in the vicinity of the threshold. This quantity for both operational regimes is illustrated in Fig. 6. It should be noted again that a detailed analysis of this problem must include more accurate considerations of quantum fluctuations in the critical ranges.

4.5 Conclusion

The results of this section demonstrate the possibility of creation of the CV entangled states of light beams in the presence of phase localizing processes. We have called such quantum states the entangled self-phase locked states of light and we have shown that they can be generated in the NOPO recently realized in the experiment [Mason 1998]. The novelty is that this device provides a high level of phase coherence between the subharmonics, which is in contrast to the case of ordinary NOPO, where the phase diffusion takes place. We have shown that both two mode squeezing and quantum phase locking phenomena are combined in such a NOPO. This development paves the way towards the generation of bright CV entangled light beams with well-localized phases. It seems that this scheme involving phase locking may potentially be useful for precise interferometric measurements and quantum communications, because it combines quantum entanglement and stability of type - II phase matching with an effective suppression of the phase noise. In practice, this phase locked scheme can effectively be used in homodyne detection setups, as well as for observation of interference with nonclassical light beams. We believe that such a source of bright entangled light providing phase coherence will also be applicable for realizing the CV quantum teleportation, as well as Heisenberg-limited interferometry [Luis 2000] with bright light beams. The price one has to pay for these advantages is the small aggravation of the degree of CV entanglement in self-phase locked NOPO, in comparison with the case of ordinary NOPO. We have studied the CV entanglement as two-mode squeezing and have shown that the entanglement is present in the entire range of pump intensities. In all cases, the maximal degree of two-mode squeezing $V \simeq 0.5$ is achieved in the vicinity of the threshold, provided that the coupling constant χ is much less than the mode detunings. We have demonstrated that, as a rule, highest degree of CV entanglement occurs for weak linear coupling strength χ. It has also been shown that the amount of entanglement can be controlled via the phase difference Φ_χ.

Acknowledgments

We thank L. Manukyan for helpful discussions. This work was supported by NFSAT PH 098-02 grant no. 12052 and ISTC grant no. A-823.

III

ADIABATIC AND NONADIABATIC PROTECTION FROM DECOHERENCE

ADIABATIC AND NONADIABATIC PROTECTION FROM DECOHERENCE

S. Pellegrin,[1] E. Brion,[2] C. Mewes,[3] L. B. Ioffe[4] and G. Kurizki,[1]

[1] *Chemical Physics Department, Weizmann Institute of Science, Rehovot 76100, Israel*

[2] *Laboratoire Aimé Cotton, Bât. 505, Campus d'Orsay, 91405 Orsay Cedex, France*

[3] *Fachbereich Physik, Technische Universität Kaiserslautern, D-67653 Kaiserslautern, Germany*

[4] *Center for Materials Theory, Department of Physics and Astronomy, Rutgers University, 136 Frelinghuysen Rd, Piscataway NJ 08854 USA*

Keywords: Decoherence, quantum information processing, errors correction codes, quantum Zeno and anti-Zeno effects, Josephson junction qubits, degenerate ground-state systems, optically-manipulated atoms, photonic band gap, nonadiabatic periodic dynamics, quantum memory, collective multiatom states, equivalent storage classes, decoherence free subspaces.

The uncontrollable interaction of an open quantum system with its environment leads to complete loss of the information initially stored in its quantum state. This phenomenon is commonly referred to as decoherence, loss of coherence, or loss of fidelity. The question of how it is possible to avoid the negative influence of this process is one of the most interesting issues in modern quantum mechanics, and concerns many different fields of physics, in particular the domains of coherent control [Shapiro 2003] and quantum information processing [Cirac 1995; Mølmer 1999; Sackett 2000].

The principal difference between quantum and classical computation is the essentially continuous nature of quantum information, which results from the superposition principle. From this characteristic follow both the amazing computational power of quantum computer (known as *quantum parallelism*) as well as its great sensitivity to the environment.

Although its actual implementation remains far away, it has been known for a long time that the hypothetical quantum computer should be a much more powerful tool than its classical analog [Ekert 1996; Steane 1998]. In particular, it might be able to simulate the behaviour of complex quantum systems far more efficiently than classical machines [Feynman 1996]. For example, to simulate an N interacting spin system classically requires about 2^N bits: therefore, such classical computations become intractable when N exceeds ~ 40 spins.

V.M. Akulin et al. (eds.), Decoherence, Entanglement and Information Protection in Complex Quantum Systems, 129–136.

By contrast, the same task could be achieved by a quantum computer comprising only N quantum bits (*qubits*), each of which represents a single spin. This substantial gain would allow us, for instance, to numerically treat frustrated quantum magnet models, which are of the utmost importance in the survey of strongly correlated electron systems and the theory of high-temperature superconductivity. But quantum simulation is not the only domain in which quantum computers are bound to surpass classical machines: actually, quantum computation can also solve classical hard problems, such as large number factorization [Shor 1994] and large list sorting [Grover 1996], parametrically faster than its classical counterpart.

On the other hand, quantum computers are particularly sensitive to the influence of their environment, which rapidly damages or destroys the very fragile correlations between qubits. In fact, the continuity of quantum information makes it much more difficult to protect from errors than binary classical information for which efficient error-correcting strategies have been developped. Quantum errors have been considered for a long time as an unavoidable impediment which would prevent quantum computation from effective implementation, until P. W. Shor [Shor 1995] established that error-correcting schemes also exist in quantum computation. Since then, a general framework of quantum error-correction has been built upon the formalism of quantum operations. The basic idea of these methods is to encode information redundantly, as in the classical case, with the hope that the errors will not corrupt it too much, so that it can be finally recovered. More precisely, the information is encoded in a subspace, called the code space, such that the effect of the errors can be detected by an appropriate measurement, and then corrected by a recovery operation [Knill 1997]. Stabilizer codes [Gottesman 1996; Gottesman 1997] and decoherence free subspaces [Lidar 1998; Zanardi 1998; Lidar 1999; Knill 2000; Viola 2000; Kempe 2001; Wu 2002] are important examples of such error-correcting codes. Generally, these methods require strong symmetry properties of the error models considered.

In the first paper of this part, a different protection method is presented which draws its inspiration from the ideas of standard classical error-correction and the quantum Zeno effect (QZE). This phenomenon introduced by Misra and Sudarshan [Misra 1977], following the early work of Khalfin [Khalfin 1968] and Fonda [Fonda 1973], takes place in a system which is subject to frequent measurements projecting it onto its (necessarily known) initial state. If the time interval between two projections is small enough the evolution of the system is nearly "frozen". This effect has been widely investigated theoretically [Khalfin 1957-58; Winter 1961; Fonda 1978; Joos 1984; Cook 1988; Itano 1990; Knight 1990; Frerichs 1991; Sakurai 1994; Panov 1996; Facchi 1998; Schulman 1998; Elattari 2000; Schmidt 2003 / 2004] as well as experimentally [Cook 1988; Itano 1990; Wilkinson 1997; Fischer 2001]. Generalizations have been proposed which employ incomplete measurements [Facchi 2002]: in this setting, the Hilbert space is split into "Zeno subspaces" (degenerate multidimensional eigenspaces of the measured observable), and the state vector of the system is compelled by frequent measurements of the physical observable to remain in its initial Zeno subspace. The dynamics of the system in the Zeno subspaces has also been studied in different specific situations [Facchi 2001 (b)]. The new protection scheme proposed by E. Brion *et al.* requires no specific symmetry of the errors. Moreover, its physical implementation in an arbitrary quantum system is suggested, and its realisation in the example of a rubidium isotope is given.

Employing QZE ideas, enriched by standard techniques from the classical coding theory [Gallager 1968], the authors propose an information protection scheme [Brion 2004] in Zanardi's spirit [Zanardi 1999], except that no symmetry assumption is done on the unitary errors considered.

The key point of the method is to entangle the initial state of an ancilla with the state of the system in such a way that the detection of the ancilla in its initial state implies that the system is also in its initial state. This is achieved through the coding procedure which has to ensure that after the exposition of the system to the action of errors, the initial state of the ancilla remains entangled only with the initial state of the system, whereas any other entanglement remains small during the time interval between consecutive measurements.

The coding procedure involves a rather complex unitary transformation in the Hilbert space of the compound system: performing such coding operations is a non-trivial quantum control problem in itself. However, this objective is achieved by employing the idea of non-holonomic control [Akulin 2001]: by applying two different Hamiltonians in sequence, one can build any desired unitary transformation.

The proposed protection scheme can be applied to various physical systems, for example to protect one qubit of information stored in the spin variable of the quantum state of a rubidium atom. The authors describe a realistic experimental setting which achieves the different steps of their scheme through the application of a sequence of laser pulses and culminates in a measurement involving spontaneous emission.

One could distinguish two types of errors that might occur in a quantum bit: an erroneous flip which is similar to the error in a classical system and a deviation of the quantum phase, $\delta\phi$, from its ideal value, the rate of these processes is known as a dephasing rate. The former is relatively easy to eliminate by ensuring that two states are separated by a significant energy barrier. The latter is much more difficult to deal with. It occurs, for instance, if the energy of two quantum states, $|0\rangle$ and $|1\rangle$ are differently affected by external noise, $V(t)$: $\delta\phi = \int (\langle 1|V(t)|1\rangle - \langle 0|V(t)|0\rangle)\, dt$. It seems very unlikely that one can screen a macroscopic system from the environment to such a degree that it is possible to achieve very low error rates required for quantum computation if the matrix element $\langle 1|V(t)|1\rangle - \langle 0|V(t)|0\rangle$ is of the same order as other energy scales characterizing the physical device. Thus, in order to achieve a low dephasing rate one needs to identify systems in which matrix elements of physical perturbations are very low, ideally, exponentially low in the system size. In other words one needs systems with exponentially large number of states which are not distinguished by physical operators. If the two states $|0\rangle$ and $|1\rangle$ have non-zero energy difference in the ideal system, fluctuations of the physical field (or device parameters) that produces this energy difference affect the energy of the states in the linear order. Further, in a solid state device with two states separated by a significant energy gap it is difficult to suppress the process of phonon emission that leads to a significant flip rate and dephasing. It seems [Ioffe 2004] that it is this phonon emission that limits the coherence of the best Josephson junction qubits [Nakamura 1999; Martinis 2002; Vion 2002; Chiorescu 2003]. Thus, a very low dephasing rate is possible only if these two states are degenerate in ideal system.

One very attractive possibility [Kitaev 1997; Kitaev 2003] involves a protected subspace [Wen 1990; Wen 1991] created by a topological degeneracy of the ground state. Typically such degeneracy happens if the system has a conservation law such as the conservation of the *parity* of the number of "particles" along some long contour. Physically, it is clear that two states that differ *only* by the parity of some big number that cannot be obtained from any local measurement are very similar to each other. The model proposed in Refs. [Kitaev 1997; Kitaev 2003] has been shown to exhibit many properties of the ideal quantum computer; however for a long time no robust and practical implementation was known. Recently in a series of papers, Josephson junction networks which implement protected degeneracy and which is possible to build in a laboratory have been proposed [Ioffe 2002 (a); Ioffe 2002 (b); Doucot 2003; Doucot 2004 (a); Doucot 2004 (b)]. In the second article of this part, L. B. Ioffe focuses on systems with degenerate ground states. He presents mathematical models which have a protected ground state and the Josephson junction arrays that implement them, and discusses the physical properties and conditions that should be satisfied by each Josephson junction.

Qualitatively, one can understand the idea of the topological protection on the example of the chessboard model shown in Fig. 1. In this model each square is either black or white, its state is described by the quantum spin $1/2$ variable. The allowed dynamics flips *simultaneously* four spins in the adjacent squares:

$$H = \sum_{i,\,j} \sigma^x_{i,j}\, \sigma^x_{i+1,j}\, \sigma^x_{i,j+1}\, \sigma^x_{i+1,j+1}. \tag{1}$$

The dynamics (1) evidently preserves the parity of the number of black (white) squares in each column. On the other hand, it is easy to verify that both even

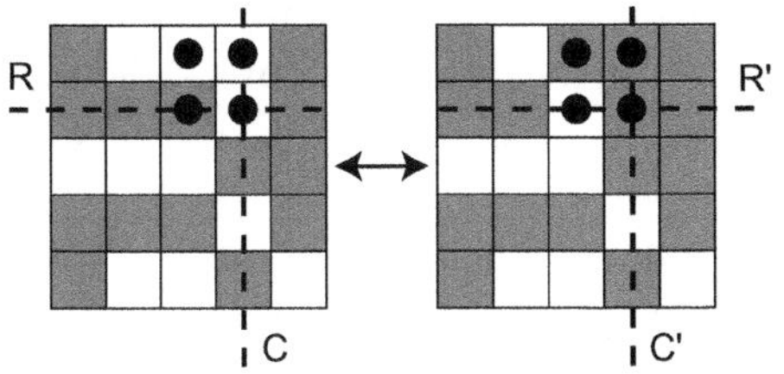

Figure 1. Formation of the protected degeneracy in the simplest quantum model. Each chessboard square is either black or white. The quantum dynamics changes white squares into black ones and vice versa simultaneously in four adjacent squares such as the ones indicated by the black dotes in the figure. This preserves the parity of the number of black squares in each column and row, *e.g.*, in column C and row R, which after the flip shown in this figure become column C' and row R'. The two states differing by the parity of the number of black squares cannot be distinguised by a local measurements.

and odd parities are in principle allowed. Furthermore, if the ground state of this model corresponds to the liquid of black (white) squares, the parity of black squares in a column is the *only* way to distinguish two states. Thus, a local operator that corresponds to a physical noise does not split the energy difference between the two states differing by their parity but otherwise similar. The model (1), however, is not perfect, because it does not ensure by itself that all columns have the same parity of black squares, without this condition the number of degenerate states is 2^{2L} where L is the linear size of the system because each column and one row can be in two different states. This would allow a small perturbation, such as $\delta H = \sigma^x_{i,j}$ to change the state of one column and one row and would lead to a large flip rate. More complicated models where such processes are expressely forbidden by the additional constraints, such as equal parities of all columns (or rows) or a local constraints imposed on each site of the (dual) lattice are also discussed.

In the domain of quantum information processing based on optically-manipulated atoms the decoherence is due to atomic spontaneous emission (SE) [Bose 1998]. A promising means of protection against such a fidelity loss is to embed the atoms in photonic crystals that possess spectrally-wide, omnidirectional photonic bandgaps (PBGs) [Yariv 1984; Joannopoulos 1995]: atomic SE would then be blocked at frequencies within the PBG [Martorell 1990; Blanco 1998; Yoshino 1998; Kurizki 2003]. Yet optical manipulations of such atoms may necessitate the consideration of atomic transition frequencies near a PBG edge (*i.e.*, the edge of the photonic mode continuum), where SE is only partially blocked, because an initially excited atom then evolves into a superposition of decaying and stable states, the stable state representing photon-atom binding [John 1990; Kofman 1994]. These PBG-edge effects may play a role if, in order to coherently manipulate an atomic transition in the PBG, one takes advantage of its proximity to the edge and couples it to a field mode outside the PBG or to a mode in the PBG created by a local defect in the photonic crystal [Kofman 1994; Joannopoulos 1997] (Fig 2).

Thus far, studies of coherent optical processes in a PBG have assumed *fixed* (static) values of the atomic transition frequency [Quang 1997]. However, in order to operate quantum logic gates, based on pairwise entanglement of atoms by field-induced dipole-dipole interactions [Brennen 1999; Petrosyan 2002; Opatrny 2003], one should be able to switch the interaction on- and off-, most conveniently by AC Stark-shifts of the transition frequency of one atom relative to the other, thereby changing its detuning from the PBG edge.
The question then arises: should such frequency shifts be performed adiabatically, in order to minimize the decoherence and maximize the quantum-gate fidelity? The answer is expected to be affirmative, based on existing treatments of adiabatic entanglement and protection from decoherence [Unanyan 2003; Calarco 2003; García-Ripoll 2003] and on the tendency of nonadiabatic effects to promote transitions to the continuum. Remarkably, the analysis developped in the third paper of this part demonstrates that periodic *"sudden"* (strongly nonadiabatic) changes of the detuning from the PBG edge may yield *higher fidelity* of qubit and quantum gate operations than their adiabatic counterparts whether for strong or weak coupling to the continuum.
This unconventional nonadiabatic protection from decoherence is attributed to the ability of the periodically alternating detuning from the PBG edge to augment the interference of the emitted and back-scattered photon amplitudes, thereby increasing the probability amplitude of the stable state. This may pave the way to new methods of controlling decay and decoherence in spectrally structured continua [Viola 1999; Facchi 2001 (c); Wu 2002; Zanardi 2003].

An important element for quantum information processing with photons [DiVincenzo 2000; Zoller 2004] is a reliable quantum memory (QM) capable of a faithful storage of their quantum state. Such a memory plays a key role in long-distance quantum communication, teleportation [Duan 2000 (d); Duan 2001; Julsgaard 2001; Kuzmich 2003; van der Wal 2003] and network quantum computing [Cirac 1997]. Photons may well be the best information carriers and, in the case of spin components of the electronic ground-state, atoms are reliable long-lived storage units. A controllable and decoherence insensitive way of coupling light to atoms is provided by Raman transitions. Although the simplest photonic qubits are individual atoms a faithful transfer of quantum states to and from the radiation field requires strongly coupling resonators. The technically difficult regime of strong-coupling cavity QED can be avoided if collective multiatom are used for coherent and reversible transfer of individual photon wave-packets [Csesznegi 1997; Fleischhauer 2000; Lukin 2000

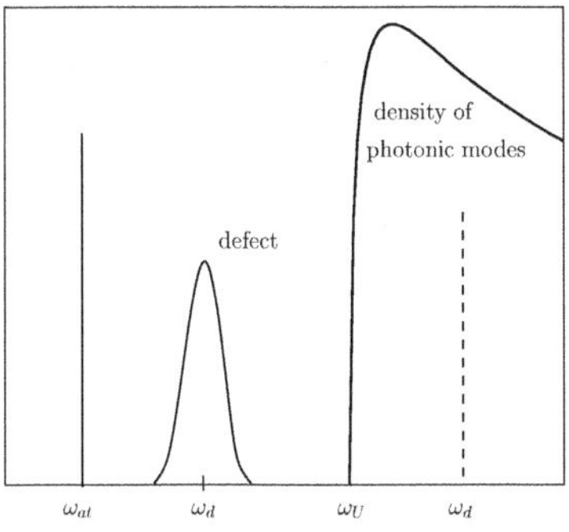

Figure 2. Mode density spectrum with a PBG. The atomic frequency ω_{at} is inside a PBG, near the cutoff frequency ω_0 and defect-mode frequency ω_d.

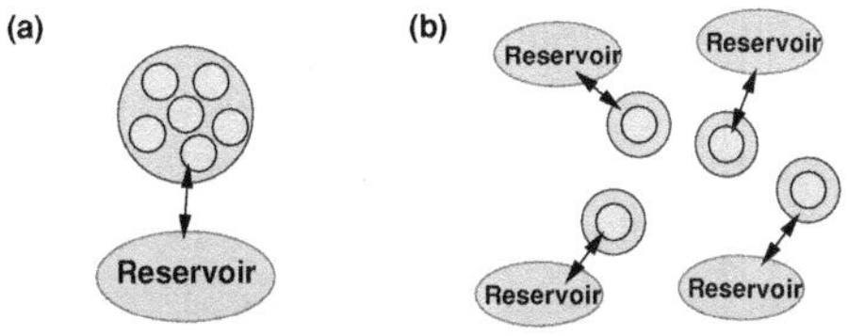

Figure 3. Collective (a) vs. individual (b) reservoir coupling.

(b); Liu 2001; Phillips 2001; Fleischhauer 2002; Lukin 2003] or cw light fields [Kuzmich 1997; Hald 1999; Kuzmich 2000; Schori 2002 (a)].

The advantages of enhanced coupling between collective many-atom states and the radiation field have to be checked against the worry that these states are highly entangled if non-classical light is stored. Entangled states are very sensitive to decoherence and one could naively expect their lifetime to decrease with the number of atoms. It is therefore important to analyze the effect of unwanted environmental influences on the fidelity of the collective QM.

Decoherence is usually modeled by coupling the system to a large reservoir of, *e.g.*, harmonic oscillators. Here two different types of couplings need to be distinguished which are indicated in Fig. 3. These are collective interactions, Fig. 3 (a), where all particles couple to the same bath and individual interactions, Fig. 3 (b), where each particle couples to its own, independent reservoir. In the third contribution by C. Mewes *et al.* only the second type of coupling is considered. It is well suited for the case of a dilute atomic vapor.

The authors show that for the retrieval of a stored field state only excitations in a certain quasi-particle mode, the dark-state polariton, are of importance. Thus all storage states with the same projection to this mode are equivalent. Due to the existence of these equivalence classes there is no enhanced sensitivity to (individual) decoherence processes in an N-atom system as compared to a single atom system. Quantum error correction requires a fidelity for all individual storage and gate operations better than one part in 10^4 [Knill 1998; Preskill 1998 (b)]. Thus it is important to develop efficient techniques to suppress the influence of decoherence. Among the currently discussed strategies are geometric quantum computation [Vedral 2003] and the use of decoherence-free subspaces [Lidar 2003]. Although a DFS of dimension one is easy to find, the interesting case $d \geq 2$ requires in general special symmetries of the system-reservoir interaction, see Fig. 3 (a). An example for this is Dicke subradiance, where a tightly confined ensemble of two-level atoms couples to the radiation vacuum [Dicke 1954]. The identical coupling of all atoms to the same vacuum modes generates a large DFS, the sub-radiant states. In many experimentally relevant circumstances, as for example in the case of a dilute atomic vapor, such symmetries do not exist however and the coupling to the environment is more accurately described by a model of individual and uncorrelated reservoirs as in Fig. 3 (b). As suggested by Kitaev for the example of Majorana fermions [Kitaev 2000; Levitov 2001] and by Dorner *et al.* for collective atom states in a 1D lattice [Dorner 2003] qubits can effectively be protected from decoherence without requiring a highly symmetric reservoir coupling, if they are stored in special collective states of interacting multi-particle systems. In the third paper of this part, it is shown that such special states can also be found in the collective photon memories. Storing photonic qubits in these states and providing for an additional nonlinear interaction a *decoherence free subspace* of dimension 2 emerges.

The prevailing view until recently has been that successive frequent measurements (interruptions of the evolution) known as the quantum Zeno effect *must slow down the decay of any unstable* system. A few years ago, Kofman *et al.* [Kofman 2000; Kofman 2001 (a)] showed that, in fact, the *opposite* is commonly true for decay into open-space continua: the anti-Zeno effect (AZE), *i.e.*, decay acceleration by frequent measurements[1], is far more ubiquitous than the QZE [Milonni 2000; Seife 2000]. How can this conclusion be understood and what was missing in standard treatments that claimed the QZE universality? The last paper of this part, by G. Kurizki *et al.* shows that:

(i) The QZE can only occur in systems with spectral width below the resonance energy.

(ii) It is *principally unattainable* in open-space radiative or nuclear β-decay, because the required measurement rates would cause the creation of new particles.

(iii) Contrary to the widespread view, frequent measurements can be chosen to *accelerate* essentially any decay process. Hence, the *anti-Zeno effect* should be *far more ubiquitous than the QZE.*

(iv) Acceleration of decay by frequent measurements can occur in a multilevel system, where interference effects (in the absence of phase-randomizing measurements) can inhibit the decay.

Notes

1. Related effects have been noted in proton decay [Lane 1983]; more recently for radiative decay in cavities [Kofman 1996], in Rabi oscillations betwen coupled quantum dots [Gurvitz 1997], in photodetachment [Lewenstein 2000]. See also [Facchi 2000 (a)].

COHERENCE PROTECTION BY THE QUANTUM ZENO EFFECT

E. Brion,[1] V. M. Akulin,[1] D. Comparat,[1] I. Dumer,[2] G. Harel,[3] N. Kébaïli,[1] G. Kurizki,[4] I. E. Mazets[4, 5] and P. Pillet[1]

[1] *Laboratoire Aimé Cotton, Bâtiment 505, Campus d'Orsay, 91405 Orsay Cedex, France*

[2] *College of Engineering, University of California, Riverside, CA 92521, USA*

[3] *Department of Computing, University of Bradford, Bradford, West Yorkshire BD7 1DP, UK*

[4] *Chemical Physics Department, Weizmann Institute of Science, Rehovot 76100, Israel*

[5] *A. F. Ioffe Physico-Technical Institute, 194021 St. Petersburg, Russia*

Abstract The protection of the coherence of open quantum systems against the influence of their environment is a very topical issue. The main features of quantum error-correction are reviewed here. Moreover, an original scheme is proposed which protects a general quantum system from the action of a set of arbitrary uncontrolled unitary evolutions. This method draws its inspiration from ideas of standard error-correction (ancilla adding, coding and decoding) and the Quantum Zeno Effect. A pedagogical demonstration of our method on a simple atomic system, namely a Rubidium isotope, is proposed.

Introduction

The uncontrollable interaction of an open quantum system with its environment leads to complete loss of the information initially stored in its quantum state. This phenomenon is commonly referred to as "loss of coherence". The question of how it is possible to avoid the negative influence of this process is one of the most interesting issues in modern quantum mechanics, and concerns many different fields of physics, in particular the domains of quantum information and computation.

In the context of quantum information, the effects of interactions with the environment, known as "quantum errors", may render information storage and processing unreliable [Preskill 1998 (c); Nielsen 2000]. Since Shor's demonstration that error-correcting schemes exist in quantum computation [Shor 1995], a general framework of quantum error-correction has been built upon the formalism of quantum operations. The idea of quantum error-cor-

V.M. Akulin et al. (eds.), Decoherence, Entanglement and Information Protection in Complex Quantum Systems, 137–174.

recting codes is to use redundancy to encode information in a subspace such that the effect of the errors can be detected by an appropriate measurement and then corrected [Knill 1997]. Stabilizer codes are an important class of quantum codes, which can be explicitly constructed using group-theoretic properties of the errors [Gottesman 1996; Gottesman 1997]. Decoherence free subspaces are another very particular type of codes: in this approach, information is stored in a subspace which is not affected by the errors [Lidar 1998; Zanardi 1998; Lidar 1999; Knill 2000; Viola 2000; Kempe 2001; Wu 2002]. As for the stabilizer codes, strong symmetry properties are required from the errors. In this paper, we present a different protection method which draws its inspiration from the ideas of standard classical error-correction and the quantum Zeno effect, and requires no specific symmetry of the errors. Moreover, we suggest its physical implementation in an arbitrary quantum system, and show how it works for the example of a Rubidium isotope.

The phenomenon known as the quantum Zeno effect takes place in a system which is subject to frequent measurements projecting it onto its (necessarily known) initial state: if the time interval between two projections is small enough the evolution of the system is nearly "frozen". This effect, and its inverse (the anti-Zeno effect), have been widely investigated theoretically [Khalfin 1957-58; Winter 1961; Misra 1977; Fonda 1978; Kofman 1996; Kofman 2000; Lewenstein 2000; Kofman 2001 (a); Schmidt 2003 / 2004] as well as experimentally [Cook 1988; Itano 1990; Wilkinson 1997; Fischer 2001]. Generalizations have been proposed which employ incomplete measurements [Facchi 2002]: in this setting, the Hilbert space is split into "Zeno subspaces" (degenerate multidimensional eigenspaces of the measured observable), and the state vector of the system is compelled by frequent measurements of the physical observable to remain in its initial Zeno subspace. The dynamics of the system in the Zeno subspaces has also been studied in different specific situations [Facchi 2001 (b)].

Employing these ideas, enriched by standard techniques from the classical coding theory [Gallager 1968], we propose an information protection scheme [Brion 2004] in Zanardi's spirit [Zanardi 1999], except that we do not make any symmetry assumption on the unitary errors we consider. We form a compound system $\mathcal{S}$ which comprises the information system $\mathcal{I}$ to be protected and an auxiliary system $\mathcal{A}$ (called ancilla). We then apply a controlled unitary operation $\widehat{C}$ (the coding matrix) which encodes the information, initially stored in $\mathcal{I}$, in an entangled state of $\mathcal{I}$ and $\mathcal{A}$. After a short time interval, during which infinitesimal errors may have occurred, we apply the unitary transformation $\widehat{C}^{-1}$ (the inverse to the initial step), which decodes information. Finally, we measure the ancilla to get rid of the infinitesimal changes that may have been caused by errors. Whereas in classical error-correction theory, the ancilla contains information about the errors allowing them to be corrected, in our quantum Zeno effect-based approach, the quantum state of the ancilla resulting from an elementary (coding-errors-decoding) sequence is close to its initial state up to second order corrections, so that the measurement of the ancilla brings it back to its initial state with a probability of nearly 1.

The key point of our method is to entangle the initial state of the ancilla with the state of the system in such a way that the detection of the ancilla in its initial state implies that the system is also in its initial state. This is achieved through the coding procedure which has to ensure that after the exposition of the system to the action of errors, the initial state of the ancilla remains entangled only with the initial state of the system, whereas any other entanglement remains small during the time interval between consecutive measurements.

The coding procedure involves a rather complex unitary transformation in the Hilbert space of the compound system: performing such coding operations is a non-trivial quantum control problem in itself. However, one can achieve this objective by employing the idea of non-holonomic control which we have previously presented [Akulin 2001]: by applying two different Hamiltonians in sequence, one can build any desired unitary transformation.

Our protection scheme can be applied to various physical systems. In this paper, we show how our method can protect one qubit of information stored in the spin variable of the quantum state of a Rubidium atom, the orbital variable playing the role of the ancilla. Here we describe a realistic experimental setting which achieves the different steps of our scheme through the application of a sequence of laser pulses and culminates in a measurement involving spontaneous emission.

The paper is organized as follows. In Sec. 1, we introduce the main features of quantum error-correction, and, particularly, we present the already well-developed theory of quantum error-correcting codes. In Sec. 2, we present a multidimensional generalization of the quantum Zeno effect and its application to the protection of the information contained in compound systems. Moreover, we suggest a universal physical implementation of the coding and decoding steps through the non-holonomic control. Finally, in Sec. 3, we focus on the application of our method to a rubidium isotope.

1. Quantum error-correction

Noise is obviously not a characteristic of quantum information, it also concerns classical devices. Indeed, if, on the one hand, components in classical computers are extremely reliable, and can almost be regarded as noiseless, systems like modems and CD players, on the contrary, do suffer from noise. To remedy this parasitic process, error-correcting codes have been well developed and are currently widely used in such classical devices.

Inspired by the existing classical error-correcting techniques, P. Shor built a code in the quantum domain [Shor 1995], which was able to protect one qubit of information against *arbitrary single qubit* errors. Following this important step, a general theory of quantum error-correcting codes has been set up, in the framework of quantum operations [Knill 1997].

In this part, we propose an overview of the field of quantum error-correction. We shall first introduce the basic concepts of error-correcting codes in the classical as well as in the quantum case. Then we shall deal with the general theory of quantum error-correction: in particular, we will present the general mathematical correction conditions, as well as the main existing technical methods to build codes explicitly.

1.1 Introduction

In this introduction, we deal with some basic and intuitive ideas of classical error-correction and show how they can adapt to the quantum case. Here, we shall focus on examples, and leave the general framework for the next paragraph.

Let us consider the following simple classical example: we want to transmit a bit of information through a noisy channel, the effect of which is to "flip" the bit with a probability p (*binary symmetric channel model*). One way to protect the initial information is to make 3 copies of the bit we have: so, if the initial bit was 0, we get the "logical bit" $0_L = 000$, and if the initial bit was 1 we get $1_L = 111$. After the transmission of the resulting string through the channel, we decide which was the original information by applying the "*majority voting*" procedure: for example, if we get 001 at the end of the channel, then we claim that the initial information was 000, that is 0_L, *etc.* It can be easily shown [Preskill 1998 (c); Nielsen 2000] that this operation leads to a new probability of error $p_C = 3p^2 - 2p^3$ which is less than p, if $p < 1/2$: so, provided that $p < 1/2$, this technique, called the *repetition code*, makes the transmission through the channel more reliable.

This simple example is very instructive and shows the basic key features of classical error-correction. First, one has to assume a particular and physically motivated *error model*: one cannot fight a *completely unknown* enemy! Then, one applies the following generic scheme. First, one *encodes* information on well-chosen states of an extended and *redundant* system of bits. For instance, in the repetition code, the original bit of information is encoded on two particular states of a three bit system. The idea is clear: redundancy prevents information from serious damage due to the errors and assures very likely recovery (let us emphasize that one uses the same kind of trick in everyday life when asking someone to repeat a sentence or a question to make sure of every word). Finally, after the transmission through the noisy channel, one *decodes* information by some appropriate procedure, as the majority voting for the repetition code: this decoding restores the initial information with a very high probability.

As we shall now see, the same ideas apply in the *quantum* domain. Nevertheless, one has to cope with typically quantum features such as the no-cloning theorem [Dieks 1982; Wootters 1982] and the continuity of errors, which make the translation from the classical to the quantum case somehow delicate. Fortunately, these quantum peculiarities can be overcome, as shown in the following paragraphs.

Let us consider the simple example, known as the *bit flip code*. The problem is the following: we want to send one qubit of information through a noisy channel which flips the qubit with a probability p: in other words, if the initial state of our qubit is $a\,|0\rangle + b\,|1\rangle$, we get $a\,|1\rangle + b\,|0\rangle$ with probability p, and $a\,|0\rangle + b\,|1\rangle$ with probability $(1-p)$. The situation is much alike the classical case we have considered above, so we could be tempted to apply the same method. But if we try to do so, we rapidly run into major quantum trouble! First, the no-cloning theorem forbids us to clone an arbitrary state. Moreover, even if cloning was possible, measurement of the qubits would completely destroy the information stored in the system. So, we have to find another way.

The appropriate technique is the following. First, one encodes the original information $a\,|0\rangle + b\,|1\rangle$ on the two logical states $|0_L\rangle = |000\rangle$ and $|1_L\rangle = |111\rangle$ of a three qubit system. This is simply achieved by adding two physical qubits, initially prepared in the state $|0\rangle$, and by applying some well-chosen unitary transformation $\widehat{C}$ to the compound system of three qubits: this operation yields the state $a\,|0_L\rangle + b\,|1_L\rangle$ which is then submitted to the action of the noisy channel. Each of the three physical qubits of the system is likely to independently undergo a bit flip (with probability p). At the end of the channel, one performs the measurement associated with the four projectors

$$\begin{aligned}
\widehat{P}_0 &= |000\rangle\langle 000| + |111\rangle\langle 111| \ \text{(no error)}\\
\widehat{P}_1 &= |100\rangle\langle 100| + |011\rangle\langle 011| \ \text{(error on the first qubit)}\\
\widehat{P}_2 &= |010\rangle\langle 010| + |101\rangle\langle 101| \ \text{(error on the second qubit)}\\
\widehat{P}_3 &= |001\rangle\langle 001| + |110\rangle\langle 110| \ \text{(error on the third qubit)}
\end{aligned} \tag{1}$$

which tells us which qubit (if any) has flipped: the result of the measurement is called the *error syndrome*. Knowing the syndrome, one can recover the initial information by applying the appropriate bit flip. For instance, if a flip occurred on the second qubit, the corrupted state at the end of the channel is

$a\ |010\rangle + b\ |101\rangle$: the measured syndrome is thus 2 and the measurement leaves the superposition intact. To recover the initial information, it remains to apply bit flip on the second qubit. Note that this is a very important feature that syndrome measurement does not give us information about a and b, which would destroy the coherent superposition, but only about which error occurred in the channel.

This procedure works perfectly provided bit flip occurs on one or fewer of the three qubits: the probability that more than one bit flip occur is $3p^2 - 2p^3$ which appears to be smaller than p, provided $p < 1/2$. In other words, when $p < 1/2$, the bit flip code decreases substantially the probability of error and thus makes the transmission through the channel sensitively more reliable.

Other simple codes exist such as the *phase flip code*, which protects information against phase flip (see below for the definition of the phase flip) and can be simply derived from the *bit flip code*. Merging these two codes, Peter Shor proposed a code which protects one qubit of information against the action of arbitrary single qubit errors (bit and phase flips): this code involves nine physical qubits and shows the same schematic structure as the previous example. Its publication renewed the interest of physicists for the domain and gave hope that quantum errors are correctable.

At the end of this brief introduction, quantum codes seem to be much alike their classical counterparts. Indeed, they are based on the same idea of redundancy, resulting from the addition of extra physical qubits. Moreover, quantum error-correcting schemes have the same frame as classical ones: after encoding the information on well chosen codewords, one sends the system through a noisy channel; then one measures the syndrome, which tells us exactly which error occurred and thus allows us to recover the original information.

In the following section, we shall set these features in the broader context of the theory of quantum error-correcting codes.

1.2 Quantum errors and correcting codes

In this section, we briefly present the general formal framework of quantum error-correction. First, we shall introduce *quantum errors* in the operator-sum formalism as the operator elements of the quantum operation describing the interaction of the computer with its environment. Then we shall review the main concepts and results of the already well developed theory of quantum error-correcting codes. Finally we will briefly present some of the most important explicit constructive methods to build quantum codes.

1.2.1 Quantum errors. A quantum computer can never be regarded as *perfectly isolated*: in all the experimental setups which have been considered (optical photon, optical cavity quantum electrodynamics, ion traps, NMR, *etc.*) the environment has some influence, which cannot be omitted. Therefore, strictly speaking, a quantum computer must be treated as an *open* system.

As a matter of fact, it must be described by a density matrix $\widehat{\rho}$ which is the partial trace over the environmental degrees of freedom of the total density matrix $\widehat{\rho}_{tot}$ of the closed compound system {quantum computer + environment}: $\widehat{\rho} = \mathrm{Tr}_{env}[\widehat{\rho}_{tot}]$. $\widehat{\rho}$ generally evolves non-unitarily: according to the operator-sum representation, the matrix $\widehat{\rho'}$ obtained after the interaction of the computer with its environment can be written under the form

$$\widehat{\rho'} = \mathcal{E}(\widehat{\rho}) = \sum_k \widehat{\mathbf{E}}_k^\dagger \, \widehat{\rho} \, \widehat{\mathbf{E}}_k, \tag{2}$$

where the operators $\left\{\widehat{\mathbf{E}}_k\right\}$, acting on the Hilbert space of the computer, are called the *operator elements*, and meet the condition[1]

$$\sum_k \widehat{\mathbf{E}}_k^\dagger \, \widehat{\mathbf{E}}_k \leq \widehat{I}. \tag{3}$$

The forms of these operators are obviously closely related to the dynamics of the compound system {quantum computer + environment}. A comprehensive discussion of this formalism, its demonstration and its physical interpretation can be found in Ref. [Nielsen 2000], Chap. 8.

In the context of quantum informatics, the $\widehat{\mathbf{E}}_k$'s are very dangerous, because they are likely to damage the information stored in the computer. They are often referred to as "*quantum errors*", and have to be corrected, in order to run a reliable computation. The main examples of quantum errors *on a single qubit* are bit-flip

$$\begin{bmatrix} 0 & 1 \\ 1 & 0 \end{bmatrix} = \widehat{\sigma}_x, \tag{4}$$

which we have already encountered in the introduction, phase flip

$$\begin{bmatrix} 1 & 0 \\ 0 & -1 \end{bmatrix} = \widehat{\sigma}_z, \tag{5}$$

bit-phase flip

$$\begin{bmatrix} 0 & -i \\ i & 0 \end{bmatrix} = \widehat{\sigma}_y, \tag{6}$$

corresponding to the three Pauli matrices, respectively. Other particular single qubit errors such as phase damping can be found in Ref. [Nielsen 2000].

1.2.2 The theory of quantum codes. Let us now show how to generalize the technique we used in the introduction in order to correct the effects of a general noise given by the quantum errors $\left\{\widehat{\mathbf{E}}_k\right\}$.

The First idea of quantum error-correction, which we have already employed in the bit flip code, is to "give space" to the system by adding extra qubits, which play the role of *ancillary* qubits; this *ancilla* adding procedure is highly related to the notion of redundancy in classical error-correction. Then, one encodes the information onto a well-chosen subspace $\mathcal{C}$, the *code space*, of the extended Hilbert space of the system comprising the initial plus extra qubits. In other words, one applies a well-chosen unitary transformation $\widehat{C}$, the *coding matrix*, which "delocalizes" information on all the qubits of the system. That is exactly what we did in the bit flip code, when encoding information onto the subspace spanned by $\{|0_L\rangle = |000\rangle\,,\ |1_L\rangle = |111\rangle\}$.

Let us now be more explicit about the properties we require from $\mathcal{C}$ to be an error-correcting code: if $\mathcal{C}_k$ denotes the subspace $\widehat{\mathbf{E}}_k\ \mathcal{C}$, then $\mathcal{C}$ must be chosen such that the different $\mathcal{C}_k$'s are orthogonal to each other and to $\mathcal{C}$; moreover, the basis vectors of $\mathcal{C}$ must be transformed by $\widehat{\mathbf{E}}_k$ into orthogonal vectors of $\mathcal{C}_k$. To be more specific, one can give the following illuminating geometric picture (taken from Ref. [Nielsen 2000]) of the properties we require from $\mathcal{C}$: if one thinks of $\mathcal{C}$ as a cube in the Hilbert space of the total system, then the different errors $\widehat{\mathbf{E}}_k$ are supposed to yield new cubes $\mathcal{C}_k$ which are disjoint. One can also easily translate these requirements into the following mathematical conditions: for any pair $\left(\widehat{\mathbf{E}}_k,\ \widehat{\mathbf{E}}_l\right)$ of quantum errors,

$$\widehat{P}\,\widehat{\mathbf{E}}_k^{\dagger}\,\widehat{\mathbf{E}}_l\,\widehat{P} = \alpha_k\,\delta_{kl}\,\widehat{P}, \tag{7}$$

where $\widehat{P}$ is the projector onto $\mathcal{C}$. These conditions can be easily checked on the example of the bit flip code. The code space $\mathcal{C} = \text{Span}\,\{|0_L\rangle\,,\ |1_L\rangle\}$ is transformed into "undeformed" and orthogonal subspaces $\{\mathcal{C}_k\}$ by the three errors

$$\widehat{\mathbf{E}}_1 = \widehat{I}\otimes\widehat{I}\otimes\widehat{\sigma}_x,\quad \widehat{\mathbf{E}}_1 = \widehat{I}\otimes\widehat{\sigma}_x\otimes\widehat{I}\ \text{ and }\ \widehat{\mathbf{E}}_1 = \widehat{\sigma}_x\otimes\widehat{I}\otimes\widehat{I}. \tag{8}$$

Let us now briefly show how error-correction works when conditions (7) are met. After transmission through the quantum noisy channel modeled by the errors $\left\{\widehat{\mathbf{E}}_k\right\}$, one diagnoses which error occurred by simply measuring in which subspace $\mathcal{C}_k$ the system is: this is achieved by measuring an appropriate observable the eigenspaces of which correspond to the $\mathcal{C}_k$'s. After this operation, and according to the assumed properties of the code space (and consequently of the related subspaces $\mathcal{C}_k$'s), the information lies undamaged in one the $\mathcal{C}_k$'s.

Finally, the initial density matrix $\widehat{\rho}$ is recovered (up to some proportionality factor) by applying the appropriate unitary transformation $\widehat{U}_k$ which brings the information back into the initial code space $\mathcal{C}$.

Formally, the whole correction procedure can be represented by a recovery quantum operation $\mathcal{R}$, such that, acting on the erroneous density matrix $\widehat{\rho'} = \mathcal{E}(\widehat{\rho}) = \sum_k \widehat{\mathbf{E}}_k^\dagger \, \widehat{\rho} \, \widehat{\mathbf{E}}_k$, it yields

$$\begin{aligned} \mathcal{R}\left(\widehat{\rho'}\right) &= \mathcal{R}\left[\mathcal{E}\left(\widehat{\rho}\right)\right] \\ &= \sum_k \widehat{U}_k^\dagger \, \widehat{P}_k \, \widehat{\mathbf{E}}_k^\dagger \, \widehat{\rho} \, \widehat{\mathbf{E}}_k \, \widehat{P}_k \, \widehat{U}_k \propto \widehat{\rho}. \end{aligned} \tag{9}$$

Actually, it is possible to extend the correction conditions (7), and show that, for a subspace $\mathcal{C}$ to be a correcting code for the set of errors $\left\{\widehat{\mathbf{E}}_k\right\}$, it is *necessary and sufficient* that there exists a complex Hermitian matrix $[\alpha_{kl}]$ such that, for any pair $\left(\widehat{\mathbf{E}}_k, \, \widehat{\mathbf{E}}_l\right)$ of quantum errors,

$$\widehat{P} \, \widehat{\mathbf{E}}_k^\dagger \, \widehat{\mathbf{E}}_l \, \widehat{P} = \alpha_{kl} \, \widehat{P}, \tag{10}$$

where $\widehat{P}$ is the projector onto $\mathcal{C}$. In other words, when relations (10) are checked, there exists a recovery quantum operation $\mathcal{R}$, mainly consisting in a syndrome measurement and a unitary transformation very similar to those we considered above (but slightly more complicated), such that $\mathcal{R}[\mathcal{E}(\widehat{\rho})] \propto \widehat{\rho}$. We shall not get into the technical (but straightforward) details of the demonstration which can be found in Ref. [Nielsen 2000] for example.

1.3 Constructing quantum codes

Until now, we have dealt with general features of quantum codes, in particular the mathematical conditions (10) they had to meet, without paying any attention to how they can be explicitly generated.

Most of the techniques used in constructing quantum codes draw their inspiration from classical error-correcting codes. In particular, the well-known class of *classical linear codes* leads to an interesting quantum generalization known as *CSS codes*, after the initials of their inventors Calderbank, Shor and Steane. We already know that coding k (classical) bits of information into n physical bits ($n > k$) consists in matching each k-bit information string with a n-bit codeword: in the particular case of classical linear codes, this encoding operation is simply performed through the application of a matrix, called the *generator matrix* G and characteristic of the code, to the k-bit string which contains the initial information. Equivalently, a classical linear code may be defined as the kernel of a matrix, called the *parity check matrix* H, which is related to the generator matrix by some duality condition. Given two classical linear codes C_1 and C_2, respectively coding k_1 bits of information into n physical bits and k_2 bits of information into n physical bits, and such that $C_2 \subset C_1$, a quantum code $\mathcal{C}$ can be constructed which encodes $(k_1 - k_2)$ qubits of information into n physical qubits; moreover, if the codes C_1 and $C_2^{\perp}$ correct up to t errors, so does the quantum code $\mathcal{C}$. The explicit construction of $\mathcal{C}$ involves simple group theoretic techniques and can be found in Ref. [Preskill 1998 (c); Nielsen 2000].

It can be shown that CSS codes are included in a more general class of quantum codes, called the *stabilizer codes*, which have been invented by D. Gottesman. Roughly speaking, the stabilizer formalism consists in characterizing state vectors by the set of Pauli operators which leave them invariant, that is by the tensor products of Pauli matrices and identity matrix which stabilize them. A quantum code, that is the vector space spanned by particular codewords, can also be defined by its "stabilizer", that is by the subgroup S comprising all the Pauli operators which leave the codewords invariant. A code $C(S)$ with a stabilizer S can correct the errors $\widehat{\mathbf{E}}_k$ of the Pauli group on n qubits which satisfy the requirement $\forall\,(j,\,k)\,,\ \widehat{\mathbf{E}}_j^{\dagger}\,\widehat{\mathbf{E}}_k \notin N(S)\backslash S$, where $N(S)$ is the *normalizer* of S, defined as the ensemble of Pauli operators $\widehat{\mathbf{E}}$ such that, for any element $\widehat{g}$ of S, $\widehat{\mathbf{E}}\,\widehat{g}\,\widehat{\mathbf{E}}^{\dagger} \in S$. The correction procedure consists in measuring all the generators $\{\widehat{g}_k\}$ of the stabilizer S, which provides the error syndrome. Having identified which error $\widehat{\mathbf{E}}_j$ has occurred one can recover information by simply applying $\widehat{\mathbf{E}}_j^{\dagger}$.

The practical techniques we have presented so far to generate quantum codes are based on the rather severe assumption that errors $\left\{\widehat{\mathbf{E}}_k\right\}$ form a subgroup of the Pauli group on n quibts. Another class of codes, known as *decoherence free subspaces* (for a review see Ref. [Lidar 2003]), exist. They can be formally considered as codes since they also meet quantum error correction conditions. However, they are rather special and constitute a completely different way to protect information against errors. The idea of these codes is to store information in a subspace $\widetilde{\mathcal{H}}$ of the total Hilbert space $\mathcal{H}$ which is not affected by the errors. In such a case, errors are not restricted to a special subgroup of the Pauli group any longer, contrary to the previous examples, but very special symmetry properties will be obviously required from the system-environment interaction for a decoherence free subspace to exist.

As we can see at the end of this short review of the main practical error-correction methods, it seems that all the ways of explicitly building quantum codes, or more generally that all the existing protection schemes require special features from the errors they combat. In the following, we address the problem of *unitary* errors and show that information can be protected from their action through a generalization of the quantum Zeno effect, without making any symmetry assumption about them.

2. Coherence protection by the quantum Zeno effect

In this section, we propose a universal scheme to protect information from the action of a given set of arbitrary unitary errors. Based upon ideas from the quantum Zeno effect, it is quite different from the regular error-correction method, even though it also draws inspiration from the conventional coding theory: roughly speaking, the point here is to prevent errors by frequently checking that information, initially encoded in a an appropriate subspace, has not been damaged.

The section starts by the presentation of a multidimensional generalization of the quantum Zeno effect, which we then employ to protect information in compound systems. Finally, we suggest the non-holonomic control technique as a physical way to implement the coding / decoding steps of our scheme.

2.1 Multidimensional quantum Zeno effect

Consider a quantum system $\mathcal{S}$, whose N-dimensional Hilbert space is denoted by $\mathcal{H}$ and whose time-dependent Hamiltonian has the form

$$\widehat{H}(\tau) = \sum_{m=1}^{M} f_m(\tau)\, \widehat{E}_m, \tag{11}$$

where $\left\{\widehat{E}_m\right\}_{m=1,\ ...,\ M}$ denote M given independent Hermitian matrices on $\mathcal{H}$ and $\{f_m(\tau)\}_{m=1,\ ...,\ M}$ denote M unknown functions of time. The Hamiltonian $\widehat{H}(\tau)$ generates an uncontrolled and undesired evolution we want to get rid of. Note that the unperturbed part of the Hamiltonian (11) is assumed to be zero (or proportional to the identity so that one can set it to zero).

The standard quantum Zeno effect [Khalfin 1957-58; Winter 1961; Misra 1977; Fonda 1978; Kofman 1996; Kofman 2000; Lewenstein 2000; Kofman 2001 (a); Schmidt 2003 / 2004] implies that we can nearly "freeze" the evolution of the system by measuring it frequently enough in one of its predetermined initial states; in other words, this effect allows us to protect the one-dimensional subspace spanned by the initial state of the system from the influence of the error-inducing Hamiltonian (11). In what follows, we generalize this effect so as to protect an arbitrary multidimensional subspace $\mathcal{C}$ from $\widehat{H}(\tau)$, under certain conditions that we will explicit in the following.

Any vector $|\psi\rangle$ of $\mathcal{C}$ evolves according to the evolution operator

$$\widehat{U}(t,\ t_0) = \mathcal{T}\left\{\exp\left[-i\int_{t_0}^{t} \widehat{H}(\tau)\, d\tau\right]\right\}, \tag{12}$$

where $\mathcal{T}$ denotes time-ordering. Hereafter we set $\hbar = 1$. For the quantum Zeno effect to hold, we shall only consider evolution during short time periods, the so-called Zeno intervals, whose duration T is such that the corresponding actions of the error Hamiltonian (11) are small, *i.e.* $\left|\widehat{E}_m \int_t^{t+T} f_m(\tau)\, d\tau\right| \ll 1$, and hence

$$\widehat{U}(t+T,\, t) = \widehat{U}_{inf} \simeq \widehat{I} - i \sum_{m=1}^{M} \left(\int_t^{t+T} f_m(\tau)\, d\tau \right) \widehat{E}_m. \tag{13}$$

This implies that after a Zeno interval T, the initial state $|\psi\rangle$ is transformed into $|\psi_e\rangle = |\psi\rangle + |\delta\psi_e\rangle$ where

$$|\delta\psi_e\rangle \simeq -i \sum_{m=1}^{M} \varepsilon_m\, \widehat{E}_m\, |\psi\rangle \text{ with } \varepsilon_m = \int f_m(\tau)\, d\tau. \tag{14}$$

Let us assume that we are physically able to perform a measurement-induced projection onto $\mathcal{C}$ in the system $\mathcal{S}$ (see the following subsection for a discussion of such projections in compound systems comprising an information subsystem and an ancilla). If we just follow the standard quantum Zeno effect procedure and merely project the state vector $|\psi_e\rangle$ (resulting from the infinitesimal evolution of the initial state $|\psi\rangle$) onto $\mathcal{C}$, we obtain a vector $|\psi_p\rangle$, which (a priori) differs from $|\psi\rangle$ (see fig. 1 a). This occurs due to the fact that usually the operators $\left\{\widehat{E}_m\right\}_{m=1,\,\dots,\,M}$ do not act orthogonally on $\mathcal{C}$, which means that the vectors $\widehat{E}_m\,|\psi\rangle$ and thus the increment vector $|\delta\psi_e\rangle$ itself are not orthogonal to $\mathcal{C}$. The standard Zeno strategy does not suffice to protect a multidimensional subspace, hence we have to adapt it in the spirit of coding theory.

To this end, we introduce a unitary matrix $\widehat{C}$ acting on $\mathcal{H}$, which we call the coding matrix, such that the Hermitian operators $\left\{\widehat{E}_m\right\}_{m=1,\,\dots,\,M}$ act orthogonally on the subspace $\widetilde{\mathcal{C}} = \widehat{C}\,\mathcal{C}$, which we call the code space. Let us denote by $I \geq 1$ the dimension of $\mathcal{C}$ and by $\{|\gamma_i\rangle\}_{i=1,\,\dots,\,I}$ one of its orthonormal bases. The set $\left\{|\widetilde{\gamma}_i\rangle = \widehat{C}\,|\gamma_i\rangle\right\}_{i=1,\,\dots,\,I}$ will denote one of the orthonormal bases of $\widetilde{\mathcal{C}}$ and the state vectors $|\widetilde{\gamma}_i\rangle$ will be called the codewords. By the definitions of $\widehat{C}$ and $\widetilde{\mathcal{C}}$, for any pair $(|\widetilde{\gamma}_s\rangle\,,\,|\widetilde{\gamma}_t\rangle)$ of codewords, and any operator $\widehat{E}_m \in \left\{\widehat{E}_m\right\}_{m=1,\,\dots,\,M}$, we have

$$\langle\widetilde{\gamma}_t|\widetilde{\gamma}_s\rangle = \delta_{st} \text{ (orthonormality condition)}, \tag{15a}$$

$$\left\langle \widetilde{\gamma}_t \middle| \widehat{E}_m \middle| \widetilde{\gamma}_s \right\rangle = 0 \text{ (orthogonality of the errors)}. \tag{15b}$$

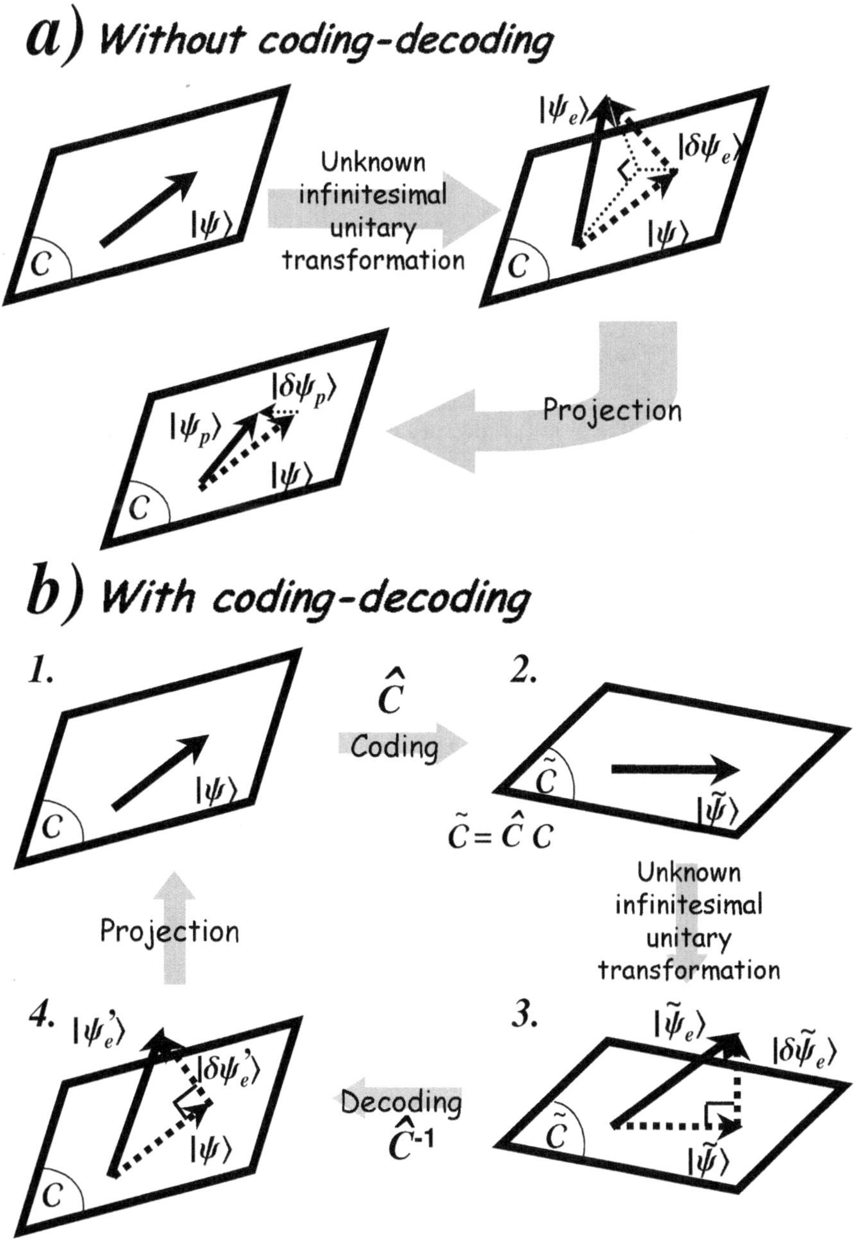

Figure 1. Multidimensional quantum Zeno effect: a) a simple projection fails to recover the initial vector, b) the sequence coding-decoding-projection protects the initial vector.

Equivalently, for any pair $(|\psi\rangle\,,\;|\chi\rangle)$ of vectors of $\mathcal{C}$ and for any operator $\widehat{E}_m \in \left\{\widehat{E}_m\right\}_{m=1,\,...,\,M}$

$$\left\langle \chi \left| \widehat{C}^{\dagger}\; \widehat{E}_m\; \widehat{C} \right| \psi \right\rangle = 0. \tag{16}$$

In particular, for any pair $(|\gamma_s\rangle\,,\;|\gamma_t\rangle)$ of basis vectors of $\mathcal{C}$ and for any operator $\widehat{E}_m \in \left\{\widehat{E}_m\right\}_{m=1,\,...,\,M}$

$$\left\langle \gamma_t \left| \widehat{C}^{\dagger}\; \widehat{E}_m\; \widehat{C} \right| \gamma_s \right\rangle = 0. \tag{17}$$

When we apply the coding matrix $\widehat{C}$ to the initial state vector $|\psi\rangle$, before exposing it to the action of the Hamiltonian (11), we obtain a new vector $|\widetilde{\psi}\rangle = \widehat{C}|\psi\rangle \in \widetilde{\mathcal{C}}$ (Figs. 1 b1 and b2) which is transformed into $|\widetilde{\psi}_e\rangle = \widehat{U}_{inf}|\widetilde{\psi}\rangle = |\widetilde{\psi}\rangle + |\delta\widetilde{\psi}_e\rangle$ after a Zeno interval T, where

$$\left|\delta\widetilde{\psi}_e\right\rangle \simeq -i \sum_{m=1}^{M} \varepsilon_m\; \widehat{E}_m \left|\widetilde{\psi}\right\rangle = -i \sum_{m=1}^{M} \varepsilon_m\; \widehat{E}_m\; \widehat{C}\,|\psi\rangle\,. \tag{18}$$

In Fig. 1 b3, we show the corresponding geometric picture. Decoding $|\widetilde{\psi}_e\rangle$ yields the vector $|\psi'_e\rangle = \widehat{C}^{-1}|\widetilde{\psi}_e\rangle = |\psi\rangle + |\delta\psi'_e\rangle$ where

$$\left|\delta\psi'_e\right\rangle \simeq -i \sum_{m=1}^{M} \varepsilon_m\; \widehat{C}^{\dagger}\; \widehat{E}_m\; \widehat{C}\,|\psi\rangle\,. \tag{19}$$

From Eq. (16) it can be seen that for any vector $|\chi\rangle \in \mathcal{C}$,

$$\langle\chi|\delta\psi'_e\rangle = -i \sum_{m=1}^{M} \varepsilon_m \left\langle \chi \left| \widehat{C}^{\dagger}\; \widehat{E}_m\; \widehat{C} \right| \psi \right\rangle = 0 \tag{20}$$

which means that $|\delta\psi'_e\rangle$ is orthogonal to $\mathcal{C}$ (Fig. 1 b4). A measurement-induced projection onto $\mathcal{C}$ finally recovers the initial vector $|\psi\rangle$ with a probability very close to 1. To be more specific, this probability coincides with unity up to second order corrections of the time dependent perturbation theory, or, in other words, the error probability is proportional to T^2. If the coding-decoding-projection sequence is frequently repeated, any vector $|\psi\rangle$ of the subspace $\mathcal{C}$ can thus be protected from the Hamiltonian (11) during the protection time T_2^2/T where T_2 is a typical period associated with this Hamiltonian. By increasing the repetition frequency $1/T$ this protection time can be done as long as needed.

Let us note that a more general version of conditions (15b) can also be considered. Indeed, if for any pair of codewords $(|\widetilde{\gamma}_s\rangle\,,\;|\widetilde{\gamma}_t\rangle)$ of $\widetilde{\mathcal{C}}$ and any error

Hamiltonian $\widehat{E}_m \in \left\{\widehat{E}_m\right\}$, $\langle\widetilde{\gamma}_t|\widehat{E}_m|\widetilde{\gamma}_s\rangle = \delta_{ts}\,\xi_m$, where δ_{ts} is the Kronecker symbol and ξ_m a real number depending only on the number m of the Hamiltonian $\widehat{E}_m$, the projection onto $\mathcal{C} = \text{Span}\{|\gamma_i\rangle,\ i = 1,\ \ldots,\ I\}$ of the state vector

$$\left|\psi_e'\right\rangle = |\psi\rangle - i \sum_{m=1}^{M} \varepsilon_m\, \widehat{E}_m\, |\psi\rangle\,, \tag{21}$$

obtained after a (coding-decoding) sequence, yields

$$\widehat{\Pi}_C \left|\psi_e'\right\rangle = |\psi\rangle - i \sum_{m=1}^{M} \varepsilon_m\, \widehat{\Pi}_C\, \widehat{E}_m\ |\psi\rangle\,, \quad \text{where} \quad \widehat{\Pi}_C = \left[\sum_{t=1}^{I} |\gamma_t\rangle\,\langle\gamma_t|\right]. \tag{22}$$

If we denote by

$$|\psi\rangle = \sum_{s=1}^{I} \alpha_s\, |\gamma_s\rangle \tag{23}$$

the decomposition of the initial information state vector, $\widehat{\Pi}_C\ \widehat{E}_m\,|\psi\rangle$ has the form

$$\begin{aligned} \widehat{\Pi}_C\, \widehat{E}_m\, |\psi\rangle \ &= \sum_{s,\ t=1}^{I} \alpha_s\, |\gamma_t\rangle \left\langle \gamma_t \left| \widehat{C}^\dagger\, \widehat{E}_m\, \widehat{C} \right| \gamma_s \right\rangle \\ &= \xi_m \sum_{s,\ t=1}^{I} \alpha_s\, \delta_{ts}\, |\gamma_t\rangle \ = \ \xi_m\, |\psi\rangle\,, \end{aligned} \tag{24}$$

which finally leads to

$$\widehat{\Pi}_C \left|\psi_e'\right\rangle = \left(1 - i \sum_{m=1}^{M} \varepsilon_m\, \xi_m\right) |\psi\rangle\,. \tag{25}$$

In other words, the Hamiltonians $\widehat{E}_m$ just introduce a global phase factor in front of the initial information state vector, but leaves its coherence intact. Obviously, the correction conditions (15b) are obtained as a particular case of the above conditions, setting $\xi_m = 0$ for all m. Yet, though less general, they will be employed in the rest of the paper for the sake of simplicity.

The multidimensional generalization of the quantum Zeno effect we have just described allows us to protect an arbitrary subspace $\mathcal{C}$ of a Hilbert space $\mathcal{H}$

against Hamiltonians of the form (11) provided the projection onto $\mathcal{C}$ is physically achievable and the coding matrix $\widehat{C}$ exists. This result is very useful in the context of information protection as we will show in the following subsection.

2.2 Information protection through the multidimensional quantum Zeno effect

Consider an information system $\mathcal{I}$ whose I-dimensional Hilbert space is denoted by $\mathcal{H}_I$ and suppose that this system is subjected to a set of M error-inducing Hamiltonians $\left\{\widehat{E}_m\right\}_{m=1,\ ...,\ M}$ which, for instance, represent interactions of the system with M uncontrolled external fields $f_m(t)$: we want to get rid of this external influence which is likely to result in the loss of the information stored in the initial state vector $|\psi_I\rangle = \sum_{i=1}^{I} c_i\,|\nu_i\rangle$, where $\{|\nu_i\rangle\}_{i=1,\ ...,\ I}$ denotes an orthonormal basis of $\mathcal{H}_I$. To this end, we will use the multidimensional Zeno effect. As the multidimensional quantum Zeno effect can only protect a subspace of the whole Hilbert space, we first have to add an A-dimensional auxiliary system $\mathcal{A}$ (called ancilla) to our system $\mathcal{I}$, so that the information is transferred from $\mathcal{H}_I$ into an I-dimensional subspace $\mathcal{C}$ of the $(N = I \times A)$-dimensional Hilbert space $\mathcal{H} = \mathcal{H}_I \otimes \mathcal{H}_A$ of the compound system $\mathcal{S} = \mathcal{I} \otimes \mathcal{A}$. Furthermore, we shall suppose that all the state vectors of the different Hilbert spaces $\mathcal{H}_I$, $\mathcal{H}_A$ and hence $\mathcal{H}$ are degenerate in energy so that the unperturbed part $\widehat{H}_0$ of the Hamiltonian of the compound system can be set to zero as in the previous subsection: the subspace $\mathcal{C}$ and the information it carries can thus be protected through the multidimensional quantum Zeno effect (in Sec. 3, on the example of *Rb*, we shall see that the multidimensional quantum Zeno effect may also be used even though $\widehat{H}_0$ is not zero, provided $\widehat{H}_0$ and the errors have some convenient properties).

To illustrate this, let us first consider the simple case in which the ancilla is initially in the pure state $|\alpha\rangle$. The information previously carried by $|\psi_I\rangle \in \mathcal{H}_I$ is then transferred into the factorized state

$$|\psi\rangle = |\psi_I\rangle \otimes |\alpha\rangle = \sum_{i=1}^{I} c_i\,|\nu_i\rangle \otimes |\alpha\rangle = \sum_{i=1}^{I} c_i\,|\gamma_i\rangle \tag{26}$$

which belongs to the tensor product subspace

$$\mathcal{C} = \mathcal{H}_I \otimes \text{Span}\,[|\alpha\rangle] = \text{Span}\left[\{|\gamma_i\rangle = |\nu_i\rangle \otimes |\alpha\rangle\}_{i=1,\ ...,\ I}\right]. \tag{27}$$

In other words, the initial density matrix of the compound system $\mathcal{S}$ is $\widehat{\rho} = (|\psi_I\rangle\langle\psi_I|) \otimes (|\alpha\rangle\langle\alpha|)$. After coding (through the matrix $\widehat{C}$) it reads $\widehat{\widehat{\rho}} = \widehat{C}^\dagger\,\widehat{\rho}\,\widehat{C}$; after the short time T (Zeno interval), infinitesimal errors have transformed it into $\widehat{\widehat{\rho}}_e = \widehat{U}_{inf}^\dagger\,\widehat{C}^\dagger\,\widehat{\rho}\,\widehat{C}\,\widehat{U}_{inf}$; finally it takes the form

$\widehat{\rho}_e = \widehat{C}\,\widehat{U}^{\dagger}_{inf}\,\widehat{C}^{\dagger}\,\widehat{\rho}\,\widehat{C}\,\widehat{U}_{inf}\,\widehat{C}^{\dagger}$ after decoding. In this setting, the projection onto $\mathcal{C}$ can be simply achieved by measuring the ancilla in its initial state $|\alpha\rangle$. As T is very short, the state of the ancilla evolves just a little within a Zeno interval, such that the probability of detecting it in its initial state $|\alpha\rangle$, and thus of projecting the state of the compound system onto $\mathcal{C}$ is very close to 1. After the projection, we trace out the ancilla to obtain the final reduced density matrix $\widehat{\rho'}_I = \langle\alpha|\,\widehat{C}\,\widehat{U}^{\dagger}_{inf}\,\widehat{C}^{\dagger}\,\widehat{\rho}\,\widehat{C}\,\widehat{U}_{inf}\,\widehat{C}^{\dagger}\,|\alpha\rangle$ for the information system $\mathcal{I}$; in the same way, one can calculate that the initial reduced density matrix is $\widehat{\rho}_I = |\psi_I\rangle\,\langle\psi_I|$. The variation $\delta\widehat{\rho}_I = \widehat{\rho'}_I - \widehat{\rho}_I$ of the information-space density matrix during the whole process can be expressed as the commutator

$$\delta\widehat{\rho}_I = -i\left[\sum_{m=1}^{M}\int f_m(\tau)\,d\tau\,\langle\alpha|\,\widehat{C}^{\dagger}\,\widehat{E}_m\,\widehat{C}\,|\alpha\rangle\,,\,\widehat{\rho}_I\right], \tag{28}$$

from which we infer that $\widehat{\rho}_I$ satisfies the equation

$$i\,\frac{d\widehat{\rho}_I}{dt} = \left[\widehat{h}_e,\,\widehat{\rho}_I\right],\quad \widehat{h}_e = \sum_{m=1}^{M} f_m\left\langle\alpha\left|\widehat{C}^{\dagger}\,\widehat{E}_m\,\widehat{C}\right|\alpha\right\rangle. \tag{29}$$

Here $\widehat{h}_e$ is an effective Hamiltonian which is determined by the error-inducing Hamiltonians transformed by the coding and decoding and projected onto the initial state of the ancilla. From Eq. (16) one can infer that $\widehat{h}_e = 0$ and hence $\widehat{\rho}_I$ remains constant in time: as long as we repeat the coding-decoding-ancilla resetting sequence, the information initially stored in $\mathcal{I}$ is protected.

It is not always feasible to directly measure the ancilla independently from the information system; in other words, it is sometimes impossible to perform a projection onto disentangled subspaces of $\mathcal{H}$ of the form $\mathcal{H}_I\otimes\text{Span}\left[|\alpha\rangle\right]$: in some cases, as for the example proposed in Sec. 3, one can only project onto entangled subspaces of the total Hilbert space $\mathcal{H}$. In such a case the information initially stored in the vector $|\psi_I\rangle = \sum_{i=1}^{I} c_i\,|\nu_i\rangle \in \mathcal{H}_I$ must be transferred into an entangled state of $\mathcal{I}$ and $\mathcal{A}$ of the form $|\psi\rangle = \sum_{i=1}^{I} c_i\,|\gamma_i\rangle$ where the I vectors $|\gamma_i\rangle$ ($i = 1,\ \ldots,\ I$) which form an orthonormal basis of the information-carrying subspace $\mathcal{C}$, are generally not factorized as earlier but entangled states. Nevertheless the same method as before can be used in that case to protect information, albeit in a different subspace $\mathcal{C}$.

To conclude this description of our method, let us now return to conditions (15a) and (15b) imposed on the codewords $\{|\widetilde{\gamma}_i\rangle\,, i = 1, ..., I\}$ and make two remarks:

A. We can establish a useful relation between the dimension of the ancilla and the number of correctable error Hamiltonians. The set of the I codewords can be seen as a collection of $2I \times N = 2I^2A$ real numbers on which $2I + 2MI = 2I^2(1+M)$ constraints, directly derived from Eqs. (15a) and

(15b), are imposed. The number of free parameters must be larger than the number of constraints, hence we necessarily have $2I^2A \geq 2I^2(1+M)$, which satisfies

$$A - 1 \geq M. \tag{30}$$

This condition gives an upper-bound on the number of independent error-inducing Hamiltonians that our method can correct simultaneously and is called the "Hamming bound".

B. We may compare our conditions (15b) with the general conditions (10) of standard quantum error-correction. First, let us introduce $\mathcal{G}\left(\left\{\widehat{E}_m\right\}\right)$, the group of all possible erroneous evolutions induced by the set $\left\{\widehat{E}_m\right\}$ of error Hamiltonians, and $\left\{\widehat{\mathbf{E}}_j\left(\left\{\widehat{E}_m\right\}\right)\right\}$, a complete basis set of operators which spans the space of erroneous evolutions induced by the $\widehat{E}_m$'s. Then, conditions (15b) can be translated, in our problem, into the following relations

$$\begin{aligned} &\forall \left(|\widetilde{\gamma}_s\rangle\,,\,|\widetilde{\gamma}_t\rangle\right) \in \widetilde{\mathcal{C}}^2,\ \forall\left(\widehat{\mathbf{E}}_k,\,\widehat{\mathbf{E}}_l\right) \in \left\{\widehat{\mathbf{E}}_j\left(\left\{\widehat{E}_m\right\}\right)\right\}, \\ &\left\langle \widetilde{\gamma}_t \left| \widehat{\mathbf{E}}_k^\dagger\, \widehat{\mathbf{E}}_l \right| \widetilde{\gamma}_s \right\rangle = \alpha_{kl} \left\langle \widetilde{\gamma}_t | \widetilde{\gamma}_s \right\rangle, \end{aligned} \tag{31}$$

where $[\alpha_{kl}]$ is a Hermitian complex matrix. The variety of all linear combinations of $\widehat{\mathbf{E}}_j$ includes not only all $\widehat{E}_m$ but also many other operators given by commutators of all orders in $\widehat{E}_m$ entering the expansion of $\widehat{U}$ for long times. The conditions (31) are therefore much more restrictive than Eq. (15b). Moreover, even for just two generic matrices $\widehat{E}_m$, the basis $\left\{\widehat{\mathbf{E}}_j\right\}$ spans the entire Hilbert space $\mathcal{H}$, yielding $\widetilde{\mathcal{C}} = \emptyset$. Only if the set $\left\{\widehat{E}_m\right\}$ belongs to an extraspecial algebra restricting the erroneous evolution operators to a subgroup $\mathcal{G}\left(\left\{\widehat{E}_m\right\}\right) \subset \mathcal{G}_U\left(\mathcal{H}\right)$ of the full unitary group in $\mathcal{H}$, a non-trivial code space $\widetilde{\mathcal{C}}$ may exist. The Zeno effect is the only way to suppress loss of coherence if it is not the case.

2.3 Physical implementation of the coding matrix

In the previous subsection, we have presented a coherence protectionxcoherence,protection method, in which the code space and the (unitary) coding matrix $\widehat{C}$ play a crucial role. This subspace and this matrix can be calculated by an appropriate algorithm, presented in Ref. [Brion 2004] and in the first appendix, whose basic idea is to approach the codewords $\{|\widetilde{\gamma}_i\rangle\,,\ i = 1,\,\ldots,\,I\}$ iteratively from a randomly picked set of orthonormal vectors. But, when determined, how can the coding matrix be *physically* implemented? The non-holonomic control technique provides a useful answer to this question.

The non-holonomic control technique has been suggested in Ref. [Akulin 2001] as a means of controlling the evolution of quantum systems. Basically, it consists in alternately applying two "well-chosen" perturbations $\widehat{V}_a$ and $\widehat{V}_b$ to the system $\mathcal{S}$ which we want to control during pulses with timings t_i. The total Hamiltonian $\widehat{H} = \widehat{H}_0 + \widehat{V}$ thus has a pulse-shaped time dependence and alternately takes the two values $\widehat{H}_a \equiv \widehat{H}_0 + \widehat{V}_a$ (during odd-numbered pulses) and $\widehat{H}_b \equiv \widehat{H}_0 + \widehat{V}_b$ (even-numbered pulses). The pulse timings t_i then play the role of free parameters one has to adjust in order to perform the desired control operation. To be more explicit, the perturbations $\widehat{V}_a$ and $\widehat{V}_b$ must be chosen such that they satisfy the *bracket generation condition, i.e.*, such that the commutators of all orders of $\widehat{H}_a \equiv \widehat{H}_0 + \widehat{V}_a$ and $\widehat{H}_b \equiv \widehat{H}_0 + \widehat{V}_b$ span the whole space of Hermitian matrices acting on the system under consideration. From the Campbell-Baker-Hausdorf formula [Cornweel 1984], it follows that this is a necessary condition of controllability. It also proves to be sufficient in all the practical cases we dealt with. For that reason, we believe that we have "good controllability conditions" as soon as the bracket generation condition is checked. The number n_C of control timings t_i depends on the particular problem to be solved. For instance, in order to impose an arbitrary evolution $\widehat{U}_{arb}$ on an N-dimensional system, we need at least $n_C = N^2$ timings t_i, since N^2 is the total number of free real parameters characterizing a $N \times N$ unitary matrix. In Ref. [Akulin 2001], a general algorithm is developed which allows us to find the appropriate timings t_i realizing

$$\widehat{U}\left(t_1,\, t_2,\, \ldots,\, t_{N^2}\right) = e^{-i\,\widehat{H}_b\, t_{N^2}}\, e^{-i\,\widehat{H}_a\, t_{N^2-1}}\,\ldots\, e^{-i\,\widehat{H}_a\, t_1} = \widehat{U}_{arb} \quad (32)$$

for any arbitrary unitary matrix $\widehat{U}_{arb}$.

We can directly apply this result to find the timings $\{t_i\}$ such that

$$\widehat{U}\left(t_1,\, t_2,\, \ldots,\, t_{N^2}\right) = e^{-i\,\widehat{H}_b\, t_{N^2}}\, e^{-i\,\widehat{H}_a\, t_{N^2-1}}\,\ldots\, e^{-i\,\widehat{H}_a\, t_1} = \widehat{C}, \quad (33)$$

where the coding matrix $\widehat{C}$ has been determined by the algorithm described in the first appendix.

However, this procedure appears to require a lot of useless numerical work. Indeed, most of the information contained in such a coding matrix is irrelevant and all the N^2 real parameters of $\widehat{C}$ do not have to be specified. It follows that the number n_C of necessary control parameters $\{t_i\}$ required for the coding is much less than N^2. Let us examine this point in more detail.

Since the coding matrix is characterized by relations (16), the problem of control reduces to finding n_C timings t_i, such that the non-holonomic evolution matrix

$$\widehat{U}\left(\vec{t}\right) = \exp\left(-i\,\widehat{H}_b\,t_{n_C}\right)\,\exp\left(-i\,\widehat{H}_a\,t_{n_C-1}\right)\,\dots\,\exp\left(-i\,\widehat{H}_a\,t_1\right) \quad (34)$$

meets conditions (16), that is for any pair $\left(|\gamma_s\rangle\,,\,|\gamma_t\rangle\right)_{1\leq s,\,t\leq I}$ of basis vectors of $\mathcal{C}$ and any operator $\widehat{E}_m \in \left\{\widehat{E}_m\right\}_{m=1,\,\dots,\,M}$

$$\left\langle\gamma_t\left|\widehat{U}^\dagger\left(\vec{t}\right)\,\widehat{E}_m\,\widehat{U}\left(\vec{t}\right)\right|\gamma_s\right\rangle = 0. \quad (35)$$

The number n_C of control parameters needs only to exceed the number of independent constraints which is clearly $\sim MI^2$, that is $n_C \sim MI^2$ which is much smaller than N^2. Therefore, here, what we really need is only a partial (and less expensive) rather than complete control of the evolution matrix. To deal with this new type of problem, we have built a new algorithm [Brion 2004], different from the complete control algorithm, which is presented in detail in the second appendix.

Note that the decoding matrix can be easily implemented when the two control Hamiltonians $\widehat{H}_a$ and $\widehat{H}_b$ can be reversed: in that case, one just needs to change the signs of the operators $\widehat{H}_a$ and $\widehat{H}_b$, and to apply the same control sequence as for the coding procedure, but backwards in time. On the contrary, if $\widehat{H}_a$ and $\widehat{H}_b$ cannot be reversed, one has to employ the general (and time-consuming) non-holonomic control technique, involving all the N^2 control parameters, to find timings which realize $\widehat{C}^{-1}$.

In this section, we have presented a method which allows one to protect information from a given set of unitary errors. Moreover, we have suggested a physical way to implement the coding and decoding steps, employing the non-holonomic control. In the last section, we propose an application of our method to a real physical system, namely a Rubidium isotope.

3. Coherence protection applied to the rubidium atom

The goal of this section is to illustrate our protection scheme by proposing its application to a real physical system, emphasizing the experimental challenges our method faces and suggesting a possible way to meet them. To this end we have chosen the following problem: we wish to protect a single qubit of information encoded on the two spin states of the ground level 5s of the isotope ^{78}Rb. As possible error-inducing perturbations $\widehat{E}_m$, we have considered 3 components of the magnetic field with the Hamiltonians $\left\{\widehat{E}_k^\beta \propto \widehat{L}_k + 2\widehat{S}_k,\ k = x,\,y,\,z\right\}$, and 3 components of the electric field, interacting with the system via the second order Hamiltonians $\left\{\widehat{E}_{k,\,l}^\varepsilon \propto \widehat{r}_k^2 - \widehat{r}_l^2,\ k,\,l = x,\,y,\,z,\ k < l\right\}$.

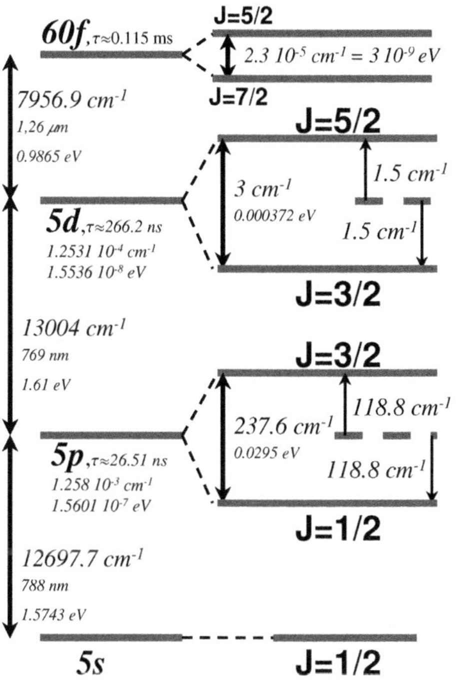

Figure 2 Spectrum of Rubidium ^{78}Rb: the useful part of the spectrum of rubidium is represented.

Before presenting the details of this physical implementation, let us motivate the choice of the atom. Alkali atoms like Rb are very interesting for the coherence-loss protection purpose because of their hydrogen-like structure. Such an atom can indeed be considered as a compound of an information subsystem, associated with the spin part of the wavefunction, and an ancilla, corresponding to the orbital part of the atomic state. As we will see, in such a setting, it is easy to manipulate the dimensionality of the ancilla by simply changing the orbital angular momentum L through controlled transitions of the atom.

We have chosen ^{78}Rb among all alkali systems because of its spectroscopic characterics (Fig. 2) [Bacher 1932; Lindgard 1977; Gallagher 1994]. In particular, ^{78}Rb has no hyperfine structure (its nuclear spin is 0) which ensures that the ground level $5s$ is degenerate: this is necessary for the projection scheme as we shall see below. Moreover it has a long enough lifetime ($\tau \simeq 17.66\,\mathrm{min}$) for the proposed experiment.

Let us now review each step of our method in detail. As mentioned above, the information to be protected is initially encoded on the two spin states

$$|\nu_1\rangle = \left|5s,\ j = \frac{1}{2},\ m_j = -\frac{1}{2}\right\rangle \text{ and } |\nu_2\rangle = \left|5s,\ j = \frac{1}{2},\ m_j = \frac{1}{2}\right\rangle \quad (36)$$

of the ground level $5s$ of the atom. These two states span the information space $\mathcal{H}_I = \text{Span}[|\nu_1\rangle,\ |\nu_2\rangle]$ whose dimension in this case is $I = 2$. The first step of our scheme consists in adding an ancilla $\mathcal{A}$ to the information system. The role of $\mathcal{A}$ is played by the orbital part of the wavefunction. In the ground state ($L = 0$), its dimension is $A = 2L + 1 = 1$ (roughly speaking, there is no ancilla). If we want to protect one qubit of information against $M = 6$ error-inducing Hamiltonians, we have to increase the dimensionality of the ancilla up to $A = M + 1 = 7$ (Eq. (30)). This can be achieved by pumping the atom to a shell nf ($L = 3$). We choose the highly excited Rydberg state $60f$ so as to make the influence of the fine structure as weak as possible (the splitting for $60f$ is approximately 10^{-5} cm^{-1} [Gallagher 1994]). We first consider the fine structure as negligible so that the $N = I \times A = 2 \times 7 = 14$ basis vectors of the total Hilbert space $\mathcal{H} = \mathcal{H}_I \otimes \mathcal{H}_A$ are almost perfectly degenerate. The validity of this approximation will be discussed at the end of this section. To be more specific, the pumping is done in such a way that

$$\begin{aligned} |\nu_1\rangle &\longrightarrow |\gamma_1\rangle = \left|60f,\ j = \frac{5}{2},\ m_j = -\frac{3}{2}\right\rangle \\ |\nu_2\rangle &\longrightarrow |\gamma_2\rangle = \left|60f,\ j = \frac{5}{2},\ m_j = -\frac{1}{2}\right\rangle . \end{aligned} \quad (37)$$

In other words, the information initially stored in $\mathcal{H}_I$ is transferred into

$$\begin{aligned} \mathcal{C} = \ \text{Span}\Bigg[&|\gamma_1\rangle = \left|60f,\ j = \frac{5}{2},\ m_j = -\frac{3}{2}\right\rangle, \\ &|\gamma_2\rangle = \left|60f,\ j = \frac{5}{2},\ m_j = -\frac{1}{2}\right\rangle\Bigg] . \end{aligned} \quad (38)$$

Such a choice of the subspace $\mathcal{C}$ will be justified later by the practical feasibility of the projection process onto $\mathcal{C}$.

Practically, the pumping can be achieved through the following multiphoton process. One applies three laser fields: the first laser is right polarized and slightly detuned from the atomic transition $(5s \longleftrightarrow 5p)$ whereas the second and third lasers are left polarized and slightly detuned from the transitions $(5p_{3/2} \longleftrightarrow 5d_{3/2})$ and $(5d_{3/2} \longleftrightarrow 60f)$ respectively. The detunings suppress real one-photon processes: the atom can only absorb three photons simultaneously and is thereby excited from the ground level $5s$ directly to the Rydberg level $60f$. With the help of the dipole selection rules, one can construct the allowed paths shown on Fig. 3: these paths only couple $|\nu_1\rangle$ and $|\nu_2\rangle$ to $|\gamma_1\rangle$ and $|\gamma_2\rangle$, respectively.

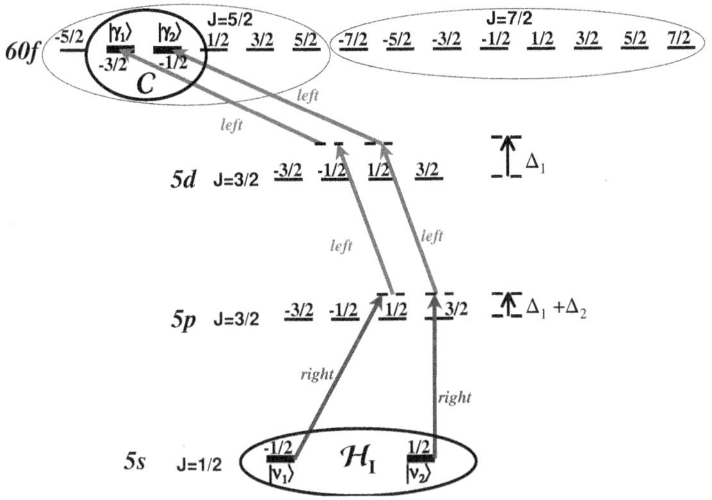

Figure 3. Ancilla adding by Pumping. Photon polarization and involved sub-Zeeman levels are represented.

The second step consists in encoding the information by the non-holonomic control technique. In order to impose the coding matrix to the excited atom, we apply $n_C = 34$ control laser pulses of timings $\{t_i\}_{i=1,\ldots,34}$ in the presence of a constant magnetic field. During the pulses, two different combinations of magnetic and Raman Hamiltonians are alternately applied to the system (see Fig. 4). To be more explicit, during odd-numbered pulses ("A" type pulses) we apply a constant magnetic field

$$\vec{B} = \begin{pmatrix} B_x = 7.10^{-3}\,T \\ \\ B_y = 8.2 \times 10^{-3}\,T \\ \\ B_z = -\,6.8 \times 10^{-3}\,T \end{pmatrix} \tag{39}$$

which is associated with the Zeeman Hamiltonian $\widehat{W}_Z$, and two sinusoidal electric laser fields

$$\vec{E}_a(t) = \text{Re}\left[\underline{\vec{E}}_a\, e^{-i\,\omega_R t}\right],$$

$$\vec{E}'_a(t) = \text{Re}\left[\underline{\vec{E}}'_a\, e^{-i\,\omega'_R t}\right],$$

$$\text{where } \underline{\vec{E}}_a = \left| \begin{array}{c} E_{x,\,a} \\ E_{y,\,a}\, e^{-i\,\varphi_{y,\,a}} \\ 0 \end{array} \right. \quad \text{and } \underline{\vec{E}}'_a = \left| \begin{array}{c} E'_{x,\,a} \\ E'_{y,\,a}\, e^{-i\,\varphi'_{y,\,a}} \\ 0 \end{array} \right. \tag{40}$$

whose frequencies ω_R and ω'_R are respectively slightly detuned from the two transitions $(60f \longleftrightarrow 5d,\ j = 3/2)$ and $(60f \longleftrightarrow 5d,\ j = 5/2)$ (detunings δ and δ'). The chosen characteristic values of these fields are

$$\begin{aligned} & E_{x,\,a} = E'_{x,\,a} = 8.5 \times 10^5\ \text{V.m}^{-1} \\ & E_{y,\,a} = E'_{y,\,a} = 5.2 \times 10^6\ \text{V.m}^{-1} \\ & \varphi_{y,\,a} = \varphi'_{y,\,a} = 2.3 \\ & \hbar\,\omega_R = 0.986324\ eV = 7955.14\ \text{cm}^{-1} \\ & \delta = -\ 0.000010\ \text{eV} = -\ 0.080654\ \text{cm}^{-1} \\ & \hbar\,\omega'_R = 0.986676\ \text{eV} = 7958.14\ \text{cm}^{-1} \\ & \delta' = 0.000010\ \text{eV} = 0.080654\ \text{cm}^{-1}. \end{aligned} \tag{41}$$

The intensity of the laser beams are typically of the order of $2.\,10^8$ W.cm^{-1}. The Raman Hamiltonian associated with these fields is

$$\begin{aligned} \widehat{W}_{R,\,a} = & \sum_{(i,\,j)\in 60f,\ k\in(5d,\ 3/2)} \frac{\left(\vec{d}_{ik}.\,\vec{E}_a\right)\left(\vec{d}_{kj}.\,\vec{E}_a\right)}{\delta}\ |i\rangle\,\langle j| \\ & + \sum_{(i,\,j)\in 60f,\ k\in(5d,\ 5/2)} \frac{\left(\vec{d}_{ik}.\,\vec{E}'_a\right)\left(\vec{d}_{kj}.\,\vec{E}'_a\right)}{\delta'}\ |i\rangle\,\langle j|\,, \end{aligned} \tag{42}$$

where $\vec{d}_{ik}$ denotes the electric dipole element $\langle i|\widehat{\vec{d}}|k\rangle$. The total perturbation is $\widehat{V}_a = \widehat{W}_Z + \widehat{W}_{R,\,a}$. During even-numbered pulses ("B" type pulses), we apply the same magnetic field as for A type pulses, which is experimentally convenient, and two other sinusoidal electric laser fields

$$\vec{E}_b(t) = \text{Re}\left[\underline{\vec{E}_b}\, e^{-i\,\omega_R t}\right],$$

$$\vec{E}'_b(t) = \text{Re}\left[\underline{\vec{E}'_b}\, e^{-i\,\omega'_R t}\right], \tag{43}$$

$$\text{where } \underline{\vec{E}_b} = \left|\begin{array}{c} E_{x,\,b} \\ E_{y,\,b}\, e^{-i\,\varphi_{y,\,b}} \\ 0 \end{array}\right. \quad \text{and } \underline{\vec{E}'_b} = \left|\begin{array}{c} E'_{x,\,b} \\ E'_{y,\,b}\, e^{-i\,\varphi'_{y,\,b}} \\ 0 \end{array}\right.$$

whose frequencies are the same as above and whose characteristics values are

$$E_{x,\,b} = E'_{x,\,b} = -\,5.2 \times 10^6\ \text{V.m}^{-1}$$

$$E_{y,\,b} = E'_{y,\,b} = 8.5 \times 10^5\ \text{V.m}^{-1} \tag{44}$$

$$\varphi_{y,\,a} = \varphi'_{y,\,a} = 2.3.$$

The Raman Hamiltonian associated with these fields is $\widehat{W}_{R,\,b}$ and can be obtained straightforwardly from the expression of $\widehat{W}_{R,\,a}$ by simply changing a's into b's.

The corresponding perturbation is $\widehat{V}_b = \widehat{W}_Z + \widehat{W}_{R,\,b}$. The fine structure of the level $60f$ is negligible, and therefore the unperturbed Hamiltonian $\widehat{H}_0$ is taken to be 0 and the total Hamiltonian has the form: $\widehat{H}_a = \widehat{V}_a$ during "A" pulses, $\widehat{H}_b = \widehat{V}_b$ during "B" pulses. The 34 different timings have been calculated so that

$$\widehat{U}(t_1,\ \ldots,\ t_{34}) = e^{-i\,\widehat{H}_b\, t_{n_C}}\, e^{-i\,\widehat{H}_a\, t_{n_C-1}} \ldots e^{-i\,\widehat{H}_a\, t_1} = \widehat{C} \tag{45}$$

meets conditions (16). At the end of the coding step the information is transferred into the code space $\widetilde{\mathcal{C}} = \widehat{C}\ \mathcal{C}$, encoded on the codewords $\left\{|\widetilde{\gamma}_i\rangle = \widehat{C}\,|\gamma_i\rangle\right\}_{i=1,\,2}$.

The total duration of the control sequence shown in Fig. 4 is $\tau \simeq 125$ ns which is approximately 10^3 times shorter than the lifetime $\tau_{life} = 0.115$ ms of the Rydberg state $60f$ as can be calculated from the spectroscopic data [Gallagher 1994]. The different pulse timings range between 2.9 ns and 7.4 ns, which are feasible. They are reproduced in Tab. 1.

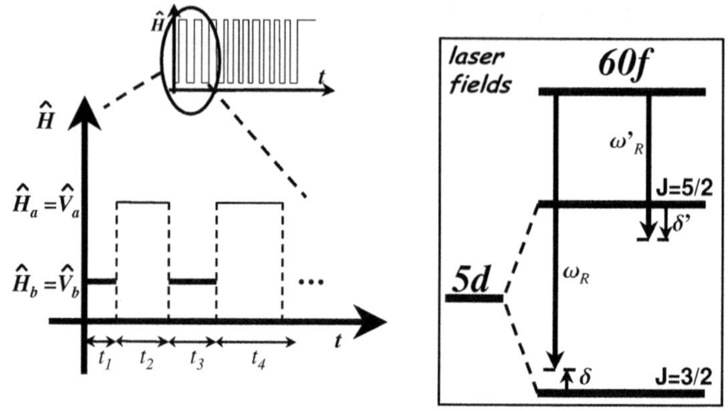

Figure 4. Coding step through the non-holonomic control technique. The two Hamiltonians $\widehat{H}_a$ and $\widehat{H}_b$ are alternately applied to the system during pulses whose timings are reproduced in Tab. 1. The pulsations of the laser fields involved in the coding step are represented on the spectrum of the rubidium atom.

Table 1. Control timings of the coding step (in nanoseconds).

n	t_n	n	t_n	n	t_n
1	32.5935	13	21.4060	25	24.8604
2	42.1221	14	34.5965	26	37.2753
3	37.3074	15	24.2740	27	26.1847
4	40.8992	16	35.7984	28	18.5771
5	38.8913	17	21.6005	29	29.1574
6	41.4479	18	30.2353	30	38.5405
7	39.8458	19	28.5720	31	29.5389
8	21.4425	20	27.9114	32	40.5240
9	30.4508	21	38.5492	33	41.5889
10	31.3759	22	36.3509	34	25.6373
11	20.8280	23	20.7170		
12	30.3198	24	41.1459		

After a short time $t < T$, the information stored in the system acquires a small erroneous component due to the action of the error Hamiltonians, which is orthogonal to the code space $\widetilde{C}$. Then, we apply the decoding matrix $\widehat{C}^{-1}$ to the atom as suggested at the end of Sec. 2. We reverse $\vec{B}$ and the detunings δ and δ', and leave all the other values unchanged (this results in taking the opposite signs of Hamiltonians $\widehat{H}_a$ and $\widehat{H}_b$), and apply the same sequence of control pulses as for the coding step, but backwards: we start with an "B" pulse whose timing is t_{n_C}, then apply an "A" pulse during t_{n_C-1}, *etc.* (see Fig. 5).

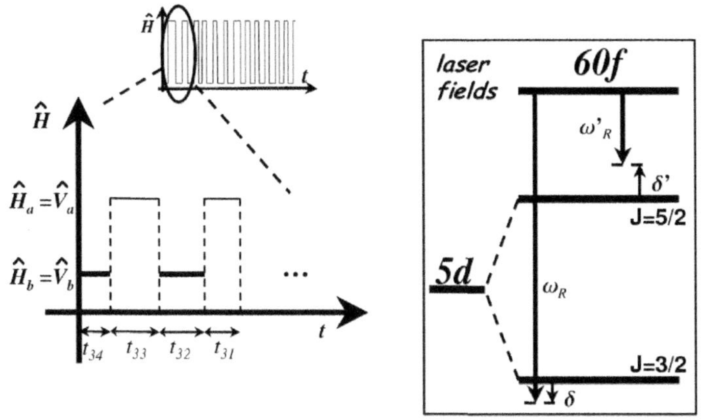

Figure 5. Decoding step through the non-holonomic control technique. We reverse the magnetic field and the detunings of electric fields, as represented on the spectrum of the Rubidium atom, and apply the same control sequence as for coding (same timings) in the reverse way.

The decoding step yields an erroneous state whose projection onto $\mathcal{C}$ is the initial information state.

In the last step the erroneous state vector is projected onto the subspace $\mathcal{C}$ to recover the initial information. Projection is a non-unitary process which cannot be achieved through a Hamiltonian process, but requires the introduction of irreversibility. To this end, we make use of a path which is symmetric with the pumping step, and consists in two stimulated and one spontaneous emissions. To be more explicit, we apply two left circularly polarized lasers slightly detuned from the transitions $(60f \longleftrightarrow 5d,\ j = 3/2)$ and $(5d,\ j = 3/2 \longleftrightarrow 5p,\ j = 3/2)$. Due to these laser fields, the atom is likely to fall towards the ground state and emit two stimulated and one spontaneous photons.

With the help of the selection rules, one can infer that, if a circularly right-polarized spontaneous photon is emitted, the only states to be coupled to the ground level are $|\gamma_1\rangle$ and $|\gamma_2\rangle$ to $|\nu_1\rangle$ and $|\nu_2\rangle$, respectively (see Fig. 6). This means that the emission of a right polarized spontaneous photon brings the "correct" part of the state vector back into $\mathcal{H}_I = \text{Span}\left[|\nu_1\rangle, |\nu_2\rangle\right]$. On the contrary, the other cases - "left polarized", "linearly polarized spontaneous photon", or "no photon at all" - do not lead to the right projection process.

The "left-polarized photon" and "no photon emitted" cases are quite unlikely: indeed the probability that they occur is proportional to the square of the error amplitude, that is to the square of the Zeno interval T, which is very short.

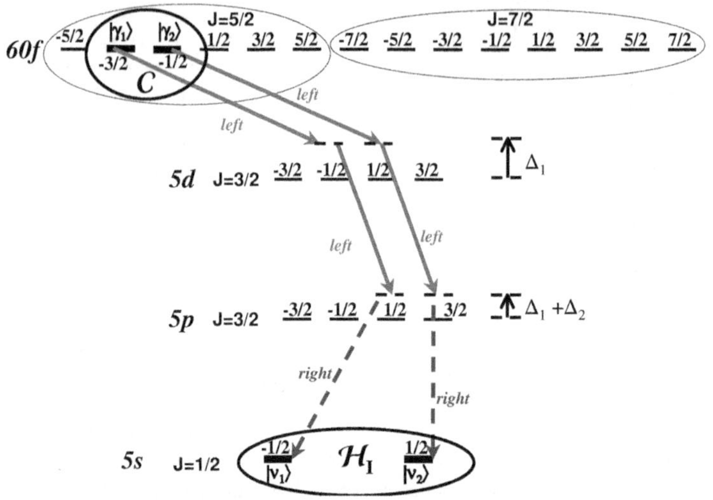

Figure 6. Projection path. The lasers involved are figured by line arrows, the spontaneous photon is represented by a dashed arrow. The different polarizations are specified.

On the contrary, the "linearly polarized photon" case is quite harmful because it mixes the two paths $|\gamma_1\rangle \longrightarrow |\nu_1\rangle$ and $|\gamma_2\rangle \longrightarrow |\nu_2\rangle$. Therefore this parasitic process and its relative probability must be suppressed with respect to the process followed by the "right-polarized" photon emission, which can be done by putting the ^{78}Rb atom into a Fabry-Perot cavity. We note that a fine tuning of the lasers driving the $60f$ - $5d$ and $5d$ - $5p$ transition is necessary in order to avoid reflection of the external laser radiation from the cavity.

The desired and unwanted projections associated with the different multi-photon processes require a finite time, which sets a constraint on the utmost efficiency of the Zeno protecting procedure. Let us evaluate the ratio of these times for the right and linearly polarized spontaneous photons. The decay rate for the 3-photon transition $|\gamma_i\rangle \longrightarrow |\nu_i\rangle$ is

$$\Gamma_{\gamma_i\,\nu_i} = 2\pi \left|\frac{d_{\gamma_i\,\lambda_i}\,E_1}{\hbar\Delta_1}\right|^2 \left|\frac{d_{\lambda_i\,\mu_i}\,E_1}{\hbar\,(\Delta_1+\Delta_2)}\right|^2 \overline{2\pi\,\hbar C\,k_s \left|\vec{d}_{\mu_i\,\nu_i}\,\vec{e}^{\,\star}_R\right|^2 \varrho\left(\vec{k}_s\right)}, \tag{46}$$

where $\vec{k}_s$ is the wave vector of the spontaneously emitted photon, $\vec{e}_R$ is the right-polarized photon polarization unit vector, $\varrho\left(\vec{k}_s\right)$ is the density of states (normalized to the cavity volume) for the cavity field at $\vec{k}_s$, and the bar denotes averaging over the directions of $\vec{k}_s$. We have used the following notations for the states

$$\left\{|\lambda_1\rangle = \left|5d,\ j=\frac{3}{2},\ m_j=-\frac{1}{2}\right\rangle,\ |\mu_1\rangle = \left|5p,\ j=\frac{3}{2},\ m_j=+\frac{1}{2}\right\rangle\right\} \tag{47}$$

and

$$\left\{|\lambda_1\rangle = \left|5d,\ j=\frac{3}{2},\ m_j=\frac{1}{2}\right\rangle,\ |\mu_2\rangle = \left|5p,\ j=\frac{3}{2},\ m_j=\frac{3}{2}\right\rangle\right\}. \tag{48}$$

coupled to $|\gamma_1\rangle$ and $|\gamma_2\rangle$, respectively, during the projective process, and the transition dipole moments are denoted by d_{ab}.

The presence of cavity results in the enhancement of the density of states for the modes propagating paraxially to the z-axis, which ensures that

$$\Gamma_{\gamma_1\,\nu_1},\, \Gamma_{\gamma_2\,\nu_2} \gg \left|\frac{d_{\gamma_i\,\lambda_j}\, E_1}{\hbar\Delta_1}\right|^2 \left|\frac{d_{\lambda_j\,\mu_k}\, E_2}{\hbar(\Delta_1+\Delta_2)}\right|^2 \gamma, \tag{49}$$

where γ is the decay rate of $|5p,\ j=3/2,\ m_j=+1/2\rangle$ into $|5s,\ j=1/2,\ m_j=+1/2\rangle$, so that the undesired process followed by the π-photon emission is relatively less important than it were in free space. For the density matrix elements ρ_{ab} the following system of equations can be written ($i=0,\ 1$):

$$\begin{aligned} \dot\rho_{\gamma_i\,\gamma_i} &= -\,\Gamma_{\gamma_i\,\nu_i}\,\rho_{\gamma_i\,\gamma_i}, \quad & \dot\rho_{\gamma_1\,\gamma_2} &= -\,\frac{\Gamma_{\gamma_1\,\nu_1}+\Gamma_{\gamma_2\,\nu_2}}{2}\,\rho_{\gamma_1\,\gamma_2}, \\ \dot\rho_{\nu_i\,\nu_i} &= \Gamma_{\gamma_i\,\nu_i}\,\rho_{\gamma_i\,\gamma_i}, & \dot\rho_{\nu_1\,\nu_2} &= \sqrt{\Gamma_{\gamma_1\,\nu_1}\,\Gamma_{\gamma_2\,\nu_2}}\,\rho_{\gamma_1\,\gamma_2}. \end{aligned} \tag{50}$$

To avoid dephasing which would corrupt the information, the coherence matrix element $\rho_{\gamma_1\,\gamma_2}$ must be transferred with the maximum efficiency into $\rho_{\nu_1\,\nu_2}$: the efficiency

$$\eta = \frac{2\sqrt{\Gamma_{\gamma_1\,\nu_1}\,\Gamma_{\gamma_2\,\nu_2}}}{\Gamma_{\gamma_1\,\nu_1}+\Gamma_{\gamma_2\,\nu_2}} \tag{51}$$

is thus a crucial quantity. According to the Wigner-Eckart theorem,

$$\frac{\Gamma_{\gamma_1\,\nu_1}}{\Gamma_{\gamma_2\,\nu_2}} = \left(\frac{C^{5/2\ -3/2}_{3/2\ -1/2\ 1\ -1}\ C^{3/2\ -1/2}_{3/2\ 1/2\ 1\ -1}\ C^{3/2\ 1/2}_{1/2\ -1/2\ 1\ 1}}{C^{5/2\ -1/2}_{3/2\ 1/2\ 1\ -1}\ C^{3/2\ 1/2}_{3/2\ 3/2\ 1\ -1}\ C^{3/2\ 3/2}_{1/2\ 1/2\ 1\ 1}}\right)^2, \tag{52}$$

where we have written the ratio of the products of the Clebsch-Jordan coefficients corresponding to the transitions in the right-hand-side. These coefficients, which can be found in Ref. [Varshalovich 1988], lead to $\eta = 12\sqrt{2}/17 \approx 0.99827$. The probability of error during the Zeno projection stage due to the small difference of the Clebsch-Gordan coefficient products for the two paths is thus equal to or less than $1 - \eta \approx 0.00173$. Note that the states $60f$, $5d$, and $5p$ have finite lifetimes τ_k (see Fig. 2). Thus the transition rates $\Gamma_{\gamma_i\ \nu_i}$ must be much larger than $1/\tau_{60f}$, $\left|d_{\gamma_i\ \lambda_j}\ E_1/(\hbar\Delta_1)\right|^2/\tau_{5d}$, and $\left|d_{\gamma_i\ \lambda_j}\ E_1/(\hbar\Delta_1)\right|^2 \left|d_{\lambda_j\ \mu_k}\ E_2/(\hbar(\Delta_1+\Delta_2))\right|^2/\tau_{5p}$, in order to minimize the influence of the errors caused by the decay of these unstable states.

To complete the projection step, one has to transfer the atom in its coherent superposition back to the $60f$ state: this is achieved by the same pumping sequence as in the first step. The mismatch of the Clebsch-Gordan coefficient products will cause again the error probability $1 - \eta$. The information is then restored with very high probability and the system is ready to undergo a new protection cycle.

To conclude this section, let us discuss some other effect which may affect the error-correction procedure. Thus far we have neglected the fine structure splitting of the level $60f$, which is approximately 2.10^{-5} cm^{-1} and corresponds to a period $\tau_f \sim 1.5$ μs. We shall now take it into account and see its effect on each step of our scheme.

Obviously the pumping and projection steps will not be affected by the fine structure, since the information-carrying vectors $\{|\gamma_1\rangle, |\gamma_2\rangle\}$ belong to the same multiplet $(J = 5/2)$.

The coding and decoding steps are neither modified by the existence of the fine structure. Indeed, since the typical period of the fine structure Hamiltonian $\tau_f \sim 1.5$ μs is more than 10 times longer than the total duration of the coding or decoding steps, it is legitimate to neglect its effect.

The influence of the fine structure on the free evolution period during which errors are likely to occur is more complicated to study in the general case. Yet, two simple limiting regimes can be considered. If the spectrum of the coupling functions $f_m(t)$'s is very narrow (*i.e.*, if the variation timescale of the $f_m(t)$'s is much longer than τ_f), one can show that our scheme applies directly as though there were no fine structure, provided the error Hamiltonians $\left\{\widehat{E}_m\right\}$ are replaced by $\left\{\widehat{E}_m^{(0)}\right\}$, where $\widehat{E}_m^{(0)}$ is obtained from $\widehat{E}_m$ by simply setting to zero the rectangular submatrices which couple the two multiplets $(J = 5/2,\ 7/2)$. The second limiting regime corresponds to a very broad spectrum for the $f_m(t)$'s (variation timescale much shorter than τ_f): in that case, one can show that our scheme applies provided one chooses a Zeno interval T multiple of τ_f.

Conclusions

In this paper, we have reviewed the main methods currently used to protect quantum information from the effects of the environment: we have presented the general idea of quantum error-correction as well as the formal theory which has recently emerged. Moreover, we have briefly introduced the most important practical methods for explicitly building quantum codes and also emphasized that they usually require strong symmetry properties from the quantum errors.

Furthermore, we have proposed an alternative scheme which allows us to protect the quantum coherence stored in an information system against the action of a set of arbitrary but known error-inducing Hamiltonians. The main recipe is the following. The information initially stored in the Hilbert space of the information system is transferred into a subspace $\mathcal{C}$ of the Hilbert space of the compound system formed by adding an ancilla to the main system. Through the application of a unitary transformation $\widehat{C}$, the information is encoded in another subspace $\widetilde{\mathcal{C}}$, called the "code space", where the error-inducing Hamiltonians $\widehat{E}_k$ act orthogonally. After a short time, the state vector acquires a small increment due to the errors, which is orthogonal to the code space. Applying the decoding transformation $\widehat{C}^{-1}$ brings the information-carrying component of the state vector back to $\mathcal{C}$, whereas the erroneous increment is transformed into a component which is orthogonal to $\mathcal{C}$. Finally, an appropriate measurement-induced projection onto $\mathcal{C}$ allows one to recover the initial state with very high probability, as it happens in the standard quantum Zeno effect. The repetition of this sequence as long as needed protects the information stored in the system. A physical achievement of the coding and decoding steps has been proposed which employs the non-holonomic control technique.

Finally, a physical application to the rubidium atom has been suggested. One qubit of information is encoded on the spin states of the atom whereas the orbital part plays the role of the ancilla. A realistic physical setting has been considered: in particular, we have suggested a projection process based on the spontaneous emission.

Acknowledgments

E. B. thanks A. Bachelier (laboratoire Aimé Cotton) for her help. The support of EU (QUACS RTN) and the computational resources of IDRIS - CNRS, Orsay, are kindly acknowledeged. I. D. was supported by NSF grant CCR-0097125.

Notes

1. The equality is checked when the dynamics of the total system is unitary; the inequality is strict if a measurement is performed on the system at some time during the interaction period.

Appendix: The code space and the code matrix

In this appendix, we describe the algorithm which allows us to explicitly construct the code space, or equivalently the coding matrix.

The problem is to find I codewords $|\widetilde{\gamma}_i\rangle$ which meet the conditions (15a) and (15b). To this end, we employ an iterative method. First, we randomly pick a set of I orthonormal state vectors which we take as the starting point. Then we repeatedly optimize a conveniently chosen functional which shows us the direction to follow at each step in the $(I \times N)$-dimensional space of parameters (coordinates of the I orthonormal vectors), in order to get arbitrarily close to the desired solution.

This appendix is organized as follows. First we present a simple iterative algorithm. Then we show that our initial multivectorial problem, set by Eqs. (15a) and (15b), can be translated into a simpler one, which may be solved by a straightforward generalization of the previous iterative algorithm. Finally, we review the technical aspects of our algorithm in detail.

Let us first present a simple algorithm which will be useful in the following. Consider a vector $|\mathsf{C}\rangle$ of some Hilbert space and a matrix $\widehat{\mathsf{E}}$ acting on this space. From the vector $|\mathsf{C}\rangle$ we want to calculate a vector $\left|\widetilde{\mathsf{C}}\right\rangle$ such that $\left\langle\widetilde{\mathsf{C}}\middle|\widehat{\mathsf{E}}\middle|\widetilde{\mathsf{C}}\right\rangle = 0$. If $\left\langle\mathsf{C}\middle|\widehat{\mathsf{E}}\middle|\mathsf{C}\right\rangle = 0$, then $|\mathsf{C}\rangle = \left|\widetilde{\mathsf{C}}\right\rangle$ and the function

$$f_{\widetilde{\mathsf{C}}}(\lambda) = \left\|\left|\widetilde{\mathsf{C}}\right\rangle + \lambda\,\widehat{\mathsf{E}}\left|\widetilde{\mathsf{C}}\right\rangle\right\|^2, \tag{A.1}$$

depending on the c-number λ, is minimal for $\lambda = 0$: indeed

$$\begin{aligned}\left\|\left|\widetilde{\mathsf{C}}\right\rangle + \lambda\,\widehat{\mathsf{E}}\left|\widetilde{\mathsf{C}}\right\rangle\right\|^2 &= \left\langle\widetilde{\mathsf{C}}\middle|\,\widetilde{\mathsf{C}}\right\rangle + \lambda\left\langle\widetilde{\mathsf{C}}\middle|\widehat{\mathsf{E}}\middle|\widetilde{\mathsf{C}}\right\rangle + \lambda^\star\left\langle\widetilde{\mathsf{C}}\middle|\widehat{\mathsf{E}}^\dagger\middle|\widetilde{\mathsf{C}}\right\rangle + |\lambda|^2\left\langle\widetilde{\mathsf{C}}\middle|\widehat{\mathsf{E}}^\dagger\,\widehat{\mathsf{E}}\middle|\widetilde{\mathsf{C}}\right\rangle \\ &= 1 + |\lambda|^2\left\langle\widetilde{\mathsf{C}}\middle|\widehat{\mathsf{E}}^\dagger\,\widehat{\mathsf{E}}\middle|\widetilde{\mathsf{C}}\right\rangle,\end{aligned} \tag{A.2}$$

and as $\left\langle\widetilde{\mathsf{C}}\middle|\widehat{\mathsf{E}}^\dagger\widehat{\mathsf{E}}\middle|\widetilde{\mathsf{C}}\right\rangle \geq 0$, $f_{\widetilde{\mathsf{C}}}(\lambda)$ is minimal for $|\lambda| = 0$, that is $\lambda = 0$. Fig. A.1 a) gives a geometric picture of this situation. The functional $f_{\mathsf{C}}(\lambda)$ can be simply interpreted in geometric terms as the square of the "length" of the vector $|\mathsf{C}\rangle + \lambda\,\widehat{\mathsf{E}}\,|\mathsf{C}\rangle$: when $|\mathsf{C}\rangle$ and $\widehat{\mathsf{E}}\,|\mathsf{C}\rangle$ are orthogonal as in Fig. A.1 a), the length of $|\mathsf{C}\rangle + \lambda\,\widehat{\mathsf{E}}\,|\mathsf{C}\rangle$ is obviously minimal for $\lambda = 0$ and so is $f_{\mathsf{C}}(\lambda)$.

But if $\left\langle\mathsf{C}\middle|\widehat{\mathsf{E}}\middle|\mathsf{C}\right\rangle \neq 0$, as in Fig. A.1 b), weapply the following "median" procedure: we minimize $f_{\mathsf{C}}(\lambda)$ with respect to λ (this is achieved when $|\mathsf{C}\rangle + \lambda\,\widehat{\mathsf{E}}\,|\mathsf{C}\rangle$ is orthogonal to $\widehat{\mathsf{E}}\,|\mathsf{C}\rangle$), and take $|\mathsf{C}'\rangle = (|\mathsf{C}\rangle + \lambda\,\widehat{\mathsf{E}}/2\,|\mathsf{C}\rangle)/\left\||\mathsf{C}\rangle + \lambda\,\widehat{\mathsf{E}}/2\,|\mathsf{C}\rangle\right\|)$ as our new vector; the angle between $|\mathsf{C}'\rangle$ and $\widehat{\mathsf{E}}\,|\mathsf{C}'\rangle$ is then closer to $\pi/2$ than the angle between $|\mathsf{C}\rangle$ and $\widehat{\mathsf{E}}\,|\mathsf{C}\rangle$. Repeating this operation yields the desired $\left|\widetilde{\mathsf{C}}\right\rangle$ such that $\left\langle\widetilde{\mathsf{C}}\middle|\widehat{\mathsf{E}}\middle|\widetilde{\mathsf{C}}\right\rangle = 0$.

Let us now show how the previous paragraph can be used to solve our initial problem. To this end, we combine the I vectors $|\widetilde{\gamma}_i\rangle$ into a $(N \times I)$ "supervector"

$$\left|\widetilde{\mathsf{C}}\right\rangle = \begin{pmatrix} |\widetilde{\gamma}_1\rangle \\ \vdots \\ |\widetilde{\gamma}_I\rangle \end{pmatrix}. \tag{A.3}$$

Then we build $E = (I(I-1)/2 + M\,I(I+1)/2)$ different $(N \times I) \times (N \times I)$-dimensional super-matrices $\widehat{\mathsf{E}}_k$ in the following way: we consider them as made of I^2 blocks of dimension $N \times N$ and we successively fill each of these blocks with the different Hamiltonians $\widehat{E}_m$ or the

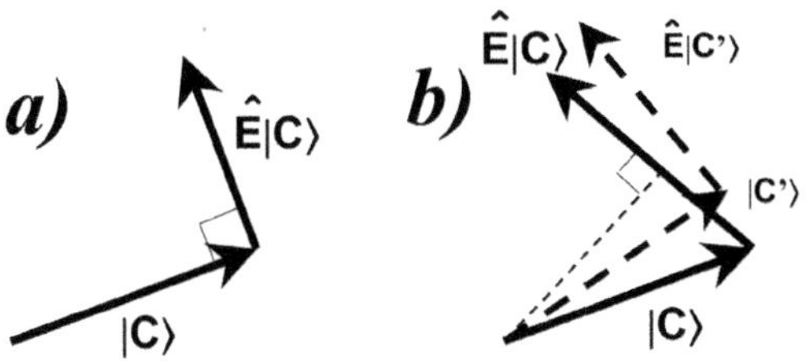

Figure A.1. Elementary step of the iterative algorithm. a) $|C\rangle$ and $\widehat{E}\,|C\rangle$ are orthogonal: the length of $|C\rangle + \lambda\,\widehat{E}\,|C\rangle$ is minimal for $\lambda = 0$ and the algorithm stops. b) $|C\rangle$ and $\widehat{E}\,|C\rangle$ are not orthogonal: one finds the new vector $|C'\rangle$ by the median procedure, the angle between $|C'\rangle$ and $\widehat{E}\,|C'\rangle$ is then closer to $\pi/2$ than the angle between $|C\rangle$ and $\widehat{E}\,|C\rangle$.

identity matrix $\widehat{I}$ or 0. To be more explicit, the first $I(I-1)/2$ matrices are built by simply placing the $N \times N$ identity matrix in each of the $I(I-1)/2$ blocks situated above the diagonal. In the last $M\ I(I+1)/2$ ones, the M operators $\widehat{E}_m$ are successively placed in each of the $I(I+1)/2$ blocks on and above the diagonal. We can thus reformulate the conditions (15a) as follows: for $1 \leq k \leq I(I-1)/2$

$$\left\langle \widetilde{C} \middle| \widehat{E}_k \middle| \widetilde{C} \right\rangle = 0. \tag{A.4}$$

Note that this form does not take the normalization condition into account, which will be imposed differently.

Similarly, the conditions (15b) are translated into the following form:

$$\text{for } \frac{I(I-1)}{2} + 1 \leq k \leq \frac{I(I-1)}{2} + \frac{M\ I(I+1)}{2},\ \left\langle \widetilde{C} \middle| \widehat{E}_k \middle| \widetilde{C} \right\rangle = 0. \tag{A.5}$$

Our initial multivectorial problem given by Eqs. (15a) and (15b) has thus been transformed into a simpler one which can be handled by the same kind of iterative algorithm as in the first paragraph: we just have to find a $(N \times I)$-dimensional supervector $\left|\widetilde{C}\right\rangle$ such that,

$$\text{for } 1 \leq k \leq \frac{I(I-1)}{2} + \frac{M\ I(I+1)}{2},\ \left\langle \widetilde{C} \middle| \widehat{E}_k \middle| \widetilde{C} \right\rangle = 0. \tag{A.6}$$

Let us now review our iterative algorithm in more detail. First we randomly pick a supervector $|C_0\rangle$ which will be the starting point of the first step: we normalize this vector by imposing to each of its I components to have norm $= 1/I$. If one of the components of $|C_0\rangle$ is non normalizable, that is equals zero, we pick up a new random supervector $|C_0\rangle$ as a starting point.

Then, as in the simple iterative algorithm we have described in the beginning of this appendix, we minimize the function

$$F_{C_0}\left(\lambda_1^{(0)}, \lambda_2^{(0)}, \ldots, \lambda_E^{(0)}\right) = \sum_{k=1}^{E} \left\| |C_0\rangle + \lambda_k^{(0)}\ \widehat{E}_k\, |C_0\rangle \right\|^2 \tag{A.7}$$

with respect to the E c-numbers $\lambda_k^{(0)}$: actually, we separate the real and imaginary parts of $\lambda_k^{(0)} = \alpha_k^{(0)} + i\,\beta_k^{(0)}$ and calculate the appropriate $\alpha_k^{(0)}$'s and $\beta_k^{(0)}$'s by solving the set of $2E$ equations

$$\frac{\partial F_{\mathsf{C}_0}}{\partial \alpha_k^{(0)}} = 0$$
$$\frac{\partial F_{\mathsf{C}_0}}{\partial \beta_k^{(0)}} = 0, \tag{A.8}$$

which can be translated into the linear system

$$\widehat{K}\left(|\mathsf{C}_0\rangle\right).\vec{\Lambda}^{(0)} = \vec{D}\left(|\mathsf{C}_0\rangle\right), \tag{A.9}$$

where $\widehat{K}\left(|\mathsf{C}_0\rangle\right)$ is a $2E \times 2E$-dimensional real matrix defined by

$$\widehat{K}_{ij}\left(|\mathsf{C}_0\rangle\right) = \begin{cases} \mathrm{Re}\left(\left\langle \mathsf{C}_0 \left| \widehat{\mathsf{E}}_i^\dagger\, \widehat{\mathsf{E}}_j \right| \mathsf{C}_0 \right\rangle\right) & \text{for } 1 \le i \le E \text{ and } 1 \le j \le E \\ -\,\mathrm{Im}\left(\left\langle \mathsf{C}_0 \left| \widehat{\mathsf{E}}_i^\dagger\, \widehat{\mathsf{E}}_{j-E} \right| \mathsf{C}_0 \right\rangle\right) & \text{for } 1 \le i \le E \text{ and } 1+E \le j \le 2E \\ \mathrm{Im}\left(\left\langle \mathsf{C}_0 \left| \widehat{\mathsf{E}}_{i-E}^\dagger\, \widehat{\mathsf{E}}_j \right| \mathsf{C}_0 \right\rangle\right) & \text{for } 1+E \le i \le 2E \text{ and } 1 \le j \le E \\ \mathrm{Re}\left(\left\langle \mathsf{C}_0 \left| \widehat{\mathsf{E}}_{i-E}^\dagger\, \widehat{\mathsf{E}}_{j-E} \right| \mathsf{C}_0 \right\rangle\right) & \text{for } 1+E \le i \le 2E \text{ and } 1+E \le j \le 2E \end{cases} \tag{A.10}$$

$\vec{D}\left(|\mathsf{C}_0\rangle\right)$ is a $2E$-dimensional real vector defined by

$$\vec{D}\left(|\mathsf{C}_0\rangle\right) = \begin{cases} -\,\mathrm{Re}\left(\left\langle \mathsf{C}_0 \left| \widehat{\mathsf{E}}_i \right| \mathsf{C}_0 \right\rangle\right) & \text{for } 1 \le i \le E \\ \mathrm{Im}\left(\left\langle \mathsf{C}_0 \left| \widehat{\mathsf{E}}_{i-E} \right| \mathsf{C}_0 \right\rangle\right) & \text{for } E+1 \le i \le 2E \end{cases} \tag{A.11}$$

and $\vec{\Lambda}^{(0)}$ is a $2E$-dimensional real vector containing the parameters $\alpha_k^{(0)}$'s and $\beta_k^{(0)}$'s

$$\vec{\Lambda}^{(0)} = \begin{pmatrix} \alpha_1^{(0)} \\ \vdots \\ \alpha_E^{(0)} \\ \beta_1^{(0)} \\ \vdots \\ \beta_E^{(0)} \end{pmatrix}. \tag{A.12}$$

Once the c-numbers $\lambda_k^{(0)} = \left(\alpha_k^{(0)} + i\,\beta_k^{(0)}\right)$'s have been found, we calculate $|\Delta\mathsf{C}_0\rangle = \sum_k \lambda_k^{(0)}\, \widehat{\mathsf{E}}_k\, |\mathsf{C}_0\rangle$ and $|\mathsf{C}_0'\rangle = |\mathsf{C}_0\rangle + |\Delta\mathsf{C}_0\rangle/2$. We normalize $|\mathsf{C}_0'\rangle$ by requiring each of its I components to have the norm $1/I$, and take the result of this operation as our new starting point $|\mathsf{C}_1\rangle$. If one of the components of $|\mathsf{C}_0'\rangle$ is non normalizable, that is equals zero, we pick up a new random supervector $|\mathsf{C}_0\rangle$ as a starting point.

We repeat this sequence of operations as long as needed. Thus, at the mth step, we minimize the function

$$F_{\mathsf{C}_{m-1}}\left(\lambda_1^{(m-1)},\, \lambda_2^{(m-1)},\, \ldots,\, \lambda_E^{(m-1)}\right) = \sum_{k=1}^{E} \left\| \left|\mathsf{C}_{m-1}\right\rangle + \lambda_k^{(m-1)}\, \widehat{\mathsf{E}}_k \left|\mathsf{C}_{m-1}\right\rangle \right\|^2 \quad \text{(A.13)}$$

by solving the real linear system

$$\widehat{K}\left(\left|\mathsf{C}_{m-1}\right\rangle\right).\vec{\Lambda}^{(m-1)} = \vec{D}\left(\left|\mathsf{C}_{m-1}\right\rangle\right). \quad \text{(A.14)}$$

This yields the $\lambda_k^{(m-1)}$'s and $\left|\Delta\mathsf{C}_{m-1}\right\rangle$ from which we calculate $\left|\mathsf{C}'_{m-1}\right\rangle = \left|\mathsf{C}_{m-1}\right\rangle + \left|\Delta\mathsf{C}_{m-1}\right\rangle/2$. If possible, we normalize $\left|\mathsf{C}'_{m-1}\right\rangle$ and take the resulting vector as the starting point $\left|\mathsf{C}_m\right\rangle$ of the $(m+1)^{th}$ step; otherwise, we pick up a new vector $\left|\mathsf{C}_0\right\rangle$ as a starting point. Finally $\left|\mathsf{C}_m\right\rangle$ tends to $\left|\widetilde{\mathsf{C}}\right\rangle$ such that $\forall k \in [1,\, I(I-1)/2]$, $\left\langle \widetilde{\mathsf{C}} \middle| \widehat{\mathsf{E}}_k \middle| \widetilde{\mathsf{C}} \right\rangle = 0$.

The codewords can be simply extracted as the I N-dimensional subcomponents of $\left|\widetilde{\mathsf{C}}\right\rangle$. Moreover, the coding matrix $\widehat{C}$ can be constructed by completing the set of I codewords $\{|\widetilde{\gamma}_i\rangle\}_{i=1,\, \ldots,\, I}$ by $(N-I)$ orthonormal vectors so as to build an orthonormal basis $\{|\widetilde{\gamma}_i\rangle\}_{i=1,\, \ldots,\, N}$ of $\widetilde{\mathcal{H}}$, and then taking the $|\widetilde{\gamma}_i\rangle$'s as columns of $\widehat{C}$.

Appendix: Physical implementation of the coding matrix by the non-holonomic control

In this second appendix, we describe the algorithm which allows us to calculate the timings $\{t_i\}_{i=1,\, \ldots,\, n_C}$ realizing the coding matrix through partial non-holonomic control of the evolution matrix, that is the timings $\{t_i\}_{i=1,\, \ldots,\, n_C}$ such that the matrix

$$\widehat{U}\left(t_1,\, t_2,\, \ldots,\, t_{n_C}\right) = e^{-i\,\widehat{H}_a\, t_{n_C}}\, e^{-i\,\widehat{H}_b\, t_{n_C-1}} \ldots e^{-i\,\widehat{H}_b\, t_1} \quad \text{(B.1)}$$

meets the conditions Eq. (35).

This algorithm mixes the iterative algorithm presented in the previous appendix and the complete control algorithm described in Ref. [Akulin 2001]. The basic idea is again to iteratively approach the solution $\vec{t}$ from a randomly picked control time vector $\vec{t_0}$. To be more explicit, at each step, one calculates the direction in the space of control parameters (timings) which allows the system to follow the direction in the Hilbert space shown by the iterative algorithm of the first appendix; then one optimizes the increment $\vec{dt}$ of the time vector such that the system gets closer to the desired solution. Repeating this procedure finally leads to the result.

This appendix is organized as follows. First, we translate our initial multivectorial problem set by Eq. (35) into a (at least formally) simpler one, involving "supervectors" and "supermatrices" as in the previous appendix. Then, we review all the technical aspects of our algorithm in detail.

Let us first put our problem under a convenient form. If we introduce the $(N \times I) \times (N \times I)$-dimensional block-diagonal matrix

$$\widehat{\mathsf{U}}\left(\vec{t}\right) = \begin{pmatrix} \widehat{U}\left(\vec{t}\right) & 0 & \ldots & 0 \\ 0 & \widehat{U}\left(\vec{t}\right) & \ldots & 0 \\ \vdots & \vdots & \vdots & \vdots \\ 0 & 0 & \ldots & \widehat{U}\left(\vec{t}\right) \end{pmatrix} \quad \text{(B.2)}$$

and the $(N \times I)$-dimensional supervector

$$|\mathsf{C}\rangle = \begin{pmatrix} |\gamma_1\rangle \\ \vdots \\ |\gamma_I\rangle \end{pmatrix} \tag{B.3}$$

composed of the coordinates of the I basis vectors of $\mathcal{C}$, we can set the problem of control Eq. (35) in the following equivalent form: we look for a time-vector $\vec{t} = \{t_i\}_{i=1,\ \dots,\ n_C}$ such that

$$\forall k,\ \left\langle \mathsf{C} \middle| \widehat{\mathsf{U}}^\dagger\left(\vec{t}\right)\ \widehat{\mathsf{E}}_k\ \widehat{\mathsf{U}}\left(\vec{t}\right) \middle| \mathsf{C} \right\rangle = 0 \tag{B.4}$$

where the matrices $\left\{\widehat{\mathsf{E}}_k\right\}_{k=1,\ \dots,\ E}$ denote E different matrices of dimension $(N \times I)\times(N \times I)$ which have been introduced in the previous appendix. In other words, we look for the time-vector $vect$ which sets to zero the test function

$$G(\vec{t}) \equiv \sum_{k=1}^{E} \left| \left\langle \mathsf{C} \middle| \widehat{\mathsf{U}}^\dagger\left(\vec{t}\right)\ \widehat{\mathsf{E}}_k\ \widehat{\mathsf{U}}\left(\vec{t}\right) \middle| \mathsf{C} \right\rangle \right|^2. \tag{B.5}$$

The idea of our algorithm is to take the super vector $|\mathsf{C}_0\rangle = \widehat{\mathsf{U}}\left(\vec{t_0}\right)|\mathsf{C}\rangle$, where $\vec{t_0}$ is a random time-vector, as the starting point for an elementary step of the iterative algorithm of the previous appendix: then, as a result, we get $|\Delta\mathsf{C}_0\rangle$ and look for the small time increment $\vec{dt_0}$ such that $\widehat{\mathsf{U}}\left(\vec{t_0} + \vec{dt_0}\right)|\mathsf{C}\rangle$ follows the direction provided by $|\mathsf{C}_0\rangle + |\Delta\mathsf{C}_0\rangle$. The repetition of this sequence finally yields $\vec{t} = \vec{t_0} + \vec{dt_0} + \vec{dt_1} + \dots$ which meets Eq. (B.4).

Let us now describe the algorithm in more detail. First, we randomly pick a set of timings $t_{0,\ i}$ in a "realistic range", dictated by the system under consideration: in particular, control-pulse timings have to be much shorter than the typical lifetime of the system and be much longer than the typical response delay required by the experiment. Then we minimize the function

$$F_{\mathsf{C}_0}\left(\lambda_1^{(0)},\ \lambda_2^{(0)},\ \dots,\ \lambda_E^{(0)}\right) = \sum_{k=1}^{E} \left\| |\mathsf{C}_0\rangle + \lambda_k^{(0)}\ \widehat{\mathsf{E}}_k\ |\mathsf{C}_0\rangle \right\|^2 \tag{B.6}$$

as we did in the iterative algorithm of the first appendix: we obtain the $\lambda_k^{(0)}$'s and $|\Delta\mathsf{C}_0\rangle = \sum_k \lambda_k\ \widehat{\mathsf{E}}_k\ |\mathsf{C}_0\rangle$. At that point, we look for the small increment $\vec{dt_0}$ of the time-vector $\vec{t_0}$ such that

$$\begin{aligned} \forall k,\quad & \left\langle \mathsf{C} \middle| \left(\frac{\partial \widehat{\mathsf{U}}^\dagger}{\partial \vec{t}}\left(\vec{t_0}\right) . \vec{dt_0} \right) \widehat{\mathsf{E}}_k\ \widehat{\mathsf{U}}\left(\vec{t_0}\right) + \widehat{\mathsf{U}}^\dagger\left(\vec{t_0}\right) \widehat{\mathsf{E}}_k \left(\frac{\partial \widehat{\mathsf{U}}}{\partial \vec{t}}\left(\vec{t_0}\right) . \vec{dt_0} \right) \middle| \mathsf{C} \right\rangle \\ & = \frac{\left\langle \mathsf{C}_0 + \frac{1}{2}\ \Delta\mathsf{C}_0 \middle| \widehat{\mathsf{E}}_k \middle| \mathsf{C}_0 + \frac{1}{2}\ \Delta\mathsf{C}_0 \right\rangle - \langle \mathsf{C}_0 | \widehat{\mathsf{E}}_k | \mathsf{C}_0 \rangle}{\left\langle \mathsf{C}_0 + \frac{1}{2}\ \Delta\mathsf{C}_0 \middle| \mathsf{C}_0 + \frac{1}{2}\ \Delta\mathsf{C}_0 \right\rangle}. \end{aligned} \tag{B.7}$$

It should be noticed that we do not consider the error super-matrices $\widehat{\mathsf{E}}_k$ corresponding to orthonormality conditions. In other words, we just take matrices $\left\{\widehat{\mathsf{E}}_k\right\}$, $k \in [I(I-1)/2 + 1,$

$I(I-1)/2+M\,I(I+1)/2]$ into account. Thus we deal with $M\,I(I+1)/2$ complex equations. This set of equations can be reduced to the real linear system

$$\widehat{S}\left(\vec{t}_0\right).\vec{dt}_0 = \vec{W}\left(|\Delta\mathsf{C}_0\rangle\right) \tag{B.8}$$

where $\widehat{S}\left(\vec{t}_0\right)$ and $\vec{W}\left(|\Delta\mathsf{C}_0\rangle\right)$ are an $MI^2 \times n_C$ real matrix and a MI^2-dimensional real vector, respectively. The technical idea is to split the set of $M\,I(I+1)/2$ complex Eqs. (B.7) into two sets of $M\,I(I+1)/2$ real equations, then reject those which are trivial $(0=0)$ or redundant to eventually obtain the form Eq. (B.8). Even though the procedure is straightforward, the different elements of $\widehat{S}$ and $\overrightarrow{W}$ are so cumbersome that we cannot reproduce them.

The linear system we have just found is a priori rectangular $(MI^2 \times n_C)$, but actually we have not fixed the number n_C yet. Previously, we stated that $n_C \geq MI^2$: we could be tempted to set $n_C = MI^2$ so as to obtain a square system, easily solvable by standard techniques of linear algebra. Yet we will proceed in a slightly different way. We set $n_C > MI^2$, say $n_C = MI^2 + \delta n$ where δn is an integer of order 1. Then we randomly pick MI^2 timings t_i among the n_C which will be considered as free parameters, whereas the other δn ones will be regarded as frozen. In other words, we randomly choose a permutation $\sigma_0 \in \mathcal{S}_{n_C}$ (symmetric group of order n_C) and take the timings $\left\{t'_i = t_{\sigma_0(i)}\right\}_{i=1,\,\ldots,\,MI^2}$ as free parameters whereas the timings $\left\{t'_i = t_{\sigma_0(i)}\right\}_{i=1+MI^2,\,n_C}$ are frozen. This leads to new versions of Eqs. (B.7) and (B.8):

$$\begin{aligned}\forall k, \quad & \left\langle \mathsf{C}\left|\left(\frac{\partial\widehat{\mathsf{U}}^\dagger}{\partial\vec{t'}}\left(\vec{t}_0\right).\vec{dt}'_0\right)\widehat{\mathsf{E}}_k\,\widehat{\mathsf{U}}\left(\vec{t}_0\right) + \widehat{\mathsf{U}}^\dagger\left(\vec{t}_0\right)\widehat{\mathsf{E}}_k\left(\frac{\partial\widehat{\mathsf{U}}}{\partial\vec{t'}}\left(\vec{t}_0\right).\vec{dt}'_0\right)\right|\mathsf{C}\right\rangle \\ & = \frac{\left\langle \mathsf{C}_0 + \frac{1}{2}\Delta\mathsf{C}_0\right|\widehat{\mathsf{E}}_k\left|\mathsf{C}_0 + \frac{1}{2}\Delta\mathsf{C}_0\right\rangle - \langle\mathsf{C}_0|\,\widehat{\mathsf{E}}_k\,|\mathsf{C}_0\rangle}{\left\langle \mathsf{C}_0 + \frac{1}{2}\Delta\mathsf{C}_0 \middle| \mathsf{C}_0 + \frac{1}{2}\Delta\mathsf{C}_0\right\rangle}.\end{aligned} \tag{B.9}$$

and

$$\widehat{S}\left(\vec{t}_0\right).\vec{dt}'_0 = \vec{W}\left(|\Delta\mathsf{C}_0\rangle\right). \tag{B.10}$$

Eq. (B.10) is now clearly a square system. Solving Eq. (B.10) yields the MI^2-dimensional increment $\vec{dt}'_0$ which we complete with δn zeros into a n_C-dimensional vector; by reordering timings we obtain the total time-vector increment $\vec{dt}_0$. Thus we have for $i \in \left[1,\,MI^2\right]$, $dt_{0,\,\sigma_0(i)} \neq 0$ (free parameters), whereas for $i \in \left[1+MI^2,\,n_C\right]$, $dt_{0,\,\sigma_0(i)} = 0$ (frozen timings). Then we set $\vec{t}_1^{\alpha} = \vec{t}_0 + \alpha\vec{dt}_0$ where α is a convergence coefficient and calculate the test function $G\left(\vec{t}\right) = \sum_k \left|\left\langle \mathsf{C}\left|\widehat{\mathsf{U}}^\dagger\left(\vec{t}\right)\widehat{\mathsf{E}}_k\,\widehat{\mathsf{U}}\left(\vec{t}\right)\right|\mathsf{C}\right\rangle\right|^2$ in $\vec{t} = \vec{t}_1^{\alpha}$ for different values of $\alpha \in [0,\,1]$. If we find an α_1 such that $G\left(\vec{t}_1^{\alpha}\right) < G\left(\vec{t}_0\right)$, we take $\vec{t}_1 \equiv \vec{t}_1^{\alpha_1}$ as our new time-vector, and keep the same free-varying timings: in other words, the permutation σ_1 governing the timings that play the role of control parameters in the second step of the algorithm remains the same, that is $\sigma_1 = \sigma_0$. If we cannot find an appropriate α_1, this means we are situated in a local minimum of G; then we set $\vec{t}_1 \equiv \vec{t}_0$ and pick a new set of free varying parameters by simply choosing a new permutation $\sigma_1 \neq \sigma_0$ randomly. This rotation procedure among control parameters allows us to avoid possible local minima of the test function G we want to cancel.

We repeat this sequence of operations as long as needed. At the mth step, we take the supervector $|\mathsf{C}_{m-1}\rangle = \widehat{\mathsf{U}}\left(\vec{t}_{m-1}\right)|\mathsf{C}\rangle$ as the starting point of an elementary step of the iterative algorithm. We calculate $|\Delta \mathsf{C}_{m-1}\rangle = \sum \lambda_k^{(m-1)} \,\widehat{\mathsf{E}}_k \,|\mathsf{C}_{m-1}\rangle$ and find MI^2-dimensional variations vector $\vec{dt}'_{m-1}$ of the MI^2 free parameters (characterized by permutation σ_{m-1}) such that

$$\forall k, \quad \left\langle \mathsf{C} \left| \left(\frac{\partial \widehat{\mathsf{U}}^\dagger}{\partial \vec{t'}} (\vec{t}_m) \,.\, \vec{dt}_m \right) \widehat{\mathsf{E}}_k \,\widehat{\mathsf{U}}\,(\vec{t}_m) + \widehat{\mathsf{U}}^\dagger\,(\vec{t}_m)\,\widehat{\mathsf{E}}_k \left(\frac{\partial \widehat{\mathsf{U}}}{\partial \vec{t'}} (\vec{t}_m) \,.\, \vec{dt}_m \right) \right| \mathsf{C} \right\rangle$$

$$= \frac{\left\langle \mathsf{C}_{m-1} + \frac{1}{2}\,\Delta \mathsf{C}_{m-1} \right| \widehat{\mathsf{E}}_k \left| \mathsf{C}_{m-1} + \frac{1}{2}\Delta \mathsf{C}_{m-1} \right\rangle - \langle \mathsf{C}_{m-1} |\, \widehat{\mathsf{E}}_k \,| \mathsf{C}_{m-1} \rangle}{\left\langle \mathsf{C}_{m-1} + \frac{1}{2}\,\Delta \mathsf{C}_{m-1} \right| \left. \mathsf{C}_{m-1} + \frac{1}{2}\,\Delta \mathsf{C}_{m-1} \right\rangle} \tag{B.11}$$

by solving the associated square linear system

$$\widehat{S}\,(\vec{t}_{m-1})\,.\,\vec{dt}'_{m-1} = \vec{W}\,(|\Delta \mathsf{C}_{m-1}\rangle)\,. \tag{B.12}$$

We complete $\vec{dt}'_{m-1}$ with δn zeros and reorder the timings so as to obtain $\vec{dt}_{m-1}$. Then we take $\vec{t}_m^{\alpha} = \vec{t}_{m-1} + \alpha\, \vec{dt}_{m-1}$. If there exists an α_m such that $F\left(\vec{t}_m^{\alpha_m}\right) < F\left(\vec{t}_{m-1}\right)$ we set $\vec{t}_m = \vec{t}_m^{\alpha_m}$ as our new time-vector, and keep the same free parameters for the $(m+1)^{th}$ step: the permutation characterizing free-varying timings in the $(m+1)^{th}$ step will be the same as in the mth step, that is $\sigma_m = \sigma_{m-1}$. Otherwise, we take $\vec{t}_m = \vec{t}_{m-1}$ as our time-vector and randomly pick up MI^2 new free parameters among the n_C timings, by choosing a new permutation σ_m for the $(m+1)^{th}$ step.

POSSIBLE IMPLEMENTATION OF TOPOLOGICALLY PROTECTED QUBITS BY A NOVEL CLASS OF JOSEPHSON JUNCTION ARRAYS

L. B. Ioffe
Center for Materials Theory, Department of Physics and Astronomy, Rutgers University, 136 Frelinghuysen Rd, Piscataway NJ 08854 USA

Abstract I review the conditions that need to be satisfied by a physical system implemeting a macroscopic quantum computer, focussing on the degree of protection of individual qubits from the noise of the environment. I argue that such degree of protection is possible in a macroscopic systems only if the information is stored very non-locally in the individual bits. I show that this can be achieved in the new class of Josephson arrays with non-trivial symmetry or topology. For such "topologically protected" arrays the effect of noise is exponentially small in the array size which allows one in principle to get extremely small error rates and extremely long dephasing times in these systems. I begin with a warm-up toy model which explains the essence of the topological protection and non-local character of information storage in such systems. I then proceed with a formulation of very general mathematical requirements on a model that ensure the appearance of the protected degenerate states and show how these conditions can be satisfied in a simple spin model. Finally I present Josephson junction array that is described by the mathematical model which satisfy these conditions and discuss its physical properties and how one can test these predictions experimentally.

Introduction.

Although it has been known for many decades [Feynman 1996] that quantum computer might be much more powerful tool than a classical one [Ekert 1996; Steane 1998] its actual implementation remains far away. In particular, it is clear that a quantum computer can solve quantum problems exponentially faster than a classical machine. A good example is provided by the models of frustrated quantum magnets important for the understanding of strongly correlated electron systems and the theory of high temperature superconductivity. These models involve quantum spins interacting with a short range but frus-

V.M. Akulin et al. (eds.), Decoherence, Entanglement and Information Protection in Complex Quantum Systems, 175–200.

trated interaction described by the Hamiltonian $H = \sum_{ij} J_{ij}\, \vec{s}_i\, \vec{s}_j$ where $\vec{s}_i$ is the quantum spin and J_{ij} is the interaction matrix. To simulate this model on a quantum computer one would need to represent each quantum spin as a quantum bit and to model the dynamics by an appropriate quantum algorithm. Roughly, in order to simulate a quantum system with N spins one needs a quantum computer with N bits while a classical computer would need to store the wave function of 2^N variables and thus would need about 2^N bits which makes such classical simulations impossible for systems containing more than $N \sim 40$ spins. Despite these obvious advantages the formidable task of building the quantum computer limited the interest to its construction. The interest to quantum computation was rekindled recently when it was proved that it can indeed solve some *classical* hard problems such as large number factorization [Shor 1994] and large list sorting [Grover 1996] parametrically faster than its classical counterpart. A few years after these pioneering works a quantum error correction was discovered [Shor 1995] showing that it is not impossible, in principle, to build the functioning quantum computer in a laboratory.

The essential difference between a quantum computer and a classical one is the continuous nature of the quantum phase characterizing each quantum bit which makes it much more difficult to protect from errors than the intrinsically discrete two state classical bit. This is why it is relatively simple to achieve error rates of $R \sim 10^{-12}$ in a classical computer (here R is the probability of the error per time needed for one operation) while the best quality factors achieved in a quantum solid state devices [Vion 2002] correspond to $R \sim 10^{-3}$. Further, the error correction codes can effectively decrease the error rates in a classical system even if the error rate in a single element is relatively high. In contrast, in a quantum system the error correction can operate only if the error rate in a single element is 10^{-3} (or lower) [Preskill 1998 (a); Dennis 2002] and even for such rates it involves a huge overhead: roughly, a single qubit is replaced by an array of $L \times L$ qubits. Further, such quantum error correction scheme requires a simultaneous operation of all L^2 qubits which is somewhat unrealistic. A more realistic alternative would be to implement quantum bits with an error rate 10^{-8} - 10^{-10} in a scalable design which allows to integrate at least $K \sim 10^4$ of such bits.

One should distinguish two types of errors that might occur in a quantum bit: an erroneous flip which is similar to the error in a classical system and a deviation of the quantum phase, $\delta\phi$, from its ideal value, the rate of these processes is known as a dephasing rate. The former is relatively easy to eliminate by ensuring that two states are separated by a significant energy barrier. The latter is much more difficult to deal with. It occurs, for instance, if the energy of two quantum states, $|0\rangle$ and $|1\rangle$ are differently affected by external noise, $V(t)$: $\delta\phi = \int (\langle 1|V(t)|1\rangle - \langle 0|V(t)|0\rangle)\, dt$. It seems very unlikely that one can screen a macroscopic system from the environment to such a degree that it is possible to achieve very low error rates required for quantum computation if the matrix element $\langle 1|V(t)|1\rangle - \langle 0|V(t)|0\rangle$ is of the same order as other energy scales characterizing the physical device. Thus, in order to achieve a low dephasing rate one needs to identify systems in which matrix elements of physical perturbations are very low, ideally, exponentially low in the system size. In other words one needs systems with exponentially large number of states which are not distinguished by physical operators. If the two states $|0\rangle$ and $|1\rangle$ have non-zero energy difference in the ideal system, fluctuations of the physical field (or device parameters) that produces this energy difference affect the energy of the states in the linear order. Further, in a solid state device with two states separated by a significant energy gap it is difficult to suppress the process of phonon emission that leads to a significant flip rate and dephasing. It seems [Ioffe 2004] that it is this phonon emission that limits the coherence of the best Josephson junction qubits [Nakamura 1999; Martinis 2002; Vion 2002; Chiorescu 2003]. Thus, a very low dephasing rate is possible only if these two states are degenerate in ideal system. In the following I shall, therefore, focus on systems with degenerate ground states. Parenthetically, I note that in physics there are many examples of systems that exhibit an exponential number of states separated by a significant barrier (for instance, spin glasses [Mezard 1997]) but the condition that these states are degenerate and are not distinguishable is very difficult to satisfy, it puts such systems in a completely new class. In philosophical terms such systems were first discussed by I. Kant more than 200 years ago (who termed them noumenons, in his thinking the noumenal world is impenetrable but contains comprehensible information) in Ref. [Kant 1781].

One very attractive possibility, proposed in an important paper by Kitaev [Kitaev 1997; Kitaev 2003] involves a protected subspace [Wen 1990; Wen 1991] created by a topological degeneracy of the ground state. Typically such degeneracy happens if the system has a conservation law such as the conservation of the *parity* of the number of "particles" along some long contour. Physically, it is clear that two states that differ *only* by the parity of some big number that can not be obtained from any local measurement are very similar to each other. The model proposed in Refs. [Kitaev 1997; Kitaev 2003] has been shown to exhibit many properties of the ideal quantum computer; however for a long time no robust and practical implementation was known. Recently in a series of papers we proposed Josephson junction networks which implement protected degeneracy and which is possible to build in a laboratory [Ioffe 2002 (a); Ioffe 2002 (b); Doucot 2003; Doucot 2004 (a); Doucot 2004 (b)]. In this paper I review our approach, present mathematical models which have a protected ground state and the Josephson junction arrays that implement them and discuss their physical properties and conditions that should be satisfied by each Josephson junction.

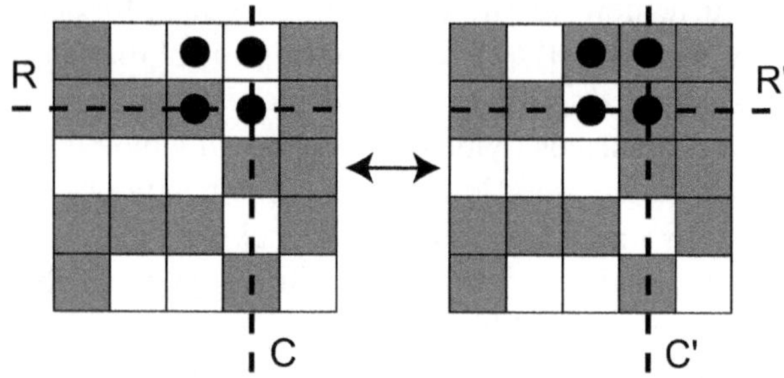

Figure 1. Formation of the protected degeneracy in the simplest quantum model. Each chessboard square is either black or white. The quantum dynamics changes white squares into black ones and vice versa simultaneously in four adjacent squares such as the ones indicated by the black dotes in the figure. This preserves the parity of the number of black squares in each column and row, *e.g.* in column C and row R, which after the flip shown in this figure become column C' and row R'. The two states differing by the parity of the number of black squares cannot be distinguised by a local measurements.

Qualitatively, one can understand the idea of the topological protection on the example of the chessboard model shown in Fig. 1. In this model each square is either black or white, its state is described by the quantum spin $1/2$ variable. The allowed dynamics flips *simultaneously* four spins in the adjacent squares:

$$H = \sum_{i,\, j} \sigma^x_{i,j}\, \sigma^x_{i+1,j}\, \sigma^x_{i,j+1}\, \sigma^x_{i+1,j+1}. \tag{1}$$

The dynamics (1) evidently preserves the parity of the number of black (white) squares in each column. On the other hand, it is easy to verify that both even and odd parities are in principle allowed. Furthermore, if the ground state of this model corresponds to the liquid of black (white) squares, the parity of the black squares in a column is the *only* way to distinguish two states. Thus, a local operator that corresponds to a physical noise does not split the energy difference between the two states differing by the parity but otherwise similar. The model (1), however, is not perfect, because it does not ensure by itself that all columns have the same parity of black squares, without this condition the number of degenerate states is 2^{2L} where L is the linear size of the system because each column and one row can be in two different states. This would allow a small perturbation, such as $\delta H = \sigma^x_{i,j}$ to change the state of one column and one row and would lead to a large flip rate. In the following I shall discuss somewhat more complicated models where such processes are expressely forbidden by the additional constraints, such as equal parities of all columns (or rows) or a local constraints imposed on each site of the (dual) lattice. I begin, however, with the general symmetry arguments for the formation of protected doublets.

1. General symmetry arguments

It is well known that the stable degeneracy of the quantum levels is almost always due to a high degree of the symmetry of the system. Examples are numerous: time inversion invariance ensures the degeneracy of the states with half-integer spin, rotational symmetry results in a degeneracy of the states with non-zero momentum, *etc.* In order for the degeneracy to be stable with respect to the local noise, one needs that the sufficient symmetry remains even if a part of the system is excluded. The simplest example is provided by the 6 Josephson junctions connecting 4 superconducting islands (so that each island is connected with every other) [Feigelman 2004 (a); Feigelman 2004 (b)] shown in Fig. 2. In this miniarray all islands are equivalent, so it is symmetric under all transformations of the permutation group S_4. This group has a two-dimensional representation and thus, pairs of the exactly degenerate states. With the appropriate choice of the parameters one can make these doublets the ground state of the system. The noise acting on one superconducting island reduces the symmetry to the permutation group of the three elements which still has two dimensional representations. So, this system is protected from the noise in the first order ($n = 2$). Here I focus on the designs providing better protection from the noise, *e.g.*, the systems that are not sensitive to the noise in the higher orders. Note that systems with higher symmetry groups, such as 5 junctions connected by 10 junctions (group S_5) typically do not have two dimensional representations, so in these systems one can typically get much higher degeneracy but not higher protection.

Generally, one gets degenerate states if there are two symmetry operations, described by the operators $\mathcal{P}$ and $\mathcal{Q}$ that commute with the Hamiltonian but do not commute with each other. If $[\mathcal{P}, \mathcal{Q}]\,|\Psi\rangle \neq 0$ for any $|\Psi\rangle$, all states are at least doubly degenerate. Local noise term is equivalent to adding other terms in the Hamiltonian which might not commute with these operators thereby lifting the degeneracy. Clearly, in order to preserve the degeneracy one needs to have two *sets* (of n elements each), $\{\mathcal{P}_i\}$ and $\{\mathcal{Q}_i\}$ of non-commuting operators, so that any given local noise field does not affect some of them; further, preferably, any given local noise should affect at most *one* $\mathcal{P}_i$ and $\mathcal{Q}_i$. In this case, the effect of the noise appears when n noise fields act *simultaneously*, *i.e.*, in the nth order in the noise strength. Another important restriction comes from the condition that these symmetry operators should not result in a higher degeneracy of the ideal system. For two operators, $\mathcal{P}$ and $\mathcal{Q}$ that implies that $[\mathcal{P}^2, \mathcal{Q}] = 0$ and $[\mathcal{P}, \mathcal{Q}^2] = 0$. Indeed, one can construct the degenerate eigenstates of the Hamiltonian starting with the eigenstate, $|0\rangle$, of H and $\mathcal{Q}$ and

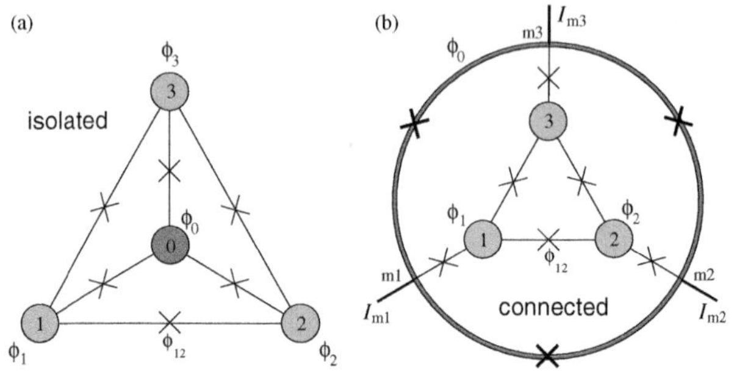

Figure 2. Miniarray containing 6 Josephson junctions that has no linear response to a local (single site) noise. The left pane shows the schematics of the isolated array which has a symmetry of the tetrahedron. The array parameters are chosen in such a way that its ground state is a doublet that transforms under a two dimensional representation of the group of tetrahedron. Namely, one has to put the array in a magnetic field so that the flux through each edge is half integer and apply potentials so that the total charge of the array is $2\ mod(4)$ in the units of Cooper pair charge ($2e$). Perturbation that affects only one site decreases the symmetry group to the one of triangle, D_3, which also has a non-trivial two dimensinal representations induced by the two-dimensional representation of the group of tetrahedron. Thus, the degeneracy of the ground state doublet is not lifted by the one site perturbation. Further, due to the time reversal symmetry the degeneracy is not lifted in the linear order by the magnetic field deviations from its ideal values. For the physical implementation it is better to use the equivalent representation shown in the right pane which allows a natural way to measure and control the states of the array. For the details see Refs.[Feigelman 2004 (a); Feigelman 2004 (b)].

acting on this state with $\mathcal{P}$. The resulting state, $|1\rangle$ should be different from the original one because $\mathcal{P}$ and $\mathcal{Q}$ do not commute: $[\mathcal{P}, \mathcal{Q}]\Psi \neq 0$ for any Ψ. In a doubly degenerate system, acting again on this state with the operator P one should get back the state $|1\rangle$, so $[\mathcal{P}^2, \mathcal{Q}] = 0$. For a set of operators, the same argument implies that in order to get a double degeneracy (and not more) one needs that $[\mathcal{P}_i\mathcal{P}_j, \mathcal{Q}] = 0$ and $[\mathcal{P}, \mathcal{Q}_i\mathcal{Q}_j] = 0$ for any i, j. Indeed, in this case one can diagonalize simultaneously the set of operators $\{\mathcal{Q}_i\}$, $\{\mathcal{Q}_i\mathcal{Q}_j\}$ and $\{\mathcal{P}_i\mathcal{P}_j\}$. Consider a ground state, $|0\rangle$, of the Hamiltonian which is also an eigenstate of all these operators. Acting on it with, say, $\mathcal{P}_1$ we get a new state, $|1\rangle$, but since $|1\rangle \propto (\mathcal{P}_i\mathcal{P}_1)\mathcal{P}_1|0\rangle = \mathcal{P}_i(\mathcal{P}_1\mathcal{P}_1)|0\rangle \propto \mathcal{P}_i|0\rangle$ all other operators of the same set would not produce a new state.

2. Spin model

The conditions derived in the previous section are fully satisfied by the spin $S = 1/2$ model on a square $n \times n$ array described by the Hamiltonian

$$H = -J_x \sum_{i,\,j} \sigma^x_{i,j}\,\sigma^x_{i,j+1} - J_z \sum_{i,\,j} \sigma^z_{i,j}\,\sigma^z_{i+1,j}. \tag{2}$$

Here σ are Pauli matrices, note that the first term couples spins in same row of the array while the second couples them along the columns. It is not important for the following discussion whether the boundary conditions are periodic or free, but since the latter are much easier to implement in a hardware we shall assume them in the following. Further, the signs of the couplings are irrelevant because for a square lattice one can always change it by choosing a different spin basis on one sublattice. For the sake of argument, we assumed that the signs of the couplings are ferromagnetic, this is also a natural sign for Josephson junction implementations of the following section.

The Hamiltonian (2) has two integrals of motion sets, $\{\mathcal{P}_i\}$ and $\{\mathcal{Q}_i\}$ with n operators each:

$$\mathcal{P}_i = \prod_j \sigma^z_{i,j} \text{ and } \mathcal{Q}_j = \prod_i \sigma^x_{i,j} \tag{3}$$

i.e., each $\mathcal{P}_i$ is the row product of $\sigma^z_{i,j}$ while $\mathcal{Q}_j$ is the column product of $\sigma^x_{i,j}$. Consider $\mathcal{P}_i$ operator first. It obviously commutes with the second term in the Hamiltonian and because the first term contains two $\sigma^z_{i,j}$ operators in the same row, $\mathcal{P}_i$ either contains none of them or both of them and since different Pauli matrices anticommute, $\mathcal{P}_i$ commutes with each term in the Hamiltonian (2). Similarly, $[\mathcal{Q}_i, H] = 0$. Clearly, different $\mathcal{P}_i$ commute with themselves, $\mathcal{P}_i^2 = 1$ and similarly $[\mathcal{Q}_i, \mathcal{Q}_j] = 0$ and $\mathcal{Q}_i^2 = 1$, but they do not commute with each other

$$\{\mathcal{P}_i,\ \mathcal{Q}_j\} = 0 \text{ and } [\mathcal{P}_i,\ \mathcal{Q}_j]^2 = 4 \tag{4}$$

so $[\mathcal{P}_i, \mathcal{Q}_j]|\Psi\rangle \neq 0$ for any $|\Psi\rangle$, thus in this model all states are at least doubly degenerate. Further, because $\mathcal{P}_i\mathcal{P}_j$ contains two $\sigma^z_{i,j}$ in any column, such product commutes with all $\mathcal{Q}_k$ operators and similarly $[\mathcal{Q}_k\mathcal{Q}_l, \mathcal{P}_i] = 0$. Thus, we conclude that in this model all states are doubly degenerate, there is no symmetry reason for larger degeneracy and that this degeneracy should be affected by the noise only in the nth order of the perturbation theory.

To estimate the effect of the noise (which appears in this high order) one needs to know the energy spectrum of the model and what are its low energy states. All states of the system can be characterized by the set, $\{\lambda_i = \pm 1\}$ of the eigenvalues of $\mathcal{P}_i$ operators (or alternatively by the eigenvalues of the $\mathcal{Q}_j$ operators). The degenerate pairs of states are formed by two sets, $\{\lambda_i\}$ and $\{-\lambda_i\}$ and each operator $\mathcal{Q}_j$ interchanges these pairs: $\mathcal{Q}_i\{\lambda_i\} = \{-\lambda_i\}$. We believe that different choices of $\{\lambda_i = \pm 1\}$ exhaust *all* low energy states in this model, *i.e.*, that there are exactly 2^n low energy states. Note that this is a somewhat unusual situation, normally one expects n^2 modes in a 2D system and thus 2^{n^2} low energy states. The number 2^n low energy states is natural for a one dimensional system and would also appear in two dimensional systems if these states are associated with the edge. Here, however, we can not associate them with the edge states because they do not disappear for the periodic boundary conditions. We can not prove our conjecture in a general case but we can see that it is true when one coupling is much larger than the other and we have verified it numerically for the couplings of the same order of magnitude. Further, one can prove that in the $J_z \gg J_x$ case the low energy states are *uniquely* characterized by the eigenvalues of $\mathcal{P}_i$ operators: the lowest energy state in $\mathcal{P}_i = 1$ sector is the ground state and the next state in this sector is separated from it by a large gap. Similarly, in the opposite limit $J_x \gg J_z$ the low energy states are characterized by the eigenvalues of $\mathcal{Q}_j$ operators. The details of this proof can be found in Ref. [Doucot 2004 (b)], here we only present the numerical results confirming this picture.

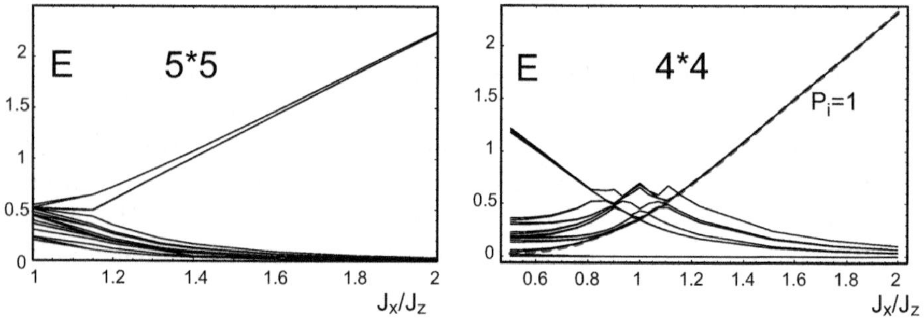

Figure 3. Energy spectrum of the 5×5 and 4×4 systems in the units of J_z coupling as a function of J_x/J_z. We show energies of the lowest 40 states for 5×5 (upper pane) and lowest 20 states for the 4×4 system (lower pane). One clearly sees that at large anisotropy a well defined low energy band is formed which contains 2^5 states for 5×5 system and 2^4 states for 4×4 one. In order to verify that low energy states are in one-to-one correspondence with P_i eigenvalues for large J_x/J_z we have calculated the second lowest eigenstate in $P_i = 1$ sector (first one is the ground state). As shown in the lower pane, this state indeed has a large gap for $J_x/J_z \gtrsim 1.2$

We have numerically diagonalized small spin systems containing up to 5 by 5 spins subjected to a small random field $h^z_{i,j}$ flatly distributed in the interval $(-\delta/2,\ \delta/2)$. We see that indeed the gap closes rather fast away from the special $J_x = J_y$ point (Fig. 3) but remains significant near $J_z = J_x$ point where it clearly has a much weaker size dependence. Interestingly, the gap between the lowest 2^n states and the rest of the spectrum expected in the limits $J_z \gg J_x$ or $J_z \ll J_x$ appears only at $J_x/J_z > j_c$ with a practically size independent $j_c \approx 1.2$. We also see that the condition $P_i = 1$ eliminates all low lying states in the $J_z \ll J_x$ limit where the lowest excited state in $P_i = 1$ sector is separated from the ground state by a large gap and in fact provides a lower bound for all high energy states. The special nature of this state appears only

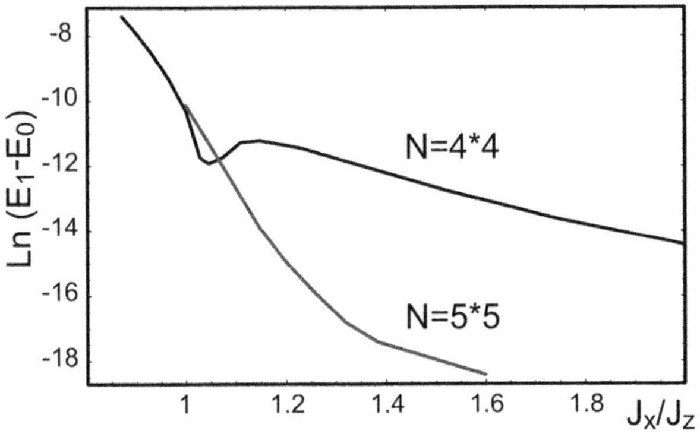

Figure 4. Ground state splitting by random field in z-direction for 5×5 and 4×4 systems. The random field acted on each spin and was randomly distributed in the interval $(-0.05, 0.05)$. Note that the effect of the random field in z-direction becomes larger for $J_x \ll J_z$ as expected (see text). Because near $J_x = J_z$ isotropic point the gap for 5×5 system is significantly smaller than the gap for the 4×4 system, this relatively large disorder has almost the same effect on these systems at $J_x \sim J_z$. We have verified numerically that decrease of the disorder by a factor of two leads to a dramatically smaller effects for 5×5 system confirming the scaling $E_1 - E_0 \propto \delta^n$ discussed in the text; for $J_x/J_z > 1.2$ the difference $E_1 - E_0$ is difficult to resolve numerically.

at $J_x/J_z > j_c$. Clearly, the system behaves quite differently near the isotropic point and away from it but the size limitations do not allow us to conclude whether these different regimes correspond to two different phases (with the "isotropic" phase restricted to a small range of parameters $j_c^{-1} < J_x/J_z < j_c$) or it is a signature of the critical region which becomes narrower as the size increases. Although we do not see any appreciable change in j_c with the system size, our numerical data do not allow us to exclude the possibility that j_c tends to unity in the thermodynamic limit. We conclude that numerical data *favor* intermediate phase scenario. In contrast to this, the analytical resuls for two and three chains indicate that the transition occurs only at $J_z = J_x$ point. Namely, both two and three chain models with periodic boundary conditions in the transverse direction can be mapped onto solvable models with transition at $J_z = J_x$ [Doucot 2004 (b)]. For a larger number of chains the number of states in each rung grows exponentially making such mappings very unlikely. In this sense two and three chain models are exceptional and it is fairly possible that the intermediate phase appears in models with larger number of chains.

Finally, we checked the effect of the $h^z_{i,j}\,\sigma^z_{i,j}$ on the ground state degeneracy splitting, our results are shown in Fig. 4. We see that, as expected, this disorder becomes relevant for $J_x \ll J_z$ while in the opposite limit its effect quickly becomes unobservable. We conclude that at (and perhaps near) isotropic point, the gap closes slowly with the system size and the effect of even significant disorder ($\delta = 0.1$) becomes extremely small for the medium sized systems.

Although it is not clear how fast the gap closes in thermodynamic limit (if it closes exponentially fast the system never becomes truly protected from the noise because the effect of the high order terms might get very large), our numerical results clearly indicate that medium size (4×4 or 5×5) systems provide an extremely good protection from the noise suppressing its effect by many orders of magnitude. This should be enough for all practical purposes. Further, it is possible [Doucot 2004 (b)] to design the physical Josephson junction arrays where $\mathcal{P}_i\mathcal{P}_j = 1$ (in other words with an additional term in the Hamiltonian $H_P = -\Delta\sum_{ij}\mathcal{P}_i\mathcal{P}_j$ with significant Δ); this eliminates the dangerous low energy states, leading to a good protection for all couplings strengths $J_x \geq J_z$. Indeed, in this case, the effects of the noise appear only in the nth order of the perturbation theory in the H^{noise}/Δ where Δ is of the order of the coefficient in the Hamiltonian H_P. Thus, in this case the effect of even a significant noise decreases exponentially fast with the array size.

3. Josephson junction arrays

3.1 Description of the array.

There are a few arrays that display topological protection, we begin with the most straighforward design (although possibly not the simplest one to implement). The basic building block of the array is a rhombus made of four Josephson junctions with each side of the rhombus containing one Josephson contact. These rhombi form a hexagonal lattice as shown in Fig. 5. We denote the centers of the hexagons by letters $a, b\ldots$ and the individual rhombi by (ab), (cd), …, because each rhombus is one-to-one correspondence with the link (ab) between the sites of the triangular lattice dual to the hexagonal lattice. The lattice is placed in a uniform magnetic field so that the flux through each rhombus is $\Phi_0/2$. The geometry is chosen in such a way that the flux, Φ_s through each David's star is a half-integer multiple of Φ_0: $\Phi = (n_s + 1/2)\Phi_0$. Finally, globally the lattice contains a number, K, of big openings (the size of the opening is much larger than the lattice constant, a lattice with $K = 1$ is shown in Fig. 5 a). The dimension of the protected space will be shown to be equal 2^K. The system is characterized by the Josephson energy, $E_J = \hbar I_c/2e$, of each contact and by the capacitance matrix of the islands (vertices of the lattice). We shall assume (as is usually the case) that the capacitance matrix is dominated by the capacitances of individual junctions, we write the charging energy as $E_C = e^2/2C$. The "phase regime" of the network mentioned above implies that $E_J > E_C$. The whole system is described by the Lagrangian

$$\mathcal{L} = \sum_{(ij)} \frac{1}{16E_C}(\dot{\phi}_i - \dot{\phi}_j)^2 + E_J \cos(\phi_i - \phi_j - a_{ij}) \tag{5}$$

where ϕ_i are the phases of individual islands and a_{ij} are chosen to produce the correct magnetic fluxes. The Lagrangian (5) contains only gauge invariant phase differences, $\phi_{ij} = \phi_i - \phi_j - a_{ij}$, so it will be convenient sometimes to treat them as independent variables satisfying the constraint $\sum_\Gamma \phi_{ij} = 2\pi\ \Phi_\Gamma/\Phi_0 + 2\pi n$ where the sum is taken over closed loop Γ and n is arbitrary integer.

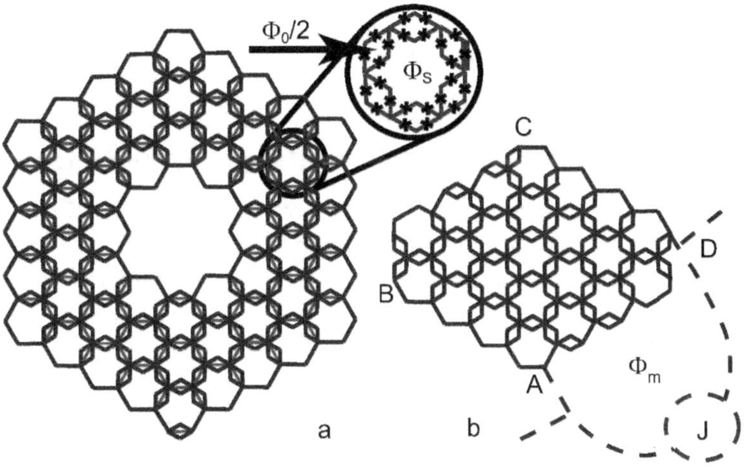

Figure 5. Example of the Josephson junction array. Thick lines show superconductive wires, each wire contains one Josephson junction as shown in detailed view of one hexagon. The width of each rhombi is such that the ratio of the area of David's star to the area of one rhombi is odd integer. The array is put in magnetic field such that the flux through each elementary rhombus and through each David's star (inscribed in each hexagon) is half integer. Thin lines show the effective bonds formed by the elementary rhombi. The Josephson coupling provided by these bonds is π-periodic. a - Array with one opening, generally the effective number of qubits, K is equal to the number of openings. The choice of boundary condition shown here makes superconducting phase unique along the entire length of the outer (inner) boundary, the state of the entire boundary is described by a single degree of freedom. The topological order parameter controls the phase difference between inner and outer boundaries. Each boundary includes one rhombus to allow experiments with flux penetration; magnetic flux through the opening is assumed to be $\Phi_0(1/2+m)/2$ with any integer m. b - With this choice of boundary circuits the phase is unique only inside the sectors AB and CD of the boundary; the topological degree of freedom controls the difference between the phases of these boundaries. This allows a simpler setup of the experimental test for the signatures of the ground state described in the text, *e.g.*, by a SQUID interference experiment sketched here that involves a measuring loop with flux Φ_m and a very weak junction J balancing the array.

As is explained below, it is crucial that the degrees of freedom at the boundary have dynamics identical to those in the bulk. To ensure this one needs to add additional superconducting wires and Josephson junctions at the boundary. There are a few ways to do this, two examples are shown in Figs. 5 a and 5 b: type I boundary (entire length of boundaries in Fig. 5 a, parts AB, CD) and type II boundary (BC, AD). For both types of boundaries one needs to include in each boundary loop the flux which is equal to $Z_b * \pi/2$ where Z_b is coordination number of the dual triangular lattice site. For instance, for the four coordinated boundary sites one needs to enclose the integer flux in these contours. In type I boundary the entire boundary corresponds to one degree of freedom (phase at some point) while type II boundary includes many rhombi so it contains many degrees of freedom.

Note that each (inner and outer) boundary shown in Fig. 5 a contains one rhombus; we included it to allow flux to enter and exit through the boundary when it is energetically favorable.

3.2 Ground state, excitations and topological order

In order to identify the relevant degrees of freedom in this highly frustrated system we consider first an individual rhombus. As a function of the gauge invariant phase difference between the far ends of the rhombus the potential energy is $U(\phi_{ij}) = -2E_J(|\cos(\phi_{ij}/2)| + |\sin(\phi_{ij}/2)|)$. This energy has two equivalent minima, at $\phi_{ij} = \pm\pi/2$ which can be used to construct elementary unprotected qubit, see Ref. [Blatter 2001]. In each of these states the phase changes by $\pm\pi/4$ in each junction clockwise around the rhombus. We denote these states as $|\uparrow\rangle$ and $|\downarrow\rangle$ respectively. In the limit of large Josephson energy the space of low energy states of the full lattice is described by these binary degrees of freedom, the set of operators acting on these states is given by Pauli matrices $\sigma_{ab}^{x,y,z}$. We now combine these rhombi into hexagons forming the lattice shown in Fig. 5. This gives another condition: the sum of phase differences around the hexagon should equal to the flux, Φ_s through each David's star inscribed in this hexagon. The choice $\phi_{ij} = \pi/2$ is consistent with flux Φ_s that is equal to a half integer number of flux quanta. This state minimizes the potential energy of the system. This is, however, not the only choice. Although flipping the phase of one dimer changes the phase flux around the star by π and thus is prohibited, flipping two, four and six rhombi is allowed; generally the low energy configurations of $U(\phi)$ satisfy the constraint

$$\hat{P}_a = \prod_b \sigma_{ab}^z = 1 \tag{6}$$

where the product runs over all neighbors, b, of site a. The number of (classical) states satisfying the constraint (6) is still huge: the corresponding configurational entropy is extensive (proportional to the number of sites). We now consider the charging energy of the contacts, which results in the quantum dynamics of the system. We show that it reduces this degeneracy to a much smaller number 2^K. The dynamics of the individual rhombus is described by a simple Hamiltonian $H = \tilde{t}\,\sigma_x$ but the dynamics of a rhombus embedded in the array is different because individual flips are not compatible with the constraint (6). The simplest dynamics compatible with (6) contains flips of three rhombi belonging to the elementary triangle, (a, b, c), $\hat{Q}_{(abc)} = \sigma^x_{ab}\,\sigma^x_{bc}\,\sigma^x_{ca}$ and therefore the simplest quantum Hamiltonian operating on the subspace defined by (6) is

$$H = -r \sum_{(abc)} Q_{(abc)}. \tag{7}$$

We first solve the simplified model (6) and (7) and show that its ground state is "protected" in the sense described above and that excitations are separated by the gap.

Clearly, it is very important that the constraint is imposed on all sites, including boundaries. Evidently, some boundary hexagons are only partially complete but the constraint should be still imposed on the corresponding sites of the corresponding triangular lattice. This is ensured by additional superconducting wires that close the boundary hexagons in Fig. 5.

We note that constraint operators commute not only with the full Hamiltonian but also with individual $\hat{Q}_{(abc)}$: $[\hat{P}_a, \hat{Q}_{(abc)}] = 0$. The Hamiltonian (7) without constraint has an obvious ground state, $|0\rangle$, in which $\sigma^x_{ab} = 1$ for all rhombi. This state, however, violates the constraint. This can be fixed by noting that since operators $\hat{P}_a$ commute with the Hamiltonian, any state obtained from $|0\rangle$ by acting on it by $\hat{P}_a$ is also a ground state. We can now construct a true ground state satisfying the constraint by

$$|G\rangle = \prod_a \frac{1 + \hat{P}_a}{\sqrt{2}}\,|0\rangle \tag{8}$$

Here $(1 + \hat{P}_a)/\sqrt{2}$ is a projector onto the subspace satisfying the constraint at site a and preserving the normalization.

Obviously, the Hamiltonian (7) commutes with any product of $\hat{P}_a$ which is equal to to the product of σ^z_{ab} operators around a set of closed loops, these products are fixed by the constraint. However, for a topologically non-trivial system there appear a number of other integrals of motion: for a system with K openings a product of σ^z_{ab} operators along contour, γ that begins at one opening and ends at another (or at the outer boundary, see Fig. 6)

$$\hat{T}_q = \prod_{(\gamma_q)} \sigma^z_{ab} \tag{9}$$

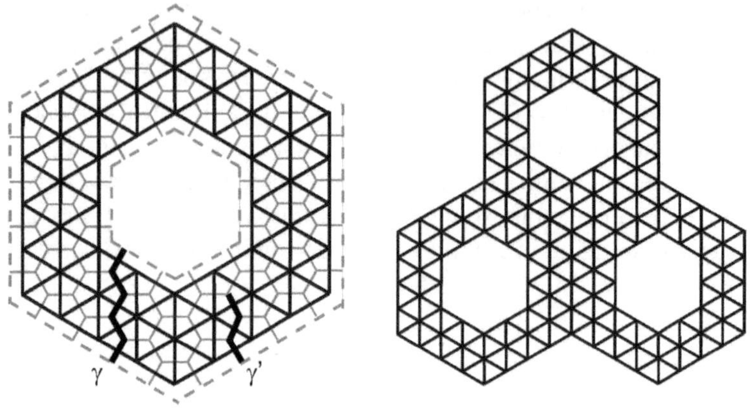

Figure 6. Left pane: Location of the discrete degrees of freedom responsible for the dynamics of the Josephson junction array shown in Fig. 5. The spin degrees of freedom describing the state of the elementary rhombi are located on the bonds of the triangular lattice (shown in thick lines) while the constraints are defined on the sites of this lattice. The dashed line indicates the boundary condition imposed by a physical circuitry shown in Fig. 5 a. Contours γ and γ' are used in the construction of topological order parameter and excitations. Right pane: The lattice with $K = 3$ openings, the ground state of Josephson junction array on this lattice is $2^k = 8$ fold degenerate.

commutes with Hamiltonian and is not fixed by the constraint. Physically these operators count the parity of "up" rhombi along such contour.

In order to establish the connection with the general symmetry theory presented in Sec. 1 we note that in the case of the constrained dynamics, the integrals of motion, $\mathcal{P}$ and $\mathcal{Q}$ introduced there should commute with both the Hamiltonian and the constraints, further they should be compatible with the boundary conditions. For the array shown in Fig. 5 the role of the operators $\mathcal{P}$ is played by $\hat{T}_q$ defined on the contours drawn on hexagonal lattice formed by rhombi and connecting the inside and the outside boundaries, see Fig. 6. The role of $\mathcal{Q}$ is played by the operators $\mathcal{Q}_\eta = \prod_{(\eta_q)} \sigma^x_{ab}$ where contour η_q is drawn on the dual (triangular) lattice and encircles one of the openings. The contour η_q crosses each hexagon twice, so operator $\mathcal{Q}_\eta$ contains exactly two operators σ^x_{ab} for all hexagons that the contour η_q crosses and thus commutes with all constraints (6). It also (obviously) commutes with the Hamiltonian (7). Further, the operators $\mathcal{Q}_\eta$ and $\mathcal{P} \equiv \hat{T}_q$ do not commute if their contours cross but their squares are trivial and commute with everything. Thus, all general conditions formulated in Sec. 1 are satisfied by this model and we expect that it has a degenerate protected doublet ground state.

Indeed, in this case the degenerate ground states can be constructed explicitly. Note that multiplying such operator by an appropriate $\hat{P}_a$ gives similar operator defined on the shifted contour so all topologically equivalent contours give one new integral of motion. Further, multiplying two operators defined along the contours beginning at the same (*e.g.*, outer) boundary and ending in different openings, A, B is equivalent to the operator defined on the contour leading from A to B, so the independent operators can be defined, *e.g.*, by the set of contours that begin at one opening and ends at the outer boundary. The state $|G\rangle$ constructed above is not an eigenstate of these operators but this can be fixed defining

$$|G_f\rangle = \prod_q \frac{1 + c_q \hat{T}_q}{\sqrt{2}} |G\rangle \tag{10}$$

where $c_q = \pm 1$ is the eigenvalue of $\hat{T}_q$ operator defined on contour γ_q. Eq. (10) is the final expression for the ground state eigenfunctions.

Construction of the excitations is similar to the construction of the ground state. first, one notices that since all operators $\hat{Q}_{abc}$ commute with each other and with the constraints, any state of the system can be characterized by the eigenvalues ($Q_{abc} = \pm 1$) of $\hat{Q}_{abc}$. The lowest excited state correspond to only one Q_{abc} being -1. Notice that a simple flip of one rhombus (by operator $\sigma^z_{(ab)}$ somewhere in the system changes the sign of *two* Q_{abc} eigenvalues corresponding to two triangles to which it belongs. To change only one Q_{abc} one needs to consider a continuous string of these flip operators starting from the boundary: $|(abc)\rangle = v_{(abc)}|0\rangle$ with $v_{(abc)} = \prod_{\gamma'} \sigma^z_{(cd)}$ where the product is over all rhombi (cd) that belong to the path, γ', that begins at the boundary and ends at (abc) (see Fig. 6 which shows one such path). This operator changes the sign of only one Q_{abc}, the one that corresponds to the "last" triangle. This construction does not satisfy the constraint, so we have to apply the same "fix" as for the ground state construction above

$$|v_{(abc)}\rangle = \prod_q \frac{1 + c_q \hat{T}_q}{\sqrt{2}} \prod_a \frac{1 + \hat{P}_a}{\sqrt{2}} v_{(abc)} |0\rangle \tag{11}$$

to get the final expression for the lowest energy excitations. The energy of each excitation is $2r$. Note that a single flip excitation at a rhombus (ab) can be viewed as a combination of two elementary excitations located at the centers of the triangles to which rhombus (ab) belongs and has twice their energy. Generally, all excited states of the model (7) can be characterized as a number of elementary excitations (11), so they give exact quasiparticle basis. Note that creation of a quasiparticle at one boundary and moving it to another is equivalent to the $\hat{T}_q$ operator, so this process acts as τ^z_q in the space of the 2^K degenerate ground states. In the physical system of Josephson junctions these excitations carry charge $2e$ so that τ_q^z process is equivalent to the charge $2e$ transfer from one boundary to another.

Consider now the matrix elements, $O_{\alpha\beta} = \langle G_\alpha|\hat{O}|G_\beta\rangle$ of a local operator, $\hat{O}$, between two ground states, *e.g.* of an operator that is composed of a small number of σ_{ab}. To evaluate this matrix element we first project a general operator onto the space that satisfies the constraint: $\hat{O} \rightarrow \mathbf{P}\hat{O}\mathbf{P}$ where $\mathbf{P} = \prod_a (1+\hat{P}_a)/2$. The new (projected) operator is also local, it has the same matrix elements between ground states but it commutes with all $\hat{P}_a$. Since it is local it can be represented as a product of σ_z and $\hat{Q}$ operators which implies that it also commutes with all $\hat{T}_q$. Thus, its matrix elements between different states are exactly zero. Further, using the fact that it commutes with $\hat{P}_a$ and $\hat{T}_q$ we write the difference between its diagonal elements evaluated between the states that differ by a parity over contour q as

$$O_+ - O_- = \langle 0| \prod_i \frac{1+\hat{P}_i}{\sqrt{2}} \hat{T}_q \hat{O}|0\rangle \tag{12}$$

This equation can be viewed as a sum of products of σ_z operators. Clearly to get a non-zero contribution each σ^z should enter even number of times. Each $\hat{P}$ contains a closed loop of six σ^z operators, so any product of these terms is also a collection of a closed loops of σ^z. In contrast to it, operator $\hat{T}_q$ contains a product of σ^z operators along the loop γ, so the product of them contains a string of σ_z operators along the contour that is topologically equivalent to γ. Thus, one gets a non-zero $O_+ - O_-$ only for the operators $\hat{O}$ which contain a string of σ^z operators along the loop that is topologically equivalent to γ which is impossible for a local operator. Thus, we conclude that for this model all non-diagonal matrix elements of a local operator are *exactly* zero while all diagonal are *exactly* equal.

3.3 Effect of physical perturbations

We now derive the parameters of the discrete model (7) starting from the original Lagrangian (5) and estimate the effect of inhomogeneity and noise on a real physical array. In the limit of small charging energy the flip of three rhombi occurs by a virtual process in which the phase, ϕ_i at one (6-coordinated) island, i, changes by π. In the quasiclassical limit the phase differences on the individual junctions are $\phi_{ind} = \pm\pi/4$; the leading quantum process changes the phase on one junction by $3\pi/2$ and on others by $-\pi/2$ changing the phase across the rhombus $\phi \rightarrow \phi + \pi$. The phase differences, ϕ, satisfy the constraint that the sum of them over the closed loops remain $2\pi(n + \Phi_s/\Phi_0)$. The simplest such process preserves the symmetry of the lattice, and changes simultaneously the phase differences on the three rhombi containing island i keeping all other phases constant. The action for such process is three times the action of elementary transitions of individual rhombi, S_0: $r \approx E_J^{3/4} E_C^{1/4} \exp(-3S_0)$, $S_0 = 1.61\sqrt{E_J/E_C}$. There are in fact many processes that contribute to this transition: the phase of island i can change by $\pm\pi$ and in addition in each rombus one can choose arbitrary the junction in which the phase changes by $\pm 3\pi/2$; the amplitude of all these processes should be added which does not change the result qualitatively. External electrical fields (created by *e.g.*, stray charges) might induce non-integer charges on each island which would lead to a randomness in the phase and amplitude of r. The phase of r can be eliminated by a proper gauge transformation $|\uparrow\rangle_{ab} \to e^{i\alpha_{ab}}|\uparrow\rangle_{ab}$ and has no effect at all. The amplitude variations result in a position dependent quasiparticle energy.

Consider now the corrections to the model (7). The most important source of corrections is the difference of the magnetic flux through each rhombus from the ideal value $\Phi_0/2$ (one can show that non-uniformity of Josephson contacts has a similar but smaller effect). If this difference is small it leads to the bias of "up" versus "down" states, their energy difference becomes $2\epsilon = 2\pi\sqrt{2}\ \delta\Phi_d\ E_J/\Phi_0$ leading to the additinal term in the Hamiltonian $\delta H_1 = \sum_{(ab)} V_{ab}\ \sigma^z_{ab}$. These terms commute with the constraint but do not commute with the main term, H, so the ground state is no longer $|G_\pm\rangle$. In other words, these terms create excitations (11) and give them kinetic energy. In the leading order of the perturbation theory the ground state becomes $|G_\pm\rangle + \epsilon\sum_{(ab)} \sigma^z_{(ab)}|G_{i\pm}\rangle/4r$. Qualitatively, it corresponds to the appearance of virtual pairs of quasiparticles in the ground state. The density of these quasiparticles is ϵ/r. As long as these quasiparticles do not form a topologically non-trivial string all previous conclusions remain valid. However, there is a non-zero amplitude to form such string - it is now exponential in the system size. With exponential accuracy this amplitude is $(\epsilon/2r)^L$ which leads to an energy splitting of the two ground state levels and the matrix elements of typical local operators of the same order $E_+ - E_- \sim O_+ - O_- \sim (\epsilon/2r)^L$.

So far we have ignored the processes that violate the constraint at each hexagon. Such processes might change the topological invariants $\hat{T}_q$ and thus are very important. In order to consider these processes we need to go back to the full description involving the continuous superconducting phases ϕ_i. Since potential energy of each rhombus is periodic in π it is convenient to separate the degrees of freedom into continuous part (defined modulus π) and discrete parts. Continuous parts have a long range order: $\langle\cos(2\phi_0 - 2\phi_r)\rangle \sim 1$. The elementary excitations of the continuous degrees of freedom are harmonic oscillations and vortices. The harmonic oscillations are small and irrelevant while vortices have an important effect. The superconducting phase changes by 0 or 2π when one moves around a closed loop. In a half vortex this is achieved if the gradual change by π is compensated (or augmented) by a discrete change by π on a string of rhombi which costs no energy. Thus, from the viewpoint of discrete degrees of freedom the position of the vortex is the hexagon where constraint (6) is violated. The energy of the vortex is found from the usual arguments $E_v(R) = \pi\ E_J(\ln R + c)/4\sqrt{6}$, $c \approx 1.2$; it is logarithmic in the vortex size, R. The process that changes the topological invariant, $\hat{T}_q$ is the one in which one half vortex completes a circle around an opening. The amplitude of such process is exponentially small: $(\tilde{t}/E_v(D))^{\Lambda}$ where $\tilde{t}$ is the amplitude to flip one rhombus and Λ is the length of the shortest path around the opening. In the quasiclassical limit one estimates: $\tilde{t} \sim \sqrt{E_J E_Q}\exp(-S_0)$.

The expression for the parameters of the model (7) is *exact* only if $E_J \gg E_C$. One expects, however, that all qualitative conclusions remain the same and the formulas derived above provide good estimates of the scales even for $E_J \sim E_C$, provided that charging energy is not so large as to result in a phase transition to a different phase. Practically, since the perturbations induced by flux deviations from Φ_0 are proportional to $\delta\Phi\ E_J/\Phi_0 r$ and r becomes exponentially small at small E_C, the optimal choice of the parameters for the physical system is $E_C \approx E_C^*$.

3.4 The most economical Josephson junction array.

We now construct with the Josephson junction array that has two sets of the integrals of motion, $\{P_i\}$ and $\{Q_i\}$ discussed above, which is shown in Fig. 7. This array contains rhombi with junctions characterized by Josephson and charging energies $E_J \gtrsim E_C$ and a weaker vertical junctions characterized by the energies ϵ_J and ϵ_C. As we explain below, although this array preserves the integrals of motion, $\{P_i\}$ and $\{Q_i\}$, it maps onto a spin model that differs from (2); we consider a more complicated arrays that are completely equivalent to spin model (2) in the next subsection. The state of the system is fully characterized by the state of each rhombi (described by the effective spin $1/2$) and by the small deviations of the continuous superconducting phase across each junction from its equilibrium (classical) values. Ignoring for the moment the continuous phase, we see that the potential energy of the array shown in Fig. 7 is given by

$$H_z = -\epsilon_J \sum_{i,\,j} \tau^z_{i,j}\,\tau^z_{i+1,j}, \quad \tau^z_{i,j} \equiv \prod_{k<j} \sigma^z_{i,k} \tag{13}$$

Physically, the variable $\tau^z_{i,j}$ describes the phase of the rightmost corner of each rhombi with respect to the left (grounded) superconducting wire modulo π. The right superconducting wire (that connects the rightmost corners of the rhombi in the last column) ensures that the phase differences along all rows are equal. In the limit of a large phase stiffness this implies that the number of the rhombi with the phase difference $\pi/2$ should be equal for all rows modulus

2. This constraint does not allow an individual rhombus flips, instead a flip of one rhombus should be always followed by a flip of another in the same row. If, however, the phase stiffness is low, the flip of one rhombus can be also compensated by the continous phase deformations in the other rhombi constituting this row, we derive the conditions at which we can exclude these processes below. The simplest allowed process is the simultaneous flip of two rhombi in one row

$$H_x = -\sum_{i,\, j,\, k} \widetilde{J_x}(j-k)\sigma^x_{i,j}\, \sigma^x_{i,k} \tag{14}$$

where $\widetilde{J_x}(k)$ is the amplitude to flip two rhombi a distance k apart. Both potential (13) and kinetic (14) energies commute with the integrals of motion, $\{P_i\}$ and $\{Q_j\}$, so that we expect that the main feature of this model, namely, the existence of the protected doublets will be preserved by this array.

As explained in the previous section, in order to achieve a really good protection one needs to eliminate all low energy states (except for the degenerate ground state) characterized by different values of the $\{P_i\}$ and $\{Q_j\}$ operators. The array shown in Fig. 7 has a boundary conditions implying $P_i P_j = 1$ for any $i,\ j$ because in this array the sum along each row of the phases across individual rhombus should be equal for all rows. Thus, for a sufficiently large tunneling amplitude $\widetilde{J_x}(k)$ this array should have two degenerate ground states separated from the rest of the spectrum by a large gap. Physically, these two states correspond to two different values of the phase difference along each row. The quantitative condition ensuring that tunneling amplitude $\widetilde{J_x}(k)$ is large enough depends on the range of $\widetilde{J_x}(k)$. The simplest situation is realized if only the nearest neighboring rhombi flip with the significant amplitude, J_x. Because flip of the two nearest rhombi is equivalent to the flip of the phase on the island between them, in this case the spin model (13) and (14) is equivalent to the collection of independent vertical Ising chains with Hamiltonian

$$H = -\sum_{i,\, j} \epsilon_J\, \tau^z_{i,j}\, \tau^z_{i+1,j} - J_x\, \tau^x_{i,j}. \tag{15}$$

For $J_x \gg 2\epsilon_J$ each chain described by this Hamiltonian has a unique ground state separated by the $\Delta = 2J_x$ from the rest of the spectrum. As the ratio ϵ_J/J_x grows, the gap decreases.

In the opposite limiting case of a very long range $\widetilde{J_x}(k) = J_x$, one can treat the interaction (14) in the mean field approximation

$$H_x = -J_x\, L_x\, \langle\sigma^x_{i,k}\rangle \sum_{i,\, j} \sigma^x_{i,j}. \tag{16}$$

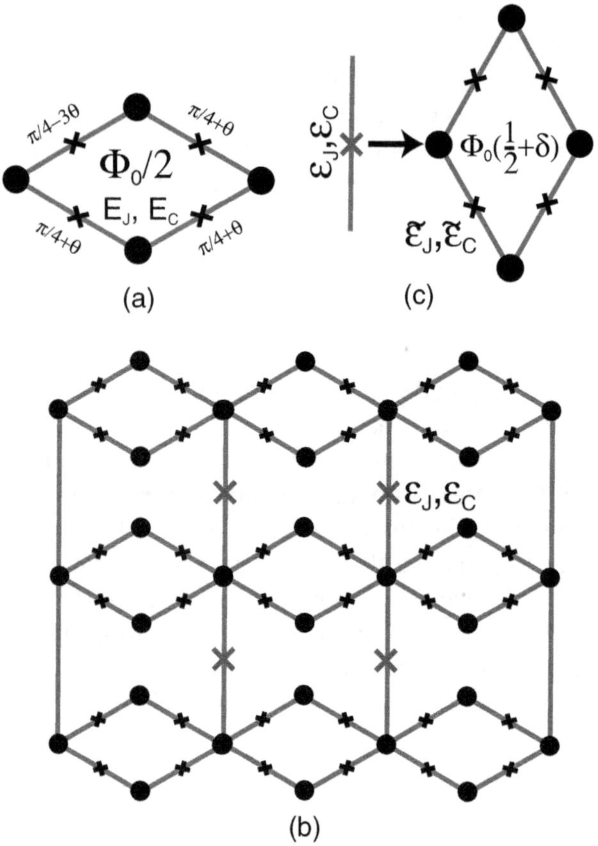

Figure 7. Schematics of the array equivalent to the spin model with the interaction (13) in the vertical direction. (a) The main element of the array, the superconducting rhombus frustrated by magnetic flux $\Phi_0/2$. Josephson energy of each rhombus is minimal for $\theta = 0$ and $\theta = -\pi/2$. Significant charging energy induces the transitions $\theta = 0 \leftrightarrow \theta = -\pi/2$ between these energy minima. (b) The array geometry. The superconducting boundary conditions chosen here ensure that $P_i P_j = 1$ thereby eliminating all low lying states in the appropriate regime (see text). (c) The requirement that continuous phase does not fluctuate much while the discrete variables have large fluctuations is easier to satisfy in a very big arrays ($L > 20$) if one replaces the vertical links by the rhombi with junctions with $\tilde{\epsilon_J}, \tilde{\epsilon_C}$ frustrated by the flux $\Phi_0(\delta + 1/2)$

At large J_x the ground state of this system is also a doublet (characterized by $\langle \sigma^x_{i,k} \rangle = \pm 1$) with all other excitations separated by the gap $\Delta = 2L_x\, J_x$ from the rest of the spectrum. As we increase the vertical coupling, ϵ_J, the gap for the excitations gets smaller. At very large ϵ_J the Hamiltonian is dominated by the ferromagnetic coupling in the vertical direction, so in this regime there are many low energy states corresponding to two possible magnetizations of each column. The magnitude of ϵ_J for which the gap decreases significantly can be estimated from the first order correction in ϵ_J. The dominant contribution comes from the transitions involving rhombi of the outmost rows. They occur with amplitude ϵ_J and lead to the states with energy Δ, so we expect that as long as $\epsilon_J \lesssim \Delta$, the system has a doubly degenerate ground state separated from the other states by gap of the order of Δ.

The amplitude, $\widetilde{J_x}(k)$, for the simultaneous flip of two rhombi can be found from the same calculation that was used in Ref. [Ioffe 2002 (b)] to calculate a single rhombus flip and the simultaneous flip of three rhombi. If $\epsilon_C \gg E_C$, the contribution of the vertical links to the total kinetic energy of the superconducting phase is small and can be treated as a small perturbation, in this case

$$\widetilde{J_x}(k) \approx E_J^{3/4}\, E_C^{1/4}\, e^{-2s\sqrt{E_J/E_C}\,\left(1+ck\,\frac{E_C}{\epsilon_C}\right)} \tag{17}$$

where $c \sim 1$. Here the factor 2 in the exponential appears because in this process one changes simultaneously the phases across two neighboring rhombi. Note that although the relative change in the action due to vertical links is always small, their contribution might suppress the flips of all rhombi except the nearest neighbors if $E_J\, E_C/\epsilon_C^2 \gg 1$. Note that even a relatively large ϵ_C (so that $E_C/\epsilon_C \ll 1$) can be sufficient to suppress the processes involving non-nearest neighbors. We conclude that the low energy states become absent as long as

$$\epsilon_J < L_{eff}\, J_x \tag{18}$$

where $L_{eff} = 1/2$ if $E_J\, E_C/\epsilon_C^2 \gg 1$ and $L_{eff} \approx \min(\epsilon_C/\sqrt{E_J\, E_C},\ L)$ if $E_J\, E_C/\epsilon_C^2 \ll 1$. These estimates assume that the main contribution to the capacitance comes from the junctions and ignores the contribution from the self-capacitance. If the self-capacitance is significant, the processes involving more than one island become quickly suppressed.

We now consider the effect of the continuous fluctuations of the superconducting phase. Generally, a finite phase rigidity allows single rhombus flip, described by the $\sum_{ij} \widetilde{t}\, \sigma_{ij}^x$ term in the effective spin Hamiltonian. This term does not commute with the integrals of motion P_i and thereby destroys the protected doublets. However, for a significant phase rigidity the energy of a state formed by a single rhombus flip, U_{sf}, is large. If, further, the amplitude $\widetilde{t}$ of these processes is small: $\widetilde{t} \ll U_{sf}$, the states corresponding to single flips can be eliminated from the effective low energy theory and the protection is restored. If $\widetilde{t} > U_{sf}$, the protection is lost.

We thus begin our analysis of the effects of the finite phase rigidity with the consideration of the dangerous single rhombus flips. Generally, the continuous phase can be represented as the sum of two parts: the one that it is due to the vortices and the spin-wave part which does not change the phase winding numbers. As usual in XY systems, it is the vortex part that is the most relevant for the physical properties. In particular, in these arrays it is the vortex part that controls the dynamics of the discrete subsystem. Notice that, unlike the conventional arrays, the arrays containing rhombi allow two types of vortices: half-vortices and full vortices because of the double periodicity of each rhombi. The flip of the individual rhombi is equivalent to the creation of the pair of half vortices. If the ground state of the system contains a liquid of half-vortices, these processes become real and the main feature of the Hamiltonian, namely the existence of two sets of anticommuting variables is lost. We now estimate the potential energy of the half vortex and of the pair associated with single flip, U_{sf}, and amplitude to create such pair, $\widetilde{t}$. We begin with the potential energy which is different in different limits. Let us consider the simpler limiting case when rhombi flips do not affect the rigidity in the vertical direction, it remains ϵ_J. Further, we have to distinguish the case of a very large size in horizontal direction and a moderate size because the contribution from the individual chains can be domininant in a moderate system if $E_J \gg \epsilon_J$. In a very large system of linear size L with rigidity ϵ_J in the vertical direction the potential energy of one vortex is

$$E_v = \pi\sqrt{E_J\,\epsilon_J}\,\ln L \tag{19}$$

while the energy of the vortex-antivortex pair at a large distance R from each other is

$$U_v(R) = \pi\sqrt{E_J\,\epsilon_J}\,\ln R \tag{20}$$

These formulas can be derived by noting that at large scales the superconducting phase changes slowly which allows one to use the continuous approximation for the energy density: $E = E_J(\partial_x\phi)^2/2 + \epsilon_J(\partial_z\phi)^2/2$. Rescaling then the x-coordinate by $x \to \widetilde{x}\sqrt{E_J/\epsilon_J}$ we get an isotropic energy density $E = \sqrt{E_J\,\epsilon_J}\,(\nabla\phi)^2/2$. The continuous approximation is valid if both rescaled coordinates $\widetilde{x},\ z \gtrsim 1$. Thus, in a system with $E_J \sim \epsilon_J$ the formulas (19, 20) remain approximately correct even at small distances $R \sim 1$, so a flip of a single rhombus creates a halfvortex - anti-halfvortex pair with energy $E_p \approx E_J$ but the formulas become parametrically different in a strongly anisotropic system. Consider first the limit $\epsilon_J = 0$. Here the chains of rhombi are completely decoupled and the energy of two half vortices separated by one rhombus in the vertical direction (the configuration created by a single flip) is due to the phase gradients in only one chain, $U_{sf}^{(0)} = \pi^2 E_J/(2L)$, which appear because the ends of the chain have the phase fixed by the boundary superconductor. A very small coupling between the chains adds $U_{sf}^{(1)} = (2\epsilon_J\,L/\pi$ to this energy, so the total potential energy of the single flip inside the array is

$$U_{sf} = \frac{\pi^2}{2L}\frac{E_J}{} + \frac{4}{\pi}\epsilon_J L \quad L \ll \sqrt{\frac{E_J}{\epsilon_J}}. \tag{21}$$

This formula is correct as long at the second term is much smaller than the first one; they become comparable at $L = \sqrt{E_J/\epsilon_J}$ and at larger L the potential energy associated with the single flip saturates at

$$U_{sf} = \gamma\sqrt{E_J\,\epsilon_J} \quad L \gg \sqrt{\frac{E_J}{\epsilon_J}} \tag{22}$$

where $\gamma \approx 3.3$. Qualitatively, a single flip leads to the continuous phase configuration where phase gradients are significant in a narrow strip in x-direction of the length $\sqrt{E_J/\epsilon_J}$ and width ~ 1. The phase configuration resulting from such process is shown in Fig. 8. These formulas assume that the rigidity of the superconducting phase remains ϵ_J which is, strictly speaking, only true if the discrete variables are perfectly ordered in the vertical direction. Indeed, the coupling in the vertical direction contains $\epsilon_J\ \cos(\phi)\ \tau^z_{i,j}\ \tau^z_{i+1,j}$ which renormalizes to $\epsilon_J \cos(\phi)\ \langle\tau^z_{i,j}\ \tau^z_{i+1,j}\rangle$ in a fluctuating system. In the opposite limit of strongly fluctuating rhombi, the average value of $\langle\tau^z_{i,j}\ \tau^z_{i+1,j}\rangle$ becomes small, we can estimate it from the perturbation theory expansion in ϵ_J which sets the lowest energy scale of the problem: $\langle\tau^z_{i,j}\ \tau^z_{i+1,j}\rangle \approx \epsilon_J/(L_{eff}\ J_x)$ which renormalizes the value of ϵ_J:

$$\epsilon_J \rightarrow \widetilde{\epsilon_J} = \frac{\epsilon_J^2}{L_{eff}\ J_x}. \tag{23}$$

This renormalized value of ϵ_J should be used in the estimates of the vortex energy (21) or (22). This does not affect much the estimates unless the system is deep in the fluctuating regime.

Unlike potential energy, the single flip processes occur with the amplitude

$$\widetilde{t} = E_J^{3/4}\ E_C^{1/4}\ e^{-s\sqrt{E_J/E_C}} \tag{24}$$

in all regimes. This formula neglects the contribution of the continuous phase to the action of the tunneling process. The reason is that both the potential energy (22) of the half-vortex and the kinetic energy required to change the continuous phase are much smaller than the corresponding energies of the individual rhombus, E_J, E_C. In order to estimate the kinetic energy, consider the contribution of the vertical links (horizontal links give equal contribution). There are roughly $\sqrt{E_J/\epsilon_J}$ such links, so their effective charging energy is about $e_C\sqrt{\epsilon_J/E_J}$. If all junctions in this array are made with the same technology their Josephson energies and capacitances are proportional to their areas, so $\epsilon_J/E_J = E_C/\epsilon_C \equiv \eta$, in the following we shall refer to such junctions as *similar*. In this case the array is characterized by two dimensionless parameters, $\eta \ll 1$ and $E_J/E_C \gg 1$ and the additional contribution to the charging

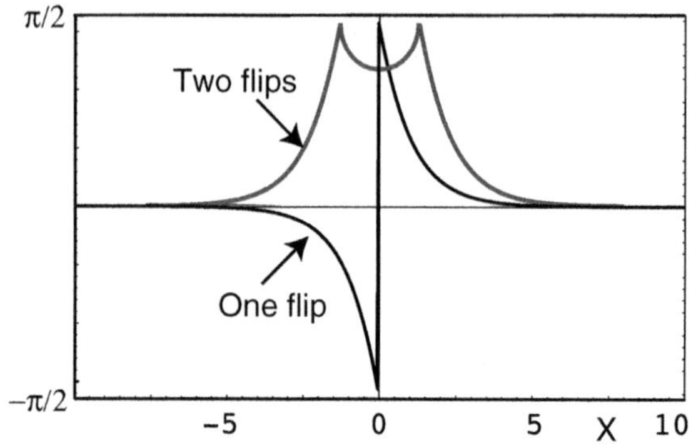

Figure 8. Phase variation along the horizontal axis after a flip of a single rhombus (solid curve) and after a consequitive flips of two rhombi located at a distance twice the core size of each rhombi, $\sqrt{E_J/\epsilon_J}$. The horizontal axis shows the distance, $X = \sqrt{\epsilon_J/E_J}x$ measured in the units of the vortex core size.

energy, $\eta^{1/2}\ E_C^{-1}$, coming from vertical links is smaller than the one of the individual rhombi, E_C^{-1} and thus do not change the dynamics.

We conclude that the dangerous real single flip processes become forbidden if $\widetilde{t} \ll U_{sf}$ where $\widetilde{t}$ is given by (24) and U_{sf} by (21) or (22). This condition is not difficult to satisfy in a real array because amplitude $\widetilde{t}$ is typically much smaller than E_J. Further, for moderately sized arrays (with $L = 5 - 10$ which already provide a very good protection) the energy of a single rhombus flip is only slightly smaller than E_J, so the condition $\widetilde{t} \ll U_{sf}$ is not really restrictive. Note, however, that in order to eliminate low energy states of the discrete subsystem we also need to satisfy the condition (18) which implies that the tunneling processes should occur with a significant amplitude. While this might be difficult in the infinite array made from the similar junctions (with the same product of charging and Josephson energies), this is not really a restrictive condition for moderately sized arrays. One can choose, for instance, for a system of $L \times L$ rhombi with $L = 5 - 10$ Josepshon contacts with $E_J = 10\ E_C$. This choice would give $\widetilde{t} \approx 0.35\ E_C$ and $J_x \approx 0.2\ E_C$ for a system of disconnected horizontal chains. The condition $\widetilde{t} \ll U_{sf}$ is well satisfied. Choosing now the vertical links with $\epsilon_J = 0.5\ E_C$ and corresponding $\epsilon_C = 20\ E_C$ we get $L_{eff} \approx 5$, so that the condition (18) is satisfied as well and there are no low energy states. It is more difficult, however, to eliminate the low energy states in the infinite array of coupled chains shown in Fig. 7 and to satisfy the condition $\widetilde{t} \ll U_{sf}$ at the same time, especially if all junctions are to be "similar" in the sense defined above. This can be achieved, however, by replacing the vertical links by rhombi frustrated by the flux $\Phi = (1/2 + \delta)\ \Phi_0$ with $\delta \ll 1$ with each junction characterized by $\widetilde{\epsilon}_J \lesssim E_J$ and $\epsilon_C \gtrsim E_C$. These rhombi would provide a significant rigidity to the continuous phase fluctuations (with effective rigidity $\widetilde{\epsilon_J}$) but only weak coupling ($\epsilon_J = \delta\widetilde{\epsilon_J}$) between discrete degrees of freedom.

Finally, we discuss the effect of the finite phase rigidity on the amplitude of the two rhombi processes, $J_x(k)$. The condition that real single flip process do not occur does not exclude the virtual processes that flip consequitively two rhombi in the same chain. This would lead to an additional contribution to $\widetilde{J_x}(k)$ (17). To estimate this contribution we note that immediately after two flips the continuous phase has a configuration shown in Fig. 8, which is associated with the energy $\sim U_{sf}$. The amplitude for such two consequitive flips is $\widetilde{t}^2/U_{sf}$; it can become of the order of $\widetilde{J_x}(k)$ in a large system (where U_{sf} is small). However, the amplitude of the full process involves additional action which further suppresses this amplitude. This happens because the two consequitive flips lead to the high energy virtual state sketched in Fig. 8 and in order to get back to the low energy state the resulting continuous phase has to evolve dynamically. To estimate the action corresponding to this evolution, we note that its dynamics is controlled by $\sqrt{E_J/\epsilon_J}$ junctions with charging energy ϵ_C. For the estimate we can replace these junctions by a single junction with capacitive energy $\epsilon_C\sqrt{\epsilon_J/E_J}$. Thus, the final stage of this process leads to the additional term in the action $\delta S \sim E_J/\epsilon_C = \eta(E_J/E_C)$. Depending on the parameter, $\sqrt{E_J\, E_C}/\epsilon_C = \eta\sqrt{E_J/E_C}$, this additional conribution to the action is smaller or bigger than the total action, but even if it is smaller, it is still large compared with unity if $E_J \gg \epsilon_C$. In this case, the processes that do not change the continuous phase dominate. We emphasize again that in any case the transitions involving two flips in the same row commute with both integrals of motion P, Q and thus do not affect the qualitative conclusions of the previous section. For practically important similar junctions, it means the following. If $\eta \gg \sqrt{E_C/E_J}$ only nearest rhombi flip with the amplitude J_x given by (17). If $\sqrt{E_C/E_J} \gg \eta \gg E_C/E_J$ the flips occur for the rhombi in the same row if they are closer than $L_{eff} < \eta^{-1}\sqrt{E_C/E_J}$. Finally for $\eta \ll E_C/E_J$ the distance between flipped rhombi exceeds the size of the half-vortex and the two rhombi flips in a large ($L \gg \sqrt{E_J/\epsilon_J}$) array happen via virtual half-vortices in the continuous phase.

In the discussion above we have implicitly assumed a superconducting boundary conditions such as shown in Fig. 7. These boundary conditions imply that in the absence of significant continuous phase fluctuations $P_iP_j = 1$. Physically, it means that if the array as a whole is a superconductor, it still has two states characterized by the phase difference, $\Delta\phi = 0$ or π between the left and the right boundaries even in the regime where individual phases in the middle fluctuate strongly between values 0 and π. In this regime of strong discrete phase fluctuations, the external fields are not coupled to the global degree of freedom $\Delta\phi$ describing the array as a whole. In principle, it is also possible to have a similar array with open boundary conditions but in this case it is more difficult to eliminate low lying states because there is no reason for the constraint $P_i P_j = 1$ in this case.

Conclusions

A realistic implementation of the quantum computer requires an extremely high degree of the protection of the constituting bits from their environment which can not be achieved in qubit implementations based on a few elements, especially in scalable solid state devices. However, the necessary degree of protection can be realized in some arrays, for instance, in Josephson junction arrays of a special geometry discussed in this review. These arrays can be built from the conventional (not even state-of-the-art) superconducting elements, the high degree of protection is due to a non-trivial collective state formed in these arrays. The elementary building blocks in these arrays are Josephson junction loops frustrated by the magnetic field. Individually, these loops have two (almost) degenerate states: in one state the current flows clockwise, in another counterclockwise. An individual qubit is an array of these loops. Even a moderately sized arrays provide an excellent protection from the environment, for instance the array of size 5×5 suppresses the 5% noise in the potential of each superconductive island to the level of 10^{-6}. Qualitatively, the protection is due to the fact that the two states of the qubit are distinguished *only* by a very non-local quantity, such as the total parity of the loops with clockwise currents along one row of the array. Such quantity does not couple to the external physical fields making these arrays optimally protected from the environment.

COHERENCE PROTECTION NEAR ENERGY GAPS

C. Mewes,[1] S. Pellegrin,[2] M. Fleischhauer,[1] and G. Kurizki[2]

[1] *Fachbereich Physik, Technische Universität Kaiserslautern, D-67653 Kaiserslautern, Germany*

[2] *Chemical Physics Department, Weizmann Institute of Science, Rehovot 76100, Israel*
gershon.kurizki@weizmann.ac.il

Abstract We compare two novel approaches to the protection of quantum states from decoherence, assuming their energies lie near forbidden energy gaps (band gaps). The first approach demonstrates that nonadiabatic, abrupt periodic changes of an atomic resonance frequency within the gap, in the neighborhood of a continuum edge can protect the atomic state from decoherence more effectively than fixing the largest possible detuning between the resonance and the continuum edge. This implies that nonadiabatic decoupling from the continuum may outperform its diabatic counterpart in maintaining a high fidelity of quantum logic operations. The other approach combats decoherence by designing multiatom quantum memories (storage) for photons, using collective atomic states near energy gaps. It is shown that despite the entanglement of the collective storage states, the insensitivity to decoherence does not scale with the number of atoms. This is due to the existence of equivalence classes of storage states. It is further shown to be possible to construct a two-dimensional decoherence free subspace if individual and independent reservoir couplings of the atoms are considered. The subspace consists of a pair of special ensemble states with a large effective distance. The probability of mutual transitions between these states due to single-atom errors scales like $1/N$, N being the total number of atoms. The existence of a sufficiently large energy gap to all other states suppresses moreover transitions out of the subspace. In this way qubits can be protected from decoherence due to both, flips and dephasing.

Introduction

Schemes of coherent control [Shapiro 2003] and quantum information processing [Cirac 1995; Mølmer 1999; Sackett 2000] that are based on optically-manipulated atoms face the challenge of protecting the quantum states of the system from decoherence, or fidelity loss, due to atomic spontaneous emis-

V.M. Akulin et al. (eds.), Decoherence, Entanglement and Information Protection in Complex Quantum Systems, 201–222.

sion (SE) [Bose 1998]. A promising means of such protection is to embed the atoms in photonic crystals (three-dimensionally periodic dielectrics) that possess spectrally-wide, omnidirectional photonic bandgaps (PBGs) [Yariv 1984; Joannopoulos 1995] atomic SE would then be blocked at frequencies within the PBG [Martorell 1990; Blanco 1998; Yoshino 1998; Kurizki 2003]. Yet optical manipulations of such atoms may necessitate the consideration of atomic transition frequencies near a PBG edge (*i.e.*, the edge of the photonic mode continuum), where SE is only partially blocked, because an initially excited atom then evolves into a superposition of decaying and stable states, the stable state representing photon-atom binding [John 1990; Kofman 1994]. These PBG-edge effects may play a role if, in order to coherently manipulate an atomic transition in the PBG, one takes advantage of its proximity to the edge and couples it to a field mode outside the PBG or to a mode in the PBG created by a local defect in the photonic crystal [Kofman 1994; Joannopoulos 1997] (Fig. 1).

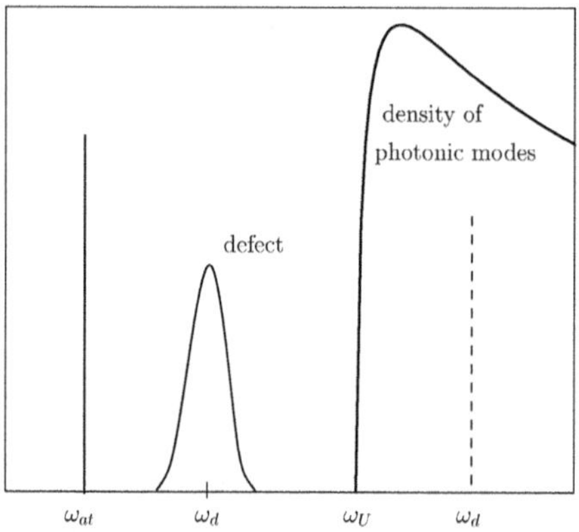

Figure 1. Mode density spectrum with a PBG. The atomic frequency ω_{at} is inside a PBG, near the cutoff frequency ω_0 and defect mode frequency ω_d.

Thus far, studies of coherent optical processes in a PBG have assumed *fixed* (static) values of the atomic transition frequency [Quang 1997]. However, in order to operate quantum logic gates, based on pairwise entanglement of atoms by field-induced dipole-dipole interactions [Brennen 1999; Petrosyan 2002; Opatrny 2003], one should be able to switch the interaction on- and off-, most conveniently by AC Stark-shifts of the transition frequency of one atom relative to the other, thereby changing its detuning from the PBG edge.
The question then arises: should such frequency shifts be performed adiabatically, in order to minimize the decoherence and maximize the quantum-gate fidelity? The answer is expected to be affirmative, based on existing treat-

ments of adiabatic entanglement and protection from decoherence [Calarco 2003; García-Ripoll 2003; Unanyan 2003] and on the tendency of nonadiabatic effects to promote transitions to the continuum. Surprisingly, the analysis presented in Sec. 1 demonstrates that periodic *"sudden"* (strongly nonadiabatic) changes of the detuning from the PBG edge may yield *higher fidelity* of qubit and quantum gate operations than their adiabatic counterparts. This unconventional nonadiabatic protection from decoherence is attributed to the ability of the periodically alternating detuning from the PBG edge to augment the interference of the emitted and back-scattered photon amplitudes, thereby increasing the probability amplitude of the stable (photon-atom bound) state. An important question which we address is the validity of the universal formula in Ref. [Kofman 2001 (a)], which presumes fast modulation of the coupling with the continuum, in the present situation. We show that this formula is valid only when the coupling is weak, *i.e.*, far enough from the PBG edge, whereas our method works equally well for strong or weak coupling to the continuum.

An important element for quantum information processing with photons [DiVincenzo 2000; Zoller 2004] is a reliable quantum memory capable of a faithful storage of their quantum state. Such a memory plays a key role in long-distance quantum communication and quantum teleportation [Duan 2000 (d); Duan 2001; Julsgaard 2001; Kuzmich 2003; van der Wal 2003] as well as in network quantum computing [Cirac 1997]. Photons may well be the best information carriers, and, in the case of spin components of the electronic ground-state atoms are reliable long lived storage units. A controllable and decoherence insensitive way of coupling light to atoms is provided by Raman transitions. Although the simplest photonic qubits are individual atoms, a faithful transfer of quantum states to and from the radiation field requires strongly coupling resonators. The technically difficult regime of strong-coupling cavity QED can be avoided if collective multiatom are used for coherent and reversible transfer of individual photon wave-packets [Csesznegi 1997; Fleischhauer 2000; Lukin 2000 (b); Liu 2001; Phillips 2001; Fleischhauer 2002; Lukin 2003] or cw light fields [Kuzmich 1997; Hald 1999; Kuzmich 2000; Schori 2002 (a)].

The advantages of enhanced coupling between collective many-atom states and the radiation field have to be checked against the worry that these states are highly entangled if non-classical light is stored. Entangled states are known to be very sensitive to decoherence and one could naively expect their lifetime to decrease with the number of atoms. It is therefore important to analyze the effect of unwanted environmental influences on the fidelity of the collective quantum memory.

Decoherence is usually modeled by coupling the system to a large reservoir of, *e.g.*, harmonic oscillators. Here two different types of couplings need to be distinguished which are indicated in Fig. 2. These are collective interactions (Fig. 2 (a)) where all particles couple to the same bath and individual interactions (Fig. 2 (b)) where each particle couples to its own, independent reservoir. In Sec. 2 of the present paper we only consider the second type of coupling which is well suited for the case of a dilute atomic vapor.

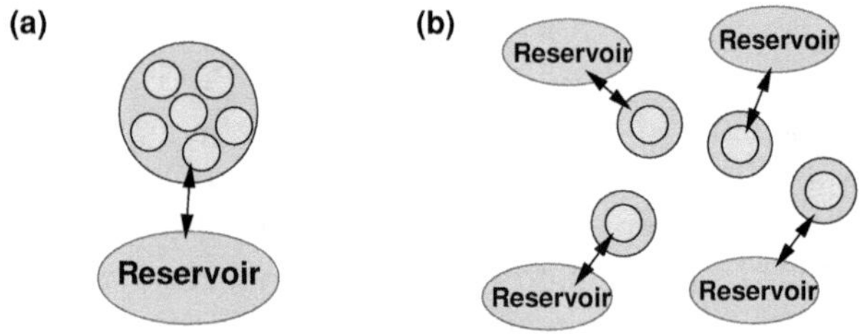

Figure 2. Collective (a) vs. individual (b) reservoir coupling.

We will show that for the retrieval of a stored field state only excitations in a certain quasi-particle mode, the dark-state polariton, are of importance. Thus all storage states with the same projection to this mode are equivalent. Due to the existence of these equivalence classes there is no enhanced sensitivity to (individual) decoherence processes in an N-atom system as compared to a single atom system.

Quantum error correction requires a fidelity for all individual storage and gate operations better than one part in 10^4 [Knill 1998; Preskill 1998 (c)]. Thus it is important to develop efficient techniques to suppress the influence of decoherence. Among the currently discussed strategies are geometric quantum computation [Vedral 2003] and the use of decoherence-free subspaces (DFS) [Lidar 2003]. Although a DFS of dimension one is easy to find, the interesting case $d \geq 2$ requires in general special symmetries of the system-reservoir interaction, see Fig. 2 (a). An example for this is Dicke subradiance, where a tightly confined ensemble of two-level atoms couples to the radiation vacuum [Dicke 1954]. The identical coupling of all atoms to the same vacuum modes generates a large DFS, the sub-radiant states. In many experimentally relevant circumstances, as for example in the case of a dilute atomic vapor, such symmetries do not exist however and the coupling to the environment is more accurately described by a model of individual and uncorrelated reservoirs as in Fig. 2 (b). As suggested by Kitaev for the example of Majorana fermions [Kitaev 2000; Levitov 2001] and by Dorner *et al.* for collective atom states in a 1D lattice [Dorner 2003] qubits can effectively be protected from decoherence without requiring a highly symmetric reservoir coupling, if they are stored in special collective states of interacting multi-particle systems. In the second section, we show that such special states can also be found in the collective photon memories. Storing photonic qubits in these states and providing for an additional nonlinear interaction a *decoherence free subspace* of dimension 2 emerges.

1. Protection from decoherence by nonadiabatic dynamics near continuum edge

1.1 Hamiltonian and equations of motion

We consider a two-level atom with excited and ground states $|e\rangle$ and $|g\rangle$ when in a photonic crystal coupled to the field of a discrete (or defect) mode and to the photonic band structure in the vacuum state. The hamiltonian of the system in the rotating-wave approximation assumes the form [Kofman 1994]

$$H = \hbar\omega_{at}\, |e\rangle\langle e| + \hbar \int_0^{+\infty} \omega\, a_\omega^\dagger\, a_\omega\, \rho(\omega)\, d\omega + \hbar\left(\kappa_d^\star\, a_d^\dagger\, |g\rangle\langle e| + h.c.\right)$$
$$+ \hbar \int_0^{+\infty} \left[\kappa^\star(\omega)\, a_\omega^\dagger\, |g\rangle\langle e| + h.c.\right] \rho(\omega)\, d\omega. \tag{1}$$

Here $\hbar\omega_{at}$ is the energy of the atomic transition frequency, $a_\omega^\dagger$ and a_ω are, respectively, the creation and annihilation operators of the field mode at frequency ω, $\rho(\omega)$ is the mode density of the PBG structure, $\kappa(\omega)$ and κ_d are the coupling rates to the atomic dipole of a mode from the continuum and the discrete mode, respectively.
Let us first consider the simple initial state obtained by absorbing a photon from the discrete mode:

$$|\Psi(0)\rangle = |e, \{0_\omega\}\rangle, \tag{2}$$

where $|\{0_\omega\}\rangle$ is the vacuum state of the field. Then the evolution of the wave-function $|\Psi(t)\rangle$ has the general form

$$|\Psi(t)\rangle = \alpha(t)\, |e, \{0_\omega\}\rangle + \beta_d(t)\, |g,\, 1_d\rangle + \int_0^{+\infty} \beta_\omega(t)\, |g, 1_\omega\rangle\, \rho(\omega)\, d\omega \tag{3}$$

where we have denoted by $|1_\omega\rangle$ and $|1_d\rangle$ the single-photon state of the relevant modes. The Schrödinger equation $i\dot{\Psi}(t) = H\ \Psi(t)$ then leads to the set of coupled differential equations

$$\dot{\alpha}(t) = -i\,\omega_{at}\,\alpha(t) - i\,\kappa_d\,\beta_d(t) - i\int_0^{+\infty} \kappa(\omega)\,\beta_\omega(t)\,\rho(\omega)\,d\omega,$$
$$\dot{\beta}_d(t) = -i\,\omega_d\,\beta_d(t) - i\,\kappa_d^\star\,\alpha(t), \tag{4}$$
$$\dot{\beta}_\omega(t) = -i\,\omega\,\beta_\omega(t) - i\,\kappa^\star(\omega)\,\alpha(t).$$

1.2 Solutions in the sudden changes approximation

Let us now introduce abrupt changes of ω_{at}, *i.e.*, of the detuning $\Delta_{at} = \omega_U - \omega_{at}$ from the upper cutoff, ω_U, of the PBG (by fast AC-Stark modulations as discussed below), at intervals τ. In the sudden-change approximation for ω_{at}, the amplitudes $(\alpha_{dyn}(t),\ \beta_{d\,dyn},\ \{\beta_{\omega\,dyn}(t)\})$ of the excited state, the discrete mode and the continuum still evolve according to Eqs. (4), except that from $t = 0$ to $t = \tau$ the atomic transition frequency is $\omega_{at} = \omega_A$, *i.e.*, the detuning $\Delta_{at} = \omega_U - \omega_A = \Delta_A$, while for $t > \tau$, we have $\omega_{at} = \omega_B$, *i.e.*, $\Delta_{at} = \Delta_B$. This dynamics leads to the relation

$$\alpha_{dyn}(t) = \alpha_A(t),\ \beta_{d\,dyn}(t) = \beta_{d,A}(t),\ \beta_{\omega\,dyn}(t) = \beta_{\omega,A}(t),\ (t \le \tau);$$
$$\alpha_{dyn}(t) = \alpha_B^{(s)}(t),\ \beta_{d\,dyn}(t) = \beta_{d,B}^{(s)}(t),\ \beta_{\omega\,dyn}(t) = \beta_{\omega,B}^{(s)}(t),\ (t > \tau). \tag{5}$$

Here both $(\alpha_A(t),\ \beta_{d,A}(t),\ \{\beta_{\omega\,A}(t)\})$ and $(\alpha_B^{(s)}(t),\ \beta_{d,B}^{(s)}(t),\ \{\beta_{\omega,B}^{(s)}(t)\})$ are solutions of Eqs. (4) with a static (fixed) atomic transition frequency, ω_A or ω_B. However, the initial condition at the instant $t = \tau$ of the frequency change from Δ_A to Δ_B is no longer the excited state (2) but the superposition:

$$|\Psi(\tau)\rangle = \alpha_A(\tau)\,|e,\,\{0_\omega\}\rangle + \beta_{d,A}(\tau)\,|g,\,1_d\rangle + \int_0^{+\infty} \beta_{\omega,A}(\tau)\,|g,\,1_\omega\rangle\,\rho(\omega)\,d\omega. \tag{6}$$

In other words, the dynamics is equivalent to two successive static evolutions, the second one starting from initial conditions $(\alpha_A(\tau),\ \beta_{d,A}(\tau),\ \{\beta_{\omega,A}(\tau)\})$. Using the Laplace transform of the system (4) with the initial condition (6), it is possible to express the dynamic amplitude of the excited state after the sudden change as

$$\alpha_{dyn}(t) = \alpha_A(\tau)\,\alpha_B(t-\tau) + \beta_{d,A}(\tau)\,\beta_{d,B}(t-\tau)$$
$$+ \int_0^{+\infty} \beta_{\omega,A}(\tau)\,\beta_{\omega,B}(t-\tau)\,\rho(\omega)\,d\omega,\ (t > \tau), \tag{7}$$

where we have used the initial conditions $(\alpha_A(\tau),\ \beta_{d,A}(\tau),\ \{\beta_{\omega,A}(\tau)\})$ and the solution $(\alpha_B(t),\ \beta_{d,B}(t),\ \{\beta_{\omega,B}(t)\})$ of Eqs. (4) for the initial condition (2).

The contribution of the first term in (7) to the excited-state population $|\alpha_{dyn}(t)|^2$ always decreases after the sudden change and then oscillates, before settling to an asymptotic non-zero value. On the other hand, the contribution of the second and third terms and their cross-product with the first term increases immediately after the sudden change. Yet whatever the time τ of the sudden change, when performing only one change, the increasing contribution is never large enough to compensate for the decreasing part. Then the dynamic population of the excited state after the sudden change always lies inbetween the two static populations obtained for Δ_A and Δ_B.

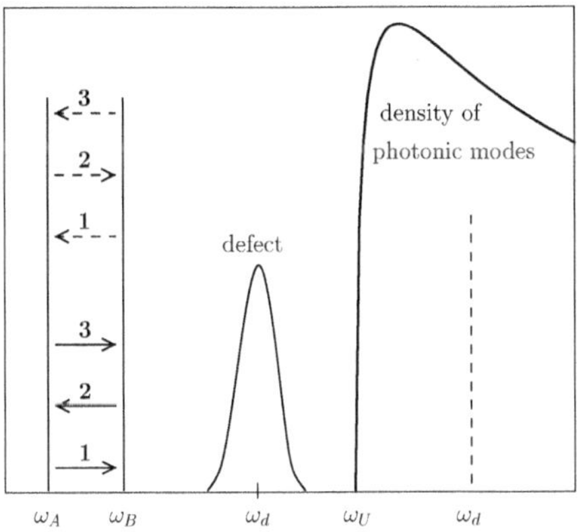

Figure 3. Left: arrows sketch the periodic alternation between two values of the atomic frequency, starting from the atomic frequency ω_A, *i.e.*, detuning $\Delta_A = \omega_U - \omega_A$ (solid line arrows) or ω_B, *i.e.*, detuning Δ_B (dashed line arrows). This frequency modulation occurs in the vicinity of a field in the discrete mode ω_d, typically in a photonic crystal with defects, and of the photonic band gap edge ω_U (right).

There is, however, an advantageous feature to the sudden change: since the time dependence of $\alpha_{dyn}(t)$ in (7) arises from the static amplitudes α_B, $\beta_{d,B}$ and $\beta_{\omega,B}$ at the *shifted* time $t - \tau$, a consequence of the sudden change is to revive the excited-state population oscillations, which tend to disappear asymptotically in the static limit. Hence, by applying several *successive* sudden changes, we should be able to maintain large-amplitude oscillations of the *coherence* between $|e\rangle$ and $|g\rangle$. The scenario leading to the largest amplitude consists in *periodic* shifts of the energy detuning from Δ_A to Δ_B. Here we have the choice between starting from $\Delta_{at} = \Delta_A$ or $\Delta_{at} = \Delta_B$ (Fig. 3). In the former case, the dynamic population experiences large amplitude oscillations but never exceeds the highest static population (Fig. 4). But when the

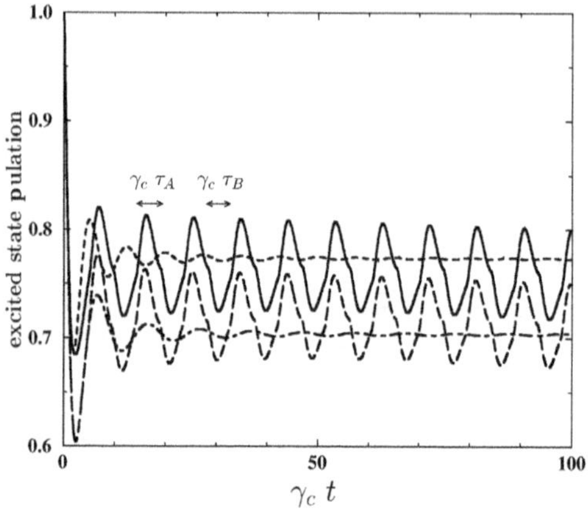

Figure 4. Excited state population as a function of dimensionless time $\gamma_c\, t$. Dashed line: static detuning $\Delta_A/\gamma_c = 0.5$. Dot-dashed line: static detuning $\Delta_B/\gamma_c = 0.25$. Solid and long-dashed lines: periodic shifts of the detuning between Δ_A and Δ_B in the sudden change approximation as shown in Fig. 3. Solid line: starting with detuning Δ_A. Long-dashed line: starting with detuning Δ_B.

initial detuning Δ_A is large and we first reduce it to Δ_B before it increases to Δ_A, the dynamic population and the $|e\rangle - |g\rangle$ coherence, thanks to the revival of oscillations, are *periodically larger* than the static ones (!) (Fig. 4).
In order to solve the system (4) the continuum is discretised, the coupling $\kappa(\omega)$ is approximated to be constant (independent of ω), and an analytic expression for the density of modes is used based on the simplified scalar, isotropic approximation [Kofman 1994]. The modification of these results to allow for the anisotropy of the density of modes can be undertaken using the results of Ref. [Woldeyohannes 2003]. We consider only the interaction of the atom with one PBG edge, assuming other edges are far enough to have negligible effects. All parameters and variables are scaled to the effective coupling γ_c, which depends on the steepness of the modes density (Fig. 3). Fig. 4 presents the excited state population for both static detunings $\Delta_{at} = \Delta_A$ and $\Delta_{at} = \Delta_B$ and for the optimized dynamics obtained in the case of periodic alternation between Δ_A and Δ_B. Here, the coupling with the discrete mode has been neglected in order to concentrate on the effects of the continuum.

The times τ_A and τ_B spent at each static detuning value Δ_A and Δ_B are proportional to the periods T_A and T_B of the static population-oscillations, respectively. The coefficients linking τ_A and τ_B to T_A and T_B depend on the difference $|\Delta_A - \Delta_B|$ but not on the initial value Δ_A or Δ_B. The dynamic asymptotic value of the excited-state population is mainly controlled by the initial value of the detuning whereas the oscillations amplitude is controlled by

the difference $|\Delta_A - \Delta_B|$. It is then possible to control the rate of spontaneous emission and the frequency of oscillations by varying only the two parameters Δ_A and Δ_B. Moreover, the time spent at one detuning or the other can be lengthened by skipping two successive changes in the periodic alternation sequence. This does not affect dramatically the oscillations amplitude.
Since the sudden change approximation is not realizable experimentally, we consider the effects of finite transition times between Δ_A and Δ_B. Although a sine modulation function gives results very close to the sudden change approximation when starting with detuning Δ_B, this is no longer the case when starting from Δ_A. Indeed, if we want to observe the periodic revival of the $|e\rangle - |g\rangle$ coherence, it is crucial that the detuning stays equal or very close to the large value before the first change. We then use a smooth function of time consisting in a sum of large pulses $\exp[-(t-\tau_n)^8/L^8]$ centered on τ_n and of half-width L of the order of $\tau_A/4 \sim \tau_B/4$. The excited state population is only slightly modified by such finite rise- and falloff-times.
We have compared our results, which allow for possibly strong coupling of $|e\rangle$ with the continuum edge, with those of the universal formula of Ref. [Kofman 2001 (a)]. This formula expresses the decay rate of $\alpha(t)$ by the convolution of the periodic modulation spectrum and the PBG coupling spectrum. We find good agreement with this formula only in the limit of *weak coupling* to the PBG edge, for which the theory in Ref. [Kofman 2001 (a)] has been developed. More quantitatively, the analytic formula well approximates the exact calculations when the dimensionless detuning parameter $\Delta_{at}/\gamma_c > 5$. When $\Delta_{at}/\gamma_c \simeq 5$, the detuning between the atomic frequency and the discrete mode becomes important for the validity of the formula, especially for the amplitude of oscillations. The larger the detuning from the discrete mode, the more accurate the formula.
Let us now consider the initial superposition

$$|\Psi(0)\rangle = \alpha(0)\,|e, \{0_\omega\}\rangle + \beta_d(0)\,|g, 1_{\omega_d}\rangle \tag{8}$$

and a non-negligible coupling constant κ_d. In this case, the periodic dynamic population of the excited state also strongly exceeds the static one. On the other hand, the discrete mode amplitude $\beta_d(t)$ diminishes as compared with the static case. Most importantly the *instantaneous dynamic fidelity* $|\langle\Psi(0)|\Psi(t)\rangle|$ is periodically enhanced as compared to the static one (see Fig. 5).

The weak coupling of the atom with the discrete mode ω_d stands in contrast to the strong reservoir-atom coupling near the PBG edge. However, the main characteristics of the results are not modified when both couplings are of the same order of magnitude, except that the excited-state and the discrete-mode amplitudes oscillate more strongly.

1.3 Application to quantum logic gates and conclusions

In order to use these results for quantum logic gates, let us consider the example of the control phase gate, which consists in shifting the phase of the target-qubit excited state by π via interaction with the control qubit. As opposed to the nonadiabatic dynamics used to inhibit spontaneous emission, the phase shift must be accumulated gradually, to preserve the coherence of the system. We have found that a single sudden shift of π or two successive sudden shifts of $\pi/2$ are too fast, but ten or twenty sudden shifts of $\pi/10$ or $\pi/20$, respectively, keep the fidelity high, with little decoherence. Without attempting to fully optimize the process, we were able to find such dynamics of the shift that preserves a high fidelity of the system state: The system begins to evolve following the detuning sequence discussed above. As soon as two sudden changes of the detuning have been performed, the conditional phase shift takes place, so that the total gate operation completed within the time-interval of maximum fidelity. Then we continue the dynamical optimization of the detuning to obtain the best protection against spontaneous emission. Fig. 5 shows the fidelity of the system relative to its initial state during the realization of a control phase gate. We can see that the fidelity is increased as compared to the static case, or even to the dynamic case without gate operations.

The following experimental scenario may be envisioned for demonstrating the proposed effect: active rare-earth (*e.g.*, Yb) dopants or quantum dots in a photonic crystal, whose transition frequency is initially detuned by $\Delta_{at} \sim 1$ MHz from the PBG edge and is abruptly modulated by fsec non-resonant laser pulses which exert ~ 10 MHz AC Stark shifts.

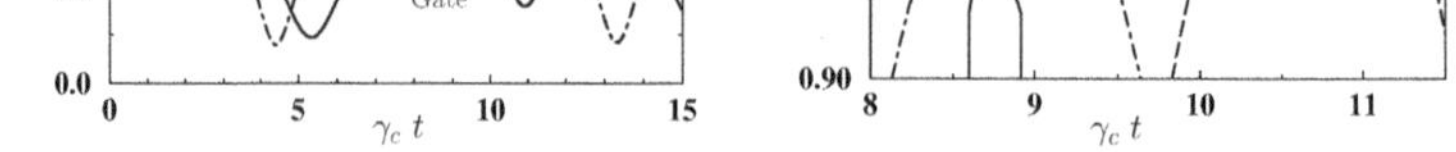

Figure 5. Fidelity of a superposition state (8) as a function of the dimensionless time $\gamma_c t$. Dot-dashed line: static detuning $\Delta_A/\gamma_c = 0.5$. Long-dashed line: periodic shifts of the detuning Δ_{at} from the large value $\Delta_A/\gamma_c = 0.5$ to the smaller one $\Delta_B/\gamma_c = 0.25$ in the sudden change approximation. Solid line: periodic shifts and control phase gates. The start and the end of the gate operation are shown by arrows. The rectangle indicates the position of the enlarged view of the right panel.

2. Decoherence and decoherence suppression in quantum memories for photons

2.1 Two-mode quantum memory

In order to simplify the discussion we will here restrict ourselves to a quantum memory for a two-mode radiation field, realized for example in a weak-coupling resonator allowing for two orthogonal polarization modes of the same frequency described by annihilation and creation operators $a_\pm$, $a_\pm^\dagger$ [Lukin 2000 (b)]. First we reexamine the adiabatic transfer scheme of Ref. [Fleischhauer 2000; Lukin 2000 (b)] for this case. We consider an ensemble of N 5-level atoms with internal states $|a_\pm\rangle$, $|b\rangle$ and $|c_\pm\rangle$ resonantly coupled to the two quantum modes $a_\pm$ and classical control fields with Rabi-frequencies $\Omega_\pm$ as shown in Fig. 6.

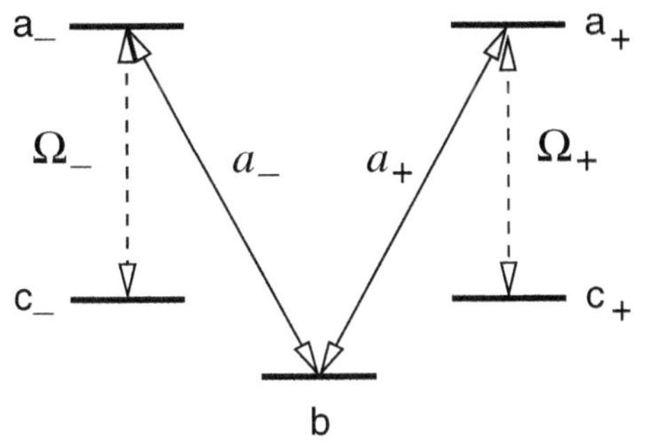

Figure 6. Scheme for parallel storage of two modes $a_\pm$ with orthogonal polarizations. $\Omega_\pm$ represent the Rabi-frequencies of coherent control fields.

The dynamics of this system is described by a non-hermitian Hamiltonian $(E_b = \hbar\,\omega_b = 0)$ $H = H_0 + H_I$, with

$$\begin{aligned} H_0 = \;& \hbar\omega\left(a_+^\dagger\, a_+ + a_-^\dagger\, a_-\right) + \hbar(\omega_a - i\gamma)\sum_{j=1}^{N}\left(\sigma_{a_+\,a_+}^j + \sigma_{a_-\,a_-}^j\right) \\ & + \hbar\,\omega_c\sum_{j=1}^{N}\left(\sigma_{c_+\,c_+}^j + \sigma_{c_-\,c_-}^j\right), \end{aligned} \tag{9}$$

and

$$H_I = \hbar \sum_{j=1}^{N} \left[g \left(a_+ \, \sigma^j_{a_+\,b} + a_- \, \sigma^j_{a_-\,b} \right) + \Omega \, e^{-i\nu t} \left(\sigma^j_{a_+\,c} + \sigma^j_{a_-\,c} \right) \right] + h.c. \tag{10}$$

Here $\sigma^j_{\mu\nu} = |\mu\rangle_j \, {}_j\langle\nu|$ is the flip operator of the jth atom and the vacuum Rabi-frequency g is assumed to be equal for all atoms. All spatial phases have been absorbed in the definition of the atomic states. We have introduced an imaginary part to the Hamiltonian to take into account losses from the excited states, *e.g.*, via spontaneous emission. The model does not take into account however relaxation from the excited states back into the lower levels.

When all atoms are initially prepared in level $|b\rangle$, the only states coupled by the interaction are the totally symmetric Dicke-states [Dicke 1954]

$$|\mathbf{b}\rangle = |b_1, \, b_2, \, \ldots, \, b_N\rangle \, , \tag{11a}$$

$$\left|\mathbf{a}^1_\pm\right\rangle = \frac{1}{\sqrt{N}} \sum_{j=1}^{N} |b_1, \, \ldots, \, a_{\pm j}, \, \ldots, \, b\rangle \, , \tag{11b}$$

$$\left|\mathbf{c}^1_\pm\right\rangle = \frac{1}{\sqrt{N}} \sum_{j=1}^{N} |b_1, \, \ldots, \, c_{\pm j}, \, \ldots, \, b\rangle \, , \quad etc. \tag{11c}$$

The couplings within the sub-systems consisting of the atomic states and the two radiation modes corresponding to at most a single excitation are shown in Fig. 7. The set of collective states can be separated into groups with specific excitation numbers $\{n_+, \, n_-\}$ in the two polarization modes.

In the following we will restrict ourselves to resonance conditions on all transitions. In this case the interaction of the N-atom system with the quan-

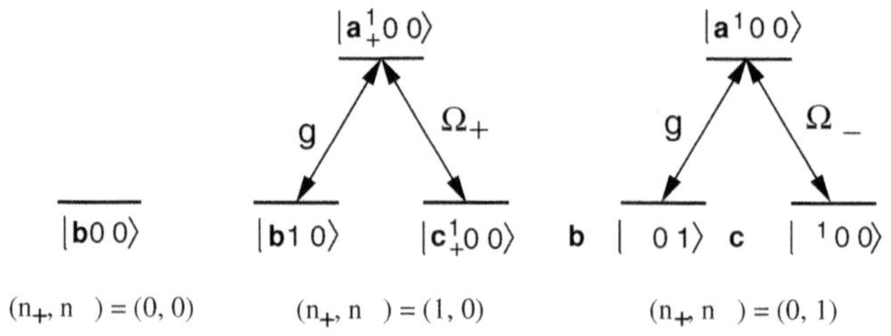

Figure 7. Coupling of bare eigenstates of atom plus cavity system for at most one photon. $|\mathbf{b}, \, 1, \, 0\rangle$ denotes collective atomic state $|\mathbf{b}\rangle$, one photon in mode a_+ and zero photons in mode a_-.

tized radiation modes has a double-family of dark-states, *i.e.*, adiabatic eigenstates with vanishing component of the excited states $|a\rangle_j$

$$|D,\, n,\, m\rangle = \sum_{k=0}^{n}\sum_{l=0}^{m} \xi_{nk}\,\xi_{ml}\,(-\sin\theta_+)^k\,(\cos\theta_+)^{n-k} \\ \times(-\sin\theta_-)^l\,(\cos\theta_-)^{m-l}\,\left|\mathbf{c}_+^k\,\mathbf{c}_-^l,\, n-k,\, m-l\right\rangle, \tag{12}$$

with $\xi_{nk} \equiv \sqrt{n!/[k!\,(n-k)!]}$. The mixing angles $\theta_\pm$ are defined as

$$\tan\theta_\pm(t) \equiv \frac{g\sqrt{N}}{\Omega_\pm(t)}, \tag{13}$$

and can be controlled by the external fields $\Omega_\pm(t)$. In (12) n and m denote the number of photons in the modes a_+ and a_- respectively. Although the dark states are degenerate, there are no transition between them even if non-adiabatic corrections are taken into account due to the symmetry of the interaction Hamiltonian.

The dark states of the N-atom system can be identified as quasi-particle excitations of the so-called dark-state polaritons $\Psi_\pm$ in the space of atoms and cavity mode [Fleischhauer 2000]

$$\Psi_\pm = \cos\theta_\pm(t)\, a_\pm - \sin\theta_\pm(t)\,\frac{1}{\sqrt{N}}\sum_{j=1}^{N}\sigma^j_{b\,c\pm}. \tag{14}$$

They are superpositions of the respective resonator mode and the collective spin corresponding to the ground-state transition $|b\rangle \longleftrightarrow |c_\pm\rangle$.

$$|D,\, n,\, m\rangle_N = \frac{1}{\sqrt{n!\,m!}}\left(\Psi_+^\dagger\right)^n\left(\Psi_-^\dagger\right)^m|\mathbf{b},\, 0,\, 0\rangle\,. \tag{15}$$

Associated with the dark polaritons are the orthogonal superpositions

$$\Phi_\pm = \sin\theta_\pm(t)\, a_\pm + \cos\theta_\pm(t)\,\frac{1}{\sqrt{N}}\sum_{j=1}^{N}\sigma^j_{b\,c\pm}, \tag{16}$$

which are called bright-polaritons. To obtain a complete set of operators we also need

$$\Phi_{\pm l} = \frac{1}{\sqrt{N}}\sum_{j=1}^{N}\sigma^j_{b\,c\pm}\,\exp\left\{2\pi\, i\,\frac{l\,j}{N}\right\},\quad l=1,\,\ldots,\,N-1, \tag{17}$$

which will be referred to as bright polaritons as well. In the limit of small atomic excitations the polariton operators obey in first order of $1/N$ bosonic commutation relations [Fleischhauer 2002].

Adiabatically rotating the mixing angles $\theta_\pm$ from 0 to $\pi/2$ leads to a complete and reversible transfer of all photonic states to a collective atomic excitation if the maximum number of photons $n+m$ is less than the number of atoms N. Let the initial quantum state of the light field be described by the density matrix

$$\hat{\rho}_f = \sum_{n,\,m}\sum_{k,\,l} \rho_{nm}^{kl}\, |n,\, k\rangle\langle m,\, l|, \tag{18}$$

where the numbers n and m correspond to mode a_+ and $k,\ l$ to mode a_-. Then the transfer process generates a quantum state of collective excitations according to

$$\begin{aligned} &\sum_{n,\,m}\sum_{k,\,l} \rho_{nm}^{kl}\, |n,\, k\rangle\langle m,\, l| \otimes |\mathbf{b}\rangle\langle\mathbf{b}| \\ &\downarrow \\ &\sum_{n,\,m}\sum_{k,\,l} \rho_{nm}^{kl}\, |D,\, n,\, k\rangle\langle D,\, m,\, l| \\ &\downarrow \\ |0\rangle\langle 0| \otimes &\sum_{n,\,m}\sum_{k,\,l} \rho_{nm}^{kl}\, |\mathbf{c}_+^n\ \mathbf{c}_-^k\rangle\langle \mathbf{c}_+^m\ \mathbf{c}_-^l|. \end{aligned} \tag{19}$$

2.2 Equivalence classes of storage states and sensitivity to decoherence

Let us consider the storage of a single photonic qubit spanned by the two polarization states $\{|1,\ 0\rangle,\ |0,\ 1\rangle\}$, *i.e.*, the initial state

$$|\psi_0\rangle = (\alpha\,|1,\ 0\rangle + \beta\,|0,\ 1\rangle) \otimes |\mathbf{b}\rangle \tag{20}$$

is mapped to the collective state $|\psi_1\rangle = |0,\ 0\rangle \otimes |\phi_1\rangle$ where

$$\begin{aligned} |\phi_1\rangle = \ &\frac{\alpha}{\sqrt{N}}\sum_j |b_1,\ \ldots,\ c_{+j},\ \ldots,\ b_N\rangle \\ &+ \frac{\beta}{\sqrt{N}}\sum_j |b_1,\ \ldots,\ c_{-j},\ \ldots,\ b_N\rangle \end{aligned} \tag{21}$$

through adiabatic rotation of the dark state

$$\begin{aligned} |D\rangle &= \alpha\,|D,\,1,\,0\rangle + \beta\,|D,\,0,\,1\rangle \\ &= \left(\alpha\,\Psi_+^\dagger + \beta\,\Psi_-^\dagger\right)|\mathbf{b},\,0,\,0\rangle \ \text{ with } \theta_+ = \theta_-. \end{aligned} \tag{22}$$

As can be seen from Eq. (21) the storage state of the photonic qubit is a maximally entangled N-atom state. These states are quite sensitive to certain decoherence processes, *e.g.*, if a single atom undergoes a spin flip from internal state $|b\rangle$ to $|c_+\rangle$ or $|c_-\rangle$ the resulting state is almost orthogonal to (21). If p denotes the probability that one atom undergoes such a transition, the total probability P_{error} to end up in an (almost) orthogonal state scales as $P_{\text{error}} \sim 1 - (1-p)^N \sim pN$. Thus one might naively expect that for the storage of a single photon the collective quantum memory will have an N times enhanced sensitivity to decoherence as compared to a single-atom device.

However, in the read-out process of the quantum memory, *i.e.*, when rotating $\theta_\pm$ from $\pi/2$ to 0, only the dark polariton excitations $\Psi_\pm$ are transferred to the cavity modes and thus only them are relevant. This can be seen upon inverting Eqs. (14) and (16):

$$a_\pm = \cos\theta_\pm(t)\;\Psi_\pm + \sin\theta_\pm(t)\;\Phi_\pm. \tag{23}$$

If W denotes the total density operator of the combined atom-cavity system after the writing process, all states W' that are identical to W when tracing out the bright polariton modes $\Phi_\pm$ and $\Phi_{\pm l}$, *i.e*, for which holds

$$\mathrm{Tr}_\Phi\left\{W'\right\} = \mathrm{Tr}_\Phi\left\{W\right\} \tag{24}$$

reproduce the same state of the radiation field. The property (24) defines an *equivalence class*. The importance of the equivalence classes stems from the fact that transitions among the states of the same class due to unwanted environmental interactions do not affect the fidelity of the quantum memory.

On the level of individual atoms the storage occurs within the three-state system consisting of $|b\rangle$, $|c_+\rangle$ and $|c_-\rangle$. We may safely neglect decoherence processes involving the excitation of other states. Then decoherence caused by individual and independent reservoir interactions can be described by the action of the two-level Pauli operators

$$X_\pm^j = \sigma^j_{b\,c\pm} + \sigma^j_{c\pm\,b},\quad Z_\pm^j = \left[\sigma^j_{b\,c\pm},\,\sigma^j_{c\pm\,b}\right],\quad Y_\pm^j = i\left(\sigma^j_{b\,c\pm} - \sigma^j_{c\pm\,b}\right). \tag{25}$$

$X_\pm^j$ describes a symmetric spin flip of the jth atom between states $|b\rangle$ and $|c_\pm\rangle$, $Z_\pm^j$ a phase flip, and $Y_\pm^j$ a combination of both. In addition there are similar processes involving states $|c_+\rangle$ and $|c_-\rangle$:

$$X_0^j = \sigma^j_{c_+\,c_-} + \sigma^j_{c_-\,c_+}, \; Z_0^j = \left[\sigma^j_{c_+\,c_-},\, \sigma^j_{c_-\,c_+}\right], \; Y_0^j = i\left(\sigma^j_{c_+\,c_-} - \sigma^j_{c_-\,c_+}\right). \tag{26}$$

Every single-atom error can be expressed in terms of the above operators by a completely positive map of the form

$$W_0 \longrightarrow W_1 = \sum_\mu p_\mu \frac{\Theta_\mu\, W_0\, \Theta_\mu}{\mathrm{Tr}\,\{\Theta_\mu\, W_0\, \Theta_\mu\}} \tag{27}$$

where the $p_\mu \geq 0$ are the probabilities for the map Θ_μ, where $\Theta_\mu \in \{\mathbf{1},\, X^j_\pm,\, Y^j_\pm,\, Z^j_\pm,\, X^j_0,\, Y^j_0,\, Z^j_0\}$ and $\sum_\mu p_\mu = 1$. We here have introduced a normalization denominator although $\mathrm{Tr}\,\{\Theta_\mu\, W_0\, \Theta_\mu\} = 1$ for all Θ_μ. This is because after expressing the Θ_μ's in terms of polariton operators we want to make explicit use of their Bose character, which is valid however only approximately.

In order to see the effect of the different decoherence processes after tracing out the bright polariton modes, we now express the operators Θ_μ in terms of polaritons in the limit $\theta_\pm = \pi/2$. This yields:

$$X^j_\pm = \frac{1}{\sqrt{N}}\left\{\sum_l \left(\eta_{jl}\, \Phi^\dagger_{\pm l} + \eta^\star_{jl}\, \Phi_{\pm l}\right) - \Psi_\pm - \Psi^\dagger_\pm\right\}, \tag{28a}$$

$$Y^j_\pm = \frac{i}{\sqrt{N}}\left\{\sum_l \left(\eta_{jl}\, \Phi^\dagger_{\pm l} - \eta^\star_{jl}\, \Phi_{\pm l}\right) - \Psi_\pm + \Psi^\dagger_\pm\right\}, \tag{28b}$$

$$Z^j_\pm = \frac{1}{N}\left[\sum_l \eta^\star_{jl}\Phi_{\pm l} - \Psi_\pm,\, \sum_m \eta_{jm}\, \Phi^\dagger_{\pm m} - \Psi^\dagger_\pm\right], \tag{28c}$$

and

$$X^j_0 = \frac{1}{N}\left(\sum_l \eta_{jl}\, \Phi^\dagger_{+l} - \Psi^\dagger_+\right)\left(\sum_m \eta^\star_{jm}\, \Phi_{-m} - \Psi_-\right) + h.a., \tag{29a}$$

$$Y^j_0 = \frac{i}{N}\left(\sum_l \eta_{jl}\, \Phi^\dagger_{+l} - \Psi^\dagger_+\right)\left(\sum_m \eta^\star_{jm}\, \Phi_{-m} - \Psi_-\right) - h.a., \tag{29b}$$

$$Z^j_0 = \frac{1}{N}\left(\sum_l \eta_{jl}\, \Phi^\dagger_{+l} - \Psi^\dagger_+\right)\left(\sum_l \eta^\star_{jl}\, \Phi_{+l} - \Psi_+\right) - (\text{terms with } + \longleftrightarrow -)\,. \tag{29c}$$

Here $\eta_{jl} = \exp(-2\pi\, l\, j/N)$. Since the $\Phi's$ and $\Psi's$ obey bosonic commutation relations up to corrections $\mathcal{O}(1/N)$, one sees from (28c) that $Z_\pm^j = 0 + \mathcal{O}(1/N)$, *i.e.*, within the bosonic quasi-particle approximation, the action of a phase flip $Z_\pm^j$ cannot be calculated. However one can draw the conclusion that a single-atom phase error only contributes in first order of $1/N$. From the other equations one recognizes an important property: if we assume that the initial state W_0 is an ideal storage state, *i.e.*, without bright polariton excitations, we find that after tracing out the bright polariton states only decoherence contributions of order $\mathcal{O}(1/N)$ survive, *e.g.*,

$$\mathrm{Tr}_\Phi\left(X_\pm^j\, W_0\, X_\pm^j\right) = \left(1 - \frac{1}{N}\right)\rho_0 + \frac{1}{N}\left(\Psi_\pm^\dagger + \Psi_\pm\right)\,\rho_0\,\left(\Psi_\pm^\dagger + \Psi_\pm\right), \tag{30}$$

where $\rho_0 = \mathrm{Tr}_\Phi(W_0)$. The factor $1/N$ in front of the second term exactly compensates for the fact that each atom can independently undergo a spin flip, which would yield a factor of N. Thus we see that due to the existence of equivalence classes, the collective quantum memory does not show an enhanced sensitivity with respect to single-atom errors.

2.3 Storage in a decoherence-free subspace

We will show in the following that the subspace of qubit storage states $\mathcal{M} \equiv \{|\mathbf{c}_+^1\rangle,\ |\mathbf{c}_-^1\rangle\}$ can form a two-dimensional decoherence-free subspace (DFS) with respect to all decoherence processes which result from individual reservoir interactions of the atoms similar to Majorana fermions as discussed in Ref. [Kitaev 2000]. We first note that it is possible to construct a one-dimensional DFS by providing an energy gap between a single non-degenerate ground state of a system and all other states. If the energy gap is sufficiently large compared to k_BT, interactions with a thermal reservoir of temperature T cannot cause transitions from the ground state to other states. If one tries to apply the same idea to a subspace with dimension $d \geq 2$ one faces the problem that an energy gap can in general not prevent transitions between the states of the subspace or dephasing of corresponding superpositions. At this point an important property of the storage states $\{|\mathbf{c}_+^1\rangle,\ |\mathbf{c}_-^1\rangle\}$ comes into play however. Calculating the transition probability amplitude from one state to the other upon a spin flip of any one atom from $|c_+\rangle$ to $|c_-\rangle$ or vice versa, one finds from Eqs. (29a) and (29b):

$$X_0^j = \frac{1}{N}\left(\Psi_+^\dagger\,\Psi_- + \Psi_-^\dagger\,\Psi_+\right) + \ldots, \tag{31}$$

where the omitted terms involve excitations or deexcitations of bright-polariton modes. One recognizes that a single-atom spin flip causes a transitions between $|\mathbf{c}_+^1\rangle$ and $|\mathbf{c}_-^1\rangle$ only with amplitude $1/N$. The corresponding transition probability scales as $1/N^2$. Thus even after taking into account that any of the N atoms can undergo a spin flip independently of all others, which leads to another factor of N, the total transition probability scales only as $1/N$. Consequently in a sufficiently large ensemble real transitions between the two states induced by single-atom spin-flips can be neglected. Similarly one finds from Eq. (29c) that for a phase flip $|c_+\rangle + |c_-\rangle \longleftrightarrow |c_+\rangle - |c_-\rangle$ holds

$$Z_0^j = \frac{1}{N}\left(\Psi_+^\dagger\, \Psi_+ - \Psi_-^\dagger\, \Psi_-\right) + \ldots, \tag{32}$$

where the omitted terms again contain excitations or deexcitations of bright polaritons. One recognizes that dephasing also leads to contributions within the subspace $\mathcal{M}$ only with amplitude $1/N$. All other terms involve the excitation or deexcitation of bright polaritons and are exponentially suppressed by the energy gap. From these arguments we arrive at the hypothesis that the subspace $\mathcal{M}$ is a decoherence-free subspace with respect to both, spin flips and dephasing processes. It is remarkable that the DFS would exist in the present system despite the fact that we have assumed individual uncorrelated bath interactions of the atoms.

To prove the above hypothesis we assume a collective interaction between the atoms of the ensemble corresponding to an effective Hamiltonian of the form

$$H_{\text{gap}} = -\hbar\, \omega_g \left(|\mathbf{c}_+^1\rangle\langle\mathbf{c}_+^1| + |\mathbf{c}_-^1\rangle\langle\mathbf{c}_-^1|\right), \tag{33}$$

which provides the necessary energy gap between the subspace $\mathcal{M}$ and all other states. Furthermore we consider a particular model for the reservoir interaction of the atoms describing both spin-flip and dephasing as indicated in Fig. 8. The atomic system is restricted to the three relevant states $|b\rangle$, $|c_+\rangle$ and $|c_-\rangle$. It is important that dissipation is not added on a phenomenological level.

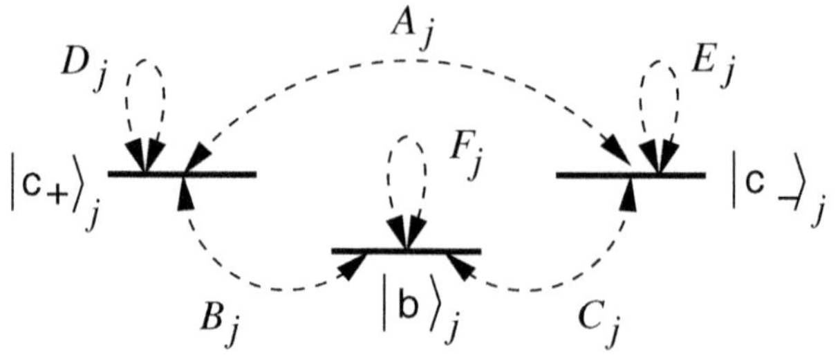

Figure 8. Model for interactions of an atom with individual and independent bosonic reservoirs decribing spin-flips (A_j, B_j, C_j) and dephasing (D_j, E_j, F_j).

The strong collective interaction (33) needs to be taken into account before including perturbative reservoir interactions and applying the standard Born-Markov approximations. We thus start from a full Hamiltonian approach of the reservoir interaction and eliminate the bath degrees of freedom later.

We consider individual bosonic reservoirs $A_j, \ldots, F_j$ with finite thermal energy $k_B T$ characterized by mode operators A_{jk}, $A^\dagger_{jk}$, *etc.* Environmentally induced spin flips can be described by the interaction Hamiltonian

$$\begin{aligned} H_{\text{spin}} = \; & \hbar \sum_{j=1}^{N} \sum_{k} \Big\{ \Big(g^k_{c_+\,c_-} A_{jk} + h.a. \Big) \Big(\sigma^j_{c_+\,c_-} + h.a. \Big) \\ & + \Big[\Big(g^k_{c_+\,b} B_{jk} + h.a. \Big) \Big(\sigma^j_{c_+\,b} + h.a. \Big) \\ & + \Big(g^k_{c_-\,b} C_{jk} + h.a. \Big) \Big(\sigma^j_{c_-\,b} + h.a. \Big) \Big] \Big\} , \end{aligned} \tag{34}$$

where the $g^j_{\mu\nu}$ denotes the coupling strength of atom j. It should be noted that the use of a rotating-wave approximation is in general not possible here due to the degeneracy of the involved states. In a similar way dephasing of the three states can be modeled by second order processes in the bosonic reservoir modes $\{D_k, E_k, F_k\}$

$$\begin{aligned} H_{\text{deph}} = \; & \hbar \sum_{j=1}^{N} \sum_{k,\,s} \Big[\kappa^j_{c_+ks} D^\dagger_{jk} D_{js} \sigma^j_{c_+\,c_+} \\ & + \kappa^j_{c_-ks} E^\dagger_{jk} E_{js} \sigma^j_{c_-\,c_-} + \kappa^j_{bks} F^\dagger_{jk} F_{js} \sigma^j_{bb} \Big] . \end{aligned} \tag{35}$$

The reservoir operators $D^\dagger_{jk}$, D_{jk}, *etc.* in (35) have a non-vanishing mean value in the thermal state. These mean values can be absorbed however in the free Hamiltonian of the atoms and will not be considered further.

Starting from the gap Hamiltonian (33) and the interactions with the reservoirs (34) and (35), eliminating the reservoir degrees of freedom within the standard Born-Markov approach we derive a master equation for the reduced density operator of the atomic ensemble. The calculation is lengthy but straight forward. Disregarding level shifts caused by the bath interaction we find for the populations in states $|\mathbf{c}^1_\pm\rangle$ the following density matrix equations in the interaction picture:

$$\frac{d}{dt}\,\rho_{\mathbf{c}_+\,\mathbf{c}_+} = -\left(1-\frac{1}{N}\right)\, f_A(\omega_g)\,\overline{n}(\omega_g)\,\rho_{\mathbf{c}_+\,\mathbf{c}_+}$$
$$-\frac{1}{N}\,f_A(0)\;[2\,\overline{n}(0)+1]\,\left(\rho_{\mathbf{c}_+\,\mathbf{c}_+}-\rho_{\mathbf{c}_-\,\mathbf{c}_-}\right)$$
$$-\,2\left(1-\frac{1}{N}\right)\,[\alpha_D(\omega_g)+\alpha_F(\omega_g)]\,\rho_{\mathbf{c}_+\,\mathbf{c}_+}$$
$$-2\,f_B(\omega_g)\,\overline{n}(\omega_g)\,\rho_{\mathbf{c}_+\,\mathbf{c}_+} + \text{decay from outside of } \mathcal{M}, \tag{36}$$

and a similar equation for the matrix element $\rho_{\mathbf{c}_-\,\mathbf{c}_-}$. Here

$$\overline{n}(\omega) = \frac{e^{-\beta\,\hbar\omega_k}}{1-e^{-\beta\,\hbar\omega_k}} \tag{37}$$

is the thermal occupation number of the reservoirs at temperature $\beta = 1/k_B T$ and energy $\hbar\omega$:

$$f_A(\omega) = 2\pi\,h_A(\omega)\,|g_{\mathbf{c}_+\,\mathbf{c}_-}(\omega)|^2, \tag{38a}$$

$$\alpha_D(\omega) = \pi\int_0^\infty d\omega'\,h_D(\omega')\,h_D(\omega+\omega')\,|\kappa_D(\omega+\omega',\,\omega)|^2$$
$$\times\,\overline{n}(\omega+\omega')\,\left[\overline{n}(\omega')+1\right], \tag{38b}$$

with $h_i(\omega)$ being the density of oscillator states at frequency ω. The terms in Eq. (36) can easily be interpreted. The first term describes the decay out of the symmetric (dark-polariton) state $|\mathbf{c}_+^1\rangle$ into non-symmetric (bright-polariton) states with the same number of excitations in internal states $|\mathbf{c}_\pm\rangle$. This process is exponentially suppressed by the term $\overline{n}(\omega_g)$ due to the presence of the energy gap. The terms on the second line correspond to an excitation exchange between the collective states $|\mathbf{c}_+^1\rangle$ and $|\mathbf{c}_-^1\rangle$. It can be seen that this process is suppressed by a factor $1/N$ as argued above. The third term is a decay of population from the symmetric (dark-polariton) collective states into non-symmetric (bright-polariton) collective states caused by dephasing of the single-atom states $|\mathbf{c}_+\rangle$ and $|\mathbf{b}\rangle$. As can be seen from Eq. (38b) this contribution is also exponentially suppressed since it contains the term $\overline{n}(\omega+\omega_g) \leq \overline{n}(\omega_g)$. Finally the third term describes spin flips to the collective ground state $|\mathbf{b}\rangle$. Also this process is exponentially suppressed by the factor $\overline{n}(\omega_g)$ due to the energy gap.

In a similar way one finds an equation for the off-diagonal density matrix element $\rho_{\mathbf{c}_+\,\mathbf{c}_-}$:

$$
\begin{aligned}
\frac{d}{dt}\,\rho_{\mathbf{c}_+\,\mathbf{c}_-} = &-\left(1-\frac{1}{N}\right)\, f_A(\omega_g)\,\overline{n}(\omega_g)\,\rho_{\mathbf{c}_+\,\mathbf{c}_-} \\
&-\frac{1}{N}\,f_A(0)\;[2\,\overline{n}(0)+1]\left(\rho_{\mathbf{c}_+\,\mathbf{c}_-}-\rho_{\mathbf{c}_-\,\mathbf{c}_+}\right) \\
&-\left(1-\frac{1}{N}\right)[\alpha_D(\omega_g)+\alpha_E(\omega_g)+2\,\alpha_F(\omega_g)]\,\rho_{\mathbf{c}_+\,\mathbf{c}_-} \\
&-3\;(f_B(\omega_g)+f_C(\omega_g))\;\overline{n}(\omega_g)\,\rho_{\mathbf{c}_+\,\mathbf{c}_-} \\
&-\frac{1}{N}\,(\alpha_D(0)+\alpha_E(0))\,\rho_{\mathbf{c}_+\,\mathbf{c}_-} + \text{decay from outside of } \mathcal{M}.
\end{aligned}
\tag{39}
$$

The first four lines of Eq. (39) are analogous to corresponding terms in Eq. (36) and can be interpreted in a similar way. They are all either exponentially small or suppressed by a factor $1/N$. The fifth line is a new contribution corresponding to collective dephasing. It is suppressed by a factor $1/N$ as expected from previous discussion, see Eq. (32).

Thus the decoherence model of Fig. 8 exactly reproduces the anticipated behavior. All decoherence processes resulting from individual and uncorrelated reservoir interactions of the atoms are either exponentially suppressed by the energy gap (33) or are proportional to $1/N$. The latter is due to the large effective distance of the collective states in state space. In this way a quasi decoherence free subspace of dimension two is generated which allows to protect a stored photonic qubit from decoherence much more efficiently than possible in quantum memories based on single particles.

Summary and conclusions

To summarize, in the first section we have reported that nonadiabatic periodic modulation of the detuning between an atomic transition and a continuum edge, whether for strong or weak coupling to the continuum, is able to protect the atomic state from spontaneous emission more effectively than fixing the largest possible detuning value. Remarkably, sudden changes of the detuning outperform its adiabatic modulation, as a means of maintaining high fidelity of quantum states and quantum-logic operations in the presence of decoherence, contrary to the prevailing approach to quantum-state control. This may pave the way to new methods of controlling decay and decoherence in spectrally structured continua [Viola 1999; Facchi 2001 (c); Wu 2002; Zanardi 2003].

In the second section of the present paper we have studied the influence of individual decoherence processes on the fidelity of a quantum memory for photons based on ensembles of atoms. Despite the fact that the atomic storage states corresponding to non-classical states of the radiation field are highly entangled, the system shows no enhanced sensitivity to decoherence as compared, *e.g.*, to single-atom storage systems, if it is caused by a coupling of the atoms to individual and independent reservoirs. This is due to the existence of equivalence classes of storage states corresponding to the excitations of only the dark-polariton modes.

We have also shown that by choosing a pair of appropriate collective storage states, which have a large effective distance in state space upon single atom errors, together with an energy gap between these states and all others a decoherence free subspaces emerges. Due to the large effective distance in state space single-atom errors cause transitions between the storage states or a relative dephasing of them only with a rate inversely proportional to the number of atoms. For a sufficiently large ensemble these processes can therefore be disregarded. Finally the energy gap, which needs to be larger than the characteristic thermal energy of the reservoir, suppresses transitions out of the subspace.

We conclude that large ensemble of atoms are well suited systems to store non-classical quantum states of light. They are not more sensitive to decoherence than single particle systems. Furthermore in the presence of a sufficiently large energy gap between appropriate collective states ensembles can lead to a suppression of decoherence much below the level possible in single-particle systems.

Acknowledgments

The support of the DFG through the SPP 1078 "Quanteninformationsverarbeitung", the EU RTN QUACS, ISF and Minerva is gratefully acknowledged. C. M. also thanks the Studienstiftung des Deutschen Volkes for financial support. S. P. acknowledges the financial support provided through the EC's Human Potential Programme under contract HPRN-CT-2002-00309, (QUACS).

ZENO AND ANTI-ZENO DYNAMICS

G. Kurizki,[1] A. G. Kofman,[1] V. M. Akulin,[2] E. Brion[2] and J. Clausen[2]

[1]*Chemical Physics Department,, Weizmann Institute of Science, Rehovot 76100, Israel*
gershon.kurizki@weizmann.ac.il
abraham.kofman@weizmann.ac.il

[2]*Laboratoire Aimé Cotton, Bât. 505, Campus d'Orsay, 91405 Orsay Cedex, France*
vladimir.akulin@lac.u-psud.fr

Abstract Decay *acceleration* by frequent measurements (interruptions of the coupling), known as the anti-Zeno effect is argued to be much more ubiquitous than its inhibition in one- or two-level systems coupled to reservoirs (continua). In multilevel systems, frequent measurements cause accelerated decay by destroying the multilevel interference, which tends to inhibit decay in the absence of measurements.

Introduction

The prevailing view until recently has been that successive frequent measurements (interruptions of the evolution) *must slow down the decay of any unstable* system [Khalfin 1968; Fonda 1973; Misra 1977; Joos 1984; Cook 1988; Itano 1990; Knight 1990; Frerichs 1991; Sakurai 1994; Panov 1996; Facchi 1998; Schulman 1998; Elattari 2000]. This is known as the quantum Zeno effect (QZE), introduced by Misra and Sudarshan [Misra 1977], following the early work of Khalfin [Khalfin 1968] and Fonda [Fonda 1973].

We have recently shown [Kofman 2000; Kofman 2001 (a)] that, in fact, the *opposite* is commonly true for decay into open-space continua: the anti-Zeno effect (AZE), *i.e.*, decay acceleration by frequent measurements[1], is far more ubiquitous than the QZE [Milonni 2000; Seife 2000]. How can this conclusion be understood and what was missing in standard treatments that claimed the QZE universality?

V.M. Akulin et al. (eds.), Decoherence, Entanglement and Information Protection in Complex Quantum Systems, 223–232.

1. Measurement-modified decay of a single state coupled to a reservoir

Consider $|e\rangle$, the measured state in a system ruled by hamiltonian $H = H_0 + V$, where V causes the coupling (decay) of $|e\rangle$ to all other eigenstates of H_0, which we refer to as the "reservoir" (Fig. 1).

The probability amplitude $\alpha(t)$ to remain in $|e\rangle$, which has the energy $\hbar\omega_a$, obeys the following *exact* integro-differential equation

$$\dot{\alpha} = -\int_0^t dt'\; e^{i\omega_a(t-t')}\; \Phi(t-t')\; \alpha(t'). \tag{1}$$

Here $\alpha(t) = \langle e|\Psi(t)\rangle\; e^{i\omega_a t}$, $\hbar\omega_a$ is the energy of $|e\rangle$, and

$$\Phi(t) = \frac{\langle e|V\, e^{-i\, H_0 t/\hbar}\; V|e\rangle}{\hbar^2} = \sum_j \frac{|V_{ej}|^2}{\hbar^2}\; e^{-i\,\omega_j t} \tag{2}$$

is the reservoir correlation function, expressed by $V_{ej} = \langle e|V|j\rangle$, where $|j\rangle$ ($\neq |e\rangle$) are H_0 eigenvectors with eigenvalues $\hbar\omega_j$.

Eq. (1) is *exactly* soluble, but it is enough to consider its short-time behavior by setting $\alpha(t) \approx \alpha(0) = 1$ in the integral of (1). This yields the expression

$$\alpha(t) = 1 - \int_0^t dt'(t-t')\; \Phi(t')\; e^{i\,\omega_a t'}, \tag{3}$$

in which *all powers of t* (phase factors!) are included and interferences between various decay channels may occur. By contrast, the standard *quadratic* expansion in t for the population [Misra 1977; Sakurai 1994] $\rho_{ee}(t) = |\alpha(t)|^2 \approx 1 - t^2/\tau_Z^2$, in which the Zeno time $\tau_Z = \hbar/(\langle e|H^2|e\rangle - \langle e|H|e\rangle^2)^{1/2}$ is the inverse variance of the energy in $|e\rangle$, may often fail, as discussed below. This is where we essentially differ from standard treatments. How does this difference show up?

Consider instantaneous measurements - projections on $|e\rangle$ interrupting its decay at intervals τ. We can use our result for $\alpha(t)$, Eq. (3), to express the population of $|e\rangle$ after n such measurements as exponentially decaying at a rate R,

$$\rho_{ee}(t = n\tau) = |\alpha(\tau)|^{2n} \approx \exp(-Rt). \tag{4}$$

The *universal form* of R is (in the frequency domain)

$$R = 2\pi \int_0^\infty d\omega\; G(\omega)\; F(\omega). \tag{5}$$

This expression is the overlap of the reservoir-coupling spectrum

$$G(\omega) = \frac{1}{\pi}\,\text{Re}\int_0^\infty dt\,\Phi(t)\,e^{i\,\omega t} = \sum_j \frac{|V_{ej}|^2}{\hbar^2}\,\delta(\omega - \omega_j) \tag{6}$$

and the measurement-induced broadening of the measured energy level

$$F(\omega) = \frac{\tau}{2\pi}\ \text{sinc}^2\,\frac{(\omega-\omega_a)\tau}{2}. \tag{7}$$

We can interpret the universal result (4) as an expression of the *time-energy uncertainty relation* for an unstable level with lifetime Δt, relating the energy broadening (uncertainty) of $|e\rangle$ to Δt, the interval between measurements (Fig. 1),

$$\text{Eq. (5)} \implies \Delta E\,\Delta t \sim \hbar. \tag{8}$$

More generally, in Eq. (8) $\Delta t = 1/\nu$, where ν is a characteristic rate of measurements. With this definition, Eq. (8) holds both for ideal and nonideal (realistic) measurements. Relation (8) comes about since measurements (projections) dephase level $|e\rangle$, analogously to phase randomization by collisions, which induce a linewidth that is equal to the collision rate ν.

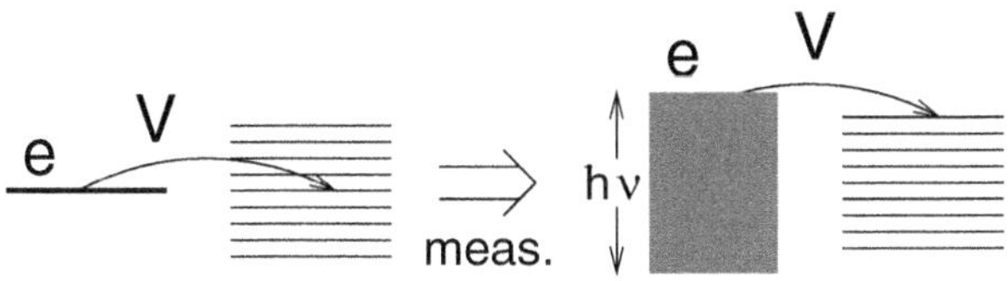

Figure 1. Left: The decay of a state into a "reservoir" via coupling. Right: Measurements broaden level $|e\rangle$, analogously to phase randomization by collisions at rate ν, drastically changing its decay into the reservoir.

A simple graphical analysis of the universal result Eq. (5) yields the main conclusions:

a) The QZE scaling, (*i.e.*, a *decrease of the decay rate R with an increase of ν*), is generally obtained when the measurement (dephasing) rate ν is much larger than the reservoir spectral width:

$$\nu \gg \Gamma_R,\ |\omega_a - \omega_M|. \tag{9}$$

Here Γ_R is the reservoir width and ω_M is the center of gravity of $G(\omega)$. In the special case of a peak-shaped $G(\omega)$, in Eq. (9) ω_M can be replaced by the position ω_m of the maximum. In the limit (9), one can approximate the spectrum $G(\omega)$ by a δ-function, with a constant C being the integrated spectrum,

$$C = \int G(\omega)\,d\omega = \langle V^2\rangle. \tag{10}$$

This approximation becomes *exact* in the case of resonant Rabi oscillations ($\Gamma_R = 0$, $\omega_a = \omega_M$), which explains why *the QZE is observable in Itano's experiment* [Itano 1990] *for any* ν. More generally, this approximation holds for any $G(\omega)$ that falls off faster than $1/\omega$ on the wings. Then our universal expression yields the most general result for the QZE, namely that R decreases with ν:

$$R \approx \frac{2C}{\nu}, \tag{11}$$

where we defined generally $\nu = [\pi \; F(\omega_a)]^{-1}$. In particular, as follows from Eq. (7), $\nu = 2/\tau$ for instantaneous projections. The *flattening of the spectral peak* of $G(\omega)$ by the broad function $F(\omega)$ in the convolution is seen to be the origin of the QZE. To put it simply, if the system is probed frequently enough, the QZE arises since *the effective decay rate is averaged over all decay channels*, many of which are weak, due to the energy uncertainty incurred by the measurements.

This result contradicts the claim of QZE universality and demonstrates the failure of the standard quadratic expansion: Eq. (8) shows that the QZE conditions can be much more stringent than the requirement to have $t \sim 1/\nu \ll \tau_Z$. The crucial point emphasized below is that Eq. (9) may be *principally impossible* to satisfy.

Under condition (9), the QZE scaling of R occurs for any $G(\omega)$ such that

$$G(\omega) \to 0 \;\; \text{at} \;\; \omega \to \infty. \tag{12}$$

Conditions (9) and (12) ensure the QZE scaling only for sufficiently large measurement rates and *do not always imply a monotonous decrease* of R as ν increases. The latter behavior, which is what one usually has in mind in discussions of the QZE, is obtained only in special situations.

b) The opposite to the QZE scaling is obtained whenever ω_a is significantly detuned from the nearest maximum of $G(\omega)$ at ω_m, so that $G(\omega_a) \ll G(\omega_m)$. In the limit (Fig. 2 - inset):

$$\nu \ll |\omega_m - \omega_a|, \tag{13}$$

the rate R *grows* with ν, since the dephasing function $F(\omega)$ is then probing more of the *rising* part of $G(\omega)$ in the convolution. This limit implies the *anti-Zeno effect* of decay acceleration by frequent measurements. Physically, this means that, as the energy uncertainty increases with the measurement rate ν, *the state decays into more and more channels, whose weight* $G(\omega)$ *is progressively larger.*

Remarkably, we may impose condition (10) in *any reservoir* that is not spectrally flat. This reveals the *universality* of the AZE, which has been noted already for radiative decay in cavities [Kofman 1996].

As an example, consider the coupling spectrum of the form

$$G(\omega) = A\,\omega^{\eta}, \;\; 0 < \omega < \omega_C,$$
$$G(\omega) = 0 \;\text{ otherwise}, \tag{14}$$

so that $\omega_a \ll \omega_C$. Using a Lorentzian $F(\omega)$, as obtained for realistic (continuous) measurements [Joos 1984; Cook 1988; Itano 1990], one obtains from Eq. (5) that for $\nu \ll \omega_C$

$$R = \frac{2\pi\; A(\omega_a^2 + \nu^2)^{\eta/2}\; \sin\eta\chi}{\sin\eta\pi} \quad (0 < \eta < 1), \tag{15a}$$

$$R = A\left[\nu\,\ln\frac{\omega_C^2}{\omega_a^2+\nu^2} + \omega_a\left(\pi + 2\,\tan^{-1}\frac{\omega_a}{\nu}\right)\right] \quad (\eta = 1) \tag{15b}$$

$$R = 2\pi\; A\,\omega_a^{\eta} + \frac{2A}{\eta - 1}\,\nu\,\omega_C^{\eta-1} \quad (\eta > 1), \tag{15c}$$

where $\tan\chi = -\nu/\omega_a$ $(0 < \chi < \pi)$. Obviously, expressions (15a) to (15c) increase with ν.

c) More subtle behavior occurs in the domain between the QZE and AZE limits. Assume, for simplicity, that $G(\omega)$ is single-peaked and satisfies condition (12). When ν increases from the limit (13) up to the range where the right-hand-side inequality is violated, then $\nu \gg |\omega_m - \omega_a|$, which is now equivalent to condition (9), implying the Zeno scaling of Eq. (11). But even in this QZE-scaling regime, *R remains larger than the Golden Rule rate*, (*i.e.*, the normal decay rate in the absence of measurements)

$$R_{GR} = 2\pi\; G(\omega_a), \tag{16}$$

up to much higher ν, as expressed by the following condition for *"genuine QZE"*

$$R < R_{GR} \;\text{ for }\; \nu > \nu_{QZE}, \tag{17}$$

where

$$\nu_{QZE} = \frac{2C}{R_{GR}} = \frac{C}{\pi\; G(\omega_a)} \tag{18}$$

in the case of a finite C.

The value of ν_{QZE} given by Eq. (18) was identified with the reciprocal "jump time", *i.e.*, the maximal time interval between measurements for which the decay rate is appreciably changed [Schulman 1997]. However, the *correct*

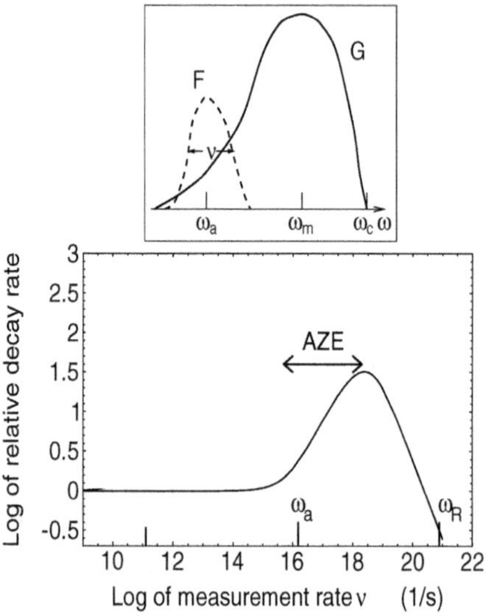

Figure 2. AZE dependence of the decay rate. Inset: conditions for the AZE [Eq. (13)]. Graph: the dependence of the logarithm of the normalized decay rate $\log_{10}(R/R_{RG})$ on $\log_{10}\nu$ for a spontaneously emitting hydrogenic state. The atomic transition frequency: $\omega_a = 1.55 \times 10^{16}\ \mathrm{s}^{-1}$, whereas the relativistic cutoff $\omega_R = 7.76 \times 10^{20}\ \mathrm{s}^{-1}$. The corresponding Bohr frequency is $\omega_B = 8.50 \times 10^{18}\ \mathrm{s}^{-1}$. The AZE range is marked.

value of the reciprocal jump time may be smaller by many orders of magnitude than ν_{QZE}. In the special case of ideal instantaneous measurements and a Lorentzian or Lorentzian-like $G(\omega)$, the genuine-QZE condition (17) reduces to that of Ref. [Facchi 2000 (b)].

These considerations apply (with some limitations) to hydrogenic radiative decay (spontaneous emission), for which $G(\omega)$ can be calculated *exactly* [Moses 1972]:

$$G(\omega) = \frac{\alpha\,\omega}{\left[1 + \left(\frac{\omega}{\omega_B}\right)^2\right]^4}. \tag{19}$$

Here $\omega_B \sim c/a_B$, where c is the vacuum light speed and a_B is the radius of the electron orbit. Then Eqs. (5) and (7) may yield the AZE trend

$$R \approx \alpha\,\nu\left[\ln\left(\frac{\omega_B}{\nu}\right) + C_1\right] \quad (\omega_a \ll \nu \ll \omega_B), \tag{20}$$

where $C_1 = 0.354$ and $\nu = 2/\tau$. The AZE trend should be observable (Fig. 2) for $\nu \gtrsim \omega_a$, *i.e.*, for microwave Rydberg transitions on a ps scale (provided we can isolate one transition) and for optical transitions on the sub-fs scale. The boundary between the AZE and QZE-scaling regions is now given by $\nu_1 \sim \omega_B$ and the genuine-QZE condition (17) by $\nu > \nu_{QZE} \sim \omega_B^2/12\pi\,\omega_a \gg \nu_1$ [*cf.* Eq. (18)], rendering $R < R_{GR} = 2\pi\,\alpha\,\omega_a$.

This analysis implies that the "genuine QZE" range $\nu > \nu_{QZE}$ is *principally unattainable*, since it requires measurement rates above the relativistic cutoff ω_R, which are *detrimental* to the system, leading to the production of new particles. A similar principal obstacle occurs for radioactive decay. By contrast, the AZE is *accessible* in decay processes, such as spontaneous emission or the nuclear β-decay, and can essentially always be imposed.

2. Decay modification by measurements in multilevel systems

Here we discuss in detail a model for measurement-induced decay modification in a multilevel system. The system with energies $\hbar\omega_n$, $1 \le n \le N$, is coupled to a zero-temperature bath of harmonic oscillators with frequencies ω_λ. The corresponding Hamiltonian, in the rotating-wave approximation, is

$$\begin{aligned} H = \;& \hbar\sum_n \omega_n\,|n\rangle\langle n| + \hbar\sum_\lambda \omega_\lambda\,a_\lambda^\dagger\,a_\lambda \\ & + \hbar\sum_{n,\,\lambda}\Big(\kappa_{n\lambda}\,a_\lambda^\dagger\,|g\rangle\langle n| + \kappa_{n\lambda}^\star\,a_\lambda\,|n\rangle\langle g|\Big)\,. \end{aligned} \tag{21}$$

We note that the same Hamiltonian decribes an n-level system coupled to a continuum, as occurs, *e.g.*, in autoionization. The both cases are described by the same equations.

The wavefunction is

$$|\Psi(t)\rangle = \sum_n \alpha_n(t)\,|n,\,\{0_\lambda\}\rangle + \sum_\lambda \beta_\lambda(t)\,|g,\,1_\lambda\rangle, \tag{22}$$

where $\{0_\lambda\}$ and 1_λ denote, respectively, the bath ground state and one quantum in the mode λ. Inserting Eqs. (21) and (22) into the Schrödinger equation yields the set of equations

$$\dot{\alpha_n} = -i\,\omega_n\,\alpha_n - i\sum_\lambda \kappa_{n\lambda}^\star\,\beta_\lambda, \tag{23a}$$

$$\dot{\beta}_\lambda = -i\,\omega_\lambda\,\beta_\lambda - i\sum_n \kappa_{n\lambda}\,\alpha_n. \tag{23b}$$

The solution of Eq. (23b) with the initial condition $\beta_\lambda(0) = 0$ is

$$\beta_\lambda(t) = -i \int_0^t dt' \, e^{-i\,\omega_\lambda (t-t')} \, i \sum_n \kappa_{n\lambda} \, \alpha_n(t'). \tag{24}$$

Inserting Eq. (24) into (23a) yields the set of integro-differential equations

$$\dot{\alpha_n} = -i\,\omega_n\,\alpha_n - \sum_{n'} \int_0^t dt' \; \Phi_{nn'}(t-t') \, \alpha_{n'}(t'). \tag{25}$$

Here

$$\Phi_{nn'}(t) = \int d\omega \, G_{nn'}(\omega) \, e^{-i\,\omega t}, \tag{26}$$

where

$$G_{nn'}(\omega) = \sum_\lambda \kappa^\star_{n\lambda} \, \kappa_{n'\lambda} \, \delta(\omega - \omega_\lambda). \tag{27}$$

The nondiagonal terms here result in *multilevel quantum interference*. One can show that

$$|G_{nn'}(\omega)|^2 \leq G_{nn}(\omega) \, G_{n'n'}(\omega). \tag{28}$$

Hence maximum quantum interference is achieved when Eq. (28) is the equality. For example, if all transition dipoles are parallel, $\vec{d_{ng}} \parallel \hat{x}$, where $\hat{x}$ is the unit vector along the x axis, then Eq. (28) holds, with

$$G_{nn'}(\omega) = d_n^\star \, d_{n'} \, g(\omega), \tag{29a}$$

$$d_n = \vec{d_{ng}} . \, \hat{x}, \;\; g(\omega) = \sum_\lambda \; |\vec{q_\lambda} . \, \hat{x}|^2 \;\; \delta(\omega - \omega_\lambda). \tag{29b}$$

Let the reservoir be specifically Lorentzian,

$$g(\omega) = \frac{1}{\pi} \frac{A\,\Gamma}{(\omega - \omega_R)^2 + \Gamma^2}, \tag{30}$$

where ω_R and Γ are the reservoir center frequency and width, respectively. Eqs. (26), (29a), and (30) yield that

$$\Phi_{nn'}(t) = d_n^\star \, d_{n'} \; A \, e^{-(i\omega_R + \Gamma)t}. \tag{31}$$

In this case the reservoir can be substituted by one level $|0\rangle$ with the energy $\hbar\omega_R$ and the decay rate 2Γ, the effective Hamiltonian being

$$H_{eff} = \hbar \sum_{n=1}^{N} \{\omega_n \, |n\rangle\langle n| + (\omega_R - i\Gamma) \, |0\rangle\langle 0| + (\kappa_n \, |0\rangle\langle n| + \kappa_n^{\star} \, |n\rangle\langle 0|)\} \,, \tag{32}$$

where κ_n is the coupling matrix element for the $|n\rangle$-$|g\rangle$ transition. The wave-function is now

$$|\Psi(t)\rangle = \sum_{n=1}^{N} \alpha_n(t) \, |n\rangle + \beta(t) \, |0\rangle, \tag{33}$$

with the initial codition $\beta(0) = 0$. The Schrödinger equation now yields

$$\dot{\alpha_n} = -i \, \omega_n \, \alpha_n - i \, \kappa_n^{\star} \, \beta, \tag{34a}$$

$$\dot{\beta} = -(i \, \omega_R + \Gamma)\beta - i \sum_{n}^{N} \kappa_n \, \alpha_n. \tag{34b}$$

The equivalence of the effective and exact Hamiltonians for the relaxation of the system follows from the fact that the exclusion of $\beta(t)$ from Eqs. (34a) and (34b) yields Eq. (25) with definition (31), if one sets

$$\kappa_n = d_n \sqrt{A}. \tag{35}$$

Thus, one can simulate relaxation of an n-level sytem, at least, in case (29a), (30), by solving the set of Eqs. (34a) and (34b). This is simpler than considering a more general case, when one should solve differential Eqs. (23a) and (23b) or, alternatively, the integro-differential Eqs. (25).

Consider now how one can modify the decoherence of the system. In order to imitate repeated instantaneous measurements, one can use a train of short, *random*, nonresonant pulses (see also Ref. [Kofman 2001 (b); Search 2000]). Let the pulses be sufficiently strong, so that the phase shift of the amplitude of state $|n\rangle$ (relative to that of $|g\rangle$) due to each pulse is large,

$$\int_{t_k}^{t_k'} dt \, (\chi_n - \chi_g) \, |E_k(t)|^2 \gg 2\pi. \tag{36}$$

Here χ_n is the polarizability of level $|n\rangle$, whereas t_k and t_k' are the beginning and end moments of the kth pulse. If these pulses slightly differ one from another (*i.e.*, in their integrated intensity), the phase shifts (27) can be considered *completely random*. For simplicity, we assume that χ_n's are the same for all levels $|n\rangle$. We also assume that the pulses are randomly distributed in time with the average interval T, take equal and real $\kappa_n = k$ and equidistant ω_n.

Then, for the above case where (32) applies one obtains the equation for the density matrix

$$\dot{\rho} = -\frac{i}{\hbar}\,[H_{eff},\,\rho] - R\,\rho. \tag{37}$$

Here R is the superoperator of measurement-dependent relaxation, whose only nonzero elements are $R_{ng,\,ng} = R_{gn,\,gn} = 1/T$, the measurement rate. With the increase of $1/T$, the relaxation dynamics should change.

Quantum interference creates a *decoherence-free subspace* and thus slows down the relaxation. Therefore, one may expect that nonzero Γ and $1/T$ can in this case *accelerate* decay. Quantum interference will be significant only for sufficiently large k, so that $\pi\ G(\omega_n) \gtrsim \omega_{n+1,\,n}$, for some n. To observe *reduced decay* with $1/T$ (the analog of QZE), one should take $\Gamma \gtrsim (\omega_n - \omega_1)/2$ and $1/T \gtrsim \Gamma$.

Conclusions

The universal formulae derived in Secs. 1 and 2 result in the following general criteria:

(i) The QZE can only occur in systems with spectral width below the resonance energy.

(ii) It is *principally unattainable* in open-space radiative or nuclear β-decay, because the required measurement rates would cause the creation of new particles.

(iii) Contrary to the widespread view, frequent measurements can be chosen to *accelerate* essentially any decay process. Hence, the *anti-Zeno effect (AZE)* should be *far more ubiquitous than the QZE.*

(iv) Acceleration of decay by frequent measurements can occur in a multilevel system, where interference effects (in the absence of phase-randomizing measurements) can inhibit the decay.

Acknowledgments

This work was supported by the EC Human Potential Programme through the Research Training Network QUACS, ISF, Minerva, and by Ministry of Absorption.

Notes

1. Related effects have been noted in proton decay [Lane 1983]; more recently (by us [Kofman 1996]) for radiative decay in cavities, in Rabi oscillations betwen coupled quantum dots [Gurvitz 1997], in photodetachment [Lewenstein 2000]. See also Ref. [Facchi 2000 (a)].

IV

NON-MARKOVIAN DECAY AND DECOHERENCE IN OPEN QUANTUM SYSTEMS

NON-MARKOVIAN DECAY AND DECOHERENCE IN OPEN QUANTUM SYSTEMS

J. Salo,[1, 2] J. Clausen[3] and I. E. Mazets[4]

[1] *Laser Physics & Quantum Optics, Royal Institute of Technology (KTH), AlbaNova, Roslagstullsbacken 21, SE-10691 Stockholm, Sweden*

[2] *Helsinki University of Technology, Materials Physics Laboratory, P.O. Box 2200, 02015 HUT, Finland*
Janne.Salo@hut.fi

[3] *Laboratoire Aimé Cotton, Bât. 505, Campus d'Orsay, 91405 Orsay Cedex, France*
jens.clausen@lac.u-psud.fr

[4] *Ioffe Physico-Technical Institute, St.Petersburg 194021, Russia*
mazets@astro.ioffe.rssi.ru

Abstract Interaction between a quantum system and its surroundings - be it another similar quantum system, a thermal reservoir, or a measurement device - breaks down the standard unitary evolution of the system alone and introduces open quantum system behaviour. Coupling to a fast-relaxing thermal reservoir is known to lead to an exponential decay of the quantum state, a process described by a Lindblad-type master equation. In modern quantum physics, however, near isolation of individual quantum objects, such as qubits, atoms, or ions, sometimes allow them only to interact with a slowly-relaxing near-environment, and the consequent decay of the atomic quantum state may become nonexponential and possibly even nonmonotonic. Here we consider different descriptions of non-Markovian evolutions and also hazards associated with them, as well as some physical situations in which the environment of a quantum system induces non-Markovian phenomena.

Keywords: Irreversible time evolution, Master Equation, non-exponential decay

Irreversibility is an everyday phenomenon in nature but it remains one of the central issues in theoretical physics. In the early days of quantum physics, open system evolution was described using the Fermi Golden Rule, which leads inherently to an exponential decay of an excited quantum state. The underlying assumption is a continuum of final states that forces all the Poincaré recurrences to infinity, and hence introduces irreversibility into the solution of an initial-value problem. While the Golden Rule yields the decay rate of the ex-

V.M. Akulin et al. (eds.), Decoherence, Entanglement and Information Protection in Complex Quantum Systems, 235–238.

citation probability, it does not, as such, describe the quantum state of the decaying system. A Liouville equation of motion for the decaying system is obtained if it is assumed to be coupled to a thermal reservoir; the implied dynamic no longer preserves the coherence of the quantum state and leads to an exponential decay.

The loss of coherence for an individual quantum system is always due to an entanglement with some other system and this holds even for a thermal reservoir. This can be seen from the following consideration: assume that the quantum system studied is coupled to the reservoir with a coupling strength α and the reservoir relaxation rate is given by κ. Standard adiabatic elimination of the reservoir then yields an atomic decay rate $\gamma = \alpha^2/\kappa$ which tends to zero for an arbitrarily fast reservoir. This is due to the fact that for such a reservoir the combined system would truly remain in a product state and no entanglement would ever be built up. This may also be taken as a variant of quantum Zeno effect, in which the decay into an infinitely fast reservoir corresponds to very rapidly repeated measurements that prevent the spontaneous decay.

Although, in principle, the description of a combined quantum system is independent of the sizes or evolution rates of the coupled subsystems, it is useful to consider different parametric regimes, yet the boundaries in-between often remain vague. Relative sizes of the parameters α and κ qualitatively determine the behaviour of the quantum system studied, illustrated schematically in Fig. 1 and studied quantitatively for a simple model below in the paper titled "*The Varieties of Master Equations*". In the intermediate range, in which the relevant parameters do not differ by much more than few orders of magnitude, we term the atom to be coupled to its surroundings or near-environment, which evolves slowly enough to respond dynamically to the decaying quantum system. This is the regime where non-Markovian and nonexponential effects are expected to play an important role.

An example of the surroundings interaction is provided by propagation of an impurity atom inside a Bose-Einstein condensate, considered here in the paper titled "*Nonexponential motional damping of impurity atoms in Bose-Einstein condensate*". Due to a long memory time of the condensate, a phonon created by a moving impurity atom may remain in the impurity's vicinity long enough for affecting its motional decay. In that sense, the condensate differs from a (thermal) environment, which is usually assumed to recover the steady state infinitely fast. A similar phenomenon is also provided by a spontaneous decay of an excited atom into a leaky cavity, in which case the photon remains within the cavity and hence close to the atom long enough for being reabsorbed.

Environment or surroundings interaction is not, however, the only way of introducing decay or decoherence into quantum evolution. Measurement processes in which the system state couples to that of the measurement device are often described in terms of instantaneous projective measurements, which also de-

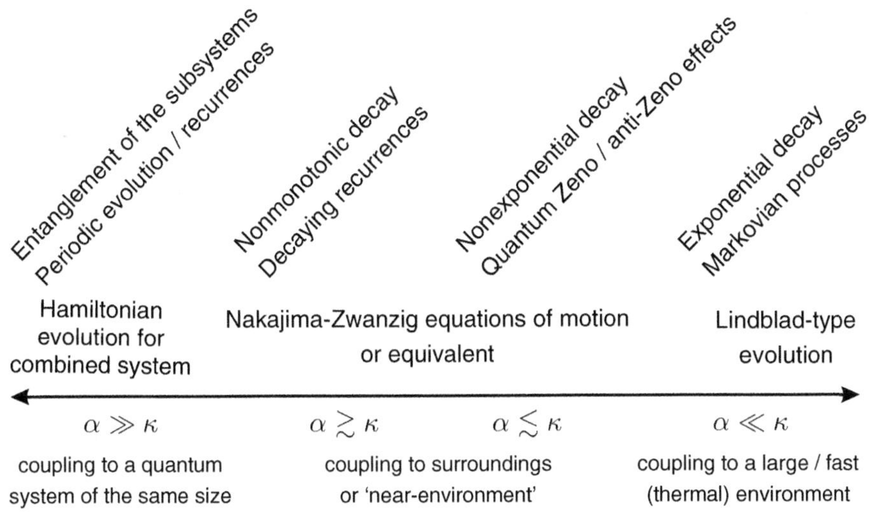

Figure 1. Different regimes of the system-environment coupling. Here α represents the coupling strength between a quantum system and its environment and κ is the environment relaxation rate. Note that, in the Markovian limit, the system decouples from the environment altogether for $\alpha/\kappa \rightarrow 0$ unless α^2/κ remain finite.

stroy quantum coherence. While individual measurements serve as projections onto the quantum states corresponding to the measurement outcome, repeated measurements allow dynamical control of the quantum evolution. Like a very fast environment, frequent measurements cause slow decoherence or no decoherence at all, while finitely spaced measurements imply more complicated dynamics, similar but not identical to those caused by surroundings interaction.

In the following three papers we consider different aspects of non-Markovian evolution of an individual quantum system. In the first paper, *The Varieties of Master Equations*, we consider different kinds of linear Master Equations and their properties, both in Markovian and non-Markovian regimes. Provided one has available an applicable reservoir model, the Master Equation may, in principle, be derived through elimination of the reservoir degrees of freedom. This is, however, often a very complicated procedure and it cannot necessarily be carried out without (or even with) approximations. Therefore we discuss possibilities of using phenomenological Master Equations inclusive of memory effects; yet this turns out to be a hazardous method since, unlike the Markovian case, no necessary requirements are known for a memory Master Equation to yield a complete positive map for time evolution. We also present two different Master Equations for a two-level atom interacting with a thermal reservoir: (i) exact Nakajima-Zwanzig equations in the case where the surroundings is represented with a thermally damped two-level atom (also called a faked-continuum model) and (ii) approximate equations of motion for a driven atom, developed to the second order in reservoir coupling constant.

The second paper, *Quantum dynamics effected by repeated measurements*, considers quantum systems coupled to an ancilla system that is subject to repeated projective measurements. Here the open quantum system behaviour arises due to coupling to a measurement device and it is intermediated to the system via the ancilla. Special emphasis is placed on the case of nondemolition interaction between the system and ancilla; this conserves the (expectation value of) system energy while allowing the phase decoherence enter in a controlled way. In this case, the system cannot be described purely with Lindblad-type operators since the projections destroy abruptly some of the non-diagonal elements of the quantum state, which would require the evolution operator to possess infinite eigenvalues, *i.e.*, decay constants. The nondemolition character of the interaction, however, allows one to find equations of motion for individual matrix elements of the system state operator, represented in the diagonal elements of the ancilla part; these equations are seen to carry memory effects and can be converted, for instance, into Nakajima-Zwanzig form. The method is applied to a harmonic oscillator coupled to a two-level ancilla. For a proper choice of parameters, an $\hat{n}^2$ effective Hamiltonian is induced on the oscillator, which repeatedly builds up a Schrödinger-cat state, eventually destroyed by decoherence.

In the third paper, *Nonexponential motional damping of impurity atoms in Bose-Einstein condensate*, the propagation of impurity atoms is considered in BEC. For atom velocities superior to that of condensate phonons, the thermalisation rate of the condesate is not high enough to allow Markovian relaxation and the motional damping of the atoms is predicted to deviate from exponential. This is also a manifestation of non-mean-field effects in the condensate since the mean-field theories yield exponential decay for the motion of impurity atoms. Measurement of the motional decay rate should, therefore, provide a way for observing non-mean-field effects in the condensate.

THE VARIETIES OF MASTER EQUATIONS

J. Salo,[1, 2] S. Stenholm,[1] G. Kurizki[3] and A. G. Kofman[3]

[1]*Laser Physics & Quantum Optics, Royal Institute of Technology (KTH), AlbaNova, Roslagstullsbacken 21, SE-10691 Stockholm, Sweden*
Janne.Salo@hut.fi, stenholm@atom.kth.se

[2]*Helsinki University of Technology, Materials Physics Laboratory, P.O. Box 2200, 02015 HUT, Finland*

[3]*Chemical Physics Department, Weizmann Institute of Science, Rehovot 76100, Israel*
gershon.kurizki@weizmann.ac.il, abraham.kofman@weizmann.ac.il

Abstract

Irreversible time evolution is in quantum systems described by Master Equations. These are usually derived by elimination of degrees of freedom belonging to an environment acting as a reservoir. When the reservoir is acting as a sink for information, it has got no memory, we obtain equations usually termed Markovian. In this limit it has been shown by semi-group arguments which form preserves all physically relevant properties of the quantum state. It has been noted that not all Markovian equations derived lead to physically meaningful conclusions. On the other hand, the derivations prove that exact equations can be obtained if the memory effects are retained. We have also observed that improper introduction of the memory features may lead to unphysical results. We pose the challenge to find the general criteria which guarantee that the time evolution described by a density matrix is physically acceptable. However, it is also possible to avoid memory problems by deriving an equation of motion with time dependent coefficients, a so called memory-less Master Equation. In order to apply the same considerations on more complicated models one runs into problems posed by the formal structure of the method. Thus we warn against ad hoc applications of irreversible Master Equations.

We explain the properties and emergence of these features with the scope of simple models which can be analysed properly; thus we hope to provide some insight into the character of irreversible time evolution. First we consider an exactly solvable model in which a two-level atom is allowed to decay into a thermally relaxing reservoir, represented by another two-level atom (so-called faked continuum model). Memory equations are, however, highly dependent on the form of memory functions and even a slight misparameterisation leads to non-positive evolution. As an alternative to the foregoing approach, we outline the derivation of a generalised non-Markovian equation for a two-level system that is coupled to a finite-temperature bath and driven by an arbitrary time-dependent

V.M. Akulin et al. (eds.), Decoherence, Entanglement and Information Protection in Complex Quantum Systems, 239–280.

field. This derivation does not invoke the rotating-wave approximation and yields dynamically-modified rates for population decay and decoherence.

Keywords: Irreversible time evolution, Master Equation, non-Markovian evolution

Introduction

The problem of irreversibility is one of the central issues in theoretical physics. From time-reversal invariant equations one has somehow to produce equations selecting a direction of time. Since the time of Boltzmann, many approaches have emerged, but the irreversibility still offers many puzzles and conflicting views. It is our purpose in this paper to summarise some of these and indicate by simple examples where the conflicts emerge. We are not able to give a comprehensive review, and the list of references has to be restricted to a few central ones.

It was early realised that the presence of a continuous spectrum would push all Poincare recurrences to infinity, and hence introduce irreversibility into the solution of an initial-value problem. Such an approach was seen in field theory in terms of simple models [Zumino 1961; Levy 1961]. The analytic behaviour of the scattering amplitude was found to be the essential issue [Peierls 1955], and the results were incorporated with the standard scattering theory [Goldberger 1964].

Many developments have emerged since, and the field has seen both successes and controversies. For a recent review of the state of the art, see Ref. [Zwanzig 2001]. In many treatments, the authors regard the issues to be resolved. We do not share this view, but want to present some of the basic questions in terms of simple models.

1. Markovian Master Equations

1.1 The universal Markovian time evolution

In quantum mechanics, the most general time evolution must be linear. That implies that having two solutions $\rho_1(t)$ and $\rho_2(t)$ of the evolution equation, we can find another solution by writing $\rho(t) = a\,\rho_1(t) + b\,\rho_2(t)$. As the density matrices are represented by ensembles of systems, it must be possible to consider the ensemble combined from $\rho_1(t)$ and $\rho_2(t)$. The linearity assumption is satisfied if we have a time evolution of the form

$$\partial_t\,\rho(t) = \mathcal{L}\,\rho(t), \tag{1}$$

where $\mathcal{L}$ is a linear operator on the set of density matrices, it is called a superoperator. The solution of Eq. (1) is the semi-group generated by the Liouvillean operator $\mathcal{L}$: $\rho(t) = e^{\mathcal{L}t}\,\rho(0)$.

An abstract approach was taken in Refs. [Gorini 1976; Sudarshan 1992]. We assume that the operators act on a Hilbert space $\mathcal{H}$. We list the most general linear transformation on a density matrix, by writing:

- the transformation $c\,\rho$ if $c \in \mathbb{C}$
- the transformation $B\rho$ for operator B on $\mathcal{H}$
- the transformation $\rho\, C$ for operators C on $\mathcal{H}$
- if $\mathcal{L}_1$ and $\mathcal{L}_2$ are linear, then so is $\mathcal{L}_1 + \mathcal{L}_2$
- if $\mathcal{L}_1$ and $\mathcal{L}_2$ are linear, then so is $\mathcal{L}_1 \circ \mathcal{L}_2$
- the identity $\mathbb{1}\rho = \rho$ is linear.

Using these rules we may construct the most general linear transformation on the density matrix by setting

$$\mathcal{L}\,\rho = c\,\rho + B\,\rho + \rho\, C + \sum_{i,\,j} \lambda_{ij}\, K_1^i\, \rho\, K_2^j. \tag{2}$$

Because the density matrix is hermitian, we may write

$$\partial_t\, \rho^\dagger(t) = [\mathcal{L}\,\rho(t)]^\dagger = \mathcal{L}\,\rho(t). \tag{3}$$

We have

$$[\mathcal{L}\,\rho(t)]^\dagger = c^\star\rho + \rho\, B^\dagger + C^\dagger\,\rho + \sum_{i,\,j} \lambda_{ij}^\star\, K_2^{j\dagger}\,\rho\, K_1^{i\dagger}. \tag{4}$$

Comparing this with (2) we find the requirements

$$\begin{gathered} c^\star = c, \qquad C^\dagger = B, \qquad B^\dagger = C \\ \lambda_{ij}^\star = \lambda_{ji}, \quad K_2^{j\dagger} = K_1^i, \quad K_1^{i\dagger} = K_2^j. \end{gathered} \tag{5}$$

Because the probability described by the density matrix must be normalised, we require

$$\partial_t\, \mathrm{Tr}\,[\rho(t)] = \mathrm{Tr}\left[\rho\left(c + B + B^\dagger + \sum_{i,\,j} \lambda_{ij}\, K^{i\dagger}\, K^j\right)\right] = 0, \tag{6}$$

giving

$$c + B + B^\dagger = -\sum_{i,\,j} \lambda_{ij}\, K^{i\dagger}\, K^j. \tag{7}$$

Separating the operator B in its hermitian components we write $B = R - i\,H$ finding

$$c\,\rho + B\,\rho + \rho\,B^\dagger = \frac{1}{2}\left[(c + 2R)\,,\;\rho\right]_+ - i\,[H,\;\rho]\,. \qquad (8)$$

Combining these, we write the equation in the form

$$\begin{aligned} \partial_t\,\rho = \;& -i\,[H,\;\rho] + \sum_{i,\,j} \lambda_{ij}\,K^i\,\rho\,K^{j\dagger} \\ & - \frac{1}{2}\sum_{i,\,j} \lambda_{ij}\,K^{i\dagger}\,K^j\,\rho - \frac{1}{2}\,\rho \sum_{i,\,j} \lambda_{ij}\,K^{i\dagger}\,K^j\,. \end{aligned} \qquad (9)$$

This form of the time evolution operator is thus the most general linear generator of a semi-group; because it depends only on the instantaneous value of the state $\rho(t)$, it is called a Markovian evolution equation. It is directly seen to preserve the trace of the density matrix. Diagonalising the hermitian matrix $[\lambda_{ij}]$, we find that its eigenvalues have to be positive if the density matrix is to remain a positive operator. In that case, it can be shown that Eq. (9) satisfies the more general condition of *complete positivity* . This states that acting on a non-interacting but possibly entangled product Hilbert space, the evolution of the density matrix stays positive definite. This is necessary to preserve the physical interpretation of an arbitrary entangled state.

In the theory of semi-groups, it is proved that the most general acceptable evolution equation has to be of the form (9), [Lindblad 1976 (b)], which is called the *Lindblad form.*

1.2 Dynamics reduced to a subsystem

We consider unitary time evolution on the direct product of two Hilbert spaces $\mathcal{H}_1 \otimes \mathcal{H}_2$. In this product space the time evolution is given by the operator $\mathcal{U}_t$ which has to be isometric, $\mathcal{U}_t^\dagger\,\mathcal{U}_t = 1$, in order to preserve the normalisation of the density matrix

$$\mathrm{Tr}\;\rho(t) \equiv \mathrm{Tr}\left[\mathcal{U}_t\,\rho(0)\,\mathcal{U}_t^\dagger\right] = \mathrm{Tr}\left[\rho(0)\,\mathcal{U}_t^\dagger\,\mathcal{U}_t\right] = \mathrm{Tr}\;\rho(0). \qquad (10)$$

This can be taken as the prototype of reduced dynamics of a subsystem #1 embedded in an environment #2, which can consist of a reservoir or a system of continuous extraction of observed information.

We use (10) to define the time evolution in the subsystem #1 by defining the linear evolution operator using the partial trace

$$U_t\,(\rho_1) = \mathrm{Tr}_2\left(\mathcal{U}_t\,\rho_1 \otimes \rho_2\,\mathcal{U}_t^\dagger\right). \qquad (11)$$

This is seen to be trace-preserving and hermitian.

We now define a set of projectors $\Pi_\alpha^{(2)}$ acting on the system #2, with the properties

$$\sum_\alpha \Pi_\alpha^{(2)} = 1, \quad \text{and} \quad \Pi_\alpha^{(2)}\, \Pi_\beta^{(2)} = \delta_{\alpha\beta}\, \Pi_\alpha^{(2)}. \tag{12}$$

With these we may expand

$$\mathcal{U}_t = \sum_\alpha \Pi_\alpha^{(2)}\, U_t^\alpha, \tag{13}$$

with $U_t^\alpha = \mathrm{Tr}_2 \left(\Pi_\alpha^{(2)}\, \mathcal{U}_t \right)$. The evolution (11) is now

$$U_t\,(\rho_1) = \sum_\alpha \eta_\alpha^2\, U_t^\alpha\, \rho_1\, U_t^{\alpha\dagger}, \tag{14}$$

where $\eta_\alpha^2 = \mathrm{Tr}_2 \left(\rho_2\, \Pi_\alpha^{(2)} \right) \geq 0$.

We now introduce a complete basis set of operators $\{K_i\}$ in the Hilbert space $\mathcal{H}_1$, where we may choose

$$K_0 = 1; \tag{15}$$

then the others can be chosen traceless. We write

$$U_t^\alpha = \sum_i u_\alpha^i(t)\, K_i. \tag{16}$$

Inserting into (14), we obtain

$$U_t\,(\rho_1) = \sum_{i,\,j} c_{ij}(t)\, K_i\, \rho_1\, K_j^\dagger, \tag{17}$$

with

$$c_{ij}(t) = \sum_\alpha \eta_\alpha^2\, u_\alpha^i(t)\, u_\alpha^{j\dagger}(t) = c_{ji}^\star(t). \tag{18}$$

We write the time evolution over a short time interval Δt and form the coarse-grained time derivative. Using (15) we separate the term with $i = 0$ to obtain

$$
\begin{aligned}
\partial_t \, \rho_1(t) \; &\equiv \frac{\rho_1(t+\Delta t) - \rho_1(t)}{\Delta t} \\
&= \frac{c_{00}(\Delta t) - 1}{\Delta t} \, \rho_1(t) + \sum_{i \neq 0} \frac{c_{i0}(\Delta t)}{\Delta t} \, K_i \, \rho_1(t) \\
&+ \sum_{j \neq 0} \frac{c_{0j}(\Delta t)}{\Delta t} \, \rho_1(t) \, K_j^{\dagger} + \sum_{i \neq 0,\, j \neq 0} \frac{c_{ij}(\Delta t)}{\Delta t} \, K_i \, \rho_1(t) \, K_j^{\dagger}.
\end{aligned} \tag{19}
$$

Letting Δt go to zero and assuming all limits exist, we re-derive a Master Equation of the Lindblad form. The problem is that the assumption about the limiting process requires some additional assumptions. These are usually taken to imply that the system #2, in some way, acts as a huge reservoir.

We see that what we need is a relation of the form

$$
c_{ij}(\Delta t) = \sum_{\alpha} \eta_\alpha^2 \, u_\alpha^i(\Delta t) \, u_\alpha^{j\dagger}(\Delta t) \sim \lambda_{ij} \, \Delta t. \tag{20}
$$

This can be achieved in two ways:

1 - If we choose to represent the environment in terms of quantum noise sources, the Ito interpretation [Gardiner 1991] allows us to write the relations

$$
u_\alpha^i(\Delta t) = \frac{\partial u_\alpha^i}{\partial t} \, \Delta t, \tag{21}
$$

which together with the relation $\Delta t^2 = \Delta t$, gives (20).

2 - In quantum theory we obtain transition rates from energy conservation relations like

$$
\begin{aligned}
\lim_{\Delta t \to \infty} \left(\int_{-\frac{\Delta t}{2}}^{\frac{\Delta t}{2}} e^{i \Delta E t} \, dt \right)^2 &\sim 2\pi \, \delta(\Delta E) \int_{-\frac{\Delta t}{2}}^{\frac{\Delta t}{2}} e^{i \Delta E t} \, dt \\
&= 2\pi \, \delta(\Delta E) \, \Delta t.
\end{aligned} \tag{22}
$$

As the S-matrix approach is based on relations like this [Goldberger 1964], we find the relation (20) again. The approximation (22) imposes restrictions on the energy spectrum encompassed by the transitions. These are essentially equivalent with the reservoir assumptions in other derivations of the Master Equations.

Again, imposing the need to conserve the trace of the density matrix, we find that the limits $\Delta t \to 0$ cannot be independent but we must have

$$\sum_{i\neq 0} \frac{c_{i0}(\Delta t)}{\Delta t} K_i \to -\frac{1}{2} \sum_{i\neq 0,\, j\neq 0} \frac{c_{ij}(\Delta t)}{\Delta t} K_j^\dagger K_i - i\, H \tag{23}$$

and

$$\sum_{j\neq 0} \frac{c_{0j}(\Delta t)}{\Delta t} K_j^\dagger \to -\frac{1}{2} \sum_{i\neq 0,\, j\neq 0} \frac{c_{ij}(\Delta t)}{\Delta t} K_j^\dagger K_i + i\, H, \tag{24}$$

where H is a hermitian part of the limit. We obtain the equation

$$\begin{aligned} \partial_t \rho \approx \;& -i\,[H,\, \rho] + \sum_{i,\, j} \lambda_{ij}\, K_i\, \rho\, K_j^\dagger \\ & -\frac{1}{2} \sum_{i,\, j} \lambda_{ij}\, K_j^\dagger\, K_i\, \rho - \frac{1}{2}\, \rho \sum_{i,\, j} \lambda_{ij}\, K_j^\dagger\, K_i. \end{aligned} \tag{25}$$

All derivations of Markovian Master Equations are based on justifying the details of the steps outlined above. Whatever models one introduces, some crucial approximations are needed to proceed from the form (19).

2. The Brownian motion

2.1 The time evolution

The standard random walk problem in physics is the Ornstein-Uhlenbeck process, which is a model of the Brownian motion in a dissipative medium. We are now looking at the possibility to generalise this to the quantum mechanical dynamics. To this end we introduce the one-dimensional canonical variables $[x,\, p] = i\hbar$, where we retain the quantum constant for dimensional reasons. We assume that these co-ordinates are physical in the sense that the laboratory positions are given by x and the physical forces are supposed to act on the momentum p only.

We now write down a Master Equation of the form

$$\partial_t\, \rho = -\frac{i}{\hbar}\,[H,\, \rho] - i\,\frac{\gamma}{2\hbar}\left[x,\, [p,\, \rho]_+\right] - \frac{D_p}{\hbar^2}\,[x,\, [x,\, \rho]]\,. \tag{26}$$

A Master Equation of this form has been derived in Refs. [Caldeira 1983; Unruh 1989]. Here we assume the Hamiltonian to be of the form

$$H = \frac{p^2}{2M} + V(x). \tag{27}$$

Using this equation we obtain the equations of motion for the first moments of the dynamical variables:

$$\partial_t \langle x \rangle = \frac{\langle p \rangle}{M}, \text{ and } \partial_t \langle p \rangle = -\left\langle \frac{dV}{dx} \right\rangle - \gamma \langle p \rangle. \tag{28}$$

This is the correct form of Ehrenfest's theorem with damping appearing only for the motion as we should expect.

For the second moments, we obtain the relations

$$\begin{aligned}
\partial_t \langle p^2 \rangle &= \left\langle \left(p \frac{dV}{dx} + \frac{dV}{dx} p \right) \right\rangle - 2\gamma \langle p^2 \rangle + 2D_p \\
\partial_t \langle (px + xp) \rangle &= 2 \frac{\langle p^2 \rangle}{M} - 2 \left\langle x \frac{dV}{dx} \right\rangle - \gamma \langle (px + xp) \rangle \\
\partial_t \langle x^2 \rangle &= \frac{\langle (px + xp) \rangle}{M}.
\end{aligned} \tag{29}$$

These are physically the correct equations of motion, only the momentum degree of freedom is spreading diffusively. This is expected classically from the medium giving the system minute impulse kicks changing its momentum only.

If we have a harmonic oscillator,

$$V(x) = \frac{1}{2} M \, \Omega^2 \, x^2, \tag{30}$$

the equations close, and we may solve for the equilibrium of the second moments. The final equilibrium energy is then given by

$$\langle H \rangle = \frac{\langle p^2 \rangle}{2M} + \frac{1}{2} M \, \Omega^2 \, \langle x^2 \rangle = \frac{D_p}{\gamma \, M}. \tag{31}$$

If we equate this with the thermal expectation value $\hbar\Omega\langle n \rangle$, we find the fluctuation-dissipation relation

$$D_p = \gamma \, M \left(\frac{\hbar\Omega}{2} \right) \cot \left(\frac{\hbar\Omega}{2 \, k_B T} \right) \to \gamma \, M \, k_B T, \tag{32}$$

which displays the correct high temperature limit.

These results seem to be fully satisfactory, both the equations of the first and second moments agree with the classical results, and the eventual equilibrium agrees with thermal considerations. However, the form (26) is not acceptable, as we can see from the following considerations [Hakim 1985; Ambegaokar 1991; Munro 1996]. For a summary of the situation see Ref. [Stenholm 1994].

In the case of the harmonic oscillator, we introduce the customary annihilation and creation operators by setting

$$x = \sqrt{\frac{\hbar}{2M\,\Omega}}\,\left(b + b^{\dagger}\right) \quad \text{and} \quad p = \sqrt{\frac{\hbar\Omega\,M}{2}}\,\left(\frac{b - b^{\dagger}}{i}\right). \tag{33}$$

Inserting this into Eq. (26) we find

$$\begin{aligned} \partial_t\,\rho = & -\frac{i}{\hbar}\,[H_{\text{eff}},\,\rho] - \frac{D_p}{\hbar^2}\,[x,\,[x,\,\rho]] - \frac{\gamma}{4}\,(b^{\dagger}\,b\,\rho + \rho\,b^{\dagger}\,b - 2b\,\rho\,b^{\dagger}) \\ & +\frac{\gamma}{4}\,(b\,b^{\dagger}\,\rho + \rho\,b\,b^{\dagger} - 2b^{\dagger}\,\rho\,b). \end{aligned} \tag{34}$$

Here not all terms are of the Lindblad form, which indicates that there may be trouble emerging. The exact nature of this will be seen in the next section. Here we only want to point out the strangeness of the situation.

Writing down the time evolution equation for the Wigner function, we find from the Master Equation (26)

$$\begin{aligned} &\frac{\partial}{\partial t}\,W(X,\,P) + \frac{P}{M}\,\frac{\partial}{\partial X}\,W(X,\,P) - \frac{dV(X)}{dX}\,\frac{\partial}{\partial P}\,W(X,\,P) \\ &\qquad = \frac{\partial}{\partial P}\,[\gamma\,W(X,\,P)] + D_p\,\frac{\partial^2}{\partial P^2}\,W(X,\,P). \end{aligned} \tag{35}$$

Solving this equation we find a kernel K defining the solution as

$$W(X,\,P,\,t) = \int_0^t K(X,\,P,\,t;\,X_0,\,P_0)\,W(X_0,\,P_0,\,0)\,dX_0\,dP_0. \tag{36}$$

The flawed time evolution tells us that, starting from some proper quantum states, we may obtain improper ones. However, Eq. (35) is the correct evolution equation for the Ornstein-Uhlenbeck process. Thus, starting from allowed classical states, it will always produce allowed classical ones. This strange situation indicates that irreversible time evolution is more intricate in quantum theory than we may guess from the classical counterpart.

Before we proceed, we should look at the Hamiltonian part of the evolution Eq. (34). The Hamiltonian turns out to be

$$H_{\text{eff}} = \hbar\Omega\,b^{\dagger}\,b + i\,\frac{\gamma}{4}\left(b^{\dagger}\,b^{\dagger} - bb\right). \tag{37}$$

This type of Hamiltonian can be diagonalized by the canonical transformation

$$b = \sqrt{i}\left[\cosh\left(\frac{\theta}{2}\right)A - \sinh\left(\frac{\theta}{2}\right)A^{\dagger}\right], \tag{38}$$

where

$$\tanh\theta = \frac{\gamma}{2\Omega}. \tag{39}$$

The result is the Hamiltonian

$$H_{\text{eff}} = \sqrt{\left(\Omega^2 - \frac{\gamma^2}{4}\right)}\, A^\dagger\, A, \tag{40}$$

which is exactly the oscillator at the frequency of the classical damped oscillator.

2.2 Non-physical features

Even if we found that the Master Equation is not of the Lindblad form, we may not actually encounter the trouble. In this section we will show that the trouble cannot be safely ignored.

Start with a real and symmetric pure state

$$\varphi(x) = \langle x|\varphi\rangle = \varphi(-x) \text{ and } \rho_0 = |\varphi\rangle\langle\varphi|. \tag{41}$$

We now consider the probability to find the system in this state at a later time by investigating its direction of change

$$\begin{aligned} \frac{\partial}{\partial t}\langle\varphi|\rho|\varphi\rangle &= -\,i\,\frac{\gamma}{2\hbar}\,\langle\varphi|\,(xp - px)\,|\varphi\rangle \\ &\quad -2\,\frac{D_p}{\hbar^2}\left(\langle\varphi|x^2|\varphi\rangle - \langle\varphi|x|\varphi\rangle^2\right) \\ &= \frac{\gamma}{2} - 2\,\frac{D_p}{\hbar^2}\,\overline{\Delta x^2}. \end{aligned} \tag{42}$$

Thus for initial symmetric states with

$$\overline{\Delta x^2} < \frac{\gamma\,\hbar^2}{4D_p} \tag{43}$$

the initial probability will grow; as it was unity to begin with, it must become larger than one violating the probability interpretation of the density matrix.

The original equation of motion preserves the trace of the density matrix. This implies that the time evolution eigenvalues sum to unity, and hence the growing state found above must be compensated by a state going negative. This is easily found to be given by

$$|\psi\rangle = p\,|\varphi\rangle. \tag{44}$$

This state is odd and orthogonal to $|\varphi\rangle$. Thus its probability is zero initially

$$\langle\psi|\rho(0)|\psi\rangle = 0. \tag{45}$$

Its evolution now follows from

$$\begin{aligned}\frac{\partial}{\partial t}\langle\psi|\rho|\psi\rangle = & -i\,\frac{\gamma}{2\hbar}\left(\langle\psi|x\,\rho\,p|\psi\rangle - \langle\psi|p\,\rho\,x|\psi\rangle\right)\\ & + 2\,\frac{D_p}{\hbar^2}\,\langle\psi|x\,\rho\,x|\psi\rangle.\end{aligned} \tag{46}$$

We calculate

$$\langle\psi|x\,\rho\,p|\psi\rangle \quad -\langle\psi|p\,\rho\,x\,|\psi\rangle = -\,i\hbar\,\langle\varphi|p^2|\varphi\rangle \tag{47a}$$

$$\langle\psi|x\,\rho\,x|\psi\rangle \quad = |\langle\varphi|xp|\varphi\rangle|^2 \tag{47b}$$

$$\langle\varphi|x\,p|\varphi\rangle \quad = i\hbar\int\varphi^2(x)\,dx + i\hbar\int\varphi(x)\,x\,\frac{d\varphi(x)}{dx}\,dx. \tag{47c}$$

With these relations (46) becomes

$$\frac{\partial}{\partial t}\langle\psi|\rho|\psi\rangle = \frac{1}{2}\left(D_p - \gamma\,\langle p^2\rangle\right). \tag{48}$$

Because the original wave function was chosen real there is no average momentum. Thus for states with

$$\overline{\Delta p^2} > \frac{D_p}{\gamma} \tag{49}$$

the initially vanishing probability grows negative. It is interesting to note that the limiting values from (43) and (49) satisfy the complementarity relation

$$\overline{\Delta p^2}\,\overline{\Delta x^2} \sim \frac{\hbar^2}{4}. \tag{50}$$

This shows that we can, indeed, violate both conditions for the states.

It is worth noticing that the problems go away if we apply the rotating-wave approximation [Stenholm 1994]. Using the relations (33) we may then write

$$\frac{D_p}{\hbar^2}\,[x,\,[x,\,\rho]] = \frac{D_p}{2M\,\hbar\Omega}\left(\left[b,\,\left[b^\dagger,\,\rho\right]\right] + \left[b^\dagger,\,[b,\,\rho]\right]\right), \tag{51}$$

and the Master Equation (34) becomes

$$\partial_t \rho = \; -\frac{i}{\hbar}[H_{\mathrm{eff}}, \rho] - D_+ (b^\dagger b \rho + \rho b^\dagger b - 2b \rho b^\dagger) \\ - D_- (b b^\dagger \rho + \rho b b^\dagger - 2b^\dagger \rho b), \tag{52}$$

Here

$$D_\pm = \frac{D_p}{2M\hbar\Omega} \pm \frac{\gamma}{4} = \frac{D_p}{2M\hbar\Omega}\left(1 \pm \frac{\hbar\Omega}{2k_BT}\right), \tag{53}$$

where (32) has been used in the high temperature limit. The ratio of the diffusion coefficients is now

$$\frac{D_+}{D_-} = \left(1 - \frac{\hbar\Omega}{2k_BT}\right)^{-1}\left(1 + \frac{\hbar\Omega}{2k_BT}\right) \approx \exp\left(\frac{\hbar\Omega}{k_BT}\right) = \frac{\overline{n}+1}{\overline{n}}, \tag{54}$$

which is what we expect from a thermal reservoir.

3. Derivations of Master Equations

3.1 Projection methods and memory kernels

One standard approach to irreversibility is separation of the system's degrees of freedom into relevant and irrelevant ones. The latter are then assumed to provide a reservoir for the relevant system of interest, which acquires irreversibility by loosing information to the environment. This environment may for instance be a thermal bath or a series of continuous observations with results not accessible in the system of interest. Based on projection operators and operator partitioning this approach was introduced in Refs. [Nakajima 1958; Zwanzig 1960; Fano 1963]. For a survey of applications, see Ref. [Argyres 1966] and more recently also Ref. [Zwanzig 2001].

The state of a quantum system is described by its density matrix ρ. The system of interest is extracted from this by introducing a projection operator $\mathcal{P}$ and its complement operator

$$\mathcal{Q} = 1 - \mathcal{P}. \tag{55}$$

If the isolated reservoir is described by the stationary density matrix ρ_0, the projector may be chosen to be

$$\mathcal{P} \equiv \rho_0 \, \mathrm{Tr}_R \otimes; \tag{56}$$

here the trace operation is over the environmental reservoir and $\mathrm{Tr}_R \otimes \rho \equiv \mathrm{Tr}_R \, \rho$. It is easy to see that $\mathcal{P}^2 = \mathcal{P}$.

With these definition the Liouville evolution equation of the combined systems

$$\partial_t \rho = \mathcal{L} \rho \tag{57}$$

can be written in the partitioned form

$$\partial_t \begin{bmatrix} \rho_P \\ \rho_Q \end{bmatrix} = \begin{bmatrix} \mathcal{L}_P & \mathcal{L}_{PQ} \\ \mathcal{L}_{QP} & \mathcal{L}_Q \end{bmatrix} \begin{bmatrix} \rho_P \\ \rho_Q \end{bmatrix}, \tag{58}$$

where

$$\rho_P \equiv \mathcal{P} \rho, \; \mathcal{L}_P \equiv \mathcal{P} \mathcal{L} \mathcal{P}, \; \mathcal{L}_{PQ} \equiv \mathcal{P} \mathcal{L} \mathcal{Q} \tag{59}$$

and so on.

We introduce the solution of the equation

$$\partial_t \mathcal{U}_Q^{\rightarrow}(t) = \mathcal{L}_Q \mathcal{U}_Q^{\rightarrow}(t), \tag{60}$$

with the initial condition $\mathcal{U}_Q^{\rightarrow}(0) = 1$. We may then write the solution to the second equation in (58) as

$$\rho_Q(t) = \mathcal{U}_Q^{\rightarrow}(t - t_0)\, \rho_Q(t_0) + \int_{t_0}^{t} \mathcal{U}_Q^{\rightarrow}(t - t')\, \mathcal{L}_{QP}\, \rho_P(t')\, dt'. \tag{61}$$

Often it is assumed that when $t_0 \rightarrow -\infty$, the memory of the initial state is forgotten, and the first term disappears. Alternatively, we may assume that it is zero. If $\rho_Q(t_0) = \rho_0$, the choice (56) makes it vanish. In the following we set $t_0 = 0$ for simplicity.

Inserting (61) into the first of Eqs. (58) we obtain for ρ_P the equation

$$\begin{aligned} \partial_t \rho_P(t) = \; & \mathcal{L}_{PQ} \mathcal{U}_Q^{\rightarrow}(t)\, \rho_Q(0) + \mathcal{L}_P\, \rho_P(t) \\ & + \int_0^t \mathcal{L}_{PQ} \mathcal{U}_Q^{\rightarrow}(t - t')\, \mathcal{L}_{QP}\, \rho_P(t')\, dt'. \end{aligned} \tag{62}$$

This equation is still exact, but it involves the history of the density matrix of the system of interest, the evolution has got memory.

It would seem tempting to obtain a memory-less equation from (62) by writing

$$\begin{aligned} \partial_t \rho_P(t) & = \mathcal{L}_P\, \rho_P(t) + \int_0^t \mathcal{L}_{PQ} \mathcal{U}_Q^{\rightarrow}(t - t')\, \mathcal{L}_{QP}\, \rho_P(t')\, dt' \\ & = \left[\mathcal{L}_P + \int_0^t \mathcal{L}_{PQ} \mathcal{U}_Q^{\rightarrow}(t - t')\, \mathcal{L}_{QP}\, \mathcal{U}_P^{\leftarrow}(t' - t)\, dt' \right] \rho_P(t) \end{aligned} \tag{63}$$

However, this cannot be done! The operator $\mathcal{U}_P^{\leftarrow}(t'-t)$ is evolution *backwards*, and it is the backwards solution of the equation we want. This just does *not exist*

The combination $\mathcal{L}_{PQ}\,\mathcal{U}_Q^{\rightarrow}(t-t')\,\mathcal{L}_{QP}$ describes the evolution in the reservoir, and if this somehow is expected to live only for short times, we may write this as

$$\begin{aligned}\partial_t\,\rho_P(t) &= \mathcal{L}_{PQ}\,\mathcal{U}_Q^{\rightarrow}(t)\,\rho_Q(0)\\ &\quad+\left[\mathcal{L}_P+\int_0^t \mathcal{L}_{PQ}\,\mathcal{U}_Q^{\rightarrow}(\tau)\,\mathcal{L}_{QP}\,d\tau\right]\rho_P(t)\\ &\equiv \mathcal{M}_P\,\rho_P(t).\end{aligned} \tag{64}$$

This is the Markovian memory-less approximation to the Master Equation. In this approximation, the effective time evolution operator becomes independent of t and the integral may be extended to infinity. It is also consistent to assume that the system lost memory of the initial state of the reservoir, whatever this was. In the limit when $\mathcal{U}_Q$ is calculated in perturbation theory and $\rho_Q(0)=0$, we obtain the conventional Born-Markov time evolution which has a long and successful history.

The problem is, however, that we need to fix the exact conditions of validity of this approximation, this was attempted already in Ref. [Fano 1954]. In particular, it has turned out that introduction of the memory effect is a very sensitive issue [Barnett 2001]. Highly reasonable but unprecise approximations may lead to non-physical time evolution. An additional problem is that the procedure does not necessarily lead to Master Equations of the Lindblad type, see above. If this is not its form, we may find well known complications, which have to be avoided if we want to escape unphysical results.

3.2 Memoryless equations

There does exist a different approach to the derivation of Master Equations using the elimination of the environmental degrees of freedom. This leads to equations which lack the memory effect. This has been formally presented in Refs. [Grabert 1977; Grabert 1978]. Occasionally the same philosophy has been applied to various problems [Alhassid 1980; Buzek 1998]. The method is presented in Sec. 9.2 of Ref. [Breuer 2002]. Here we formulate the principle in terms of the technique presented above.

We assume that we may solve the full equation of evolution (58) in the form of a partitioned unitary evolution

$$\begin{bmatrix} \rho_P(t) \\ \rho_Q(t) \end{bmatrix} = \begin{bmatrix} \mathcal{U}_P(t) & \mathcal{U}_{PQ}(t) \\ \mathcal{U}_{QP}(t) & \mathcal{U}_Q(t) \end{bmatrix} \begin{bmatrix} \rho_P(0) \\ \rho_Q(0) \end{bmatrix}. \tag{65}$$

From this we obtain the formal solutions

$$\rho_P(0) = \mathcal{U}_P^{-1}(t)\, \rho_P(t) - \mathcal{U}_P^{-1}(t)\, \mathcal{U}_{PQ}(t)\, \rho_Q(0) \tag{66}$$

and

$$\begin{aligned} \rho_Q(t) &= \mathcal{U}_{QP}(t)\, \rho_P(0) + \mathcal{U}_Q(t)\, \rho_Q(0) \\ &= \mathcal{U}_{QP}(t)\, \mathcal{U}_P^{-1}(t)\, \rho_P(t) + \left[\mathcal{U}_Q(t) - \mathcal{U}_{QP}(t)\, \mathcal{U}_P^{-1}(t)\, \mathcal{U}_{PQ}(t)\right] \rho_Q(0). \end{aligned} \tag{67}$$

Inserting this into the equation of motion from (58) we obtain

$$\begin{aligned} \partial_t\, \rho_P(t) = & \left[\mathcal{L}_P + \mathcal{L}_{PQ}\, \mathcal{U}_{QP}(t)\, \mathcal{U}_P^{-1}(t)\right] \rho_P(t) \\ & + \mathcal{L}_{PQ} \left[\mathcal{U}_Q(t) - \mathcal{U}_{QP}(t)\, \mathcal{U}_P^{-1}(t)\, \mathcal{U}_{PQ}(t)\right] \rho_Q(0). \end{aligned} \tag{68}$$

This equation is still exact, and in the case of no contribution from the initial state of the reservoir, we obtain an equation of the form

$$\partial_t\, \rho_P(t) = \mathcal{K}_P(t)\, \rho_P(t). \tag{69}$$

This is the memory-less exact time evolution equation. In the limit when the Markovian property is expected to hold, we should obtain by comparison with (64) the result

$$\lim_{t\to\infty} \mathcal{K}_P(t) = \mathcal{M}_P. \tag{70}$$

In this limit, the two approaches should be equivalent.

The problem is, however, the very existence of the limit

$$\lim_{t\to\infty} \left(\mathcal{L}_{PQ}\, \mathcal{U}_{QP}(t)\, \mathcal{U}_P^{-1}(t)\right) = \mathcal{M}_P - \mathcal{L}_P. \tag{71}$$

From (66) we find that, ignoring the initial state of the reservoir, we have

$$\rho_P(0) = \mathcal{U}_P^{-1}(t)\, \rho_P(t), \tag{72}$$

which says that the initial state of the system of interest is uniquely obtainable from the present state of the system. This is far from the case in most physical situations; it is quite conceivable that the state $\rho_P(t)$ is recurring. In fact, most quantum systems are periodic or quasi-periodic, and then the operator $\mathcal{U}_P^{-1}(t)$

becomes singular. Then the operator does not exist across such a point in time, and the formal expression looses its useful meaning. This is the problem expected to occur in the memory-less situation, see Ref. [Breuer 2002]. The result is going to be very sensitive to any approximation near such singularities. Incorrect approximations will render the result entirely useless.

4. Simple models

4.1 The decay model

There are but few models of irreversibility which can be solved exactly. Such a model was originally introduced by Ref. [Friedrichs 1948] but it also goes under the name the Fano model, [Fano 1961; Cohen-Tannoudji 1992]. It offers the theory of a single quantum state imbedded in a continuum or a quasi-continuum. It has been used in molecular physics as the Bixon-Jortner model [Bixon 1968] and in solid state physics as an Anderson model [Anderson 1961].

The relevant Hamiltonian is

$$H = \omega_0 \,|0\rangle\langle 0| + \sum_\epsilon \epsilon \,|\epsilon\rangle\langle\epsilon| + \sum_\epsilon \left(V_\epsilon \,|\epsilon\rangle\langle 0| + V_\epsilon^\star \,|0\rangle\langle\epsilon|\right). \tag{73}$$

Expanding the state as

$$|\psi\rangle = a_0 \,|0\rangle + \sum_\epsilon c_\epsilon \,|\epsilon\rangle, \tag{74}$$

we obtain the equations

$$i\,\partial_t\, a_0 = \omega_0\, a_0 + \sum_\epsilon V_\epsilon^\star\, c_\epsilon \;\text{ and }\; i\,\partial_t\, c_\epsilon = \epsilon\, c_\epsilon + V_\epsilon\, a_0. \tag{75}$$

It is now practical to introduce the Laplace-Fourier transform defined by

$$\tilde{\psi}(\omega) = -i \int_{-\infty}^{\infty} e^{i\omega t}\, \psi(t)\, dt, \tag{76}$$

with the property

$$\partial_t\, \psi(t) \to -i \left[\omega\, \tilde{\psi}(\omega) - \psi(t=0)\right]. \tag{77}$$

With this definition we retain the interpretation of ω as a frequency, and the inverse transform

$$\psi(t) = \frac{i}{2\pi} \int_{-\infty+i\eta}^{\infty+i\eta} e^{-i\omega t}\, \tilde{\psi}(\omega)\, d\omega, \tag{78}$$

shows that ω must approach real values from above. The Laplace-Fourier transform is convenient in solving initial-value problems, because it offers simpler ways to honour causality than the double frequency Fourier transform.

We assume the system to be in the state $|0\rangle$. Transforming Eq. (75) we find

$$\tilde{c}_\epsilon = \frac{V_\epsilon}{\omega - \epsilon}\,\tilde{a}_0 \text{ and } \tilde{a}_0 = \frac{1}{\omega - \omega_0 - \sum(\omega)}, \tag{79}$$

where

$$\sum(\omega) = \sum_\epsilon \frac{|V_\epsilon|^2}{\omega - \epsilon}. \tag{80}$$

The inversion Eq. (78) shows that the time dependence is determined by the singularities of the function $\tilde{a}_0$ in the complex ω-plane. To this end we write $\omega = \omega' - i\,\omega''$ and find

$$\sum(\omega' - i\,\omega'') = \sum_\epsilon \frac{|V_\epsilon|^2\,(\omega' - \epsilon + i\,\omega'')}{(\omega' - \epsilon)^2 + \omega''^2}. \tag{81}$$

The imaginary part of the denominator in (79) gives

$$\omega'' \left[1 + \sum_\epsilon \frac{|V_\epsilon|^2}{(\omega' - \epsilon)^2 + \omega''^2} \right] = 0. \tag{82}$$

It is obvious that this relation has no solution for ω'' and hence it seems that no damping can emerge from this approach.

However, the emergence of irreversibility follows only in the limit when the ϵ-spectrum becomes genuinely continuous [Zumino 1961; Levy 1961]. We assume that the strength of this is determined by a density of states $D(\epsilon)$. Then the real axis becomes a branch cut for the function $\sum(\omega)$ and the calculation has to be performed on the correct sheet of the analytic function. The inversion (78) shows that the contour has to be closed in the unphysical half-plane $\mathrm{Im}\omega < 0$, and hence the singularities have to be found in the analytic continuation of the function from the upper half-plane through the branch cut. We find that this can most easily be carried out by setting in the lower half-plane

$$\sum^+(\omega) \to \int \frac{|V_\epsilon|^2\,D(\epsilon)}{\omega - \epsilon}\,d\epsilon - 2i\pi\,|V_\omega|^2\,D(\omega). \tag{83}$$

It is easy to see that this function at $\omega = \omega' - i\eta$ gives the same value as the original function at $\omega = \omega' + i\eta$.

Now the imaginary part of the denominator in (79) becomes

$$\omega'' = \mathrm{Im} \int \frac{|V_\epsilon|^2\,D(\epsilon)}{\omega' - \epsilon - i\,\omega''}\,d\epsilon \approx \pi\,|V_{\omega_0}|^2\,D(\omega_0). \tag{84}$$

Here we have assumed that the energy shift $\propto \mathrm{Re}\sum(\omega)$ is negligible and ω'' is small enough to allow the continuation according to (83). Now introducing

$$\gamma = 2\pi \left|V_{\omega_0}\right|^2 D(\omega_0) = -2\,\mathrm{Im}\sum{}^{+}(\omega_0 - i\eta), \tag{85}$$

we find from (78) the result

$$\psi(t) = \frac{i}{2\pi}\int_{-\infty}^{\infty} \frac{e^{-i\omega t}\,d\omega}{\omega - \omega_0 + i\,\mathrm{Im}\sum^{+}(\omega_0 - i\eta)}, \tag{86}$$

which by closing the contour in the lower half plane gives

$$\psi(t) = e^{-i\omega_0 t} \times e^{-\gamma t/2}. \tag{87}$$

Thus we obtain the correct decay into the continuum and the decay rate (85) is the correct Weisskopf-Wigner expression. We note that our procedure of analytic continuation directly introduces the correct decay in the forward direction of time. This requirement by causality follows uniquely from the way we perform the analytic continuation.

4.1.1 The analytic continuation. In order to see how the analytic continuation uniquely determines the damping rate in the model, we introduce the simplest possible branch cut in the spectrum [Stenholm 1997]

$$\begin{aligned} |V_\epsilon|^2\, D(\epsilon) &= A^2,\ |\epsilon| < L; \\ |V_\epsilon|^2\, D(\epsilon) &= 0,\ |\epsilon| > L. \end{aligned} \tag{88}$$

The relation (80) then becomes

$$\sum(\omega) = \int \frac{|V_\epsilon|^2 D(\epsilon)}{\omega - \epsilon}\, d\epsilon = A^2 \log\left[-\left(\frac{L+\omega}{L-\omega}\right)\right]. \tag{89}$$

The sign of the imaginary part of this function clearly depends on how the branch of the logarithm function is chosen. To find the correct answer, we use the continuation according to (83) and we obtain with $\omega = \omega' - i\,\omega''$

$$\sum{}^{+}(\omega) = A^2 \left\{ \log\left[\frac{-\left(1+\dfrac{\omega'}{L}\right) + i\,\dfrac{\omega''}{L}}{\left(1-\dfrac{\omega'}{L}\right) + i\,\dfrac{\omega''}{L}}\right] - 2i\pi \right\}. \tag{90}$$

Assuming the branch cut along the real axis and letting $L \to \infty$, we find the result

$$\gamma = -2\,\mathrm{Im}\sum{}^{+}\left(\omega_0 - i\,\frac{\omega''}{L}\right) = 2\pi\, A^2. \tag{91}$$

This simple result shows how the procedure avoids all ambiguity in the sign of the decay and enforces the causal condition of forward time. Essential is, however, the existence of the continuous spectrum. As long as the spectrum consists of discrete poles, no branches emerge and the system remains reversible but generally only quasi-periodic.

4.1.2 The faked continuum. In this model, we consider only one term in the ϵ-summations in (73). A single state does not, however, offer a continuum to decay into, and thus we must modify the model. This can be achieved by letting the mode simulate a continuum by having a fast decay. This may well derive from some further system coupled to a genuine continuum. From the view of the initially populated state, the second system seems like a continuum because of its broad lineshape. We call this a "faked continuum" [Stenholm 1986].

The equations of motion are now modified to

$$i\,\partial_t\,a_0 = \omega_0\,a_0 + V^\star\,c \text{ and } i\,\partial_t\,c = -i\,\Gamma\,c + \epsilon\,c + V\,a_0. \tag{92}$$

Introducing the Laplace-Fourier transform we now obtain

$$\tilde{c} = \left(\frac{V}{\omega - \epsilon + i\,\Gamma}\right)\tilde{a}_0 \tag{93}$$

and

$$\tilde{a}_0 = \left[\omega - \omega_0 - \frac{|V|^2}{\omega - \epsilon + i\,\Gamma}\right]^{-1}. \tag{94}$$

Again introducing $\omega = \omega' - i\,\omega''$ we obtain with $\omega' \approx \omega_0$

$$\omega'' = \pi|V|^2\left(\frac{\Gamma/\pi}{(\omega_0 - \epsilon)^2 + \Gamma^2}\right). \tag{95}$$

This corresponds exactly to the Weisskopf-Wigner result (85); at resonance $\omega_0 = \epsilon$, the density of state seen by the single initial state is

$$D(\omega_0) = \frac{1}{\pi\Gamma}, \tag{96}$$

as expected from a Lorentzian line shape. Our approach with a faked continuum thus produces physically reasonable results. There is also a shift of the resonance but, as usual, its significance becomes model dependent; the renormalisation of energy is sensitive to the whole spectral profile not only to the density of states near resonance.

In particular, the sign of the damping comes out in a trivial manner. This derives from the fact that the sign of the damping term in (92) is dictated by physics. The correct causal behaviour follows then automatically.

If the decay continuum has got some structure, it may be approximated by a set of poles on the second sheet of the lower plane. In that case, the decay can be described in terms of pseudo-modes similar to our faked continuum [Garraway 1997 (a); Garraway 1997 (b)].

4.2 Phase-diffused Bloch equations

4.2.1 The adiabatic limit. In the rotating-wave approximation, the most general version of the Bloch equations, [Allen 1975], can be written as

$$\frac{d}{dt}\begin{bmatrix} u \\ v \\ w \end{bmatrix} = \begin{bmatrix} -\gamma & \Delta\omega & 0 \\ -\Delta\omega & -\gamma & \Omega \\ 0 & -\Omega & -\gamma_0 \end{bmatrix}\begin{bmatrix} u \\ v \\ w \end{bmatrix} + \gamma_0 \begin{bmatrix} 0 \\ 0 \\ w_0 \end{bmatrix}. \tag{97}$$

Here γ and γ_0 are the transverse and longitudinal relaxation rates respectively; Ω is the Rabi frequency and w_0 is the equilibrium population difference. In the limit of strong phase diffusion, we may assume

$$\gamma \gg \gamma_0; \tag{98}$$

we take this to justify the neglect of the terms proportional to γ_0. We then loose the information about the ultimate final state of the system, but over a time $t \ll \gamma_0^{-1}$ this treatment holds. Here we consider such dynamics as a model to discuss the emergence of the Markovian limit.

In order to simplify the treatment, we look at the resonant case $\Delta\omega = 0$. The equations are then given by

$$\partial_t\, v = \Omega\, w - \gamma\, v \text{ and } \partial_t\, w = -\Omega\, v. \tag{99}$$

For the case of no damping, $\gamma = 0$, this describes a harmonic motion $\ddot{w}(t) = -\Omega^2\, w(t)$. The damping can be made to simulate a continuous spectrum of the second mode, a faked reservoir.

In the adiabatic limit $\gamma \to \infty$, the elimination gives the equation of motion

$$\partial_t\, w(t) = -\frac{\Omega^2}{\gamma}\, w(t), \tag{100}$$

with the solution

$$w(t) = \exp\left(-\frac{\Omega^2}{\gamma}\, t\right) w(0). \tag{101}$$

Note that this follows from the solution

$$v(t) = e^{-\gamma t}\, v(0) + \Omega \int_0^t e^{-\gamma(t-t')}\, w(t')\, dt' \tag{102}$$

in the limit $t \gg \gamma^{-1}$. If we perform the Markovian replacement $w(t') \to w(t)$ here, we find from (99) the equation

$$\partial_t\, w(t) = -\Omega\, e^{-\gamma t}\, v(0) - \frac{\Omega^2}{\gamma}(1 - e^{-\gamma t})\, w(t). \tag{103}$$

For $t \gg \gamma^{-1}$ this gives the right behaviour. Look at the solution obtained by setting $v(0) = 0$ and, according to Ref. [Haake 1994],

$$\frac{d\tau}{dt} = (1 - e^{-\gamma t}), \tag{104}$$

which gives from (103) the solution

$$w(\tau) = \exp\left(-\frac{\Omega^2}{\gamma}\tau\right) w(0), \tag{105}$$

where

$$\tau = \frac{\gamma t - \left(1 - e^{-\gamma t}\right)}{\gamma}. \tag{106}$$

This is the function known from the Ornstein-Uhlenbeck process; for short times it goes like $\sim t^2$ but for long times it gives

$$\tau \approx t - \frac{1}{2\gamma}. \tag{107}$$

This gives the adiabatic results (101) with the initial condition

$$w_{\text{eff}}\,(0) = \exp\left(\frac{\Omega^2}{\gamma^2}\right)\, w(0). \tag{108}$$

This is the initial slip, which may give an unphysical initial value to $w(t)$ [Haake 1983; Gnutzmann 1996]. The exponentially decaying solution is compared with the exact one in Fig. 1. The role of the initial slip to correct for the short-time behaviour is clearly seen.

The eigenvalues of the coupled equations of motion are

$$\lambda_\pm = -\frac{\gamma}{2} \pm i\, W, \tag{109}$$

where

$$W = \sqrt{\Omega^2 - \frac{\gamma^2}{4}}. \tag{110}$$

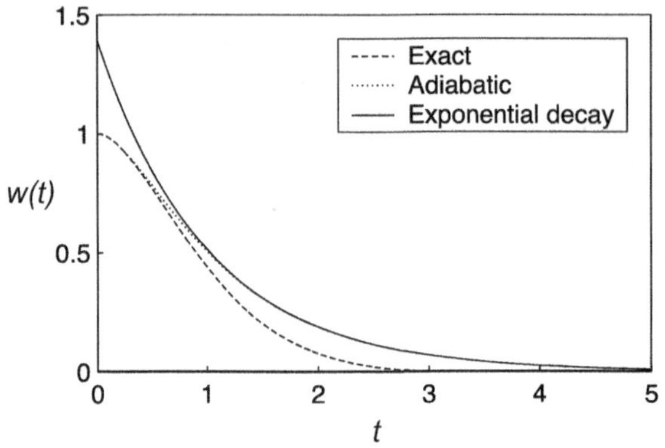

Figure 1 Exact evolution (Eq. (113)), adiabatic evolution (Eq. (105)), and corresponding exponential decay with non-physical initial slip (Eqs. (107) and (108)). Here the parameters $\gamma = 3$ and $\Omega = \sqrt{3}$ have been chosen such that the three curves remain illustrative. In the MarkovMarkovian regime limit $\Omega/\gamma \to 0$, $\Omega^2/\gamma =$ constant the curves become indistinguishable.

This works also for over-damped cases $\gamma \gg \Omega$. In that case, we can expand

$$i\,W = \frac{\gamma}{2} - \frac{\Omega^2}{\gamma}. \tag{111}$$

This gives the two decay time scales

$$\lambda_+ = -\,\gamma;\ \lambda_- = -\,\frac{\Omega^2}{\gamma}. \tag{112}$$

This is the limit when adiabatic elimination is expected to work. Because $|\lambda_-| \ll |\lambda_+|$ the time-scales separate, and the adiabatic limit is expected to work. In this limit, also the initial slip (108) becomes insignificant.

The limit of strong damping is, actually, the proper limit to consider the damped component $v(t)$ as offering a faked continuum. Without damping the effective level splitting is given by the Rabi frequency Ω. In order that the damping may look as a broad continuum, the line width γ must exceed this splitting, *i.e.* $\gamma \gg \Omega$, see also Ref. [Cohen-Tannoudji 1992]. Thus the faked continuum emerges correctly in the adiabatic limit.

4.2.2 Exact results.

The exact solutions for the Eqs. (99) are

$$w(t) = e^{-\gamma t/2}\left[\left(\cos Wt + \frac{\gamma}{2W}\,\sin Wt\right) w(0) - \frac{\Omega}{W}\,\sin Wt\, v(0)\right]$$

$$v(t) = e^{-\gamma t/2}\left[\left(\cos Wt - \frac{\gamma}{2W}\,\sin Wt\right) v(0) + \frac{\Omega}{W}\,\sin Wt\, w(0)\right]. \tag{113}$$

Because of the linearity of the equations, we can solve the first equation with respect to $w(0)$

$$w(0) = \frac{e^{\gamma t/2}}{\left(\cos Wt + \frac{\gamma}{2W}\sin Wt\right)}\, w(t) + \frac{\Omega}{W}\frac{\sin Wt}{\left(\cos Wt + \frac{\gamma}{2W}\sin Wt\right)}\, v(0). \tag{114}$$

We insert this into the second equation and then obtain from $\dot{w} = -\Omega v$ the exact equation of motion for $w(t)$ in terms of $w(t)$ and $v(0)$. We find

$$\dot{w}(t) = -\frac{\Omega^2}{W}\left(\frac{\sin Wt}{\cos Wt + \frac{\gamma}{2W}\sin Wt}\right) w(t) - \Omega\left(\frac{e^{-\gamma t/2}}{\cos Wt + \frac{\gamma}{2W}\sin Wt}\right) v(0). \tag{115}$$

This is the exact memory-less equation of motion. Its solution should be identical with the original solution. We can see that after a time $t \gg \gamma^{-1}$ we loose the influence of the initial value for v but it is needed if the exact solution is required.

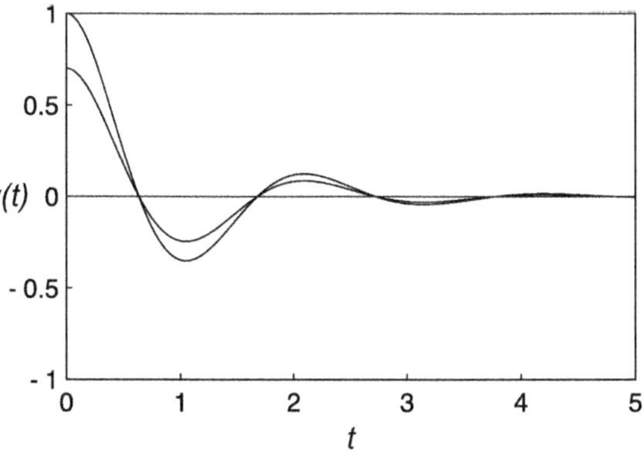

Figure 2 Time evolution for two different initial conditions in the oscillating (underdamped) situation with $\gamma = 1$ and $\Omega = 3$. The time evolution is not invertible when $w(t) = 0$ and, consequently, the memory-less Master Equation does not have a unique solution.

We have derived exact memory-less differential equations by eliminating $w(0)$ in terms of the value $w(t)$. This has been carried out analytically, and this has given formally correct equations. However, this process is not unique, because we get the same $w(t)$ for several $w(0)$. As long as $\gamma \leq \Omega$, the inversion from (114)

$$w(0) = \frac{e^{\gamma t/2}}{\left(\cos Wt + \frac{\gamma}{\Omega}\sin Wt\right)}\, w(t) \tag{116}$$

is not unique for all t.

The situation is illustrated in Fig. 2, which shows how a zero result emerges independently of the initial value. Thus the procedure is expected to be singular. If the exact expression is known, the evolution equation can be integrated across this singularity as we will find to be possible even in the undamped case, see also Ref. [Buzek 1998]. However, any approximation here may introduce uncontrolled errors.

However, in the over-damped case $\gamma > 2\Omega$, we have a different situation, compare Ref. [Cohen-Tannoudji 1992]. Now we see that the inverse

$$w(0) = \frac{e^{\gamma t/2}}{\left(\cosh \Theta t + \dfrac{\gamma}{2\Theta} \sinh \Theta t\right)} w(t) \tag{117}$$

exists and the inversion is possible. In this, over-damped limit, it is seen that the elimination procedure works, see Fig. 3.

However, in the general case it is assumed that the elimination can be carried out also in the weakly coupled limit. This seems to be dangerous, and may lead to incorrect results. The same conclusion is given in Ref. [Breuer 2002].

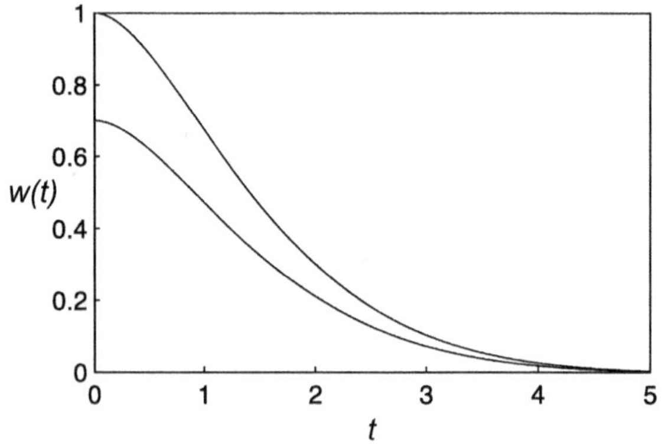

Figure 3 Time evolution for two different initial conditions for over-damped situation with $\gamma = 3$ and $\Omega = 1$. Now $w(t) \neq 0$ for all finite t and the memory-less Master Equation has a unique solution.

4.2.3 No damping. In this limit $\gamma = 0$ and $\Omega = W$. Then the equation of motion becomes

$$\dot{w}(t) = -\,\Omega\, \tan \Omega t\, w(t) - \Omega\, \frac{v(0)}{\cos Wt}. \tag{118}$$

If $v(0) = 0$, we find directly the appropriate solution

$$w(t) = \cos \Omega t\, w(0). \tag{119}$$

With the inhomogeneous term we find

$$
\begin{aligned}
w(t) &= \cos\Omega t\, w(0) - \Omega\, \cos\Omega t \int_0^t \frac{dt'}{\cos^2 \Omega t'}\, v(0) \\
w(t) &= \cos\Omega t\, w(0) - \sin\Omega t\, v(0),
\end{aligned} \tag{120}
$$

which is the correct limit of (113). Thus we have obtained the correct memory-less equation for this case. In spite of the singular behaviour of the inversion, analytic integration of the equations does give the correct solutions. All time dependence is given by the Rabi frequency Ω, which gives the splitting of the driven spectrum in the resonant case.

4.2.4 The over-damped case. We now look at the limit $\gamma \gg 2\Omega$. This gives both the limit of faked continuum and justification of the adiabatic approximation. We may set $i\,W = \Theta \approx \gamma/2$. In this case we have

$$
\begin{aligned}
&\sin Wt = i\, \sinh\Theta \sim \frac{i}{2}\, e^{\Theta t}, \quad \cos\Omega t = \cosh\Theta \sim \frac{1}{2}\, e^{\Theta t}, \\
&\text{and} \quad i\,W = \Theta \approx \frac{\gamma}{2} - \frac{\Omega^2}{\gamma} + \dots .
\end{aligned} \tag{121}
$$

In the exact solutions (113) we now introduce

$$
\left(\cos Wt + \frac{\gamma}{2W}\, \sin Wt\right) = \left(\cosh\Theta t + \frac{\gamma}{2\Theta}\, \sinh\Theta t\right) \to e^{\Theta t} \tag{122}
$$

and

$$
e^{-\gamma t/2}\, e^{\Theta t} = \exp\left[-\frac{\gamma}{2}\, t + \left(\frac{\gamma}{2} - \frac{\Omega^2}{\gamma} + \dots\right) t\right] \to \exp\left(-\frac{\Omega^2}{\gamma}\, t\right). \tag{123}
$$

Thus the solution becomes

$$
\begin{aligned}
w(t) &= \exp\left(-\frac{\Omega^2}{\gamma}\, t\right)\, w(0) - \frac{\Omega}{\gamma}\, \exp\left(-\frac{\Omega^2}{\gamma}\, t\right)\, v(0) \\
&\approx \exp\left(-\frac{\Omega^2}{\gamma}\, t\right)\, w(0)
\end{aligned} \tag{124}
$$

in full agreement with the result of the adiabatic elimination (101). However, the initial slip is not emerging in this approach. However, if we look at the exact equation, with $y(0) = 0$, we have

$$
\begin{aligned}
w(t) &= e^{-\gamma t/2}\left(\cosh\Theta t + \frac{\gamma}{2\Theta}\sinh\Theta t\right) w(0) \\
&\approx \frac{e^{-\gamma t/2}\, e^{i\Theta t}}{2}\left(1+\frac{\gamma}{2\Theta}\right) w(0) \\
&= \exp\left(-\frac{\Omega^2}{\gamma}\, t\right) w_{\text{eff}}(0),
\end{aligned}
\tag{125}
$$

where

$$
w_{\text{eff}}(0) = \frac{\left(\Theta+\frac{\gamma}{2}\right)}{2\Theta}\, w(0). \tag{126}
$$

Expanding

$$
\begin{aligned}
\frac{\left(\Theta+\frac{\gamma}{2}\right)}{2\Theta} &= \frac{1}{2}\left(\frac{\gamma-\frac{\Omega^2}{\gamma}}{\frac{\gamma}{2}-\frac{\Omega^2}{\gamma}}\right) \approx \left(1-\frac{\Omega^2}{\gamma^2}\right)\left(1+\frac{2\Omega^2}{\gamma^2}\right) \\
&\approx \left(1+\frac{\Omega^2}{\gamma^2}\right) \approx \exp\left(\frac{\Omega^2}{\gamma^2}\right)\ldots .
\end{aligned}
\tag{127}
$$

Thus (125) is fully equivalent with the earlier result (108). Consequently we have recovered all the results obtained in the adiabatic limit. Not only do we get the adiabatically obtained decay rate but we also identify the initial slip.

4.3 Hazards of memory functions

We consider a most simplified damping by introducing for a two-level system the Lindblad type relaxation operator

$$
\begin{aligned}
\mathcal{L}\,\rho = \;& \Gamma\;(\bar{n}+1)\,(2\sigma^-\rho\,\sigma^+ - \rho\,\sigma^+\sigma^- - \sigma^+\sigma^-\,\rho) \\
& -\Gamma\,\bar{n}\;(2\sigma^+\rho\,\sigma^- - \rho\,\sigma^-\sigma^+ - \sigma^-\sigma^+\,\rho)\,,
\end{aligned}
\tag{128}
$$

where the σs are the Pauli operators $[\sigma^+,\,\sigma^-] = \sigma_3$ and $\bar{n}$ is the bosonic population function.

We test the consequences of introducing memory effects in an ad hoc manner [Barnett 2001; Barnett 2004] by writing the equation of motion in the form

$$
\partial_t\,\rho(t) = \int_0^t K(t-\tau)\,\mathcal{L}\,\rho(\tau)\,d\tau. \tag{129}
$$

If the equation is derived by eliminating a reservoir, we know that there exists an exact expression of this form from Eq. (61). However, in the general physical situation this is not known and we may have to be satisfied with a physically motivated guess.

In the present case we may choose the apparently reasonable form

$$K(t) = V^2 \, e^{-\gamma t}. \tag{130}$$

The existence of an exponentially decaying memory is expected to emerge from most physically motivated derivations. For an initially excited atom we may solve the equation of motion using the ansatz

$$\rho(t) = \begin{bmatrix} \dfrac{\bar{n}}{2\bar{n}+1} & 0 \\ 0 & \dfrac{\bar{n}+1}{2\bar{n}+1} \end{bmatrix} + A(t) \begin{bmatrix} 1 & 0 \\ 0 & -1 \end{bmatrix}, \tag{131}$$

with $A(0) = \dfrac{\bar{n}+1}{2\bar{n}+1}$. We find

$$\mathcal{L}\,\rho = -\,2\Gamma\ (2\bar{n}+1)\,A(t) \begin{bmatrix} 1 & 0 \\ 0 & -1 \end{bmatrix}. \tag{132}$$

The time evolution equation now reduces to

$$\frac{dA(t)}{dt} = -\,2\Gamma\,(2\bar{n}+1) \int_0^t K(t-\tau)\,A(\tau)\,d\tau \ldots . \tag{133}$$

This equation can be solved by the Laplace transform method to give the solution for $\rho(t)$. See Fig. 4.

We look at the probability of excitation of the upper level

$$\begin{aligned} \rho_{11}(t) \ &= \frac{\bar{n}}{2\bar{n}+1} + \frac{\bar{n}+1}{2\bar{n}+1} \left\{ \frac{1}{2}\left(1 - i\,\frac{\gamma}{2W}\right) \exp\left[-\left(\frac{\gamma}{2} - i\,W\right)t\right] \right. \\ &\left. + \frac{1}{2}\left(1 + i\,\frac{\gamma}{2W}\right) \exp\left[-\left(\frac{\gamma}{2} + i\,W\right)t\right] \right\}, \end{aligned} \tag{134}$$

where we have

$$\begin{aligned} \Omega^2 &= 2V^2\,\Gamma\,(2\bar{n}+1) = 2V^2\,\Gamma \coth\left(\frac{\hbar\omega}{2\,k_B T}\right) \\ W &= \sqrt{\Omega^2 - \frac{\gamma^2}{4}}, \end{aligned} \tag{135}$$

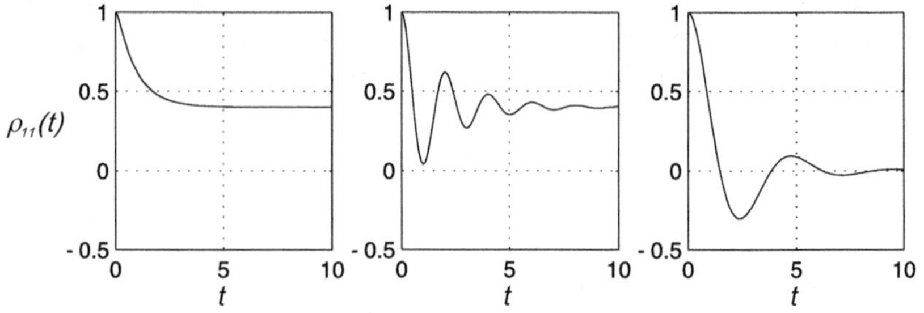

Figure 4. Three different parameter regions. Left: damping without oscillations ($\bar{n} = 2$, $\gamma = 10$). Centre: damping with oscillations ($\bar{n} = 2$, $\gamma = 1$). Right: damping with oscillations that violate the state operator properties ($\bar{n} = 0$, $\gamma = 1$). The other parameters are $\Gamma = V = 1$. In the high-temperature limit ($\bar{n} \to \infty$), there are always oscillations but their amplitude is smaller than the steady-state excitation and, consequently, the probability never goes negative.

and where the ordinary thermal relationship has been introduced. When $\gamma \geq 2\Omega$, the excitation decreases monotonically from unity to its steady state value $\rho_{11}(\infty) = \bar{n}/(2\bar{n}+1)$.

On the other hand, when $\gamma \leq 2\Omega$ the excitation displays damped oscillation around its eventual steady state value. The minimum of the excitation is given when $Wt_1 = \pi$ giving the value

$$\rho_{11}(t_1) = \frac{\bar{n}}{2\bar{n}+1} - \frac{\bar{n}+1}{2\bar{n}+1} \exp\left(-\frac{\gamma\,\pi}{2W}\right). \tag{136}$$

This is negative if

$$\exp\left(-\frac{\gamma\,\pi}{2W}\right) > \frac{\bar{n}}{\bar{n}+1} = \exp\left(-\frac{\hbar\omega}{k_B T}\right), \tag{137}$$

For temperatures

$$k_B T < \frac{2W}{\gamma\,\pi}\,\hbar\omega \tag{138}$$

a negative population value emerges, and the physical interpretation of the density matrix breaks down, see Fig. 4. Thus for any finite memory time, the function (130) leads to unphysical consequences, in spite of its seemingly natural form. In the limit $\gamma \to \infty$, it seems that the condition (138) can never be satisfied. Thus, in this limit of vanishing memory range, no unphysical features emerge. A more detailed investigation, however, shows how this situation comes about in our simple model.

For physical reasons we have to assume that

$$\int_0^\infty K(t)\,dt = v^2 \neq 0 \tag{139}$$

giving that in the memory-less limit

$$V^2 = v^2\,\gamma. \tag{140}$$

From (145) we find that then

$$\Omega^2 \equiv \Lambda^2\,\gamma \tag{141}$$

and

$$W = \sqrt{\Lambda^2\,\gamma - \frac{\gamma^2}{4}}. \tag{142}$$

Thus W cannot stay real in the limit $\gamma \to \infty$, when the memory effects disappear. In case that $W = i\Theta$ and Eq. (134) becomes

$$\begin{aligned}\rho_{11}(t) = \;& \left\{ \frac{1}{2}\left(1 - \frac{\gamma}{2\Theta}\right) \exp\left[-\left(\frac{\gamma}{2} + \Theta\right) t\right] \right. \\ & \left. + \frac{1}{2}\left(1 + \frac{\gamma}{2\Theta}\right) \exp\left[-\left(\frac{\gamma}{2} - \Theta\right) t\right] \right\}.\end{aligned} \tag{143}$$

In this form we may now consider the Markovian limit $\gamma \to \infty$ using the relation

$$\frac{\gamma}{2\Theta} = \left(1 - 4\,\frac{\Lambda^2}{\gamma}\right)^{-1/2} \approx 1 + 2\,\frac{\Lambda^2}{\gamma} + \ldots . \tag{144}$$

In (143) this gives

$$\rho_{11}(t) = \frac{\bar{n}}{2\bar{n}+1} + \frac{\bar{n}+1}{2\bar{n}+1}\left[-\frac{\Lambda^2}{\gamma}\,e^{-\gamma t} + e^{-\Lambda^2 t}\right]. \tag{145}$$

Here the two time-scales, the rapid and the slow one, are clearly seen. In the limit $\gamma \to \infty$, the physical Markov expression is obtained.

In this model, we have seen how careless introduction of a memory function may lead to unphysical behaviour. In the present simple case, we have seen how the physical situation forces us to work with a parameter choice that removes the difficulty. However, as we do not, in general, have a criterion for acceptable memory functions, an ad hoc modification of the equations may easily lead to unexpected and absurd results. Approximative memory functions are hazardous, [Barnett 2001].

5. The damped qubit system

5.1 The model

In quantum information applications one often treats the system of two qubits being manipulated as part of some information processing. This is modelled by a couple of interacting two-level systems. Following the approach in the present paper, we consider the one system to be strongly damped. In that case it serves as a faked continuum for the other one, and we desire to derive an equation of motion for the originally undamped system. The damping is described by a Markovian term of the Lindblad type.

The system of interest is labelled as #1 and the damped one as #2. We assume the systems initially uncorrelated and further that the system #2 initially is in a thermal state

$$\bar{\rho}_2 = \begin{bmatrix} \dfrac{\bar{n}}{2\bar{n}+1} & 0 \\ 0 & \dfrac{\bar{n}+1}{2\bar{n}+1} \end{bmatrix}, \tag{146}$$

where $\bar{n}$ is the thermal bosonic function. The time evolution is given by the equation

$$\partial_t \, \rho = -i \, \left[(H_1 + H_2 + H_{int}) \, , \, \rho \right] + \mathcal{L}_2 \, \rho \equiv \mathcal{L} \, \rho, \tag{147}$$

where

$$\begin{aligned} H_1 &= \omega_1 |1\rangle_1 \langle 1| \\ H_2 &= \omega_2 |1\rangle_2 \langle 1| \\ H_{int} &= \alpha \left(\sigma_1^+ \, \sigma_2^- + \sigma_1^- \, \sigma_2^+ \right) \\ \mathcal{L}_2 \, \rho &= \kappa \, (n+1) \left(2\sigma_2^- \rho \, \sigma_2^+ - \sigma_2^+ \sigma_2^- \rho - \rho \, \sigma_2^+ \sigma_2^- \right) \\ &\quad + \kappa \, n \left(2\sigma_2^+ \rho \, \sigma_2^- - \sigma_2^- \sigma_2^+ \rho - \rho \, \sigma_2^- \sigma_2^+ \right) . \end{aligned} \tag{148}$$

We introduce the thermal projectors as in (56) and find the equation of motion for the reduced density matrix ρ^S of system #1

$$\partial_t \, \rho^S(t) = -i \left[H_1, \, \rho^S(t) \right] + \mathcal{W} \left(\rho^S(t) \right) , \tag{149}$$

where we have

$$\mathcal{W}\left(\rho^S(t)\right) = -\int_0^t d\tau \begin{bmatrix} f_1(t-\tau)\, R_3(\tau) & f_2(t-\tau)\, \rho_{10}^S(\tau) \\ f_2(t-\tau)\, \rho_{01}^S(\tau) & -f_1(t-\tau)\, R_3(\tau) \end{bmatrix} \tag{150}$$

and

$$R_3 \equiv \frac{\bar{n}+1}{2\bar{n}+1}\, \rho_{11}^S - \frac{\bar{n}}{2\bar{n}+1}\, \rho_{00}^S. \tag{151}$$

The memory functions $f_i(t)$ are calculated from the model in a straightforward way; for resonance $\omega_1 = \omega_2$ they become simply

$$\begin{aligned} f_1(t) = \ & \alpha^2 \left(1 - \frac{\kappa_{\text{eff}}}{2\Theta}\right) \exp\left[-\left(\frac{3}{2}\,\kappa_{\text{eff}} + \Theta\right) t\right] \\ & + \alpha^2 \left(1 + \frac{\kappa_{\text{eff}}}{2\Theta}\right) \exp\left[-\left(\frac{3}{2}\,\kappa_{\text{eff}} - \Theta\right) t\right] \\ f_2(t) = \ & \alpha^2 \exp\left[-\left(\kappa_{\text{eff}} + i\,\omega\right) t\right]. \end{aligned} \tag{152}$$

Here we have introduced the notation

$$\kappa_{\text{eff}} \equiv \kappa\ (2\bar{n}+1) \quad \text{and} \quad \Theta = \frac{1}{2}\sqrt{\kappa_{\text{eff}}^2 - 8\,\alpha^2}. \tag{153}$$

In the strong damping region $\kappa_{\text{eff}} > 2\sqrt{2}\,\alpha$, the memory functions decay in a monotonic manner. However, even in the limit $\alpha \to 0$, we have two different decay constants: $2\kappa_{\text{eff}}$ and κ_{eff}. As we wish the system #2 to provide a reservoir, we must assume that the amplitude of the Rabi flopping induced by the coupling α is not extending the spectrum outside the effective bandwidth of the faked continuum; this is given by κ_{eff}. This is thus the proper regime for the application of the model.

However, the solution of the system does also apply in the strong driving limit $\kappa_{\text{eff}} < 2\sqrt{2}\,\alpha$. As the original equation of motion is of the Lindblad form, no unphysical features should emerge. In the case of no damping $\kappa \to 0$, we obtain the simple result

$$f_1(t) = 2\alpha^2 \cos\left(\sqrt{2}\,\alpha t\right), \tag{154}$$

and the off-diagonal function reduces to simple exponential form.

5.2 Solution of the equations

The equations of motion (149) are easily handled in the case of no coherent coupling between the states of the oscillators. We make an ansatz for the density matrix of system #1 in the form

$$\rho^S(t) = \bar{\rho}_2 + A_3(t)\,\sigma^3 + A_-(t)\,\sigma^+ + A_+(t)\,\sigma^-, \tag{155}$$

where the σ:s are the Pauli matrices. The ensuing equations of motion are

$$\frac{d}{dt}\,A_3(t) = -\int_0^t d\tau\; f_1(t-\tau)\,A_3(\tau)$$

$$\frac{d}{dt}\,A_-(t) = -i\,\omega\,A_-(t) - \int_0^t d\tau\; f_2(t-\tau)\,A_-(\tau) \tag{156}$$

$$\frac{d}{dt}\,A_+(t) = i\,\omega\,A_+(t) - \int_0^t d\tau\; f_2^\star(t-\tau)\,A_+(\tau).$$

These equations are in the form of convolutions and can be solved straightforwardly by the Laplace transform. The ensuing time evolution is consistent with our physical expectations in all parameter regions; some results are shown in Fig. 5.

5.2.1 Hazards of parameter mismatch. In the present model all results are solvable. We know the exact memory kernels in (152) and we can solve for the time evolution from (156). However, in realistic applications, one must ordinarily resort to approximations. In this situation, we get equations of the type (150) but with incorrect values of the parameters in the memory function. In order to display the possible disasters that may ensue, we modify the functions f_1 and f_2 arbitrarily and look at the time evolution of the density matrix.

Instead of (152), we write the memory functions as

$$\begin{aligned} f_1(t) = &\; \alpha_1^2\left(1-\frac{\kappa_1}{2\Theta}\right)\,\exp\left[-\left(\frac{3}{2}\,\kappa_1+\Theta\right)t\right] \\ &+\alpha_1^2\left(1+\frac{\kappa_1}{2\Theta}\right)\,\exp\left[-\left(\frac{3}{2}\,\kappa_1-\Theta\right)t\right] \\ f_2(t) = &\; \alpha_2^2\,\exp\left[-\left(\kappa_2+i\omega\right)t\right], \end{aligned} \tag{157}$$

where

$$\Theta = \frac{1}{2}\sqrt{\kappa_1^2\,(2\bar{n}+1)^2 - 8\,\alpha_1^2}. \tag{158}$$

Here the parameters α_1, α_2, κ_1 and κ_2 are free to be chosen. For any choice, we may solve Eqs. (156) and construct the time dependent 2×2 density matrix. In order to test if this is physically acceptable, we calculate its eigenvalues. The

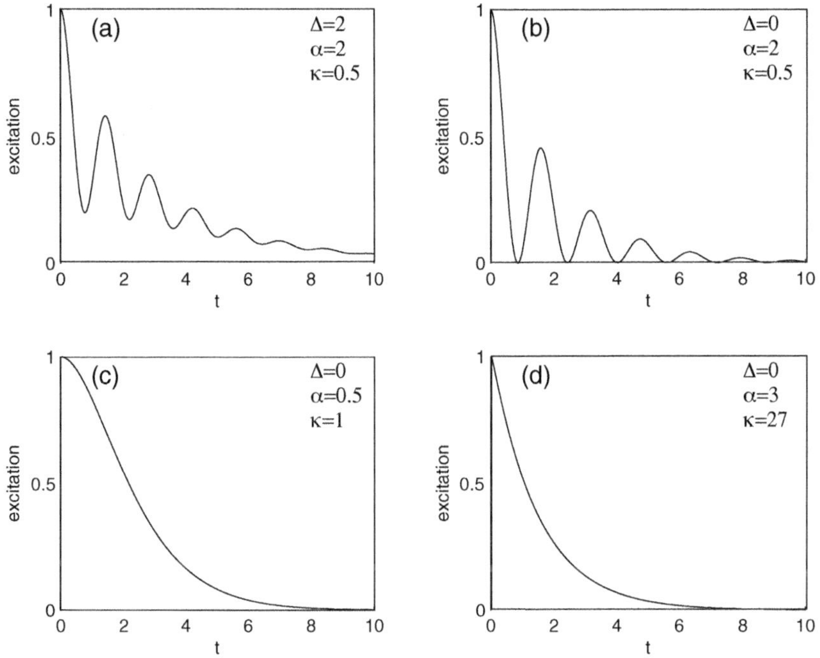

Figure 5. Probability of the excited state; evolution of an initially excited state due to faked-continuum dissipation. In this figure, we consider the limit $T = 0$ only. (a) Detuning-dominated, (b) weak-damping, (c) strong-damping, and (d) Markovian limit. Due to the finite values of all parameters, even the case of the Markovian limit has a non exponential decay within a very short initial slip, although it is hardly visible here.

probability interpretation requires these to be in the interval $[0, 1]$; if not the result is unphysical.

We first choose a rather weak coupling, $\alpha_1 = \alpha_2 = 1$ but with $\kappa_1 = 3$ and $\kappa_2 = 5$. As we can see from the top left-hand part of Fig. 6, the excitation remains positive, but in the top right-hand part of the figure, we find that the eigenvalues of the density matrix become unphysical albeit only very little. We then increase the coupling to $\alpha_1 = \alpha_2 = 3$ but keep the damping low, $\kappa_1 = 1$ and $\kappa_2 = 0.4$. As we can see in middle part of the same figure, the density matrix still behaves well, but the eigenvalues repeatedly assume unphysical magnitudes. Finally we investigate whether we can achieve the same behaviour by using different coupling strengths only. We set $\alpha_1 = 3, \alpha_2 = 2$ but with $\kappa_1 = \kappa_2 = 1$. In Fig. 6 bottom we display the result: the density matrix seems regular enough, but the eigenvalues are strongly unphysical.

As the product of the eigenvalues is the determinant of the density matrix, we find that our time evolution in Figs. 6 violate the condition

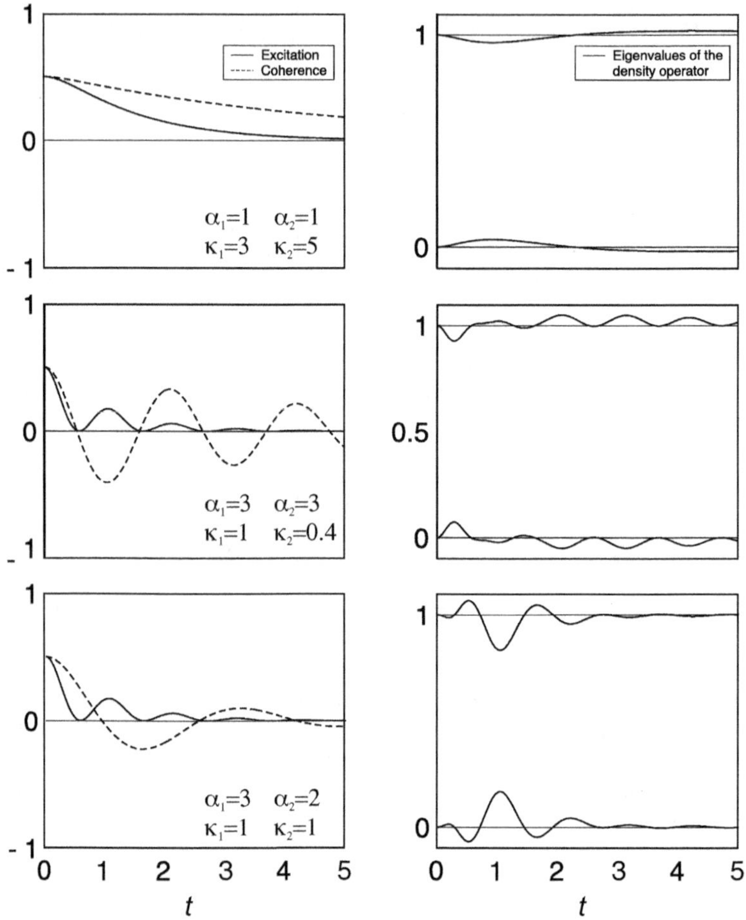

Figure 6. Examples on the breakdown of the positivity. Top: off-diagonal memory is too short compared to the diagonal memory and the coherences decay too slowly. Middle: coherences oscillate too much. Bottom: diagonal and off-diagonal oscillations have different frequencies and the zeros do not occur simultaneously.

$$\rho_{11}\,\rho_{22} \geq \rho_{12}\,\rho_{21}; \tag{159}$$

this causes the breakdown of the probability interpretation. We thus learn that memory kernels are fragile; introducing the wrong ones may lead to unphysical consequences. In the present case, we have the exact solution available, and we may thus resolve the problem explicitly. We do not, however, have a universal criterion, which from the form of the Master Equation allows us to judge if the time evolution preserves physical sense or not. We need a generalisation of the Lindblad condition to Master Equations with memory.

5.2.2 The Markovian limit. The correct memory-less Markovian limit is the following

$$\frac{\alpha}{\kappa_{\text{eff}}} \to 0; \quad \frac{\alpha}{\kappa_{\text{eff}}} \to \text{finite}$$
$$\frac{\omega}{\kappa_{\text{eff}}} \to 0; \quad \Theta \to \frac{1}{2}\,\kappa_{\text{eff}}. \tag{160}$$

Then the memory functions can be integrated to

$$\int_0^\infty f_1(\tau)\,d\tau = \frac{2\alpha^2}{\kappa_{\text{eff}}} \quad \text{and} \quad \int_0^\infty f_2(\tau)\,d\tau = \frac{\alpha^2}{\kappa_{\text{eff}}}. \tag{161}$$

The equation of motion becomes

$$\begin{aligned} \partial_t\,\rho^S(t) = & -i\,\left[H_1,\,\rho^S(t)\right] + \Gamma\,(\bar{n}+1)\left(2\sigma_2^-\rho\,\sigma_2^+ - \sigma_2^+\sigma_2^-\rho - \rho\,\sigma_2^+\sigma_2^-\right) \\ & + \Gamma\,\bar{n}\left(2\sigma_2^+\rho\,\sigma_2^- - \sigma_2^-\sigma_2^+\rho - \rho\,\sigma_2^-\sigma_2^+\right). \end{aligned} \tag{162}$$

The effective decay rate is now given by

$$\begin{aligned} \Gamma &= \frac{\alpha^2}{(2\bar{n}+1)\,\kappa_{\text{eff}}} = \frac{\alpha^2}{(2\bar{n}+1)^2\,\kappa} \\ &= \coth^2\left(\frac{\hbar\omega}{2\,k_B T}\right)\frac{\alpha^2}{\kappa} \to \left(\frac{2\,k_B T}{\hbar\omega}\right)^2\frac{\alpha^2}{\kappa}. \end{aligned} \tag{163}$$

In this model we thus find the high-temperature result $\Gamma \propto T^2$.

5.3 The memory-less formulation

Using the approach indicated in the Sec. 3.2, we can derive an equation of motion in the form

$$\partial_t \begin{bmatrix} \rho^S_{11}(t) & \rho^S_{10}(t) \\ \rho^S_{01}(t) & \rho^S_{00}(t) \end{bmatrix} = \begin{bmatrix} \mathcal{K}_1(t)\,R_3(t) & \mathcal{K}_-(t)\,\rho^S_{10}(t) \\ \mathcal{K}_+(t)\,\rho^S_{01}(t) & -\,\mathcal{K}_1(t)\,R_3(t) \end{bmatrix}, \tag{164}$$

where as before

$$R_3(t) = \left(\frac{\bar{n}+1}{2\bar{n}+1}\,\rho^S_{11}(t) - \frac{\bar{n}}{2\bar{n}+1}\,\rho^S_{00}(t)\right). \tag{165}$$

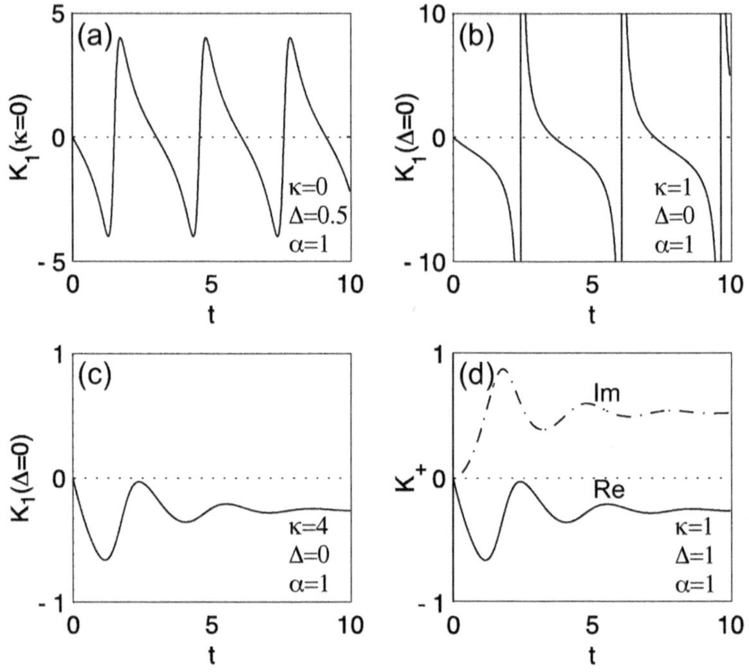

Figure 7. (a) $\mathcal{K}_\infty$ in the absence of decay; any nonzero detuning removes the divergences. (b) $\mathcal{K}_\infty$ for under-damped case in the absence of detuning. (c) $\mathcal{K}_\infty$ for over-damped case in the absence of detuning. (d) $\mathcal{K}_+$ for an arbitrary case.

The expressions for the coefficients $\mathcal{K}$ are rather complicated. In Fig. 7 we show how they behave as functions of time. As expected, in the case when we have repeated zeros of the function $\rho_{11}^S(t)$ we find that $\mathcal{K}_1(t)$ blows up. Across these points, it is dangerous to apply the Eq. (164).

6. Dynamically modified decay and decoherence in two-level systems coupled to baths (reservoirs)

Our previously elaborated approach, [Kofman 2000; Kofman 2001 (a)], to dynamical control of states coupled to an arbitrary *zero-temperature* "bath" or continuum has reaffirmed the intuitive anticipation that, in order to suppress their decay, we must modulate the system-bath coupling at a rate exceeding the spectral interval over which the coupling is significant. The spectra of baths (continua) corresponding to vibrational or collisional decay or decoherence typically allow *dynamical suppression*, using realistic rates of modulation, [Kofman 2000; Kofman 2001 (a)].

These results leave several basic questions open: How to derive a non-Markovian master equation (ME) for *arbitrary* time-dependent driving and modulation of a thermally relaxing two-level system? Would *the two-level system (TLS) model hold* at all for modulation rates, that are comparable to the TLS transition frequency ω_a (between its states $|e\rangle$ and $|g\rangle$) which may invalidate the standard rotating-wave approximation (RWA), [Cohen-Tannoudji 1992]? Would *temperature effects*, which are known to incur *upward* $|g\rangle \rightarrow |e\rangle$ transitions, [Lifshitz 1980], further complicate the dynamics and perhaps hinder the suppression of decay? How to control decay in an efficient, *optimal* fashion? We address these questions by outlining the derivation of a ME of a TLS that is coupled to an *arbitrary bath* and is driven by an *arbitrary* time-dependent field.
The Hamiltonian in question is the sum of the system (S), reservoir bath (B) and system-bath interaction (I) terms,

$$H = H_S(t) + H_B + H_I(t), \; H_I(t) = \mathcal{S}(t)\; B(t). \tag{166}$$

Here $H_S(t)$ is the driven (and modulated) system Hamiltonian, $\mathcal{S}(t)$ is a system operator and $B(t)$ is a bath operator, whose choice depends on the system-bath coupling (linear or quadratic, diagonal or off-diagonal). These operators vary with time due to the external fields. This general form of $H_I(t)$, unlike common treatments, *does not invoke the RWA*, [Cohen-Tannoudji 1992], which may fail for ultrafast modulation. The combined state of the system and the bath is described by the density matrix $\rho_{S+B}(t)$.
Let $\rho_{S+B}(0) = \rho_S(0) \otimes \rho_B$, $\rho_S(t)$ being the density matrix of the system and

$$\rho_B = Z^{-1} \exp[-\frac{\beta}{\hbar}\, H_B] \tag{167}$$

the density matrix of the bath in equilibrium, with Z as the normalization factor, $\beta = \hbar/k_B T$ the inverse temperature (in frequency units), and k_B the Boltzmann constant. Using the projection-operator technique, we have derived, to *second order* in the coupling, the quantum ME in the following *differential* form

$$\dot{\rho} = -\frac{i}{\hbar}\,[H_S(t),\, \rho] + \int_0^t dt' \Big\{\Phi(t,\, t')\, \Big[\tilde{\mathcal{S}}(t',\, t)\, \rho\, \mathcal{S}(t) - \mathcal{S}(t)\, \tilde{\mathcal{S}}(t',\, t)\, \rho\Big] + \text{h.c.}\Big\}. \tag{168}$$

Here

$$\Phi(t,\, t') = \langle U_B^{\dagger}(t-t')\; B(t)\; U_B(t-t')\; B(t')\rangle \tag{169}$$

is the bath "memory" (correlation) function (CF), $\langle \ldots \rangle = \mathrm{Tr}\,(\ldots \rho_B)$ and $\tilde{\mathcal{S}}(t',\, t) = U_S(t,\, t')\; \mathcal{S}(t')\; U_S^\dagger(t,\, t')$, written using the evolution operators $U_B(t) = \exp[-i\, H_B\, t/\hbar]$ and

$$U_S(t,\, t') = T_+\, \exp\left[-\frac{i}{\hbar}\int_{t'}^{t} H_S(\tau)\, d\tau\right], \tag{170}$$

T_+ being the time-ordering operator. In the derivation of Eq. (168) we assumed that $\langle B(t)\rangle = 0$. It needs to be stressed that Eq. (168) *generalizes* previously known master equations to *arbitrary* time-dependent hamiltonians, $H_S(t)$ for the system and $H_I(t)$ for system-bath coupling, [Cohen-Tannoudji 1992].

Henceforth, we explicitly consider a driven TLS undergoing decay, whose resonant frequency and *dipolar* coupling to the reservoir are dynamically modulated, so that

$$\begin{aligned} H_S(t) &= \hbar\,[\omega_a + \delta_a(t) + \delta_r(t)]\; |e\rangle\langle e|, \\ H_I(t) &= \mathcal{S}(t)\; B = \tilde{\epsilon}(t)\; \sigma_x\; B. \end{aligned} \tag{171}$$

Here $\delta_a(t)$ is the dynamically imposed Stark shift of the TLS resonance frequency, $\delta_r(t)$ is its *random* counterpart representing proper dephasing, $\sigma_x = |e\rangle\langle g| + |g\rangle\langle e|$ is the dipole-transition operator, whose time-modulated form is given by $\mathcal{S}(t)$, with the *real* amplitude $\tilde{\epsilon}(t)$. If the bath consists of oscillators, then

$$H_B = \sum_\lambda \hbar\omega_\lambda\, a_\lambda^\dagger\, a_\lambda, \; B = \hbar \sum_\lambda (\kappa_\lambda\, a_\lambda + \kappa_\lambda^\star\, a_\lambda^\dagger), \tag{172}$$

where ω_λ and a_λ are the frequency and annihilation operator, respectively, of the mode λ and κ_λ is the coupling amplitude. Clearly, terms such as $|e\rangle\langle g|\; \kappa_\lambda^\star\; a_\lambda^\dagger$ or $|g\rangle\langle e|\; \kappa_\lambda\; a_\lambda$ in the system-bath interaction $H_I(t)$ are *anti-resonant*, in violation of the RWA.

Upon using Eq. (171) in (168), we obtain our generalized Bloch equations for the components of the TLS density matrix (compare with [Cohen-Tannoudji 1992])

$$\dot{\rho}_{ee} = -\,\dot{\rho}_{gg} = i\,V(t)\,(\rho_{eg} - \rho_{ge}) - R_e(t)\,\rho_{ee} + R_g(t)\,\rho_{gg}, \tag{173a}$$

$$\begin{aligned} \dot{\rho}_{eg} = \dot{\rho}_{ge}^\star &= -\,\{R(t) + i\,[\tilde{\omega}_a(t) + \delta_a(t) + \delta_r(t)]\}\;\rho_{eg} \\ &+ i\,V(t)\,(\rho_{ee} - \rho_{gg}) + [R(t) - i\,\Delta_a(t)]\;\rho_{ge}. \end{aligned} \tag{173b}$$

Eqs. (173a) and (173b) account for the presence of *upward transitions* $|g\rangle \to |e\rangle$ (caused by either temperature or anti-resonant effects - see below) at a rate

$R_g(t)$, in addition to *downward* decay $|e\rangle \to |g\rangle$ at a rate $R_e(t)$. Their half-sum $R(t) = [R_e(t) + R_g(t)]/2$ contributes to the decoherence rate, which is further augmented by the random shift $\delta_r(t)$ (see below). The resonance frequency is dynamically shifted by $\tilde{\omega}_a(t) - \omega_a = \Delta_a(t) = \Delta_e(t) - \Delta_g(t)$, where $\hbar\, \Delta_{e(g)}(t)$ is the Lamb shift of $|e\rangle$ ($|g\rangle$), caused by the dynamically modified coupling to the bath. The last term on the right-hand side of Eq. (173b) is known as "non-secular", [Cohen-Tannoudji 1992]; though usually negligible, it can be important if the modulated resonant frequency $\omega_a + \delta_a(t)$ can vanish or be comparable to $R(t) + |\Delta_a(t)|$.

Here we consider situations wherein $R_{e(g)}(t)$ and $R(t)$, *the rates of decay and decoherence are dominant compared to the proper-dephasing rate* [determined by $\delta_r(t)$], so that the latter may be neglected in Eq. (173b). The dynamically affected transition rates and shifts, obtained from Eqs. (171), (173a) and (173b), are then given by the real and imaginary parts of the expression

$$\frac{R_{e(g)}(t)}{2} + i\, \Delta_{e(g)}(t) = \int_0^t dt'\; K_{e(g)}(t,\, t')\; \Phi_T(t - t'). \tag{174}$$

Here $\Phi_T(t) = \langle \tilde{B}(t)\;\tilde{B}(0)\rangle$ is the bath CF at temperature T, where $\tilde{B}(t) = U_B^\dagger(t)\; B\; U_B(t)$ is the operator B in the interaction representation, and $K_i(t,\, t') = \langle i|S(t)\; \tilde{S}(t',\, t)|i\rangle$ $(i = e,\, g)$ is the correlation function (CF) of the dipole moment in the state $|i\rangle$. One can show that

$$K_e(t,\, t') = K_g^\star(t,\, t') = \epsilon(t)\; \epsilon^\star(t'), \tag{175}$$

where $\epsilon(t) = \tilde{\epsilon}(t)\, \exp\left[i \int_0^t \delta_a(\tau)\, d\tau + i\, \omega_a\, t\right]$ is the coupling-modulation function, allowing for both amplitude and phase modulations.

Since we are interested here in dynamical control of relaxation, we shall concentrate on the transition rates $R_{e(g)}(t)$ rather than the level shifts. Using Eq. (174), one obtains that, for sufficiently short times, the probability of the $|e\rangle \to |g\rangle$ transition $\int_0^t dt'\; R_e(t')$ and its $|g\rangle \to |e\rangle$ counterpart $\int_0^t dt'\; R_g(t')$ are given by

$$\int_0^t dt'\; R_{e(g)}(t') = 2\pi\; Q(t) \int_{-\infty}^{\infty} d\omega\; F_t(\omega)\; G_T(\pm\omega). \tag{176}$$

Here the upper (lower) sign corresponds to the subscript e (g),

$$Q(t) = \int_0^t d\tau\; |\tilde{\epsilon}(\tau)|^2, \tag{177}$$

and

$$F_t(\omega) = [2\pi\; Q(t)]^{-1} \int_0^t dt' \int_0^t dt''\; K_e(t',\, t'')\; e^{i\omega(t''-t')} \tag{178}$$

is the (normalized to unity) spectral density (SD) describing the modulation-induced splitting / shifting of the spectral line at the TLS transition frequency in the interval $(0,\ t)$. The SD of the bath CF

$$G_T(\omega) = (2\pi)^{-1} \int_{-\infty}^{\infty} \Phi_T(t)\, e^{i\omega t}\, dt \tag{179}$$

can be shown to be nonnegative, [Lifshitz 1980], with $G_T(-\omega) = e^{-\beta\omega} G_T(\omega)$, and vanish for $\omega < 0$ at $T = 0$: $G_0(\omega) = 0\ \ (\omega < 0)$. For the oscillator bath (172) one finds that

$$G_T(\omega) = [n(\omega) + 1]\, G_0(\omega) + n(-\omega)\, G_0(-\omega), \tag{180}$$

where $G_0(\omega) = \sum_\lambda |\kappa_\lambda|^2\, \delta(\omega - \omega_\lambda)$ and $n(\omega) = (e^{\beta\omega} - 1)^{-1}$ is the average number of quanta in the oscillator (bath mode) with frequency ω.

We apply Eq. (176) to the case of *coherent modulation of quasiperiodic form*, $\epsilon(t) = \sum_k \epsilon_k\, e^{i\omega_k t}$, where ω_k $(k = 0,\ \pm 1,\ \ldots)$ are arbitrary discrete frequencies with the minimal spectral distance Ω. Without a limitation of the generality, we can assume that $\sum_k |\epsilon_k|^2 = 1$. We then find, using Eq. (176), that the rates $R_{e(g)}(t)$ tend to the long-time limits

$$R_{e(g)} = 2\pi \int_{-\infty}^{\infty} d\omega\, F(\omega)\, G_T(\pm\omega), \tag{181}$$

where

$$F(\omega) = \lim_{t\to\infty} F_t(\omega) = \sum_k |\epsilon_k|^2\, \delta(\omega - \omega_a - \omega_k) \tag{182}$$

or

$$R_{e(g)} = 2\pi \sum_k |\epsilon_k|^2\, G_T[\pm(\omega_a + \omega_k)]. \tag{183}$$

Eq. (181) is the *pivotal general expression* derived here: it shows that R_e (R_g) *is given by the overlap of the modulation spectrum* $F(\omega)$ *with the bath-CF spectrum* $G_T(\omega)$ $[G_T(-\omega)]$. The limits (183) are approached when

$$\Omega t \gg 1 \text{ and } t \gg t_c \equiv \max_k \left\{ \frac{1}{\xi \pm (\omega_a + \omega_k)} \right\}. \tag{184}$$

Here t_c is the bath memory (correlation) time, defined as the inverse of $\xi(\omega)$, the spectral interval over which $G_T(\omega)$ changes around the relevant frequencies.

Had we used the standard dipolar RWA hamiltonian in the case of an oscillator bath, dropping the antiresonant terms in $H_I(t)$ (Eqs. (171) and (172)), we would have arrived at the transition rates

$$R_{e(g)}^{\mathrm{RWA}} = 2\pi \int_0^\infty d\omega\, F(\omega)\, G_T(\pm\omega), \tag{185}$$

wherein the integration is performed from 0 to ∞, rather than from $-\infty$ to ∞, as in (181). This means that the RWA transition rates hold for a slow modulation, when $F(\omega) \simeq 0$ at $\omega < 0$, being peaked near ω_a. However, whenever the suppression of $R_{e(g)}$ requires modulation at a rate comparable to ω_a, the RWA is inadequate. For instance, Eqs. (180) and (185) imply that, at $T = 0$, the rate R_g^{RWA} vanishes identically, irrespective of $F(\omega)$, in contrast to the true upward-transition rate R_g in Eq. (181), which may be comparable to R_e for ultrafast modulation. The difference between the RWA and non-RWA decay rates stems from the fact that the RWA implies that a downward (upward) transition is accompanied by emission (absorption) of a bath quantum, whereas the non-RWA (negative-frequency) contribution to $R_{e(g)}$ in Eq. (181) allows for just the opposite: downward (upward) transitions that are accompanied by absorption (emission). The latter processes are possible since the modulation may cause *level* $|e\rangle$ *to be shifted below* $|g\rangle$.

The validity of the (decohering) TLS model in the presence of modulation at a rate $\gtrsim \omega_a$ is now elucidated: it requires that $R_{e(g)j}\, t \ll 1$, $R_{e(g)j}$ being the effective transition rate from level e (g) to *any other* level j, and, in particular, $R_{e(g)}\, t \ll 1$. If $R_{e(g)}$ are strongly suppressed by the modulation, the TLS model holds for long times.

Conclusions

In the first part of this paper we have reviewed the main concepts and properties of various Master Equations. We have illuminated the behaviour within the framework of simple models, some would say too simple ones. The advantage with these is that the results can be worked out in detail and the validity and the consequences of the approaches can be surveyed.

We have found that even in these simple cases things can go wrong. A physically reasonable Master Equation may not be of the Lindblad form. The corresponding Lindblad form may, on the other hand, violate simple rules like the fluctuation-dissipation theorem.

Another problem is that memory kernels seem to be delicate entities. Erroneous kernels can destroy the physical sense of the time evolution of an initially acceptable density matrix. We do not have a general criterion to help us judge from the Master Equation with memory if the evolution is acceptable. In the Markovian limit, we know that the Lindblad form is certain to preserve the physical interpretation. It is a challenge for the theory of irreversibility in quantum systems to find such a criterion when memory effects are important.

As an alternative to Master Equations with memory, many authors suggest memory-less equations with time dependent coefficients. This approach seems to avoid some of the hazards of the memory-approach, but it offers its own dangers, and we expect approximate expressions for the coefficients to lead to disasters like the ones we encountered in the memory-formulation.

A solvable model which we have not investigated is the one of coupled harmonic oscillators. This was introduced by Ford and his collaborators [Ford 1965] and also by Ref. [Ullersma 1966]. This model provides a formally exact derivation of the Master Equation. Many features of irreversible evolution can be investigated exactly within this model; for example see Ref. [Haake 1985; Strunz 2003]. The result is also equivalent with the approaches in Refs. [Caldeira 1983; Unruh 1989].

In all approaches, it is usually assumed that the irreversible behaviour is generic. If we have calculated the dissipative terms in a given environment, we may change the potential part in the system of interest and retain the dissipative form. This seems justifiable in the weak coupling Born-Markov limit, but it is far from obvious that it is universally valid. With memory effects or strong coupling we have found indications that changing the Hamiltonian of the system, this will affect the influence of the environment on the time evolution. It may also seem that the dynamic behaviour of the reservoir may be crucial. Nonlinearities in its dynamics might give rise to recurrences, which will complicate the memory influence on the system. All such considerations are usually hidden in the assumption of a universal dissipative part of the Master Equation. We think that this is an oversimplification; only in the Born-Markov limit can we really claim to understand the intricacies of irreversible time evolution.

In the second part, we have presented a unified, comprehensive anlysis of dynamically suppressed decay and decoherence in driven TLS coupled to finite-temperature baths and undergoing random frequency fluctuations. This treatment has resulted in both principal and practical general conclusions:
(a) anti-resonant (non-RWA) effects that cause dynamically modified "upward" ($|g\rangle \rightarrow |e\rangle$) transitions may dominate over their temperature-activated counterparts.
(b) control (qubit-flipping) fields can yield resonantly suppressed dephasing as effectively as the "bang-bang" technique.
(c) the dependence of dynamically suppressed decay on ω_a and T [Eqs. (178) and (180)] allows us to design the optimal phase jumps or Stark shifts Δ, *i.e.*, the largest τ or the smallest Δ that can effect the suppression.

QUANTUM DYNAMICS EFFECTED BY REPEATED MEASUREMENTS

J. Clausen,[1] V. M. Akulin,[1], J. Salo[2, 3] and S. Stenholm[2]

[1]*Laboratoire Aimé Cotton, Bât. 505, Campus d'Orsay, 91405 Orsay Cedex, France*
jens.clausen@lac.u-psud.fr, vladimir.akulin@lac.u-psud.fr

[2]*Laser Physics & Quantum Optics, Royal Institute of Technology (KTH), AlbaNova, Roslagstullsbacken 21, SE-10691 Stockholm, Sweden*
Janne.Salo@hut.fi, stenholm@atom.kth.se

[3]*Helsinki University of Technology, Materials Physics Laboratory, P.O. Box 2200, 02015 HUT, Finland*

Abstract We consider the time evolution of a dynamic quantum system coupled to a repeatedly measured ancilla. Given the time lapse Δt between two subsequent measurements, the combined system may be described using a difference master equation whereas, in the Zeno-limit $\Delta t \to 0$, the evolution of the dynamic system is unitary and defined by the state of the ancilla. For an arbitrary Δt, we also formulate a master equation that interpolates smoothly the exact evolution given by the difference equation.

In the special case of a nondemolition interaction Hamiltonian, the master equation of the total system reduces to uncoupled systems of first order differential equations, whose dimensions are the same as the dimension N_a of the ancilla Hilbert space. After having traced over the ancilla state, the master equation of the dynamic system can be expressed either as ${N_a}^{\mathrm{th}}$ order differential equations in time or, equivalently, as Zwanzig equations with an explicit memory over the system evolution.

The above results are applied to a harmonic oscillator coupled to a two-level system, that serves as the repeatedly measured ancilla. Relatively sparse measurements are shown to destroy the coherence of the oscillator whereas, in the Zeno-limit, the coherence is preserved for all times. This is demonstrated by a periodic generation of a Schrödinger cat-like state. The decoherence process is highly nonlinear in the initial state amplitude and the decoherence time decreases rapidly for increasing amplitude.

Keywords: Zeno effect, measurement-induced dynamics, nondemolition measurement

V.M. Akulin et al. (eds.), Decoherence, Entanglement and Information Protection in Complex Quantum Systems, 281–306.

Introduction

Dynamics of an open system interacting with the environment considered as a thermostate may be formulated in terms of a master equation with an integral operator allowing for the relaxation process, [Zwanzig 1960]. In some particular cases this operator has a short-lasting kernel that enables one to consider the relaxation as a Markovian process and to obtain the master equation in the Lindblad form, [Lindblad 1976 (a)]. In some situations the memory effects become, however, important and the dynamics of the system gets much more involved, [Barnett 2001]. A similar situation arises in the case where a set of consecutive or continuous measurements is performed. The purpose of this article is to consider a situation where some simplification of the general form of the master equation with memory is still possible and the result is a simpler master equation. In particular, we consider the case of a dynamic system coupled to a measured ancilla via a nondemolition interaction, [Caves 1980]. This simplifies the consideration essentially whereas providing an important special case in which the energy of the dynamic part is conserved.

We consider a composite quantum system consisting of a dynamic part interacting with an ancillary part, the latter being subject to repeated projective measurements. The entire quantum system is assumed to evolve unitarily during time Δt between the measurements. As a specific example, we analyze a harmonic oscillator coupled to a two-level ancilla that is subject to measurements. It may serve as a model for different physical situations, such as a spin-1/2 particle moving in a one-dimensional harmonic potential or a two-level atomic system interacting with a single cavity mode. Such a model is the simplest possible non-trivial problem of interesting system, each of which has a symmetry, SU(2) and SU(1, 1) for the spin and the oscillator, respectively. These both processes have been studied intensively in the context of atoms interacting with a maser cavity [Raithel 1995; Raimond 2001], as well as in the context of a cold ion moving in an optical trap [Cirac 1995].

1. Effects of measured ancilla

We consider a coupled quantum system consisting of two subsystems, one of which is referred to as dynamic part and the other one as ancilla. The evolution of the system is governed by the Schrödinger equation

$$\frac{d}{dt}\,\varrho = -\,\frac{i}{\hbar}\,[\hat{H},\,\varrho]. \tag{1}$$

We assume that the Hamiltonian $\hat{H}$ has no explicit time-dependence, and, therefore, the unitary evolution operator over Δt may be written as

$$\varrho(t+\Delta t) = e^{-\frac{i}{\hbar}\,\hat{H}\,\Delta t}\,\varrho(t)\,e^{\frac{i}{\hbar}\,\hat{H}\,\Delta t} = \hat{U}_{\Delta t}\,\varrho(t)\,\hat{U}^{\dagger}_{\Delta t} = \mathcal{U}_{\Delta t}\{\varrho(t)\}. \tag{2}$$

We denote linear operators acting on the Hilbert space by capital letters with a hat, such as $\hat{U}_{\Delta t}$, whereas linear maps on operators (operators in the Liouville space) are denoted by calligraphic letters such as $\mathcal{U}_{\Delta t}$, where the argument is often placed within braces in order to avoid ambiguity.

The unitary evolution of the combined system is interrupted repeatedly with a measurement performed on the ancilla. Formally, the effect of a projective measurement is described with an operator $\mathcal{V}$ that acts only in the Hilbert space of the ancilla, $\varrho \to \mathcal{V}\,\varrho$, and the effect of $\mathcal{V}$ is taken to be instantaneous. Therefore, two immediately subsequent measurements yield the same outcome and $\mathcal{V}^2 = \mathcal{V}$ is a projector.

The evolution of the combined system over one measurement cycle is given by $\mathcal{V}\,\mathcal{U}_{\Delta t}$, where the cycle is taken to begin immediately after one measurement and to end after the next. Since the one-cycle evolution always begins with a state that already satisfies $\varrho = \mathcal{V}\,\varrho$, we may write the one-cycle evolution operator as $\mathcal{Y} = \mathcal{V}\,\mathcal{U}_{\Delta t}\,\mathcal{V}$.

1.1 Three measurement descriptions

In order to develop the mathematical framework for the master equation, we consider three descriptions of the system time evolution.

1.1.1 Finite difference equation. In the general case of an arbitrary Δt, the evolution of the coupled system is given by the difference equation

$$\varrho(t_j) = \mathcal{Y}\,\varrho(t_{j-1}) = \mathcal{Y}^j\,\varrho(t_0), \tag{3}$$

where j enumerates the state at times $t_j = t_0 + j\,\Delta t$. This is, however, only a 'stroboscopic' description of the system evolution since the state is obtained only for discrete times immediately after each measurement. The intermediate times are then obtained using the unitary evolution as

$$\varrho(t_j + \tau) = (\mathcal{U}_{\Delta t})^{\tau/\Delta t}\,\varrho(t_j), \tag{4}$$

where $\tau \in [0,\ \Delta t)$. In the frequency domain, the quantum state is given by

$$\tilde{\varrho}(\varepsilon) = \int_{-\infty}^{\infty} \varrho(t)\,e^{i\,\varepsilon t}\,dt, \tag{5}$$

and the Fourier transform of the time evolution equation

$$\varrho(t + \Delta t) = \mathcal{Y}\,[\varrho(t) + \delta(t - t_0)\,r_0] \tag{6}$$

where $r_0 = \varrho(t_0)$ represents the initial condition at $t = t_0]$ is

$$e^{-i\varepsilon\,\Delta t}\,\tilde{\varrho}(\varepsilon) = \mathcal{Y}\,[\tilde{\varrho}(\varepsilon) + e^{i\varepsilon\,t_0}\,r_0]. \tag{7}$$

In principle, this allows one to solve for

$$\tilde{\varrho}(\varepsilon) = \frac{e^{i\varepsilon\, t_0}}{e^{-i\varepsilon\,\Delta t} - \mathcal{Y}}\, \mathcal{Y}\, r_0, \tag{8}$$

and for $t_k = t_0 + k\,\Delta t$,

$$\begin{aligned} \varrho(t_k) &= \frac{1}{2\pi}\int_{-\infty}^{\infty} \frac{e^{-i\varepsilon\, t_k}\, e^{i\varepsilon\, t_0}}{e^{-i\varepsilon\,\Delta t} - \mathcal{Y}}\, \mathcal{Y}\, r_0\, d\varepsilon \\ &= \frac{1}{2\pi}\int_{-\infty}^{\infty} \frac{\left(e^{-i\varepsilon\,\Delta t}\right)^k}{e^{-i\varepsilon\,\Delta t} - \mathcal{Y}}\, \mathcal{Y}\, r_0\, d\varepsilon \\ &= s\,\frac{\Delta t}{2\pi}\int_{\varepsilon_0}^{\varepsilon_0+\frac{2\pi}{\Delta t}} \frac{\left(e^{-i\varepsilon\,\Delta t}\right)^k}{e^{-i\varepsilon\,\Delta t} - \mathcal{Y}}\, \mathcal{Y}\, d\varepsilon\, r_0. \end{aligned} \tag{9}$$

The factor

$$s = \sum_{j=-\infty}^{\infty} \delta(t_k - j\,\Delta t - t_0) \tag{10}$$

is the δ-inhomogeneity added in (6) together with its periodic copies due to the periodicity of the integrand and reflects the fact that the original difference equation has a solution only for periodic moments of time. With the substitution $z = e^{-i\,\varepsilon\,\Delta t}$ the integral may be transformed into the contour integral

$$\frac{\Delta t}{2\pi}\int_{\varepsilon_0}^{\varepsilon_0+\frac{2\pi}{\Delta t}} \frac{\left(e^{-i\varepsilon\,\Delta t}\right)^k}{e^{-i\varepsilon\,\Delta t} - \mathcal{Y}}\, \mathcal{Y}\, d\varepsilon = \frac{1}{2i\pi}\oint \frac{z^{k-1}}{z - \mathcal{Y}}\, \mathcal{Y}\, dz = \mathcal{Y}^k, \tag{11}$$

provided that the eigenvalues of $\mathcal{Y}$ lie inside the complex unit circle. Since the time evolution consists of unitary operations combined with projections, this condition will naturally be met.

1.1.2 Quasi-continuous analogy. We assume now that it is possible to adapt a continuous-time picture given by

$$\varrho(t) = \mathcal{Y}^{\frac{t-t_0}{\Delta t}}\, \varrho(t_0) = e^{\frac{t-t_0}{\Delta t}\ln \mathcal{Y}}\, \varrho(t_0). \tag{12}$$

Here, $e^{\frac{t-t_0}{\Delta t}\ln \mathcal{Y}}$ is a one-parameter semi-group element and the state obeys the master equation given by its generator $\mathcal{L}$,

$$\frac{d}{dt}\,\varrho(t) = \frac{1}{\Delta t}\ln \mathcal{Y}\, \varrho(t) = \mathcal{L}\,\varrho(t), \tag{13}$$

where the generator may also be obtained as

$$\begin{aligned} \frac{d}{dt}\,\varrho(t) &= \left(\frac{d}{dt}\,\mathcal{Y}^{\frac{t-t_0}{\Delta t}}\right)\varrho(t_0) \\ &= \underbrace{\left(\frac{d}{dt}\,\mathcal{Y}^{\frac{t-t_0}{\Delta t}}\right)\left(\mathcal{Y}^{\frac{t-t_0}{\Delta t}}\right)^{-1}}_{\equiv\,\mathcal{L}}\;\varrho(t). \end{aligned} \tag{14}$$

The generator $\mathcal{L}$ yields a smooth time evolution that interpolates the exact evolution given by Eq. (3) and coincides with it for all $t = t_j = t_0 + j\,\Delta t$. The smaller, however, Δt is, the more accurate the intermediate solution becomes. In the limit where Δt is the smallest time scale relevant for the system evolution, the measurements are considered quasi-continuous and, formally, the master equation is determined by the generator $\mathcal{L}$.

The quasi-continuous approach carries, however, a potentially dangerous hazard: the time-evolution $\mathcal{Y}$ contains projections and, therefore, it has zero eigenvalues. This renders it, in principle, impossible to find an inverse or logarithm operator. Sometimes it will be, however, possible to exclude the null-space of $\mathcal{Y}$ from the consideration, which enables one to define these operators. This will be the case in our example below.

1.1.3 Zeno-limit.

In the limit $\Delta t \to 0$, the time evolution needs to be considered only to the first order in Δt, leading to

$$\varrho(t+\Delta t) = \mathcal{V}\left\{\varrho(t) - \Delta t\,\frac{i}{\hbar}\,\hat{H}\,\mathcal{V}\{\varrho(t)\} + \Delta t\,\frac{i}{\hbar}\,\mathcal{V}\{\varrho(t)\}\,\hat{H}\right\} \tag{15}$$

since, because of the effect of the previous measurement, $\varrho(t) = \mathcal{V}\,\varrho(t)$. This leads to

$$\frac{\varrho(t+\Delta t) - \varrho(t)}{\Delta t} = -\frac{i}{\hbar}\,\mathcal{V}\left\{\hat{H}\,\mathcal{V}\{\varrho(t)\}\right\} + \frac{i}{\hbar}\,\mathcal{V}\left\{\mathcal{V}\{\varrho(t)\}\,\hat{H}\right\} \tag{16}$$

and, for $\Delta t \to 0$, we obtain a master equation

$$\frac{d}{dt}\varrho(t) = -\frac{i}{\hbar}\,\mathcal{V}\left\{\left[\hat{H},\,\mathcal{V}\{\varrho(t)\}\right]\right\}. \tag{17}$$

If the operator $\mathcal{V}$ represents a complete measurement on the ancilla, its reduced state $\varrho_2 = \mathrm{Tr}_1(\varrho)$ is frozen to satisfy $\varrho_2 = \mathcal{V}\{\varrho_2\}$.

1.2 Measurement schemes

Here we consider different alternatives for the measurement processes. In all of them, the system is taken to consist of two parts: dynamic part, the evolution of which is of eventual interest in the experiment, coupled to an ancilla through which the non-unitary effects are produced. The ancilla measurements are carried out in some orthogonal meter basis $\{|\Psi_k\rangle_2\}$, which does not necessarily consist of energy eigenstates, but of some other measurable states. This is often favorable since projections onto the energy states do not yield all the variety of possible effects.

We propose five different schemes.

1 No measurements are carried out; the total system evolves unitarily all the time, see Fig. 1 (a). The measurement projection is, naturally, a unit operator, $\mathcal{V}\{\varrho\} = \varrho$.

2 The entire system evolves unitarily but a noise term is present in the ancilla Hamiltonian, see Fig. 1 (b). This is not, in fact, a measurement in the sense considered in this work; nonetheless, it provides similar physics since a stochastic noise acting on the ancilla produces coherence loss to the combined system.

3 A complete projective measurement is carried out repeatedly on the ancilla but the outcome is ignored, see Fig. 1 (c), whence

$$\mathcal{V}\{\varrho\} = \sum_k |\Psi_k\rangle_2\langle\Psi_k|\varrho|\Psi_k\rangle_2\langle\Psi_k|. \tag{18}$$

The system state may be expanded at all times as

$$\varrho(t) = \sum_{i,\,j} |\Psi_i\rangle_2 \underbrace{{}_2\langle\Psi_i|\varrho|\Psi_j\rangle_2}_{\varrho_{ij}(t)} {}_2\langle\Psi_j|, \tag{19}$$

where $\varrho_{ij}(t)$ are operators for the dynamic part, and the reduced state for the dynamic part is given by $\varrho_{\text{dyn}}(t) = \sum_i \varrho_{ii}(t)$. In the Zeno-limit, the equation of motion (17) yields

$$\begin{aligned}\frac{d}{dt}\varrho_{ii}(t) &= -\frac{i}{\hbar}\left[{}_2\langle\Psi_i|\hat{H}|\Psi_i\rangle_2,\ {}_2\langle\Psi_i|\varrho|\Psi_i\rangle_2\right]\\ &= -\frac{i}{\hbar}\left[\hat{H}_{ii},\ \varrho_{ii}\right], \qquad (20)\\ \frac{d}{dt}\varrho_{ij}(t) &= 0 \ \text{ for } i \neq j.\end{aligned}$$

Thus, the evolutions of the conditional reduced states $\varrho_{ii}(t)$ decouple and they evolve independently. If, in particular, at least one measurement outcome is known, the corresponding reduced state has a probability one and the others disappear. In such a case, the dynamic system evolves unitarily and the reduced Hamiltonian $\hat{H}_{ii}$ is fixed by the outcome of the Zeno measurement.

4 State of the ancilla is repeatedly reset to a fixed initial state $|\Psi_0\rangle_2$, see Fig. 1 (d). Here,

$$\begin{aligned} \mathcal{V}\{\varrho\} &= |\Psi_0\rangle_2 \left(\sum_k {}_2\langle\Psi_k|\varrho|\Psi_k\rangle_2 \right) {}_2\langle\Psi_0| \\ &= \mathrm{Tr}_2\{\varrho\}\, |\Psi_0\rangle_2\langle\Psi_0|. \end{aligned} \tag{21}$$

In the Zeno-limit,

$$\frac{d}{dt}\,\varrho(t) = -\frac{i}{\hbar}\, |\Psi_0\rangle_2 \left[{}_2\langle\Psi_0|\hat{H}|\Psi_0\rangle_2,\ \mathrm{Tr}_2\{\varrho(t)\} \right] \langle\Psi_0|, \tag{22}$$

whence

$$\begin{gathered} \frac{d}{dt}\,\varrho_{00}(t) = -\frac{i}{\hbar} \left[{}_2\langle\Psi_0|\hat{H}|\Psi_0\rangle_2,\ \mathrm{Tr}_2\{\varrho(t)\} \right], \\ \text{and} \\ \frac{d}{dt}\,\varrho_{ij}(t) = 0 \ \text{ for } i \text{ or } j \neq 0. \end{gathered} \tag{23}$$

Therefore, the reduced state ϱ_1 of the dynamic system obeys the equation of motion

$$\frac{d}{dt}\,\varrho_1(t) = -\frac{i}{\hbar} \left[{}_2\langle\Psi_0|\hat{H}|\Psi_0\rangle_2,\ \varrho_1(t) \right], \tag{24}$$

and the dynamic system evolution is governed by H_{00}, similarly with the previous case, when the measurement outcome is known to have been $|\Psi_0\rangle$ even once.

5 Only the special case is considered, in which the measurement outcome always corresponds to a fixed state $|\Psi_0\rangle_2$, see Fig. 1 (e). Now

$$\mathcal{V}\{\varrho\} = \frac{|\Psi_0\rangle_2\langle\Psi_0|}{P_0}\, {}_2\langle\Psi_0|\varrho|\Psi_0\rangle_2 = \frac{\varrho_{00}}{P_0}\, |\Psi_0\rangle_2\langle\Psi_0|, \tag{25}$$

where $P_0 =_2 \langle\Psi_0|\mathrm{Tr}_1\{\varrho\}|\Psi_0\rangle_2$ is the probability of the measurement outcome corresponding to $|\Psi_0\rangle_2$.

In the Zeno-limit,

$$\frac{d}{dt}\,\varrho(t) = -\,\frac{i}{\hbar}\,|\Psi_0\rangle_2\langle\Psi_0|\,\left[\,{}_2\langle\Psi_0|\hat{H}|\Psi_0\rangle_2,\ \varrho_{00}\right] \tag{26}$$

(since $P_0 = 1$ after an infinitesimal evolution), and the dynamic system is again governed by H_{00}, as before.

2. Master equation for nondemolition interaction

The master equation for the combined system may be represented in a simplified way provided that the Hamiltonian of the dynamic part alone commutes with the interaction Hamiltonian. In this case, the matrix elements of the dynamic system decouple and a master equation may be written to each of them separately. In this section we present three different kinds of master equations.

The unitary evolution of the combined system is now described by

$$\frac{d}{dt}\,\varrho = -\,\frac{i}{\hbar}\,\left[\hat{H}_1 + \hat{H}_{\text{int}} + \hat{H}_2,\ \varrho\right], \tag{27}$$

where $[\hat{H}_1,\ \hat{H}_{\text{int}}] = 0$. We call this nondemolition interaction because it does not induce any transitions between eigenstates of different energies. The relative phases do, however, change, which leads to pure phase decoherence of the reduced state of the dynamic system.

The dynamic part may be represented with the help of the energy eigenstates $\hat{H}_1\,|n\rangle_1 = E_n\,|n\rangle_1$, for which $\hat{H}_{\text{int}}\,|n\rangle_1 = \hat{F}_n\,|n\rangle_1$. Note that the quantities $\varrho^{(m,\,n)} =_1 \langle m|\varrho|n\rangle_1$ and $\hat{F}_n$ are operators in the ancilla Hilbert space. The equation of motion for $\varrho^{(m,n)}$ may now be written as

$$\frac{d}{dt}\,\varrho^{(m,\,n)} = -\,\frac{i}{\hbar}\left[(E_m + \hat{F}_m + \hat{H}_2)\,\varrho^{(m,\,n)} - \varrho^{(m,\,n)}(E_n + \hat{F}_n + \hat{H}_2)\right], \tag{28}$$

and the corresponding unitary evolution over Δt is

$$\begin{aligned}\varrho^{(m,\,n)}(t+\Delta t) &= e^{-\frac{i}{\hbar}(E_m+\hat{F}_m+\hat{H}_2)\Delta t}\,\varrho^{(m,\,n)}(t)\,e^{\frac{i}{\hbar}(E_n+\hat{F}_n+\hat{H}_2)\Delta t}\\ &= \hat{U}^m_{\Delta t}\,\varrho^{(m,\,n)}(t)\,\hat{U}^{n\dagger}_{\Delta t}.\end{aligned} \tag{29}$$

We choose the measurement scheme where the outcome is ignored, and the measurement projection is therefore given by

$$\mathcal{V}\{\varrho\} = \sum_k |\Psi_k\rangle_2\langle\Psi_k|\varrho|\Psi_k\rangle_2\langle\Psi_k|. \tag{30}$$

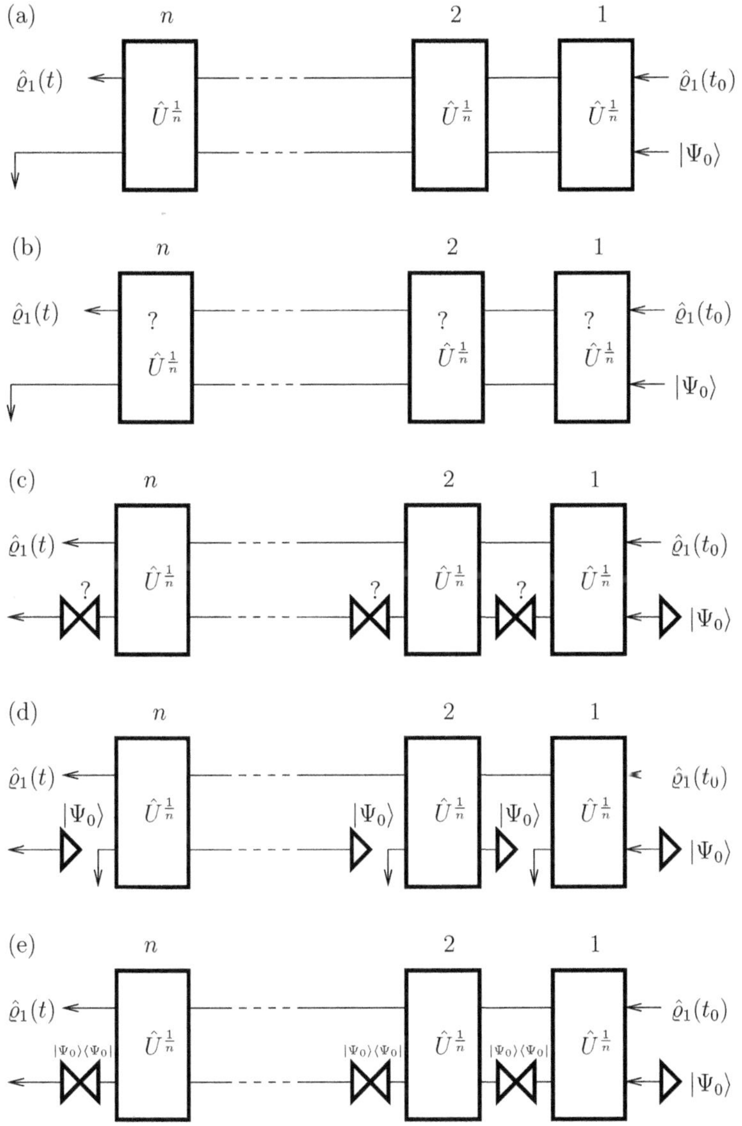

Figure 1. (a) No measurements are carried out; the total system evolves unitarily all the time. (b) Combined system evolves unitarily but a noise term is included in the ancilla Hamiltonian. (c) A complete projective measurement is carried out repeatedly on the ancilla but the outcome is ignored. (d) State of the ancilla is repeatedly reset to a fixed initial state $|\Psi_0\rangle_2$. (e) Only the special case is considered, in which the measurement outcome always corresponds to a fixed state $|\Psi_0\rangle_2$.

The nonunitary evolution over one measurement cycle is (note that after the precedent measurement the state satisfies $\varrho(t) = \mathcal{V}\{\varrho(t)\}$)

$$\begin{aligned}
\varrho^{(m,\,n)}(t+\Delta t) &= \mathcal{V}\left\{\hat{U}^{m}_{\Delta t}\,\mathcal{V}\{\varrho^{(m,\,n)}(t)\}\,\hat{U}^{n\dagger}_{\Delta t}\right\} \\
&= \sum_{k,\,l} |\Psi_k\rangle_2 \langle\Psi_k|\hat{U}^{m}_{\Delta t}|\Psi_l\rangle_2 \langle\Psi_l|\varrho^{(m,\,n)}(t)|\Psi_l\rangle_2 \\
&\quad \times {}_2\langle\Psi_l|\hat{U}^{n\dagger}_{\Delta t}|\Psi_k\rangle_2 \langle\Psi_k| \\
&= \sum_{k,l} |\Psi_k\rangle_2\, U^{m}_{\Delta t,\,kl}\, \varrho^{(m,\,n)}_{ll}(t)\, U^{n\star}_{\Delta t,\,kl}\, {}_2\langle\Psi_k|,
\end{aligned} \tag{31}$$

whence

$$\varrho^{(m,\,n)}_{kk'}(t+\Delta t) = \delta_{kk'} \sum_{l} U^{m}_{\Delta t,\,kl}\, \varrho^{(m,\,n)}_{ll}(t)\, U^{n\star}_{\Delta t,\,kl}. \tag{32}$$

The exact difference equation of motion for the matrix element (m, n) in the dynamic part Hilbert space is given as

$$\varrho^{(m,\,n)}_{kk}(t+\Delta t) = \sum_{l} \mathcal{U}^{(m,\,n)}_{\Delta t,\,kl}\, \varrho^{(m,\,n)}_{ll}(t), \tag{33}$$

where the evolution is defined by the matrix $\mathcal{U}^{(m,\,n)}_{\Delta t}$ whose elements are

$$\mathcal{U}^{(m,\,n)}_{\Delta t,\,kl} = U^{m}_{\Delta t,\,kl}\, U^{n\star}_{\Delta t,\,kl}. \tag{34}$$

For quasi-continuous measurements, the time evolution is generated by the operator

$$\mathcal{L}^{(m,\,n)}_{\Delta t} = \frac{1}{\Delta t} \ln\left[\mathcal{U}^{(m,\,n)}_{\Delta t}\right], \tag{35}$$

and the Zeno limit is obtained for $\Delta t \to 0$.

We examine now the special case of a two-level ancilla. For fixed (m, n) the equation of motion is now

$$\frac{d}{dt}\begin{bmatrix} \varrho^{(m,\,n)}_{00}(t) \\ \varrho^{(m,\,n)}_{11}(t) \end{bmatrix} = \begin{bmatrix} \mathcal{L}^{(m,\,n)}_{\Delta t,\,00} & \mathcal{L}^{(m,\,n)}_{\Delta t,\,01} \\ \mathcal{L}^{(m,\,n)}_{\Delta t,\,10} & \mathcal{L}^{(m,\,n)}_{\Delta t,\,11} \end{bmatrix} \begin{bmatrix} \varrho^{(m,\,n)}_{00}(t) \\ \varrho^{(m,\,n)}_{11}(t) \end{bmatrix}, \tag{36}$$

where, in the time evolution generator matrix, the elements $\mathcal{L}^{(m,\,n)}_{\Delta t,\,kl}$ are expected to be time-independent. For the sake of simplicity, the superscripts ^{mn} and the subscript ${}_{\Delta t}$ are omitted from now on.

In order to describe the time-evolution of the dynamic system alone, we need to find an evolution equation for $\rho = \mathrm{Tr}_2[\varrho] = \varrho_{00}(t) + \varrho_{11}(t)$. For that purpose, using the abbreviations

$$\begin{bmatrix} \rho \\ \eta \end{bmatrix} = \begin{bmatrix} \varrho_{00}(t) + \varrho_{11}(t) \\ \varrho_{00}(t) - \varrho_{11}(t) \end{bmatrix} \tag{37}$$

and

$$\begin{aligned} \mathbb{B} &= \begin{bmatrix} B_{00} & B_{01} \\ B_{10} & B_{11} \end{bmatrix} \\ &= \begin{bmatrix} \dfrac{\mathcal{L}_{00} + \mathcal{L}_{10} + \mathcal{L}_{01} + \mathcal{L}_{11}}{2} & \dfrac{\mathcal{L}_{00} + \mathcal{L}_{10} - \mathcal{L}_{01} - \mathcal{L}_{11}}{2} \\ \dfrac{\mathcal{L}_{00} - \mathcal{L}_{10} + \mathcal{L}_{01} - \mathcal{L}_{11}}{2} & \dfrac{\mathcal{L}_{00} - \mathcal{L}_{10} - \mathcal{L}_{01} + \mathcal{L}_{11}}{2} \end{bmatrix}, \end{aligned} \tag{38}$$

Eq. (36) is further written as

$$\frac{d}{dt}\begin{bmatrix} \rho \\ \eta \end{bmatrix} = \mathbb{B}\begin{bmatrix} \rho \\ \eta \end{bmatrix}, \tag{39}$$

where $\mathbb{B}$ is the time-evolution generator (*i.e.*, Liouvillian) in the basis chosen above.

We present here three different methods for finding an equation of motion for ρ alone.

1 Second-order differential equation in time.

From the above Eqs. (39), we obtain

$$\rho = \frac{1}{B_{10}}\left(\dot{\eta} - B_{11}\,\eta\right) \text{ and } \eta = \frac{1}{B_{01}}\left(\dot{\rho} - B_{00}\,\rho\right). \tag{40}$$

Further, differentiation of Eq. (39) a second time yields

$$\begin{aligned} \ddot{\rho} &= \left(B_{00} + B_{01}\,B_{11}\,\frac{1}{B_{01}}\right)\dot{\rho} + \left(B_{01}\,B_{10} - B_{01}\,B_{11}\,\frac{1}{B_{01}}\,B_{00}\right)\rho \\ &= \left(B_{00} + B_{11}\right)\dot{\rho} + \left(B_{01}\,B_{10} - B_{11}\,B_{00}\right)\rho, \end{aligned} \tag{41}$$

which may also be written as

$$\left(\frac{d^2}{dt^2} - \text{Tr}\,\mathbb{B}\,\frac{d}{dt} + \text{Det}\,\mathbb{B}\right)\rho = 0. \tag{42}$$

Note that this is a second-order differential equation since the ancilla is chosen to be a two-level system. For an N-level ancilla, the equation would be of Nth order and read

$$\text{Det}\left(\mathbb{B} - \mathbb{I}\,\frac{d}{dt}\right)\rho = 0. \tag{43}$$

The initial value for $\dot{\rho}(0)$ is obtained as

$$\frac{d}{dt}\,\rho(0) = B_{00}\rho(0) + B_{01}\eta(0). \tag{44}$$

Note that the initial condition depends on the initial ancilla state and, therefore, the time evolution has a memory.

2 Nakajima-Zwanzig equation with memory, see, *e.g.*, [Breuer 2002]. Introducing the Nakajima-Zwanzig projectors

$$\mathcal{P} = \begin{bmatrix} 1 & 0 \\ 0 & 0 \end{bmatrix} \text{ and } \mathcal{Q} = \begin{bmatrix} 0 & 0 \\ 0 & 1 \end{bmatrix}, \tag{45}$$

the equation of motion for

$$\mathcal{P}\begin{bmatrix} \rho \\ \eta \end{bmatrix} = \begin{bmatrix} \rho \\ 0 \end{bmatrix} \tag{46}$$

is

$$\frac{d}{dt}\,\mathcal{P}\begin{bmatrix} \rho(t) \\ \eta(t) \end{bmatrix} = \mathcal{P}\,\mathcal{L}\,\mathcal{P}\begin{bmatrix} \rho(t) \\ \eta(t) \end{bmatrix} + \mathcal{P}\,\mathcal{L}\,\mathcal{G}(t-0)\,\mathcal{Q}\begin{bmatrix} \rho(0) \\ \eta(0) \end{bmatrix} + \int_0^t \mathcal{P}\,\mathcal{L}\,\mathcal{G}(t-s)\,\mathcal{Q}\,\mathcal{L}\,\mathcal{P}\begin{bmatrix} \rho(s) \\ \eta(s) \end{bmatrix} ds, \tag{47}$$

with

$$\mathcal{G}(t-s) = e^{(t-s)\mathcal{QL}} = \begin{bmatrix} 1 & 0 \\ \frac{B_{10}}{B_{11}}\left(e^{B_{11}(t-s)} - 1\right) & e^{B_{11}(t-s)} \end{bmatrix}. \tag{48}$$

This can be written out explicitly as

$$\frac{d}{dt}\,\rho(t) = B_{00}\,\rho(t) + \int_0^t B_{01}\,e^{B_{11}(t-s)}\,B_{10}\,\rho(s)\,ds + B_{01}\,e^{B_{11}(t-0)}\,\eta(0). \tag{49}$$

The influence of the initial spin state is carried out by the last term but it decays exponentially, since the real part of B_{11} is expected to be negative.

This equation can also be used for obtaining the master equation in the Markovian limit. In that case, the exponential decays faster than the state ρ evolves; $\rho(s)$ may therefore be replaced with $\rho(t)$ and taken out of the integral. Assuming still $\mathrm{Re}(B_{11}) < 0$, the integral may be performed explicitly and the master equation is

$$\frac{d}{dt}\,\rho(t) = \left(B_{00} - B_{01}\,\frac{1}{B_{11}}\,B_{10}\right)\rho(t). \tag{50}$$

3 The Nakajima-Zwanzig equation containing an explicit time convolution can also be transformed into a time-convolutionless form, in which the equation of

motion is

$$\frac{d}{dt}\,\mathcal{P}\begin{bmatrix}\rho(t)\\ \eta(t)\end{bmatrix} = \mathcal{H}(t)\,\mathcal{P}\begin{bmatrix}\rho(t)\\ \eta(t)\end{bmatrix} + \mathcal{I}(t)\,\mathcal{Q}\begin{bmatrix}\rho(0)\\ \eta(0)\end{bmatrix}, \tag{51}$$

with the following definitions:

$$\begin{aligned} \Sigma(t) &= \int_0^t e^{(t-s)\mathcal{QL}}\,\mathcal{Q}\,\mathcal{L}\,\mathcal{P}\,e^{-(t-s)\mathcal{L}}\,ds, \\ \mathcal{H}(t) &= \mathcal{P}\,\mathcal{L}\,\frac{1}{1-\Sigma(t)}\,\mathcal{P}, \\ \mathcal{I}(t) &= \mathcal{P}\,\mathcal{L}\,\frac{1}{1-\Sigma(t)}\,e^{(t-0)\mathcal{QL}}\,\mathcal{Q}. \end{aligned} \tag{52}$$

For the case of a two-level ancilla, these quantities can be evaluated in a closed form, yielding

$$\mathcal{H}_{ij}(t) = \left[B_{00} + \frac{2B_{01}\, B_{10}}{B_{00} - B_{11} + \beta \,\coth\left(\frac{1}{2}\,\beta t\right)} \right] \delta_{i0}\delta_{j0},$$

$$\mathcal{I}_{ij}(t) = \frac{2B_{01}\,\beta\, e^{\frac{1}{2}(B_{00}+B_{11}+\beta)t}}{(-B_{00}+B_{11}+\beta) + (B_{00}-B_{11}+\beta)\,e^{\beta t}} \delta_{i0}\delta_{j1}, \tag{53}$$

with $\beta = \sqrt{(B_{00} - B_{11})^2 + 4B_{01}\,B_{10}}$, and the time-convolutionless equation of motion reads explicitly

$$\frac{d}{dt}\,\rho(t) = \mathcal{H}_{00}(t)\,\rho(t) + \mathcal{I}_{01}(t)\,\eta(0). \tag{54}$$

We want to emphasize here that the choice of the measurement period Δt, as well as of the measurement scheme and of the ancilla meter basis allows one to modify these parameters to yield a desired evolution of the dynamic system.

3. Harmonic oscillator coupled to a spin

The general scheme presented in the previous sections is now applied to the special case of a harmonic oscillator coupled to a two-level system (spin) by means of a time-independent interaction. In the absence of measurements, the time evolution is generated by the Hamiltonian

$$\hat{H} = \hbar\omega\left(\hat{n} + \frac{1}{2}\right) + \hbar\beta\,\hat{\sigma}_x + \alpha\, m\,\omega\,\hat{x}^2\,\hat{\sigma}_z. \tag{55}$$

This model may describe, for example, a spin $\frac{1}{2}$-particle confined to move in a one-dimensional harmonic potential whose spin is subject to a harmonic magnetic field or a two-level atomic system interacting with a single mode of a cavity field. It is of interest here as example of an interaction between a discrete- and a continuous-variable system.

3.1 Free evolution

We assume a parametric region where the internal frequencies of the oscillator and of the spin yield the fastest time scales,

$$\omega^{-1},\ \beta^{-1} \ll \Delta t, \tag{56}$$

where Δt denotes the evolution period, *i.e.*, time between measurements. Provided that there are no resonances between the oscillator and the spin, *i.e.*, if $\Gamma = \omega/\beta$ is not close to some fraction p/q with small integers p and q, the Hamiltonian (55) may be in the Dirac (interaction) picture approximated by an effective nondemolition Hamiltonian. To see this, we apply

$$\hat{U}_{\text{free}} = \exp\left\{-i\left[\omega\left(\hat{n}+\frac{1}{2}\right)+\beta\,\hat{\sigma}_x\right]\Delta t\right\} \tag{57}$$

to the interaction part of the Hamiltonian (55),

$$\begin{aligned}
\hat{H}_D^{\text{int}}(t) &= \hat{U}_{\text{free}}^{\dagger}\,\alpha\,m\,\omega\,\hat{x}^2\,\hat{\sigma}_z\,\hat{U}_{\text{free}} \\
&= \alpha\,m\,\omega\left(e^{i\omega\,\Delta t\,\hat{n}}\,\hat{x}^2\,e^{-i\omega\,\Delta t\,\hat{n}}\right)\left(e^{i\beta\,\Delta t\,\hat{\sigma}_x}\,\hat{\sigma}_z\,e^{-i\beta\,\Delta t\,\hat{\sigma}_x}\right) \\
&= \frac{\hbar\,\alpha}{2}\left(e^{2i\,\omega\,\Delta t}\,\hat{a}^{\dagger 2}+e^{-2i\,\omega\,\Delta t}\,\hat{a}^2+2\hat{n}+1\right) \\
&\quad\times\left[\hat{\sigma}_z\,\cos(2\beta\,\Delta t)+\hat{\sigma}_y\,\sin(2\beta\,\Delta t)\right].
\end{aligned} \tag{58}$$

Here, we substituted $\hat{x} = \sqrt{\hbar/(2m\,\omega)}\,(\hat{a}+\hat{a}^{\dagger})$ and applied the relations

$$\begin{aligned}
\alpha^{\hat{n}}\,f(\hat{a},\,\hat{a}^{\dagger})\,\alpha^{-\hat{n}} &= f(\alpha^{-1}\hat{a},\,\alpha\,\hat{a}^{\dagger}), \\
e^{\alpha\hat{\sigma}_x}\,\hat{\sigma}_z\,e^{-\alpha\,\hat{\sigma}_x} &= \cosh(2\alpha)\,\hat{\sigma}_z - i\,\sinh(2\alpha)\,\hat{\sigma}_y
\end{aligned} \tag{59}$$

for the oscillator and spin part, respectively. From the periodic second factor in (58) we see immediately that in the limit (56), the first order average vanishes,

$$\frac{1}{\Delta t}\int_t^{t+\Delta t} dt_1\;\hat{H}_{\text{D}}^{\text{int}}(t_1) \approx 0, \tag{60}$$

whereas the second order average yields a new Hamiltonian $\hat{H}$, which replaces (55),

$$\begin{aligned}
&\frac{1}{i\hbar\,\Delta t}\int_t^{t+\Delta t} dt_1 \int_t^{t_1} dt_2\;\hat{H}_D^{\text{int}}(t_1)\;\hat{H}_D^{\text{int}}(t_2) \\
&\approx \frac{\hbar\,\alpha^2}{2\beta}\left[\bar{\omega}\left(\hat{n}+\frac{1}{2}\right)+\bar{\beta}\hat{\sigma}_x+\bar{\alpha}\hat{n}(\hat{n}+1)\hat{\sigma}_x\right] = \hat{H},
\end{aligned} \tag{61}$$

where

$$\bar{\omega} = \left(\Gamma^{-1}-\Gamma\right)^{-1},\ \bar{\beta} = \frac{3-\Gamma^2}{4-4\Gamma^2}\ \text{ and }\ \bar{\alpha} = \frac{3-2\Gamma^2}{2-2\Gamma^2}. \tag{62}$$

Physically, we consider the parametric regime where coupling is so weak that it hardly induces any transition during Δt and, consequently, the effective Hamiltonian has a nondemolition form in the sense defined in the previous section. Since the interaction term commutes with both system Hamiltonians, the (expectation values of the) subsystem energies are constants of time. This is, in fact, a consequence of the assumed lack of resonances between the oscillator and the spin.

In contrast to the original Hamiltonian given by Eq. (55), in Eq. (61) we observe a functional dependence of the form $\hat{H} = f_1(\hat{H}_1) + f_2(\hat{H}_2) + f_{12}(\hat{H}_1,\ \hat{H}_2)$, where $\hat{H}_{1,\ 2}$ are the original (commuting) Hamiltonians of the free harmonic oscillator and spin, respectively, as given in Eq. (55). Of particular importance for us is the third term $f_{12}(\hat{H}_1,\ \hat{H}_2) \sim \hat{n}(\hat{n}+1)\hat{\sigma}_x$, describing the interaction between the spatial and the spin part, as seen by an observer co-rotating appropriately with the subsystems. In the semiclassical limit, *i.e.*, for large photon numbers, it may be approximated by $f_{12}(\hat{H}_1,\ \hat{H}_2) \sim \hat{n}^2\ \hat{\sigma}_x$.

For notational simplicity we consider from now on a Hamiltonian

$$\hat{H} = \hbar\left[a\,\hat{n} + b\,\hat{\sigma}_x + c\,\hat{n}(\hat{n}+1)\,\hat{\sigma}_x\right], \tag{63}$$

where a, b, and c are arbitrary real constants. Over time interval Δt it generates the unitary evolution $\mathcal{U}_{\Delta t}\ \varrho = \hat{U}_{\Delta t}\ \varrho\ \hat{U}^{\dagger}_{\Delta t}$, where

$$\hat{U}_{\Delta t} = e^{\frac{\hat{H}\ \Delta t}{i\hbar}} = \frac{\cos\left[(b + c\,\hat{N})\Delta t\right] - i\,\hat{\sigma}_x\ \sin\left[(b + c\,\hat{N})\Delta t\right]}{e^{ia\ \hat{n}\ \Delta t}} \tag{64}$$

with $\hat{N} = \hat{n}(\hat{n}+1)$. The state-operator elements $\varrho^{(m,\ n)} = {}_1\langle m|\varrho|n\rangle_1$ are decoupled due to the nondemolition nature of the interaction, and they evolve as

$$\begin{aligned}
\varrho^{(m,\ n)}(t+\Delta t) &= {}_1\langle m|\hat{U}_{\Delta t}\ \varrho(t)\ \hat{U}^{\dagger}_{\Delta t}|n\rangle_1 \\
&= \sum_{p,\ q} {}_1\langle m|\hat{U}_{\Delta t}|p\rangle_1 \langle p|\varrho(t)|q\rangle_1 \langle q|\hat{U}^{\dagger}_{\Delta t}|n\rangle_1 \\
&= \hat{U}^{m}_{\Delta t}\ \varrho^{(m,\ n)}(t)\ \hat{U}^{n\dagger}_{\Delta t} = \mathcal{U}^{(m,\ n)}_{\Delta t}\{\varrho^{(m,\ n)}(t)\},
\end{aligned} \tag{65}$$

where

$$
\begin{aligned}
\hat{U}^m_{\Delta t} &= {}_1\langle m|\hat{U}(\Delta t)|n\rangle_1 \\
&= \delta_{mn}\, \frac{\cos\left[(b + c\, m(m+1))\Delta t\right] - i\, \hat{\sigma}_x\, \sin\left[(b + c\, m(m+1))\Delta t\right]}{e^{ia\, m\, \Delta t}} .
\end{aligned}
\tag{66}
$$

3.2 Evolution within one measurement cycle

We consider the state ϱ of the total system and assume that in time intervals Δt a given meter basis $\{|\Psi_k\rangle\}$, $k = 0,\ 1$ is measured on the discrete subsystem without knowing the measurement result. We may now apply the methods developed in Sec. 2. Evolution over one cycle is then obtained from Eq. (33)

$$
\varrho^{(m,\, n)}_{kk}(t + \Delta t) = \sum_l \mathcal{U}^{(m,\, n)}_{\Delta t,\, kl}\; \varrho^{(m,\, n)}_{ll}(t), \tag{67}
$$

with

$$
\mathcal{U}^{(m,\, n)}_{\Delta t,\, kl} = {}_2\langle\Psi_k|\hat{U}^m_{\Delta t}|\Psi_l\rangle_2 \langle\Psi_l|\hat{U}^{n\dagger}_{\Delta t}|\Psi_k\rangle_2 = U^m_{\Delta t,\, kl}\, U^{n\star}_{\Delta t,\, kl},
$$

and (68)

$$
U^m_{\Delta t,\, kl} = e^{-ia\, m\, \Delta t}\left(\delta_{kl}\, \cos\zeta_m - i\, \gamma_{kl}\, \sin\zeta_m\right).
$$

Here, $\zeta_m = (b + c\, m(m+1))\Delta t$ and $\gamma_{kl} = {}_2\langle\Psi_k|\hat{\sigma}_x|\Psi_l\rangle_2$. We may write $\gamma_{00} = \cos\theta$, $\gamma_{01} = e^{i\varphi}\, \sin\theta$, $\gamma_{10} = e^{-i\varphi}\, \sin\theta$ and $\gamma_{11} = -\cos\theta$ where θ is the angle between the measurement (meter) and quantization axes (here chosen to be the x-axis) and φ is the azimuthal angle around the measurement axis.

We can readily evaluate the matrix elements of $\mathcal{U}^{(m,\, n)}_{\Delta t,\, kl}$, and the diagonal and off-diagonal elements are, respectively,

$$
\begin{aligned}
\mathcal{U}^{(m,\, n)}_{\Delta t,\, kk} &= e^{-i\, a(m-n)\, \Delta t}\left[\cos\zeta_m\, \cos\zeta_n + |\gamma_{kk}|^2\, \sin\zeta_m\, \sin\zeta_n \right. \\
&\qquad \left. -i\, \gamma_{kk}\, \sin(\zeta_m - \zeta_n)\right], \\
\mathcal{U}^{(m,\, n)}_{\Delta t,\, k\neq l} &= e^{-i\, a(m-n)\, \Delta t}\, |\gamma_{kl}|^2\, \sin\zeta_m\, \sin\zeta_n,
\end{aligned}
\tag{69}
$$

and the exact difference equation for the time evolution is given by

$$\begin{bmatrix} \varrho_{00}^{(m,\,n)}(t+\Delta t) \\ \\ \varrho_{11}^{(m,\,n)}(t+\Delta t) \end{bmatrix} = \begin{bmatrix} \mathcal{U}_{\Delta t,\,00}^{(m,\,n)} & \mathcal{U}_{\Delta t,\,01}^{(m,\,n)} \\ \\ \mathcal{U}_{\Delta t,\,10}^{(m,\,n)} & \mathcal{U}_{\Delta t,\,11}^{(m,\,n)} \end{bmatrix} \begin{bmatrix} \varrho_{00}^{(m,\,n)}(t) \\ \\ \varrho_{11}^{(m,\,n)}(t) \end{bmatrix}. \tag{70}$$

Note that if the measurement basis coincides with the eigenbasis of the spin Hamiltonian, we have $\gamma_{kl} = \pm\delta_{kl}$, whence

$$\mathcal{U}_{\Delta t,\,kl}^{(m,\,n)} = \delta_{kl}\; e^{-i\;[a\;(m-n)\;\Delta t\;\pm\;(\zeta_m-\zeta_n)]}, \tag{71}$$

and the different k and l states decouple. If, on the other hand, the meter axis is orthogonal to the quantization axis, we have $\gamma_{kk} = 0$, and the evolution simplifies to

$$\begin{aligned} \mathcal{U}_{\Delta t,\,kk}^{(m,\,n)} &= e^{-i\;a(m-n)\;\Delta t}\;\cos\zeta_m\;\cos\zeta_n, \\ \mathcal{U}_{\Delta t,\,k\neq l}^{(m,\,n)} &= e^{-i\;a(m-n)\;\Delta t}\;\sin\zeta_m\;\sin\zeta_n. \end{aligned} \tag{72}$$

Therefore, the measurement axis should be chosen neither parallel nor orthogonal to the quantization axis in order to obtain a nontrivial contribution to the dynamic system.

Similarly to Eq. (39), we change into the basis

$$\begin{bmatrix} \rho^{(m,\,n)}(t) \\ \\ \eta^{(m,\,n)}(t) \end{bmatrix} = \begin{bmatrix} \varrho_{00}^{(m,\,n)}(t) + \varrho_{11}^{(m,\,n)}(t) \\ \\ \varrho_{00}^{(m,\,n)}(t) - \varrho_{11}^{(m,\,n)}(t) \end{bmatrix}, \tag{73}$$

which obeys the equation of motion

$$\begin{bmatrix} \rho^{(m,\,n)}(t+\Delta t) \\ \\ \eta^{(m,\,n)}(t+\Delta t) \end{bmatrix} = \begin{bmatrix} \mathcal{A}_{\Delta t,\,00}^{(m,\,n)} & \mathcal{A}_{\Delta t,\,01}^{(m,\,n)} \\ \\ \mathcal{A}_{\Delta t,\,10}^{(m,\,n)} & \mathcal{A}_{\Delta t,\,11}^{(m,\,n)} \end{bmatrix} \begin{bmatrix} \rho^{(m,\,n)}(t) \\ \\ \eta^{(m,\,n)}(t) \end{bmatrix} \tag{74}$$

with

$$
\begin{aligned}
\mathcal{A}^{(m,\,n)}_{\Delta t,\,00} &= e^{-i\,a(m-n)\,\Delta t}\,\frac{\mathcal{U}^{(m,\,n)}_{\Delta t,\,00}+\mathcal{U}^{(m,\,n)}_{\Delta t,\,01}+\mathcal{U}^{(m,\,n)}_{\Delta t,\,10}+\mathcal{U}^{(m,\,n)}_{\Delta t,\,11}}{2}\\
&= e^{-i\,a(m-n)\,\Delta t}\,\cos(\zeta_m-\zeta_n),\\
\mathcal{A}^{(m,\,n)}_{\Delta t,\,01} &= e^{-i\,a(m-n)\,\Delta t}\,\frac{\mathcal{U}^{(m,\,n)}_{\Delta t,\,00}-\mathcal{U}^{(m,\,n)}_{\Delta t,\,01}+\mathcal{U}^{(m,\,n)}_{\Delta t,\,10}-\mathcal{U}^{(m,\,n)}_{\Delta t,\,11}}{2}\\
&= -i\,e^{-i\,a(m-n)\,\Delta t}\,\cos\theta\,\sin(\zeta_m-\zeta_n),\\
\mathcal{A}^{(m,\,n)}_{\Delta t,\,10} &= e^{-i\,a(m-n)\,\Delta t}\,\frac{\mathcal{U}^{(m,\,n)}_{\Delta t,\,00}+\mathcal{U}^{(m,\,n)}_{\Delta t,\,01}-\mathcal{U}^{(m,\,n)}_{\Delta t,\,10}-\mathcal{U}^{(m,\,n)}_{\Delta t,\,11}}{2}\\
&= -i\,e^{-i\,a(m-n)\,\Delta t}\,\cos\theta\,\sin(\zeta_m-\zeta_n),\\
\mathcal{A}^{(m,\,n)}_{\Delta t,\,11} &= e^{-i\,a(m-n)\,\Delta t}\,\frac{\mathcal{U}^{(m,\,n)}_{\Delta t,\,00}-\mathcal{U}^{(m,\,n)}_{\Delta t,\,01}-\mathcal{U}^{(m,\,n)}_{\Delta t,\,10}+\mathcal{U}^{(m,\,n)}_{\Delta t,\,11}}{2}\\
&= e^{-i\,a(m-n)\,\Delta t}\left(\cos\zeta_m\,\cos\zeta_n+\cos 2\theta\,\sin\zeta_m\,\sin\zeta_n\right),
\end{aligned}
\tag{75}
$$

where we have used

$$
\begin{aligned}
&\frac{|\gamma_{00}|^2+|\gamma_{01}|^2+|\gamma_{10}|^2+|\gamma_{11}|^2}{2}=1,\\
&\frac{|\gamma_{00}|^2-|\gamma_{01}|^2-|\gamma_{10}|^2+|\gamma_{11}|^2}{2}=\cos 2\theta,\\
&\text{and}\quad \gamma_{11}=-\,\gamma_{00}=-\,\cos\theta.
\end{aligned}
\tag{76}
$$

3.3 Quasi-continuous description

Parallel to Eqs. (35) and (36), the differential equation for the system evolution is obtained as

$$
\frac{d}{dt}\begin{bmatrix}\rho^{(m,\,n)}(t)\\ \eta^{(m,\,n)}(t)\end{bmatrix}=\begin{bmatrix}\mathcal{L}^{(m,\,n)}_{\Delta t,\,00} & \mathcal{L}^{(m,\,n)}_{\Delta t,\,01}\\ \mathcal{L}^{(m,\,n)}_{\Delta t,\,10} & \mathcal{L}^{(m,\,n)}_{\Delta t,\,11}\end{bmatrix}\begin{bmatrix}\rho^{(m,\,n)}(t)\\ \eta^{(m,\,n)}(t)\end{bmatrix},
\tag{77}
$$

where the generator is resolved to be

$$\mathcal{L}_{\Delta t}^{(m,\,n)} = \begin{bmatrix} \mathcal{L}_{\Delta t,\,00}^{(m,\,n)} & \mathcal{L}_{\Delta t,\,01}^{(m,\,n)} \\ \mathcal{L}_{\Delta t,\,10}^{(m,\,n)} & \mathcal{L}_{\Delta t,\,11}^{(m,\,n)} \end{bmatrix} = \frac{1}{\Delta t} \ln \begin{bmatrix} \mathcal{A}_{\Delta t,\,00}^{(m,\,n)} & \mathcal{A}_{\Delta t,\,01}^{(m,\,n)} \\ \mathcal{A}_{\Delta t,\,10}^{(m,\,n)} & \mathcal{A}_{\Delta t,\,11}^{(m,\,n)} \end{bmatrix}$$
$$= \frac{1}{\Delta t} \left(\frac{\mathcal{A}_{\Delta t}^{(m,\,n)} - \lambda_+ \mathbb{I}}{\lambda_- - \lambda_+} \ln \lambda_- + \frac{\mathcal{A}_{\Delta t}^{(m,\,n)} - \lambda_- \mathbb{I}}{\lambda_+ - \lambda_-} \ln \lambda_+ \right). \tag{78}$$

Here, $\mathcal{A}_{\Delta t}^{(m,\,n)}$ (assumed to be regular) is the 2×2 matrix consisting of elements $\mathcal{A}_{\Delta t,\,ij}^{(m,\,n)}$ and $\lambda_\pm$ are its eigenvalues,

$$\lambda_\pm = e^{-ia(m-n)\Delta t} \left[\frac{\cos(\zeta_m - \zeta_n) + \cos\zeta_m \cos\zeta_n + \cos 2\theta \sin\zeta_m \sin\zeta_n}{2} \right.$$
$$\left. \pm \sqrt{\sin^4\theta \sin^2\zeta_m \sin^2\zeta_n - \cos^2\theta \sin^2(\zeta_m - \zeta_n)} \right]. \tag{79}$$

Absolute values of $\lambda_\pm$ are smaller or equal to one by construction (they are eigenvalues of an operator consisting of unitary transformations and projections), and, therefore, the logarithm operator has eigenvalues with real part smaller than or equal to zero.

3.4 Zeno-limit of the evolution

In order to find the Zeno-limit evolution, time evolution need to be resolved to the first order in Δt. The evolution-matrix elements are then

$$\begin{aligned} \mathcal{A}_{\Delta t,\,00}^{(m,\,n)} &= 1 - i\,a(m-n)\,\Delta t + O(\Delta t^2), \\ \mathcal{A}_{\Delta t,\,01}^{(m,\,n)} &= \frac{i}{2}(-\gamma_{00} + \gamma_{11})\,(\zeta_m - \zeta_n) + O(\Delta t^2), \\ \mathcal{A}_{\Delta t,\,10}^{(m,\,n)} &= \frac{i}{2}(-\gamma_{00} + \gamma_{11})\,(\zeta_m - \zeta_n) + O(\Delta t^2), \\ \mathcal{A}_{\Delta t,\,11}^{(m,\,n)} &= 1 - i\,a(m-n)\,\Delta t + O(\Delta t^2), \end{aligned} \tag{80}$$

and the generator can be found to be

$$\mathcal{L} = \lim_{\Delta t \to 0} \mathcal{L}_{\Delta t}^{(m,\,n)} = i \begin{bmatrix} -a(m-n) & C^{(m,\,n)} \\ C^{(m,\,n)} & -a(m-n) \end{bmatrix}, \tag{81}$$

where $C^{(m,\,n)} = -\cos\theta\; c\,[m(m+1) - n(n+1)]$. The equation of motion (77) can be solved straightforwardly to yield

$$\begin{bmatrix} \rho^{(m,\,n)} \\ \eta^{(m,\,n)} \end{bmatrix} = e^{-i\,a(m-n)\,t} \times \begin{bmatrix} \cos(C^{(m,\,n)}t) & i\,\sin(C^{(m,\,n)}\,t) \\ i\,\sin(C^{(m,\,n)}\,t) & \cos(C^{(m,\,n)}\,t) \end{bmatrix} \begin{bmatrix} \rho^{(m,\,n)}(0) \\ \eta^{(m,\,n)}(0) \end{bmatrix}. \tag{82}$$

We note that, using the above definitions, the system evolution in the Zeno limit is identically obtained from Eq. (20). Now

$$\hat{H}_{ii} = {}_2\langle\Psi_i|\hat{H}|\Psi_i\rangle_2 = \hbar\,[a\,\hat{n} + b\,\gamma_{ii} + c\,\hat{n}(\hat{n}+1)\,\gamma_{ii}]\,, \tag{83}$$

and

$$\frac{d}{dt}\,\varrho_{ii}^{(m,\,n)} = -i\,\{a(m-n) + c\,\gamma_{ii}\,[m(m+1) - n(n+1)]\}\;\varrho_{ii}^{(m,\,n)}. \tag{84}$$

The evolution generator is diagonal in the meter basis whereas in the $[\rho,\ \eta]^T$ basis it is given by Eq. (81).

3.5 Generation of a Schrödinger cat-like state

A Schrödinger cat-like state is a superposition of two macroscopically distinguishable classical states, [Schrödinger 1935 (a)], which for the harmonic oscillator are represented by strongly excited and sufficiently well separated (thus orthogonal) coherent states. To evolve a coherent state into a superposition, we may apply a unitary operator

$$\exp\left\{i\,\frac{\pi}{2}\,k\,\hat{n}^2\right\} = \frac{1+i^k}{2}(+1)^{\hat{n}} + \frac{1-i^k}{2}(-1)^{\hat{n}} \tag{85}$$

($k = 0, \pm1, \ldots$), since $e^{i\varphi\,\hat{n}}\,|\alpha\rangle = |e^{i\varphi}\,\alpha\rangle$. This possibility has been considered theoretically in the example of optical fibers and electrical circuits, [Milburn 1986 (b); Milburn 1986 (a); Yurke 1986]. In an optical medium, a Hamiltonian $\hbar\,\kappa\,\hat{n}^2$ describes in lowest order the dependence of the refractive index on the electric field strength (optical Kerr effect). Since κ is typically many orders of magnitude smaller than unity, an accumulation over large optical path lengths is required and decoherence has so far prevented the experimental observation of a Schrödinger cat-like state in an optical fiber. It has however been successfully prepared and its decoherence observed for a mesoscopic cavity field, [Brune 1996; Bertet 2002; Auffeves 2003].

Here we analyze a measurement-assisted generation of a superposition state. This is in line with the intentions of earlier proposals such as preparation by means of continuous photodetection, [Ogawa 1991], or via state reduction in a Mach-Zehnder interferometer containing a Kerr medium, [Gerry 1999]. In order to implement the Zeno-limit evolution given by Eq. (85), we assume in the Hamiltonian (63) for simplicity $b = 0$ and the parameter c to be sufficiently large such that for some detected ancilla state $|\Psi_0\rangle$ the condition $\langle\Psi_0|\hat{\sigma}_x|\Psi_0\rangle = \gamma_{00} = \cos\theta = -\,c^{-1}\,a$ can be fulfilled. This yields for

$$t_k = \left(k + \frac{1}{4}\right)\frac{2\pi}{\gamma_{00}\;c} = -\,\frac{2\pi}{a}\left(k + \frac{1}{4}\right) \tag{86}$$

a unitary evolution given by

$$\begin{aligned} \hat{U} &= \exp\left\{\frac{\langle\Psi_0|\hat{H}|\Psi_0\rangle\;t_k}{i\hbar}\right\} = \exp\left\{-i\,c\,\hat{n}^2\,\gamma_{00}\,t_k\right\} \\ &= \exp\left\{-i\,\hat{n}^2\left(k + \frac{1}{4}\right)2\pi\right\}. \end{aligned} \tag{87}$$

Consequently, at times t_k the desired superposition is obtained, where the index $k = 0,\ 1,\ \ldots$ marks the periodicity intervals of (87). This defines the evolution period

$$T = |t_{k+1} - t_k| = \frac{2\pi}{|\gamma_{00}\;c|} = \frac{2\pi}{|a|} \tag{88}$$

as given by Eq. (86). The Zeno-limit is however only a mathematical idealization. In practice, the time interval Δt between two subsequent measurements always remains finite and, therefore, the system evolution is subject to decoherence. As a consequence, the state begins to depart from the superposition as k increases. The duration $\Delta t = T/N$ of a measurement cycle can be tuned by choosing the number N of measurements per evolution period (88) to be sufficiently large. A simulation has been carried out to estimate the size of Δt tolerated to prevent decoherence over the minimal time t_0 required to build up a superposition.

We choose the meter states $|\Psi_0\rangle$ and $|\Psi_1\rangle$ to be defined in terms of a complex parameter z, so that (in the $\hat{\sigma}_z$-representation)

$$\Psi_0 = \frac{1}{\sqrt{1+|z|^2}}\begin{pmatrix} 1 \\ z \end{pmatrix} \quad \text{and} \quad \Psi_1 = \frac{1}{\sqrt{1+|z|^2}}\begin{pmatrix} z^\star \\ -\,1 \end{pmatrix}. \tag{89}$$

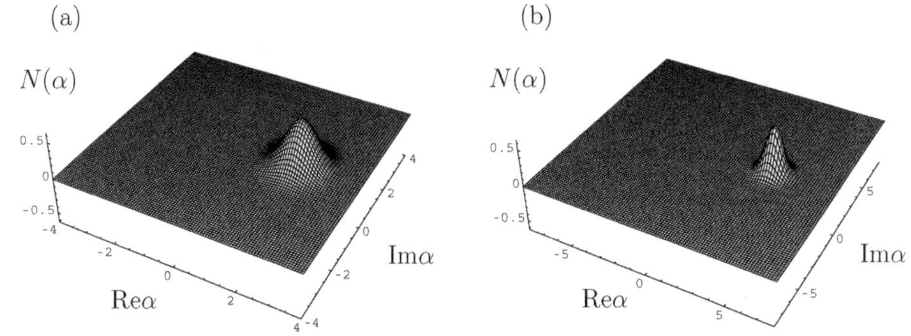

Figure 2. Wigner function $N(\alpha)$ of the initial coherent states $|\beta\rangle$ of the oscillator for coherent amplitudes (a) $\beta = 2$ and (b) $\beta = 4$.

Note that for $z = \pm 1$ these are eigenstates of $\hat{\sigma}_x$, for $z = \pm i$ of $\hat{\sigma}_y$, and for $z = 0, \infty$ of $\hat{\sigma}_z$, whereas the expectation value of the spin is given by

$$\gamma_{00} = \frac{z + z^\star}{1 + |z|^2}. \tag{90}$$

We consider the initial state $\varrho_{12}(t = 0) = |\beta\rangle_1\langle\beta| \otimes |\Psi_0\rangle_2\langle\Psi_0|$, where $|\beta\rangle$ is a coherent state, so that the initial condition for individual matrix elements becomes

$$\varrho_{kl}^{(m,\,n)}(0) = \frac{\beta^m \beta^{\star\,n}}{\sqrt{m!\,n!}} e^{-|\beta|^2}\,\delta_{k0}\delta_{l0}. \tag{91}$$

For simulations, we have chosen coherent amplitudes $\beta = 2$ and $\beta = 4$, see Fig. 2, in order to compare the decoherence rate, and an initial ancilla state given by $z = 0.5$, thus avoiding eigenstates of Pauli matrices. Figs. 2 to 4 show the Wigner function

$$\begin{aligned} N(\alpha) = {} & \tfrac{2}{\pi} \sum_{m,\,n} \rho^{(m,n)} \left(\frac{\nu!}{\mu!}\right)^{1/2} (-1)^\nu (2\,|\alpha|)^{\mu-\nu} L_\nu^{\mu-\nu}(4\,|\alpha|^2) \\ & \times \exp\left\{i(n-m)\,\varphi_\alpha - 2\,|\alpha|^2\right\} \end{aligned} \tag{92}$$

of the respective states, where $\mu = \max[m,\, n]$, $\nu = \min[m,\, n]$, and L is the generalized Laguerre polynomial.

We consider the generation of the Schrödinger cat-like state and the coherence loss as function of both the initial coherent amplitude β and the measurement cycle time Δt. We study the cases with $\Delta t = 10^{-4}\,T$ and $\Delta t = 10^{-9}\,T$, shown in Figs. 3 and 4, respectively. 10^4 measurement cycles per evolution period is hardly enough to produce the cat-like state for $\beta = 2$ during the first period and the loss of coherencexdecoherence is obvious for the second period. For $\beta = 4$ the cat-state decoheres much before it is even created. For the case of 10^9 measurements per period the cat-state is created in both cases during the first period. After 10^4 periods the coherence is still preserved for $\beta = 2$ whereas it is lost for $\beta = 4$. The rate of coherence loss is therefore seen to be highly nonlinear in the coherent amplitude.

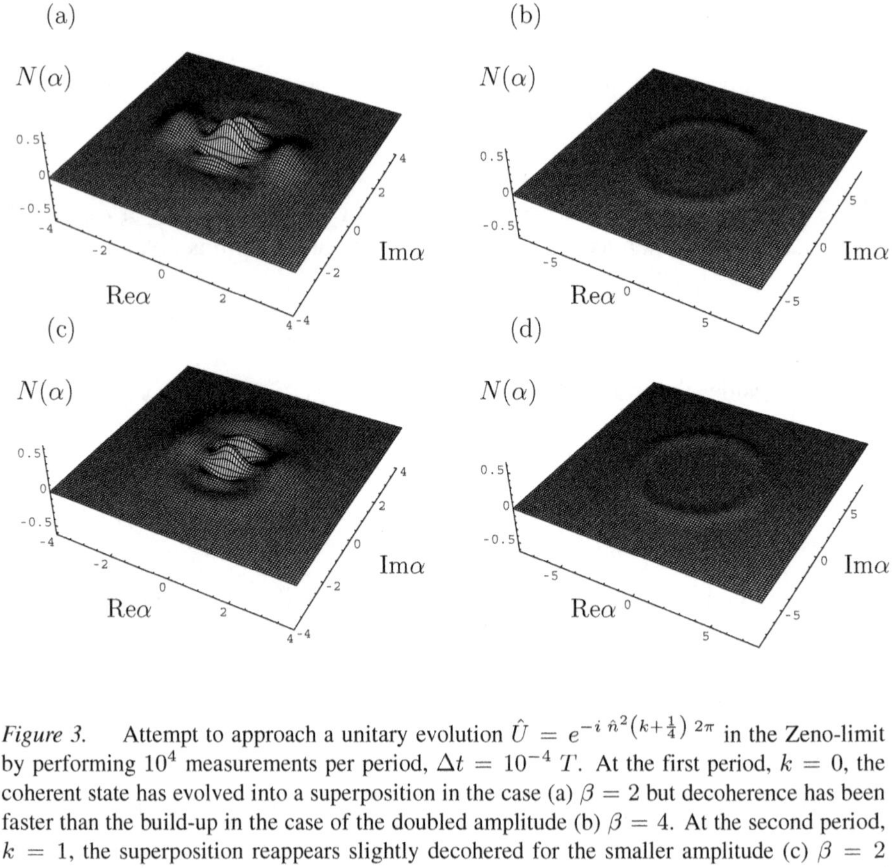

Figure 3. Attempt to approach a unitary evolution $\hat{U} = e^{-i\,\hat{n}^2\left(k+\frac{1}{4}\right)\,2\pi}$ in the Zeno-limit by performing 10^4 measurements per period, $\Delta t = 10^{-4}\,T$. At the first period, $k = 0$, the coherent state has evolved into a superposition in the case (a) $\beta = 2$ but decoherence has been faster than the build-up in the case of the doubled amplitude (b) $\beta = 4$. At the second period, $k = 1$, the superposition reappears slightly decohered for the smaller amplitude (c) $\beta = 2$ whereas no change has occurred for (d) $\beta = 4$ since in this case the asymptotic state had already been obtained during the previous period.

The eventual steady state is obtained directly from Eq. (70). Since, for $m \neq n$ (and the measurement angle $\theta \neq 0$) the eigenvalues of the time-step operator are smaller than one but not equal to zero, the steady state is given by

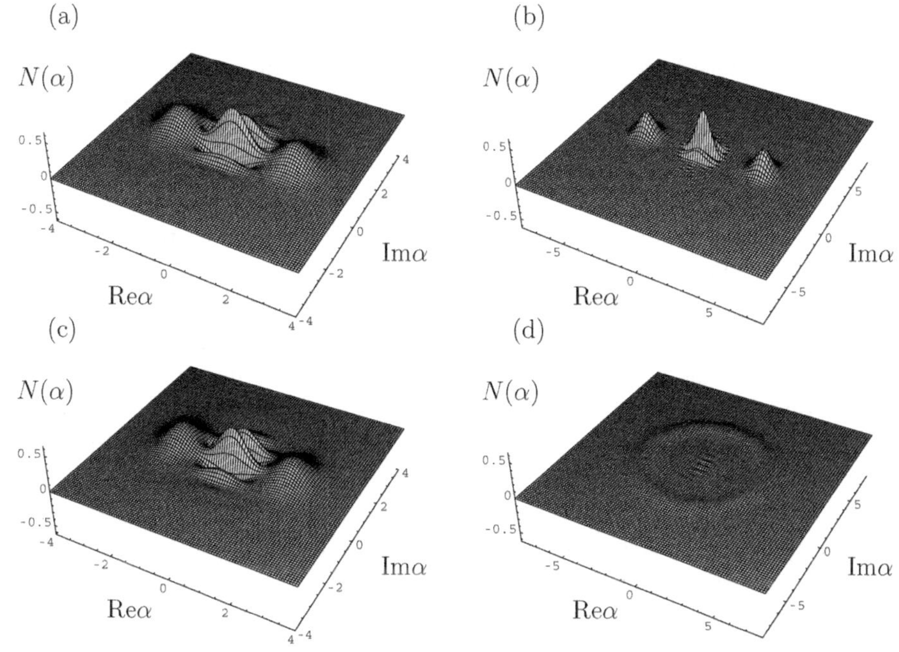

Figure 4. In order to obtain a superposition even for the doubled amplitude, we have to increase the number of measurements per period by five orders of magnitude, so that we now realize $\Delta t = 10^{-9}T$. At the first period, $k = 0$, the coherent state has evolved into a superposition for both (a) $\beta = 2$ and (b) $\beta = 4$. After 10^4 periods, $k = 10^4$, the superposition is still present for (c) $\beta = 2$ but is decohered for the (d) $\beta = 4$ case.

$[0\ 0]^T$. Therefore, the coherences tend to zero whereas the diagonal elements of the oscillator stay constant, and the initial coherent state $|\beta\rangle$ evolves into a Poissonian mixture of Fock states

$$\begin{aligned} \rho_{\text{asympt}} &= \frac{1}{2\pi}\int_0^{2\pi} d\varphi_\beta\, |\beta\rangle\langle\beta| = e^{-|\beta|^2}\,\frac{|\beta|^{2\hat{n}}}{\hat{n}!} \\ &\overset{|\beta|\gg 1}{\approx} \frac{1}{\sqrt{2\pi\,|\beta|^2}}\exp\left\{-\frac{\left(\hat{n}-|\beta|^2\right)^2}{2|\beta|^2}\right\}, \end{aligned} \tag{93}$$

which, in the semiclassical limit, resembles a Gaussian mixture.

Conclusions

We have considered repeated projective measurements on an ancilla as a tool for manipulating the evolution of a dynamic quantum system of interest. Due to an interaction between the dynamic system and the ancilla, the nonunitary evolution of the ancilla extends equally to the dynamic system, but close to the Zeno-limit the coherence of the dynamic system may still be preserved. Of particular interest here are systems coupled with a nondemolition interaction, since they can be described in an essentially simplified manner. Depending on the dimension N_{a} of the ancilla, individual elements of the reduced state of the dynamic part obey master equations that are $N_{\mathrm{a}}^{\mathrm{th}}$ order differential equations in time. Equivalently, the master equations can be written in the Nakajima-Zwanzig or time convolutionless form.

We have applied the above approach to a harmonic oscillator coupled to a spin by means of a photon number - nondemolition Hamiltonian. The spin is being measured periodically, whereas the measurement outcome is ignored. For a sufficiently high measurement frequency, the state of the harmonic oscillator evolves in a unitary manner which can be influenced by a choice of the meter basis. In practice however, the time interval Δt between two subsequent measurements always remains finite and, therefore, the system evolution is subject to decoherence. As an example of application, we have simulated the evolution of an initially coherent state of the harmonic oscillator into a Schrödinger cat-like superposition state. The state departs from the superposition as time increases. The simulations confirm that the decoherence rate increases dramatically with the amplitude of the initial coherent state, thus destroying very rapidly all macroscopic superposition states.

NON-EXPONENTIAL MOTIONAL DAMPING OF IMPURITY ATOMS IN BOSE-EINSTEIN CONDENSATES

I. E. Mazets[1, 2] and G. Kurizki[2]

[1] *Ioffe Physico-Technical Institute, St.Petersburg 194021, Russia*
mazets@astro.ioffe.rssi.ru

[2] *Chemical Physics Department, Weizmann Institute of Science, Rehovot 76100, Israel*

Abstract We demonstrate that the damping of the motion of an impurity atom injected at a supercritical velocity into a Bose-Einstein condensate can exhibit appreciable deviations from the exponential law on time scales of 10^{-5} s.

Keywords: Non-exponential decay, Bose-Einstein condensate, non-mean-field effects

The decay of an unstable quantum system has long been known to deviate from exponential law for both very short and very long times. The short-time deviation from the exponential decay gives rise to either slowdown or speedup of the decay by frequently repeated measurements, known, respectively, as the quantum Zeno effect (QZE), [Khalfin 1957-58; Winter 1961; Misra 1977; Fonda 1978; Itano 1990], and the anti-Zeno effect (AZE), [Lane 1983; Schieve 1989; Kofman 1996; Facchi 2001 (a)]. A general unified theory of the QZE and AZE has been given in Refs. [Kofman 2000; Kofman 2001 (a)]. Among the proposed realizations of the QZE / AZE are photodetachement of negative ions, [Rzazewski 1982], and radiative decay of excited atoms in cavities, [Kofman 1996], photonic band gap structures, [Kofman 1994], or in the presense of a Bose-Einstein condensate (BEC), [Mazets 1997]. The first unambiguous observation of non exponential decay in an unstable quantum system has been reported in Ref. [Wilkinson 1997], followed by the demonstration of the QZE and AZE in Ref. [Fischer 2001]. These experiments have measured the escape of cold atoms from wells of an accelerating periodic potential induced by a standing light wave with varying frequency. Qualitatively, escape from a trapping potential resembles nuclear alpha-decay, [Gamow 1928; Condon 1928], rather than decay via quanta emission into a bath, as in radiative decay, [Heitler 1954] or in beta-decay, [Fermi 1950].

V.M. Akulin et al. (eds.), Decoherence, Entanglement and Information Protection in Complex Quantum Systems, 307–314.

The difficulty impeding the demonstration of non exponential decay via quanta emission has been the short non-Markovian (memory) time of the relevant (electromagnetic or leptonic) bath, which is the time-scale of the effect, [Kofman 2000; Kofman 2001 (a)]. Here we propose the emission of sound quanta (phonons) by atoms into a BEC as a candidate process for the observation of non-exponential relaxation, taking advantage of the rather long memory time of the condensate. The detailed understanding of such decoherence processes is important for the envisaged use of atomic BECs in metrology and interferometry, [Choi 2001; Gupta 2002; Dunningham 2002; Gorlitz 2003].

Another intriguing point is that non-exponential damping of impurity motional states in a BEC provides an experimentally feasible example of manifestation of non-mean-field effects in a BEC. Up to now, only few experimental works on non-mean-field effects in BECs are available: influence of a large thermal fraction of a bosonic system not far below the condensation point on damping and frequency shifts of elementary excitations was studied in Ref. [Jin 1997], and the Belyaev coupling in a trapped BEC under resonant conditions was demonstrated in Ref. [Hodby 2001]. It is worth to note that the critical temperature shift due to atomic interactions with respect to the case of an indeal Bose gas can be accounted for within a mean-field model, [Davies 1997]. Our system seems advantageous for detecting the non-mean-field effects via observation of deviations from the exponential decay law.

Let us consider an "impurity" atom moving in an atomic BEC. Atoms of another isotope or the same isotope but in a different internal (hyperfine) state can be viewed as impurities as long as their density is small enough not to modify considerably the BEC excitation spectrum. At "supercritical" velocities, namely, above the speed of sound in the BEC, the impurity atom is decelerated due to phonon creation in the BEC. The rate of such a process according to the standard Fermi golden rule (*i.e.*, assuming exponential decay of the amplitude of the initial state) has been calculated for both a uniform BEC, [Timmermans 1998], and a harmonically trapped BEC, [Idziaszek 1999], and has been determined experimentally, [Chikkatur 2000]. We shall step beyond this approach and calculate more generally the time evolution of the impurity-atom motion in such a system.

The state vector of the system can be written as (we set $\hbar = 1$)

$$|\psi(t)\rangle = \alpha_{in}(t)\ |in\rangle + \int \frac{d^3q}{(2\pi)^3}\, \beta_q(t)\ |\vec{q}\rangle\,, \tag{1}$$

where the initial state $|in\rangle$ corresponds to an impurity atom of mass m_1 moving at the velocity $\vec{V}_1$ in a BEC with no elementary excitation, and $|\vec{q}\rangle$ denotes the state where one elementary excitation of the BEC with the momentum $\vec{q}$ is present (correspondingly, the impurity momentum is changed to $m_1\vec{V}_1 - \vec{q}$). The initial conditions are, naturally,

$$\alpha_{in}(0) = 1, \ \beta_q(0) = 0. \tag{2}$$

The set of equations describing motional damping of the impurity atom is

$$i\,\dot{\alpha}_{in} = \left(\frac{m_1\,V_1^2}{2} + \tilde{g}_{12}\,n\right)\alpha_{in} + \tilde{g}_{12}\sqrt{n}\int\frac{d^3q}{(2\pi)^3}\,(u_q - v_q)\,\beta_q, \tag{3a}$$

$$i\,\dot{\beta}_q = \left[\frac{(m_1\,\vec{V}_1 - \vec{q})^2}{2m_1} + \epsilon(q) + \tilde{g}_{12}\,n\right]\beta_q + \tilde{g}_{12}\,\sqrt{n}\,(u_q - v_q)\,\alpha_{in}. \tag{3b}$$

Here n is the uniform BEC density, $\tilde{g}_{12}$ is the effective interspecies coupling constant, $u_q = \{[E_{HF}(q)/\epsilon(q)+1]/2\}^{1/2}$ and $v_q = (u_q^2-1)^{1/2}$ are the Bogoliubov transformation coefficients, and $\epsilon(q) = [E_{HF}^2(q)-\mu^2]^{1/2}$ is the energy of the elementary excitation with momentum $\vec{q}$ [Bogoliubov 1947]. Here we have introduced the Hartree-Fock excitation energy $E_{HF} = q^2/(2m_2) + \mu$ and the chemical potential of the BEC $\mu = 4\pi\,a_{22}\,n/m_2$, a_{22} being the intraspecies s-wave scattering length for the condensed atoms of mass m_2. Correspondingly, the speed of sound in the BEC is $c_s = \sqrt{\mu/m_2}$.

To proceed, we have to reckon with the renormalization of the interspecies coupling constant, [Lifshitz 1980; Belyaev 1958]. The renormalized constant $\tilde{g}_{12}$ is expressed in terms of the bare coupling constant g_{12} as

$$\tilde{g}_{12} = g_{12}\left[1 + 2m\,g_{12}\,(2\pi)^{-3}\int d^3q\,q^{-2}\right], \tag{4}$$

with $g_{12} = 2\pi\,a_{12}/m$, a_{12} being the interspecies s-wave scattering length and $m = m_1\,m_2/(m_1 + m_2)$ being the reduced mass. In the approximate (perturbative) solution of Eqs. (3a) and (3b) we shall keep the terms up to the second order in the *bare* constant g_{12}.

The easiest way to solve Eqs. (3a) and (3b) is by the Laplace transformation (*cf.* Ref. [Kofman 1994]). We adopt the interaction representation, wherein the probability amplitude of the initial state is

$$\alpha(t) = \alpha_{in}(t)\,\exp[i\,(\frac{m_1\,V_1^2}{2} + g_{12}\,n)\,t]. \tag{5}$$

The algebraic solution for the Laplace transform of $\alpha(t)$,

$$\overline{\alpha}(s) = \int_0^\infty dt\,\exp(-st)\,\alpha(t), \tag{6}$$

has the form

$$\overline{\alpha}(s) = \left[s + \overline{\Omega}(s)\right]^{-1}, \tag{7a}$$

$$\overline{\Omega}(s) = g_{12}^2 n \int \frac{d^3q}{(2\pi)^3} \left[\frac{(u_q - v_q)^2}{s + i\,\Delta(q)} + \frac{2i\,m}{q^2}\right], \tag{7b}$$

$\Delta(q) = \epsilon(q) + q^2/(2m_1) - \vec{q}.\vec{V_1}$ being the energy mismatch between the the states $|in\rangle$ and $|\vec{q}\rangle$. The second term in the square brackets in Eq. (7b) arises from the coupling-constant renormalization in Eq. (4) and compensates for the ultraviolet divergence of the first term. This compensation is completely analogous to that of the electron mass renormalization in calculations of the radiative shift of an atomic optical transition [Bethe 1947; Cohen-Tannoudji 1992].

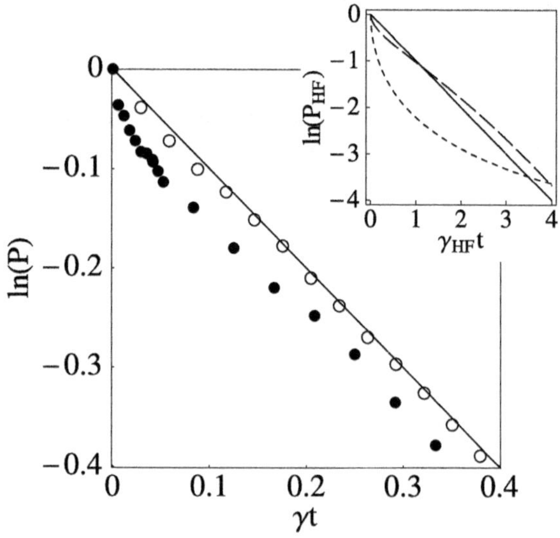

Figure 1. Numerically calculated logarithm of survival probability $P(t)$ of the initial state of impurity atoms in a ^{87}Rb BEC plotted versus dimensionless time γt for $n = 10^{14}$ cm^{-3} ($c_s = 0.2$ cm / s); $m_1 = m_2, a_{12} = 3\,a_{22}$. filled circles: $V_1 = 3\,c_s, \gamma = 1.1 \times 10^3$ s^{-1}. Open circles: $V_1 = 7\,c_s, \gamma = 4.4 \times 10^3$ s^{-1}. Solid line: exponential law $\exp(-\,\gamma t)$. Inset: the survival probability of the initial state (on a logarithmic scale) calculated in the HF approximation, plotted versus $\gamma_{HF}t$ for $K = 10$ (straight line, indistinguishable from the exponential decay), $K = 1$ (long-dashed line), and $K = 0.1$ (short-dashed line). Note that the inset horizontal axis is scaled by the HF value of the exponential decay rate, γ_{HF}, unlike the horizontal axis of the main plot, which is scaled by the *exact* value of γ.

Eqs. (3a) and (3b) yield over a broad time interval, excluding very short times, exponential decay of $\alpha(t) \propto \exp[-(\gamma/2 + i\omega_s)\,t]$, with the rate γ and the frequency shift ω_s

$$\gamma = \lim_{s\to 0}\, 2\,\mathrm{Re}\,\overline{\Omega}(s),\ \omega_s = \lim_{s\to 0}\, \mathrm{Im}\,\overline{\Omega}(s). \tag{8}$$

The relaxation rate γ can be calculated within the exact Bogoliubov theory using Fermi's golden rule, [Timmermans 1998; Chikkatur 2000].

It is possible to obtain a non-exponential analytical solution for $\alpha(t)$ in the Hartree-Fock (HF) approximation, [Esry 1997] (hereafter we label all the quantities in this approximation by HF): $\epsilon_{HF}(q) = E_{HF}(q)$, $u_{q\,HF} = 1$, $v_{q\,HF} = 0$. Then the integral in Eq. (7b) can be evaluated analytically, yielding in Eq. (8) the exponential decay rate $\gamma_{HF} = \gamma_0\,[1 - c_{HF}^2\,/\,V_1^2]^{1/2}$ if $V_1 > c_{HF} \equiv \sqrt{2\mu/m_2}$ and zero otherwise. Here $\gamma_0 = 4\pi\, a_{12}^2\, n\, V_1$ is the collison rate calculated for the impurity atom using the Fermi golden rule in the limit $a_{22} \to 0$ (an ideal BEC). Finally, in the case $V_1 > c_{HF}$ we obtain

$$\overline{\alpha}_{HF}(s) = \left[\gamma_{HF}\left(\frac{s}{\gamma_{HF}} + \frac{1}{2}\sqrt{1 + \frac{i\,s}{\gamma_{HF}\,K}}\right)\right]^{-1}, \tag{9}$$

where $K = (m\,V_1^2/2 - \mu)/\gamma_{HF}$ is the energy of collision of the impurity atom with a condensate atom in their center-of-mass frame (including the correction due to the BEC chemical potential μ) scaled to γ_{HF}. By inverting such a Laplace transform, [Abramowitz 1964], we get

$$\alpha_{HF}(t) = \frac{e^{i\,\Lambda t}}{\xi_1 - \xi_2}\left[\xi_1\,\varphi(\xi_1\sqrt{t}) - \xi_2\,\varphi(\xi_2\sqrt{t})\right], \tag{10}$$

where $\varphi(z) = \exp(z^2)\mathrm{erfc}(-z)$, $\mathrm{erfc}(z)$ being the complementary error function, and $\xi_{1,\,2} = -\,\Xi/2 \pm \sqrt{\Xi^2/4 - i\,\Lambda}$ are the roots of the quadratic equation $\xi^2 + \Xi\,\xi + i\,\Lambda = 0$. The coefficients of the latter equation are $\Xi = \gamma_0\,\sqrt{i/(2m\,V_1^2)}$ and $\Lambda = m\,V_1^2/2 - \mu$. The survival probability $P_{HF}(t) = |\alpha_{HF}(t)|^2$ is displayed in the inset of Fig. 1.

Although our numerical results (Fig. 1) show that the HF approximated solution Eq. (10) is rather crude, it nonetheless provides a qualitative guidance to the physical behavior. Simple scaling considerations lead us to the conclusion that α_{HF} depends on two parameters: the dimensionless time variable $\gamma_{HF}\,t$ and K [Eq. (9)]. Eq. (10) predicts that at $t \to 0$ the decay is more rapid than exponential, so the survival probability $P(t) = |\alpha(t)|^2$ behaves in the HF approximation as $P_{HF}(t) \approx 1 - 4\,\mathrm{Re}\,(\xi_1 + \xi_2)\sqrt{t/\pi}$, becoming exponential at larger times, $P_{HF}(t) \approx \exp(-\,\gamma_{HF}\,t)$. The deviation from exponential decay is appreciable only for $K \lesssim 1$. There are two ways to attain $K < 1$. One is to take a small difference between the impurity atom velocity V_1 and the critical velocity, but this would reduce the damping rate, which may be experimentally inconvenient. A much better way is to strive for a large interspecies scattering length a_{12}, as discussed below.

The results of our numerical calculations based on Eqs. (7a) and (7b), which use the *exact* expressions for $\epsilon(q)$, u_q and v_q instead of the HF approximation, clearly reveal a deviation from exponential decay for small times. Under such conditions, frequent measurements would accelerate the decay, causing the anti-Zeno effect (AZE). Alternatively, one may accelerate the decay by periodically modulating the coupling of the initial state to the continuum, [Kofman 2000; Kofman 2001 (a)], instead of repeated projective measurements. This can be done by changing the impurity velocity using a sequence of Bragg or Raman laser pulses [Stenger 1999; Steinhauer 2002].

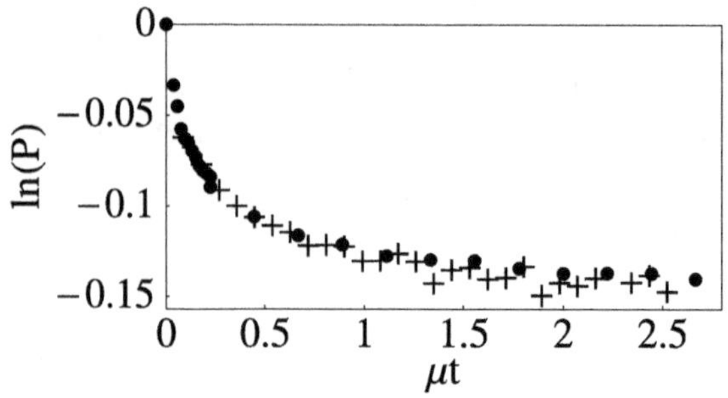

Figure 2 The numerically calculated logarithm of survival probability versus the time scaled by the BEC chemical potential μ. The parameters of the BEC and impurity atoms are the same as in the main plot of Fig. 1, except the impurity velocity: $V_1 = 0.3\ c_s$ (filled circles) and $V_1 = 0.7\ c_s$ (crosses).

Fig. 2 shows that if $V_1 < c_s$, the survival probability first decreases and then approaches the constant value of about 0.85, almost independent on V_1. This behavior reveals the physical reason for the short-time non exponential decay: the initial conditions Eq. (2) imply that, initially, the impurity atom is surrounded by *no virtual phonons*, while in the steady state, the impurity atom must be surrounded by a cloud of virtual phonons (*cf.* the polaronic effect for electrons in a crystal, [Isihara 1971]). Thus the non-exponential stage of the decay is associated with the formation of such a phonon cloud.

The faster the impurity atom moves, the weaker its coupling to the phonon cloud is. Therefore the decrease of ω_s corresponds to the vanishing of non-exponential decay effects as V_1 increases. This behavior is displayed in Fig. 3 by the numerically calculated [from Eq. (8)] decay rate γ and frequency shift ω_s.

Our numerical studies of Eqs. (3a) and (3b) always yield *decay acceleration* at short times, typically on the scale of 10^{-5} s. But should one not expect, from general considerations [Khalfin 1957-58; Winter 1961; Misra 1977; Fonda 1978; Itano 1990], $P(t) = 1 - \text{const} \cdot t^2$ at $t \to 0$, in accordance with the QZE? To answer this question, we should apply the general theory developed in Refs. [Kofman 2000; Kofman 2001 (a)], whereby the short-time behavior is determined by the spectrum (*i.e.*, the dependence on the emitted quantum energy ϵ) of the reservoir response $G(\epsilon)$. This spectrum is given by the interaction matrix element squared multiplied by the density of the reservoir states. In our case, we find that

$$G(\epsilon) = \left[\frac{2\pi\, a_{12}\,(u_q - v_q)}{m}\right]^2 \frac{n\, q^2}{2\pi^2}\, \frac{dq}{d\epsilon} \tag{11}$$

monotonously increases with the emitted phonon energy ϵ. According to Refs. [Kofman 2000; Kofman 2001 (a)], if the energy uncertainty $\sim t^{-1}$ associated with the finite observation time t covers the energy range where $G(\epsilon)$ increases, then decay acceleration (AZE) takes place.

However, our approach [Eqs. (3a) and (3b)] leading to Eq. (11) is valid only for small transferred momenta. If $q \gtrsim r_0^{-1}$, where $r_0 \sim 10^{-7}$ cm is the characteristic radius of the interatomic potential, we cannot consider a_{12} as constant any more. Instead, the interaction matrix element decreases with q in this range. In the inset of Fig. 3 we schematically display the spectrum of $G(\epsilon)$, including its decreasing part, whose detailed calculation is beyond the scope of the present work. This spectrum implies that at very short times ($\lesssim 10^{-9}$ s), the energy uncertainty broadening covers the whole profile of $G(\epsilon)$, thereby giving rise to the QZE.

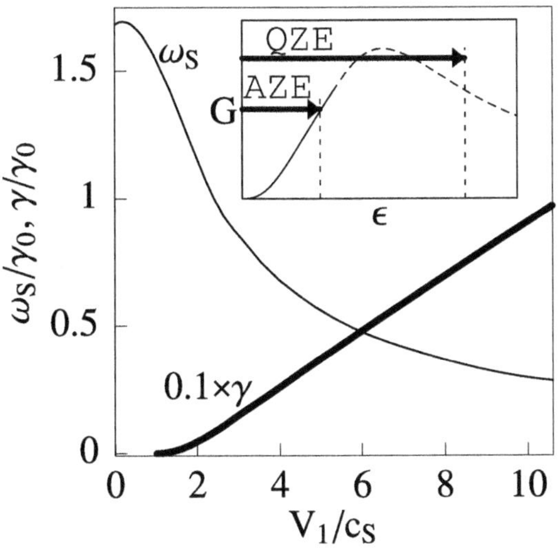

Figure 3 Numerically calculated frequency shift ω_s and the exponential decay rate γ (scaled to γ_0) versus the impurity dimensionless velocity V_1/c_s. Both ω_s/γ_0 and γ/γ_0 display a universal behaviour, independent of the BECBose-Einstein condensate density and the atomic species. Inset: schematic representation of the spectrum $G(\epsilon)$; solid line: the increasing part given by Eq. (11); dashed line: the remaining part, which decreases at $\epsilon \gtrsim 1/(m\ r_0^2)$. The arrows indicate ranges of the energy uncertainty corresponding to AZE and QZE.

The parameters used in Fig. 1 for a BECBose-Einstein condensate of ^{87}Rb may correspond to impurity atoms of the same isotope but in a different hyperfine state, obtained by a short Raman pulse. This method of impurity atom admixing conforms to the initial conditions of Eq. (2). However, to reach appreciable non-exponentiality by this method, one has to enhance interspecies scattering either by means of interspecies Feshbach resonance or via laser-induced dipole-dipole interactions [Kurizki 2002]. Another possibility is the use of two-isotope mixture, for example, a BEC of ^{87}Rb atoms with admixed fermionic ^{40}K atoms. Such a choice is of particular interest, since the large interspecies scattering length in this mixture [Roati 2002], as well as the mass ratio between these two elements seem very promising for experimental search of non exponential decay effects. We note that the condition on the impurity velocity in this case is opposite to that of the aforementioned case of impurities *generated* from a BEC by a Raman pulse. Indeed, the ^{40}K atoms at rest co-exist with the ^{87}Rb BEC for a time long enough to form virtual phonon clouds around them. If then the ^{40}K atoms suddenly acquire, by the action of a Bragg or Raman pulse, the velocity $V_1 \gg c_s$, the initial conditions, instead of Eq. (2), should assume the pre-existence of the phonon cloud (prior to the pulse), which vanishes when the impurities attain high (supercritical) velocities. Thus we expect that in the case of a two-element ultracold mixture, non exponential decay features are most pronounced for $V_1 \gg c_s$.

Now can estimate the influence of the finite BEC temperature T on the non-exponential decay of impurity wave packets. Let us consider $\mu/k_B \lesssim T \ll T_c$, where k_B is the Boltzmann constant, $T = 3.31\, n_t^{2/3}/(m_2\, k_B)$, [Isihara 1971], and $n_t = n + n_a$ is the total number density of the degenerate atomic sample. The number density of the above-condensate (thermal) fraction is $n_a = n_t\,(T/T_c)^{3/2}$, [Isihara 1971]. The characteristic time scale τ_{th}, on which an impurity atom experiences a collision with the atom belonging to the above-condensate fraction, is given by

$$\tau_{th} = \left[4\pi\, a_{12}^2\, n_a\, \max(\sqrt{k_B T/m},\, V_1)\right]^{-1}. \tag{12}$$

If τ_{th} is longer than the characteristic time scale of the non-exponential decay of the initial state of the impurity motion, then the thermal effects can be neglected, *e.g.*, we find for a BEC with parameters given in the caption of Fig. 1 and $V_1 < 3c_s$ that if T is lower than 100 nK that comprises $0.25\ T_c$ then τ_{th} is greater by an order of magintude than the characteristic time of non-exponentiality, which is about 5.10^{-5} s.

To conclude, we have outlined the possibility of observing deviations from exponential decay for unstable momentum states of impurity atoms moving in a BEC with a supercritical velocity on time scales of 10^{-5} s.

Acknowledgments

We thank Dr A. G. Kofman for helpful discussions. The support of the German-Israeli Foundation, the EC (QUACS RTN), ISF and Minerva is acknowledged. I. E. M. also thanks the RFBR (projects 02-02-17686, 03-02-17522) and the program Leading Russian Scientific Schools (grant 1115.2003.2).

V

INTERNAL-TRANSLATIONAL ENTANGLEMENT AND INTERFERENCE IN ATOMS AND MOLECULES

INTERNAL-TRANSLATIONAL ENTANGLEMENT AND INTERFERENCE IN ATOMS AND MOLECULES

T. Opatrný,[1] M. Arndt,[2] T. F. Gallagher,[3] R. Garcia-Fernandez,[4] S. Haroche,[5] M. Leibscher,[6] P. Pillet,[7] and J. Sherson[8]

[1]*Department of Theoretical Physics, Palacký University, 17. listopadu 50, 77146 Olomouc, Czech Republic*

[2]*Institut für Experimentalphysik, Universität Wien, Boltzmanngasse 5, A-1090 Wien*

[3]*Department of Physics, University of Virginia, Charlottesville, VA 22901, U.S.A.*

[4]*Department of Physics, University of Kaiserslautern, Erwin-Schrödinger Str., D-67663 Kaiserslautern, Germany*

[5]*Laboratoire Kastler Brossel, Département de Physique, Ecole Normale Supérieure, 24 rue Lhomond, and Collège de France, 11 Place Marcellin Berthelot, F-75005 Paris, France*

[6]*Royal Institute of Technology, Department of Physics, Alba Nova, Roslagstullbacken 21, 10691 Stockholm Sweden and Weizmann Institute of Science, Chemical Physics Department, Rehovot 76100, Israel*

[7]*Laboratoire Aimé Cotton, CNRS, Bât. 505, Campus d'Orsay, 91405 Orsay cedex, France*

[8]*Danish National Research Foundation, Center for Quantum Optics - QUANTOP, Department of Physics and Astronomy, University of Aarhus, Ny Munkegade bygning 520, 8000 Aarhus, Denmark*

Introduction

Entanglement in the dynamics of atoms and molecules, especially entanglement between their internal and external degrees of freedom is one of the key issues in modern quantum research. A better understanding of this most remarkable feature of physics is hoped to enable us, to asses and to control decoherence and to better exploit various quantum interference phenomena at different scales. This part contains both theoretical and experimental contributions studying entanglement, interference and decoherence phenomena in different atomic and molecular systems. In particular, we report experimental results for flying fullerene molecules, sodium dimers in strong laser fields, single Rydberg atoms in high Q microwave cavities, cold Rydberg gases of both atomic cesium and rubidium, and for distant macroscopic samples of cesium

V.M. Akulin et al. (eds.), Decoherence, Entanglement and Information Protection in Complex Quantum Systems, 317–324.

atoms. We present theoretical investigations on how to steer quantum states of diatomic molecules beyond the Franck-Condon approximation, and we discuss the possibility of generating translationally entangled pairs of atoms in optical lattices.

1. Atom-mesoscopic field entanglement in microwave cavities

Very interesting questions concern coherence properties of a coupled system consisting of a single atom and the electromagnetic field. What are the consequences of the quantum nature of the field? What happens if the field contains more and more photons and its features are more classical? Can one entangle and disentangle the atom and the field on demand and to which extent?

Two experiments in microwave cavity QED are reviewed in the first article by S. Haroche *et al.* in which entanglement between a two-level atom and a mesoscopic field has been demonstrated and the coherence of the field state superpositions produced in the process has been probed. The experiments involve circular Rydberg states of ^{85}Rb atoms; it was observed that a mesoscopic field made of several tens of microwave photons exhibits quantum features when interacting with such an atom in a high Q cavity. In the off-resonant regime [Brune 1996] the atom shifts the phase of the field depending on its internal state. Vice versa, the phase of the atomic state is shifted depending on the field state, which can be observed using Ramsey interferometry. In the resonant regime [Auffeves 2003] one can observe Rabi oscillations during which the field and the atom periodically exchange energy. Interesting correlations between the field and the atom can be observed. The field is split into two components whose phases differ by an angle inversely proportional to the square root of the average photon number. The field and the atomic dipole are phase entangled. These manifestations of photon graininess vanish at the classical limit. These studies constitute an exploration of the quantum-classical boundary and open the way to studies of non-local mesoscopic state superpositions.

2. Fullerene interference

A key motivation for macromolecule interferometry is the question up to which molecular size and complexity non-classical features of quantum mechanics are still observable. How far can the limits of the wave-particle duality be pushed for massive particles, first predicted by de Broglie [de Broglie 1923] and demonstrated by Davisson and Germer [Davisson 1927] for electrons.

Formally, we describe the state of the particle during the propagation as a coherent superposition of states, in particular of position states, that are classically mutually exclusive. A classical object will either take one or the other path with certainty. A quantum object cannot be said to do that since the intrinsic information content of the quantum system is insufficient to allow such a description [Brukner 2002]. Matter wave interferometers prove this experimentally. The intriguing part is that a full interference visibility can only be obtained if we exclude all possibilities of detecting the path the object has taken. This seems to imply the philosophically challenging statement that in a wave-particle experiment the *"possibility of having a position"* is often the only objective reality in contrast to the property of *"having a well-defined position"*.

In addition to the duality between being localized and showing interference, we also find the duality between determinism and objective randomness. The entire pattern on the screen is well determined for the ensemble, but the detection of a single object is completely unpredictable in all experiments. This is why Richard Feynman said the double-slit experiment to be at the heart of quantum mechanics [Feynman 1965]. One might argue that another central issue of quantum physics, namely entanglement, is missing in this example. However, it turns out to be an essential ingredient if we consider how which-path information is diffused into the environment, *i.e.*, if we include decoherence, as will be shown in the second contribution.

One of the remarkable properties of quantum mechanics is that the wave nature of matter completely escapes perception in our everyday life, although this feature is a cornerstone of the theory. The smallness of Planck's constant and therefore of the de Broglie wavelength of a macroscopic object is certainly largely responsible for the non-observability of quantum effects in the classical world. However, it is important to ask whether there are fundamental limits to quantum physics and how far we can push the experimental techniques to visualize quantum effects in the mesoscopic world for objects of increasing size, mass and complexity. Where are the fundamental limits on the way towards larger objects?

The work by M. Arndt *et al.* starts with the description of far-field diffraction experiments with fullerene molecules C_{60}. Since for larger objects the far-field observations become much more challenging, the authors also study the feasibility of near-field Talbot Lau interferometry with C_{70}. This particular technique allows them to work with a spatially incoherent beam and thus with a much increased count rate. Moreover it has a better wavelength scaling in the grating constant and can potentially be applied to much smaller wavelengths at a reasonable grating constant.

Decoherence theory explains that a quantum object may loose some of its particular quantum properties due to the interaction with its environment. This phenomenon is investigated along two lines. First the authors study the influence of small angle collisions exerted by a dilute thermal gas on the molecules in the interferometer. They also focus on a novel property which is unique to large objects with many internal degrees of freedom, namely the thermal emission of photons. If sufficiently many photons of sufficiently short wavelength are emitted, they can diffuse which-path information from the molecule to the environment and thus lead to an effective loss of interference fringe visibility.

All reported experiments so far indicate that the experimental limits of molecule interferometry are still far out. And M. Arndt *et al.* discuss some open questions in the quest for quantum experiments with macromolecules.

3. Distant entanglement of macroscopic samples of atomic cesium

One of the main ingredients in most quantum information protocols is a reliable source of two entangled systems. Such systems have been generated experimentally several years ago for light [Aspect 1982 (b); Shih 1988; Ou 1992; Kwiat 1995; Schori 2002 (b)] but has only in the past few years been demonstrated for atomic systems [Hagley 1997; Sackett 2000; Julsgaard 2001; Roos 2004]. None of these approaches however involve two atomic systems situated in separate environments. This is necessary for the creation of entanglement over arbitrary distances which is required for many quantum information protocols such as atomic teleportation [Bennett 1993; Kuzmich 2000]. Experimental realization of entanglement of atomic samples is reported in the third contribution by J. Sherson *et al.*, with a separation of 0.35 m (arbitrarily scalable) and a typical number of atoms of 10^{11}. The entanglement is generated via the off-resonant Kerr interaction between the atoms and a pulse of light, similarly as in Ref. [Julsgaard 2001]. Because the entangled objects are macroscopic and the distance straightforward scalable, the ability to implement various quantum information protocols using macroscopic samples of atoms has therefore been greatly increased. The authors describe the essential parts of the experimental setup and present a theoretical modeling in terms of canonical position and momentum operators $\hat{X}$ and $\hat{P}$ describing the entanglement generation and verification in presence of decoherence mechanisms.

4. Translational entanglement of atoms in optical lattices

Einstein, Podolsky and Rosen (EPR) [Einstein 1935] asked the question of whether the quantum mechanical description of physical world is complete, giving the following example. Two-particles are in the quantum state showing strange correlations: if one measures the position or momentum of one particle, one can predict with certainty the result of measuring their counterpart for the second particle. Thus, depending on which measurement is chosen for the first particle, the value of either the momentum or position can be predicted with arbitrary precision for the other particle. The later discussion has concerned the interpretation of the EPR paradox and its implications on quantum theory [Bohr 1935]. Later, Bohm considered [Bohm 1951] two entangled spin-$1/2$ particles, which have become the center of attention on this EPR issue: their *discrete-variable* entangled states have served to demonstrate the incompatibility of quantum mechanics with local realism, by the Bell's inequality violation.

However, the original EPR state of two translationally entangled particles is hardly realizable experimentally. It would occupy infinite space and have infinite kinetic energy. One can consider more realistic variants of this state, *e.g.*, a Gaussian state where the particles' positions and momenta are correlated within some finite widths Δx_- and Δp_+. After measuring the position of particle 1 (projection onto $x_1 = \text{const}$), the position of particle 2 is centered close to x_1 with the uncertainty $\approx \Delta x_-$. Similar relations hold also for the momentum of particle 2 after the momentum of particle 1 is measured. Since Δx_- and Δp_+ can be, in principle, arbitrarily small, either of the two conjugate quantities of particle 2 can be predicted with arbitrarily high precision. Of course, the Heisenberg uncertainty relation is not violated, since for a single system one can measure only one of the two conjugate quantities.

Approximate versions of the translational EPR state, wherein the δ-function correlations are replaced by finite-width (Gaussian) distributions, have been shown to characterize the quadratures of the two optical-field outputs of parametric down-conversion, or of a fiber interferometer with Kerr nonlinearity. Such states allow for various schemes of continuous-variable quantum information processing such as quantum teleportation [Braunstein 1998 (b); Furusawa 1998] or quantum cryptography [Silberhorn 2002]. A similar state has also been predicted and realized using collective spins of large atomic samples [Polzik 1999; Julsgaard 2001]. It has been shown that if suitable interaction schemes can be realized, continuous-variable quantum states of the original EPR type could even serve for quantum computation.

However, the realization and measurement of the EPR translational correlations of *material particles* appears to be very difficult. In order to generate the translational EPR entanglement between interacting material particles, one must be able to accomplish several challenging tasks:
(a) switch on and off the entangling interaction;
(b) confine their motion to single dimension;
(c) infer and verify the dynamical variables of particle 2 *at the time of measurement* of particle 1.
The latter requirement is particularly hard for free particles, since by the time we complete the prediction for particle 2, its position will have changed. The scheme proposed in the fourth contribution by T. Opatrný *et al.* is based on the following elements:
(i) controlling the diatom formation and dissociation in an optical lattice by switching on and off a laser-induced dipole-dipole interaction;
(ii) controlling the motion and effective masses of the atoms and the diatom by changing the intensities of the lattice fields.

The authors discuss the different regimes of the diatomic systems and study the evolution of their entanglement.

5. Autler-Townes effect in sodium dimers

R. Garcia-Fernandez *et al.* present the study of decoherence of internal degrees of freedom of molecules, which starts with the preparation of a coherent ensemble of states using coherent laser radiation. The Autler-Townes (AT) effect, known since 1955 [Autler 1955], is at the heart of such preparation schemes. Decoherence in molecules is of particular interest, and significant new insights may be gained from studies of the AT effect. Although it has been widely studied for atoms [Fisk 1986], few high-resolution studies have been reported for molecules [Qi 2000]. It is in this context that the authors have started a program within QUACS to better characterize the AT-effect under varying conditions (coupling schemes, intensities, and detunings).

In its simplest form, the AT effect can be regarded as originating in the coherent interaction of a strong laser field with a two-state quantum system. The resulting alteration of the atomic or molecular energies can be probed in various ways, such as through fluorescence excitation spectrum obtained by scanning the frequency of a weak laser field across the resonance with some third level. The resulting signals, typically visible as a doublet structure, incorporate not only the physics of the coherent interaction revealed in the separation of the two doublet components, but also aspects of decoherence, such as spontaneous emission lifetimes, as revealed by the widths and intensities of the two components. The experiments reported in the fifth contribution study with high-resolution the Autler-Townes effect in molecules, observed as fluorescence from both upper and intermediate levels in a three-level ladder coupled by two laser fields. They indicate the possibility of novel implementation of the AT effect for the characterization of excited molecular states.

6. Molecular states steered via space-dependent coupling

Can one prepare a molecule in a particular quantum mechanical state of its vibrational and rotational degrees of freedom? Controlling atomic and molecular quantum states with the help of laser light has been an intensively studied field in the last years. The ability to produce and shape femtosecond laser pulses made it possible to create specially tailored wave packets and to manipulate their dynamics in order to reach pre-assigned goals. The techniques of quantum control have been applied to the problem of steering chemical reactions as well as intramolecular dynamics. Quantum control methods make use of the time- and frequency dependence of the external laser field, usually assuming that the spatial dependence of the coupling between two electronic energy surfaces of a molecule is constant. In the sixth contribution, M. Leibscher *et al.* ask what may be the influence of the spatial dependence of the coupling and wether it can also be used for steering molecular transitions.

The dipole couplings between molecular levels is often weak enough to allow the transfer of only a minute fraction of the ground state population into excited levels. As a consequence, working within the perturbation approximation is often justified and make easier the discussion of laser-induced molecular transitions. To this end, the authors assume some spatial dependence of the coupling between two electronic energy surfaces of a molecule. The usual assumption of a constant coupling leads to the well known Franck-Condon effect: the ground state wave function is lifted up to the excited state without distortion. This can also be seen as the molecular equivalent of momentum conservation. The initial wave function is lifted to the spatial turning points of the motion, where it moves with nearly zero momentum. This argument changes when the levels coupling depends on position along the energy surface. This spatial dependence imposes a modification of the initial momentum distribution, which may have a profound effect on the transition probabilities. The influence of space-dependent couplings for rather simple model systems is discussed. The authors consider one-dimensional systems, that might describe a diatomic molecule, or the reaction path of some simple reaction. Furthermore, they neglect all influence of the rotational motion. They assume a model space-dependence for the coupling and discuss how the transfer of a wave packet is affected by the various parameters of the coupling. They choose the space-dependence such that the parameters have direct physical interpretation within the model system. In this way, they can discuss possible consequences within a general framework. A model allowing to steer intramolecular dynamics by varying the spatial dependence of the coupling is also proposed.

7. Coherence and decoherence in Rydberg gases

Shining properly chosen short laser pulses onto cold clouds of atomic Cs or Rb can reveal a fascinating world where microsecond is as long as eternity for one kind of motion and extremely short for another: whereas the highly excited atoms almost do not move, excitations exchanged via their dipole-dipole interactions can jump from one atom to another and travel over the sample. How do coherent phenomena occur here and what are the effects of decoherence?

Dipole-dipole interactions (DDIs) between cold Rydberg atoms have recently attracted much interest since they play a central role in proposed quantum logic gates [Jaksch 2000; Lukin 2001]. While DDIs between cold, or stationary Rydberg atoms, are only beginning to be explored [Anderson 1998 (b); Mourachko 1998], resonant dipole-dipole collisions between Rydberg atoms have been studied extensively [Gallagher 1992]. The connection between the two phenomena is that the interactions between cold stationary atoms correspond to freezing a pair of colliding atoms at their point of closest approach.

An ensemble of cold Rydberg atoms is easily obtained after laser excitation of a cold atomic cloud, as those performed in a Cs or Rb vapor-cell magneto-optical trap, at temperatures of 135 μK or 300 μK respectively. In the case of Cs or Rb (experiments performed at Laboratoire Aimé Cotton or at the University of Virginia respectively), the atoms p-excited by cooling lasers are Rydberg-excited by a pulse provided by a dye laser pumped by the third harmonic of a $Nd{:}YAG$ laser: $Cs(6p_{3/2}) + h\nu_L \rightarrow Cs(nl)$ or $Rb(5p_{3/2}) + h\nu_L \rightarrow Rb(nl)$ with laser wavelengths ($\lambda_L = c/\nu_L$) of $\sim$ 515 nm or 480 nm, for Cs or Rb respectively [Anderson 1998 (b); Mourachko 1998]. The principal quantum number n ranges from 20 to 50 and $l = s, p, d$. Typically, the number of Rydberg atoms ranges from 10^3 to 10^6 in a volume corresponding to the MOT cloud volume $\sim (0.5 \text{ mm})^3$, meaning a density ranging from 10^7 to 10^{10} cm^{-3}.

The time scale of interest in the experiments discussed by P. Pillet, T. F. Gallagher *et al.* (of the order of a few μs) is smaller than the Rydberg lifetime. As a consequence, the Rydberg atoms move only small distances relative to their average separation ($\overline{R} \approx 4.5 - 45$ μm). For a typical velocity around $v \sim 10$ cm/s, the displacement $\Delta R = v\tau$ is $\sim$ 100 nm during a time $\tau = 1$ μs. Such a displacement is much closer to the size of the Rydberg atoms $R_0 \sim 4n^2$ (in a. u.) ($\sim$ 80 nm for $n = 20$) than their average distance. In a first approximation, it is therefore possible to ignore the motion of the Rydberg and to consider the atomic ensemble as a frozen gas. Frozen gases are difficult to realize with neutral atoms, and elegant examples are the well defined spatial structure of cold ions observed in ion traps [Diedrich 1987; Wineland 1987]. The long-range Coulomb interaction plays a key role in the observation of the crystalline phase. Cold Rydberg atoms, with their large sizes and dipole moments, appear to be natural candidates for studying frozen neutral gas expected to behave as an amorphous solid.

At large density ($\sim 10^{10}$ cm^{-3}) and large $n \gtrsim 40$, the authors observe the spontaneous evolution of a cold Rydberg gas in an ultracold plasma [Robinson 2000]. Such spectacular behavior is first the result of ionization processes due to blackbody radiation and hot Rydberg - cold Rydberg collisions. After the first electrons leave the Rydberg cloud, the formation of positive space charge due to an ion cloud traps subsequent electrons leading finally to the rapid avalanche ionization of the ensemble of the Rydberg gas. The resulting cold plasma expands slowly and persists for tens of microseconds. Such a dramatic behavior does not occur at lower densities and for low n. Due to the long-range interaction between Rydberg atoms, the ensemble of cold Rydberg atoms can exhibit many-body phenomena, offering an interesting quantum mesoscopic system for studying the properties of coherence and decoherence.

ATOM-MESOSCOPIC FIELD ENTANGLEMENT

S. Haroche,[1, 2] M. Brune,[2] and J. M. Raimond[2]

[1] *Laboratoire Kastler Brossel, Département de Physique, Ecole Normale Supérieure, 24 rue Lhomond, F-75005 Paris, France*

haroche@lkb.ens.fr

[2] *Collège de France, 11 Place Marcellin Berthelot, F-75005 Paris, France*

Abstract We review two experiments in microwave cavity QED where we have demonstrated entanglement between a two-level atom and a mesoscopic field and probed the coherence of the field state superpositions produced in the process. These studies constitute an exploration of the quantum-classical boundary and open the way to studies of non-local mesoscopic state superpositions.

Keywords: Cavity QED, mesoscopic state superpositions.

The entanglement of a microscopic object with a large system, made of many particles, is a central issue in quantum physics, related to the problem of macroscopic state superpositions and decoherence [Joos 2003]. When the number of degrees of freedom in the large system increases, the state superpositions tend to be rapidly replaced by statistical mixtures and the entanglement is lost. Preserving macroscopic superpositions from decoherence is obviously a central problem in quantum information processing [Bouwmeester 2000].

Cavity quantum electrodynamics in the microwave domain is an ideal testing ground for the study of entanglement at the quantum-classical boundary [Raimond 2001]. Here, circular state Rydberg atoms interact with a coherent field in a high Q cavity, generated by a classical source. The average number of photons in this field can be continuously varied, from the vacuum to arbitrarily large values. When the field is small, it behaves as a quantum system and can easily be entangled with the atom. When the number of photons is increased, the field becomes progressively more and more classical. Its entanglement with the atom takes longer and longer times, and becomes more and more sensitive to decoherence processes. We have studied this phenomenon in two series of experiments, which have constituted an exploration of the quantum to classical frontier [Brennen 1999; Auffeves 2003].

V.M. Akulin et al. (eds.), Decoherence, Entanglement and Information Protection in Complex Quantum Systems, 325–328.

In a first experiment [Brennen 1999], we have used a dispersive non-resonant interaction to entangle the atom and the field. The two systems were then unable to exchange energy. The atom was shifting the frequency of the field, the sign of the effect depending upon the energy state of the atom. By preparing initially the atom in a state superposition, the field acquired two phase shifts of opposite signs at once and evolved into a superposition of coherent states correlated to the two atomic energies. By performing a Ramsey interferometric measurement on the atom crossing the cavity, we have been able to monitor the separation of the field components, which produced a collapse of the Ramsey fringe contrast. This was a manifestation of quantum mechanical complementarity. By analyzing the correlations between two successive atoms, we could also detect the coherence of the field state superposition. We observed in this way the decoherence process, as a decay of the correlation signal when the time interval between the two atoms was increased. This experiment was limited to fields containing, on average, three to five photons.

Recently, we have replaced the dispersive interaction by a resonant one [Auffeves 2003]. The atom and the field then reversibly exchange energy. This is the well known Rabi oscillation phenomenon, resulting in a periodic variation of the probability to find the atom in its initial energy eigenstate. During this process, the field is slowly split into two components with opposite phases, each being correlated to an atomic state, superposition with equal weights of the two energy states pertaining to the atomic transition resonant with the field [Gea-Banacloche 1990; Buzek 1992]. As the field components separate in phase space, the contrast of the Rabi oscillation collapses, again a manifestation of complementarity. The atom and the field are then entangled. When the field component merge again, after each has rotated by half a turn in phase space, the Rabi signal is expected to revive [Eberly 1980].

We have directly observed the separation of the field components by performing a homodyne phase measurement. After interacting with the atom, the field is mixed again with a reference coherent field of same amplitude and variable phase. This phase is swept until one component of the field is canceled, bringing in this component the total field back to vacuum. This cancelation is checked by having the resulting field absorbed with probe atoms sent in the cavity after a delay following the first phase-shifter atom. In this way, the phase distribution of the phase-shifted field is measured, as the phase of the reference field is continuously swept. We observe a distribution presenting two symmetrical peaks, with a separation increasing with time. Figs. 1 (a) and (b) show the phase distributions for various photon numbers and two interaction times. Fig. 1 (c) shows the two component field phases as a function of the dimensionless parameter $\Omega\, t/4\sqrt{\overline{n}}$ (Ω: Rabi frequency in vacuum, t: atom-field interaction time, $\overline{n}$: average field photon number). There is a very good agreement between the experimental points and the theoretical lines. The data

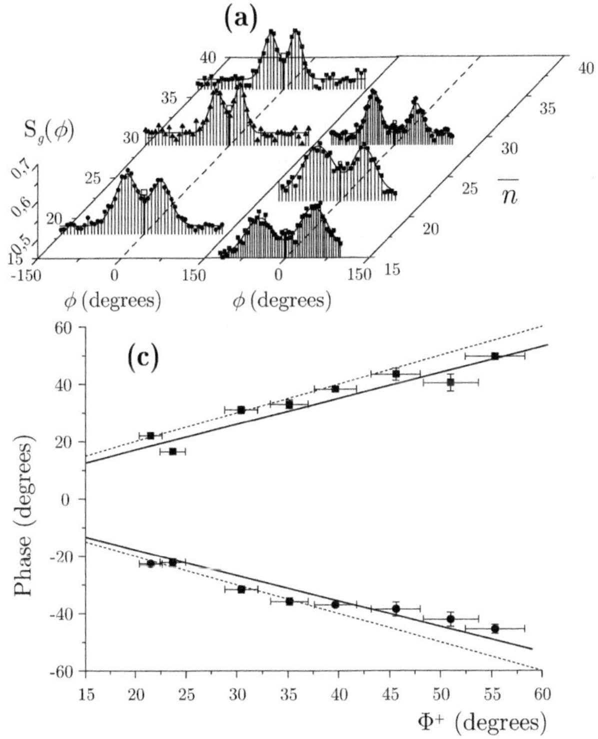

Figure 1. (a) Field phase distribution [probability $S_g(\phi)$ that the probe atom has not absorbed a photon as a function of the phase ϕ of the reference homodyning field] for three different average photon numbers $\overline{n}$ (18, 29 and 36) and an effective atom-cavity interaction time t = 32 μs. The points are experimental and the curves are fits on sums of gaussian shapes. The thick vertical line indicates the zero phase reference. (b) Field phase distribution for $\overline{n} = 15, 22$ and 29 and for $t = 53$ μs. The field phase splitting is clearly apparent in (a) and (b). (c) Phases of the two field components versus the dimensionless parameter $\Omega\ t/4\sqrt{\overline{n}}$. Dotted lines give the predictions of a simple analytical model and solid lines result from a numerical simulation (from Ref. [Auffeves 2003]).

show that the speed of the phase splitting is inversely proportional to the field amplitude in the cavity. It becomes smaller and smaller as this amplitude is increased, revealing the onset of classical behaviour when the average number of photons becomes large.

By applying an echo technique proposed in Ref. [Morigi 2002], we have also been able to reverse the field separation and to induce an early Rabi revival signal as the two field components are refocused back into their initial position [Meunier 2004]. These studies have been made with fields containing up to 40 photons on average, a noticeable improvement over the first generation of mesoscopic state superposition experiments [Brennen 1999]. The resonant interaction is more efficient than the dispersive one to couple the atom and the field and can produce larger superpositions in a given time.

We plan to extend these experiments to a set-up with two identical cavities separated by several centimeters. In this way, we hope to study mesoscopic field superpositions having a non-local character. Tens of photons could be put in state superpositions involving the two cavities, revealing many situations unusual to a classical mind. Teleportation of the quantum state of a material particle at macroscopic distances could be achieved in this way, as well as other demonstrations of nonlocality [Milman 2004].

Acknowledgments

Laboratoire Kastler Brossel is a laboratory of Université Pierre et Marie Curie and ENS, associated to CNRS (UMR 8552). We acknowledge support of the European Community, of the Japan Science and Technology corporation (International Cooperative Research Project: Quantum Entanglement).

COHERENCE AND DECOHERENCE EXPERIMENTS WITH FULLERENES

M. Arndt,[1] L. Hackermüller,[1] K. Hornberger,[1] A. Zeilinger[1, 2]

[1] *Institut für Experimentalphysik, Universität Wien, Boltzmanngasse 5, A-1090 Wien*

[2] *Institut für Quantenoptik & Quanteninformation, Austrian Academy of Sciences, Boltzmanngasse 3, A-1090 Wien*

zeilinger-office@exp.univie.ac.at

Abstract We present a review of recent experiments on molecular coherence and decoherence with fullerene molecules. Nearly perfect quantum interference with high fringe contrast can be observed in far-field diffraction as well as in near-field interferometry, when the molecules are sufficiently well isolated from their environment. This is true for ambient pressures below 10^{-7} mbar and internal temperatures below 1000 K. The fringe contrast decreases gradually as the interaction with the environment is smoothly turned on by either increasing the ambient pressure or by heating the molecules.

Keywords: Quantum optics, matter waves, molecule interferometry, decoherence, fullerenes

Introduction

At the beginning of the 20th century several fundamental discoveries were made in physics which lead to mind-boggling experiments and notions that seemed to escape any interpretation in terms of classical, pre-quantum physics. One key result of that time, which is closely connected to the development of Schrödinger's wave-equation, was the prediction by de Broglie [de Broglie 1923] and the experimental demonstration by Davisson and Germer [Davisson 1927] that massive particles also propagate in a wave-like manner.

Formally, we describe the state of the particle during the propagation as a coherent superposition of states, in particular of position states, that are classically mutually exclusive. A classical object will either take one or the other path for sure. A quantum object cannot be said to do that since the intrinsic information content of the quantum system is insufficient to allow such a description [Brukner 2002]. Matter wave interferometers prove this experimentally. The intriguing part is that a full interference visibility can only be obtained if we exclude all possibilities of detecting, even in principle, the

V.M. Akulin et al. (eds.), Decoherence, Entanglement and Information Protection in Complex Quantum Systems, 329–352.

path the object has taken. This seems to imply the philosophically challenging statement that in a wave-particle experiment the *"possibility of having a position"* is often the only objective reality in contrast to the property of *"having a well-defined position"*.

In addition to the duality between being localized and showing interference, we also find the duality between determinism and objective randomness. The entire pattern on the screen is well determined for the ensemble, but the detection of a single object is completely unpredictable in all experiments.

This is why Richard Feynman said the double-slit experiment to be at the heart of quantum mechanics [Feynman 1965]. One might argue that another central issue of quantum physics, namely entanglement, is missing in this example. However, it turns out to be an essential ingredient if we consider how which-path information is diffused into the environment, *i.e.*, if we include decoherence, as will be done in Sec. 3 and Sec. 4.

One of the remarkable properties of quantum mechanics is that the wave nature of matter completely escapes perception in our everyday life, although this feature is a cornerstone of the theory. The smallness of Planck's constant and therefore of the de Broglie wavelength of a macroscopic object is certainly largely responsible for the non-observability of quantum effects in the classical world. However, it is important to ask whether there are fundamental limits to quantum physics and how far we can push the experimental techniques to visualize quantum effects in the mesoscopic world for objects of increasing size, mass and complexity.

We shall therefore briefly review the experimental efforts in this field throughout the last century. Soon after Louis de Broglie proposed his wave hypothesis for material particles matter wave phenomena were experimentally verified for electrons [Davisson 1927], atoms and dimers [Estermann 1930], and neutrons [Halban 1936; Gähler 1991]. A replica of Young's double-slit experiment with matter waves was demonstrated by Jönsson for electrons [Jönsson 1974], by Zeilinger *et al.* for neutrons [Zeilinger 1988], by Carnal and Mlynek for atoms [Carnal 1991] and by Schöllkopf and Toennies for small molecules and noble gas clusters [Schöllkopf 1994; Schöllkopf 1996; Bruch 2002].

Further advances in matter wave physics with atoms were made possible by sophisticated techniques exploiting the interaction between atoms and light. Already in 1975 ideas were put forward for slowing and cooling of atoms using light scattering [Hänsch 1975; Wineland 1975]. The rapid progress of this field was recognized by the fact that the most important developments were recently awarded the Nobel prize for laser cooling [Chu 1998; Cohen-Tannoudji 1998; Phillips 1998] in 1997 and for the experimental realization of Bose-Einstein condensates with dilute atomic vapor [Anderson 1995; Davis 1995] in 2001. In Bose-Einstein condensates all atoms have extremely long de Broglie wavelengths and they are coherent over macroscopic distances up to a millimeter. However, similar to light quanta in a laser beam, the atoms in a Bose-Einstein condensate occupy a single-particle state with a macroscopic population and are weakly interacting. Therefore, in spite of their large number, the interfering object is still of relatively small mass and complexity. Experiments demonstrating interference between two Bose-Einstein condensates [Andrews 1997] can be viewed as a double-slit experiment with many individual atoms. This is witnessed by the fact that the fringe spacing is determined by the de Broglie wavelength which is derived from the mass of an individual atom, and not from the total mass of the condensate.

Different questions and new experimental challenges arise if we study particles in the opposite parameter regime where the interaction among the particles is strong. Covalently bound atoms form a new entity, a molecule or cluster, and the de Broglie wavelength of this system is determined by the total mass of all the atoms and by the center-of-mass velocity of the bound system. In the following we shall focus on these complex objects.

The very first demonstration of molecule interference dates back to the days of Estermann and Stern [Estermann 1930] who demonstrated experimentally diffraction of H_2 at a LiF crystal surface in 1930. Further experiments with diatomic molecules had to await progress and interest in atom optics. A Ramsey-Bordé interferometer was realized for the iodine dimer in 1994 [Bordé 1994] and was recently used [Lisdat 2000] with K_2. Similarly, a Mach-Zehnder interferometer was demonstrated [Chapman 1995 (a)] for Na_2. The near-field analog to the Mach-Zehnder interferometer, a Talbot-Lau interferometer, was recently applied to experiments with Li_2 [Berman 1997]. Diffraction at nanofabricated gratings also turned out to be the most effective way to prove the existence of weakly bound helium dimer [Schöllkopf 1996] and to measure its binding energy [Grisenti 2000].

Based on these historical achievements one may ask how far one might be able to extend such quantum experiments and for what kind of objects one might still be able to show the wave-particle duality. Recently, a new set of experiments exceeding the mass and complexity of the previously used objects by about an order of magnitude has been developed in our laboratory. These far-field diffraction experiments with the fullerene molecule C_{60} will be shown in Sec. 1.

Given the currently available techniques it seems however that far-field experiments are rather challenging when it comes to objects in the mass range of beyond 2000 amu and velocities still beyond 200 $\mathrm{m/s}$. This is why we investigated the feasibility of near-field Talbot-Lau interferometry with C_{70} as described in Sec. 2. This particular technique allows to work with a spatially incoherent beam and thus with a much increased count rate. Moreover it has a better wavelength scaling in the grating constant and can potentially be applied to much smaller wavelengths at a reasonable grating constant.

Once the high de Broglie interference visibility with large molecules has been established, it is straight forward to ask if there are experimental or fundamental limits on the way towards larger objects. Decoherence theory explains that a quantum object may loose some of its particular quantum properties due to the interaction with its environment. We investigate this phenomenon along two lines. Sec. 3 is dedicated to the influence of small angle collisions exerted by a dilute thermal gas on the molecules in the interferometer. Here we establish a quantitative limit for future interference experiments and we find that masses of up to 10^7 amu may still exhibit their wave-particle duality in an appropriate experiment. Sec. 4 is focused on a novel property which is unique to large objects with many internal degrees of freedom, namely the thermal emission of photons. If sufficiently many photons of sufficiently short wavelength are emitted they can spread which-path information about the molecule and thus lead to an effective loss of interference fringe visibility. Sec. 5 is finally dedicated to open questions in the quest for quantum experiments with macromolecules.

1. Far-field diffraction with C_{60}

The cage-like carbon molecules earned their name "buckminster fullerenes" or simply "fullerenes" because of their close resemblance to geodesic structures that were first discussed by Leonardo da Vinci [Saffaro 1992] and implemented in buildings by the architect Buckminster Fuller [Marks 1960]. This new modification of pure carbon was discovered in 1985 by Kroto *et al.* [Kroto 1985] and it is particularly stable when exactly sixty carbon atoms are arranged in one molecule to form the smallest natural soccer ball, the buckyball. From a practical point of view, fullerenes are well suited macromolecules for beam experiments, since they are available to a sufficiently large amount, and they can be sublimated to produce an effusive beam. Their particle nature is undoubted in all standard experiments, in particular when they are attached to a surface where one can see them as well localized balls in scanning tunneling microscopy [Chen 1991]. And yet, as discussed in the following, their wave-nature is equally well proven. Indeed, these molecules can be delocalized by more than one thousand times their own diameter.

Fullerenes and their derivatives not only represent the most massive and most complex single particles in interference experiments until recently, they also mark a *qualitative* step towards the mesoscopic world. In many aspects they resemble rather small solids than simple quantum systems: they possess collective many-particle states like plasmons and excitons, and they exhibit thermionic electron, photon and particle emission [Mitzner 1995; Hansen 1998] - which may be regarded as microscopic analogs of glow emission, blackbody radiation and thermal evaporation. Fullerenes contain about two

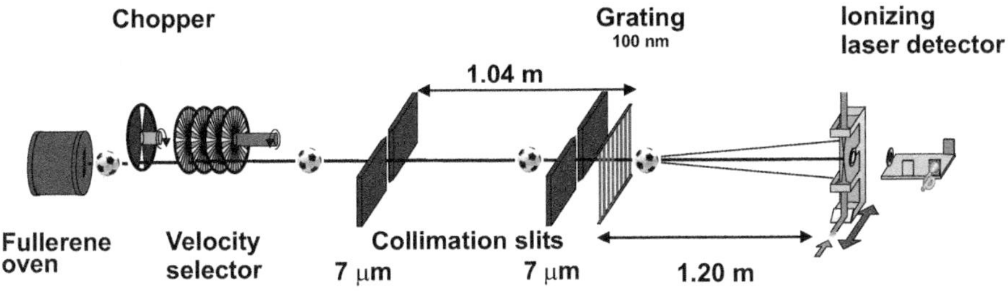

Figure 1. Setup of the far-field diffraction experiment. Fullerenes are sublimated, collimated by two narrow slits and diffracted at a nanofabricated SiN grating. The ionizing laser is tightly focused and scans over the molecular density distribution.

hundred different vibrational modes and highly excited rotational states which lead to broad optical lines, and both non-radiative and radiative transitions between many of them. It is therefore only natural to ask whether this internal complexity can can give rise to new decoherence phenomena.

To answer this question, we have set up the experiment as shown in Fig. 1. It resembles very much the standard Young's double-slit experiment. Like its historical counterpart, our setup also consists of four main parts: the source, the collimation, the diffraction grating, and the detector.

To bring the buckyballs into the gas phase, fullerene powder is sublimated in a ceramic oven at a temperature of about 900 K. The vapor pressure is then sufficient to eject molecules, in a statistical sequence, one by one through a small slit in the oven. The molecules have a most probable velocity v_{mp} of about 200 $\mathrm{m/s}$ and a nearly thermal velocity spread of $\Delta v/v_{\mathrm{mp}} \simeq 60\%$. Here Δv is the full width of the distribution at half height.

To calculate the expected diffraction angles, we first need to know the de Broglie wavelength which is uniquely determined by the momentum of the molecule $\lambda = h/mv$, where h is Planck's constant. Accordingly, for a C_{60} fullerene with a mass of $m = 1.2\times10^{-24}$ kg and a velocity of $v = 200\,\mathrm{m/s}$, we find a wavelength of $\lambda = 2.8$ pm. This is about 350 times smaller than the size of each molecule and about five orders of magnitude smaller than any realistic free-standing mechanical structure. We therefore expect the characteristic size of the interference phenomena to be small and a sophisticated machinery is needed to see the effect.

We used a free standing silicon nitride grating as the diffracting element with a nominal grating constant of $d = 100$ nm, slit openings of $s = 55\pm5$ nm and thickness of only 200 nm along the beam trajectory. These gratings are at the cutting edge of current technology and were provided by H. Smith and T. Savas, at MIT, Cambridge [Savas 1995].

We can now calculate the deflection angle to the first diffraction order in the small angle approximation as the ratio of the wavelength and the grating constant,

$$\theta = \frac{\lambda}{d} = \frac{2.8 \times 10^{-12}\ \mathrm{m}}{10^{-7}\ \mathrm{m}} = 28\ \mu\mathrm{rad}. \quad (1)$$

Since our detector is placed at 1.2 m downstream from the grating, the separation between the interference peaks at the detector amounts then to only $L \times \theta = 1.2\ \mathrm{m} \times 28\ \mu\mathrm{rad} = 34\ \mu\mathrm{m}$.

The detector has to comply with three different criteria: firstly, the spatial resolution must be sufficient to distinguish the diffraction orders, secondly, it has to be efficient and thirdly, it has to be selective. It must not detect any molecule in the vacuum chamber but the fullerenes. This is a challenging task without the addition of a mass spectrometer, since even at a background gas pressure of typically 10^{-8} mbar the density of air molecules exceeds the density of fullerenes by far.

We find that laser induced thermionic emission of electrons is a very well suited method for our purposes. The idea consists simply of irradiating the molecules by a tightly focused laser beam. When many photons are absorbed by a single molecule, the latter is heated such that it emits an electron. The resulting positive ion is then extracted and detected.

Thermionic detection of fullerenes has been studied by several groups with short pulse [Campbell 1991] and cw lasers [Helden 1998]. In contrast to coherent multi-photon schemes one has to deposit roughly five to ten times more energy than one would expect from the ionization potential at 7.6 eV [Steger 1992]. The reason for this is the fact that any stored energy rapidly dissipates into the various internal degrees of freedom. This heating eventually leads to the emission of an electron when the required ionization energy of 7.6 eV is concentrated on one of them. The mechanism has the great advantage to work for a broad range of laser wavelengths, in particular also in the blue / green spectrum, while a single-photon process would require a vacuum-UV light source. Moreover the intensity requirements are much less demanding than in a coherent multi-photon scheme since the photons do not have to arrive at the molecule simultaneously.

A continuous-wave green laser beam (argon ion laser, all lines) with a maximum power of up to 28 W is focused to the beam width of only 4 μm. As shown in Fig. 1, the vertically aligned laser beam runs orthogonal to the molecular beam. All molecules that pass the laser beam at or very close to the focus are heated to an internal temperature above 3000 K and ionize. The positive fullerene ions are then accelerated towards an electrode at 10 kV where they induce the emission of electrons. The electrons in turn are again multiplied and the charge pulses are subsequently counted. The overall molecule detection efficiency is about 10% and thus about two orders of magnitude higher than, for example, in electron beam impact ionization - which is used in many mass spectrometers. We find that our laser detector is sensitive only to fullerenes. This is due to their particular level scheme and their high stability against fragmentation. The counting background of 0.2 counts per second is solely due to electronic noise of the detector. Because of the tight focusing of the laser beam, the effective width of our detector [Nairz 2000] amounts to 8 μm, which is sufficient to resolve the individual diffraction orders. To record a diffraction pattern, we scan the laser across the molecular beam in steps of 2 μm. The interferograms shown in Figs. 2 and 4 represent molecule ion counts as a function of the transverse laser position.

Let us now turn to the coherence properties of our molecular beam. In general, coherence means that there is a fixed and well-defined phase relation in space and time between two or more wavefronts.

The spatial (transverse) coherence of our source is almost negligible right after the oven. Inside the source, the coherence width is actually only of the order of the thermal de Broglie wavelength. As is true in general for extended sources with uncorrelated emitters, the visibility is reduced by the fact that the many partial interferometers - each originating at one point in the source and forming two trajectories through the double-slit towards a point in the detector - acquire different phase differences along their path to a given spot on the screen.

Behind the oven, we therefore need to enlarge the *spatial coherence* width by about five orders of magnitude in order to illuminate at least two neighboring slits in the 100 nm grating coherently. The spatial coherence is essentially determined by the geometry of the experiment and grows linearly with increasing distance from the source and with decreasing size of the first collimation slit. This general rule for the influence of collimating elements on transverse coherence is commonly known as the van Cittert-Zernike theorem[1]. It states that the spatial coherence function of a spatially incoherent beam can be obtained from the diffraction curve of a coherent beam which is prepared by the same apertures along the beam path. The limiting and preparing element in our case is in particular the first collimation slit.

Obviously the gain in coherence has to be paid for by a dramatic drop in the count rate because the signal decreases quadratically with the distance from the source and linearly with the size of each collimation slit. Although the first collimating slit alone already provides coherence, we still have to introduce a second collimating slit - in our case also 7 μm wide and about 1 m downstream from the first slit. The reason for this is the requirement that the collimated beam width needs to be significantly smaller than the separation between the diffraction orders behind the grating in order to clearly resolve the diffraction peaks in Fig. 2.

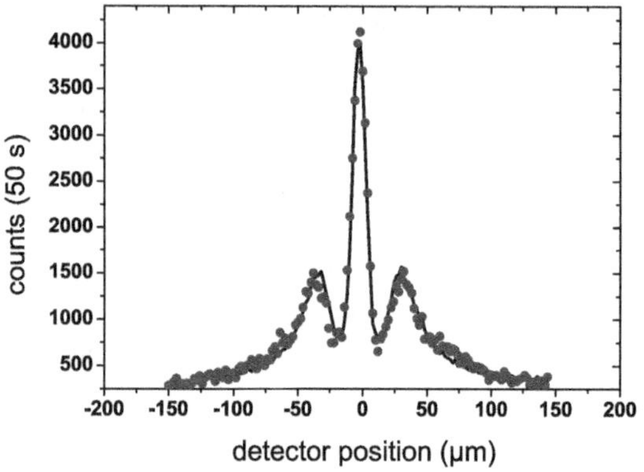

Figure 2. Far-field diffraction curve recorded with the full thermal distribution of the C_{60} beam [Nairz 2003]. The velocity is centered around $v = 200\ \mathrm{m/s}$ and has a FWHM of $\Delta v/v = 0.6$.

The *spectral coherence* of the source enters as well because molecules with different velocities and therefore different de Broglie wavelengths follow different diffraction angles. The detector records the sum of the correspondingly stretched or compressed diffraction pictures so that the interference pattern gets washed out. In contrast to the spatial contribution, there is no gain in longitudinal (spectral) coherence during free flight. This is due to the fact that different velocity classes will evolve separately. In a pulsed beam experiment one would observe a chirped packet, that is, a wave packet with short wavelengths in the pulse lead and long wavelengths in its tail. Even though the packet spreads out in the course of its evolution, the coherence does not grow since the momentum distribution remains unchanged.

The picture of a single wave packet is in any case inappropriate for the description of a continuous source because it implies a well-defined internal phase structure. This is not provided in our experiment, and the beam can only be regarded as a statistical, and therefore incoherent, mixture of the various momenta. Nevertheless, the beam can operationally be characterized by a coherence length, which is the length that measures the fall-off of the interference visibility when the difference between two interfering paths increases. The longitudinal coherence length is determined by the Fourier transformation of the velocity distribution and it is of the order $L_c \simeq \lambda^2/\Delta\lambda = \lambda v/\Delta v$.

In elementary textbooks the diffraction equation Eq. (1) is usually derived by noting that the first constructive interference occurs when the difference between two neighboring paths is equal to one de Broglie wavelength, see

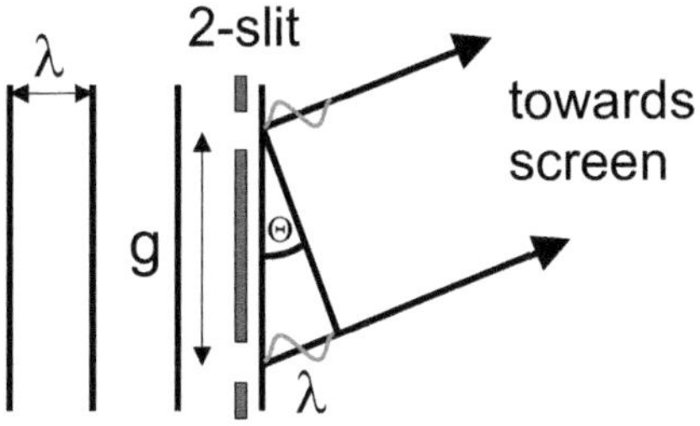

Figure 3. First order far field diffraction at a two slit grating.

Fig. 3. For our thermal beam of Fig. 2 with $\Delta v/v \simeq 0.6$ we find $L_c \simeq 1.7\lambda$, which is just enough to guarantee the existence of the first order interference fringes. We can clearly discern the first interference orders on both sides of the central peak. But the limited coherence is reflected by the fact that we cannot see any second or higher order peaks in the interferogram of Fig. 2.

To see more fringes we have to increase the coherence length and therefore decrease the velocity spread. For this purpose we employ a mechanical velocity selector, as shown after the oven in Fig. 1. It consists of four slotted disks that rotate around a common axis. The first disk chops the fullerene beam. Only those molecules are transmitted which traverse the distance from one disk to the next in the same time that the disks rotate from one open slot to the next. Although two disks would suffice for this purpose, the additional disks decrease the velocity spread even further and help to eliminate velocity sidebands. By varying the rotation frequency of the selector, the desired velocity class of the transmitted molecules can be adjusted. To measure the time of flight distribution we chopped the fullerene beam with the chopper right behind the source (see Fig. 1). The selection is of course accompanied by a significant loss in count rate, but we can still retain about 7% of the unselected molecules.

In contrast to the unselected case, the velocity spread now amounts to $\Delta v/v = 17\%$. This increase in longitudinal coherence by a factor of more than three allows for the observation of higher order diffraction peaks, as can be seen in Fig. 4.

It should also be pointed out that by using the velocity selector we can now choose a slower mean velocity centered about $136\ \mathrm{m/s}$, which corresponds to a de Broglie wavelength of 4.1 pm. It is obvious that this increase in wavelength results in a wider separation of the diffraction peaks, which can be seen by comparing Figs. 2 and 4.

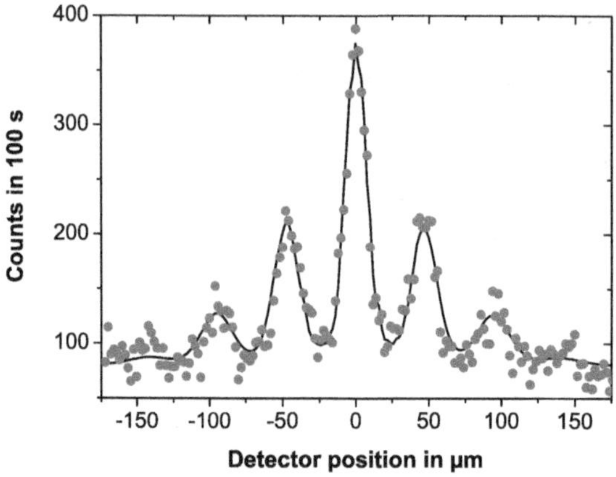

Figure 4 Far-field diffraction curve recorded with a velocity selected C_{60} beam, where the longitudinal coherence length has been improved by more than a factor of three [Nairz 2003]. The velocity is centered at $v = 136$ m/s and has a FWHM of $\Delta v/v = 0.17$.

In principle, the diffraction patterns can be quantitatively understood within the Fraunhofer approximation of Kirchhoff's diffraction theory as described in any optics textbook (*e.g.*, [Hecht 1994]). However, Fraunhofer's optical diffraction theory misses an important point of our experiments with matter waves and material gratings: the attractive interaction between the molecule and the wall results in an additional phase of the molecular wavefunction [Grisenti 1999]. Although the details of the calculations are somewhat involved[2], the qualitative effect of this attractive force on far-field diffraction can be understood as a narrowing of the real slit width to an *effective* slit width [Brühl 2002]. For our fullerene molecules the reduction can be as big as 20 nm for the unselected molecular beam and almost 30 nm for the slower, velocity selected beam. The stronger effect on slower molecules is due to the longer and therefore more influential interaction between the molecules and the wall.

The full lines in Figs. 2 and 4 are fits of our data to this modified Kirchhoff-Fresnel theory. To obtain such a good fit we also have to take into account an enhanced contribution in the zeroth order which we attribute to mechanical defects (holes) of the grating which are significantly larger than the grating period.

It is important to note that the interference pattern is built up from single, separate particles. There is no interference between two or more particles during their evolution in the apparatus. *Single particle interference* is evidenced in our case by two independent arguments.

The first argument is based on the spatial separation between the molecules. The molecular flux at an average speed of 200 m/s is about 3×10^9 cm^{-2}.s^{-1} at the plane of the detector. This flux corresponds to an average molecular density of 1.7×10^{11} m^{-3} or an average molecular distance of 200 μm. This is three orders of magnitude wider than any realistic range of molecular (van der Waals) forces, which are typically confined to below 100 nm.

The second argument is based on the fact that interference occurs only between indistinguishable states. All molecules in the beam are in different states. There are 174 different vibrational modes and also the rotational modes are populated at various energies with a Boltzmann distribution. The chance of having two subsequent molecules in exactly the same state of all internal modes is vanishingly small. Therefore, interference in our experiments really is a single particle phenomenon.

2. Near-field interferometry with C_{70}

Several challenges arise when one moves from small to large systems. At a given velocity the de Broglie wavelength $\lambda = h/(mv)$ shrinks with increasing mass, and in addition it becomes increasingly difficult to slow down massive systems which have a large kinetic energy. The requirements on interferometer technology for de Broglie wavelengths in the sub-picometer range are therefore rather demanding.

One can conceive various experimental arrangements to demonstrate the wave-nature of material particles and many interferometers have already been built for molecules as mentioned in Sec. 1. However, all these arrangements needed well collimated beams or experimentally distinguishable internal states in order to separate the various diffraction orders. This requirement makes them less suitable for large clusters and molecules for which brilliant sources and highly efficient detection schemes still have to be developed.

In contrast to these near field setups, a near-field interferometer of the Talbot-Lau type, does away with the collimation requirement and accepts spatially incoherent molecular beams. Such a scheme is much more compact, rugged and allows a much higher transmission [Clauser 1992].

The basic idea of this device, the lens-less periodic imaging through the Talbot effect [Talbot 1836] of molecular density distributions, has been frequently applied in optics with light, X-rays and ultra-sound [Patorski 1989] and it was also recently applied to massive objects [Chapman 1995 (c)]. The general idea can be understood from the rule that the separation D of neighboring diffraction peaks in the image plane at distance L_T from the grating should equal the grating constant g:

$$\Theta \simeq \sin\Theta = \frac{\lambda}{g} \simeq \tan\Theta = \frac{D}{L_T} \stackrel{!}{=} \frac{g}{L_T} \Rightarrow L_T = \frac{g^2}{\lambda}. \tag{2}$$

However, this simple explanation of Talbot images still requires plane waves corresponding to a highly collimated and therefore weak input beam. The full intensity gain of the Talbot effect is only deployed when it is applied to uncollimated and therefore much more intense molecular beams [Clauser 1994; Brezger 2002]. This is realized if the single diffraction grating is re-

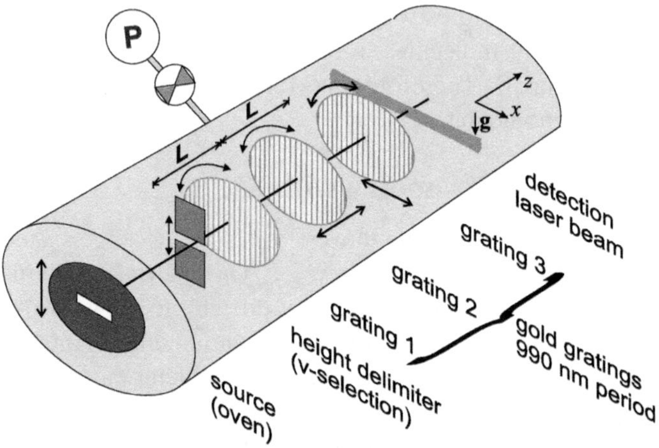

Figure 5. Artist's view of the Talbot-Lau interferometer. A beam of C_{70} molecules is generated from fullerene powder at 900 K. The beam passes a series of three gold gratings each with a grating constant of $d = 990$ nm, an open width of 480 nm and a grating thickness of 500 nm. The grating separation L was set to 38 cm. The whole vacuum chamber is evacuated to 2×10^{-8} mbar and can then be pressurized with different gases, typically up to 10^{-6} mbar. The fullerenes are again detected using laser ionization. The molecule interferogram is scanned by shifting the third grating along the grating vector [Brezger 2002; Hornberger 2003 (a)].

placed by three gratings which act as a multiplexing collimator, a diffraction grating and a detection mask (for details see, *e.g.*, [Brezger 2003]). Each point in the grating then acts as the source of an interference pattern and, even though there is no coherence between different source points, the independent interference patterns originating from each of them overlap in a position-synchronized manner to form a pattern of high fringe visibility. One may also regard the first grating as a tool to impose some coherence on the uncollimated molecular beam. The finite width of each opening in the first grating induces then a lateral coherence at the second grating which is of the order of 2 - 3 grating periods.

Our experimental setup is based on this idea, and a sketch of it is shown in Fig. 5. The effusive C_{70} beam is produced in the same source as the C_{60} in Sec. 1 but this time the beam is essentially uncollimated. However, it is still vertically selected by three spatially separated height delimiters, namely the oven aperture (200 μm), a central height delimiter (between 50 and 150 μm), and the detector laser beam with a gaussian beam waist of 8 μm, which is now horizontally oriented and fixed in position. By shifting the oven up and down

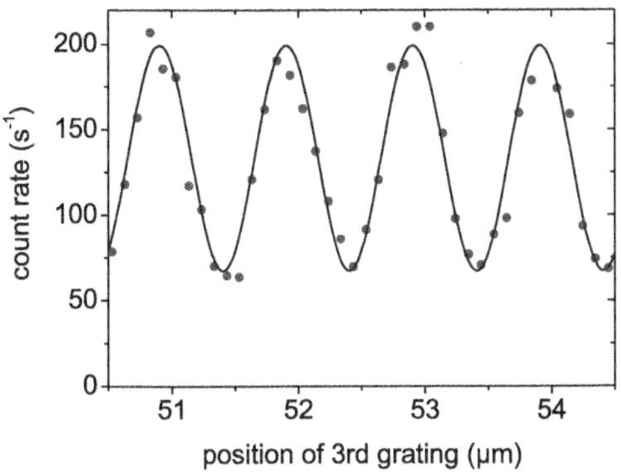

Figure 6 Interference fringes in C_{70} Talbot-Lau interferometry. The periodicity of the signal corresponds to the grating constant [Hornberger 2003 (a)].

one can then select a well-determined free-flight parabola and thus a certain velocity class. This way, mean velocities could be chosen from $90\ \mathrm{m/s}$ to $220\ \mathrm{m/s}$ with the full width at half maximum of their distributions ranging from 7% to 17% of the mean velocity. The gravitational method is superior to the set of rotating slotted disks of Sec. 1 because it avoids additional vibrations and allows up to 100% throughput at the central selected velocity.

The transverse coherence of the beam is unprepared until the molecules pass the first grating. The three gold gratings have a period of 990 nm, a nominal open fraction of 0.48 ± 0.02 (as specified by the manufacturer Heidenhain, Traunreut) and a flat open field of roughly 16 mm diameter. We limit the lateral width of the molecular beam to 1 mm which is also comparable to the width of the ionization range of the detecting laser beam [Nairz 2000]. The distance between consecutive gratings was set to $L = 0.38$ m. This corresponds to roughly twice the Talbot length and enables us to observe two Talbot recurrences (see Fig. 7). All gratings can be rotated around the molecular beam to align them with an accuracy of about 1 mrad both with respect to each other and to the direction of earth's acceleration. The third grating G_3 masks the molecular density pattern behind the second grating. G_3 is mounted on a piezo translation stage (Piezosystem Jena) and is scanned perpendicular to the molecular beam in steps of 100 nm. Those molecules which are in phase with the openings of G_3 pass the grating and are heated by the crossing Ar^+ laser beam (488 nm, 15 W). The positive ions which are generated by the laser induced thermionic emission are counted as a function of the lateral position x_s of G_3. For our particular experiment we expect an almost sine-wave shaped molecule interferogram $S(x_s)$. This is indeed observed in our experiment, and the fringe contrast $V_\lambda = (S_{max} - S_{min})/(S_{max} + S_{min})$ serves to characterize the interference pattern as shown in Fig. 6. Before taking the experimental signal as evidence for quantum interference one must discuss the potential classical explanation of periodic fringes due to a Moiré-like effect. The classical expectation can be calculated by propagating the classical phase space density of an uncollimated particle stream through the interferometer subjected to the same forces and approximations as in the quantum case [Hornberger 2004]. In the presence of van der Waals interactions the deflection tends to be underestimated in the classical analog of the eikonal approximation, so that our classical calculation gives an upper limit for the classical visibility. Hence, whenever the experimental contrast is significantly greater than the classical value one has evidence that quantum interference took place. An even stronger proof is the characteristic wavelength dependence of the fringe visibility. Varying the mean molecular velocity corresponds to changing the mean molecular de Broglie wavelength. This dependence is used to scan the Talbot length L_λ and to demonstrate the periodic wave nature of the molecular Talbot Lau effect in Fig. 7.

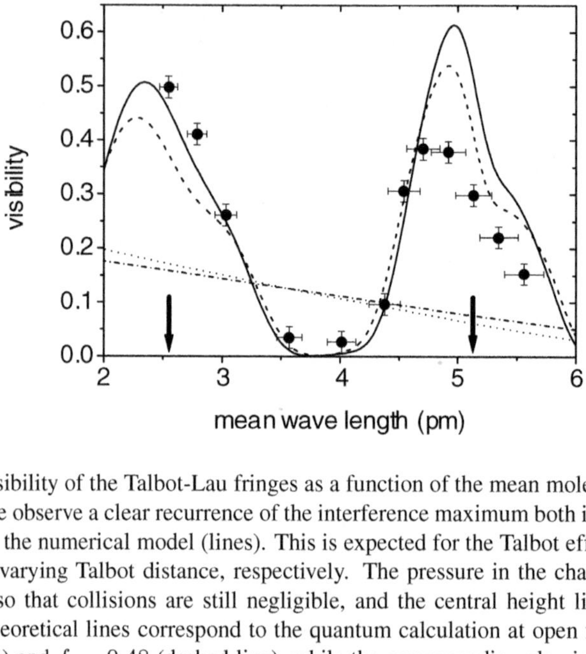

Figure 7. Visibility of the Talbot-Lau fringes as a function of the mean molecular de Broglie wavelength. We observe a clear recurrence of the interference maximum both in the experiment (circles) and in the numerical model (lines). This is expected for the Talbot effect with varying wavelength or varying Talbot distance, respectively. The pressure in the chamber was below $3.\,10^{-8}$ mbar so that collisions are still negligible, and the central height limiter was set to $50\ \mu$m. The theoretical lines correspond to the quantum calculation at open fractions of $f = 0.45$ (solid line) and $f = 0.48$ (dashed line), while the corresponding classical expectation is shown by the dotted ($f = 0.45$) and dash-dotted line ($f = 0.48$). The arrows indicate the wavelengths where the Talbot criterion $L = m$, $L_\lambda = mg^2/\lambda$, $m \in \mathbb{N}$, is met with $m = 1$ for $\lambda = 2.58$ pm and $m = 2$ for $\lambda = 5.14$ pm. With respect to ideal gratings the true maxima are slightly shifted to smaller wavelengths and the two maxima have different heights. This is due to the interaction between the molecule and the wall [Hackermüller 2003].

The experimental data are shown as full circles and are generally well represented by the quantum theoretical calculation (solid line). Both clearly show the expected recurrence of the visibility with λ_{dB}. The classical expectation (dash-dotted line) completely fails to reproduce the observed effects. Note that the visibility peaks are neither equally high nor symmetric around their maxima. Also, the peaks do not occur exactly at the Talbot length (indicated by arrows in Fig. 7). Instead the maxima are shifted to shorter wavelengths and the peaks have a broad shoulder towards the longer wavelengths. These deviations from the simple optical Talbot-Lau effect appearing in the experiment are reproduced in the quantum calculation once we take into account the retarded van der Waals interaction [Casimir 1948] between the polarizable molecules and the gold bars of the gratings. This interaction reduces the effective slit width. At fixed grating distance this corresponds to a shift of the maxima towards smaller wavelengths.

Fig. 7 shows two theoretical predictions which represent the experimental data. The dashed line assumes a grating with an open fraction, *i.e.*, a ratio of opening to grating constant, of 0.48 as originally specified when the gratings were purchased and mounted about two years ago. The solid line assumes an open fraction of 0.45. A possible explanation for the apparent shrinking of the openings might be a deposited layer of fullerenes which are also visible to the unaided eye, at least on the first grating. But also tiny mechanical grating deformations might be a reason.

We observe a notable difference between theory and observation in the peak height at a wavelength around 5 pm, corresponding to molecules with a mean velocity of around $100\ \mathrm{m/s}$. We have evidence for the hypothesis that the experimental reduction of the interference contrast is mainly due to remaining vibrations of the setup with oscillation amplitudes of a few ten nanometers. Further investigations of this effect are still under way.

3. Collisional decoherence

The technical challenges in establishing macromolecule interferometry are already rather demanding. But even more important are the potential mechanisms which may lead to decoherence, that is to a loss of visibility in the interference pattern due to the coupling of the quantum system to its environment (see, *e.g.*, [Joos 1985; Zurek 1991; Joos 2003]).

The investigation of this seeming loss of quantum properties has become an important corner stone of modern quantum physics because of its fundamental role in mesoscopic physics and its importance for the understanding of the quantum-classical transition.

Any increase in size and complexity generally opens new decoherence channels, and for large molecules one can think of many interactions with the environment, either by scattered radiation [Chapman 1995 (b)], by collisions with particles [Hornberger 2003 (a)], by an interaction with fluctuating quasi-static electro-magnetic fields [Myatt 2000] or even by the interaction with gravitational waves [Reynaud 2002].

While it is impossible to manipulate and track the details of the perturbations for truly macroscopic systems, the environment of isolated mesoscopic quantum systems can still be efficiently controlled. The present section focuses on one particular interaction between large molecules and the environment, namely on collisions between the coherently propagating molecules and various background gases [Hornberger 2003 (a)].

We use the same near-field interferometer as in the previous section and introduce various gases at low, controllable pressure (between 5.10^{-8} mbar and 2.5×10^{-6} mbar) into the vacuum chamber. Each collision between a fullerene molecule and a gas particle entangles their motional states. In order to obtain the properties of the isolated quantum system one has to trace over the state of the scattered molecule. We assume that the mass of the fullerene molecule is much greater than the mass m_g of the gas particle and find that the density operator for the fullerene molecule alone, $\rho_0(\vec{r}, \vec{r}')$, is modified by a multiplicative factor because of the collision

$$\rho\left(\vec{r}, \vec{r}'\right) = \rho_0\left(\vec{r}, \vec{r}'\right) \eta\left(\left|\vec{r} - \vec{r}'\right|\right). \tag{3}$$

This factor η may be called the decoherence function since it describes the effective loss of coherence in the fullerene state. For elastic scattering with an isotropic potential and the gas initially in a thermal state it reads [Hornberger 2003 (b)]

$$\eta(\delta r) = \int_0^\infty dv_\mathrm{g} \frac{g(v_\mathrm{g})}{\sigma(v_\mathrm{g})} \int d\Omega \, |f(\cos\theta)|^2 \,\mathrm{sinc}\left[\sin\left(\frac{\theta}{2}\right) \frac{2m_\mathrm{g}\, v_\mathrm{g}\, \delta r}{\hbar}\right]. \tag{4}$$

This expression involves an integration over the thermal distribution $g(v_\mathrm{g})$ of the gas velocities and an integral over the scattering angle $\Omega = (\theta, \phi)$. In the argument of the sinc function one finds the distance between the two superposed position states times the momentum change in units of $\hbar$. Hence, the sinc function suppresses the integrand whenever the change in the state of the gas particle during a collision is able to resolve the distance δr. This leads to a reduction of the corresponding off-diagonal elements in Eq. (3) when the gas particle transmits (partial) position information about the molecule to the environment. For small distances the sinc approaches unity so that the angular integral yields the total scattering cross section $\sigma(v_\mathrm{g})$. Hence, for $\delta r \to 0$ the decoherence function approaches unity as required from the conservation of the trace in Eq. (3).

As described above, the Talbot-Lau interferometer is essentially a lens-less imaging device which produces a strictly periodic molecular density pattern immediately in front of the third grating. This pattern can therefore be decomposed in a Fourier series. It turns out [Hornberger 2004] that the effect of collisional decoherence can be treated analytically and that it is completely described by a multiplication of the even Fourier coefficients (B_{2m}) by the factor

$$B_{2m}^{(\lambda)} \to B_{2m}^{(\lambda)} \exp\left(- n\,\sigma_{\rm eff} \int_0^{2L} \left[1 - \eta \left(m\, \frac{L - |z-L|}{L_\lambda}\, d \right) \right] dz \right). \quad (5)$$

Here $n\,\sigma_{\rm eff}$ is the number density of gas particles times the effective total cross section defined below in Eq. 6, describing the number of collisions per unit length. The integral in the exponent covers the various positions in the interferometer where a collision may occur. As discussed above we have $\eta(0) = 1$, and the function decreases to zero for increasing arguments. It follows that the $m = 0$ component, related to the mean flux through the interferometer, is not affected by decoherence. The other components of the interference signal are most sensitive to collisions occurring close to the second grating, at $z = L$, where the path separation is greatest. Indeed, if the Talbot criterion is met, $L = lL_\lambda$, $l \in \mathbb{N}$, the distances entering the decoherence function are integer multiples of the grating period d. For the other z positions the sensitivity decreases according to the path separation and, as one expects, a collision event will not contribute to decoherence directly at the first or at the third grating, at $z = 0,\ 2L$.

In our experiment each collision with a background gas particle is so strong that it localizes the fullerene to within a few nanometers. This is very small compared to the typical path separation of 1 μm. One can therefore safely approximate the integral in Eq. (5) by $2L$ if $m \neq 0$. On the other hand, the effective scattering cross section must be evaluated with care, since it must account for the longitudinal velocity $v_{\rm m}\,\vec{e}_z$ of the fullerene and for the thermal distribution $\mu(\vec{v}_{\rm g})$ of the gas particle velocities. The general expression reads

$$\sigma_{\rm eff}(v_{\rm m}) = \int \mu(\vec{v}_{\rm g})\, \sigma\left(|v_{\rm m}\,\vec{e}_z - \vec{v}_{\rm g}|\right) \frac{|v_{\rm m}\,\vec{e}_z - \vec{v}_{\rm g}|}{v_{\rm m}}\, d\vec{v}_{\rm g}. \quad (6)$$

In our experiment the interaction potential is well described by the isotropic London dispersion force, *i.e.*, the van der Waals force between polarizable molecules. The corresponding potential $U(r) = -\,C_6/r^6$ has a single parameter C_6 that can be found in [Hornberger 2003 (a)] for a number of gases. The cross section $\sigma(v)$ for a fixed relative velocity follows from a semiclassical calculation [Maitland 1981] and the remaining integration in Eq. (6) can be performed asymptotically. This effective cross section exceeds the geometric one by two orders of magnitude at the velocities of our experiment ($v_{\rm m} = 80 \ldots 240$ m/s).

A further simplification comes from the fact that for our experimental setup the visibility of the interference signal is essentially determined by the zeroth and first Fourier components only, which means that we expect a quasi sinusoidal interference pattern. The expected reduction of the contrast is therefore easily evaluated:

$$V_\lambda(p) = V_\lambda(0)\ \exp\left(-\frac{2L\,\sigma_{\text{eff}}}{k_B T}\,p\right) = V_\lambda(0)\ \exp\left(-\frac{p}{p_0}\right). \tag{7}$$

The exponential decrease of the visibility as a function of the gas pressure p is the expected experimental signature of collisional decoherence. Although similar to Beer's law for absorption, it should not be confused with it since absorption alone would lead to an exponential decrease of the mean signal at *constant* visibility.

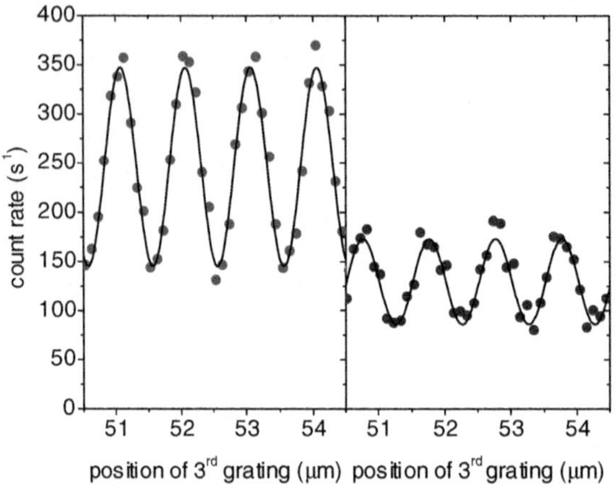

Figure 8 Left: C_{70} interference fringes at a pressure of 3.10^{-8} mbar (residual background gases) shown as full circles. Right: the same signal in the presence of argon gas, at a pressure of 5.10^{-7} mbar. The lines are fits of a sine function. The mean velocity of the fullerene molecules was $189\ \text{m/s}$.

The decohering influence of collisions can be seen in Fig. 8. It shows the change in the interference pattern of C_{70} if a small amount of argon gas is added to the vacuum chamber. We observe a significant reduction in visibility from 42% to 34% if the pressure in the chamber is increased from 3.10^{-8} mbar to 5.10^{-7} mbar. The horizontal shift between the two curves is not significant since it can be explained by thermal drifts of the setup between the two recordings. In contrast to that, the values of the visibilities are significant and reproducible within $\pm\ 2\%$.

We have recorded the visibility for a series of interferograms at different gas pressures and we observe the expected exponential decay. A good quantitative agreement with decoherence theory is obtained after taking into account the details of the velocity selection in our experiment. For example our predictions for the visibility are obtained by weighting Eq. (7) with the classical velocity distribution in the detector - which corresponds to an averaging over

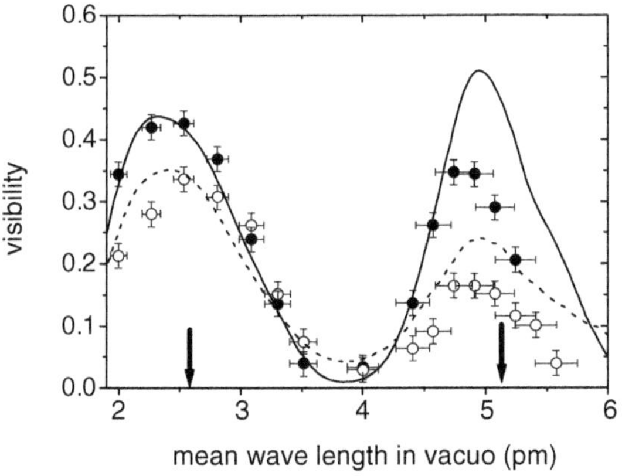

Figure 9 Talbot-Lau interference visibility for different pressures [Hackermüller 2003]. The reduction of the visibility with respect to the theoretical expectation at $\lambda_{\rm dB} = 5$ pm can be attributed to grating vibrations.

a distribution of de Broglie wavelengths. Also the reduction of the mean count rate found in Fig. 8 is well reproduced by our calculation.

The loss of coherence with increasing pressure is conveniently described by the "decoherence pressure" p_0 defined in Eq. (7). Its value depends on the specific nature of the colliding gases and has to be computed quantum mechanically. The experimental values vary between $5.\,10^{-7}$ and 1.3×10^{-6} mbar among He, Ne, Ar, Kr, Xe, air and methane. They are all well described by our theoretical model [Hornberger 2003 (a)].

Note that decoherence pressure depends rather weakly on the specific collision gas partner. This can be explained by assuming that the polarizability of the gas particle is proportional to its mass m_g. Then also C_6 is proportional to m_g, and the mass dependencies of the interaction constant and of the most probable gas velocity almost cancel out leaving $\sigma_{\rm eff} \propto m_g^{1/10}$.

It should also be mentioned that the total time of flight plays an important role for the absolute value of the fringe contrast. Clearly, for slow molecules the interaction time with the gratings is longer, and vibrations will be more detrimental. Also, the effective cross section (6) increases in general for decreasing molecular velocities. In Fig. 9 we observe the increasing influence of collisions in an interferometer filled with argon, for various fullerene velocities, *i.e.*, various interaction times.

Fig. 9 shows the experimental visibility curves for $p_{\rm low} = 3.\,10^{-8}$ mbar (full circles) and $p_{\rm high} = 5.\,10^{-7}$ mbar (hollow circles) and compares them to the quantum calculation (solid and dashed line, respectively) with the same model parameters as already used for Fig. 7. The remarks of the discussion of Fig. 7 apply also here. This holds for the reduction of the visibility at long wavelengths due to vibrations, the shift of the maxima with respect to the Talbot length and the asymmetric line shapes - caused by the molecule-wall interaction. The new feature in this graph is the contrast reduction due to the scattering events in the argon atmosphere and their dependence on the interaction time. For wavelengths in the 2.5 pm regime ($v \simeq 200$ m/s) the pressure increase leads to the ratio $V_\lambda(p_{\text{high}})/V_\lambda(p_{\text{low}}) = 0.8$ whereas for wavelengths in the 5 pm regime ($v \simeq 100$ m/s) the increased pressure results in $V_\lambda(p_{\text{high}})/V_\lambda(p_{\text{low}}) = 0.5$. These ratios are identical for theory and experiment, both for the slow and for the fast molecules. However, the *absolute* values still differ at long wavelengths. Again this shows that the different decoherence mechanisms (collisions and vibrations) are independent of each other and their effects can be considered as separate multiplicative factors to the visibility.

4. Thermal decoherence

Collisional decoherence is a universal mechanism which occurs for objects of any size. In contrast to that, the following section describes a mechanism which only appears for complex quantum systems, namely decoherence,thermal due to the emission of their own heat radiation.

All large objects, but also molecules of sufficient complexity, are able to store energy and to interact with their environment via thermal emission of photons. It is generally believed that warm macroscopic bodies emit by far too many photons to allow the observation of de Broglie interferences, whereas individual atoms or molecules can be sufficiently well isolated to exhibit their quantum nature. Obviously there must be a transition region between these two limiting cases. Interestingly, as we have shown recently [Hackermüller 2004], C_{70} molecules have just the right amount of complexity to exhibit high quantum interference contrast at temperatures below 1000 K, as shown in the previous sections, and to gradually lose all their quantum behaviour when the internal temperature is increased up to 3000 K. One can thus trace the quantum-to-classical transition in a controlled and quantitative way.

The complexity of large molecules adds a novel quality with respect to previously performed experiments with atoms [Pfau 1994; Chapman 1995 (b); Kokorowski 2001] in that the energy in molecules may be equilibrated in many internal degrees of freedom during the free flight and a fraction of the vibrational energy will eventually be reconverted into emitted photons. Therefore the internal dynamics of the molecule is also relevant for the quantum behaviour of the center-of-mass state.

The basic setup of our experiment is sketched in Fig. 10. In variance to the collisional coherence experiment the vacuum is pumped to 2.10^{-8} mbar, but the molecules now pass a heating stage where they cross a multiply folded and focused argon ion laser beam. The laser heating increases the molecular temperature by 140 K per absorbed photon up to temperatures of the order of

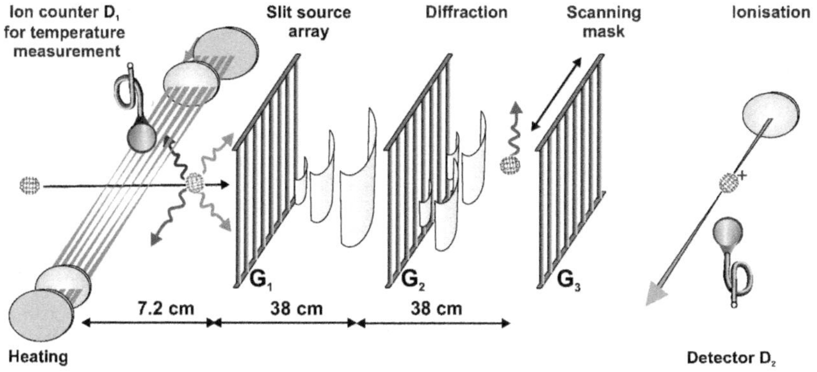

Figure 10. Setup for the investigation of thermal decoherence. Up to 16 laser beams are used to heat the fullerenes before they enter the Talbot-Lau interferometer.

5000 K. Reemission of thermal photons is very efficient so that even the hottest molecules are cooled to below 3000 K before they cross the first interferometer grating 7.2 cm downstream of the heating stage.

The essence of the experiment is now to measure the variation of the interference fringe visibility. Fig. 11 clearly shows a non-trivial monotonic decrease of the interference contrast with increasing laser heating power. This is the unambiguous signature of decoherence which we attribute to the enhanced probability for the emission of thermal photons that carry which-path information.

We also note that the count rate varies considerably as the heating is varied. This can be understood by the dependence of the final ionization efficiency on the internal energy of the fullerenes. It proves that much internal energy remains in the molecules during their flight through the apparatus.

We model the evolution of the distribution of the internal energies on their way through the interferometer. The temperature dependence of the spectral photon emission rate then yields the loss of fringe visibility as predicted by decoherence theory. The first photon absorption populates the electronic triplet state T_1 via the excited singlet S_1. Given the known C_{70} triplet life-times and non-radiative transition rates [Dresselhaus 1998] we can assume both that all further excitation occurs in the triplet system and that the absorbed excess energy is rapidly transferred to the vibrational levels. It is known that fullerenes may store more than 100 eV for a very short time, and it was observed that at high temperatures three different cooling mechanisms start to compete, which are the thermal emission of either photons, electrons, or C_2 dimers [Mitzner 1995]. These processes are the molecular analogues of the bulk phenomena known as blackbody radiation, thermionic emission and evaporative cooling

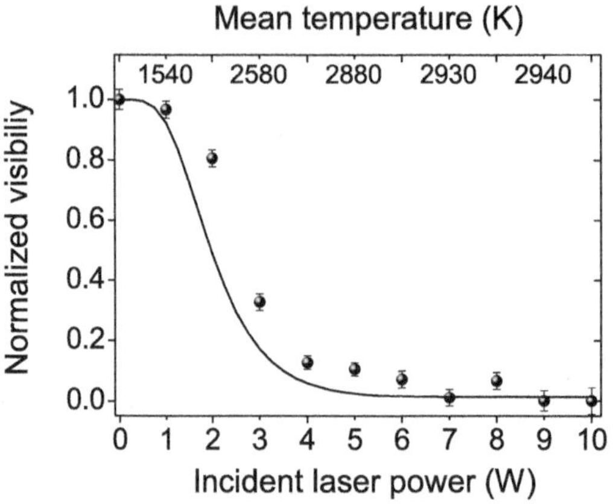

Figure 11 Decrease of the fringe visibility with increasing laser heating power. The upper scale gives the corresponding mean fullerene temperature at the first grating [Hackermüller 2004].

already mentioned in Sec. 1. Following the most recent experimental data [Matt 1999] we assume that fragmentation is the least efficient mechanism. In contrast to that, thermally activated ionization is an important mechanism which we use both in our fullerene detector [Nairz 2000] and for molecule thermometry, as discussed below. Nevertheless, we can safely neglect both delayed ionization and fragmentation for the discussion of the fringe contrast since the recoil upon fragmentation and ionization is generally so large that the affected molecules will miss the narrow detector. We have also experimentally confirmed that neither ions nor C_{68} or smaller fragments from the heating region are recorded by the detector D_2. However, ions - and potentially ionized fragments - can be detected immediately above the heating stage by the electron multiplier D_1 (Fig. 10). To get an estimate of the molecular temperature distribution we record the number of ions as a function of the heating power and of the fullerene velocity. By comparing the data to a model calculation we can extract the parameters which govern the molecular heating. The same calculation also describes the heating dependent increase in count rate at the detector D_2 and yields independent information on the temperature distribution in the molecular beam.

The mean temperature in the beam drops rapidly behind the heating stage through the emission of thermal photons. The emission of a continuous photon spectrum has already been observed for fullerenes in other experiments [Mitzner 1995; Heszler 1997]. In our calculation we take into account that the functional form of the photon spectrum differs from the simple Planck law. Firstly, the thermal wavelengths are much larger than the size of the fullerene and usual mode density factor has to be multiplied by the known, frequency dependent absorption cross section. Secondly, the particle is much hotter than its environment and stimulated emission does not occur. Finally, C_{70} has a finite heat capacity so that its internal energy changes during a photon emission. Nonetheless the internal energy is conveniently characterized by a (micro-canonical) temperature.

We find that at temperatures below 2000 K the emission rate is negligible, while at higher temperatures the molecules may emit photons whose wavelengths are comparable to or even smaller than the maximum path separation of 1 μm. Photons with wavelengths below 2 μm then transmit (partial) which-path information to the environment and lead to a reduction of the fullerene interference fringe contrast. Around 3000 K the molecules have a high probability to emit several visible photons. This leads to a complete vanishing of the fringe visibility in our interferometer. A formal description of this qualitative picture can be given by decoherence theory. It considers the entanglement of the molecule with the emitted photon and shows how coherences vanish once a trace over the photon state is performed. Since only the center of mass motion is concerned the predictions of decoherence theory are quantitatively consistent with the Heisenberg's argument that any observation will necessarily lead to a perturbation of the quantum system. The isotropically emitted photons thus lead to a different fringe shift for each molecule, thus averaging out the interference visibility when we average over many molecules.

In Fig. 11 we compare our decoherence calculation with the experiments by plotting the interference fringe visibility as a function of the laser power. We observe a good agreement between decoherence theory (solid line) and the experiment (circles). The experiment is reproducible within the indicated error bars for a given laser alignment, but small displacements of the laser focus will influence the shape and slope of the observed decoherence curve. The difference between the theoretical and the experimental curve is of the order of this variation.

This auto-localization of an object by its own heat radiation is a fundamental process limiting the ultimate observability of quantum effects in macroscopic objects. However, for nanometer sized systems [Arndt 1999; Hackermüller 2004; Clauser 1997] this mechanism becomes only relevant at high temperatures and it is not expected to be a limitation for the interference of objects which are even considerably larger than the fullerenes, such as proteins.

5. Summary and perspectives for quantum interferometry with complex systems

Our experiments demonstrate that the observation of high contrast quantum interference is possible for large, complex and hot molecules. With the fullerenes both far-field and near-field interference could be established, but the near-field interferometer seems to be better suited for large objects because of its higher molecule throughput, its better scaling behavior with decreasing de Broglie wavelength and its high spatial resolution.

If one wants to observe the wave nature of even more massive molecules brilliant beams are needed that have a low molecular velocity. In practice one will hardly be able to work with de Broglie wavelengths less than 100 fm. For particles in the mass range of 10^5 amu this requires velocities of the order of $v_m = 10\ rmm/\mathrm{s}$. Although this is a rather demanding requirement it seems not impossible to develop appropriate sources in the future. Moreover, a realistic earth-bound interferometer would be limited to a Talbot length of the order of one meter.

The large grating separation in the present Talbot-Lau interferometer permitted a detailed quantitative study of decoherence due to collisions with the background gas and due to the thermal emission of photons. Our experiments on collisional decoherence show that decoherence by scattering of small particles, which is ubiquitous in our macroscopic world, can be understood and well controlled under high vacuum conditions. Based on the good agreement that we found by comparing our experiments with numerical simulations we estimate that the residual gas pressures required to observe the quantum nature of much larger objects are still within the range of modern technologies.

Thermal decoherence seems to be equally well understood and we conclude that one should be able to suppress this mechanism in future experiments with cooled particles. Internal temperatures of 77 K instead of the thousands of Kelvins in the described study appear to be reachable in future experiments.

From an experimental perspective the major challenge for future quantum interferometry with proteins or nanocrystals is the development of a suitable source, which provides a sufficiently intense beam of slow, massive, well-collimated, internally cold, neutral particles. Also novel detection schemes for such objects are currently being investigated, since the current technique of thermal ionization cannot be applied to thermolabile molecules.

Once these techniques are established another key research issue will be to investigate controlled entanglement. This field is still essentially an empty plain open to be discovered by both theorists and experimentalists.

Acknowledgments

The authors wish to acknowledge the contributions by Olaf Nairz who was the first PhD student to see Fullerene interference and by Björn Brezger who worked as a Marie Curie postdoc in our laboratory. Our work has been supported by the Austrian science foundation (FWF) in the programs START Y177 and SFB F1505, as well as the European research and training networks (HPRN-CT-2002-00309 and HPRN-CT-2000-00125), as well as through an Emmy-Noether grant by the DFG (K. H.).

DISTANT ENTANGLEMENT OF MACROSCOPIC GAS SAMPLES

J. Sherson,[1] B. Julsgaard[2] and E. S. Polzik[2]
Danish National Research Foundation, Center for Quantum Optics - QUANTOP

[1]*Department of Physics and Astronomy, University of Aarhus, Ny Munkegade bygning 520, 8000 Aarhus, Denmark*

[2]*Niels Bohr Institute, Blegdamsvej 17, 2100 Copenhagen, Denmark*

Abstract One of the main ingredients in most quantum information protocols is a reliable source of two entangled systems. Such systems have been generated experimentally several years ago for light [Aspect 1982 (b); Shih 1988; Ou 1992; Kwiat 1995; Schori 2002 (b)] but has only in the past few years been demonstrated for atomic systems [Hagley 1997; Sackett 2000; Julsgaard 2001; Roos 2004]. None of these approaches however involve two atomic systems situated in separate environments. This is necessary for the creation of entanglement over arbitrary distances which is required for many quantum information protocols such as atomic teleportation [Bennett 1993; Kuzmich 2000]. We present an experimental realization of such distant entanglement based on an adaptation of the entanglement of macroscopic gas samples containing about 10^{11} cesium atoms shown in Ref. [Julsgaard 2001]. The entanglement is generated via the off-resonant Kerr interaction between the atomic samples and a pulse of light. The achieved entanglement distance is 0.35 m but can be scaled arbitrarily. The feasibility of an implementation of various quantum information protocols using macroscopic samples of atoms has therefore been greatly increased. We also present a theoretical modeling in terms of canonical position and momentum operators $\hat{X}$ and $\hat{P}$ describing the entanglement generation and verification in presence of decoherence mechanisms.

Introduction

Ever since Einstein, Podolsky, and Rosen in their seminal paper from 1935 [Einstein 1935] introduced the possibility of entangling two quantum system, entanglement has been viewed as one the most curious and spectacular phenomena in quantum mechanics. In the past few years the role of entanglement in quantum mechanics has shifted dramatically from being a fundamental test of the foundation of the entire quantum mechanical theory to being a techni-

V.M. Akulin et al. (eds.), Decoherence, Entanglement and Information Protection in Complex Quantum Systems, 353–372.

cal resource in the rapidly developing field of quantum information. Thus, the hunt is on for reliable sources of entanglement. These have been available for discrete as well as continuous states of the electromagnetic field. However, entangled states of material particles have presented a greater experimental challenge.

Macroscopic samples of atoms as a resource of entanglement have attracted a lot of attention in recent years because of their relatively simple experimental realization (works at room temperature) and robustness to single particle decoherence. In Ref. [Julsgaard 2001] entanglement of this kind was accomplished. However, the two samples were located only 1 cm apart in the same shielded environment. This meant that this implementation did not incorporate the important feature in, *e.g.*, the teleportation protocol, that the distance of teleportation given by the separation of the entangled systems could be arbitrary. We have created two separate environments each containing a gas sample of cesium atoms at room temperature. As we will see, we have successfully created entangled states between two such systems being 0.35 m apart. In the present paper we will also focus on a better method to verify the generation of entangled states as compared to the experiment in Ref. [Julsgaard 2001].

1. Light-atom interaction

In this section we introduce the physical systems involved in the experiment, *i.e.*, we introduce the atomic spin samples and the polarization state of laser pulses interacting with each other. Based on the equations of motion we will explain how the interaction can be utilized for entanglement generation in Sec. 2.

1.1 Atomic system

Our atomic system is composed of two separate samples of spin polarized cesium vapour placed in paraffin coated glass cells at room temperature. Cesium has a hyperfine split ground state with total angular momentum $F = 3$ and $F = 4$, the latter being our atomic quantum system of interest. Having a macroscopic ensemble of atoms (around 10^{11}) we will define the theoretically discrete but effectively continuous collective spin variables $\hat{J}_k = \sum_i \hat{j}_k^i$, where $k = x, y, z$ and i denotes the individual atom. These will retain regular angular momentum commutation relations, $[\hat{J}_y, \hat{J}_z] = i\ \hat{J}_x$. The Heisenberg uncertainty relation will then lead to $\text{Var}(\hat{J}_y)\ \text{Var}(\hat{J}_z) \geq |\langle \hat{J}_x \rangle|^2/4$. We will be interested in the state in which practically all atoms are in the $F = 4, m_F = 4$ state with x as quantization axis. All the relevant interactions only involve minute changes in the macroscopic spin orientation. $\hat{J}_x$ can then be regarded as a constant classical number. If all atoms are independent this will give rise to a minimum uncertainty state called the coherent spin state (CSS) with:

$$\mathrm{Var}(\hat{J}_y) = \mathrm{Var}(\hat{J}_z) = \frac{J_x}{2} = 2\, N_{\mathrm{atoms}}. \quad \text{(CSS)} \tag{1}$$

We therefore see that the variance, often referred to as the projection noise, of the CSS will grow proportionally to the number of atoms. This scaling is manifestly quantum and will be important for the verification of the entanglement.

1.2 Light system

In complete analogy to the atomic collective spin variables, the polarization state of a pulse of light can be described by a vector, the so-called Stokes vector. For light propagating along the z-axis we define

$$\hat{S}_x(t) = \frac{1}{2}\left(\phi_x - \phi_y\right),\ \hat{S}_y(t) = \frac{1}{2}\left(\phi_{45} - \phi_{135}\right),\ \hat{S}_z(t) = \frac{1}{2}\left(\phi_+ - \phi_-\right), \tag{2}$$

where ϕ_x, ϕ_y are the photon fluxes of x and y-polarized photons, ϕ_{45}, ϕ_{135} are photon fluxes measured in a basis rotated 45° with respect to the x-, y-axes, and ϕ_+, ϕ_- refer to photons with σ_+ and σ_--polarization. In our experiments the light will with very good approximation be linearly polarized along the x-axis. Then $\hat{S}_x(t)$ can be described by a classical c-number. The $\hat{S}_y(t)$ and $\hat{S}_z(t)$ operators will contain the interesting quantum variables. The Stokes operators defined above have dimension time^{-1}. This is convenient for describing light / matter interactions as we will see below. But for entanglement generation or quantum information protocols in general it is more convenient to consider entire pulses which are time integrated versions of the above. It can be shown that

$$\left[\int_0^T \hat{S}_y(t)\, dt,\ \int_0^T \hat{S}_z(t)\, dt\right] = i\int_0^T \hat{S}_x(t)\, dt = \frac{n_{\mathrm{ph}}}{2}, \tag{3}$$

where the last equality holds in our case for strong linear polarization along the x-axis. From this commutator we derive the variance of the minimum uncertainty state called the coherent state:

$$\mathrm{Var}\left(\int_0^T \hat{S}_y(t)\, dt\right) = \mathrm{Var}\left(\int_0^T \hat{S}_z(t)\, dt\right) = \frac{n_{\mathrm{ph}}}{4}. \quad \text{(shot noise)} \tag{4}$$

Note again the characteristic linear quantum scaling with the number of particles n_{ph}. We often refer to the variance of the coherent light state as the shot noise level.

For completeness we will note that for coherent light states it can be shown [Julsgaard 2003] that

$$\left\langle \hat{S}_y(t)\, \hat{S}_y(t')\right\rangle = \left\langle \hat{S}_z(t)\, \hat{S}_z(t')\right\rangle = \frac{S_x}{2}\, \delta(t-t'). \tag{5}$$

The usefulness of this will be shown below. In our experiment we will use $\hat{S}_y$ detections to create entanglement between atomic samples. If initially x-polarized light is rotated a small angle θ around the axis of propagation (z) we get $\langle \hat{S}_y \rangle = 2\, S_x\, \theta$. This is why an $\hat{S}_y$ detection is sometimes referred to as a polarization rotation measurement.

1.3 Interaction

We couple our light and atomic system by tuning a laser beam off-resonantly to the $6S_{1/2} \rightarrow 6P_{3/2}$ dipole transition in cesium. This leads to the following equations of interaction:

$$\hat{S}_y^{\mathrm{out}}(t) = \hat{S}_y^{\mathrm{in}}(t) + a\, S_x\, \hat{J}_z(t), \tag{6a}$$

$$\hat{S}_z^{\mathrm{out}}(t) = \hat{S}_z^{\mathrm{in}}(t), \tag{6b}$$

$$\frac{\partial}{\partial t}\, \hat{J}_y(t) = a\, J_x\, \hat{S}_z^{\mathrm{in}}(t), \tag{6c}$$

$$\frac{\partial}{\partial t}\, \hat{J}_z(t) = 0, \tag{6d}$$

where $a = -\gamma\, \lambda^2/(8\pi\, A\, \Delta)$, A is the beam cross section, Δ is the detuning (red positive), λ is the optical wavelength, and γ is the natural linewidth of the excited state $6P_{3/2}$. In and out refer to light before and after passing the atomic sample, respectively. The above equations have been derived carefully in Ref. [Julsgaard 2003] but we will give a short physical explanation here.

First of all, the interaction is refractive in nature (the absorption of the off-resonant light is negligible). It is convenient to consider the incoming linearly polarized light in the σ_+ and σ_- basis. The phase shift of a σ_+ and a σ_- photon propagating through atoms will be different if there is a spin component $\hat{J}_z$ along the propagation direction. For instance (quantized along z) an atom in the $m_F = F$ magnetic sub-level couples strongly to σ_+ photons and weakly to σ_- photons. For the $m_F = -F$ sub-level the situation is reversed. The differential phase shift of σ_+, σ_- photons turns out to depend linearly on $\hat{J}_z$ which leads to a polarization rotation of the incoming linearly polarized light proportionally to $\hat{J}_z$ (also known as Faraday rotation). Eq. (6a) is a first order approximation of this effect.

The different coupling strengths for different sub-levels also lead to a Stark shift of atomic levels depending on the sub-level quantum number m_F and the incoming light polarization. Integrated over time, the Stark shifts lead to different phase changes of the magnetic sub-levels which changes the spin state. If for instance there are more σ_+ than σ_- photons, the $m_F = F$ sub-state will be affected more than the $m_F = -F$ state. The amount of σ_+ and σ_- photons is measured by the $\hat{S}_z$ operator. It turns out that the spin state evolution can be described as a rotation of the spin around the z-axis by an amount proportional to $\hat{S}_z$. This is given to first order in Eq. (6c) (the rotation is so small that J_x is unaffected).

In the interaction process the σ_+ and σ_- photons experience phase shifts but are not absorbed. In the off-resonant limit the flux of σ_+ and σ_- photons are individually conserved leading to Eq. (6b). By conservation of angular momentum along the z-direction this leads to the constancy of $\hat{J}_z$ expressed by Eq. (6d).

We see from Eqs. (6a) and (6d) that in the case of a large interaction strength (*i.e.*, if $a\ S_x\ \hat{J}_z$ dominates $\hat{S}_y^{\mathrm{in}}$) a measurement on $\hat{S}_y^{\mathrm{out}}$ amounts to a measurement of $\hat{J}_z$ without destroying the state of $\hat{J}_z$. This is termed a quantum non demolition (QND) measurement of $\hat{J}_z$. Using off-resonant light for QND measurements of spins has also been discussed in Refs. [Kuzmich 1998; Takahashi 1999]. We note that Eq. (6c) implies that a part of the state of light is also mapped onto the atoms. This opens up the possibility of using this sort of system for quantum memory. One step in this direction is discussed in Refs. [Schori 2002 (a); Julsgaard 2003].

1.4 Adding a magnetic field

In the experiment a constant and homogeneous magnetic field is added in the x-direction. We discuss the experimental reason for this below. For our modeling the magnetic field adds a term $H_B = \Omega\ J_x$ to the Hamiltonian. This makes the transverse spin components precess at the Larmor frequency Ω depending on the strength of the field. Introducing the rotating frame coordinates:

$$\begin{pmatrix} \hat{J}_y' \\ \hat{J}_z' \end{pmatrix} = \begin{pmatrix} \cos\Omega t & \sin\Omega t \\ -\sin\Omega t & \cos\Omega t \end{pmatrix} \begin{pmatrix} \hat{J}_y \\ \hat{J}_z \end{pmatrix} \tag{7}$$

we can easily show that Eqs. (6a) to (6d) will transform into:

$$\hat{S}_y^{\rm out}(t) = \hat{S}_y^{\rm in}(t) + a\, S_x \left(\hat{J}_y'(t)\ \sin\Omega t + \hat{J}_z'(t)\ \cos\Omega t\right), \qquad (8a)$$

$$\hat{S}_z^{\rm out}(t) = \hat{S}_z^{\rm in}(t), \qquad (8b)$$

$$\frac{\partial}{\partial t}\ \hat{J}_y'(t) = a\ J_x\ \hat{S}_z^{\rm in}(t)\ \cos\Omega t, \qquad (9a)$$

$$\frac{\partial}{\partial t}\ \hat{J}_z'(t) = a\ J_x\ \hat{S}_z^{\rm in}(t)\ \sin\Omega t. \qquad (9b)$$

Thus, the atomic imprint on the light is encoded in the Ω-sideband instead of at the carrier frequency. The advantage of this added feature is threefold. The first and perhaps most important advantage is that lasers are generally a lot more quiet at high sideband frequencies compared to the carrier. A measurement without a magnetic field will be a DC measurement and the technical noise would dominate the subtle quantum signal. Secondly, the B-field introduces a Larmor splitting of the magnetic sublevels of the hyperfine ground state multiplet, thus lifting the degeneracy. This will introduce an energy barrier strongly suppressing spin flipping collision. The lifetime of the atomic spin state is consequently greatly increased. The last advantage is that as long as the measurement time is longer than $1/\Omega$ Eq. (8a) enables us to access both J_y' and J_z' at the same time. We are of course not allowed to perform non-destructive measurements on these two operators simultaneously since they are non-commuting. This is also reflected by the fact that neither $\hat{J}_y$ nor $\hat{J}_z$ are constant in Eqs. (9a) and (9b). In Sec. 2 we shall consider two atomic samples and the third advantage becomes evident.

2. Entanglement creation and modeling

In this section we define what is meant by entangled states of atomic samples and we adapt the equations of motion from previous sections for this purpose. We will derive a simple model for the entanglement creation and describe how to verify that the states created really are entangled. In Ref. [Sherson 2004] a much more elaborate description of entanglement creation is given.

2.1 Entanglement criterion

Let us here state the criterion to fulfil in order to prove the generation of entangled states. Since entanglement is the nonlocal interconnection of two systems we need to have two atomic samples A and B. Entanglement is usually defined in terms of density matrices so that A and B are entangled if they are connected in such a way that it is impossible to write the total density matrix as a product, $\rho_{tot} = \sum_i p_i \, \rho_{Ai} \, \rho_{Bi}$. For our continuous variable system we have the experimentally practical criterion derived from the above definition in Ref. [Duan 2000 (b)]:

$$\mathrm{Var}(\hat{J}'_{y1} + \hat{J}'_{y2}) + \mathrm{Var}(\hat{J}'_{z1} + \hat{J}'_{z2}) < 2\, J_x \tag{10}$$

where we have assumed both samples to be macroscopically oriented with same magnitude J_x. The two samples are indexed by 1 and 2. Comparing to Eq. (1) we get an equality for two independent atomic samples in the CSS. To have entanglement we thus need to know the sums of the spin components along the y- and the z-directions better than we could ever know each of the spin projections by itself. It is now interesting to examine the commutator between the sums, $\left[\hat{J}'_{y1} + \hat{J}'_{y2},\, \hat{J}'_{z1} + \hat{J}'_{z2}\right] = i(J_{x1} + J_{x2})$. A non-zero commutator means that increasing our knowledge of one component will automatically decrease our knowledge of the other, thus making our attempts to break the inequality of Eq. (10) futile. Now comes the trick that makes entanglement generation possible in our experiment. Assume J_{x1} and J_{x2} to be equal in magnitude but opposite in direction. The commutator will then become zero and we can at least theoretically measure both components with arbitrary precision, thereby satisfying the entanglement criterion of Eq. (10).

2.2 Two oppositely oriented spins

Inspired by the above we will from now on assume $J_{x1} = -J_{x2} \equiv J_x$. We will re-express the equations of motion (8a) to (9b) for two samples in a way which is much more convenient for the understanding of our entanglement creation and verification procedure. We introduce position and momentum like operators $\hat{X}$ and $\hat{P}$ to describe pulses of light and the atomic systems. This is a more abstract but hopefully also more well known and intuitive way to express the interactions creating the entangled states.

For two atomic samples we write equations of motion:

$$\hat{S}_y^{\mathrm{out}}(t) = \hat{S}_y^{\mathrm{in}}(t) + a\, S_x \left\{ \left[\hat{J}'_{y1}(t) + \hat{J}'_{y2}(t)\right] \sin\Omega t \right. \tag{11a}$$

$$\left. + \left[\hat{J}'_{z1}(t) + \hat{J}'_{z2}(t)\right] \cos\Omega t \right\},$$

$$\frac{\partial}{\partial t}\left(\hat{J}'_{y1}(t) + \hat{J}'_{y2}(t)\right) = a\,(J_{x1} + J_{x2})\; \hat{S}_z^{\mathrm{in}}(t)\; \cos\Omega t = 0, \tag{11b}$$

$$\frac{\partial}{\partial t}\left(\hat{J}'_{z1}(t) + \hat{J}'_{z2}(t)\right) = a\,(J_{x1} + J_{x2})\; \hat{S}_z^{\mathrm{in}}(t)\; \sin\Omega t = 0. \tag{11c}$$

The fact that the sums $\hat{J}'_{y1}(t) + \hat{J}'_{y2}(t)$ and $\hat{J}'_{z1}(t) + \hat{J}'_{z2}(t)$ have zero time derivative relies on the assumption of opposite spins of equal magnitude. The constancy of these terms together with Eq. (11a) allows us to perform QND measurements on the two sums. We note that each of the sums can be accessed by considering the two operators

$$\int_0^T \hat{S}_y^{\text{out}} \cos\Omega t\, dt = \int_0^T \hat{S}_y^{\text{in}} \cos\Omega t\, dt + \frac{a\, S_x}{2}\left(\hat{J}'_{z1}(t) + \hat{J}'_{z2}(t)\right),$$
$$\int_0^T \hat{S}_y^{\text{out}} \sin\Omega t\, dt = \int_0^T \hat{S}_y^{\text{in}} \sin\Omega t\, dt + \frac{a\, S_x}{2}\left(\hat{J}'_{y1}(t) + \hat{J}'_{y2}(t)\right). \quad (12)$$

We have used the fact that $\int_0^T \cos^2\Omega t\, dt \approx \int_0^T \sin^2\Omega t\, dt \approx 1/2$ and that $\int_0^T \cos\Omega t\, \sin\Omega t\, dt \approx 0$. Each of the operators on the left hand side can be measured simultaneously by making a $\hat{S}_y$-measurement and multiplying the photocurrent by $\cos\Omega t$ or $\sin\Omega t$ followed by integration over the duration T. The possibility to gain information about $\hat{J}'_{y1}(t) + \hat{J}'_{y2}(t)$ and $\hat{J}'_{z1}(t) + \hat{J}'_{z2}(t)$ enables us to break the inequality (10). At the same time we must loose information about some other physical variable. This is indeed true, the conjugate variables to these sums are $\hat{J}'_{z2}(t) - \hat{J}'_{z1}(t)$ and $\hat{J}'_{y1}(t) - \hat{J}'_{y2}(t)$, respectively. These have the time evolution

$$\frac{\partial}{\partial t}\left(\hat{J}'_{y1}(t) - \hat{J}'_{y2}(t)\right) = 2a\, J_x\, \hat{S}_z^{\text{in}}(t)\, \cos\Omega t,$$
$$\frac{\partial}{\partial t}\left(\hat{J}'_{z1}(t) - \hat{J}'_{z2}(t)\right) = 2a\, J_x\, \hat{S}_z^{\text{in}}(t)\, \sin\Omega t. \quad (13)$$

We see how noise from the input $\hat{S}_z$-variable is piling up in the difference components while we are allowed to learn about the sum components via $\hat{S}_y$ measurements. The above equations clearly describe the physical ingredients in play but the notation is cumbersome. Therefore we define new operators. For the atomic system we take

$$\hat{X}_{A1} = \frac{\hat{J}'_{y1} - \hat{J}'_{y2}}{\sqrt{2\, J_x}}, \quad \hat{P}_{A1} = \frac{\hat{J}'_{z1} + \hat{J}'_{z2}}{\sqrt{2\, J_x}},$$
$$\hat{X}_{A2} = -\frac{\hat{J}'_{z1} - \hat{J}'_{z2}}{\sqrt{2\, J_x}}, \quad \hat{P}_{A2} = \frac{\hat{J}'_{y1} + \hat{J}'_{y2}}{\sqrt{2\, J_x}}. \quad (14)$$

New light operators will be

$$\hat{X}_{L1} = \sqrt{\frac{2}{S_x\,T}}\int_0^T \hat{S}_y(t)\;\cos\Omega t\;dt,\;\hat{P}_{L1} = \sqrt{\frac{2}{S_x\,T}}\int_0^T \hat{S}_z(t)\;\cos\Omega t\;dt,$$

$$\hat{X}_{L2} = \sqrt{\frac{2}{S_x\,T}}\int_0^T \hat{S}_y(t)\;\sin\Omega t\;dt,\;\hat{P}_{L2} = \sqrt{\frac{2}{S_x\,T}}\int_0^T \hat{S}_z(t)\;\sin\Omega t\;dt. \tag{15}$$

Each pair of $\hat{X}$, $\hat{P}$ operators satisfy the usual commutation relation, *e.g.*, we have $\left[\hat{X}_{L1},\,\hat{P}_{L1}\right] = i$. All previous equations now translate into

$$\hat{X}^{\text{out}}_{Li} = \hat{X}^{\text{in}}_{Li} + \kappa\,\hat{P}^{\text{in}}_{Ai}, \tag{16a}$$

$$\hat{P}^{\text{out}}_{Li} = \hat{P}^{\text{in}}_{Li}, \tag{16b}$$

$$\hat{X}^{\text{out}}_{Ai} = \hat{X}^{\text{in}}_{Ai} + \kappa\,\hat{P}^{\text{in}}_{Li}, \tag{16c}$$

$$\hat{P}^{\text{out}}_{Ai} = \hat{P}^{\text{in}}_{Ai}, \tag{16d}$$

where we remember $i = 1,\ 2$ refer to the definitions above and not the two samples. The parameter describing the strength of light / matter-interactions is given by $\kappa = a\sqrt{J_x\,S_x\,T}$. The limit to strong coupling is around $\kappa \approx 1$. Note, we have two decoupled sets of interacting light and atomic operators.

In the transition from Stokes operators to canonical variables in Eqs. (15) the result (5) is a convenient tool for calculating variances. If for instance the input light state is the coherent vacuum state we have

$$\begin{aligned}\text{Var}(\hat{X}_{L1}) &= \left\langle \hat{X}^2_{L1}\right\rangle \\ &= \frac{2}{S_x\,T}\int_0^T \cos\Omega t\;dt\int_0^T \cos\Omega t'\;dt'\left\langle \hat{S}^{\text{in}}_y(t)\;\hat{S}^{\text{in}}_y(t')\right\rangle \\ &= \frac{1}{2}\end{aligned} \tag{17}$$

which is as expected. Likewise, if the two atomic samples are each in the coherent state we will derive, *e.g.*, $\text{Var}(\hat{X}_{A1}) = 1/2$. Coherent states of atomic or light systems as defined above correspond to what is known as coherent states of the $\hat{X}$-, $\hat{P}$-operators.

The entanglement criterion (10) written in $\hat{X}$, $\hat{P}$-language is

$$\text{Var}(\hat{P}_{A1}) + \text{Var}(\hat{P}_{A2}) < 1. \tag{18}$$

We see that entanglement of the two atomic samples can be considered as so-called two mode squeezing. The uncertainty in the uncoupled pair of operators $\hat{P}_{A1}$ and $\hat{P}_{A2}$ is reduced on the expense of the increased noise in the operators $\hat{X}_{A1}$ and $\hat{X}_{A2}$.

2.3 Entanglement generation and verification

Now we turn to the actual understanding of entanglement generation and verification. Experimentally we perform the following steps (more details will be given in Sec. 3). First the atoms are prepared in the oppositely oriented coherent states corresponding to creating the vacuum states of the two modes $(\hat{X}_{A1},\ \hat{P}_{A1})$ and $(\hat{X}_{A2},\ \hat{P}_{A2})$. Next a pulse of light called the *entangling pulse* is sent through atoms and we measure the two operators $\hat{X}_{L1}^{\text{out}}$ and $\hat{X}_{L2}^{\text{out}}$ with outcomes A_1 and B_1, respectively. These results bear information about the atomic operators $\hat{P}_{A1}$ and $\hat{P}_{A2}$ and hence we reduce variances $\text{Var}(\hat{P}_{A1})$ and $\text{Var}(\hat{P}_{A2})$. To prove we have an entangled state we must confirm that the variances of $\hat{P}_{A1}$ and $\hat{P}_{A2}$ fulfil the criterion (18). That is we need to know the mean values of $\hat{P}_{A1}$ and $\hat{P}_{A2}$ with a total precision better than unity. For this demonstration we send a second *verifying pulse* through the atomic samples again measuring $\hat{X}_{L1}^{\text{out}}$ and $\hat{X}_{L2}^{\text{out}}$ with outcomes A_2 and B_2. Now it is a matter of comparing A_1 with A_2 and B_1 with B_2. If the results are sufficiently close the state created by the first pulse was entangled.

Now let us be more quantitative. The interaction (16a) mapping the atomic operators $\hat{P}_{Ai}$ out on light is very useful for a strong κ and useless if $\kappa \ll 1$. We will describe in detail the role of κ for all values. To this end we first describe the natural way to determine κ experimentally. If we repeatedly perform the first two steps of the measurement cycle, *i.e.*, prepare coherent states of the atomic spins and performing the first measurement pulse with outcomes A_1 and B_1, we may deduce the statistical properties of the measurement outcomes. Theoretically we expect from Eq. (16a)

$$\langle A_1 \rangle = \langle B_1 \rangle = 0 \text{ and } \text{Var}(A_1) = \text{Var}(B_1) = \frac{1+\kappa^2}{2}. \tag{19}$$

The first term in the variances is the shot noise (SN) of light. This can be measured in absence of the interaction where $\kappa = 0$. The quantum nature of the shot noise level is confirmed by checking the linear scaling with photon number of the pulse, see Eq. (4). The second term arises from the projection noise (PN) of atoms. Hence, we may calibrate κ^2 to be the ratio $\kappa^2 = \text{PN}/\text{SN}$ of atomic projection noise to shot noise of light. Theoretically κ^2 has the linear scaling $\kappa^2 = a\, J_x\, S_x\, T$ with the macroscopic spin size J_x which must be confirmed in the experiment.

Next we describe how to deduce the statistical properties of the state created by the *entangling pulse*. Based on the measurement results A_1 and B_1 of this pulse we must predict the mean value of the second measurement outcome. If $\kappa \to \infty$ we ought to trust the first measurement completely since the initial noise of $\hat{X}_{Li}^{\text{in}}$ is negligible, *i.e.*, $\langle A_2 \rangle = A_1$ and $\langle B_2 \rangle = B_1$. On the other hand, if $\kappa = 0$ we know that atoms must still be in the vacuum state such that $\langle A_2 \rangle = \langle B_2 \rangle = 0$. It is natural to take in general $\langle A_2 \rangle = \alpha\, A_1$ and $\langle B_2 \rangle = \alpha\, B_1$. We need not know a theoretical value for α. The actual value can be deduced from the data. If we repeat the measurement cycle N times with outcomes $A_1^{(i)}$, $B_1^{(i)}$, $A_2^{(i)}$, and $B_2^{(i)}$, the correct α is found by minimizing the conditional variance

$$\begin{aligned}\mathrm{Var}(A_2|A_1) + \mathrm{Var}(B_2|B_1) = \; & \min_{\alpha} \frac{1}{N-1} \sum_{i}^{N} \left[\left(A_2^{(i)} - \alpha\, A_1^{(i)} \right)^2 \right. \\ & \left. + \left(B_2^{(i)} - \alpha\, B_1^{(i)} \right)^2 \right] . \end{aligned} \tag{20}$$

In order to deduce whether we fulfil the entanglement criterion (18) we compare the above to our expectation from Eq. (16a). For the verifying pulse we get

$$\begin{aligned}\left\langle \left(\hat{X}_{Li}^{\text{out}} - \left\langle \hat{X}_{Li}^{\text{out}} \right\rangle \right)^2 \right\rangle & = \left\langle \left(\hat{X}_{Li}^{\text{in, 2nd}} + \kappa \left[\hat{P}_{Ai}^{\text{ent}} - \left\langle \hat{P}_{Ai}^{\text{ent}} \right\rangle \right] \right)^2 \right\rangle \\ & = \frac{1}{2} + \kappa^2\, \mathrm{Var}(\hat{P}_{Ai}^{\text{ent}}), \end{aligned} \tag{21}$$

where $\hat{X}_{Li}^{\text{in, 2nd}}$ refers to the incoming light of the *verifying pulse* which has zero mean. $\hat{P}_{Ai}^{\text{ent}}$ refers to the atoms after being entangled. We see that the practical entanglement criterion becomes

$$\begin{aligned}\mathrm{Var}(A_2|A_1) + \mathrm{Var}(B_2|B_1) & = 1 + \kappa^2 \left(\mathrm{Var}(\hat{P}_{A1}^{\text{ent}}) + \mathrm{Var}(\hat{P}_{A2}^{\text{ent}}) \right) \\ & < 1 + \kappa^2 = \mathrm{Var}(A_1) + \mathrm{Var}(B_1). \end{aligned} \tag{22}$$

In plain English, we must predict the outcomes A_2 and B_2 with a precision better than the statistical spreading of the outcomes A_1 and B_1 with the additional constraint that A_1 and B_1 are outcomes of quantum noise limited measurements.

2.4 Theoretical entanglement modeling

Above we described the experimental procedure for generating and verifying the entangled states. Here we present a simple way to derive what we expect for the mean values (*i.e.*, the α-parameter) and for the variances $\mathrm{Var}(\hat{P}_{Ai}^{\mathrm{ent}})$.

We calculate directly the expected conditional variance of A_2 based on A_1:

$$\begin{aligned}\left\langle\left(\hat{X}_{L1}^{\mathrm{out,\ 2nd}} - \alpha\,\hat{X}_{L1}^{\mathrm{out,\ 1st}}\right)^2\right\rangle &= \left\langle\left(\hat{X}_{L1}^{\mathrm{in,\ 2nd}} - \alpha\,\hat{X}_{L1}^{\mathrm{in,\ 1st}}\right.\right. \\ &\quad \left.\left. + \kappa\left[\hat{P}_{A1}^{\mathrm{in}} - \alpha\,\hat{P}_{A1}^{\mathrm{ent}}\right]\right)^2\right\rangle = \frac{1}{2}\left[1+\alpha^2+\kappa^2(1-\alpha)^2\right].\end{aligned} \tag{23}$$

In the second step we assumed that the measurement is perfectly QND and without any decoherence, *i.e.*, $\hat{P}_{A1}^{\mathrm{ent}} = \hat{P}_{A1}^{\mathrm{in}}$. By taking the derivative with respect to α we obtain the theoretical minimum

$$\begin{aligned}\mathrm{Var}(A_2|A_1) + \mathrm{Var}(B_2|B_1) &= 1 + \frac{\kappa^2}{1+\kappa^2} \\ \Rightarrow \mathrm{Var}(\hat{P}_{A1}^{\mathrm{ent}}) + \mathrm{Var}(\hat{P}_{A2}^{\mathrm{ent}}) &= \frac{1}{1+\kappa^2}\end{aligned} \tag{24}$$

obtained with the α-parameter

$$\alpha = \frac{\kappa^2}{1+\kappa^2}. \tag{25}$$

It is interesting that in principle any value of κ will lead to creation of entanglement. The reason for this is our prior knowledge to the entangling pulse. Here the atoms are in the coherent state which is as well defined in terms of variances as possible for separable states. We only need an "infinitesimal" extra knowledge about the spin state to go into the entangled regime.

It is interesting to see what happens to the conjugate variables $\hat{X}_{Ai}$ in the entangling process. This is governed by Eq. (16c). We do not perform measurements of the light operator $\hat{P}_{Li}^{\mathrm{in}}$ so all we know is that both $\hat{X}_{Ai}^{\mathrm{in}}$ and $\hat{P}_{Li}^{\mathrm{in}}$ are in the vacuum state. Hence $\mathrm{Var}(\hat{X}_{Ai}^{\mathrm{ent}}) = (1+\kappa^2)/2$ and we preserve the minimum uncertainty relation $\mathrm{Var}(\hat{X}_{Ai}^{\mathrm{ent}})\mathrm{Var}(\hat{P}_{Ai}^{\mathrm{ent}}) = 1/4$.

2.5 Entanglement model with decoherence

Practically our spin states decohere between the light pulses and also in the presence of the light. We model this decoherence naively by putting the entire effect between the two pulses, *i.e.*, we assume there is no decoherence in presence of the light but a larger decoherence between the pulses. We may then perform an analysis in complete analogy with the above with the only difference that $\hat{P}_{A1}^{\text{ent}} = \beta\, \hat{P}_{A1}^{\text{in}} + \sqrt{1-\beta^2}\, \hat{V}_p$ where $\hat{V}_p$ is a vacuum operator admixed such that $\beta = 0$ corresponds to a complete decay to the vacuum state and $\beta = 1$ corresponds to no decoherence. Completing the analysis we find the theoretical conditional variances

$$\begin{aligned} \mathrm{Var}(A_2|A_1) + \mathrm{Var}(B_2|B_1) &= 1 + \kappa^2\, \frac{1 + (1-\beta^2)\kappa^2}{1+\kappa^2} \\ \Rightarrow \mathrm{Var}(\hat{P}_{A1}^{\text{ent}}) + \mathrm{Var}(\hat{P}_{A2}^{\text{ent}}) &= \frac{1 + (1-\beta^2)\kappa^2}{1+\kappa^2} \end{aligned} \tag{26}$$

obtained with α-parameter

$$\alpha = \frac{\beta\, \kappa^2}{1+\kappa^2}. \tag{27}$$

In the limit $\beta \to 1$ these results agree with (24) and (25). For $\beta \to 0$ we have $\alpha \to 0$ (outcomes A_1 and B_1 are useless) and the variance approaches that of the vacuum state which is a separable state.

3. Experimental setup

In this section we describe the details of the experimental setup, *e.g.*, laser settings, pulse lengths, detection systems, *etc.* A picture of the experimental setup is shown in Fig. 1. In part (a) we see two cylindrical magnetic shields which each contain a paraffin coated vapour cell with cesium. The distance between the two cells is 35 cm. In part (b) of the figure we show schematically the timing of laser pulses and the detection system setup.

3.1 Laser settings and pulse timing

In one measurement cycle the first step is to create the coherent spin state of the atomic samples. To this end we have two diffraction grating stabilized diode lasers, one at the 894 nm $D1$ transition $6S_{1/2} \to 6P_{1/2}$ and one at the 852 nm $D2$ transition $6S_{1/2} \to 6P_{3/2}$. We call these the optical pump and the repump, respectively. Both are sent through the first gas sample along the x-axis with σ_+ polarization, thus driving $\Delta m = +1$ transitions only. In the second gas sample the polarization is σ_-. The main pumping is done by the optical pump laser, which drives the atoms towards the $F = 4$, $m_F = 4$ state (for σ_+-polarization). This state will be unaffected by the optical pumping laser (a dark state) because of the absence of an $F = 5$ state in the $6P_{1/2}$ multiplet. In the pumping process some of the atoms will decay into the $F = 3$ ground state, which is why we need the repumping laser from this state to the $6P_{3/2}$ state to

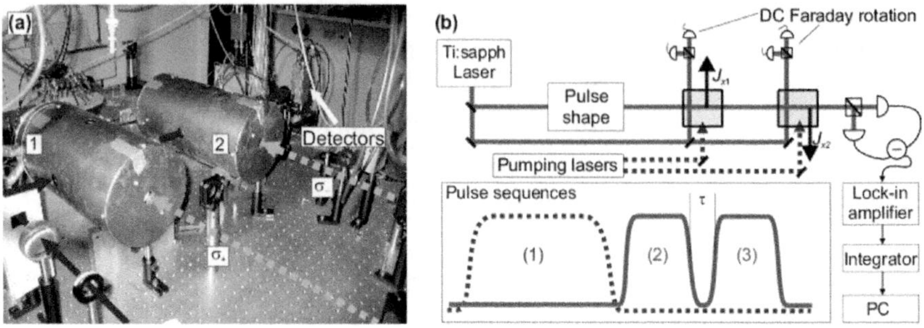

Figure 1. (a) A photographic view of the experimental setup. Atomic vapour cells are placed inside the cylindrical magnetic shields. The pumping beams are indicated with dashed arrows and the path of the entangling and verifying pulse is marked with the solid arrows. (b) A schematic view of the setup. The pulses reach a detection system measuring $\hat{S}_y(t)$. The photocurrent is sent to a lock-in amplifier which singles out the $\sin \Omega t$ and $\cos \Omega t$ parts. These are integrated and stored in a PC. The pulse sequence consists of a (1) 4 ms pumping pulse, (2) 2 ms entangling pulse, a small delay $\tau = 0.25$ ms, and (3) 2 ms verifying pulse. In addition to the pumping lasers and the entangling and verifying pulse, a laser beam is sent through each sample to measure the magnitude J_x of the macroscopic spins by Faraday rotation measurements.

return them to the pumping cycle. Note that we can to some extent control the number of atoms in the $F = 4$ ground state with the power of the repump laser. With this optical pumping scheme we can obtain spin polarization above 99% (measured by methods similar to Ref. [Julsgaard 2004]). The pumping pulses last for 4ms and are shaped by acousto optical modulators.

The atomic samples are now prepared in the CSS with anti-parallel macroscopic orientation. Next the off-resonant entangling pulse shaped by an electro optical modulator of duration 2 ms is sent through both cells to the $\hat{S}_y$ detection. The difference signal is fed into a lock-in amplifier and the result is two numbers A_1 and B_1 corresponding to the integrated $\hat{J}'_{y1} + \hat{J}'_{y2}$ and $\hat{J}'_{z1} + \hat{J}'_{z2}$ signals with an additional light contribution from the incoming $\hat{S}_y^{\rm in}(t)$. The entangling (and verifying) pulses have power of $P = 4.5$mW and are blue detuned by 700 MHz compared to the $6S_{1/2},\ F = 4 \rightarrow 6P_{3/2},\ F' = 5$ transition. They emerge from a Microlase Ti:sapphire laser which is pumped by an 8 W Verdi laser.

After the entangling pulse is a short $\tau = 0.25$ ms delay before the verifying pulse is sent through the atoms. The entangling and verifying pulses have exactly the same shape and duration. The verifying pulse will again result in two numbers A_2 and B_2 stored in a PC. We now must predict the outcomes A_2 and B_2 based on A_1 and B_1 as described in Sec. 2. In order to calculate the statistics properly we repeat the measurement cycle 10.000 times.

3.2 Measuring the macroscopic spin

In addition to the entanglement creation and verification we also measure the macroscopic value J_x of the spin samples by sending linearly polarized light along the direction of optical pumping in both samples. This light experiences polarization rotation proportional to J_x. As already noted we have for small angles $\theta = S_y/2\, S_x$. We also know $S_y = a\, S_x\, J_x$ (for small angles, S_y^{in} is zero classically) when light is propagating along the x-direction. Holding these together we find (this holds for all angles)

$$\theta[\text{rad}] = \frac{a\, J_x}{2} = -\frac{\gamma\, \lambda^2\, J_x}{16\pi\, A_{\text{eff}}\, \Delta}. \tag{28}$$

Instead of the beam cross section A we need to use the *effective* cross section $A_{\text{eff}} = 6.0\,\text{cm}^2$ which is such that the volume of the vapour cell is $V = A_{\text{eff}}.\, L$ where L is the length traversed by the laser beam (we call it "effective" since the vapour cell is not exactly box like). We only measure polarization rotation from atoms inside the beam cross section (which does not fill the whole volume) and the equation must be scaled in order to count J_x for all atoms (we have $J_x = J_x^{\text{in beam}}.\, A_{\text{eff}}/A$).

The DC Faraday rotation angle θ is a practical handle on the macroscopic spin size J_x but we may in addition calculate a theoretical level for the projection noise to shot noise ratio $\kappa^2 = \text{PN/SN}$. Theoretically $\kappa^2 = a^2\, J_x\, S_x\, T$ where again it is a question which cross section A to insert in a. The entangling and verifying pulses do also not fill the entire vapour cell volume. We assume that the measurement results are not depending much in the real cross section for the following reason. If A is small the interaction with the atoms inside the beam is stronger, but each atom will spend less time inside the beam leading to a correspondingly shorter effective interaction time T (atoms are at room temperature and move in and out of the beam). If this simple consideration has some validity we may assume that the light fills the whole cell volume and the correct cross section to insert is the effective area A_{eff}. Now this can be hold up against (28). Inserting $\lambda = 852$ nm, $\gamma = 5$ MHz and expressing the photon flux $\phi = 2S_x$ in terms of probe power we may derive

$$\kappa^2_{\text{theory}} = \frac{18.6\, P[\text{mW}]\, T[\text{ms}]\, \theta[\text{deg}]}{\Delta[\text{MHz}]}. \tag{29}$$

We remember the theory is a bit crude but we will see in Sec. 4 that the model holds pretty well.

4. Experimental results

The experimental data is shown in Fig. 2, in the following we carefully explain the details of this graph.

4.1 The projection noise level

The graph in Fig. 2 has on the abscissa the measured DC Faraday rotation angle which is proportional to J_x. The squares are the variances $\mathrm{Var}(A_1) + \mathrm{Var}(B_1)$ for the entangling pulses with shot noise (and electronics noise) subtracted. Also, the results are normalized to shot noise. The circles are the variances $\mathrm{Var}(A_2) + \mathrm{Var}(B_2)$ for the verifying pulses with the same normalization. According to Eq. (19) we have plotted the experimental ratio $\kappa^2 = \mathrm{PN}/\mathrm{SN}$ (the unity term in (19) is the subtracted shot noise). The linearity with J_x confirms that we measure projection noise of atoms and not extra classical noise. We have roughly $\kappa^2_{\mathrm{exp}} = 0.10\ \theta[\mathrm{deg}]$ which should be compared to the prediction (29) $\kappa^2_{\mathrm{theory}} = 0.24\ \theta[\mathrm{deg}]$. The discrepancy is a little more than a factor of two but this is acceptable for our quite simple modeling. Note that the noise of the verifying pulse is the same as that of the entangling pulse. This is expected since we are performing a QND measurement. It is only when we remember the information given to us by the measurement results A_1 and B_1 that we can tell more about the state created by the entangling pulse.

4.2 Conditional variances and entanglement

The tip down triangles in Fig. 2 is the conditional variance $\mathrm{Var}(A_2|A_1) + \mathrm{Var}(B_2|B_1)$ normalized to shot noise and with shot and electronics noise subtracted. According to Eq. (22) we thus plot $\kappa^2(\mathrm{Var}(\hat{P}^{\mathrm{ent}}_{A1}) + \mathrm{Var}(\hat{P}^{\mathrm{ent}}_{A2}))$. The fact that the points are lower than the straight line (κ^2) is a direct indication that the entanglement criterion (18) is fulfilled. For the higher densities the reduction is 25% but we note that entanglement is also observed for smaller densities with $\kappa^2 < 1$. The latter was impossible with the older methods applied in Ref. [Julsgaard 2001]. The corresponding α-parameters from the minimization procedure (20) are plotted in Fig. 2 with tip up triangles.

The expected entangled noise level in the ideal case is given by Eq. (24). This is drawn as the dash-dotted curve (κ^2 times $1/(1 + \kappa^2)$). We see the conditional variance lies higher than this curve and hence the entanglement is worse than expected. According to (25) we also would expect the α-parameters to lie on the same dash-dotted curve in the ideal case. It is clearly not the case, the experimental α-parameters are lower which indicates that the results A_1 and B_1 can not be trusted to as high a degree as expected.

Let us try to apply the simple decoherence model given by Eqs. (26) and (27). Taking the decoherence parameter $\beta = 0.65$ we get the dashed lines in the figure. These match nicely the experimental data. We conclude that the simple decoherence model has some truth in it and we must accept that the entangled state created can only be verified to be around "65% as good" as expected in an ideal world.

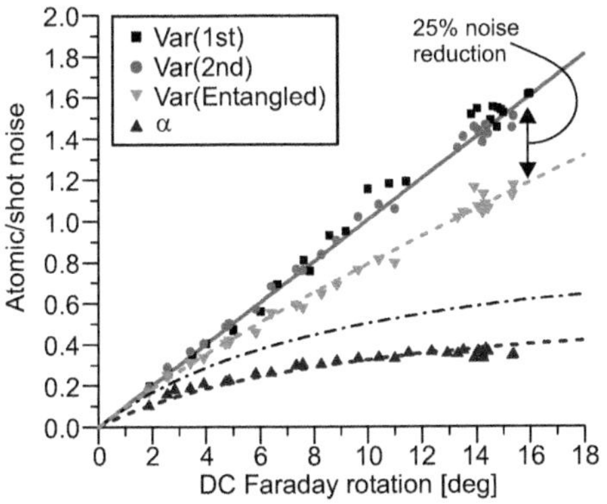

Figure 2. Atomic noise in units of shot noise as a function of atomic density (measured by DC Faraday rotation). Shot and electronics noise has been subtracted. Squares show the 1st pulse noise, circles the 2nd pulse noise. The linearity of these data is a finger print of the projection noise level which is then given by the solid line linear fit. Tip down triangles show the noise of the entangled states. Tip up triangles show the weight factor α. The two dashed curves trough triangles is the model described in Eqs. (26) and (27) with $\beta = 0.65$. The dash-dotted curve is the theoretically best for the triangles ($\beta = 1$). Note that the states created by the first pulse measurements are really entangled states (according to the criterion (22)) since the noise is clearly below the PN. We observe up to 25% noise reduction. Note, entanglement is observed for low densities also with $\kappa^2 < 1$.

4.3 Physical decoherence processes

Above we quantify the observed decoherence with the β-parameter. Here we comment on the physical grounds for the decoherence. A well known parameter for describing the decay of the *transverse* spin components J_y and J_z is the T_2-time defined by

$$\frac{\partial J_i}{\partial t} = -\frac{J_i}{T_2} \tag{30}$$

where $i = y, z$. We have studied the T_2-time extensively. A very good method for this is to create (by applying an RF-magnetic pulse) a displaced version of the coherent spin state with, *e.g.*, $\langle \hat{J}'_y \rangle \neq 0$. This non-zero mean value can be detected by our standard $\hat{S}_y$-detection method and the decay following Eq. (30) may be observed.

Our experience tells us that power broadening by the laser pulses combined with light assisted atomic collisions play the important role in the decoherence processes. For high densities and high optical powers we may find T_2 as low as 5 ms. A fair guess for β is an exponential decay over a typical time scale of 2 ms (the time between the central parts of the entangling and verifying pulses). This yields $\beta \approx \exp(-2/5) \approx 0.67$. This is not far from the observed

β but we should say here that the 5 ms is a typical value not directly measured in the case of the data given in Fig. 2. It is our experience that the observed decoherence in the entanglement experiments is stronger than that expected from the processes mentioned above. This is indeed true for shorter pulses. We believe that the atomic motion in and out of the laser beam combined with inhomogeneous light / atom coupling due to the Doppler effect may also play a role, but we need further investigation to confirm this completely.

4.4 Conclusion of experimental results

To conclude the experimental section we emphasize that we have generated entangled states between distant atomic samples in the sense that each vapour cell sits in its own magnetic shield. The two shields can in principle be moved as far apart as is practical, our experiment was performed with a distance of 35 cm. In the future we hope to extend this distance further. The noise reduction below the level set by separable states was measured up to 25%. This number is mainly limited by power broadening and light assisted collisional relaxation of the atomic spins. The decoherence is successfully modeled by a single parameter β. A much more elaborate theory on entanglement generation in presence of decoherence and losses exists [Sherson 2004]. We should note, that the generated entangled states have random but known (based on A_1 and B_1) mean values. It is possible by applying RF-magnetic fields to shift the entangled states to having zero mean value while preserving the reduced variance. Experimental demonstration of such will be considered elsewhere.

5. Perspectives

We will now present a brief overview of some of the interesting applications in the field of quantum information of our reliable source of distant atomic entanglement.

5.1 Atomic teleportation

Quantum teleportation was first proposed in 1993 [Bennett 1993] and the year after for the special case of continuous variables [Vaidman 1994]. Teleportation is extremely important since direct transport of physical states is often hindered by exponential decoherence. With quantum teleportation the information is cleanly separated into a classical part, which can be transmitted over arbitrary distances, and a quantum mechanical part, which only needs to interact locally.

A proposal for spin state teleportation was given in Ref. [Duan 2000 (d)]. Three atomic samples are needed as shown in Fig. 3 (a). Adjacent samples are oriented oppositely along the x-axis so that both collective measurements on cells 1 and 2 and on cells 1 and 3 will be regular entangling interactions as

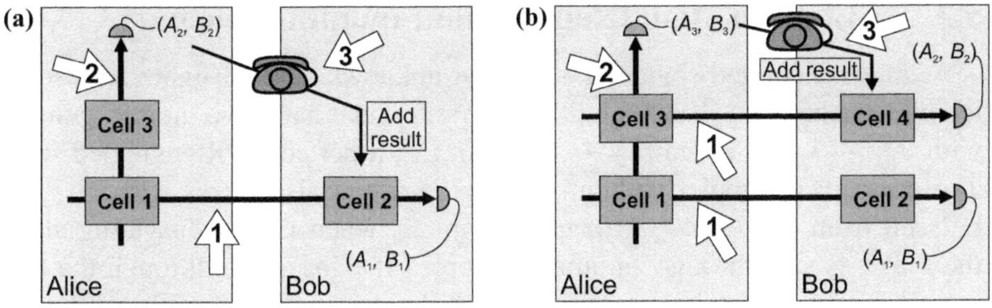

Figure 3. (a) Teleporting an unknown quantum state: first cells 1 and 2 are entangled. Then Alice sends a light pulse through cell 1 and the unknown quantum state in cell 3. The measurement results are communicated classically to Bob who by applying a displacement to his system based on the results of the two measurement recreates the unknown quantum state in his cell. (b) Entanglement swapping: the same procedure as in (a) but with two sets of entangled states initially. After a displacement based on the measurement results cells 2 and 4 are entangled even though they have never interacted directly.

discussed earlier. Cells 1 and 3 are located at Alice's site and cell 2 at Bob's site. The goal is now to teleport an unknown state in cell 3 onto cell 2. First cells 1 and 2 are entangled giving the measurement results $(A_1,\ B_1)$. Next a pulse is sent through cells 1 and 3 and the measurement results $(A_2,\ B_2)$ are communicated classically to Bob. He can now by applying a rotation to his system based on the outcomes of the two measurements recreate the unknown state in his cell. This of course only works with perfect fidelity in the large interaction regime ($\kappa^2 \gg 1$).

5.2 Teleporting an entangled state: entanglement swapping

As described in Ref. [Polzik 2003] an entangled state can also be teleported using macroscopic samples of atoms. In this case Alice and Bob each have two samples (Fig. 3 (b)). First each one of Alice's samples are entangled with one of Bob's. Then a pulse of light is sent through Alice's two samples making it an entangled state. Alice now sends the result of a measurement on this last entangling pulse to Bob. Using this and the results of the two primary entangling pulses he can displace one of his states. This entangles his two samples without ever bringing the two into direct contact. Had Alice shifted one of her spin states prior to creating the entanglement between her two cells the exact same protocol would allow Bob to recover this shift in his cells. This allows for secret quantum communication.

5.3 Light to atom teleportation: quantum memory

With an entangled source of atoms an unknown state of light can also be teleported onto this [Kuzmich 2000]. Assume we have two atomic samples with $\hat{J}_{y1} + \hat{J}_{y2} = 0$ and $\hat{J}_{z1} + \hat{J}_{z2} = 0$, *i.e.*, a perfect EPR-entangled state. The protocol is simpler without rotating spins but also works with. As can be seen from Eq. (6a) $\hat{J}_z$ is mapped onto $\hat{S}_y$ when light propagating along the z axis is sent through an atomic sample. If light is sent through the first atomic sample and subsequently detected the measurement result can be fed back into $\hat{J}_{z2}$ such that the original atomic variables exactly cancel. $\hat{S}_y$ has now been mapped perfectly onto $\hat{J}_{z2}$. Another effect of the light pulse is seen from Eq. (6c). $\hat{S}_z$ is mapped onto $\hat{J}_{y1}$. If the transverse spins of sample 1 are rotated 90° and a new strong light pulse ($\kappa^2 \gg 1$) is sent through this sample $\hat{J}_{y1}^{\text{out}}$ is measured. This result can be fed back onto $\hat{J}_{y2}$ in such a way that the two original $\hat{J}_y$'s cancel. $\hat{S}_z$ is now stored in $\hat{J}_{y2}$ and the teleportation is complete. This process could also be reversed so that an unknown atomic state is mapped onto a light pulse via two EPR-entangled light beams. This shows that macroscopic samples of atoms offers a very feasible protocol for complete quantum memory.

Summary and conclusion

We have presented the first experimental realization of distant atomic entanglement in the sense that the two atomic systems are placed in separate environments, thus enabling entanglement between system separated by arbitrary distances. Given the abundance of available quantum information protocols for this type of continuous variable atomic system the importance of this achievement is evident. Although quantum teleportation of atomic states has been achieved recently [Barett 2004; Riebe 2004] none of these approaches display directly scalable distance between the unknown quantum state and the target system. Our system must therefore be considered an important candidate for achieving the long standing goal of high fidelity transfer of atomic states over great distances upon which many of the proposed technical applications of entanglement and teleportation critically depend. The existence of several quantum information protocols for our physical system is based on the simplicity of the interaction between light and atoms as expressed by Eqs. (16a) to 16d).

Acknowledgments

This work is supported by the Danish National Research Foundation and the European Union through the grant QUICOV.

POSITION AND MOMENTUM ENTANGLEMENT OF DIPOLE-DIPOLE INTERACTING ATOMS IN OPTICAL LATTICES

T. Opatrný,[1] M. Kolář[1] and G. Kurizki[2]

[1]*Department of Theoretical Physics, Palacký University, 77146 Olomouc, Czech Republic*

[2]*Chemical Physics Department, Weizmann Institute of Science, Rehovot 76100, Israel*

Abstract We consider a possible realization of the position- and momentum-correlated atomic pairs that are confined to adjacent sites of two mutually shifted optical lattices and are entangled via laser-induced dipole-dipole interactions. The Einstein-Podolsky-Rosen (EPR) "paradox" [Einstein 1935] with translational variables is then modified by lattice-diffraction effects. We study a possible mechanism of creating such diatom entangled states by varying the effective mass of the atoms.

Keywords: Entanglement, optical lattice, dipole-dipole interaction, effective mass.

Introduction

Einstein, Podolsky and Rosen (EPR) [Einstein 1935] asked the question of whether the quantum mechanical description of physical world is complete, giving the following example. Two-particles are in the quantum state showing strange correlations (dubbed "entanglement" or "Verschränkung" by Schrödinger [Schrödinger 1935 (b)]): if one measures the position or momentum of one particle, one can predict with certainty the result of measuring their counterpart for the second particle. Thus, depending on which measurement is chosen for the first particle, the value of either the momentum or position can be predicted with arbitrary precision for the other particle. The later discussion has concerned the interpretation of the EPR paradox and its implications on quantum theory [Bohr 1935]. Later, Bohm considered [Bohm 1951] two entangled spin-1/2 particles, which have become the center of attention on this EPR issue: their *discrete-variable* entangled states have served to demonstrate the incompatibility of quantum mechanics with local realism (advocated by EPR supporters), by the Bell's inequality violation [Bell 1964; Clauser 1969; Aspect 1982 (b); Perrie 1985; Ou 1998; Tittel 1998; Weihs 1998; Rowe

V.M. Akulin et al. (eds.), Decoherence, Entanglement and Information Protection in Complex Quantum Systems, 373–390.

2001; Hasegawa 2003]. In recent years, there has been revival of interest in *continuous-variable* entanglement, in the spirit of the original EPR problem [Reid 1988; Gisin 1991; Gisin 1992; Ou 1992; Braunstein 1998 (b); Braunstein 1998 (a); Furusawa 1998; Lloyd 1998; Lloyd 1999; Polzik 1999; Parkins 2000; Julsgaard 2001; Opatrný 2001; Silberhorn 2001; Silberhorn 2002; Opatrny 2003; Zhang 2003].

The original *ideal* EPR [Einstein 1935] state of two particles 1 and 2 (distinguishable by the spatial part of the wavefunction), is, respectively, represented as follows in their coordinates or momenta (in one dimension),

$$\begin{aligned} \langle x_1,\, x_2|\psi_{\rm EPR}\rangle &= \delta(x_1 - x_2), \\ \langle p_1,\, p_2|\psi_{\rm EPR}\rangle &= \delta(p_1 + p_2). \end{aligned} \tag{1}$$

If two particles are prepared in such a state, and one measures the value of x (or p) of one particle (*e.g.*, 1), one can predict the outcome of measuring x (or p, respectively) on the other particle with perfect precision. However, the state of Eq. (1) is hardly realizable experimentally. It would occupy infinite space and have infinite kinetic energy. One can consider more realistic variants of this state, *e.g.*, a Gaussian state given by

$$\begin{aligned} \langle x_1,\, x_2|\psi_{\rm EPR}\rangle &= \frac{\exp\left(-\dfrac{(x_1 - x_2)^2}{4\,\Delta x_-^2}\right)\,\exp\left(-\dfrac{(x_1 + x_2)^2}{4\,\Delta x_+^2}\right)}{\sqrt{\pi \Delta x_-\, \Delta x_+}}, \\ \langle p_1,\, p_2|\psi_{\rm EPR}\rangle &= \frac{\exp\left(-\dfrac{(p_1 - p_2)^2}{4\,\Delta p_-^2}\right)\,\exp\left(-\dfrac{(p_1 + p_2)^2}{4\,\Delta p_+^2}\right)}{\sqrt{\pi \Delta p_-\, \Delta p_+}}, \end{aligned} \tag{2}$$

where $\Delta p_\pm \equiv \hbar/\Delta x_\pm$ (see Fig. 1). We approach the original EPR state as the ratio $\Delta x_-/\Delta x_+ \;\to\; 0$. After measuring the position of particle 1 (projection onto $x_1 = $ const.), the position of particle 2 is centered at

$$\bar{x}_2 = x_1 \frac{1 - \left(\frac{\Delta x_-}{\Delta x_+}\right)^2}{1 + \left(\frac{\Delta x_-}{\Delta x_+}\right)^2}, \tag{3}$$

with the uncertainty $\Delta x_- \left[1 + (\Delta x_-/\Delta x_+)^2\right]^{-1/2}$. In the limit of $\Delta x_-/\Delta x_+ \ll 1$, the position of particle 2 is centered at x_1, and its uncertainty is $\approx \Delta x_-$. Similar relations hold also for the momentum of particle 2 after the momentum of particle 1 is measured. Thus, either of the two conjugate quantities of particle 2 can be predicted with arbitrarily high precision. Of

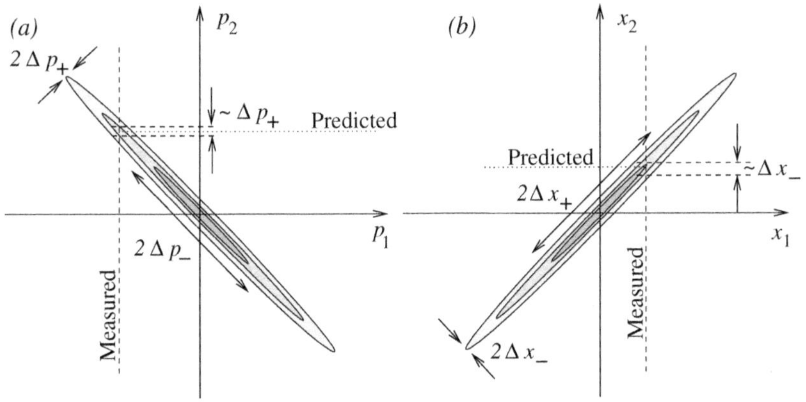

Figure 1. Joint probability distribution of (a) momenta and (b) position of EPR-pair ensembles. If one measures the position of particle 1, one can predict the position of particle 2 with uncertainty $\approx \Delta x_-$, whereas if one measures momentum of particle 1, one can predict the momentum of particle 2 with uncertainty $\approx \Delta p_+$.

course, the Heisenberg uncertainty relation is not violated, since for a single system one can measure only one of the two conjugate quantities.

Approximate versions of the translational EPR state, wherein the δ-function correlations are replaced by finite-width (Gaussian) distributions, have been shown to characterize the quadratures of the two optical-field outputs of parametric down-conversion [Reid 1988; Ou 1992], or of a fiber interferometer with Kerr nonlinearity [Silberhorn 2001]. Such states allow for various schemes of continuous-variable quantum information processing such as quantum teleportation [Braunstein 1998 (b); Furusawa 1998] or quantum cryptography [Silberhorn 2002]. A similar state has also been predicted and realized using collective spins of large atomic samples [Polzik 1999; Julsgaard 2001]. It has been shown that if suitable interaction schemes can be realized, continuous-variable quantum states of the original EPR type could even serve for quantum computation [Braunstein 1998 (a); Lloyd 1998; Lloyd 1999].

Notwithstanding its applications to quantum information processing, the translational variables of the EPR state (2) do not violate local realism: such a state has a *non-negative* Wigner function, controlling the position and momentum distribution of each particle. Nevertheless, there exist measurement schemes in which an analog of Bell's inequality is violated [Gisin 1991; Gisin 1992] for such a state - as for any pure entangled state. The point is that other variables have to be measured, *e.g.*, displaced parity operators as in Refs. [Banaszek 1998; Banaszek 1999].

The realization and measurement of the EPR translational correlations of *material particles* appears to be very difficult. There have been suggestions

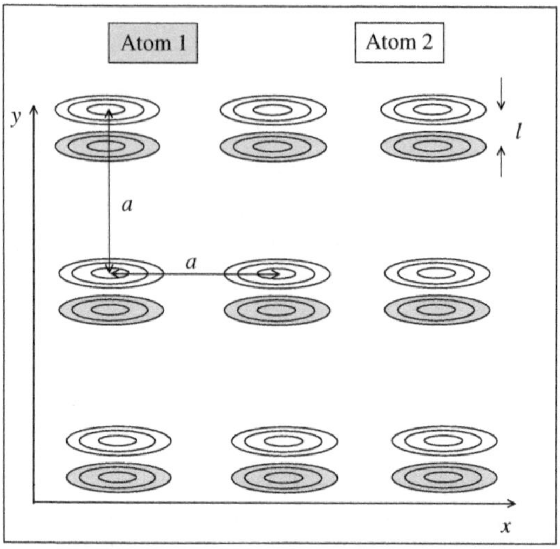

Figure 2 Proposed scheme of two kinds of mutually shifted overlapping optical lattices used to create the translational EPR state. The lattices are displaced from each other in the y direction by l. They are sparsely occupied by two different kinds of atoms. Each kind of atoms interacts with a different lattice. The oval regions depict the energy minima (potential wells) of the lattices.

to start with entangled light fields and to transfer their quantum state into the state of trapped ions in optical cavities [Parkins 2000] or of vibrating mirrors [Zhang 2003]. We have proposed to realize translational EPR states by taking advantage of interatom correlations in a dissociating diatom [Opatrný 2001]. More recently, we have considered dipole-dipole coupled cold atoms in an optical lattice as a source of translational EPR states [Opatrny 2003].

In order to generate the translational EPR entanglement between interacting material particles, one must be able to accomplish several challenging tasks:
(a) switch on and off the entangling interaction;
(b) confine their motion to single dimension;
(c) infer and verify the dynamical variables of particle 2 *at the time of measurement* of particle 1.
The latter requirement is particularly hard for free particles, since by the time we complete the prediction for particle 2, its position will have changed. In Ref. [Opatrný 2001] we suggested to overcome these hurdles by transforming the wavefunction of flying (ionized) atoms emerging from diatom dissociation by an electrostatic / magnetic lens onto the image plane, where its position corresponds to what it was at the time of the diatom dissociation. In Ref. [Opatrny 2003] we have proposed a solution based on the following steps:
(i) controlling the diatom formation and dissociation in an optical lattice by switching on and off a laser-induced dipole-dipole interaction;
(ii) controlling the motion and effective masses of the atoms and the diatom by changing the intensities of the lattice fields.
Here we discuss our proposal in more detail and elaborate on its principles.

Our aim here is to demonstrate the feasibility of preparing a momentum- and position-entangled state of atom pairs in optical lattices, which would be a vari-

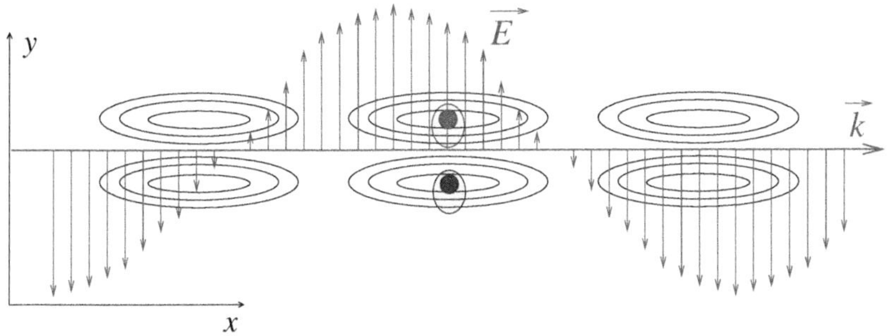

Figure 3. Scheme of the LIDDI interaction: a traveling laser field induces dipole moments in the atoms, thereby causing the interatomic interaction. The laser field is propagating in the direction x along which the atoms are weakly confined. The electric field vector is in the xy plane.

ant of the original EPR state, owing to lattice diffraction. In Sec. 1 we specify the physical system under study. In Sec. 2 the basic properties of single-atom states in optical lattices are discussed. In Sec. 3 we discuss the binding effect of the dipole-dipole interaction. Sec. 4 deals with the preparation of EPR states by manipulation of the effective masses of the atoms. In Sec. 5 we discuss experimental demonstration possibilities of measuring the EPR.

1. System description

Let us consider two overlapping optical lattices with the same lattice constant a (see Fig. 2). The lattices are very sparsely occupied by two kinds of atoms, each kind interacting with only one of the two lattices. We can achieve this, *e.g.*, by using two different internal (say, hyperfine) states of the atoms [Brennen 1999; Deutsch 2000; Mandel 2003]. The potential in each lattice is strongly confining in the y and z directions (realized by strong laser fields), whereas in the x direction the lattice potential is only moderately to weakly confining (weaker laser field). Thus, the motion of each particle is restricted to the x direction. In each direction we assume that only the lowest vibrational energy band is occupied. Initially, the potential minima of the lattices are displaced from each other by an amount $l \ll a$ in the y direction. An auxiliary laser produces mutual interaction between the atoms, so called laser-induced dipole-dipole interaction (LIDDI). It is linearly polarized along the y axis, traveling in the x direction and has a wavelength λ_C, moderately detuned from an atomic transition that *is different* from the transition used to trap the atoms in the lattice. If we assume two kinds of atoms with identical polarizabilities in the geometry of Fig. 3, the interatomic LIDDI potential induced by a linearly polarized laser is of the form [Thirunamachandran 1980]

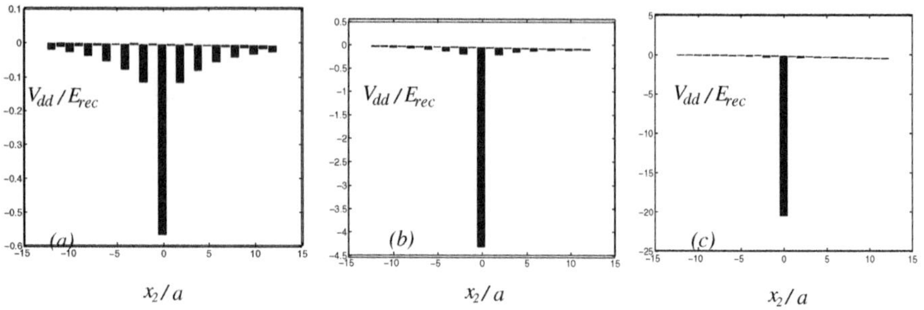

Figure 4. LIDDI potential as a function of the position of atom 2, given that atom 1 occupies site 0, for different separations of the two lattices: (a) $l = 150$ nm, (b) $l = 70$ nm, (c) $l = 40$ nm. The other parameters are specified in Sec. 4.

$$V_{dd} = -\, V_C \, F_\theta(kR), \tag{4}$$

where

$$F_\theta(kR) = \cos(kR\,\cos\theta) \times \left\{ (2 - 3\cos^2\theta) \left[\frac{\cos kR}{(kR)^3} + \frac{\sin kR}{(kR)^2} \right] + \cos^2\theta \, \frac{\cos kR}{kR} \right\}, \tag{5}$$

and

$$V_C = \frac{\alpha^2\, k^3\, I_C}{4\pi\, \epsilon_0^2\, c}. \tag{6}$$

Here $k = 2\pi/\lambda_C$ is the wavenumber, I_C is the coupling laser intensity, and the atomic dynamic polarizability α is

$$\alpha = \frac{2\omega_A\, |\mu|^2}{\hbar(\omega_A^2 - \omega^2)}, \tag{7}$$

μ being the dipole moment element of the used transition, ω_A the atomic transition frequency, and $\omega = kc$. The part $F_\theta(kR)$ is position-dependent function of the distance R between the atoms, and the angle θ between the interatomic axis and the wavevector of the coupling laser. Since $l \ll 2a$, $V_{dd}(R)$ has a pronounced minimum for atoms located at the nearest sites, $R \simeq l$, where

$$V_{dd}^{min}(l) \simeq -\, \frac{V_C}{4\pi^3} \left(\frac{\lambda_C}{l} \right)^3. \tag{8}$$

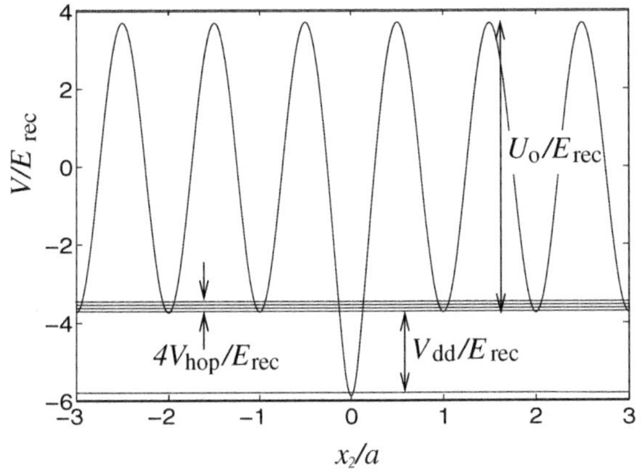

Figure 5 Position dependence of the potential energy of atom 2, under condition that atom 1 is located at site 0. The horizontal lines denote the lowest band of single atoms (bandwidth $4V_{hop}$) and the narrower band for diatoms V_{dd} (below the band of uncorrelated atoms).

The LIDDI energy as a function of l and the relative position of the atoms is shown in Figs. 4 and 5. Under the above assumptions, we can treat the system as consisting of pairs of "tubes", either empty or occupied, that are oriented along x. Only atoms within adjacent tubes are appreciably attracted to each other along y, due to the LIDDI.

2. Single-atom states in the Wannier basis

Let us focus on the subensemble of tube-pairs in which each tube is occupied by exactly one atom. In the 1D optical lattice, the single-atom Hamiltonian in x representation is

$$\hat{H}_{lat} = \frac{U_0}{2} \cos\left(\frac{2\pi\, x}{a}\right) + \frac{\hat{p}_x^2}{2m}. \tag{9}$$

Here m is the atomic mass, $\hat{p}_x$ is the momentum operator, U_0 is the maximum potential energy due to the interaction of the atomic dipole with the laser field,

$$U_0 = \frac{4\,|\mu_L|^2}{\epsilon_0\, \hbar c\, \delta_L}\, I_L, \tag{10}$$

where μ_L is the dipole matrix element of the lattice transition, δ_L is the detuning of the lattice field from this transition, and I_L is the intensity of the lattice field. The Hamiltonian (9) describes a quantum pendulum. The eigenfunctions of the Schrödinger equation governed by this Hamiltonian are the Mathieu functions [Slater 1952]. The eigenvalues form bands, whose spectrum depends on the ratio of U_0 to the recoil energy,

$$E_{rec} = \frac{2\pi^2\hbar^2}{m\,\lambda_L^2}, \tag{11}$$

so that one can distinguish between strongly binding ($U_0 \gg E_{\rm rec}$) and weakly binding ($U_0 \sim E_{\rm rec}$) lattice potentials. We assume that all the atoms are cooled down to the lowest energy band of the lattice (at beginning of the preparation), in the absence of LIDDI.

It is convenient to describe the state of each atom in terms of Wannier functions $|\chi_j\rangle$ [Wannier 1937] that are *localized* at lattice sites labeled by index j. The Wannier functions are superpositions of the delocalized Bloch eigenfunctions $|\phi_k\rangle$ of the same band,

$$|\chi_j\rangle = \frac{1}{\sqrt{N}} \sum_k \exp\left(-i\, k\, x_j\right) |\phi_k\rangle, \tag{12}$$

where N is the number of lattice sites, and x_j is the position of the jth site. Since the Wannier functions are not eigenfunctions of the single-particle Hamiltonian, this Hamiltonian will have off diagonal elements, that causes following: if an atom is initially prepared in a Wannier state that is localized at one site, there is non zero probability of its tunneling to the neighboring sites. Nevertheless, if the tunneling rate is sufficiently slow we can neglect the "far" off diagonal elements and the single-particle Hamiltonian (9) in the Wannier basis has a relatively simple form:

$$H_{lat} \approx \begin{pmatrix} \cdots & \cdots & \cdots & \cdots & \cdots & \cdots \\ \cdots & H_0 & V_{hop} & 0 & 0 & \cdots \\ \cdots & V_{hop} & H_0 & V_{hop} & 0 & \cdots \\ \cdots & 0 & V_{hop} & H_0 & V_{hop} & \cdots \\ \cdots & 0 & 0 & V_{hop} & H_0 & \cdots \\ \cdots & \cdots & \cdots & \cdots & \cdots & \cdots \end{pmatrix}. \tag{13}$$

Here the diagonal elements H_0 are equal to the energy at the center of the band, and off-diagonal elements V_{hop} express hopping between the neighboring sites.

$$H_0 = \left\langle \chi_j \left| \hat{H}_{lat} \right| \chi_j \right\rangle, \; V_{hop} = \left\langle \chi_j \left| \hat{H}_{lat} \right| \chi_{j+1} \right\rangle. \tag{14}$$

The hopping rate is related to the energy bandwidth of the lowest lattice band V_B by $V_B \approx 4\, |V_{hop}|$ (for exact expressions see Ref. [Slater 1952]).

For a relatively shallow lattice potential ($U_0 \lesssim 15\ E_{rec}$), it is possible to derive the approximate formulae for V_{hop} and the single-atom effective mass m_{eff} from the quantum-pendulum Schrödinger equation:

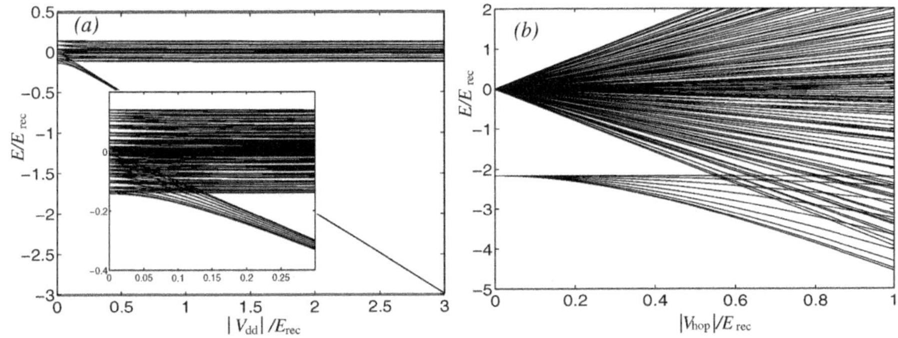

Figure 6. Eigenvalues of the two-atom Hamiltonian as a function of (a) the LIDDI coupling $|V_{dd}|$ (for a constant hopping potential $|V_{hop}| = 0.0355\ E_{rec}$), and (b) the hopping potential V_{hop} (for a constant dipole-dipole coupling potential $V_{dd} = 2.16\ E_{rec}$).

$$V_{hop} \approx \frac{1}{4}\, E_{rec}\, \exp\left(-\, 0.26\, \frac{U_0}{E_{rec}}\right), \tag{15a}$$

$$m_{eff} = \frac{2\hbar^2}{a^2\, V_B} \approx \frac{\hbar^2}{2a^2\, |V_{hop}|}. \tag{15b}$$

We can use Gaussians as a relatively accurate approximation of the lowest band Wannier functions:

$$\langle x|\chi_j\rangle \approx \langle x|\psi_j^{\rm Gauss}\rangle = \frac{1}{\sqrt{\sigma_G\, \sqrt{2\pi}}}\, \exp\left(-\, \frac{(x-x_j)^2}{4\, \sigma_G^2}\right), \tag{16}$$

where

$$\sigma_G^2 = \frac{\lambda_L^2}{4\pi^2}\, \sqrt{\frac{E_{rec}}{2U_0}}. \tag{17}$$

This approximation yields for $U_0 \gtrsim 6\ E_{rec}$ (see Fig. 11) the overlap (fidelity) $|\langle\psi_j^{\rm Gauss}|\chi_j\rangle|^2 > 98\%$. Let us stress, however, that although Wannier functions are sufficiently well approximated by Gaussians, we may not use Gaussians to calculate hopping potential (14), because of its sensitivity to non-Gaussian tails of the Wannier functions.

3. Diatom binding and translational EPR states

Let us now assume two neighboring tubes close to each other ($l \ll \lambda_C$ as in Fig. 3), occupied by one atom each. The LIDDI depends on the atomic positions as in Fig. 4 (c). Under these conditions, the interaction Hamiltonian

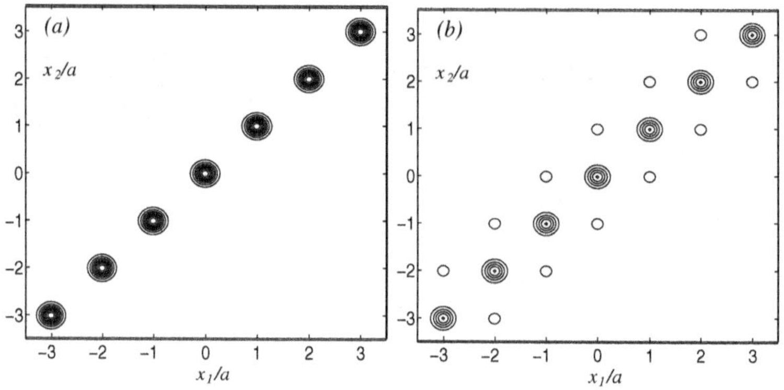

Figure 7. Joint probability distribution of the positions of two atoms in the ground state of Hamiltonian (19) for $|V_{hop}| = 0.0355E_{rec}$: (*a*) $|V_{dd}| = 1.0\ E_{rec}$, (*b*) $|V_{dd}| = 0.10\ E_{rec}$.

in the Wannier basis has nonzero elements only for atoms placed at the nearest sites,

$$\hat{H}_{int} \approx V_{dd} \sum_j \left|\chi_j^{(1)}\right\rangle \left|\chi_j^{(2)}\right\rangle \left\langle \chi_j^{(1)}\right| \left\langle \chi_j^{(2)}\right|, \tag{18}$$

and the total two-system Hamiltonian is

$$\hat{H}^{(2\ at)} = \hat{H}_{lat}^{(1)} \otimes \hat{1}^{(2)} + \hat{1}^{(1)} \otimes \hat{H}_{lat}^{(2)} + \hat{H}_{int}. \tag{19}$$

Hamiltonian (19) has been diagonalized numerically, and its energy spectrum is shown in Fig. 6, as a function of the binding and hopping potential strengths. From Fig. 6 (a) we can see that for an increasing ratio $|V_{dd}|/|V_{hop}|$ a band of diatomic energies is split off the band of single atom energies towards lower values.

For a strong LIDDI coupling field, $|V_{hop}| \ll |V_{dd}|$, the ground state of the Hamiltonian (19) corresponds to a tightly bound diatom which can be approximated by

$$|\psi_0\rangle \approx \frac{1}{\sqrt{N}} \sum_j \left|\chi_j^{(1)}\right\rangle \left|\chi_j^{(2)}\right\rangle. \tag{20}$$

This is a highly correlated state: when particle 1 is found at the jth site of lattice 1, then particle 2 is found at the jth site of lattice 2, with position dispersion given by the half-width σ of the atomic Wannier function in the lowest band, $\sigma \approx \sigma_G$ (Fig. 7 (a)). The Fourier transform of this wave function yields its momentum representation. The corresponding momentum probability dis-

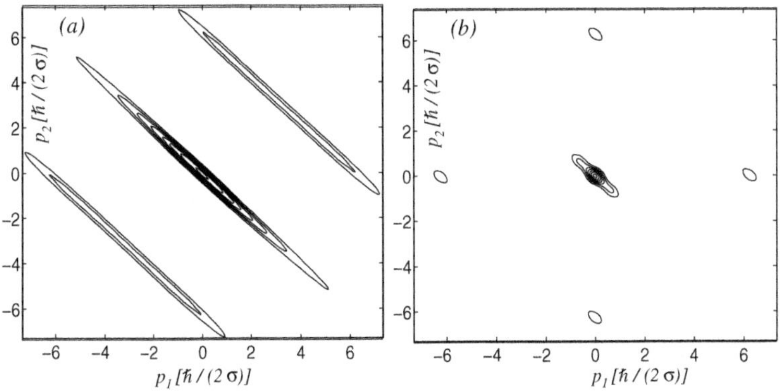

Figure 8. Joint probability distribution of the momenta of two atoms in the ground state of Hamiltonian (19) for $|V_{hop}| = 0.0355\ E_{rec}$: (a) $|V_{dd}| = 1.0\ E_{rec}$, (b) $|V_{dd}| = 0.10\ E_{rec}$.

tribution exhibits anti-correlation similarly to the EPR states (1) or (2), but it reflects the lattice periodicity (Fig. 8 (a)). In momentum space, the state occupies a region of half-width $\hbar/(2\sigma)$, and the probability distribution has narrow ridges along $p_2 = -p_1$. The width of the ridges is inversely proportional to the lattice size, $\Delta p_+ \sim \hbar/(Na)$, and they are shifted by $2\pi\hbar/a$ from each other.

The probability of atoms to escape their EPR partners "over the next" sites increases with the ratio $|V_{hop}/V_{dd}|$. This leads to an increase of the position dispersion which can be estimated by first-order perturbation theory: let atoms 1 and 2 occupy the jth site in the absence of V_{hop}. With the perturbation V_{hop} on, atom 2 can occupy also sites $j \pm 1$, which have energies $|V_{dd}|$ above the unperturbed state, with the probability $\approx |V_{hop}|^2/|V_{dd}|^2$. This contributes to an increase in the diatomic separation dispersion,

$$\Delta x_-^2 \approx \sigma^2 + 2a^2 \left(\frac{|V_{hop}|}{|V_{dd}|} \right)^2, \tag{21}$$

resulting in the joint probability distribution of the atomic positions and momenta as a function of $|V_{hop}|/|V_{dd}|$, shown in Figs. 7 (b) and 8 (b), respectively.

The states of the tightly bound diatom form a separate band whose bandwidth is

$$V_B^{(2\ at)} \approx 4 \left| V_{hop}^{(2\ at)} \right|, \tag{22}$$

below the lowest atomic vibrational band. We can estimate the *diatomic* hopping potential $V_{hop}^{(2\ at)}$ if we realize, that there are two ways for the diatom to hop to the neighboring site, *i.e.*, the change

$$\left|\chi_j^{(1)}\right\rangle \left|\chi_j^{(2)}\right\rangle \to \left|\chi_{j+1}^{(1)}\right\rangle \left|\chi_{j+1}^{(2)}\right\rangle \tag{23}$$

is realized either via

$$\left|\chi_j^{(1)}\right\rangle \left|\chi_j^{(2)}\right\rangle \to \left|\chi_{j+1}^{(1)}\right\rangle \left|\chi_j^{(2)}\right\rangle \to \left|\chi_{j+1}^{(1)}\right\rangle \left|\chi_{j+1}^{(2)}\right\rangle, \tag{24}$$

or via

$$\left|\chi_j^{(1)}\right\rangle \left|\chi_j^{(2)}\right\rangle \to \left|\chi_j^{(1)}\right\rangle \left|\chi_{j+1}^{(2)}\right\rangle \to \left|\chi_{j+1}^{(1)}\right\rangle \left|\chi_{j+1}^{(2)}\right\rangle. \tag{25}$$

By adiabatic elimination of the higher-energy intermediate states, one obtains

$$V_{hop}^{(2\,at)} \approx 2\,\frac{|V_{hop}|^2}{V_{dd}}. \tag{26}$$

All the states of the diatomic band have correlated positions. However, the momenta are not anti-correlated in all these states in the same way as in the diatomic ground state. To achieve highly anti-correlated momentum state, we have to prepare a state that predominantly originates from the bottom of the diatomic band. If we work with thermal states this means that the temperature of the system must satisfy

$$k_B T \ll V_B^{(2\,at)}. \tag{27}$$

Near the bottom of the band, the diatomic dynamics can be described by means of the 2-atom effective mass given by

$$m_{eff}^{(2\,at)} = \frac{2\hbar^2}{V_B^{(2\,at)}\,a^2} \approx \frac{\hbar^2\,|V_{dd}|}{4\,V_{hop}^2\,a^2}. \tag{28}$$

The thermal (kinetic) energy of the diatom is then related to the degree of momentum anti-correlation through the sum-momentum spread $\Delta p_+ = p_{x1} + p_{x2}$,

$$\Delta p_+^2 \approx k_B T\, 2m_{eff}^{(2\,at)} \approx \frac{\hbar^2\,|V_{dd}|}{4\,V_{hop}^2\,a^2}\,k_B T. \tag{29}$$

To determine how "strong" the EPR effect is, we compare the product of the half-widths of the position and momentum peaks in the tightly bound diatom state with the Heisenberg uncertainty limit through the parameter [Opatrný 2001; Opatrny 2003]:

$$s = \frac{\hbar}{2\,\Delta x_-\,\Delta p_+}. \tag{30}$$

A value of s higher than 1 indicates the occurrence of the EPR effect; the higher the value of s, the stronger the effect.

Strictly speaking, for the multi-peak momentum distribution, one should use a more general uncertainty relation, as discussed, *e.g.*, in Refs. [Hilgevoord 1983; Uffink 1984; Uffink 1985], that distinguishes the uncertainty of multiple narrow peaks from that of a single broad peak. However, even the simple half-width of the peaks is a useful measure of the EPR effect. In order to maximize s, we must adhere to the trade-off between reducing either Δx_-, by decreasing $|V_{hop}/V_{dd}|$, or Δp_+, by increasing $|V_{hop}/V_{dd}|$. The optimum value of s generally depends on the temperature of the diatom, as detailed below.

4. EPR state preparation

Cooling down the diatomic system is a quite non-trivial task, similar to cooling down a molecule. We suggest, in our case, a method which takes advantage of the different effective masses of diatoms in comparison to uncorrelated single atoms. The scheme has three basic steps:

(1) We switch on the shallow external harmonic 1D potential, that has its minimum in the lattice area (all other potentials are off), and cool down the atomic motion in the x direction to its ground state. The width σ_E is several times the lattice constant a and it is related to the desired momentum anti-correlation by $\sigma_E \approx \hbar/(\sqrt{2}\,\Delta p_+)$. The temperature necessary to achieve this is

$$T \ll \frac{\hbar^2}{4m\, k_B\, \sigma_E^2}. \tag{31}$$

(2) A weak lattice potential in the x-direction is then adiabatically switched on, so that the overall state is a product state

$$\left(\sum_j \alpha_j \left|\chi_j^{(1)}\right\rangle\right)\left(\sum_l \alpha_l \left|\chi_l^{(2)}\right\rangle\right) = \sum_j \alpha_j^2 \left|\chi_j^{(1)}\right\rangle \left|\chi_j^{(2)}\right\rangle + \sum_{j\neq l} \alpha_j\, \alpha_l \left|\chi_j^{(1)}\right\rangle \left|\chi_l^{(2)}\right\rangle, \tag{32}$$

where the coefficients

$$\alpha_j \sim \exp\left[-\frac{(j-j_0)^2\, a^2}{4\,\sigma_E^2}\right] \tag{33}$$

are Gaussians localized around the minimum of the external harmonic potential.

(3) We switch on the LIDDI and change the external potential, from an attractive (harmonic) well to a repulsive linear potential, that acts to remove the

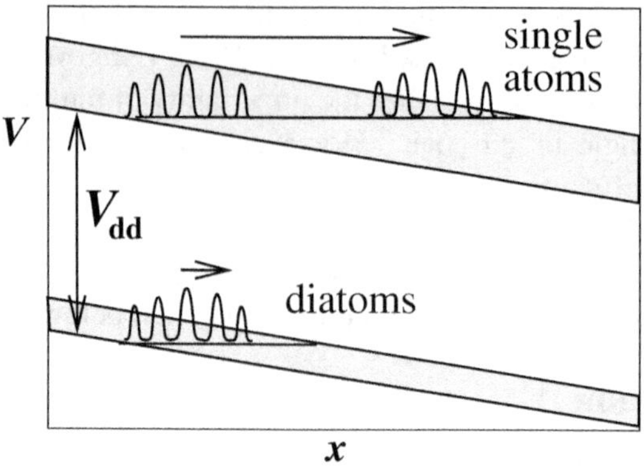

Figure 9 Separating single (unpaired) atoms from diatoms (compare to Fig. 5): an external repulsive potential causes both the single atoms and the diatoms to move on a surface of constant energy. The diatoms hit the top of the energy band after being displaced by a much shorter length than the unpaired atoms.

particles out of the lattice. The two parts of the wavefunction (32) will respond differently: the motion of the paired atoms (corresponding to $\sum_j \alpha_j^2 |\chi_j^{(1)}\rangle |\chi_j^{(2)}\rangle$) will remain in the vicinity of the initial position because their energy band is narrow and during the motion the system hits the upper edge (Fig. 9). Single (unpaired) atoms, whose bandwidth is substantially larger, will travel a much longer distance. Thus, after a properly chosen time t_c, the unpaired atoms could be removed from the x-region of interest, which is $\approx |V_{hop}/V_{hop}^{(2\ at)}|$ longer for single atoms than for diatoms [see Eq. (22)].

Provided the time t_c is short enough for the system to remain near the bottom of these two bands, the dynamics can be interpreted in terms of the appropriate effective masses: the "heavy" diatoms with mass (28) move much slower than the "light" single atoms with mass (15b). Thus, after changing the external potential from harmonic to linear repulsive, the unpaired atoms will be ejected out of the lattice and separated from the diatoms as glumes from grains.

This effect is illustrated by the results of the numerical model in Fig. 10 for two lithium atoms in two lattices with $\lambda_L = 323$ nm (corresponding to the transition $2s$ - $3p$) and a dipole-dipole coupling field of $\lambda_C = 670.8$ nm (transition $2s$ - $2p$). The dipole moment element of the lattice transition is 1.26×10^{-30} cm, while the LIDDI coupling dipole element is 2.7×10^{-29} cm. From these values we get the recoil energy $E_{rec} = 1.85 \times 10^{-28}$ J. The lattice and LIDDI field intensities are $I_L = 0.186\ \mathrm{W/cm^2}$ and $I_C = 0.023\ \mathrm{W/cm^2}$. The corresponding field detunings are $\delta_L = 50\ \gamma_L$, $\delta_C = 100\gamma_C$, the respective decay rates being $\gamma_L = 1.2 \times 10^6\ \mathrm{s^{-1}}$, and $\gamma_C = 3.7 \times 10^7\ \mathrm{s^{-1}}$. The two lattices are displaced by $l = 40$ nm. From these values we get the lattice potential $U_0 = 3.93\ E_{rec}$, the LIDDI potential of the nearest atoms $V_{dd}^{min} = -0.5\ E_{rec}$, and the hopping potential $V_{hop} = -0.09\ E_{rec}$. The two-particle hopping potential is then $V_{hop}^{(2\ at)} \approx -0.0324\ E_{rec}$. The correlated pairs are prepared by first cooling independent atoms in an external harmonic potential with the ground-state half-width of $\sigma_E = 5a$ (frequency of

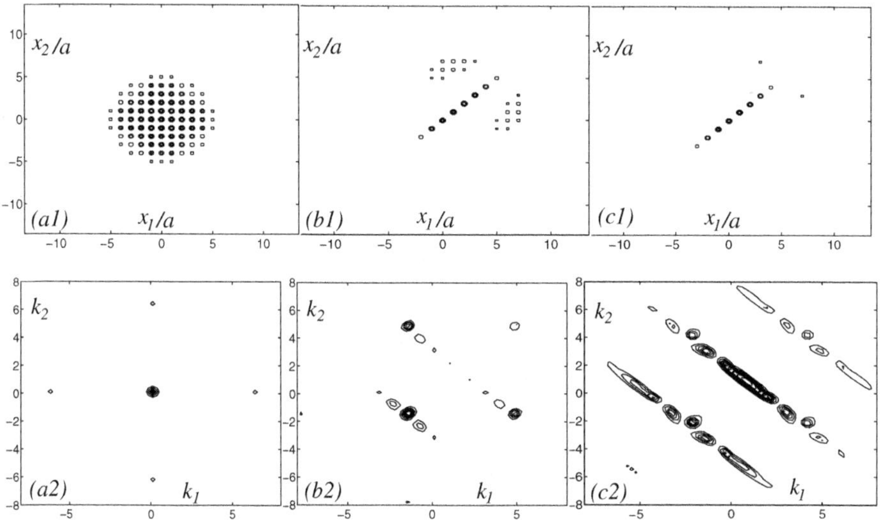

Figure 10. Simulation of the EPR state preparation in an optical lattice with 25 sites, at three consecutive times. First row shows the joint probability distribution in x representation, the second one in p representation. $(a1)$ and $(a2)$: initially ($t = 0$), the atoms are cooled down to the external harmonic potential ground state, whereas the LIDDI is off. $(b1)$ and $(b2)$: at $t = 1.4 \times 10^{-4}$ s LIDDI and the repulsive linear potential (with the slope $0.04\ E_{rec}$ per lattice site) are on, whereas the harmonic potential is off. The diatoms are moving through the lattice *very* slowly in comparison to the single atoms. $(c1)$ and $(c2)$: at $t = 2.16 \times 10^{-4}$ s single atoms are ejected out of the lattice and discarded and the diatoms are separated out.

1 kHz $\sim$ 30 nK). When the linear external potential is switched on, the atoms start moving in the direction of decreasing potential energy. Fig. 10, which captures the situation at three consecutive times, shows that unpaired atoms (off-diagonal peaks) gain with the same potential significantly higher velocity than the diatoms (diagonal peaks).

The paired atoms remaining in the lattice are then in the state $\sim \exp[-(j-j_0)^2 a^2/(2\sigma_0^2)]\ |\chi_j^{(1)}\rangle\ |\chi_j^{(2)}\rangle$ wherein positions and momenta are correlated with the uncertainties $\Delta x_+ \approx \sigma_E/\sqrt{2}$ and $\Delta p_+ \approx \hbar/\Delta x_+$, respectively. At higher temperatures the atoms are not cooled to the ground state of the external potential and the momentum anti-correlation has the spread

$$\Delta p_+ \approx \frac{\hbar}{\sqrt{2}\,\sigma_E\ \tanh\left[\frac{\hbar^2}{2\ \sigma_E^2\ m\ k_B T}\right]}. \tag{34}$$

The parameter s of Eq. (30) can then be estimated as

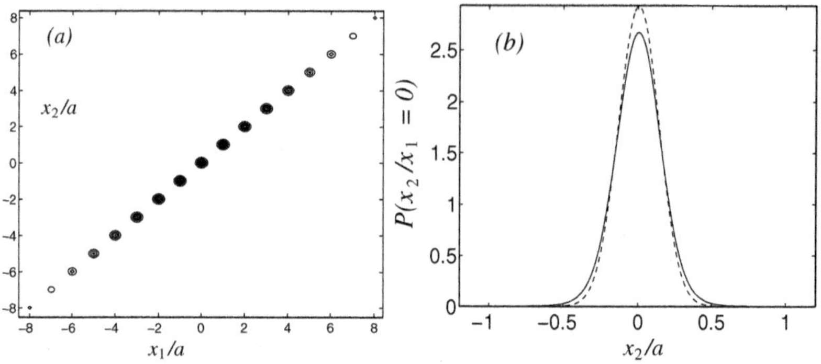

Figure 11. (a) Joint probability distribution of the positions of two lithium atoms in adjacent optical lattices, prepared in a diatom state as specified in the text, using the ground state of the external harmonic potential with half-width $\sigma_E = 6a$ and temperature of 10 nK. (b) Solid line: position probability of atom 2 in the state above, conditional on atom 1 being measured at site 0. Dashed line: Gaussian approximation of the Wannier function with the half-width $\sigma = 0.14\ a$.

$$s \approx \frac{\sigma_E}{\sqrt{2}\,\sigma}\ \tanh\left[\frac{1}{\pi^2}\left(\frac{a}{\sigma_E}\right)^2\frac{E_{rec}}{k_B T}\right]. \tag{35}$$

This relation enables us to select the optimal external harmonic potential (specified here by σ_E) such that the parameter s is maximized, for a given temperature T.

The small effective mass of unpaired atoms allows us to cool them to temperatures higher than that corresponding to the bottom of the diatomic band. The price is, however, that most of the atoms are discarded and only a small fraction of $\sim a/\sigma_E$ will remain in the bound diatom state. The different behavior of the paired and unpaired atoms in a periodic potential is a sparse-lattice analogy of the transition from Mott-insulator to a superfluid state in the fully occupied lattice, recently observed in Ref. [Greiner 2002].

The two-particle joint position distribution of the ground state is a chain of peaks of half-width σ separated by a that are located along the line $x_2 = x_1$ (Fig. 11). The corresponding joint momentum distribution spreads over an area of half-width $\hbar/(2\sigma)$ and consists of ridges in the direction $p_2 = -p_1$, that are separated by $2\pi\hbar/a$, and have the half-width $\pi\hbar/(Na)$ for a lattice of N sites (Fig. 12).

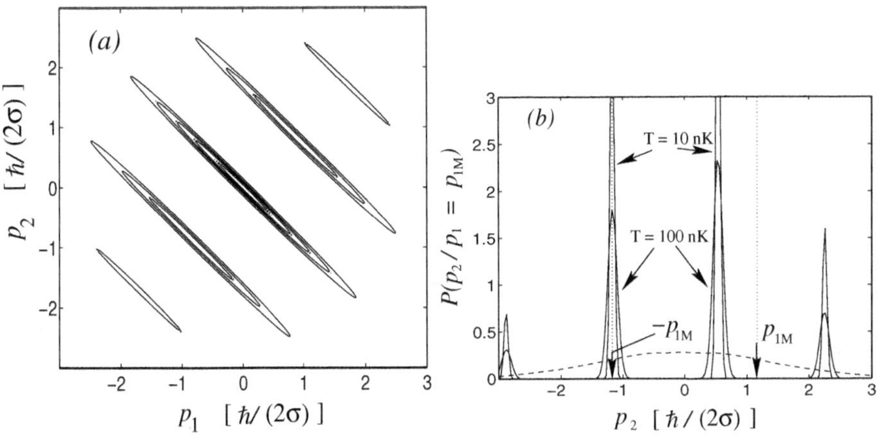

Figure 12. (a) Joint probability distribution of the atomic momenta in the aforementioned state with $T = 100$ nK. (b) Conditional probability of the momentum of atom 2, given that the momentum of atom 1 has been measured (the measured value $p_1 = p_{1M}$ is indicated by an arrow) for lithium diatoms prepared as in the text. The dashed line corresponds to the marginal probability distribution of momentum p_2 irrespective of the momentum of atom 1 at the temperature $T = 100$ nK. The half-width of each peak is equal to $1/s$ of Eq. (30).

5. Measurements

After preparing the system in the EPR state, how can one test its properties experimentally? To this end we may increase the lattice potential U_0, switch off the field inducing the LIDDI, and separate the two lattices by changing the laser-beam angles. By increasing U_0, the atoms lose their hopping ability and their quantum state is "frozen" with a large effective mass: the bandwidth V_B decreases exponentially with U_0 and the effective mass increases exponentially, so that the atoms become too "heavy" to move. One has then enough time to perform measurements on each atom:

a) The atomic position can be measured by detecting its resonance fluorescence. After finding the site occupied by atom 1, one can infer the position of atom 2. If this inference is confirmed in a large ensemble of measurements, it would suggest that there is an "element of reality" [Einstein 1935] corresponding to the position of particle 2. An example of the conditional probability of position of particle 2 after measuring the position of particle 1 is given in Fig. 11.

b) The momentum can be measured by switching off the x-lattice potential of the atom (thus bringing it back to its "normal" mass m). The distance traversed by the atom during a fixed time is proportional to its momentum. An example of the conditional probability of the momentum of particle 2 given the momentum of particle 1 is shown in Fig. 12.

c) One can test the EPR correlations between the atomic *ensembles* occupying the two lattices, testing large number of pairs in a single run. The correlations in x and anti-correlations in p would be observed by matching the *distribution histograms* measured on atoms from the two lattices.

Conclusions

We have discussed a scheme which can be used to prepare a translationally entangled pair of massive particles in a state analogous to the original EPR state [Einstein 1935]. A novel element of the present scheme is the extension of the EPR correlations to account for lattice-diffraction effects. Their momentum and position correlations principally differ from those of free particles [Eqs. (1) and (2)]: due to the lattice periodicity, the *position and momentum distributions have generally a multi-peak structure.*

The realization of the proposed scheme is expected to be based on the adaptation of existing techniques (optical trapping, cooling, controlled dipole-dipole interaction), to the requirements spelled out in Secs. 4 and 5. The most important ingredient of the scheme is the manipulation of the effective mass, for EPR-pairs preparation (by separating the "light" unpaired atoms from the "heavy" diatoms) and for their detection (by "freezing" the atoms in their initial state so that their EPR correlations are preserved long enough).

One may envision extensions of the present approach to matter teleportation [Opatrný 2001] and quantum computation based on continuous variables [Braunstein 1998 (a); Lloyd 1998; Lloyd 1999]. Such extensions may involve the coupling of entangled atomic ensembles in optical lattices by photons carrying quantum information.

Acknowledgments

We acknowledge the support of ISF, Minerva and the EU Networks QUACS and ATESIT.

INVESTIGATION OF AUTLER-TOWNES EFFECT IN SODIUM DIMERS

R. Garcia-Fernandez, A. Ekers, B. W. Shore*, J. Klavins†, L. P. Yatsenko‡ and K. Bergmann
Department of Physics, University of Kaiserslautern,
Erwin-Schrödinger Str., D-67663 Kaiserslautern, Germany

Abstract Experiments under way will provide high-resolution studies of the Autler-Townes effect in molecules, observed as fluorescence from both upper and intermediate levels in a three-level ladder coupled by two laser fields.

Keywords: Autler-Townes effect, dynamic Stark shift, molecular spectra

Introduction

The study of decoherence starts with the preparation of a coherent ensemble of states using coherent laser radiation. The Autler-Townes (AT) effect, known since 1955 [Autler 1955], is at the heart of such preparation schemes. Decoherence in molecules is of particular interest, and significant new insights may be gained from studies of the AT effect. Although it has been widely studied for atoms [Fisk 1986], few high-resolution studies have been reported for molecules [Qi 2000]. It is in this context that we have started a program within QUACS to better characterize the AT-effect under varying conditions (coupling schemes, intensities, and detunings).

In its simplest form, the AT effect can be regarded as originating in the coherent interaction of a strong laser field with a two-state quantum system. The resulting alteration of the atomic or molecular energies can be probed in various ways, such as through fluorescence excitation spectrum obtained by scanning a weak laser field across the resonance with some third level. The resulting signals, typically visible as a doublet structure, incorporate not only the

*Permanent address: 618 Escondido Cir., Livermore CA 94550, USA
†Permanent address: Institute of Atomic Physics and Spectroscopy, University of Latvia, Raina bulv. 19, LV-1586 Riga, Latvia
‡Permanent address: Institute of Physics, Ukrainian Academy of Sciences, prospect Nauki 46, Kiev-39, 03650, Ukraine

V.M. Akulin et al. (eds.), Decoherence, Entanglement and Information Protection in Complex Quantum Systems, 391–394.

physics of the coherent interaction revealed in the separation of the two doublet components, but also aspects of decoherence, such as spontaneous emission lifetimes, as revealed by the widths and intensities of the two components.

1. Molecular system

Our experiments use a weak probe field P to excite molecular levels (possibly degenerate) that are coupled by a stronger field S to more highly excited states in a three-level ladder linkage of energies $E_g < E_e < E_f$. Fig. 1 shows the states involved, and the spontaneous emission which is detected, as discussed below.

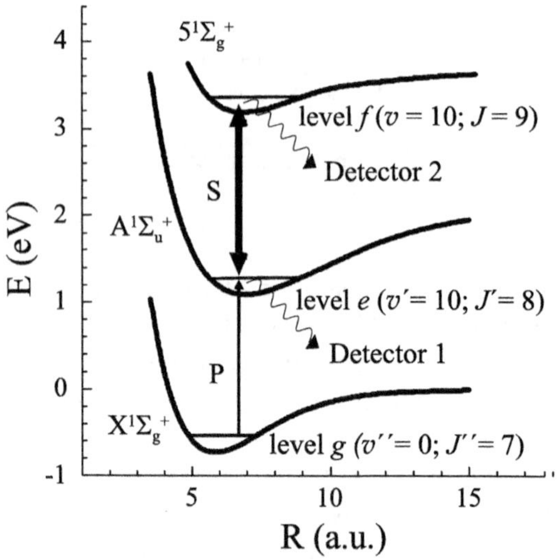

Figure 1 Excitation scheme of Na_2 involving the electronic and rovibrational levels of interest, showing weak P field, strong S field, and spontaneous emission from levels e and f.

2. Experimental arrangement

Fig. 2 shows the arrangement of the experiment. The molecular beam is generated in a source oven at a temperature of 900 K, from which vapor emerges through a 0.4 mm diameter nozzle. This produces a supersonic beam of Na atoms and Na_2 molecules, cooled such that 99% of the molecules are in the ground vibrational level, $v'' = 0$, and the distribution over rotational levels peaks at $J'' = 7$. The flow velocity in the beam is $v_f = 1340$ m/s, and the $1/e$ width of the longitudinal velocity distribution is $\Delta v_{mol} = 260$ m/s for the molecules.

The beam is collimated by two skimmers and an aperture at the entrance of the interaction region to a divergence of $1°$, resulting in a transverse velocity spread $\leq \pm 10$ m/s. The concentration of molecules in the reaction region 20 cm downstream from the nozzle is $2. 10^{10}$ cm^{-3}.

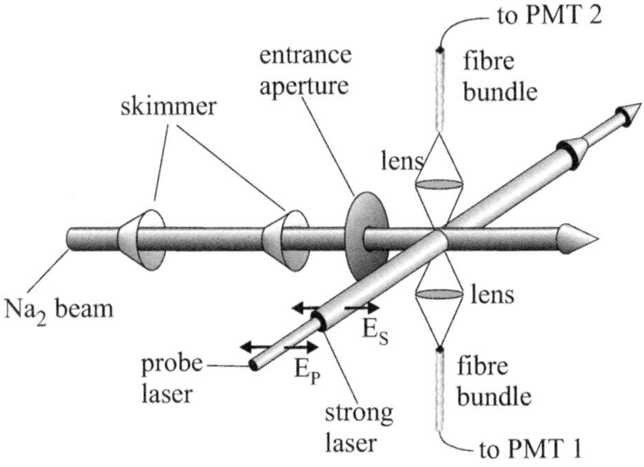

Figure 2 Experimental arrangement of the molecular and laser beams.

The molecular beam is crossed at right angles by two co-propagating cw laser beams (both from coherent CR-699-21 dye lasers, with bandwidth $\Delta\nu_L =$ 1 MHz) which are focused onto the molecular beam axis by cylindrical lenses. Both laser beams have the same linear polarization which is held parallel to the molecular beam axis.

The weaker laser P, driving the first step, is tuned near the frequency of the the g - e transition $X^1\Sigma_g^+,\ v'' = 0,\ J'' = 7 \rightarrow A^1\Sigma_u^+,\ v' = 10,\ J' = 8$, while the stronger laser S, coupling the intermediate and upper level is tuned near the e - f transition $A^1\Sigma_u^+,\ v' = 10,\ J' = 8 \rightarrow 5^1\Sigma_g^+,\ v = 10,\ J = 9$.

Although the axis of the two lasers coincide, the S beam waist (typically a few hundred μm) is about twice the size of the P waist. Thus the S field can be considered as constant, and maximal, across the extension of the P field.

2.1 The signals

We observe the fluorescence collected into two optical fibers fixed at right angles to the molecular beam and to the laser beams. The light passes from the fibers, through filters, into photomultiplier detectors. One detector channel is equipped with a cut-off filter that removes light with wavelengths longer than 600 nm. This detector registers the total fluorescence, from level f to the $A^1\Sigma_u^+$ state (see Fig. 1). The second detector, accepting only light with wavelength longer than 620 nm, registers the total fluorescence from the $A^1\Sigma_u^+$ state.

The excitation spectrum signals are obtained by fixing the frequency of the S laser and sweeping the frequency of the P laser through resonance with the $X^1\Sigma_g^+,\ v'' = 0,\ J'' = 7 \rightarrow A^1\Sigma_u^+,\ v' = 10,\ J' = 8$ transition, while monitoring separately the fluorescence from the $A^1\Sigma_u^+$ and $5^1\Sigma_g^+$ state.

2.2 Observations

Fig. 3 shows results of scanning the probe-field detuning for several choices of S field detuning, (a) for the $5^1\Sigma_g^+$ state and (b) for the $A^1\Sigma_u^+$ state.

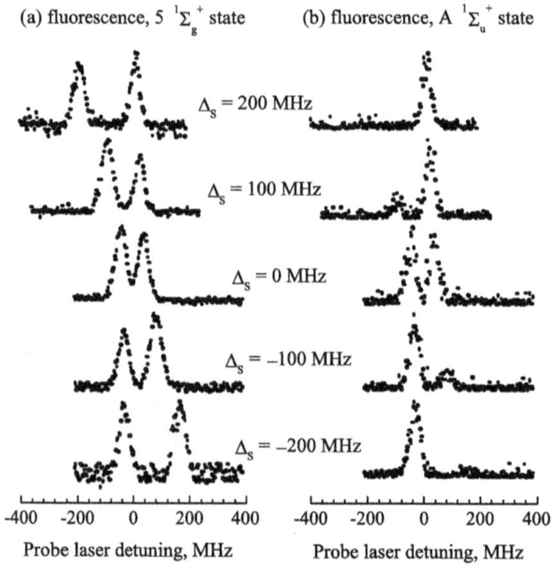

Figure 3 Fluorescence signals as a function of P field detuning, for several values of the S field detuning. (a) Fluorescence from the $5^1\Sigma_g^+$ state. (b) Fluorescence from the $A^1\Sigma_u^+$ state.

As can be seen, the observations of Fig. 3 fit the pattern of a well resolved Autler-Townes doublet. The separation of the two components is smallest when the detuning is smallest; it then gives a direct measure of the Rabi frequency of the S-field transition.

Some patterns of Fig. 3 (b) differ significantly from those of Fig. 3 (a). Both the component separation and the relative heights are strongly dependent on the detuning of the S-field for the e level doublets. However, the peak heights from the f level patterns are nearly unaffected by the S-field detuning.

The patterns shown here can be simulated, using density matrix equations, and from such modeling the relative lifetimes of states e and f can be deduced.

We have obtained a large number of data sets of the AT effect, with several choices of the molecular states, over a range of detunings and intensities of the two lasers. These reveal the influence of these various parameters on the pattern, which under some conditions differs significantly from the simple doublet pattern. Analysis of these data is currently under way.

Acknowledgments

In addition to support by the QUACS RTN we also acknowledge support by the Deutsche Forschungs Gemeinschaft and INTAS-2001-155.

TRANSITION STEERING VIA SPACE-DEPENDENT COUPLING

M. Leibscher[1, 2] and S. Stenholm[1]

[1] *Royal Institute of Technology, Department of Physics, Alba Nova, Roslagstullbacken 21, 10691 Stockholm, Sweden*

[2] *Weizmann Institute of Science, Chemical Physics Department, 76199 Rehovot, Israel*

monika@chemie.fu-berlin.de

stenholm@atom.kth.se

Abstract The transition between electronic energy surfaces of molecules is usually described in the Franck-Condon approximation where the spatial variation of the coupling matrix elements is neglected. In this work we go beyond this approximation and explore the effects of such a variation using a simply parameterized interaction instead of the usually poorly known realistic variations. Moreover, we propose a model that allows us to steer molecular transitions by shaping the space-dependence of the coupling.

Keywords: Transition dipole moment, quantum control

Introduction

Controlling atomic and molecular quantum states with the help of laser light has been an intensively studied field in the last years (see, *e.g.*, [Tannor 1985; Judson 1992; Rice 2000; Shapiro 2003]). The ability to produce and shape femtosecond laser pulses [Zewail 1994; Manz 1995] made it possible to create specially tailored wave packets and to manipulate their dynamics in order to reach pre-assigned goals. The techniques of quantum control have been applied to the problem of steering chemical reactions [Tannor 1985; Assion 1998; Shapiro 2003] as well as intramolecular dynamics [Averbukh 1993; Goodson 2000].

Quantum control methods make use of the time- and frequency dependence of the external laser field, usually assuming that the spatial dependence of the coupling between two electronic energy surfaces of a molecule is constant. In this work we ask what may be the influence of the spatial dependence of the coupling and can it be also used for steering molecular transitions?

V.M. Akulin et al. (eds.), Decoherence, Entanglement and Information Protection in Complex Quantum Systems, 395–410.

The effects of a laser-induced molecular transition is most easily discussed within the perturbation approximation. Since the dipole couplings between molecular levels is often weak enough to allow the transfer of only a minute fraction of the ground state population into excited levels, working within the perturbative frame is often justified.

To this end, we assume some spatial dependence of the coupling between two electronic energy surfaces of a molecule. The usual assumption of a constant coupling leads to the well known Franck-Condon effect: the ground state wave function is lifted up to the excited state without distortion. This can also be seen as the molecular equivalent of momentum conservation. The initial wave function is lifted to the spatial turning points of the motion, where it moves with nearly zero momentum. This argument changes when the coupling between the levels depends on position along the energy surface. This spatial dependence imposes a modification of the initial momentum distribution, which may have a profound effect on the transition probabilities.

Any spatial dependence of the coupling between two molecular potentials derives from the spatial dependence of the transition matrix element. Quantum chemical calculations of the transition dipole moment tend to be rather difficult and require considerable effort in order to be reliable. Different approaches to calculate the transition dipole moments lead to contradictory results [Hessel 1974; Akopyan 1999; Magnier 2000], and no general trend seems to emerge. We found only few attempts to determine the generic behavior that governs such effect from semiempirical considerations [Woerdman 1981; Akopyan 1999; Johnston 2003].

In this work, we discuss the influence of space-dependent couplings for rather simple model systems. We consider one-dimensional systems, that might describe a diatomic molecule, or the reaction path of some simple reaction. Furthermore, we neglect all influence of the rotational motion.

The motivation of this work is twofold. In the first part, we assume a model space-dependence for the coupling and discuss how the transfer of a wave packet is affected by the various parameters of the coupling. Here, we choose the space-dependence such that the parameters have direct physical interpretation within the model system. In this way, we can discuss possible consequences within a general framework. To see the effects in a realistic system, the system would have to be re-investigated in order. Such investigations are however beyond the scope of this work. In the second part of this work, we propose a model that allows us to steer intramolecular dynamics by varying the spatial dependence of the coupling.

The paper is organized as follows. In Sec. 1, we define the problem of space-dependent coupling between two molecular potentials in the perturbative framework. The outcome for transition between two harmonic oscillators coupled by a space-dependent interaction is discussed in Sec. 2. In Sec. 3, we

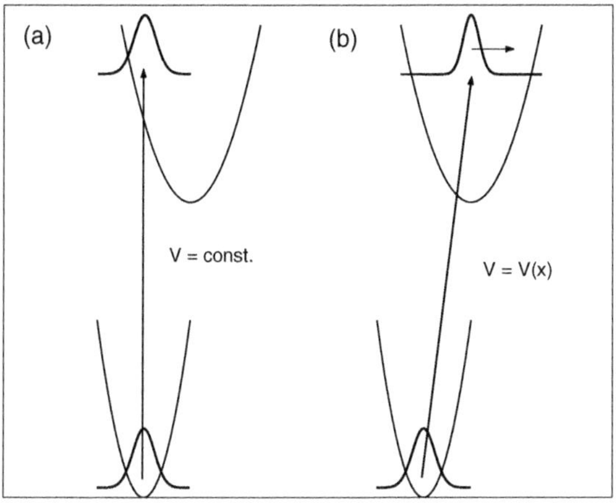

Figure 1 Transition between two electronic states of a molecule (a) in Franck-Condon approximation, and (b) with a coordinate-dependent interaction.

propose a system consisting of three coupled potentials that allows to induce a space-dependence coupling. The prospects of steering the transfer of a wave packet by varying the space-dependent coupling is discussed in Sec. 4. We finally summarize our results in the concluding Sec. 4.

1. Electronic transitions in momentum space

In the dipole approximation, the interaction between molecules and a laser light is given by $V_{int}(\vec{r},\, t) = -\,\vec{E}(t).\,\vec{\mu}(\vec{r})$, where $E(t)$ is the electric field and $\vec{\mu}(\vec{r})$ is the dipole moment. Here, we will consider only one-dimensional systems, *i.e.*, diatomic molecules, with the coordinate x denoting the internuclear distance between the two atoms. In first order perturbation theory, the probability for a transition between the initial state $\varphi_i(x)$ on the ground potential $U_1(x)$ and the nth vibrational eigenstate $\varphi_n(x)$ of the excited potential $U_2(x)$ is proportional to

$$\mathcal{P}_n = \left| \int dx\; \varphi_n^\star(x)\; V(x)\; \varphi_i(x) \right|^2 . \tag{1}$$

Here, we assumed that the interaction with the laser pulse is so short that the motion of the molecule is "frozen" during the interaction, and $V_{int}(x,\, t) = V(x)\, \delta(t)$. Note, that we use dimensionless variables throughout the paper. Since we will be mainly concerned with transitions between harmonic oscillators, we scale $x \to x\sqrt{m\,\omega/\hbar}$ and $p \to p/\sqrt{\hbar\omega\, m}$, where m is the reduced mass of the molecule and ω is the oscillator frequency. Moreover, we set $t \to t/\omega$ and $U_i \to U_i/(\hbar\omega)$ as well as $V \to V/(\hbar\omega)$.

If the interaction $V(x)$ is constant, Eq. (1) describes the well known Franck-Condon factors that induce a copy of the ground state wave function in the excited potential via a vertical transition, as depicted in Fig. 1 (a). In this case we can write Eq. (1) as

$$\mathcal{P}_n \propto \left| \int dx\, \varphi_n^\star(x)\, \varphi_i(x) \right|^2 = \left| \int dp\, \tilde{\varphi}_n^\star(p)\, \tilde{\varphi}_i(p) \right|^2, \qquad (2)$$

where we have expressed the vibrational eigenstates in terms of the momentum representation. This shows that the Franck-Condon principle is the best a molecular transition can do to satisfy the conservation of momentum. However, the introduction of spatial dependence of the coupling influences the selection between initial and final momentum distributions. The interaction gives the nuclei a kick during interaction. Thus, if $V(x)$ does depend on the internuclear distance, also non-vertical transitions may occur, see Fig. 1 (b). In order to study such transitions more systematically, we look at the transition probability in the momentum representation,

$$\mathcal{P}_n = \frac{1}{2\pi} \left| \int dp \int dp'\, \tilde{\varphi}_f^\star(p)\, \tilde{\varphi}_i(p')\, \tilde{V}(p'-p) \right|^2, \qquad (3)$$

where

$$\tilde{V}(p'-p) = \frac{1}{\sqrt{2\pi}} \int dx\, \exp[-i(p'-p)x]\, V(x). \qquad (4)$$

Note, that the interaction in the momentum representation depends only on the difference between the momenta p' and p. Therefore, we can introduce the coordinate $P = p' - p$. Then, the transition probability can be written as

$$\mathcal{P}_n = \frac{1}{2\pi} \left| \int dP\, \tilde{V}(P)\, I(P) \right|^2. \qquad (5)$$

Here, we have introduced the notation

$$I(P) = \int dp\, \tilde{\varphi}_n^\star(p)\, \tilde{\varphi}_i(p-P). \qquad (6)$$

The function $I(P)$ can be regarded as a kind of generalized Franck-Condon factor. It describes the momentum mismatch between the initial state and an excited one. Eq. (5) tells how efficiently the potential V is able to compensate for this mismatch.

In perturbation theory only a fraction of the population is transferred during the interaction. In order to compare the transition probabilities for different forms of the interaction, it is essential to normalize the excited wavefunction. Therefore, we assume that

$$\int_{-\infty}^{\infty} |V(x)|^2\, dx = V_0^2 \qquad (7)$$

is constant. This form of normalization can be seen as an analog to exciting wave packets at constant laser intensity, a constraint that is frequently used, *e.g.*, in the field of laser control of molecular dynamics. In many cases, the pulse area $\int_{-\infty}^{\infty} V(x)dx$ is regarded as fixed. The normalization according to Eq. (7) has also the advantage that the momentum interaction is normalized too,

$$\int_{-\infty}^{\infty} \left|\tilde{V}(P)\right|^2 dP = V_0^2. \tag{8}$$

2. Transition between harmonic oscillators

In order to investigate the momentum effects, we consider simple model systems. We assume that the spatial dependence of the coupling is of the form

$$V(x) = V_0 \left(\frac{\sigma}{\sqrt{\pi}}\right)^{1/2} \exp\left[-\frac{\sigma^2(x-x_1)^2}{2} + i\,p_1(x-x_1)\right], \tag{9}$$

or, in the momentum representation,

$$\tilde{V}(P) = V_0 \left(\frac{1}{\sqrt{\pi}\,\sigma}\right)^{1/2} \exp\left[-\frac{(P-p_1)^2}{2\sigma^2} - i\,P\,x_1\right], \tag{10}$$

where σ is the half-widths of the momentum distribution, p_1 is the transition-induced shift in momentum, and x_1 is the shift in the x-direction. For an interaction of such form, we can easily consider the limit $\sigma \to 0$ to obtain the limit of constant interaction. Moreover, such a shape allows us to study the effects of broadening the momentum distribution, of having an additional average momentum p_1 transferred during the interaction, and of shifting the maximum of the interaction out of the equilibrium distance $x = 0$. To this end, we consider the transition between two harmonic oscillators

$$U_1(x) = \frac{x^2}{2} \text{ and } U_2(x) = \frac{s^2}{2}(x-x_0)^2 + D, \tag{11}$$

where D is the energy difference between the electronic states. The excited electronic states are usually broader than the ground state, a situation we will model by choosing $s < 1$. We assume that the system is initially in the ground vibrational state of $U_1(x)$. In Ref. [Leibscher 2004] we have shown that the probability for a transition to the nth vibrational state of U_2 is given by

$$\mathcal{P}_n = \frac{\mathcal{C}}{2^n\, n!} \left|\frac{s^2-1-\sigma^2}{s^2+1+\sigma^2}\right|^n \left|H_n\left[-\frac{s\left(p_1 + i\,\sigma^2(x_0-x_1) + i\,x_0\right)}{\sqrt{s^4-(1+\sigma^2)^2}}\right]\right|^2, \tag{12}$$

where

$$\mathcal{C} = \frac{|V_0|^2}{\sqrt{\pi}} \frac{2s\,\sigma}{1+s^2+\sigma^2} \times \exp\left[-\frac{p_1^2 + x_0^2\,s^2(1+\sigma^2) + x_1^2\,\sigma^2(1+s^2) - 2x_0\,x_1\,s^2\,\sigma^2}{1+s^2+\sigma^2}\right] \tag{13}$$

is the normalization constant that depends on the parameters of the interaction, V_0, σ, x_1 and p_1, as well as on the parameters of the excited potential, s and x_0. For a constant interaction, $V(x) = V_0$, Eq. (12) reduces to the well-known probability distribution for the Franck-Condon transition between two harmonic oscillators [Stenholm 1994]. In the following, we present the various effects the space-dependent interaction Eq.(9) has on the transition probabilities.

2.1 Results

In order to demonstrate the effects a coupling of the form Eq. (9), we consider the transition between two harmonic oscillators shifted by $x_0 = 4$. The frequency of the excited potential U_2 is given by $s = 0.7$.

In the example, we choose $x_1 = 0$ and $p_1 = 0$ and consider only the broadening of the momentum distribution $\tilde{V}(p)$ by varying its width σ. The results can be seen in Fig. 2 The transition probability for $\sigma = 0.1$ is shown in Fig. 2 (a). The distribution is centered around $n = 5$ and has the Poisson-like shape of a Franck-Condon type of interaction, as is expected for a quasi-constant interaction with small σ. With increasing σ, the maximum is shifted slightly towards higher values of n. But the most obvious effect is the broadening of the probability distribution. For $\sigma \to \infty$, an extremely narrow position wave packet with its center at $x = 0$ is excited on the upper potential. In this limit, the probability distribution is

$$\mathcal{P}_n \propto \frac{1}{2^n\,n!}\,|H_n(-\,s\,x_0)|^2\,. \tag{14}$$

It can be seen in Fig. 2 (c), that such a distribution is very different from the Poisson-like distribution of a Franck-Condon transition, Fig. 2 (a). Certain levels, determined by the zeros of the Hermite polynomials, are not excited at all. Therefore, although it is not possible to selectively excite pre-selected eigenstates by varying σ, it is possible to suppress the excitation of some of the eigenstates.

Next, we consider a coupling with $x_1 = 0$ but $p_1 \neq 0$. In this case, we still have a vertical transition, but a non-zero average momentum is transferred during the interaction, as indicated in Fig. 1. Fig. 3 shows the resulting excitation probability for different values of p_1 for the case of transitions between two oscillators shifted by $x_0 = 4$. It can be seen that with increasing values of p_1, the maximum of the transition is shifted towards larger values of n, and the distribution becomes broader. We note that the shift in the probability distribution depends only on $|p_1|$. In Fig. 3, we have set $\sigma = 0.1$. For larger values of σ, the shift in the probability distribution is superimposed on the general broadening of the distribution observed above. In the limit $\sigma \to \infty$, the probability distribution does not depend on p_1 at all. We note that this effect can be used to select the range of excited states which is populated by the process.

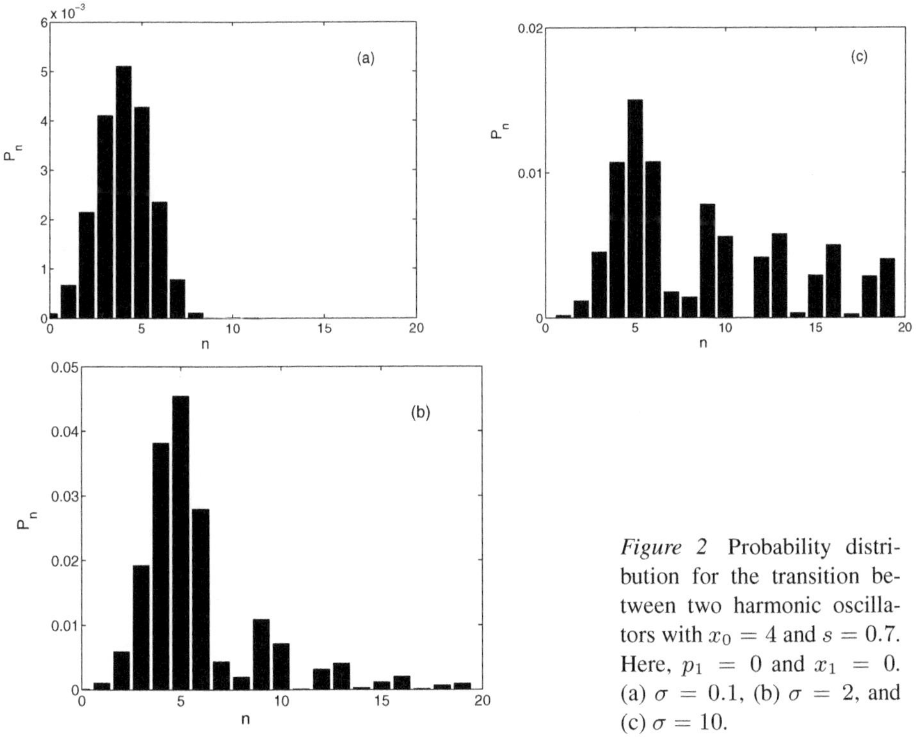

Figure 2 Probability distribution for the transition between two harmonic oscillators with $x_0 = 4$ and $s = 0.7$. Here, $p_1 = 0$ and $x_1 = 0$. (a) $\sigma = 0.1$, (b) $\sigma = 2$, and (c) $\sigma = 10$.

If the maximum of the transition $V(x)$ is displaced from the equilibrium position ($x_1 \neq 0$), the interaction induces a non-vertical transition. The excited wave packet $\psi(x, 0) = V(x)\,\varphi_i(x)$ is centered around

$$x = x_1 \frac{\sigma^2}{1 + \sigma^2}. \tag{15}$$

For $\sigma \to 0$, the displacement is zero (Franck-Condon type of excitation). The displacement increases with σ to $x = x_1$ for $\sigma \to \infty$. Fig. 4 shows the probability distribution for a transition between two oscillator states shifted

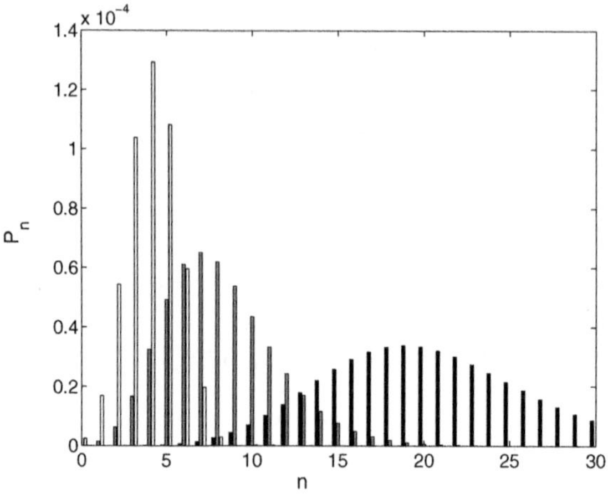

Figure 3 Probability distribution for $x_0 = 4$ and $\sigma = 0.1$. Here, $p_1 = 0$ (white bars), $p_1 = 2$ (grey bars), and $p_1 = 4$ (black bars). The frequency of the excited potential is $s = 0.7$.

by $x_0 = 4$ for a displacement $x_1 = \pm 1$. Compared with the vertical transition which is shown by the white bars in Fig. 4, the transition probability for $|x_1| = 1$ is shifted towards larger or smaller values of n, depending on the sign of x_1. For positive values of x_1, the probability distribution is shifted towards smaller values of n until, for

$$x_1 = x_0 \frac{\sigma^2 + 1}{\sigma^2}, \tag{16}$$

it is maximal at $n = 0$. For this value of x_1, the center of the excited wave packet is at the equilibrium position of the excited potential. In other words, the argument of the Hermite polynomial in Eq. (12) is zero, and therefore, the probability distribution is proportional to that of two unshifted oscillators. That implies, for example, that only states with even n become excited. If the shift x_1 of the interaction becomes even larger than in Eq. (16), the average value $\bar{n}$ of the probability distribution increases again. Since the average momentum of the interaction was chosen to be zero, the average energy of the excited wave packet is equal to the potential energy it posses at the position where it is placed during the interaction. It has been shown recently in Ref. [Henrikson 2003] that instantaneous non-vertical transitions cannot be achieved with a shaped femtosecond laser pulse and constant dipole moment. It can be obtained, as we saw, with a spatial shaping of the interaction.

3. Inducing a space-dependent interaction

In the previous sections, we demonstrated how a space-dependent coupling between two electronic potentials influences the transition probabilities between molecular energy levels. In the following, we ask whether the space-dependence of the coupling can be used to *steer* molecular transitions. There-

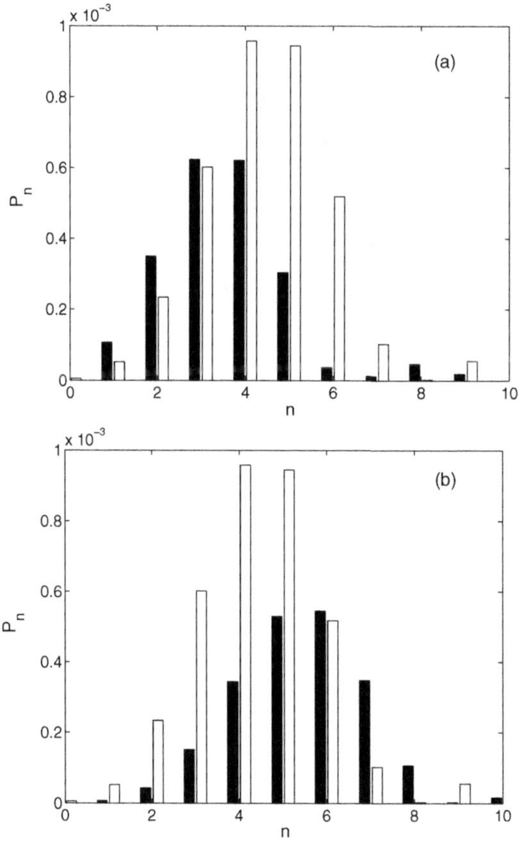

Figure 4. Probability distribution for $x_0 = 4$, $s = 0.7$, $\sigma = 1$, and $p_1 = 0$. White bars corresponds to $x_1 = 0$ and black bars to (a) $x_1 = 1$ and (b) $x_1 = -1$.

fore, we consider a model system of three molecular potentials, as shown in Fig. 5. A steady-state laser with frequency ω_l couples the potentials $V_{22}(x)$ and $V_{33}(x)$, and a short laser pulse $\mathcal{E}(t)$ induces the transfer of a wave packet between $V_{22}(x)$ and $V_{11}(x)$. The dynamics of the system is described by the Schrödinger equation

$$i\hbar \frac{\partial}{\partial t} |\psi\rangle = H |\psi\rangle, \tag{17}$$

with

$$H = \begin{bmatrix} H_1 & V_{12} & 0 \\ V_{21} & H_2 & V_{23} \\ 0 & V_{32} & H_3 \end{bmatrix}, \tag{18}$$

where

$$H_j = \frac{p^2}{2} + V_{jj}(x). \tag{19}$$

We use dimensionless variables scaled as in the previous section. The coupling between V_{22} and V_{33} is described by

$$V_{23} = -\, \Omega \, \cos(\omega_l \, t), \tag{20}$$

where $\Omega = \mu_{23}\, \mathcal{E}_0/(\hbar\omega)$ is the dimensionless interaction strength with μ_{23} being the transition dipole moment and $\mathcal{E}_0$ the field amplitude. Moreover,

$$V_{12} = -\, \frac{\mu_{12}\, \mathcal{E}(t)}{\hbar\omega}. \tag{21}$$

Here, μ_{12} is the transition dipole moment for the transition between V_{22} and V_{11}. In the adiabatic approximation (see, *e.g.*, Ref. [Stenholm 1994]), the coupled bare potentials V_{22} and V_{33} can be replaced by uncoupled adiabatic potentials. Using the rotating-wave approximation, the dynamics of the system is then described by the Schrödinger equation

$$i\hbar \frac{\partial}{\partial t} |\chi\rangle = U \; H \; U^{\dagger} \, |\chi\rangle, \tag{22}$$

with

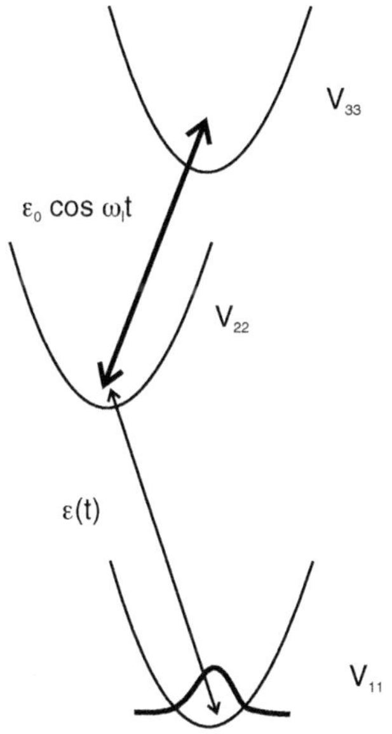

Figure 5. The potentials V_{22} and V_{33} are coupled by a steady-state laser with frequency ω_l. A short laser pulse $\mathcal{E}(t)$ couples V_{22} with V_{11}.

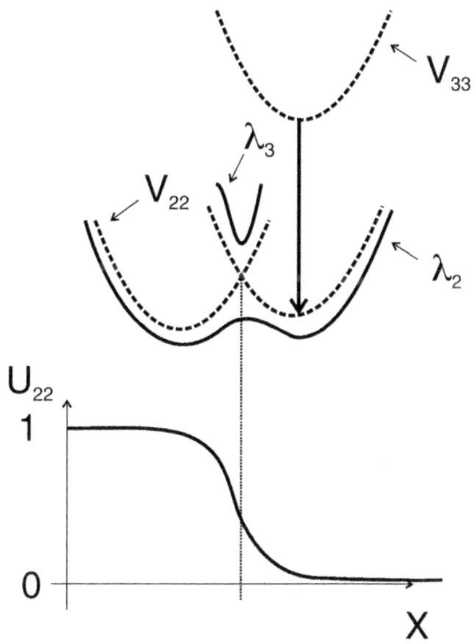

Figure 6. Schematic picture of the original molecular potentials $V_{22}(x)$ and $V_{33}(x)$ (dashed lines), the adiabatic potentials $\lambda_2(x)$ and $\lambda_3(x)$ (solid lines), and the matrix element $U_{22}(x)$.

$$U\,H\,U^{\dagger} = \begin{bmatrix} \frac{p^2}{2} + V_{11}(x) & V_{12}\,U^{\star}_{22}(x) & V_{12}\,U^{\star}_{32}(x) \\ V_{21}\,U_{22}(x) & \frac{p^2}{2} + \lambda_2(x) & 0 \\ V_{21}\,U_{32}(x) & 0 & \frac{p^2}{2} + \lambda_3(x) \end{bmatrix}. \tag{23}$$

Here, $|\chi\rangle = U\,|\psi\rangle$. The matrix

$$U(x) = \begin{bmatrix} 1 & 0 & 0 \\ 0 & U_{22}(x) & U_{23}(x) \\ 0 & U_{32}(x) & U_{33}(x) \end{bmatrix}, \tag{24}$$

with

$$U_{22}(x) = U_{33}(x) = \left\{1 + \frac{4}{|\Omega|^2}\,[\lambda_2(x) - V_{22}(x)]^2\right\}^{-\frac{1}{2}} \tag{25}$$

and

$$U_{32}(x) = -U_{23}(x) = \left\{1 + \frac{4}{|\Omega|^2}\,[\lambda_3(x) - V_{22}(x)]^2\right\}^{-\frac{1}{2}}. \tag{26}$$

is a unitary matrix that decouples V_{22} and V_{33}. Note, that these matrix elements are real functions. Instead of the coupled potentials $V_{22}(x)$ and $V_{33}(x)$, we have now the two uncoupled adiabatic potentials

$$\lambda_{2(3)}(x) = \frac{1}{2}\left(V_{22}(x) + V_{33}(x) - w_l \mp \sqrt{|\Omega|^2 + [V_{22}(x) - V_{33}(x) + w_l]^2}\right). \tag{27}$$

As we can see in Eq. (23), the off-diagonal elements that couple V_{11} to the other potentials are now coordinate-dependent. We further assume that the interaction V_{12} is far off resonant to the transition between V_{11} and λ_3. Then, we have a transition between the states described by the potentials λ_2 and V_{11}, driven by the coordinate dependent interaction

$$V_{int}(x,\,t) = V_{12}(t)\,U_{22}(x). \tag{28}$$

We have now two parameters which allow us to control the coordinate dependence of $V_{int}(x)$: the frequency of the steady state laser, ω_l, and the interaction strength, $|\Omega|$. It can be seen in Fig. 6 that the coordinate-dependence of the interaction $V_{int}(x)$ can be easily determined by the geometry of the potentials $V_{22}(x)$ and $V_{33}(x)$. In the rotating-wave approximation, the potential V_{33} is shifted by ω_l. The avoided crossing is represented by the adiabatic potentials $\lambda_2(x)$ and $\lambda_3(x)$. For sufficiently small $|\Omega|$, the potential $\lambda_2(x)$ is approximately the same as $V_{22}(x)$ on the left side of the crossing, and approximately the same as $V_{33}(x) - \omega_l$ on the right side. The energy gap between λ_2 and λ_3 at the crossing point is determined by the interaction strength: it increases with increasing $|\Omega|$. According to Eq. (25), $U_{22} \approx 1$ if $\lambda_2 \approx V_{22}$. If the difference between λ_2 and V_{22} is large, U_{22} is approximately zero. The matrix element U_{22} changes from one to zero at the crossing point of the potentials V_{22} and $V_{33} - \omega_l$. The position of the avoided crossing can be controlled by changing the laser frequency ω_l. The interaction strength $|\Omega|$ determines the slope of U_{22}. A weak interaction causes a steep change of $U_{22}(x)$, while for a strong interaction, $U_{22}(x)$ is a slowly varying function of x.

In Sec. 2 we discussed how the shape of a space-dependent interaction affects the transfer of a wave packet to an excited molecular potential. With this knowledge, the pictorial considerations about the shape of $V_{int}(x)$ allow us an easy, qualitative estimation of how the transition to V_{11} can be steered by the coupling to a third potential $V_{33}(x)$. In the following, we present an example of such a transition steering.

4. An example for transition steering

We assume that the molecular potentials V_{ii} can be described by harmonic oscillators

$$\begin{aligned} V_{11}(x) &= \frac{x^2}{2}, \\ V_{22}(x) &= \frac{s_2^2}{2}\,(x - x_2)^2 + D_2, \\ V_{33}(x) &= \frac{s_3^2}{2}\,(x - x_3)^2 + D_3, \end{aligned} \tag{29}$$

where D_i is the energy difference between the excited electronic states and the ground state, x_i are the shifts with respect to V_{11}, and the squeezing parameters s_i determine the frequencies of the oscillators. We further consider a system that is initially in its ground state so that

$$\psi_1(x, 0) = \pi^{1/4} \exp\left(-\frac{x^2}{2}\right) \tag{30}$$

is the ground vibrational state of V_{11} and $\psi_2(x, 0) = \psi_3(x, 0) = 0$, as shown in Fig. 5. Here, ψ_i denotes the component of the wavefunction on the molec-

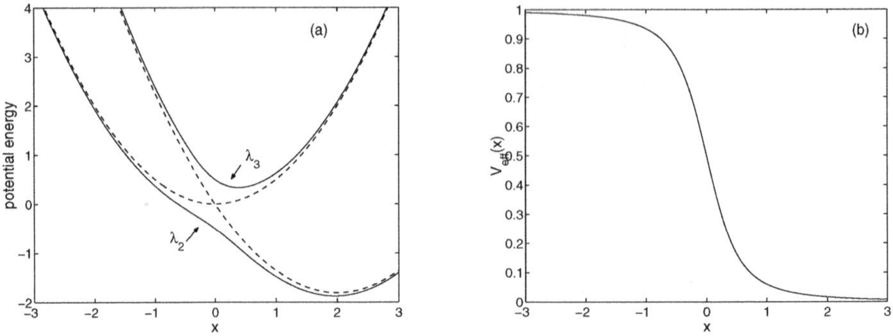

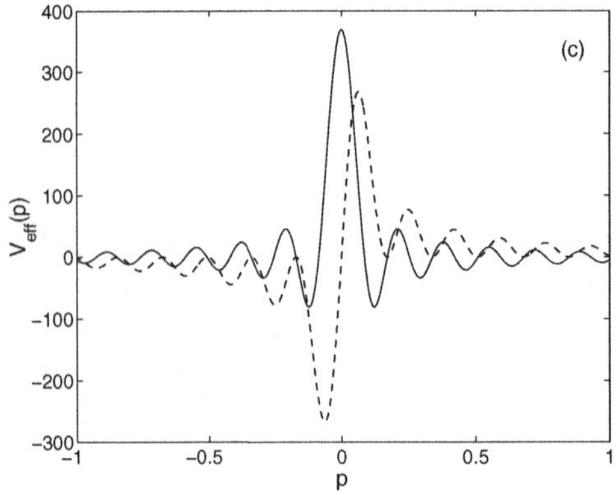

Figure 7 (a) Original potentials V_{22} and $V_{33} - \omega_l$ (dashed lines) compared to the adiabatic potentials λ_2 and λ_3 (solid lines). The parameters are $\Omega = 1$, $\Delta_l \equiv D_3 - D_2 - \omega_l = 1.81$, $s_2 = 1$, $x_2 = 0$, $s_3 = 0.95$ and $x_3 = 2$. (b) $V_{eff}(x)$, measured in units of V_0. (c) Real (solid line) and imaginary (dashed line) parts of $\tilde{V}_{eff}(p)$.

ular potential labeled with i. Moreover, we assume that the interaction V_{12} is weak and can be treated in the perturbation limit. Furthermore, the field $\mathcal{E}(t)$ shall be a short pulse such that the interaction occurs instantaneously. Immediately after the interaction with $\mathcal{E}(t)$,

$$\chi_2(x,\,0) = V_0\, U_{22}(x)\, \psi_1(x,\,0), \tag{31}$$

where

$$V_0 = \int_{-\infty}^{\infty} V_{12}(t)\, dt \tag{32}$$

is the integrated interaction strength. If the steady state laser is switched off during a time $t \ll 1/|\Omega|$, the excited wave packets evolve on the bare potentials V_{22} and V_{33}. To this end, consider only the wave function on V_{22}, which is, according to Eq. (24)

$$\psi_2(x,\,0) = V_0\, U_{22}^2(x)\, \psi_1(x,\,0). \tag{33}$$

An example of transition steering can be seen in Fig. 7. Here, the potentials V_{22} and $V_{33} - \omega_l$ have one point of intersection in the relevant coordinate range.

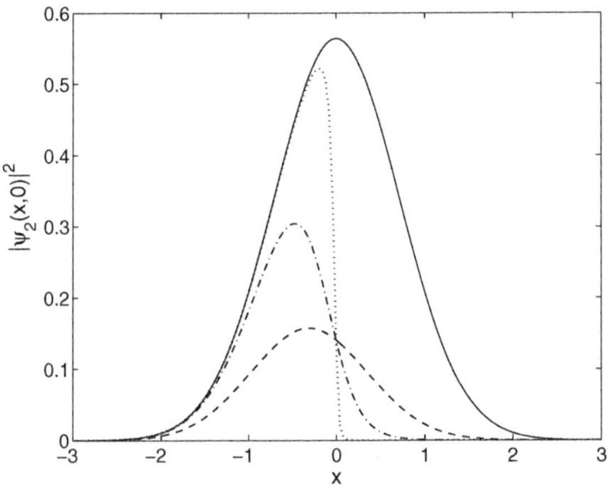

Figure 8 Probability distribution of the wave packet excited on V_{22} for the same parameters as in Fig. 7. The solid line represents the excited wave packet without interaction between V_{22} and V_{33}. The dotted line corresponds to $\Omega = 0.1$, the dash-dotted line corresponds to $\Omega = 1$, and for the dashed line, $\Omega = 5$.

The resulting effective interaction $V_{eff}(x) \equiv V_0\, U_{22}^2(x)$, Fig. 7 (b), switches from one to zero at $x_c = 0$. The Fourier transformation of the interaction, $\tilde{V}_{eff}(p)$ can be seen in Fig. 7 (c). The probability density $|\psi_2(x,\,0)|^2$ is shown in Fig. 8. Without the interaction between V_{22} and V_{33}, we have a usual Franck-Condon-type of interaction, where the short laser pulse transfers a copy of the ground state to V_{22}, and

$$\psi_2^{FC}(x,\,0) = V_0\,\psi_1(x,\,0). \tag{34}$$

The probability distribution $|\psi_2^{FC}(x,\,0)|^2$ is represented by the solid line in Fig. 8. The other curves show the probability distribution for different values of the interaction strength $|\Omega|$. One can clearly see the effect of coordinate-dependent interaction. It induces a shift of the center of the wavefunction and reduces the widths of the excited wave packet.

By changing the frequency and the intensity of the steady state laser, we can vary the width and the displacement of the excited wave packet. With this example, we want to demonstrate how the space-dependence of a dipole coupling can be used to steer the transfer of a wave packet to an excited molecular potential. However, this is only one example out of many possible. This method is by no means restricted to a model system consisting of harmonic oscillators but can be easily applied to any form of one-dimensional potential curves. Possible applications might be steering of a reaction to one side of a potential barrier by displacing the excited wave packet to the desired side, or coupling to a dissociative state in order to steer the dissociation of a molecule.

Summary

In this work, we investigated the possibilities to utilize the spatial dependence of the dipole-coupling to influence the transition probability between molecular energy levels. In this case, we go beyond the standard Frack-Condon results based on constant coupling.

In the first part of our work, we consider a simple form of the space dependent coupling and compare the transition probabilities to those of a Franck-Condon transition. Our model is based on an interaction characterized by a few parameters only. The coupling is displaced from the equilibrium position of the lower level potential, it delivers an average kick of momentum, and it encompasses a range of momenta with given width. The Franck-Condon transition favors only direct vertical transitions to states with turning points above the initial state. An interaction with displaced peak can push the initial state to correspond more exactly with the momentum of a desired final state. The transfer of the initial state to a desired excited vibrational state can be facilitated by a proper momentum kick. A coupling with a broad momentum distribution can cover a larger range of final states than would otherwise be possible.

In the second part of this work, we addressed the problem of how to use the above described effects of a space-dependent interaction to steer molecular transition. We proposed a model that allows us to induce a space-dependent coupling between two molecular potentials via a steady-state coupling to a third potential surface. By changing the frequency and intensity of the steady state laser, we can shape the space-dependence of the coupling. We illustrated the method with an example of three coupled harmonic oscillators and showed how displacement and width of the excited wave packet can be controlled.

We point out that the steering of transition between two harmonic oscillators is only one example for the proposed model. Other possible applications are the control of a chemical reaction, *i.e.*, the steering of a wave packet to one or the other side of a potential barrier, or the control of molecular dissociation.

Acknowledgments

The contributions from M. L. have been supported by the European Union Research and Training Network QUACS, contract No. HPRN-CT-2002-00309.

COHERENCE AND DECOHERENCE IN RYDBERG GASES

P. Pillet, D. Comparat, M. Muldrich, T. Vogt, N. Zahzam, V. M. Akulin
Laboratoire Aimé Cotton, CNRS, Bât. 505, Campus d'Orsay, 91405 Orsay cedex, France

T. F. Gallagher, W. Li, P. Tanner, M. W. Noel, and I. Mourachko
Department of Physics, University of Virginia, Charlottesville, VA 22901, U.S.A.

Introduction

Dipole-dipole interactions between cold Rydberg atoms have recently attracted much interest since they play a central role in proposed quantum logic gates [Jaksch 2000; Lukin 2001]. While dipole-dipole interactions between cold, or stationary Rydberg atoms, are only beginning to be explored [Anderson 1998 (b); Mourachko 1998], resonant dipole-dipole collisions between Rydberg atoms have been studied extensively [Gallagher 1992]. The connection between the two phenomena is that the interactions between cold stationary atoms correspond to freezing a pair of colliding atoms at their point of closest approach.

In the first section of this article we describe resonant dipole-dipole energy transfer collisions between Rydberg atoms. As we shall see, very subtle features can be discerned in these collisions, and they are a good starting point for the development of an understanding of dipole-dipole interactions in a frozen Rydberg gas.

An ensemble of cold Rydberg atoms is easily obtained after laser excitation of a cold atomic cloud, as those performed in a Cs or Rb vapor-cell magneto-optical trap, at a temperature of 135 μK or 300 μK respectively. In the case of cesium (for the experiments performed at Laboratoire Aimé Cotton) or rubidium (for the experiment performed at the University of Virginia), the atoms p-excited by the cooling lasers are Rydberg-excited by using a laser pulse provided by a dye laser pumped by the third harmonic of a $Nd:YAG$ laser

$$Cs(6\,p_{3/2}) + h\nu_L \rightarrow Cs(nl) \tag{1}$$

or

V.M. Akulin et al. (eds.), Decoherence, Entanglement and Information Protection in Complex Quantum Systems, 411–436.

$$Rb(5\,p_{3/2}) + h\nu_L \rightarrow Rb(nl) \tag{2}$$

where the laser wave length ($\lambda_L = c/\nu_L$) is $\sim$ 515 nm or 480 nm, for the cesium or rubidium respectively [Anderson 1998 (b); Mourachko 1998]. The principal quantum number n ranges from 20 to 50 and $l = s, p, d$. Typically, the number of Rydberg atoms ranges 10^3 - 10^6 in a volume corresponding to the MOT cloud volume $\sim$ $(0.5\ \mathrm{mm})^3$, meaning a density ranging 10^7 - 10^{10} cm^{-3}.

Over the time scale of interest of the experiments in the order of a few microseconds, smaller than the Rydberg lifetime, the Rydberg atoms move only small distances relative to their average separation ($\overline{R} \approx 4.5 - 45\ \mu$m). For a typical velocity around $v \sim 10$ cm/s, the displacement $\Delta R = v\tau$ is $\sim$ 100 nm during a time $\tau = 1\ \mu$s. Such a displacement is much more closer to the size of the Rydberg atoms $R_0 \sim 4n^2$ atomic units ($\sim$ 80 nm for $n = 20$) as their average distance. In a first approximation, it is therefore possible to ignore the motion of the Rydberg and to consider the atomic ensemble as a frozen gas. Frozen gases are difficult to realize with neutral atoms, and elegant examples are the well defined spatial structure of cold ions observed in ion traps [Diedrich 1987; Wineland 1987]. The long-range Coulomb interaction plays a key role in the observation of the crystalline phase. Cold Rydberg atoms, with their large sizes and dipole moments, appear to be natural candidates for studying frozen neutral gas expected to behave as an amorphous solid.

At large density ($\sim 10^{10}$ cm^{-3}) and large $n \gtrsim 40$, we observe the spontaneous evolution of a cold Rydberg gas in an ultracold plasma [Robinson 2000]. Such spectacular behavior is first the result of ionization processes due to blackbody radiation and hot Rydberg - cold Rydberg collisions. After the first electrons leave the Rydberg cloud region, the formation of positive space charge due to an ion cloud traps subsequent electrons leading finally to the rapid avalanche ionization of the ensemble of the Rydberg gas. The resulting cold plasma expands slowly and persists for tens of microseconds. Such a dramatic behavior does not occur at lower densities and for low n. Due to the long-range interaction between Rydberg atoms, the ensemble of cold Rydberg atoms can exhibit many-body phenomena, offering an interesting quantum mesoscopic system for studying the properties of coherence and decoherence.

1. Resonant dipole-dipole collisions

The first experiments were carried out using Rydberg atoms in the Na energy levels shown in Fig. 1. The $17s$ level lies roughly halfway between the $16p$ and $17p$ levels, and an electric field shifts the p states to higher energy, so that at $E = 540$ V/cm the $17s$ state if halfway between the two p states. In other words, we can use the E field tuning to shift the process

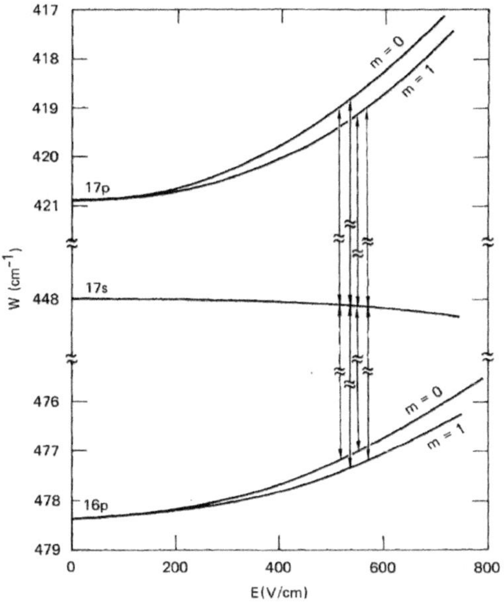

Figure 1. Na energy levels showing the locations of the collisional resonances for the process $Na(17s) + Na(17s) \rightarrow Na(17p) + Na(16p)$.

$$Na(17s) + Na(17s) \rightarrow Na(16p) + Na(17p) \tag{3}$$

into resonance.

The experiments are done by exciting Na atoms in an atomic beam to the $17s$ state, by the route $3s \rightarrow 3p \rightarrow 17s$, using two 5 ns dye laser pulses. The atoms are allowed to collide for $\sim 3\ \mu$s, and then the final states are analyzed by selective field ionization. In particular, atoms in the $17p$ state are detected as the tuning electric field is slowly swept over many shots of the laser. In Fig. 2 we show the resonances observed in this way. There are four resonances corresponding to the two possible $|m_1|$ values of the np states, $|m_1| = 0$ and 1. It is convenient to label the four resonances by the $|m_1|$ values of the upper and lower p states, *e.g.*, (0,0) has $|m_1| = 0$ in both.

An interesting aspect of the collisional resonances shown in Fig. 2 is that they are quite easy to observe. In fact, most of the Rydberg atoms undergo collisions, and we can estimate the cross section rather easily using

$$\Gamma = N\sigma\, v, \tag{4}$$

where Γ is the collision rate, N the atomic number density, and v the typical collision velocity. Since $\Gamma \sim 10^6\ \mathrm{s}^{-1}$, $N \approx 10^8\ \mathrm{cm}^{-3}$, and $v = 10^5\ \mathrm{cm/s}$, we quickly arrive at $\sigma \sim 10^{-7}\ \mathrm{cm}^2$, which is a little hard to believe. However, it is easy to show that this cross section is precisely what is expected for resonant dipole-dipole energy transfer collisions. Imagine that atom A passes atom B with velocity v and impact parameter b. If atom A produces an oscillating field at atom B of magnitude $E' \approx \mu'/R^3$, where R is the distance between the two atoms and μ' is the dipole moment from the initial state to the upper final state, for example the $17s$ - $17p$ dipole moment. On resonance this oscillating field can drive atom B down to the lower final state if

$$\mu\, E'\tau = 1 \tag{5}$$

where μ is the dipole matrix element connecting the initial state to the lower final state, *e.g.*, the $17s$ - $16p$ dipole moment, and τ is the duration of the collision. E' is most important when $R \sim b$ so we set $E' \approx \mu'/b^3$ and correspondingly, set $\tau = b/v$. With these substitutions we find

$$\sigma \approx b^2 = \frac{\mu\,\mu'}{v} \tag{6}$$

and

$$\tau = \sqrt{\frac{\mu\,\mu'}{v^{3/2}}}. \tag{7}$$

Since $\mu \approx \mu' = n^2$ case we find

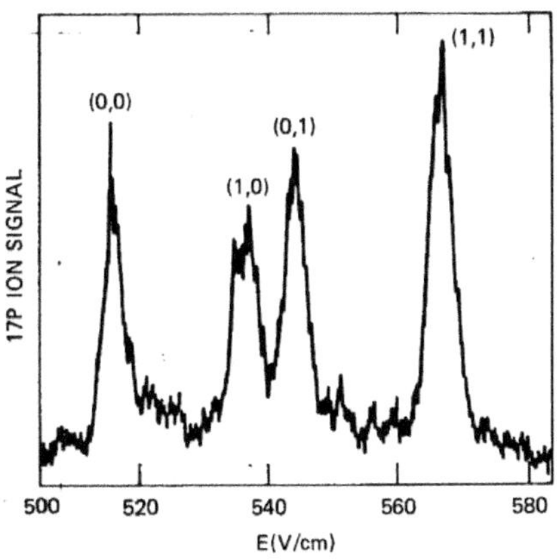

Figure 2. Observed collisional resonances for the process $Na(17s) + Na(17s) \rightarrow Na(17p) + Na(16p)$. The resonances are labeled by the $|m_l|$ values of the (lower, upper) final p states, as shown in Fig. 1.

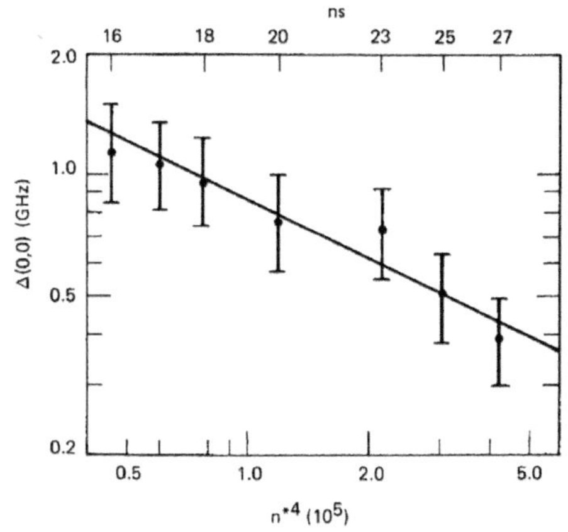

Figure 3. Frequency widths of the $Na(ns){+}Na(ns) \rightarrow Na(np){+}Na((n{-}1)p)$ $(0,0)$ resonances vs. n.

$$\sigma = \frac{n^4}{v} \tag{8}$$

and

$$\tau = \frac{n^2}{v^{3/2}}. \tag{9}$$

Eq. (8) is given in atomic units, and if we re-express it in laboratory units, for $n = 20$, we find $\sigma = 3.\,10^{-8}$ cm^2 and $\tau = 10^{-9}$ s. This value of the cross section is in accord with our earlier rough estimate, and the inverse of the collision time is consistent with the linewidths of the collisional resonances shown in Fig. 2. The n scaling of both the cross section and the resonance width has been verified, and in Fig. 3 we show the observed dependence of the width Δ of the $(0,0)$ resonance on n [Gallagher 1982].

One of the more interesting aspects of the collisional resonances is how sharp they are. Most atomic collision processes have durations of 10^{-12} s, and these collsions last for times in excess of a nanosecond. Furthermore, the collisional resonances should become even narrower as the collision velocity is reduced. To explore this issue we used the K system

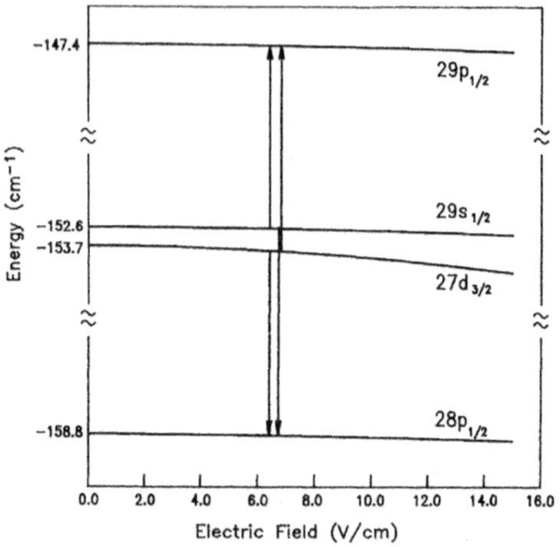

Figure 4 K energy levels for the resonant collision $K(29s) + K(27d) \rightarrow K(29p) + K(28p)$

$$K(29s) + K(27d) \rightarrow K(29p) + K(28p) \qquad (10)$$

which can be tuned into resonance at the field 6 $\mathrm{V/cm}$ as shown by Fig. 4.

These experiments are slightly more difficult in that both the $29s$ and $27d$ states must be excited. In this case the atoms are allowed to collide for a few microseconds after pulsed laser excitation, and the $29p$ atoms are detected as the tuning field is slowly swept over many shots of the laser. The important difference from the Na experiment is that the K atoms can be velocity selected. As shown in Fig. 5, there is a slotted disc in front of the oven, and if the disc is turning, a burst of the atoms passes through the slit, and only those within a narrow velocity range are excited by the 5 ns dye laser pulse 200 μs after the burst of atoms passes through the slit.

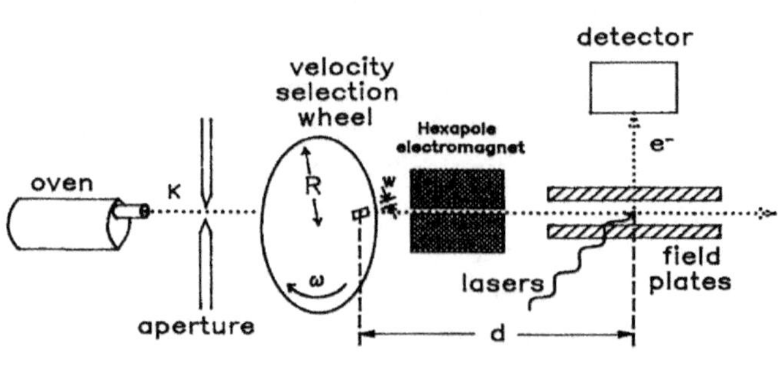

Figure 5. Apparatus for producing velocity selected K beams

In Fig. 6 we show data collected in three different ways. First we show the collisional resonance observed when the liquid nitrogen trap is not filled. The apparatus becomes a cell, with atoms moving in all directions, and the collisional resonance is quite broad, 240 MHz. If the trap is filled K atoms in the beam are condensed when they strike the trap, and the only K atoms present are in the beam and moving in the same direction but with a wide velocity distribution. Nonetheless, the resonance is 57 MHz wide, narrower than observed with a cell. Finally, when the velocity selector is used the resonance is narrowed to 6 MHz, a width which is due primarily to the earth's magnetic field. Shielding the interaction region from the earth's field reduces the width of the resonance to 1 MHz, implying that the duration of a collision is 1 μs [Thomson 1990]. To observe 1 MHz wide collisional resonances the atoms are allowed to collide for several microseconds, so it is not clear when individual collisions begin and end. However, if the time the atoms are allowed to collide is reduced to below 1 μs, the resonances become transform broadened, and we see transform limited collisions, as shown in Fig 7.

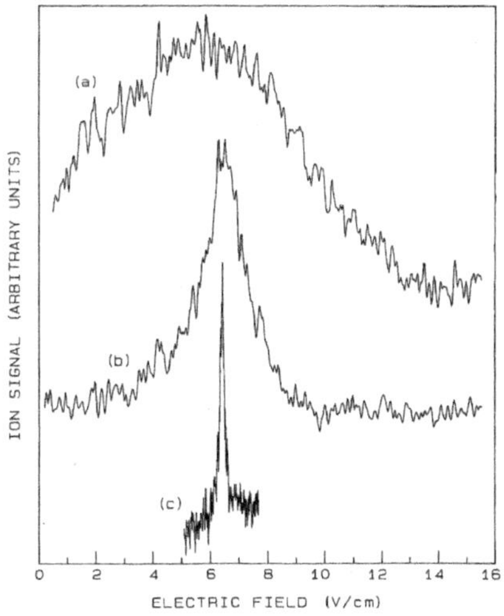

Figure 6 Resonances of $K(29s) + K(27d) \rightarrow K(29p) + K(28p)$. (a) 240 MHz wide resonance obtained with a cell of K atoms with no liquid nitrogen in the trap. (b) 51 MHz wide resonance obtained a thermal beam K. (c) 6 MHz wide resonance obtained with a velocity selected beam.

The notion of transform broadening is, of course, quite general, but in this case the transform broadening means that we know when an atomic collision begins and ends [Thomson 1990]. Later we shall take advantage of this unusual feature of these collisions.

As we have already stated, a Rydberg-Rydberg resonant dipole-dipole collision lasts for a long time compared to most atomic collisions. As a result, any process which happens during the collision can be carried out at a slower rate and observed more easily. An excellent example is a radiatively assisted collision in which a photon is absorbed or emitted during the collision to balance the energy transfer. In particular, we have examined the process [Pillet 1983; Pillet 1987],

$$Na(ns) + Na(ns) + m\ h\nu \rightarrow Na(np) + Na((n-1)p) \qquad (11)$$

which becomes resonant at slightly different fields than does the resonant process of Eq. (3).

Experiments to probe this system were done by passing an atomic beam of Na through a 15 GHz microwave cavity where it is excited by two counter propagating laser beams tuned to the $3s$ - $3p$ and $3p$ - ns transitions. There is a septum in the waveguide which enables us to apply a static field to tune the energy levels and to field ionize the atoms. There is a 1 mm diameter hole in the center of the top of the waveguide through which the ions resulting from field ionization pass. Due to the location of the hole we only detect ions from those atoms at an antinode of the microwave field. In Fig. 8 we show the collisional resonances observed when $Na(18s)$ state is populated and the static field scanned in the presence of several microwave field amplitudes.

As shown by Fig. 8 we can see collisions in which stimulated emission of up to three microwave photons occurs using very low microwave powers. It is interesting to note that laser assisted collisions typically require optical intensities of $\mathrm{MW/cm^2}$ to be visible at all [Falcone 1977]. In Fig. 8 it appears that the one photon assisted collision disappears only to reappear at higher microwave fields. This unexpected feature becomes more apparent if we plot the resonance signal *vs* the microwave field, as shown in Fig. 9. Also shown by the lines in Fig. 9 are the results of a model describing collisions in which from zero to three photons are emitted.

The model is based on a molecular description of the resonant collisions which we outline here, first for the case of no microwave field. Briefly, the two molecular states are $ss = ns \otimes ns$ and $pp' = np \otimes (n-1)p$. The p state energy can be tuned with an electric field due to the Stark shifts of the two p states, and at infinite internuclear separation the ss and pp' levels cross at the resonance field. At any finite separation there is an avoided crossing of magnitude $2\mu\mu'/R^3$ due to the dipole-dipole coupling. Using this picture we can derive the size of the cross section as follows. Imagine that the field is set to the resonance field, so that ss and pp' are degenerate at infinite separation. During the collision the atoms come closer together, and the dipole-dipole coupling lifts the degeneracy of the ss and pp' states so that the good eigenstates are $\psi_\pm = (ss \pm pp')/\sqrt{2}$, which are separated by $2\mu\mu'/R^3$. During a collision R varies with time, and the energy spacing varies with time. By virtue of the laser excitation the initial population is in the ss state, or in the coherent superposition $(\psi_+ + \psi_-)/\sqrt{2}$. During the collision the amplitudes follow both the ψ_+ and ψ_- curves which separate. If the area between the ψ_+ and

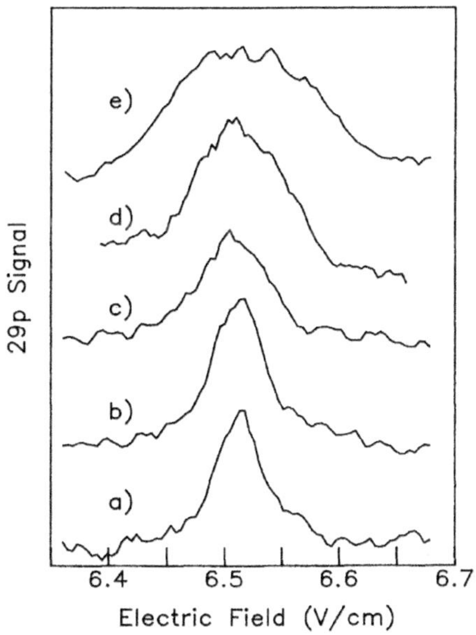

Figure 7. Resonances of $K(29s) + K(27d) \rightarrow K(29p) + K(28p)$ obtained with allowed collision times of (a) 3.0 μs, (b) 2.0 μs, (c) 1.0 μs, (d) 0.4 μs, and (e) 0.2 μs. The FWHM of the resonances are (a) 1.4 MHz, (b) 1.8 MHz, (c) 2.2 MHz, (d) 3.1 MHz, and (e) 5.2 MHz. For (d) and (e) the resonances are transform broadened, and we know when the collisions start and end.

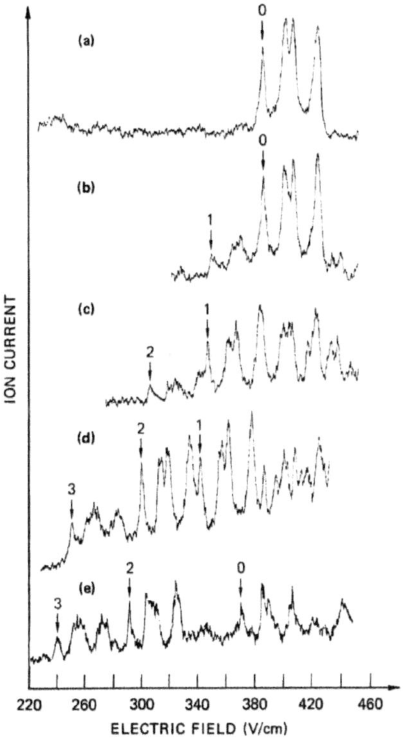

Figure 8. $Na(18s) + Na(18s) \rightarrow Na(18p) + Na(17p)$ collisions assisted by zero to three 15.4 GHz microwave photons. Above the $(0,0)$ resonances are indicated how many photons have been emitted. In each trace there are obvious sets of four resonances corresponding to those of Fig. 2. Microwave fields (a) 0 V/cm, (b) 13.5 V/cm, (c) 50 V/cm, (d) 105 V/cm, and (e) 165 V/cm.

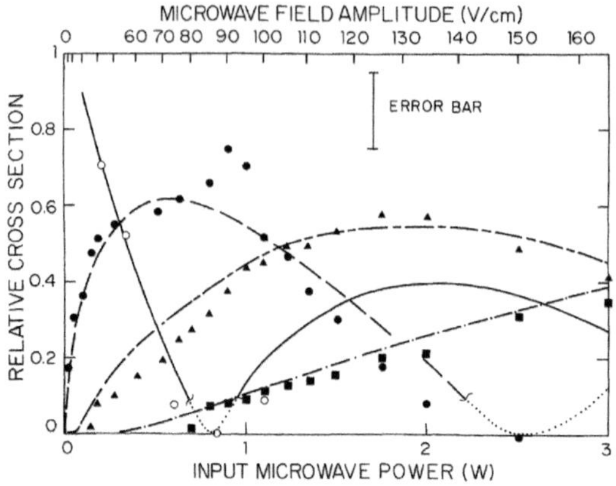

Figure 9 Relative cross sections for collisions assisted by zero photons (○ and solid line), one photon (● and dashed line), two photons (▲ and long-short dashed line), three photons (■, dot-dashed line). The lines are calculated from Eq. (14).

ψ_- curves integrated over the collision is π, the initial superposition evolves to $(\psi_+ - \psi_-)/\sqrt{2}$ or pp' after the collision. Since the area is: $\mu\mu'/R^3 . R/v$ we arrive at the same conclusion as before, that $\sigma = \mu\mu'/v$. More important, the cross section is proportional to $\chi_0 = \mu\mu'/R_0^3$, the size of the avoided crossing of the ss and pp' levels at an arbitrary internuclear separation R_0.

Now let us consider what happens when we add a microwave field. The ss state has no Stark shift and is unaffected. In the presence of an existing static field the pp' state has a linear differential Stark shift due to an added field, such as the microwave field. Consequently the microwave field $E \sin\omega t$ modulates the energy of the pp' state and breaks it into a carrier and sidebands just as modulating a laser beam or radio wave produces sidebands. To represent the carrier and sidebands the pp' wave function is replaced by a Bessel function expansion, *i.e.* [Pillet 1987],

$$\psi_{pp'} \rightarrow \psi_{pp'} \sum_m J_m\left(\frac{kE}{\omega}\right) e^{-i\, m\, \omega t}, \tag{12}$$

where k is the dipole moment of the pp' state. Now the single avoided crossing of the pp' state with the ss state is replaced by a set of avoided crossings, between the ss state and the carrier and sidebands of the pp' state. The mth sideband state of the pp' state has an avoided crossing with the ss state of magnitude

$$\chi(E)_m = \chi\, J_m\left(\frac{kE}{\omega}\right). \tag{13}$$

Consequently the cross section for the j photon assisted collision is

$$\sigma(E)_m = \sigma \; J_m \left(\frac{kE}{\omega} \right). \tag{14}$$

As before σ is the cross section in the absence of a microwave field. Naturally $\sigma_0(0) = \sigma$. This expression is the origin of the lines drawn in Fig. 9, which evidently match the experimental cross sections. Similar results have also been observed with the K system of Eq. (10) using velocity selected beams to obtain narrower collisional linewidths allowing, the use of rf frequencies of 4 MHz, instead of 15 GHz [Thomson 1992]. Since the collisions last longer, the rf fields can be very weak, < 0.1 V/cm. An interesting aspect of both the 15 GHz and the 4 MHz measurements is that the Bessel function oscillation of the cross section with microwave or rf field amplitude is observed indicating that the coherence of the colliding atoms is maintained over multiple field cycles.

The radiatively assisted collisional resonances shown in Fig. 8 are taken under the condition that the microwave frequency far exceeds the linewidth of the collisions. If the frequency is far less than the linewidth, or equivalently the duration of the collision is short compared to the rf period, then the rf field simply adds to the static field. If in a static field the collisional resonance occurs at the field E_R, then if the collision occurs at time $t = 0$ in the combined field

$$E = E_S + E_{rf} \; \cos\left(\omega t + \phi\right) \tag{15}$$

the resonance will be observed at the static field $E_S = E_R - E \; \cos\phi$ [Renn 1994]. Unlike the high frequency regime, the phase ϕ matters.

The two extreme cases $\omega \gg 1/\tau$ and $\omega \ll 1/\tau$ are easily understood, but in very different terms, so it is less clear how to think about the case $\omega \approx 1/\tau$. Nonetheless, the low frequency regime suggests that the phase of the rf field relative to when the collision occurs is likely to be important. To control this phase we use transform limited collisions of $\mathcal{K}$ atoms in a velocity selected beam so that we know when the collisions begin and end, and we synchronize the rf field to the laser pulse initiating the collisions [Renn 1991]. Specifically, we have studied the process of Eq. (10) in rf fields phase locked to the collisions.

In particular, the collisions occur in a $0.7\ \mu$s, interval centered at $t = 0$, in the presence of an rf field $E = E_{rf} \; \cos(\omega t + \phi)$. In Fig. 10 we show the population observed in the $\mathcal{K}$ $29p$ state as a function of static field for an rf amplitude of 0.21 V/cm, frequency $\omega/2\pi = 0.75$ MHz, and the phases 0, and $\pi/2$.

Not shown in Fig. 10, the $\phi = \pi$ trace is approximately a mirror image of the $\phi = 0$ trace, with a main peak and subsidiary peaks. The $\phi = \pi/2$ pulse on the other hand leads to a resonance which is almost a rectangular shape. The

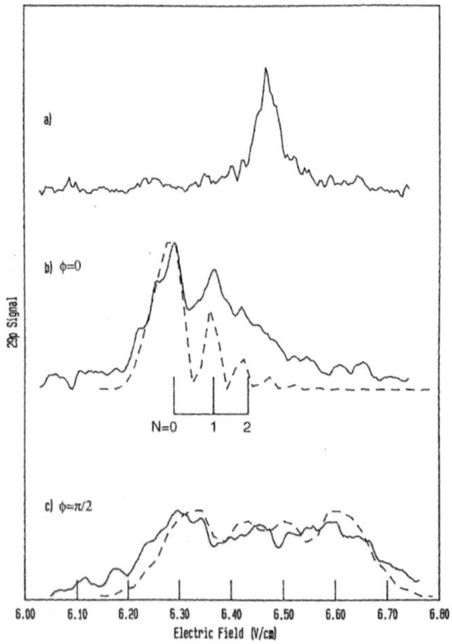

Figure 10 Resonance of $K(29s) + K(27d) \rightarrow K(29p) + K(28p)$ observed in the presence of a 0.21 V/cm 0.75 MHz field $E\cos(\omega t + \phi)$. The allowed collision time of 0.7 μs is symmetric about $t = 0$. (a) $\phi = 0$ (b) $\phi = \pi/2$. The dotted lines are the result of numerical integration of the Schrödinger equation, and the stick figures are the resonances due to integral numbers of 2π phase shifts from Fig. 11.

origin of the differences in the lineshapes can be understood by examining the energy levels shown in Fig. 11, which shows the energies W_i of the initial state $(29s + 27d)$, and W_f of the final state, $29p + 28p$, as functions of time for $\phi = 0$ and $\pi/2$.

For the $\phi = 0$ case shown in Fig. 11 (a), as the static field is scanned through the resonance the two levels come into resonance once at the field $E_S = E_R - E_{rf}$, leading to the peak at $E_S = 6.3$ V/cm in Fig. 10 (a). As the static field is raised the levels come into resonance twice, on the rising and falling edge of the pulse. There are two avoided level crossings, and at the first of these two avoided crossings the amplitude is split into two coherent paths, and depending on the phase accumulation at the second avoided crossing the second crossing increases of decreases the transition probability. The subsidiary peaks at 6.4 and 6.5 V/cm in Fig. 10 (a) are cases in which the phase accumulation is 2π and 4π. Such oscillations are observed in differential scattering where they are termed Stuckelberg oscillations. In contrast to Fig. 10 (a), in Fig. 10 (b) there is essentially no interference structure, and the reason is evident in Fig. 11 (b). The atoms only come into resonance once, so there is no possibility of interference.

The above description of the collisions in the regime $\omega = 1/\tau$ is an extension of how we viewed the low frequency regime. It can be extended to the high frequency regime to make the connection with the Bessel function description of Eq. (14) [Renn 1994]. Imagine that, instead of having the ini-

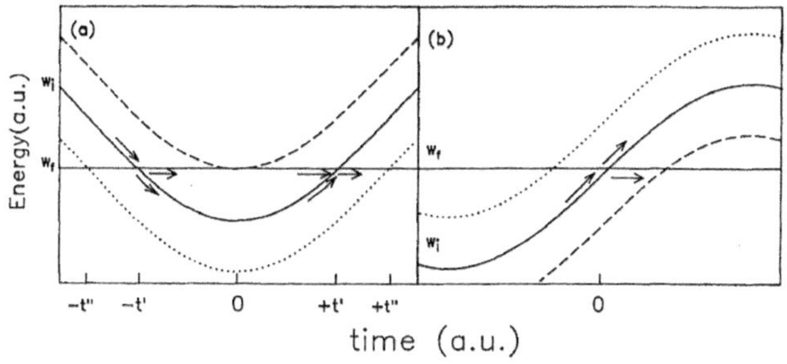

Figure 11. The initial- and final- state energies, W_i and W_f in a 0.75 MHz rf field with phases (a) $\phi = 0$ and (b) $\phi = \pi/2$. In both cases, the allowed collision time is 0.7 μs. As the static electric field is scanned, the single interaction period at $t = 0$, when $E_S = E_R - E_{rf}$ (dashed line), evolves into two interaction periods at $t = \pm t'$ (solid line) and $\pm t''$ (dotted line) when $E_S > E_R - E_{rf}$, as shown in (a). The transition amplitudes interfere constructively when the total phase shift of Φ between $-t'$ and t' is $2\pi N$. In (b) only one interaction period occurs irrespective of E_S.

tial and final energy levels cross many times, not once or twice, as shown in Fig. 11. For example, in Fig. 12 we show the energies of the initial and final states, W_i and W_f, over two cycles of the microwave field.

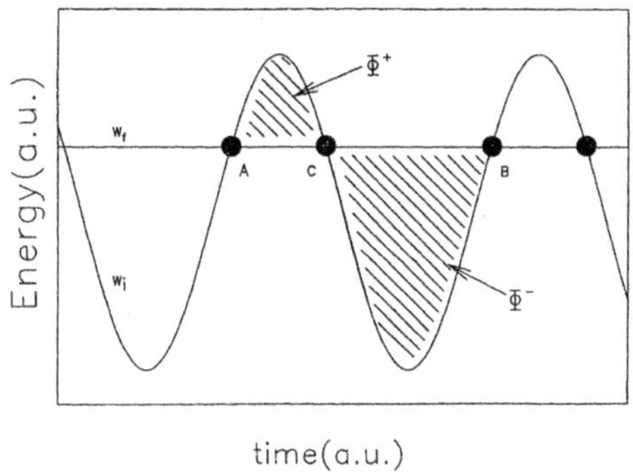

Figure 12. The initial- and final- state energies, W_i and W_f in a high frequency rf field. Only two cycles of the rf field are shown. Each time $W_i = W_f$ a collisional interaction occurs and the relative amplitude of ψ_i and ψ_f change. If $\Phi_- - \Phi_+ = 2\pi\ N$, then the transition amplitudes over successive cycles add constructively and a resonance in the cross section is observed. For the maxima in the cross section Φ_- and Φ_+ must separately be equal to integral multiples of 2π.

For the transition amplitudes at two points one cycle apart, such as A and B in Fig. 12, to add constructively we require that

$$\Phi_- - \Phi_+ = \int_{t_0}^{t_0 - 2\pi/\omega} (W_f - W_i)dt = 2\pi\, m \tag{16}$$

where m is an integer. Since W_i has a sinusoidal dependence on the microwave field, Eq. (16) reduces to

$$W_f - W_i = m\omega \tag{17}$$

which is simply the resonance condition for the m photon assisted collision. For the transition amplitudes from point A and C to add constructively we impose the additional requirement that Φ_+, and therefore Φ_-, be an integral multiple of 2π. To see how this requirement leads to the Bessel function dependence we now imagine that the static field is chosen so that the system is at the four photon assisted resonance, *i.e.*, $m = 4$. As the microwave field is increased from zero, the levels will first reach the field where they only come into resonance once per cycle (leading to the maximum cross section). In Fig. 12 this condition corresponds to $\Phi_+ = 0$ and $\Phi_- = 8\pi$. As the field is further increased, both Φ_+ and Φ_- increase, and when they are π and 9π, respectively, there is a zero in the cross section. A further increase in the field leads to $\Phi_+ = 2\pi$ and $\Phi = 10\pi$ with a maximum in the cross section. As the field continues to increase there is an oscillation in the cross section which i simply the Bessel function oscillation of Eq. (14) and Fig. 10.

2. Cold resonant Rydberg atom - Rydberg atom collisions

The dipole-dipole collisional energy exchange between two Rydberg atoms can be experimentally studied for many configurations. We can consider here the case of the cesium atom where the $np_{3/2}$ level is located in the energy diagram midway between the two ns and $(n+1)s$ states. Adding a small electric field displace the p state and allows us to match the resonance for the reaction with exchange of internal energy between both atoms (see Fig. 13)

$$2\, Cs(n\, p_{3/2}) \rightarrow Cs(ns) + Cs((n+1)s) \tag{18}$$

Due to the degeneracy of the $n\, p_{3/2}$ level three resonance are expected

$$
\begin{aligned}
A: &\quad 2\,Cs\left(n\,p_{3/2},\ |m|=\frac{1}{2}\right) \rightarrow Cs(ns)+Cs((n+1)s), \\
B: &\quad Cs\left(n\,p_{3/2},\ |m|=\frac{1}{2}\right)+Cs\left(n\,p_{3/2},\ |m|=\frac{3}{2}\right) \\
&\qquad \rightarrow Cs(ns)+Cs((n+1)s), \\
C: &\quad 2\,Cs\left(n\,p_{3/2},\ |m|=\frac{3}{2}\right) \rightarrow Cs(ns)+Cs((n+1)s)
\end{aligned}
\tag{19}
$$

for three different values of the electric field as shown in Fig. 14 for the example of $n = 23$.

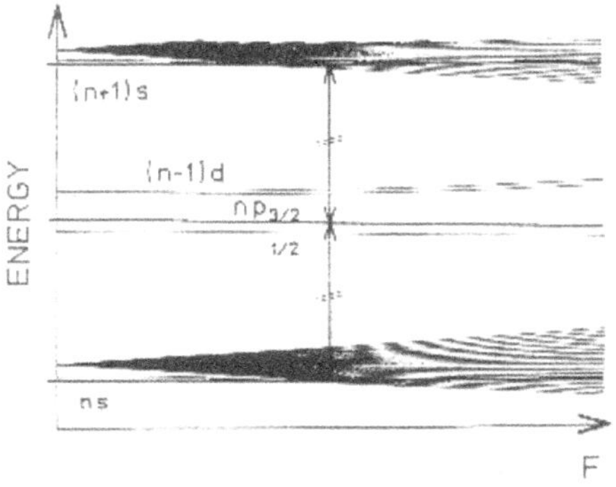

Figure 13. Stark diagram of cesium atom in the vicinity of the $(n-3)$ and $(n-2)$ manifolds.The arrows indicate the energy resonance for the internal energy exchange of the reaction (18).

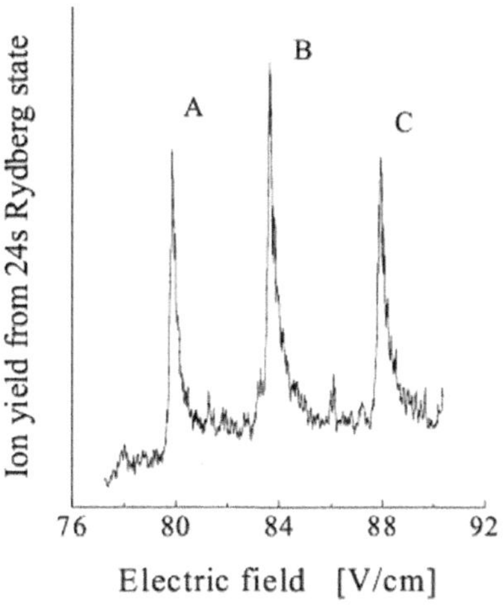

Figure 14. Signal of resonant Rydberg-Rydberg collisions for the reaction (19), in a Cs atomic beam.

The theoretical treatment of the long-range dipole-dipole collisions can be found in Refs. [Gallagher 1982; Pillet 1987]. Considering two colliding atoms as shown in Fig. 15, one of the atom, let say 1, is assumed to be stationary at the origin and the other, 2, passes the first atom with an impact parameter b at $x = 0$ with a relative velocity v parallel to the axis x (see Fig. 15). We construct the product states

$$|\Psi_P\rangle = |n\, p_{3/2}\rangle_1 \otimes |n\, p_{3/2}\rangle_2$$

$$|\Psi_{SS}\rangle = [|ns\rangle_1 \otimes |(n+1)s\rangle_2 + |(n+1)s\rangle_1 \otimes |ns\rangle_2]\, \sqrt{2} \tag{20}$$

$$|\Psi_{AS}\rangle = [|ns\rangle_1 \otimes |(n+1)s\rangle_2 - |(n+1)s\rangle_1 \otimes |ns\rangle_2]\, \sqrt{2}.$$

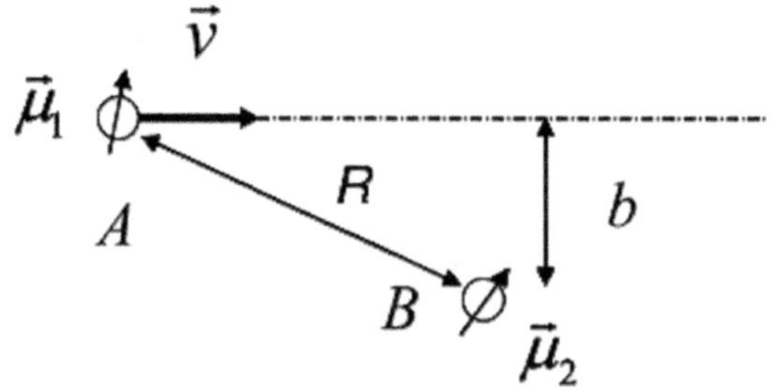

Figure 15 Geometry of the collision of two dipoles. μ_2 is at rest and μ_1 passes with a velocity v and impact parameter b. The dipoles are separated by R.

The long-range dipole-dipole interaction, V, couples the state $|\Psi_P\rangle$ with the symmetric state $|\Psi_{SS}\rangle$ and not with the antisymmetric one $|\Psi_{AS}\rangle$, which can be ignored in the following of the treatment

$$\langle\Psi_P|V|\Psi_{SS}\rangle = \left\langle \Psi_P \left| \frac{\vec{\mu}_1.\vec{\mu}_2}{R^3} - \frac{3\left(\vec{\mu}_1.\vec{R}\right)\left(\vec{\mu}_2.\vec{R}\right)}{R^5} \right| \Psi_{SS} \right\rangle \tag{21}$$

$$\langle\Psi_P|V|\Psi_{AS}\rangle = 0$$

where $\vec{R}$ is the vector between the two atoms and $\vec{\mu}_{1(2)}$ the dipole vector of the atom $1(2)$. It is interesting to notice that the state of the pair of atoms is an entangled state. We neglect in this treatment the nonresonant couplings to other molecular states which lead to variations in energy with the interatomic distance R. The total wave function for the system may be written as

$$|\Psi(t)\rangle = C_P(t)\ |\Psi_P\rangle + C_{SS}(t)\ |\Psi_{SS}\rangle\,. \tag{22}$$

The Hamiltonian of the system is given by

$$H = H_0 + V, \tag{23}$$

where H_0 is the Hamiltonian of the two atoms at $R \to \infty$, and can be written

$$H_0 = \begin{pmatrix} E_P(F) & 0 \\ 0 & E_{SS}(F) \end{pmatrix}. \tag{24}$$

$E_P(F)$ is the Stark energy of the pair of atoms in the state Ψ_P, and $E_{SS}(F) \simeq 0$ in the state Ψ_{SS}. The resonance, $E_P(F) = 0$, is matched for the value F_0 of the static electric field F. The interaction matrix elements of V can be simplified by taking the rms value

$$\langle \Psi_P | \, V \, | \Psi_{SS} \rangle = \sqrt{2} \, \frac{\langle n \, p_{3/2} \, |\mu| \, ns \rangle \, \langle n \, p_{3/2} \, |\mu| \, (n+1)s \rangle}{R^3} = \frac{\hbar \chi}{R^3} \tag{25}$$

with $\langle n \ p_{3/2}|\mu|ns\rangle \simeq \langle np_{3/2}|\mu|(n+1)s\rangle \approx n^2$ a.u.. At resonance, $E_P(F) = 0$, we obtain

$$C_{SS}(t \to \infty) = -i \, \sin \left[\int_{-\infty}^{+\infty} \frac{\chi}{(b^2 + v^2 t^2)^{3/2}} \, dt \right] = -i \, \sin \left(\frac{2\chi}{v \, b^2} \right). \tag{26}$$

For experiments in an atomic beams where faster atoms collide the slower ones, the cross section is given by

$$\sigma = 2\pi \int_0^{+\infty} |C_{SS}(t = \infty)|^2 \, b \, db, \tag{27}$$

which gives at resonance

$$\sigma = 2\pi \int_0^{+\infty} \sin^2 \left(\frac{2\chi}{v \, b^2} \right) b \, db = \frac{\pi^2 \chi}{v}. \tag{28}$$

By assuming for an atomic beam that the average on the relative velocity can in a first approximation be written as $\langle 1/v \rangle \approx 1/\langle v \rangle \approx \sqrt{M/(2 \, k_B T)}$, we obtain

$$\sigma \approx \frac{\pi^2 \chi}{\langle v \rangle}. \tag{29}$$

We can define the impact parameter at resonance, b_0, for $|C_{SS}(t \to \infty)| = 1$, leading to $2\chi/(v \, b_0^2) = \pi/2$ and

$$b_0 \approx \frac{2}{\sqrt{\pi}} \sqrt{\frac{\chi}{\langle v \rangle}}, \tag{30}$$

and we define the characteristic time for the collision as

$$\tau = \frac{b_0}{\langle v \rangle} \approx \frac{2}{\sqrt{\pi}} \frac{\sqrt{\chi}}{\langle v \rangle^{3/2}}. \tag{31}$$

Using these expressions we obtain the following orders of magnitude at room temperature for the above considered collisional configuration for cesium. We

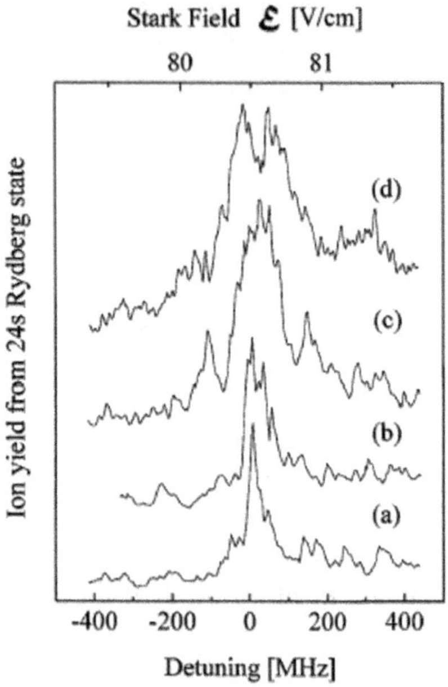

Figure 16 Energy tranfer resonance of Eq. (19) for 23 $p_{3/2}$, $|m| = 1/2$. Estimated densities: (a) 4, (b) 7, (c) 40, (d) 100×10^8 cm^{-3}.

have $\langle n\, p_{3/2}|\mu|ns\rangle \simeq \langle np_{3/2}|\mu|(n+1)s\rangle \simeq 200$ a.u. for $n = 23$. At a temperature $T = 300$ K, we take $\langle v\rangle \sim 200$ m/s, and we obtain $b_0 \sim 1.5$ μm, $\tau \sim 7.5$ ns. Considering the Doppler temperature for the cesium atom $T \sim 125$ μK, we obtain a huge impact parameter $b_0 \sim 60$ μm, comparable with the size ~ 300 μm of the atomic cloud. Clearly, the impact parameter is here larger than the average interatomic distance, R, meaning three-body and many-body collisional processes should be present. However, we have also to consider here the characteristic collisional time $\tau \sim 500$ μs, which becomes much larger than the lifetime of the Rydberg. We are here clearly in the sit-

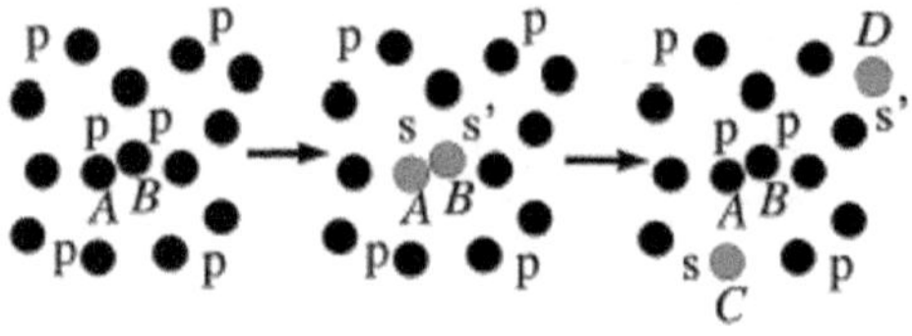

Figure 17. Illustration of elementary act of excitation exchange, which includes creation of an $ss\prime = ns + (n+1)s$ couple for a pair of p-Rydberg atoms $(A + B)$, embedded in the ensemble of other atoms. Evacuation of s-excitation towards C or D atoms allow the pair of atoms $(A + B)$ to react again.

uation where we have no longer a collisional problem, but an ensemble of interacting cold Rydberg atoms, where many-body processes could appear. In a first approximation, we can ignore the motion of the atoms and consider the Rydberg sample as a frozen Rydberg gas.

3. Many-body effects in a frozen Rydberg gas

The basic scheme of the experiment is the same as that followed for resonant collisions in an atomic beam, as for the reactions (19) of Fig. 14. At the trap position a static field, F, and a pulsed high voltage field can be applied by means of a pair of electric field grids spaced by 15 mm. The 6 $p_{3/2}$atoms are pulsed-laser excited in the presence of the static field 6 $p_{3/2} \rightarrow n\ p_{3/2}$. After the excitation the atoms evolve during a few microseconds (typically 2 μs) and when a quasi-steady-state regime attains, we apply a high voltage pulse of a 300 ns risetime and selectively ionize the ns upper state. The ions are expelled out of the interaction region and detected by a pair of microchannel plates. By slowly sweeping the field, F, we chirp the detuning of the reaction (18) and we observe ion peaks corresponding to resonances for the different components m. The control of the Rydberg atomic density is performed by attenuating the dye laser. Fig. 16 shows typical resonance for $|m| = 1/2 + |m| = 1/2$.

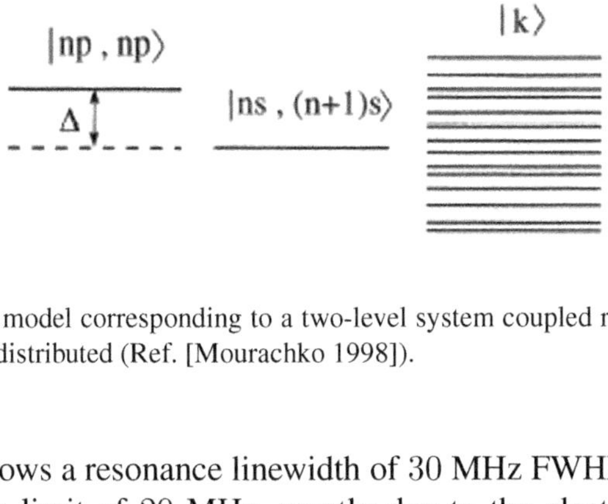

Figure 18. Band model corresponding to a two-level system coupled randomly with a band of levels randomly distributed (Ref. [Mourachko 1998]).

Fig. 16 (a) shows a resonance linewidth of 30 MHz FWHM, which is close to the resolution limit of 20 MHz, mostly due to the electric field inhomogeneities. At higher densities, Figs. 16 (b)-(d) show a progressive broadening of the lines about 70, 120, 180 MHz, respectively. The width of the resonance are no longer dependent of the velocity, because the Rydberg atoms are not moving. It will be dependent of the interaction of each atom with the ensemble of the other atoms. In a two-body framework, one may expect the linewidth to be of the order of typical size of the interaction given by Eq. (25) for $R = \overline{R}$, meaning due to the interaction between neighboring Rydberg atoms. Clearly, the observed lines are two orders of magnitude larger [Anderson 1998 (b); Mourachko 1998]. The observed density dependence of the lines is the signature of many-body effects. When the reaction (18) is tuned in resonance by a Stark shift, aside from the creation of ns, $(n+1)s$ pair, the s-excitation of the products of the reaction can migrate by excitation exchange with other atoms (see Fig. 17)

$$Cs_A(ns) + Cs_C(n\,p_{3/2}) \rightarrow Cs_A(n\,p_{3/2}) + Cs_C(ns)$$

$$\text{or} \qquad (32)$$

$$Cs_A((n+1)s) + Cs_D(n\,p_{3/2}) \rightarrow Cs_A(n\,p_{3/2}) + Cs_D((n+1)s).$$

The interpretation of the evolution of the total system has been proposed in several references [Mourachko 1998; Akulin 1999; Frasier 1999]. We do not develop in this article the details of the different theoretical approaches. It is clear that the pairs of closest neighboring Rydberg atoms play a particular role in the reaction. They can be considered as two-level systems coupled with a

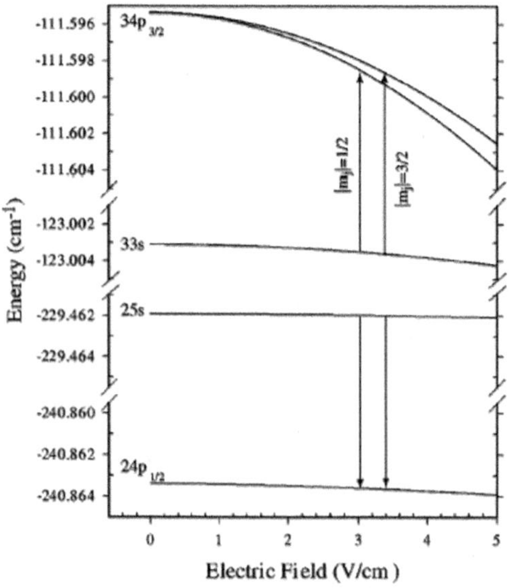

Figure 19. Energy levels of the relevant states for ^{85}Rb in an electric field showing the resonant transitions at 3.0 and 3.4 V/cm.

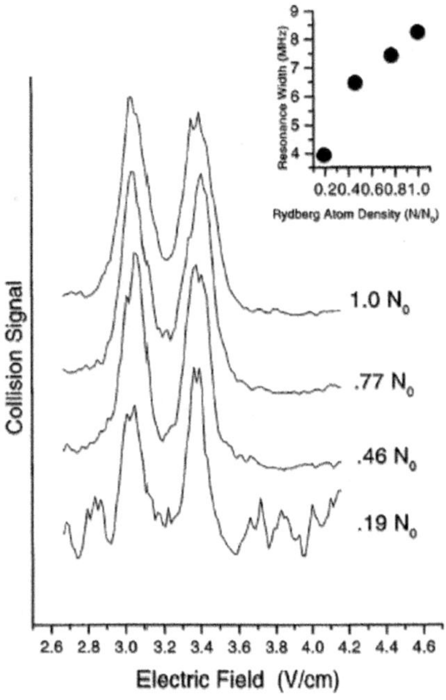

Figure 20. Resonances corresponding to the diagram of Fig. 19. The four densities are given for $N_0 = 10^9$ cm^{-3}. The inset shows the width of the observed resonances.

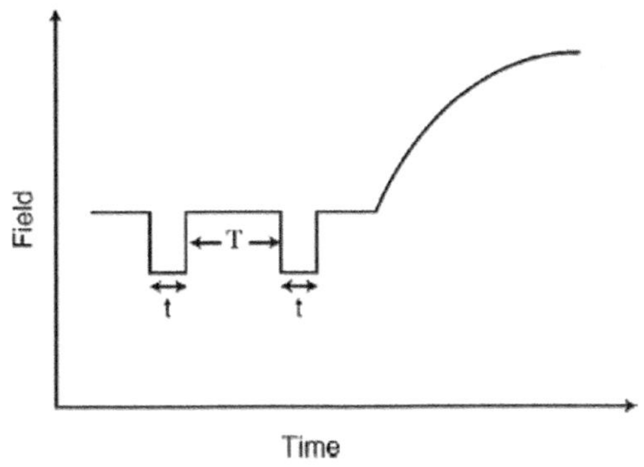

Figure 21. Timing diagram for the Ramsey double field pulse.

reservoir (the ensemble of other atoms), which can be described as a band of levels (see Fig. 18).

The frozen Rydberg gas corresponds to a quantum mesoscopic system, where the coherence of the two-level system is shared with the ensemble of the other atoms, offering an interesting example of decoherence through interaction with the environment [Joos 2003].

4. Coherence through a Ramsey scheme

To study the coherence and the decoherence problem in a mesoscopic frozen Rydberg gas, a variant of the Ramsey interference method has been proposed. The experiment has been performed for the rubidium atom by considering the reaction studied in reference [Anderson 1998 (b)],

$$Rb(25\,s_{1/2}) + Rb(33\,s_{1/2}) \rightarrow Rb(24\,p_{1/2}) + Rb(34\,p_{3/2}) \qquad (33)$$

which is resonant as shown in Fig. 19. To observe the resonances, we detect selectively the $34p$ level by field-ionization.

Fig. 20 shows the resonances observed in a magneto-optical trap at $300\ \mu$K by scanning the static electric field for four different densities. As in the previous case of the cesium atom, the resonances are broader than expected for the dipole-dipole interaction alone. We have to take into account the migration processes of the excitation of the reaction products

$$A\ :\ Rb(25\,s_{1/2}) + Rb(24\,p_{1/2}) \rightarrow Rb(24\,p_{1/2}) + Rb(25\,s_{1/2}) \qquad (34a)$$

$$B\ :\ Rb(33\,s_{1/2}) + Rb(34\,p_{3/2}) \rightarrow Rb(34\,p_{3/2}) + Rb(33\,s_{1/2}) \qquad (34b)$$

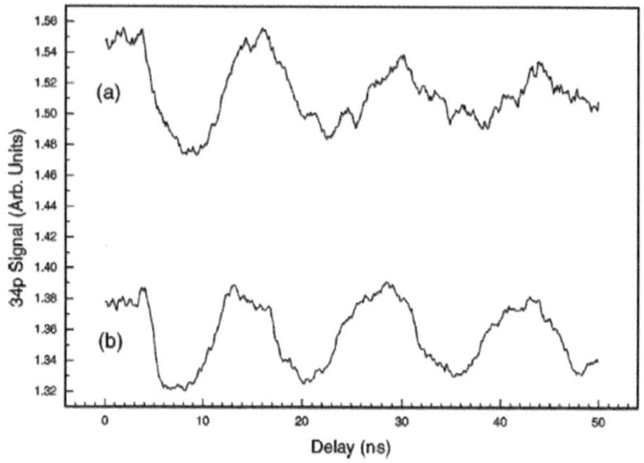

Figure 22. Ramsey fringes showing dephasing for two densities: (a) 7.6 and (b) 1.9 × 10^9 cm^{-3}.

We consider now the Ramsey scheme given by Fig. 21. The atoms are first Rydberg-excited out of resonance for the dipole-dipole energy transfer. After the excitation, the system is now tuned to resonance during two periods $t = 50$ ns, separated by a delay $T = 0 - 50$ ns. At resonance, the $34p$ ion signal should present Ramsey fringes versus the delay T. In fact the migration of the products (here p-excitation) of the reaction leads to the decoherence of the system and to the damping of the Ramsey-signal (see Fig. 22).

There is no observable dependence of the beat frequency on the density of Rydberg atoms in the trap, while the decay rate depends on the Rydberg density. For the high and low-density traces of the Figs. 22 (a) and (b), $1/\tau = 33.9$ and 11.4×10^9 s^{-1}, respectively. The damping rates are roughly comparable to the width of the observed energy transfer resonance, which implies that the widths are linked with the processes of decoherence in the frozen Rydberg gases.

5. Above the frozen Rydberg gas

In the frozen Rydberg gas, we neglect the motion of the atoms and any force due to the dipole-dipole interaction between Rydberg atoms. If the frozen Rydberg gas is a valid picture in a first approximation, we can show that dipole-dipole forces affect the dynamics at long-range distance and should be taken into account in resonant transfer mechanism. To evaluate the role of the long-range forces between cold Rydberg atoms, we have considered the following experiment. We consider a pair of two cesium Rydberg atoms in the initial state $|i\rangle = |23\ p_{3/2},\ |m| = 3/2\rangle \otimes |23\ p_{3/2},\ |m| = 1/2\rangle$, products of two

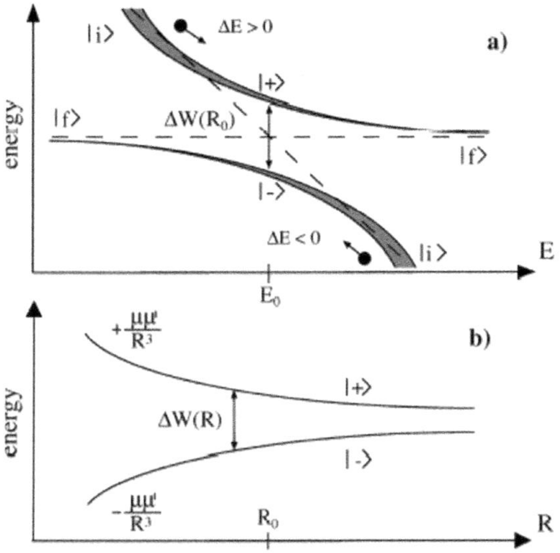

Figure 23 Energy diagram for Rydberg diatomic levels. (a) fixed internuclear distance $R = R_0$. (b) Fixed electric field. The spacing between the $|+\rangle$ and the levels at resonance is $\Delta W(R)$.

Rydberg states, evolving towards the final state $|f\rangle = |23s\rangle \otimes |24s\rangle$. For large interatomic distance, R, $|i\rangle$ and $|f\rangle$ are good eigenstates and their energy are the sum of the Stark energy of the Rydberg atom in their respective state nl, as shown by the broken line of Fig. 23 (a). At a smaller distance R the dipole-dipole interaction couples $|i\rangle$ and $|f\rangle$, giving eigenstates $|+\rangle$ and $|-\rangle$, as shown by the solid line of Fig. 23 (a). In Fig. 23 (b), we show the energy of the $|+\rangle$ and $|f\rangle$ states versus R at the electric field value $E = E_0$.

To observe effects due to the fact that the upper curve $|+\rangle$ of Fig. 23 (b) is repulsive while the lower one $|-\rangle$ is attractive, we excite first the initial diatomic state $|i\rangle$ and we detect the final state $|f\rangle$ as we scan the static electric field. We observe that the obtained resonance is asymmetric (see Fig. 24 (a)). The asymmetry arises because, for $E > E_0$, the atoms are preferentially excited on the attractive curve thus enhancing the energy transfer when the atoms move closer together. For $E < E_0$, the reverse is true. To verify our interpretation, we carried out what might loosely be called adiabatic rapid passage experiments. We have added, 2 μs after the Rydberg laser excitation a slowly varying field pulse $\Delta E = \pm\, 2.4$ V/cm, which rose or fell to its final value with a 1 μs time constant. The field ionization pulse comes 6 μs after the laser pulse, and we record the final state signal from the $24s$ state as the static field is scanned. Figs. 24 (b) and (c) show typical results for negative and positive field pulses, respectively. Clearly the results show that the experiments (b) and (c) are not mirror images altogether. First, resonances at $E = E_0$ in Fig. 24 (b) and (c) are due to the energy transfer which occurs before the ΔE field. The signal in (b) is clearly larger than in (c). The disparity is due to the difference

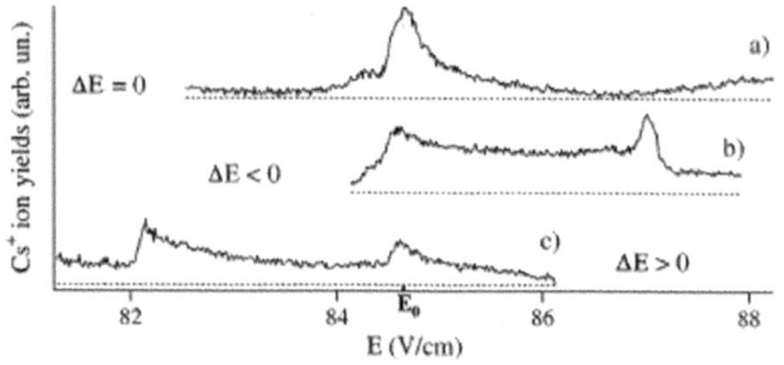

Figure 24 Asymmetric profiles of resonant energy transfer among Rydberg atoms. In (a) a normal field scan and, in (b) and (c), with adiabatic rapid passage.The dashed lines are the zeros of the ionic signals.

in how adiabatically the pair of atoms pass from dipole-dipole coupled states at resonance to the uncoupled states away from resonance. The resonances for $E > E_0$, Fig. 24 (b), and $E < E_0$, Fig. 24 (c) correspond to energy transfer during the ΔE field. If we assume an adiabatic following of the curves, atoms are prepared to the attractive lower curve $|-\rangle$ (respectively, to repulsive upper one $|+\rangle$) and move toward each together (respectively, move apart) leading to increasing of the dipole-dipole interaction (respectively, decreasing). The adiabatic assumption should be more valid is we consider the attractive state $|-\rangle$. Clearly the resonance in Fig. 24 (b) is more intense than in (c), and more details of the discussion can be found in Ref. [Fioretti 1999].

6. Prospectives and conclusion

In the first part of this article we have briefly summarized the study of resonant dipole-dipole energy transfer collisions between Rydberg atoms. Due to the long range of the dipole-dipole interaction the collision process can be understood with nearly spectroscopic accuracy, and this understanding forms the basis for understanding dipole-dipole interactions in a frozen Rydberg gas.

Frozen Rydberg gases offer a wide field of possible experiment to test coherence and decoherence in a mesoscopic quantum system. Clearly, such an ensemble in the resonant energy transfer configuration looks as a kind of amorphous solid, the properties of which can be properly tested. Adding a microwave field can increase the field of the possibilities as shown in Fig. 25.

From the top to the bottom of Fig. 25, the curves correspond to no microwave (upper curve), and with microwave field with frequency $\nu_{\mu w} = 1$, 20, 40, 165 and 400 MHz, and with field amplitude of $F_{\mu w} = 0.5$, 1.0, 1.0, 1.4 and 7.1 $\mathrm{V/cm}$, respectively. We observe here the clear evolution of the microwave assisted energy transfer when the microwave frequency, $\nu_{\mu w}$, becomes larger than the width of the resonance in absence of microwave. For frequency $\nu_{\mu w} \lesssim 40$ MHz, we have a broadening of the resonance corresponding to twice the microwave field amplitude, $2F_{\mu w}$. For larger frequency, we observe the appearance of sidebands around the resonance corresponding to

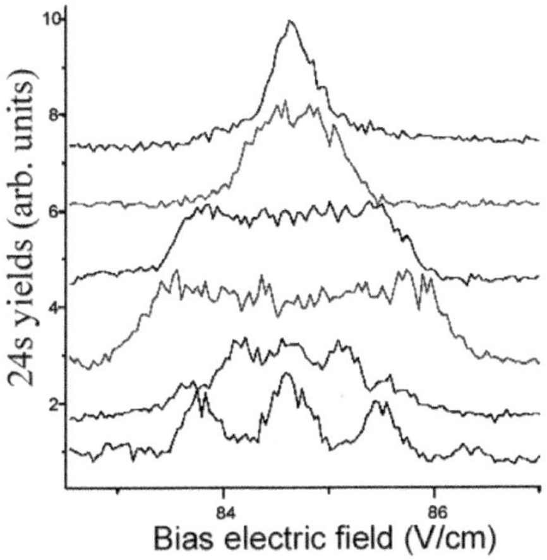

Figure 25. Microwave assisted resonant transfer in a frozen Rydberg gas (Fig. 19 B).

the emission or the absorption of one (or more) microwave photon. In the case of one photon, the resonance are located to position corresponding to a frequency of $\pm\nu_{\mu w}$. These results are quite analogue to those observed in the radiative Rydberg-atom - Rydberg-atom collisions experiments [Pillet 1987].

An interesting approach to test the decoherence of an assembly of cold Rydberg atoms is to increase the rate for the migration of the products of the reactions of energy transfer (18) or (33). One experiment (not yet published) has recently been performed at the University of Virginia considering the reaction (33). It consists in transferring a part of the population from the $33s$ level to the $34s$ one, with a two-photon microwave excitation on the transition $33s \rightarrow 34s$. The products $34\,p_{3/2}$ of the reaction (33) can now migrate with the exchange equation

$$Rb(34\,s_{1/2}) + Rb(34\,p_{3/2}) \rightarrow Rb(34\,p_{3/2}) + Rb(34\,s_{1/2}). \qquad (35)$$

By comparing to the migration due to Eq. (34b), we can increase the rate of migration up to a factor 50, because the electric dipole momenta are $\langle 34\,p_{3/2}|\,\mu\,|34\,s_{1/2}\rangle = 930$ a.u. and $\langle 33\,p_{3/2}|\,\mu\,|33\,s_{1/2}\rangle = 126$ a.u.. We observe in the experiment a broadening of the resonance depending on the population in the $34\,s_{1/2}$ state.

In a near future the experiments will evolve by considering a cw laser excitation of the Rydberg atoms. One of the reason is to avoid the strong fluctuations of the number of excited atoms from shot to shot, due to the frequency fluctuations of the multi-mode dye laser. But the main reason is to have a good control of the excitation and to be able to prepare pairs of Rydberg atoms A

and B with a given dipole-dipole interaction coupling and in a symmetric or anti-symmetric superposition of states, for instance

$$\frac{1}{\sqrt{2}}\left|23\,p_{3/2},\ |m|=\frac{3}{2}\right\rangle_A \otimes \left|23\,p_{3/2},\ |m|=\frac{1}{2}\right\rangle_B \pm \frac{1}{2\sqrt{2}}\left[\left(|23s\rangle_A+|24s\rangle_A\right)\otimes\left(|23s\rangle_B+|24s\rangle_B\right)\right] \tag{36}$$

as shown in Fig. 23. Such a configuration offers the possibility to demonstrate the dipole blockade of the Rydberg excitation, which can be used for coherent manipulation and entanglement of collective excitation in mesoscopic ensembles and lead to the realization of scalable quantum logic gates [Lukin 2001].

The study of elementary processes of interatomic interaction in a frozen Rydberg gas can also lead to a better understanding in the evolution of a cold Rydberg sample with $n \sim 70$ towards an ultra-cold plasma, where for so high n excitation, the blackbody radiation is not enough to explain the initial formation of the ionic space charge.

Acknowledgments

M. M. is postdoctoral fellow in the European Network COMOL. Laboratoire Aimé Cotton is associated to the University Paris-Sud and is member of the CNRS Research Federation "LUMAT". The work performed at University of Virginia has been supported by the U. S. Air Force Office of Scientific Research.

VI

PROTON ENTANGLEMENT AND DECOHERENCE IN SOLIDS

SCHRÖDINGER'S CAT STATES OF PROTONS IN CONDENSED MATTER

M. Krzystyniak,[1] T. Abdul-Redah,[2] C. A. Chatzidimitriou-Dreismann,[1] F. Fillaux,[3] E. B. Karlsson,[4] J. Mayers,[2] I. E. Mazets,[5] H. Naumann[6] and S. Stenholm[7]

[1] *Institut für Chemie, Stranski Laboratorium, Technische Universität Berlin, Straße des 17. Juni 112, D-10623 Berlin, Germany*

[2] *ISIS Facility, Rutherford Appleton Laboratory, Chilton / Didcot, OX11 0QX, UK*

[3] *LADIR-CNRS and Université Pierre et Marie Curie, UMR 7075, 2 rue H. Dunant, 94320 Thiais, France*

[4] *Dept. of Physics, Uppsala University, P. O. Box 530, SE-75121 Uppsala, Sweden*

[5] *Ioffe Physico-Technical Institute, St.Petersburg 194021, Russia*

[6] *Institut für Chemie, Vollmer Laboratorium, PC 14, Technische Universität Berlin, Straße des 17. Juni 135, D-10623 Berlin Germany*

[7] *Laser Physics & Quantum Optics, Royal Institute of Technology (KTH), AlbaNova, Roslagstullsbacken 21, SE-10691 Stockholm, Sweden*

The development of the experimental and theoretical work on *Proton entanglement and decoherence in condensed matter* is recent but yet has a rich history starting with the pioneering work of C. A. Chatzidimitriou-Dreismann *et al.* [Chatzidimitriou-Dreismann 1997 (a)]. There, the authors presented experimental results of neutron Compton scattering (NCS) from H_2O / D_2O mixtures showing a striking effect manifested by anomalous shortfall of the scattering cross section of the protons. It has been claimed [Chatzidimitriou-Dreismann 1997 (a)] that these results provided for the first time direct evidence for the existence of short-lived nuclear quantum entanglement,proton / neutron. This experiment was followed by another one on a biologically important material, *i.e.*, urea dissolved in H_2O / D_2O mixtures, which showed results different than those of the pure solvent [Chatzidimitriou-Dreismann 1999].

A new dynamical aspect based on previous work by V. F. Sears [Sears 1984] and G. I. Watson [Watson 1996] was introduced to the data interpretation by the work of E. B. Karlsson *et al.* [Karlsson 1999] on niobium hydride. While the cross section anomalies found in the previous work [Chatzidimitriou-Dreismann 1997 (a); Chatzidimitriou-Dreismann 1999] were independent of the ex-

V.M. Akulin et al. (eds.), Decoherence, Entanglement and Information Protection in Complex Quantum Systems, 439–444.

perimentally available scattering time τ_s, niobium hydride represented a system in which a transition could be observed between a dynamical region where quantum entanglement of protons did exist thus exhibiting large proton cross section anomalies and a dynamical region where the proton cross section takes its conventionally expected value.

These NCS experiments [Chatzidimitriou-Dreismann 1997 (a); Chatzidimitriou-Dreismann 1999; Karlsson 1999], which were motivated by the theoretical work of C. A. Chatzidimitriou-Dreismann [Chatzidimitriou-Dreismann 1991; Chatzidimitriou-Dreismann 1997 (b)] on protonic quantum entanglement in condensed systems and by the results of a an earlier Raman experiment on liquid H_2O / D_2O mixtures [Chatzidimitriou-Dreismann 1995], were followed by a series of other experiments on liquid and solid organic materials [Chatzidimitriou-Dreismann 2000 (b); Chatzidimitriou-Dreismann 2001; Chatzidimitriou-Dreismann 2002 (a)], various metallic hydrides [Abdul-Redah 2000; Karlsson 2002 (b); Karlsson 2003 (b)], liquid hydrogen [Chatzidimitriou-Dreismann 2004 (b)] and among others an ionic solid [Abdul-Redah 2004] using the same experimental technique, *i.e.*, neutron Compton scattering. All these experiments confirmed the anomalous results found earlier and also revealed certain new aspects of the considered effect.

In the course of time (since 1995), the data analysis has been subject to various criticisms [Blostein 2001; Blostein 2003 (a); Blostein 2003 (b); Cowley 2003] which, however, have been refuted by several authors [Chatzidimitriou-Dreismann 2002 (b); Abdul-Redah 2003; Mayers 2004]. A further criticism was that all these effects were found on one instrument only, *i.e.*, VESUVIO spectrometer of the ISIS neutron spallation source at the Rutherford Appleton Laboratory in UK. There was no possibility to confirm this effect on a different apparatus. However, very recently, this effect was confirmed using an independent experimental method, electron-proton Compton scattering at the Australian National University, Canberra [Chatzidimitriou-Dreismann 2003 (a)]. This experiment has attracted a vast attention by the scientific community [Physics News 2003; Physics Today 2003; Scientific American 2003].

As already mentioned above the experimental work on short-lived protonic quantum entanglement in condensed matter has been triggered by earlier theoretical and experimental work of C. A. Chatzidimitriou-Dreismann [Chatzidimitriou-Dreismann 1991; Chatzidimitriou-Dreismann 1997 (b)]. One model for the theoretical interpretation of the NCS results has been put forward by C. A. Chatzidimitriou-Dreismann *et al.* starting from the van Hove formalism for the scattering process and explicitly taking into account the irreversible dynamics of the decoherence process [Chatzidimitriou-Dreismann 2000 (b)]. This model has experienced a further development and extension [Chatzidimitriou-Dreismann 2003 (b)]. E. B. Karlsson and S. W. Lovesey have proposed another model [Karlsson 2002 (a)] relying on the existence of exchange correlations between proton pairs. This model has been also developed further [Karlsson 2002 (c)] and applied [Karlsson 2003 (a)] to the first NCS experimental results on H_2O / D_2O mixtures [Chatzidimitriou-Dreismann 1997 (a)]. Another important theoretical work by F. Fillaux [Fillaux 1998] introduces an entanglement mechanism arising from symmetry-related normal coordinates of quantum oscillators. This type of entanglement gives rise to long-lived and spatially extended quantum interference of many protons and finds applicatiom to diffraction experiments on $KHCO_3$ and other crystals [Fillaux 2003 (a)]

In the following a brief overview of the contributions to this part of the book are presented. First the experimental papers are described showing the striking results of neutron and electron Compton scattering where the shortfall of the total scattering intensity is observed. In a separate contribution also the elastic neutron scattering and its application to the investigation of the long-lived quantum entanglement will be presented. The novel technique of inelastic X-ray scattering, lately applied to the investigation of short-lived quantum entanglement of vibrations in condensed matter, is also discussed. Second, the theoretical papers are presented. These are three different contributions focusing on slightly different aspects of quantum entanglement in condensed matter.

In the first experimental paper entitled *Anomalous neutron inelastic cross sections at eV energy transfers* J. Mayers and T. Abdul-Redah give an overview of the NCS technique and a full description of the data analysis procedure followed to extract the required information about the scattering cross sections. Due to the various criticisms [Blostein 2001; Blostein 2003 (a); Blostein 2003 (b); Cowley 2003] of the interpretation of the results and the NCS data treatment, the VESUVIO instrument as well as the data treatment procedure has been subject to a thorough scrutiny by the authors. They present the results both from experiments as well as from Monte Carlo simulations showing that none of the criticisms can account for the observed anomalies in the neutron scattering cross section of hydrogen.

This paper is followed by two further experimental papers dealing with NCS experiments. The paper by T. Abdul-Redah *et al.*, *Quantum entanglement and decoherence due to coupling of protons to electronic environment* deals with the question of the involvement of the electronic degrees of freedom in the anomalous scattering behavior and in the decoherence process. The influence of the electronic structure surrounding the H atoms in different materials on the neutron scattering cross section of hydrogen is investigated. It is found that different electronic structures cause different H cross section anomalies. Furthermore, the comparison of water and LiH data shows different temperature dependence behavior. In addition, results of recent experimental tests of the data analysis procedure are presented showing its correctness and refuting the

criticisms [Blostein 2001; Blostein 2003 (a); Blostein 2003 (b); Cowley 2003] already mentioned above.
In the second one entitled *Attosecond effects in scattering of neutrons and electrons from protons* C. A. Chatzidimitriou-Dreismann *et al.* present NCS results from organic materials and from liquid hydrogen and H_2 - D_2 mixtures and results from both NCS and electron-proton Compton scattering (ECS) on a solid polymer (formvar). The proton momentum distributions measured in NCS and ECS on formvar turn out to be the same and the ECS data reveal also the anomalous decrease of the total proton scattering intensity. Moreover, the good agreement between the NCS and ECS results from a solid polymer demonstrates the "general" character of the considered effect because it appears to be independent of the fundamental interaction (strong for NCS, electromagnetic for ECS) involved [Chatzidimitriou-Dreismann 2003 (a); Physics News 2003; Physics Today 2003; Scientific American 2003].

This part continues with the paper *Macroscopic quantum entanglement in the $KHCO_3$ crystal* by F. Fillaux which is different from the preceding contributions in that it does not deal with neutron Compton scattering but rather with elastic neutron scattering techniques. Novel aspects of proton dynamics in $KHCO_3$ are presented. This crystal, called by the author a "quantal-crystal", is an example of yet a different entanglement mechanism. Quantum correlation arises exclusively from normal coordinates representing dynamics of indistinguishable particles. A detailed theoretical interpretation is given in the framework of the symmetry-adapted wave function of protons in the crystal that takes into account the fermionic and / or bosonic character of the involved particles.

Another novel technique applied to the investigation of short-lived proton entanglement in condensed matter is the inelastic X-ray scattering (IXS). The paper *Probing short lived entanglement with inelastic X-ray scattering* by H. Naumann *et al.* presents preliminary IXS experimental results of molecular vibrations of liquid light and heavy water and the equimolar light-heavy water mixture. The results indicate again the presence of an anomalous shortfall of scattering intensity from the OH-stretching vibrational modes. This effect is very similar to the earlier one observed with Raman light scattering [Chatzidimitriou-Dreismann 1995]. The possible connection of these observations with the results of neutron and electron Compton scattering from protons in condensed matter is mentioned, as well as their interpretation in terms of attosecond entanglement.

In addition to the papers presenting the results of the various experimental techniques (*e.g.*, NCS, ECS, elastic neutron scattering, IXS, *etc.*) applied to investigate protonic quantum entanglement in condensed matter, there are also papers in this part aiming at the theoretical interpretation of the experimental results. Although the theoretical models presented are different from each other, they all have in common the existence of quantum entanglement of the particles involved in the scattering processes.

The theoretical part opens with the contribution *Proton-proton correlations in condensed matter* by E. B. Karlsson in which a model of proton-proton correlations is presented. This model which has been proposed earlier by E. B. Karlsson and S. W. Lovesey [Karlsson 2002 (a); Karlsson 2002 (c)] makes use of quantum exchange correlations between proton (or deuteron) pairs and is inspired by a theoretical work of L. Pitaevskii and S. Stringari [Pitaevskii 1999] dealing with the interference of spatially separated Bose-Einstein condensates in momentum space. It is based on the idea that proton pairs, or larger clusters, may stay quantum entangled in condensed matter for measurable times. Mechanisms leading to local proton entanglement and natural decoherence is discussed. Then the model is presented in which the scattering on two indistinguishable particles is considered leading to a decrease in the proton cross-section as a result of destructive interference.

Another theoretical paper presenting a different approach than the work of Karlsson and making use of the generalized Fermi's Golden Rule approach to the problem of scattering is presented by I. E. Mazets *et al.*. The paper *Is Fermi's Golden Rule always true for Compton scattering?* describes a simple situation where Fermi's Golden Rule in its original formulation does not apply, even if an unstable quantum system decays exponentially. The authors present the most general expression for the double-differential cross-section taking into account the environment-induced broadening of the lineshape recorded in the NCS. This expression leads, for a certain set of experimental parameters, to a modification of the total proton cross-section. In this model, entanglement and its decoherence is implicitly connected with the assumed "struck particle-environment" relaxation.

This part closes with the contribution *On correlation approach to scattering in the decoherence timescale* by C. A. Chatzidimitriou-Dreismann and S. Stenholm. The authors provide a first principles description of the scattering process taking into account the irreversible dynamics of an open quantum system. The relevant quantum system which contains a struck proton and its adjacent electrons (and perhaps also other nuclei) obeys a Lindblad-type equation [Breuer 2002] that is specifically rewritten for the case of scattering. This is done using the so-called preferred basis as introduced in general decoherence theory [Breuer 2002]. The basis vectors in this treatment are not the simple momentum eigenstates of the scattering particle and target. The authors consider in particular the case when the duration of the scattering process is of similar order as the decoherence time of the scatterer. The conclusion from the derivations presented is that the irreversible time-evolution may cause a reduction of the system's transition rate (which is tantamount to total scattering intensity).

To sum up, the contributions described above represent a few chosen examples of the experimental and theoretical work on the subject of quantum entanglement and coherence in condensed matter. This fascinating new field of science has not yet been given a common name. With no doubt it is, however, science "in statu nascendi" and as such needs further development. From the experimental point of view this new field needs new techniques involved. Some examples (NCS, neutron diffraction, ECS, IXS) have already been mentioned above. Further work is also clearly needed to make the theoretical models mentioned above more specific and directly applicable to the experiment.

ANOMALOUS NEUTRON INELASTIC CROSS SECTIONS AT EV ENERGY TRANSFERS

J. Mayers and T. Abdul-Redah
ISIS Facility, Rutherford Appleton Laboratory, Chilton / Didcot, OX11 0QX, UK
j.mayers@rl.ac.uk, t.abdul-redah@rl.ac.uk

Abstract It has been proposed that short-lived quantum entanglement of protons in condensed matter systems would result in anomalous inelastic scattering cross sections at eV energy transfers. This proposal seems to be confirmed by neutron measurements on the VESUVIO spectrometer at ISIS and by measurements using other techniques. However, there have been a number of published suggestions of ways in which the observed effects on VESUVIO could be introduced by assumptions used in the data analysis. In this paper it is shown using experimental data and Monte Carlo simulations, that these suggestions cannot explain the observed cross section anomalies. The other assumptions of the data analysis are also examined. It is shown that the assumption of a Gaussian peak shape for the neutron Compton profile can introduce significant errors into the determination of cross section ratios but also cannot explain the observed anomalies.

Keywords: Quantum entanglement, sub-femtosecond dynamics, neutron Compton scattering

Introduction

The neutron scattering experiments discussed in this paper were motivated by the proposal of C. A. Chatzidimitrou-Dreismann [Chatzidimitriou-Dreismann 1997 (b)], that if the time interval during which an incident particle may interact with the target system is comparable to the quantum decoherence time, anomalous inelastic scattering cross sections would be observed. It was suggested that quantum entanglement would give different ratios of cross sections to those calculated from the sample composition and standard theory. This proposal obtained initial experimental support from experiments on liquid H_2O - D_2O mixtures using Raman light-scattering [Chatzidimitriou-Dreismann 1995] and eV neutron scattering [Chatzidimitriou-Dreismann 1997 (a)], performed on the VESUVIO spectrometer at the ISIS pulsed neutron source. These measurements showed anomalous ratios of the hydro-

V.M. Akulin et al. (eds.), Decoherence, Entanglement and Information Protection in Complex Quantum Systems, 445–468.

gen and deuterium cross sections. Subsequent VESUVIO measurements have been performed by Karlsson *et al.* on niobium and palladium hydrides and deuterides [Karlsson 1999; Abdul-Redah 2000; Karlsson 2003 (b)] and by Chatzidimitrou-Dreismann *et al.* on polystyrene [Chatzidimitriou-Dreismann 2000 (b)], benzene [Chatzidimitriou-Dreismann 2002 (a)] and other sytems [Chatzidimitriou-Dreismann 2001; Abdul-Redah 2003] . These measurements all exhibited anomalous ratios of the proton and deuterium inelastic cross sections. Recently Chatzidimitrou-Dreismann *et al.* [Chatzidimitriou-Dreismann 2003 (a)] compared neutron Compton scattering (NCS) results obtained on the amorphous polymer "Formvar" ($C_8H_{14}O_2$) with results of electron-proton scattering from the same material. The two techniques showed the same anomalous ratio of H and C inelastic cross sections as a function of momentum transfer, despite the very different experimental methods used in the two cases.

Many authors [Chatzidimitriou-Dreismann 2000 (b); Karlsson 2000; Karlsson 2002 (c); Chatzidimitriou-Dreismann 2003 (b); Karlsson 2003 (a)] have presented different theoretical treatments of this effect. Both treatments share the assumption that the observations are due to quantum effects, which exist over the time scales $\sim 10^{-15}$ seconds accessed by measurements using eV neutrons. However, these theoretical interpretations have been questioned by Cowley and Colognesi [Colognesi 2003; Cowley 2003]. The accuracy of the experimental results has also been questioned. It has been argued by Blostein *et al.* [Blostein 2001; Blostein 2003 (a); Blostein 2003 (b); Blostein 2004] that the way in which the instrument resolution function is incorporated into the data fitting routines, introduces serious errors into the data analysis and that this could account for the observed anomalies. It has also been stressed by Cowley, that an accurate correction of the measured data for the variation of the incident neutron intensity with energy is necessary and suggested that any errors in this correction could also account for the observed anomalies. In the present paper these two points are specifically addressed. It is shown using measured data and Monte Carlo simulations of data that neither can explain the anomalies observed. The other assumptions of the data analysis are also critically examined.

The outline of the paper is as follows. In Sec. 1, the expression used to fit the VESUVIO data is derived. Sec. 2 describes the way in which the instrument resolution is incorporated into the data analysis. Sec. 3 describes the Monte Carlo (MC) procedure, used to test the fitting programs. In Secs. 4 to 6 the influence of instrumental effects on the results are evaluated, using MC simulations and also by comparing data taken under different experimental conditions. Sec. 4 considers the effects of the correction for the incident beam intensity. Sec. 5 considers the systematic errors generated by approximations made to incorporate the instrument resolution in the fitting programs. Sec. 6 discusses effects dependent on sample size, such as attenuation, multiple scattering and detector dead time effects. Secs. 7 and 8 discuss the validity of the assumptions (a) that scattering can be described within the Impulse Approximation (IA) and (b) that the atomic momentum distributions have Gaussian peak shapes.

1. Theory of the data analysis procedure

1.1 Count-rates in time of flight neutron scattering experiments

We first consider a system of N identical atoms, scattering neutrons into a detector subtending solid angle $d\Omega$, at scattering angle θ. It follows from the definition [Lovesey 1984] of the partial differential scattering cross section $d^2\sigma/d\Omega dE$, that the number of neutrons with incident energies in the range E_0 to $E_0 + dE_0$, detected with final energies between E_1 and $E_1 + dE_1$ is

$$C_D(E_0,\, E_1)\, dE_0\, dE_1 = I(E_0)\, D(E_1)\, \frac{d^2\sigma(E_0,\, E_1,\, \theta)}{d\Omega\, dE_1}\, d\Omega\, dE_0\, dE_1 \quad (1)$$

where $I(E_0)\, dE_0$ is the number of incident neutrons per unit area with energies between E_0 and $E_0 + dE_0$ and $D(E_1)$ is the probability that a neutron of energy E_1 is detected. It follows from standard theory [Lovesey 1984] that for isotropic scattering,

$$\frac{d^2\sigma(E_0,\, E_1,\, \theta)}{d\Omega\, dE_1} = |b|^2 \sqrt{\frac{E_1}{E_0}}\, S(q,\, \omega) \quad (2)$$

where b is the nuclear scattering length, the energy transfer in the measurements is $\omega = m(v_0^2 - v_1^2)/2$ and the momentum transfer $q = m(v_1^2 + v_0^2 - 2\, v_0\, v_1\, \cos\theta)^{1/2}$. The velocity of the scattered neutron is

$$v_1 = \sqrt{\frac{2E_1}{m}} \quad (3)$$

with a similar expression for the velocity v_0 of the incident neutron, where m is the neutron mass. The neutron time of flight t is thus

$$t = \frac{L_0}{v_0} + \frac{L_1}{v_1} \quad (4)$$

where L_0 is the incident flight path and L_1 is the final flight path (Fig. 1). Eq. (3) and (4) can be used to define E_0 in terms of E_1 and t

$$E_0(E_1,\, t) = \frac{m}{2}\left(\frac{L_0\, v_1}{v_1\, t - L_1}\right)^2. \quad (5)$$

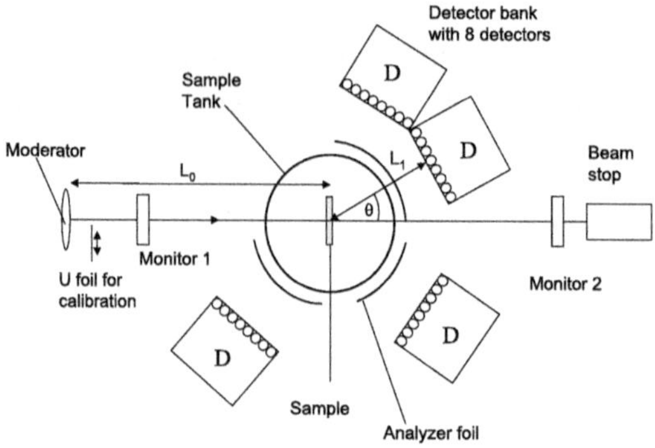

Figure 1 A schematic representation of the VESUVIO spectrometer at ISIS

The total number of neutrons detected in a time channel between t and $t + dt$ can be expressed as

$$C(t)\,dt = \left[\int C_D\,[E_0(t,\,E_1),\,E_1]\,\frac{dE_0(t,\,E_1)}{dt}\,dE_1\right]\,dt. \qquad (6)$$

An alternative and equally valid approach, used by Blostein *et al.* [Blostein 2001], is to calculate $C(t)$ by expressing E_1 as a function of t and E_0 and integrating over E_0. However, it is more convenient to use Eq. (6), when discussing an "inverse geometry" instrument such as VESUVIO, where the energy of the scattered neutron is analysed. It follows from (3) and (5) that

$$\frac{dE_0}{dt} = -\frac{2^{3/2}}{L_0\,m^{1/2}}\,E_0^{3/2}. \qquad (7)$$

For an ideal inverse geometry instrument, in which L_0, L_1, θ are precisely known and only neutrons of a precisely defined energy E_R are detected, *i.e.*,

$$D(E_1) = D(E_R)\,\delta(E_1 - E_R) \qquad (8)$$

it follows from (1) and (6) to (8) that

$$C(t) = 2\left(\frac{2}{m}\right)^{1/2}\frac{E_0^{3/2}}{L_0}\,I(E_0)\,D(E_R)\,N\,\frac{d^2\sigma}{d\Omega\,dE_1}\,d\Omega \qquad (9)$$

where $E_0(E_R,\,t)$ is defined via (5). Eq. (9) is the standard expression for the count rate in an inverse geometry time of flight spectrometer [Windsor 1981].

1.2 The impulse approximation

The VESUVIO spectrometer is mainly used to determine atomic momentum distributions in condensed matter systems, by "deep inelastic neutron scattering" (DINS). DINS relies upon the fact that, at sufficiently high momentum transfer, the IA can be used to interpret data. The validity of the IA in neutron scattering has been discussed by many authors [Sears 1984; Mayers 1989; Mayers 1990; Glyde 1994]. At the energy and momentum transfers attained on VESUVIO is accurate to within $\sim$ 5% in hydrogeneous samples [Evans 1996]. A basic assumption of the IA is that for neutron wavelengths much less than the inter-atomic spacing, atoms scatter incoherently. Thus if atoms of different mass M are present in the sample, it follows from (9) that the count rate is

$$C(t) = 2\left(\frac{2}{m}\right)^{1/2} \frac{E_0^{3/2}}{L_0}\, I(E_0)\, D(E_R) \sum_M N_M \frac{d^2\sigma_M}{d\Omega\, dE_1}\, d\Omega \tag{10}$$

where N_M is the number of atoms of mass M and $d^2\sigma_M/d\Omega\, dE_1$ is the partial differential cross section for mass M. The IA effectively treats the scattering as single atom "billiard ball" scattering with conservation of momentum and kinetic energy of the neutron + target atom. The dynamic structure factor for atoms of mass M is thus [Lovesey 1984]

$$S_M(\vec{q},\, \omega) = \int n_M(\vec{p})\; \delta\left(\omega + \frac{p^2}{2M} - \frac{(\vec{p}+\vec{q})^2}{2M}\right) d\vec{p} \tag{11}$$

where $n_M(\vec{p})$ is the atomic momentum distribution for mass M. It is important to understand that the total scattering cross section given by the IA is the "free atom" value, which is not the same as the cross section in the neutron-nucleus centre of mass frame. If $n_M(\vec{p}) = \delta(\vec{p})$ it follows from (11) and (2) that

$$\frac{d^2\sigma_M}{d\Omega\, dE_1} = b_M^2 \sqrt{\frac{E_1}{E_0}}\; \delta\left(\omega - \frac{q^2}{2M}\right) \tag{12}$$

where b_M is the "bound" scattering length for atoms of mass M. Integrating Eq. (12) over the solid angle $d\Omega$ and final energies E_1 gives [Lovesey 1984] the "free atom" cross section.

$$\int \frac{d^2\sigma_M}{d\Omega\, dE_1}\, d\Omega\, dE_1 = \frac{4\pi\, b_M^2}{\left(1+\dfrac{M}{m}\right)^2} = \frac{\sigma_M^2}{\left(1+\dfrac{M}{m}\right)^2} \tag{13}$$

where σ_M is the standard tabulated "bound" total scattering cross section for mass M.

It follows from (11) that

$$S_M(q,\, \omega) = \frac{M}{q}\, J_M(y_M) \tag{14}$$

where

$$y_M = \frac{M}{q}\left(\omega - \frac{q^2}{2M}\right) \quad \text{and} \quad J_M(y_M) = \int n_M(\vec{p})\ \delta\left(y_M - \frac{\vec{p}.\vec{q}}{q}\right)\ d\vec{p}. \tag{15}$$

The "neutron Compton profile" $J_M(y_M)$ is the probability distribution of the momentum component of mass M along the direction of $\vec{q}$ and is analogous to the "Compton profile", measured in Compton scattering of photons from electrons.

It follows from Eq. (14) and (2) that,

$$\frac{d^2\sigma_M}{d\Omega\, dE_1} = b_M^2 \sqrt{\frac{E_1}{E_0}}\, \frac{M}{q}\ J_M(y_M). \tag{16}$$

Combining Eqs. (10) and (16) we get

$$C(t) = \frac{E_0\ I(E_0)}{q} \sum_M A_M\ M\ J_M(y_M) \tag{17}$$

where

$$A_M = \frac{2}{L_0}\ D(E_R)\sqrt{\frac{2\ E_R}{m}}\ \Delta\Omega\ N_M\ b_M^2 \tag{18}$$

is proportional to the scattering intensity from mass M.

1.3 Fitting expression

In the derivation of Eq. (17) it is assumed that the "instrument parameters" L_0, L_1, θ and E_1 are known exactly. In reality these quantities can be determined only according to some probability distribution $P(L_0,\ L_1,\ \theta,\ E_1)$, which determines the instrument resolution. The measured count rate $C_m(t)$ is an average over the possible values of these parameters, weighted by their probability of occurrence

$$C_m(t) = \int C(t) P(L_0,\ L_1,\ \theta,\ E_1)\ dL_0\ dL_1\ d\theta\ dE_1. \tag{19}$$

Thus the exact incorporation of the instrument resolution function would entail the evaluation of this four dimensional integral for each data point, in addition to the convolution in t, required to incorporate the uncertainty in the measurement of time of flight. To reduce data processing times, the approximation is made in the data analysis that the resolution can be incorporated as a single convolution in t space, with a different resolution function $R_M(t)$ for each mass. Thus (17) is modified to

$$C_m(t) = \left[\frac{E_0\, I(E_0)}{q}\right] \sum_M A_M\, M\, J_M(y_M) \otimes R_M(t). \tag{20}$$

The approximation of replacing the exact expression (19) by (20) is referred to in the rest of the paper as the "convolution approximation" (CA).

A second approximation of the data analysis is that $J_M(y_M)$ is assumed to have a normalised Gaussian form

$$J_M(y_M) = \frac{1}{\sqrt{2\pi\, w_M^2}} \exp\left(-\frac{y_M^2}{2\, w_M^2}\right). \tag{21}$$

The data analysis consists of fitting Eqs. (20) and (21) to the data with two fitting parameters for each atomic mass, A_M and σ_M. A_M determines the integrated peak intensity corresponding to a mass M and σ_M determines the peak width. It follows from (18) that

$$\frac{A_M}{A_{M'}} = \frac{N_M\, b_M^2}{N_{M'}\, b_{M'}^2} = \frac{N_M\, \sigma_M}{N_{M'}\, \sigma_{M'}} \tag{22}$$

where σ_M is the "bound" cross section for mass M and σ'_M that for mass M'. Thus, if the sample composition (and hence N_M/N'_M) is known, the ratios of cross sections for atoms of different masses can be determined from the ratio of the fitted parameters A_M and A'_M. All results discussed in this paper were obtained in this way.

There are a number of possible sources of error in the determination of cross section ratios from the fitting expression defined by (20) and (21), which will be examined in this paper.

(1) The incident spectrum intensity $I(E_0)$ must be accurately known.

(2) The approximation of the resolution function components as Gaussian or Lorentzian functions is an approximation. Furthermore incorporation of the resolution function as a convolution in t is an approximation.

(3) Multiple scattering and sample attenuation effects may change the fitted cross section ratios.

(4) An implicit assumption in Eqs. (17) to (19) is that the probability $D(E_1)$ that a neutron of energy E_1 is detected, is independent of t, which is not true if detector dead time effects are significant.

(5) The impulse approximation is strictly valid only as $q \to \infty$ and corrections at the finite q of measurements must therefore be evaluated.

(6) The functions $J_M(y_M)$ may have non-Gaussian peak shapes.

2. The VESUVIO resolution function

It is assumed in the data analysis that the distributions of instrument parameters L_0, L_1, θ and E_1 are statistically independent and that these distributions can be defined in terms of mean values $\bar{L}_0$, $\bar{L}_1$, $\bar{\theta}$, $\bar{E}_1$ and widths of the distributions, ΔL_0, ΔL_1, $\Delta\theta$, ΔE_1. Both mean values and widths are determined by calibration measurements as described previously [Fielding 2002]. For the calculation of E_0, q, y_M in terms of t, in the fitting expression (20), the mean values $\bar{L}_0$, $\bar{L}_1$, $\bar{\theta}$ and $\bar{E}_1$ are used. The widths are used to determine the mass dependent resolution widths of the functions $R_M(t)$ as follows.

For a given M, the position $t_M(L_0,\ L_1,\ \theta,\ E_1)$ of the peak centre in time of flight is determined by Eq. (4) and the relation,

$$\frac{v_1}{v_0} = \frac{\cos\theta + \sqrt{\left(\frac{M}{m}\right)^2 - \sin^2\theta}}{\frac{M}{m} + 1}. \tag{23}$$

Eq. (23) follows from the assumption of the IA that the peak centre corresponds to scattering from an atom with zero momentum, with conservation of momentum and kinetic energy. The width in t due to the uncertainty in for example E_1, is calculated as $\Delta t_{ME_1} = (\partial t_M/\partial E_1)\,\Delta E_1$ with similar expressions for the other resolution components in L_0, L_1 and θ. All resolution components other than the energy resolution are assumed to have a Gaussian peak shape in t and their widths are therefore added in quadrature

$$\Delta t_{MI}^2 = \left(\frac{\partial t_M}{\partial L_0}\,\Delta L_0\right)^2 + \left(\frac{\partial t_M}{\partial L_1}\,\Delta L_1\right)^2 + \left(\frac{\partial t_M}{\partial \theta}\,\Delta\theta\right)^2. \tag{24}$$

The resolution function for these components is represented as a Gaussian in t, with standard deviation Δt_{MI}.

On VESUVIO the final energy is determined using the filter difference method [Seeger 1985] where two measurements are taken, one with an absorbing filter placed between the sample and detector and one with the filter removed. The difference between these two measurements is determined by the probability that the filter absorbs a neutron. This is

$$A(E_1) = 1 - \exp[-N\,d\sigma(E_1)] \tag{25}$$

where N is the number of filter atoms / cm^3 the filter thickness is d and the filter total cross section is $\sigma(E_1)$. The effective detection probability $D(E_1) = A(E_1)\,\eta(E_1)$ is equal to the product of the filter absorption with the detector efficiency $\eta(E_1)$.

Filters with resonance absorption peaks, which absorb neutrons only in a narrow range of energies are used to determine E_1. Two filters have been used in the measurements, either a gold filter with $Nd = 7.35 \times 10^{19}\text{cm}^{-2}$, which defines a final energy $E_1 = 4908$ meV, with an approximately Lorentzian shape of half width at half maximum (HWHM) $\Delta E_1 \sim 140$ meV, or a uranium filter with $Nd = 1.46 \times 10^{20}$ cm^{-2}, $E_1 = 6671$ and an approximately

Gaussian shape with standard deviation $\Delta E_1 \sim 63$ meV. For the gold filter, the total resolution function $R_M(t)$ is therefore represented as a convolution of a Lorentzian of HWHM Δt_{ME_1} and a Gaussian of standard deviation Δt_{MI}, whereas with the U filter analyser $R_M(t)$ is represented as a Gaussian function of standard deviation $\sqrt{\Delta t_{ME_1}^2 + \Delta t_{MI}^2}$.

3. Monte Carlo simulations of VESUVIO

Monte Carlo (MC) simulations of VESUVIO data can be performed using the computer code DINSMS, which has been described previously [Mayers 2002]. The MC program follows individual neutron histories through the spectrometer and then bins them in t, according to the time they have taken to travel between moderator and detector. The input to the program is

(1) the mean flight paths $\bar{L}_0$, $\bar{L}_1$ and the scattering angle $\bar{\theta}$ for each detector, defined by the position of the sample and detector centres;

(2) the incident neutron spectrum intensity $I(E_0)$, the moderator size and the probability distribution of times the neutron leaves the moderator;

(3) the definition of the sample in terms of (a) the sample size, geometry and composition N_M and (b) $d^2\sigma_M/d\Omega\, dE_1$ for each mass - this is defined by Eqs. (15) and (16), and input values of σ_M and $J_M(y_M)$;

(4) the detector size, geometry and efficiency $\eta(E_1)$, calculated from the known thickness and doping levels of the ^6Li doped glass scintillator detectors, used on VESUVIO;

(5) the filter absorption $A(E_1)$ - this is calculated from the known filter thickness and the tabulated neutron resonance parameters for the gold and uranium resonances [Mughabgab 1984]. All final energies between 0.1 and 60 eV are included in the definition of the filter resolution, for both Au and U filters.

DINSMS exactly incorporates resolution effects, multiple scattering and sample attenuation and is used in this paper to assess systematic errors introduced into cross section ratios, derived by fitting data with Eq. (20).

The procedure followed is

(a) simulate a lead calibration measurement and use the standard instrument calibration routines [Fielding 2002] to determine the energy width ΔE_1 for each detector;

(b) generate simulated data sets for the sample and use the standard instrument fitting routines, incorporating Eqs. (20) and (21) and the fitted value of ΔE_1 from step (a), to determine the cross section ratios;

(c) compare the fitted cross section ratios σ_M/σ'_M with those input to the MC simulation.

Data is simulated for all detectors used in measurements and the simulated data is analysed in precisely the same way as real data. Simulations with "perfect resolution" can also be made by setting the moderator, sample and detector sizes set to $\sim 10^{-6}$ cm and all final energies to exactly 4.908 eV, so that L_0, L_1, θ and E_1 are precisely defined.

4. Correction for incident neutron intensity

4.1 Determination of $I(E_0)$

It has been stressed [Blostein 2001; Blostein 2003 (a); Blostein 2003 (b); Cowley 2003] that an accurate determination of the incident neutron intensity $I(E_0)$ is essential for the determination of cross section ratios on VESUVIO. $I(E_0)$ was measured using the VESUVIO incident beam monitor 1 (see Fig. 1). The incident energy of the neutrons is related to their time of flight measured in the monitor via

$$E_0(t) = \frac{1}{2}\, m \left(\frac{L}{t}\right)^2 . \tag{26}$$

The incident beam intensity is related to the measured monitor counts $C_m(t)\,dt$ via

$$I(E_0)\, dE_0 = \frac{C_m(t)}{\eta_m(E_0)} \frac{dt}{dE_0}\, dE_0 \tag{27}$$

where $\eta_m(E_0)$ is the monitor efficiency at energy E_0. An example of a measurement of $I(E_0)$, determined from (27), is shown in Fig. 2, together with a fit to the function

$$I(E_0) = \frac{B}{E^\gamma} \tag{28}$$

with B and γ as adjustable parameters.

The fit gave $\gamma = 0.90$ with statistical errors at the $\sim 10^{-5}$ level, in agreement with Monte Carlo calculations of the moderator performance [Taylor 1984], which predict that $\gamma = 0.9$. A large number of such data sets, collected over the past 10 years, all give consistent values $\gamma = 0.90 \pm 0.01$.

As a further test on the accuracy of the measurement of $I(E_0)$, a second procedure was used to determine γ. A uranium filter was cycled in the incident beam (see Fig. 1) and the difference between the counts with the foil in and foil out was calculated. An example of such a difference measurement is shown in Fig. 3.

The difference foil out - foil in is given by

$$\Delta(t)\, dt = I(E_0) \frac{dE_0}{dt}\, dt\, \eta_m(E_0)\, A(E_0) \tag{29}$$

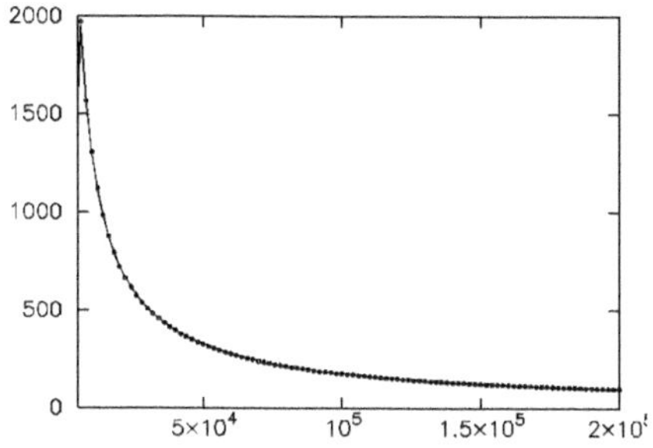

Figure 2. The points are the incident neutron spectrum after correction for the monitor efficiency . The solid line is a fit to Eq. (28).

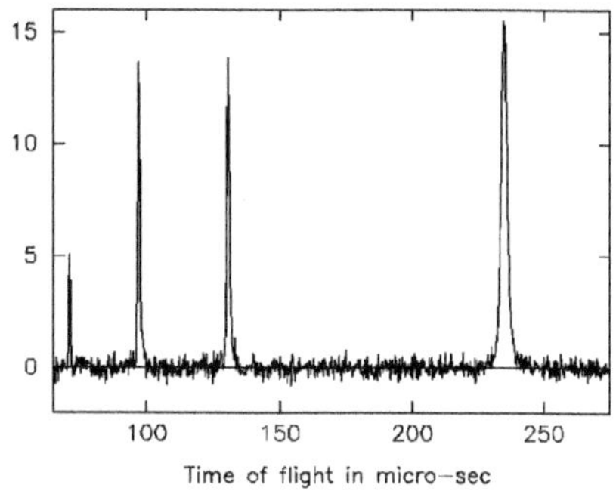

Figure 3. Difference between monitor spectra obtained with a uranium foil in the incident beam and with no foil in the beam. The four peaks correspond to resonances at 6.7, 22, 37 and 66 eV.

where $A(E_0)$ is the filter absorption, defined in Eq. (25). It follows from (29) that the sum of counts in time of flight, over the area of a single resonance peak, centred at E_R, is

$$\int_{t_1}^{t_2} \Delta(t)\, dt = \int I(E_0)\, \eta_m(E_0)\, A(E_0)\, dE_0 \approx I(E_R)\, \eta_m(E_R)\, \alpha_R \qquad (30)$$

where t_1 and t_2 are chosen to include only a single resonance peak and

$$\alpha_R = \int A(E_0)\, dE_0 \qquad (31)$$

is the total absorption corresponding to the resonance peak chosen. $I(E_0)$ and $\eta(E_0)$ can be taken outside the integral in (30) due to the narrowness of the peaks. The absorption factor α_R was also calculated from (25) and (31) for uranium resonances at 6.7 eV, 21 eV, 37 eV and 66 eV, using tabulated resonance parameters for uranium [Mughabgab 1984], which are known with very high accuracy. The quantity Nd was determined by weighing the filter, which has uniform thickness over a 10 by 10 cm area. The detector efficiency $\eta_m(E_R)$ can also be calculated and for the 0.5 mm thick glass beads used in standard ISIS monitors is $\propto 1/\sqrt{E_R}$ to $\sim$1%. Using calculated values of α_R and $\eta_m(E_R)$, $I(E_R)$ was calculated from S_R for each of the four resonance peaks. These values of $I(E_R)$ were then fitted to the function $I(E_0)$ defined in Eq. (27), to obtain γ. The advantage of this measurement over the direct determination of γ from the full monitor spectra, is that γ is determined only by the neutrons absorbed by the foil. Thus for example, any delayed neutron background, has little effect on γ values obtained in this way. The results obtained are shown in column 2 of Tab. 1. For comparison, values of γ obtained by direct fitting of the spectra as in Fig. 2 are also given. There is a small systematic difference in γ values obtained by the two different methods, but as will be shown in the following section, this difference is much too small to explain the anomalies observed.

Table 1. Lists measurements of γ obtained by the methods described in Sec. 4. The means with standard deviations are also given.

Run	*U Foil Transmission*	*Direct Measurement*
10539	0.93	0.89
10540	0.90	0.89
10541	0.94	0.89
10545	0.90	0.89
10550	0.96	0.90
Mean	0.93 ± 0.01	0.892 ± 0.02

4.2 Effects of errors in $I(E_0)$

The measurements described in Ref. [Chatzidimitriou-Dreismann 1997 (a)], were of the ratio of the H and D cross sections, σ_H/σ_D, in mixtures of D_2O and H_2O as a function of D_2O concentration x_D. In order to test how sensitive these measurements are to the accuracy with which γ is known, complete data sets were simulated by DINSMS, as described in Sec. 3, using perfect

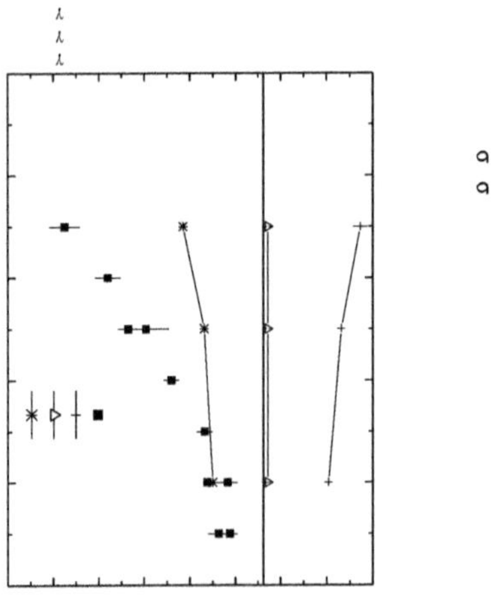

Figure 4 Crosses were obtained from fits to simulations for H_2O / D_2O mixtures with $\gamma = 0.8$, uptriangles are for $\gamma = 0.9$ and stars for $\gamma = 1.0$. Full squares show the results from VESUVIO data collected using an Au filter.

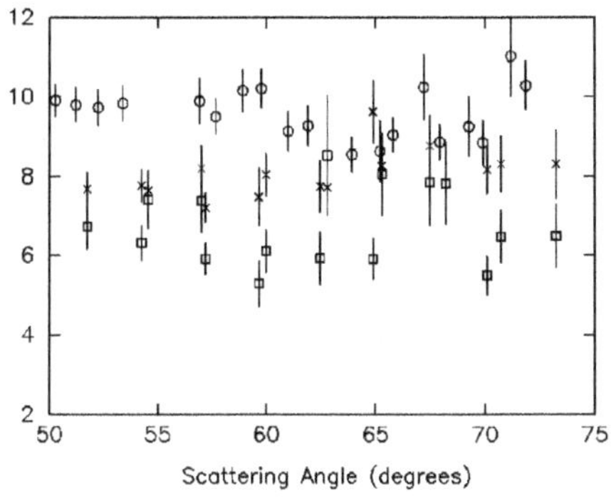

Figure 5 Values of σ_H/σ_D obtained from fitting individual time of flight spectra for the H_2O / D_2O data published in Ref. [Chatzidimitriou-Dreismann 1997 (a)] as a function of scattering angle. Circles are for $x_D = 0.9$, crosses for $x_D = 0.5$ and squares for $x_D = 0.3$. Within error at a given x_D, there is no angular dependence

resolution. Incident intensities of the form in (28), with $\gamma = 0.8$, $\gamma = 0.9$ and $\gamma = 1.0$ were input to three different simulations. These three simulations were then fitted using the standard data analysis routines, which assume $\gamma = 0.9$. Values of σ_H/σ_D were calculated from the fitted parameters, as an average over the angular range 50 - 75°, following exactly the same procedure used for real data. Fig. 4 shows values of σ_H/σ_D as a function of x_D, obtained from for thc three different values of γ input to the simulation. Also shown are the data measured in Ref. [Chatzidimitriou-Dreismann 1997 (a)].

With an input value to DINSMS of $\gamma = 0.9$, the fitted parameters were identical within statistical error to the values input to the simulation. As γ increases above this value, the values of σ_H/σ_D obtained from the fit decrease, but it is clear from Fig. 4 that γ would have to be ~1.1 to account for the large anomalies observed in the data. Similar comments apply to anomalies observed in other systems. This is well outside the errors in the measurement of γ given in Tab. 1.

4.3 Jacobians

Cowley [Cowley 2003] has also pointed out that there is a large Jacobian factor involved in any conversion between a VESUVIO time of flight scan in q, ω space and a constant q scan and suggested that any errors in the incorporation of this factor, could seriously affect the peak areas obtained from the fitting. However we note that the only Jacobian, dE_0/dt, involved in the fitting expression is well known [Windsor 1981]. Furthermore, neglecting resolution effects, if the IA is valid, Eq. (17) is an *exact* expression for the count rate as a function of t and is true for any point in q, ω space accessed by the spectrometer. Thus the exact line of the scan in q, ω space is immaterial, since in principle every scan will give the same values for the fitting parameters, whether it is at constant $\vec{q}$ or constant θ. For example fitting DINSMS simulations with perfect resolution to Eq. (17) recovers the cross section ratios input to the simulation to within a statistical error $\sim 1\%$, at any scattering angle as can be seen in the $\gamma = 0.9$ simulation shown in Fig. 4.

It should also be noted that any errors in either the assumed Jacobian dE_0/dt, or $I(E_0)$, would produce a consistent angular dependence in the cross section ratios of H to heavier atoms in all samples, which is not observed. The ratio σ_H/σ_D obtained from measurements on H_2O / D_2O mixtures is essentially independent of angle. This is illustrated in Fig. 5, where this ratio is shown as a function of angle for three different values of D_2O concentration x_D, measured in Ref. [Chatzidimitriou-Dreismann 1997 (a)].

VESUVIO measurements on NbD [Karlsson 1999] and polystyrene [Karlsson 2003 (b)], also show ratios of H to heavier atom cross sections which are independent of angle. In contrast measurements on Formvar [Chatzidimitriou-Dreismann 2003 (a)], shown in Fig. 6, give a marked fall off in the H to heavy atom cross section ratio as the scattering angle and hence q is increased. Measurements on NbH [Karlsson 1999] have an even stronger angular dependence. Thus no single γ value can explain anomalies observed in different samples.

5. The convolution approximation

5.1 Tests using experimental data and DINSMS simulations

The importance of assessing the inaccuracies introduced by the "convolution approximation" (CA), defined in Sec. 1.3, was discussed by Chatzidimitrou-Dreismann *et al.* [Chatzidimitriou-Dreismann 1997 (a)] in the first published measurement of cross section anomalies on VESUVIO in H_2O / D_2O mixtures. In order to eliminate the possibility that the CA could be responsible for the observed anomalies in this system, two independent checks were made.

(1) A DINSMS simulation of the measurements was made following the procedure outlined in Sec. 3. Although a simulation can never fully reproduce experimental data, it should provide a useful test, since the simulation does not incorporate the CA. An example of a simulation is compared with real data in Fig. 7 for a 50% mixture of H_2O / D_2O. The ratio σ_H/σ_D, obtained from fitting simulated data, is shown in Fig. 8, for both Au and U filters, as a function of D concentration x_D.

The solid line is the "conventional" cross section ratio, input to the simulation. It can be seen that with a Au analyser, the DINSMS simulation predicts that the CA introduces a systematic reduction of the observed ratio σ_H/σ_D by between 5 and 7%. However with a U analyser, fitting simulated data with Eq. (19) recovers the cross section ratios input to the simulations to within $\sim 1\%$, over the entire range of concentrations x_D.

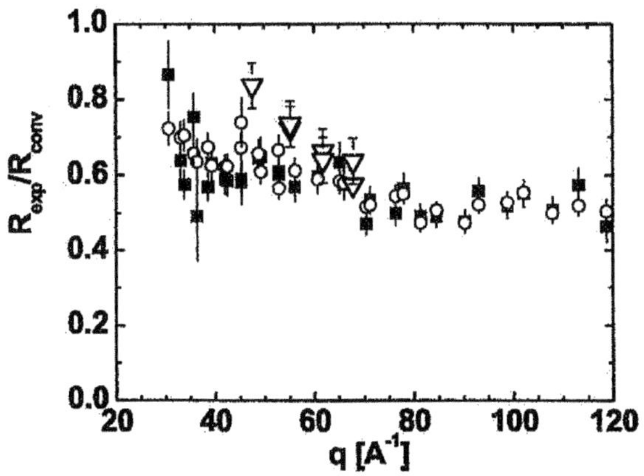

Figure 6. Anomalous reduction of scattering intensity from H of Formvar ($C_8H_{14}O_2$), as a function of momentum transfer q [Chatzidimitriou-Dreismann 2003 (a)]. The results are normalized by dividing the measured cross section ratios of H to $C + O$ by the "conventional" ratio 21.6. VESUVIO results for full squares: $0:1$ mm formvar foils and open circles: $0:2$ mm foils. Large open triangles: high-energy electron-proton scattering measurement from Formvar films of $50 - 100$ Å thickness, as described in Ref. [Chatzidimitriou-Dreismann 2003 (a)].

(2) The VESUVIO measurement was repeated with a U filter, where the MC simulation indicated that the effects of the CA were negligible. The results of the Au and U filter measurements are also shown in Fig. 8. It can be seen that the difference between the data with the two filters is consistent with the MC calculation, with an offset of the values obtained from the different filters, but that the slope of the experimental curve is unchanged. Furthermore the size of the observed anomaly is much larger than the systematic errors introduced by the CA for both Au and U filters.

5.2 Criticisms

It was concluded [Chatzidimitriou-Dreismann 1997 (a)] on the basis of these tests that the CA had no significant effect on the observed anomalous cross section ratios in H_2O / D_2O mixtures. Simulations of measurements on other systems also indicated that in all cases that the effects of the CA are small and do not significantly affect the results obtained. In contrast to this conclusion Blostein *et al.* have produced a series of papers [Blostein 2001; Blostein 2003 (a); Blostein 2003 (b)], arguing that the CA does introduce serious errors into cross section ratios and that this could explain the anomalies observed in H_2O / D_2O mixtures. They base this conclusion on their own numerical simulations of the experiments. Their simulations indicate that the systematic errors introduced by the CA are comparable with the observed anomalies, with both

Au and U filters. The results in Fig. 8 should be compared with the calculations of Blostein *et al.*, illustrated in Fig. 4 of Ref. [Blostein 2001]. For an Au analyser, we calculate a maximum reduction in σ_H/σ_D of $\sim 7\%$ at $x_D = 0.3$ compared with the 25% reduction calculated by Blostein *et al.*. With a U filter analyser, we calculate a maximum $\sim 1\%$ reduction in σ_H/σ_D compared with the $\sim$40% reduction calculated by Blostein *et al.*.

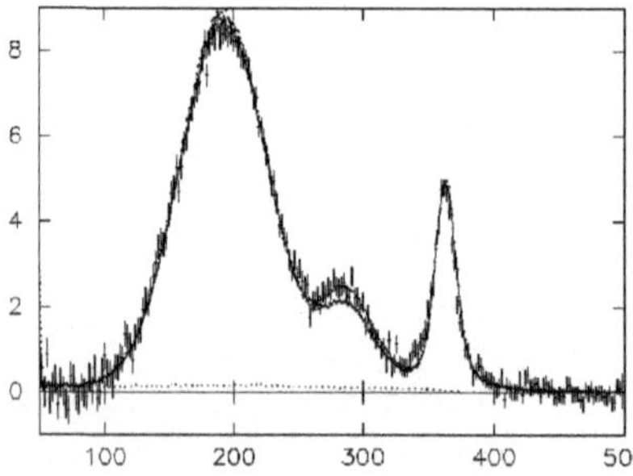

Figure 7. Data (points) and DINSMS simulations for a 0.17 mm thick sample of a 50% mixture of H_2O / D_2O. The solid line was calculated using cross sections determined from the fitting procedure. The second solid line shows a simulation using the tabulated values for cross sections and illustrates the larger D cross section obtained from the data. The dotted line shows the multiple scattering contribution. Three detectors with angles between 58° - 65° were summed to improve the statistical accuracy of the data.

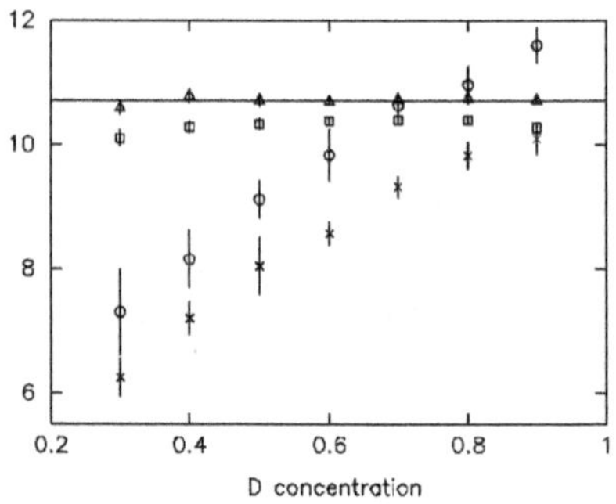

Figure 8. Ratio σ_H/σ_D in H_2O / D_2O mixtures as a function of D_2O concentration. Squares: MC simulation with a gold filter. Triangles: MC simulation with a uranium filter. The error bars shown are the standard deviation from the mean over the angular range 50° - 75°. The solid line is the ratio 10.7 expected from tabulated values of σ_H and σ_D, input to the MC simulation. Crosses: data [Chatzidimitriou-Dreismann 1997 (a)] taken with a Au filter and circles: data with a U filter.

One possible explanation of the discrepancy between our and the Blostein *et al.* calculations is that the latter incorporate only the energy resolution, whereas all resolution components are included in the DINSMS simulations. With a U filter for example, the main resolution component is due to the angular rather than the energy resolution[1]. However if all resolution widths other than the energy resolution are set to zero in DINSMS, this has little effect on the results obtained from the simulations. The systematic errors in the fit parameters are, somewhat surprisingly, increased by $\sim$ 1 - 2% when other resolution component are neglected, but this cannot explain the large discrepancy between the results of our calculation and those of Blostein *et al.*.

The result of Blostein *et al.* that the errors introduced by the CA with a uranium filter are comparable with those introduced with a gold filter, are particularly puzzling and are crucial to their argument that the anomalies are an artefact of the CA. In Fig. 9 we show the filter absorption $A(E_1)$ calculated from tabulated nuclear resonance parameters for Au (solid line) and U (dotted line) filters of the thicknesses used in the experiments. Fig. 10 shows measured data from lead, which is used to determine the energy resolution widths for the two filters [Fielding 2002]. Both calculation and measurements indicate that a factor 2 - 3 improvement in resolution is obtained with a U filter, compared with a Au filter. A rough estimate of the errors introduced by the instrument resolution function can be obtained by assuming that the intrinsic peak shape in time of flight of width w and the resolution function of width r, are both Gaussians. Adding the widths in quadrature thus implies that the measured width w_m is

$$w_m = \left(w^2 + r^2\right)^{1/2} \sim w\left(1 + \frac{r^2}{2w^2}\right) \tag{32}$$

With a Au filter, the energy resolution width is typically $\sim 1/4$ that of the intrinsic peak width of H and D peaks and this introduces a $\sim 3\%$ increase in the peak width due to resolution effects. With a U filter the resolution width is $\sim 1/10$ that of the peak width and introduces a $\sim 0.5\%$ increase in peak width.

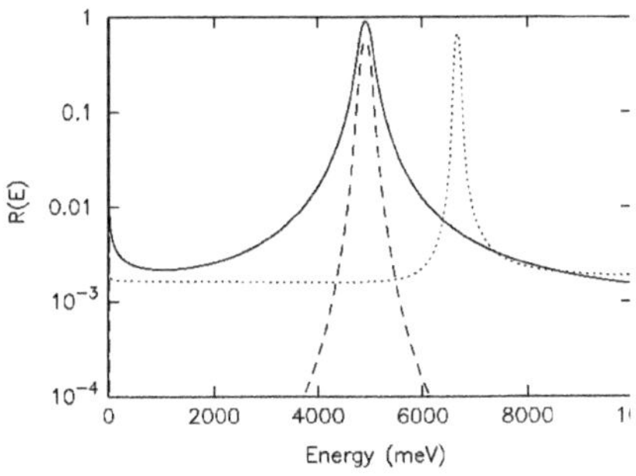

Figure 9. Calculated absorption $A(E_1)$ of the analyser filters as a function of E_1. Solid line: Au filter in single difference. Dotted line: U filter in single difference. Dashed line: Au filter in double difference.

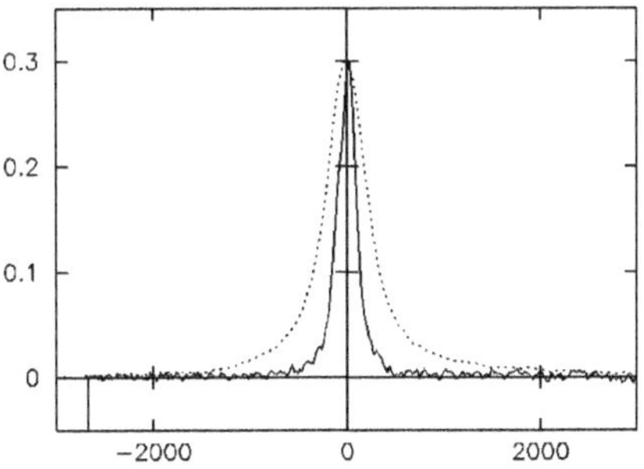

Figure 10. Measured data from lead for a gold filter (dotted line) and for a U filter (solid line) in momentum space. The measurement is resolution dominated and is used to determine the energy resolution function $D(E_1)$.

With the assumption that errors introduced by the CA, also scale roughly in the same way, one would expect that the effects of the CA would therefore be less by a factor $\sim$ 6 with U than Au. This agrees with the factor $\sim$ 7 decrease in errors indicated by the MC calculation shown in Fig. 8. In contrast Blostein *et al.* calculate that the error introduced by the CA for a uranium filter is almost twice that for a gold filter, despite the much reduced width of the uranium resolution function, a result which appears to contradict basic physical considerations.

Blostein *et al.* have also argued that the large errors they calculate in fitted cross section ratios, are introduced by the wings in the filter absorption $A(E_1)$, which are not properly accounted for by the approximation of the resolution function as a Lorentzian, or by the approximation of incorporating the resolution function as a convolution. However the wings in the Au resolution function are much more significant than those in the U resolution function. Thus, one would expect that if long wings in the resolution function were responsible for the reduced values of σ_H/σ_D, this effect should be larger with a Au filter than a U filter, whereas the Blostein *et al.* calculation indicates that it is smaller.

5.3 Double difference measurements

A further experimental check on the possibility that long wings in $A(E_1)$ could affect the cross section ratios derived from fitting, has recently become possible with the installation of the "double difference" (DD) technique [Seeger 1985] on VESUVIO. This consists of taking three measurements, with no filter, a filter of thickness d_1 and absorption $A_1(E_1)$ and a filter of thickness d_2 and absorption $A_2(E_1)$. The "double difference" of the three measurements is

$$R_{DD}(E_1) = A_1(E_1) - \frac{d_1}{d_2} A_2(E_1). \tag{33}$$

The DD technique relies upon the fact that when $\sigma(E)$ is small,

$$A_1(E_1) = 1 - \exp[-N\, d_1\, \sigma(E_1)] \sim N\, d_1\, \sigma(E_1) \tag{34}$$

with a similar expression for $A_2(E_1)$. Thus, when $\sigma(E)$ is small $R_{DD}(E) = 0$ and the wings of the function $A_1(E_1)$in single difference (SD) are removed, whatever their functional form. This is illustrated in Fig. 10 where the calculated energy resolution function $R_{DD}(E_1)$ for a Au analyser is also shown.

Figs. 11 and 12 show recent data collected from a 50:50 mixture of H_2O / D_2O using the SD and DD methods, respectively.

The improved resolution of the DD data is most obvious in the narrowing of the width of the peak at $\sim$ 370 μs, which is a combination of scattering from

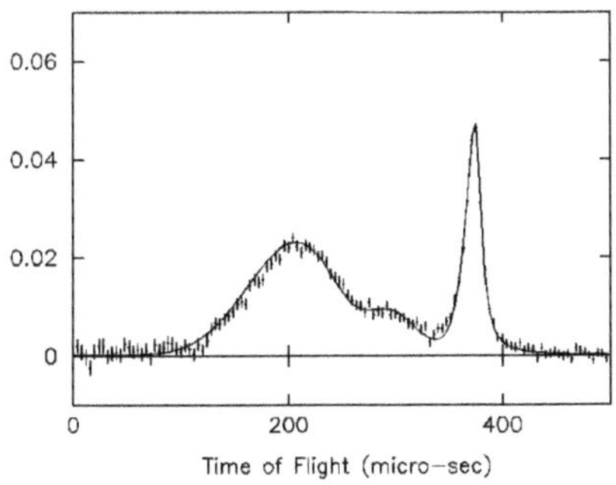

Figure 11. Sum of data and fits from 8 detectors in the angular range 53° - 68°, for a 50:50 mixture of H_2O / D_2O, using the single difference technique.

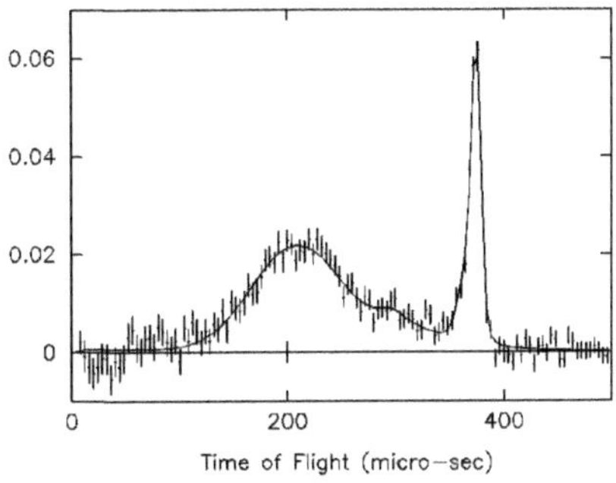

Figure 12. Sum of data and fits from the same 8 detectors in the angular range 53° - 68°, for a 50 : 50 mixture of H_2O / D_2O, using the double difference technique.

the O atom and the niobium container. The mean of σ_H/σ_D over 16 detectors in the angular range 50° - 80° was 8.20 ± 0.09 for the SD data and 7.70 ± 0.20 for the DD data. Both results are in good agreement with the results given in Ref. [Chatzidimitriou-Dreismann 1997 (a)] at the same concentration. Fig. 13 shows the results of fitting to the SD and DD data for Formvar $(C_8H_{14}O_2)^2$.

Again there is a slight increase in the anomalies when the DD method is used. This is the opposite trend to that which would be expected, if the anomalies were due to the method used to incorporate the resolution function.

To summarise, the calculations of Blostein *et al.* disagree with our MC simulations. Their results are also counter-intuitive, with a factor ~ 2 increase in the errors introduced by the CA, when the resolution width is decreased by a factor 2 to 3. Whereas their simulations have not been tested against real data, it has been shown that DINSMS accurately reproduces both single and multiple scattering in VESUVIO data for atomic masses ranging between 1 and 207 [Mayers 2002]. We also stress that DINSMS incorporates the energy resolution of VESUVIO accurately. To within 1 - 2%, simulated lead calibration measurements give the same widths for the energy resolution function as real calibrations, for both Au and U filters. The fact that the anomalies observed on VESUVIO are essentially the same for energy resolution functions varying in width by a factor 2 - 3, whether or not significant wings are present, also provides very strong experimental evidence that the effects of the convolution approximation are small and cannot explain the observed anomalies.

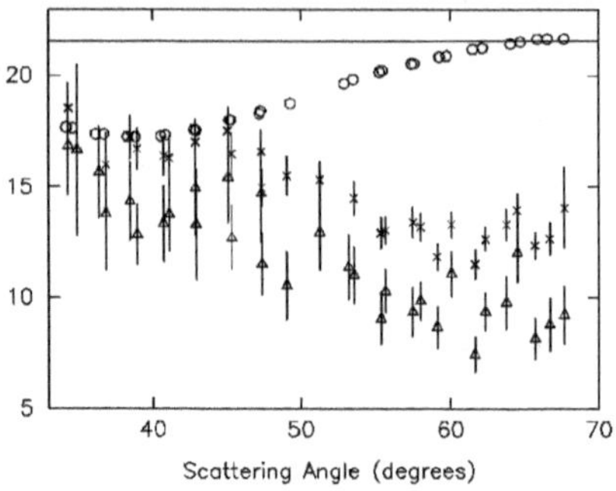

Figure 13. Data for Formvar [Chatzidimitriou-Dreismann 2003 (a)]. The solid line is the calculated ratio of $H/(C+O) = 21.6$. Crosses: single difference results and triangles: double difference results. Circles: DINSMS simulation described in Sec. 8.

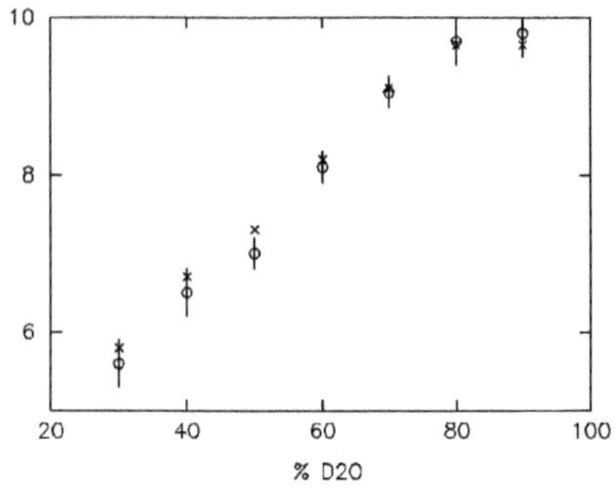

Figure 14. Ratio σ_H/σ_D with (crosses) and without (circles) a correction for multiple scattering, for data taken with a gold analyser filter.

6. Sample size effects

A series of measurements on H_2O / D_2O mixtures with $x_D = 0.5$, listed in Tab. 2, with a variety of sample geometries and scattering intensities varying by a factor $\sim$ 5, have given the same ratio σ_H/σ_D, within error. Similarly in NbH [Karlsson 1999] and Formvar [Chatzidimitriou-Dreismann 2003 (a)] (see Fig. 6), varying the sample thickness by a factor two made no difference to the cross section ratios obtained.

Table 2. Ratio σ_H/σ_D for 50:50 H_2O / D_2O mixtures measures with different materials for the sample can and different scattering geometries and scattering powers. The scattering power was determined by comparing the sample scattering at 60°, with that from a 1mm thick lead sample.

Date	*Can Material*	*Geometry*	*Thickness*	*Scattering Power*	σ_H/σ_D
March 1995	Al	Flat	0.5 mm	6.5 %	8.0 ± 0.5
July 1995	V	Flat	0.2 mm	3.3 %	7.6 ± 0.2
May 1997	Al	Flat	0.5 mm	15.0 %	7.4 ± 0.2
Aug 1997	V	Flat	0.2 mm	7.5 %	7.8 ± 0.5
June 1998	Al	Annular	0.5 mm	12.0 %	7.5 ± 0.5
June 1998	Nb	Annular	0.5 mm	12.0 %	7.0 ± 0.4
July 2003	Nb	Annular	0.5 mm	8.4 %	8.2 ± 0.1

The fact that the results are independent of the sample size is a strong evidence that sample attenuation effects and multiple scattering plays no significant role in the observed anomalies. Multiple scattering corrections can also be calculated using DINSMS, as described in Ref. [Mayers 2002]. This was done for the thickest sample used in the H_2O / D_2O experiments (0.5 mm) and the ratio σ_H/σ_D, obtained from fitting data, with and without a correction for multiple scattering, are shown in Fig. 14. The form of the multiple scattering contribution is shown in Fig. 7. It can be seen that calculated multiple scattering effects are essentially negligible. Another possible cause of the reduction in the intensity of H peaks, observed in data as the scattering angle is increased in NbH [Karlsson 1999] and Formvar [Chatzidimitriou-Dreismann 2003 (a)], is the presence of dead time effects in the VESUVIO detectors. This would have the effect of making the detector efficiency $\eta(E_1)$ a function of time of flight t, with lower $\eta(E_1)$ at short t, where the count rate is largest. Since the H peak moves to lower t as the scattering angle increases, this would introduce a reduction in intensity of the H peak with increasing scattering angle, similar to that observed in Formvar and metal hydride systems. A number of independent checks on the electronic counting chain, detailed in Ref. [Karlsson 2003 (b)], have been made to eliminate this possibility. Most conclusively, the fact that the effects observed on VESUVIO are essentially independent of count rates varying by a factors of up to ~ 5 demonstrates that dead time effects have no significant influence on the results.

7. Deviations from the IA

The corrections to the IA for the finite q of measurement, known as "final state effects" (FSE), have been extensively discussed in the literature [Sears 1984; Mayers 1989; Mayers 1990; Glyde 1994; Evans 1996]. The method of Sears [Sears 1984] is incorporated in standard VESUVIO data analysis routines. He showed that the effects of finite $\vec{q}$ and ω can be accounted for by expressing the neutron Compton profile $J_M(y_M)$ as,

$$\begin{aligned} J(y) = \quad & J_{IA}(y) - \left(\frac{M\langle\nabla^2 V\rangle}{36\hbar^2 q}\right)\frac{\partial^3 J_{IA}(y)}{\partial y^3} \\ & + \left(\frac{M^2\langle F^2\rangle}{72\hbar^4 q^2}\right)\frac{\partial^4 J_{IA}(y)}{\partial y^4} - \dots \end{aligned} \tag{35}$$

where $J_{IA}(y)$ is the IA result. $\langle\nabla^2 V\rangle$ is the mean value of the Laplacian of the potential energy of the atom and F is the force on the atom. At the q values observed on VESUVIO, it is only necessary to include the first correction term in (35) in the analysis. It is assumed that FSE in H and D atoms are identical to those observed in an isotropic harmonic potential, implying that

$\langle \nabla^2 V \rangle = 12\hbar^2 w_M^4 / M$, where w_M is the Gaussian width defined in Eq. (21). Within this approximation FSE can be incorporated in to the fitting expression for $J_M(y_M)$, without increasing the number of fitting parameters. It has been shown previously [Evans 1996] that this parameterisation of FSE is broadly consistent with deviations from the IA observed on VESUVIO.

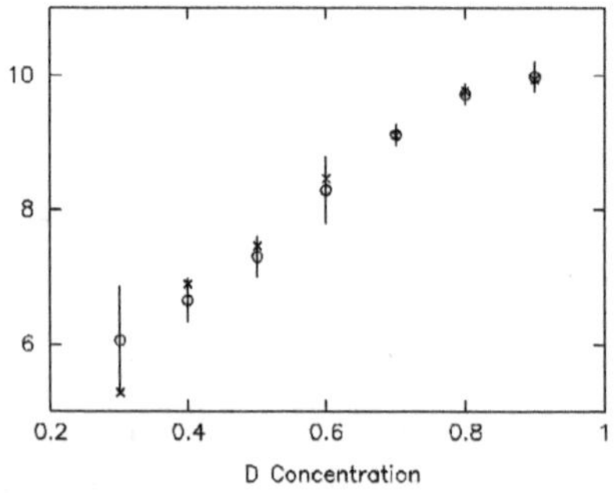

Figure 15. σ_H/σ_D: circles include a correction for deviations from the IA, while crosses do not.

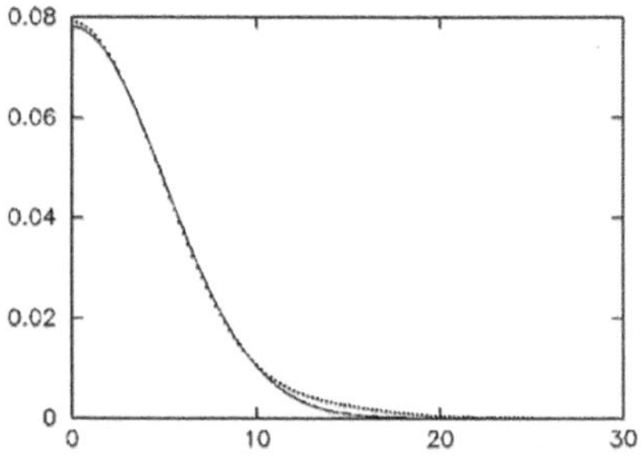

Figure 16. Dots: J(y) determined for H in Formvar. Solid line: the best Gaussian fit to J(y).

The effects of FSE were evaluated by fitting time of flight data from the different mixtures of D_2O and H_2O, described in Ref. [Chatzidimitriou-Dreismann 1997 (a)], with and without an FSE correction of the form given in Eq. (35), in the fitting expression. The mean values of σ_H/σ_D over the angular range 50 - 75°, obtained by the two procedures, are shown in Fig. 15. It can be seen that differences are small and that the trend to smaller values of σ_H/σ_D with decreasing x_D is essentially the same. This indicates that FSE on VESUVIO have little effect on the measured ratio σ_H/σ_D in this system. Similar comments apply to other systems studied.

8. Peak shapes

The approximation that $J(y)$ for all masses is a Gaussian function has also been questioned [Cowley 2003]. Recently it has become possible to fit the peak shapes of VESUVIO data exactly and hence to measure $J(y)$, without any assumptions about the peak shape [Reiter 2002]. In Fig. 16 we show the $J(y)$ derived for the H atom in Formvar using the procedure given in Ref. [Reiter 2002], together with the best Gaussian fit.

It can be seen that there are wings in $J(y)$, which are not well described by a Gaussian function. An MC simulation of Formvar data was made with

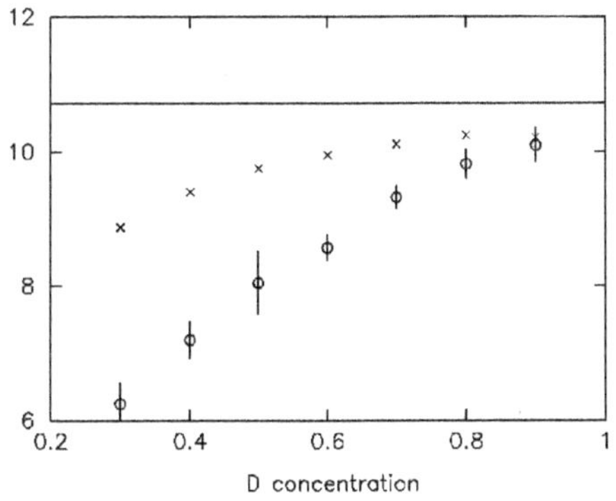

Figure 17 Crosses show σ_H/σ_D obtained by fitting simulated data with non-Gaussian peak shapes for H and D, with the standard fitting programs, which assume that the peak shapes are Gaussian. The circles show values of σ_H/σ_D obtained by fitting experimental data.

the measured $J(y)$ for H in Fig. 16, input to the simulation. The C and O peaks were represented as Gaussians. The latter assumption should be a good one since the widths of these peaks are resolution dominated. The results of analysing the simulated data, with the standard fitting programs, assuming Gaussian peak shapes for all atoms are shown in Fig. 13. It can be seen that the simulation gives an increase in σ_H/σ_C with increasing scattering angle. This can be understood in terms of the decreasing overlap between the H and heavy atom peaks as the scattering angle is increased and is the opposite trend to that observed in the data. Simulations for NbH [Karlsson 1999] gave similar results to those for Formvar with an increase of $\sim 7\%$ in the ratio σ_H/σ_{Nb} as the angle was increased between $30°$ and $80°$, compared with the $\sim 40\%$ decrease observed in real data. A similar procedure was followed to simulate data for H_2O / D_2O mixtures as a function of x_D, using measured non-Gaussian $J(y)$'s for the H and D peak shapes. The results of fitting the simulated data are shown in Fig. 17. It can be seen that in this case, the non-Gaussian peak shapes do reduce the fitted ratio σ_H/σ_D, in a similar way to that observed in data, although only by $\sim 30\%$ of the observed value.

Summary

We have demonstrated that the following considerations cannot explain the observed cross section anomalies.

Inaccuracies in the incident intensity, as suggested by Cowley [Cowley 2003]. The Jacobian dE_0/dt is a standard textbook expression [Windsor 1981] and is simply calculated. The incident beam spectrum $I(E_0)$ has been calculated [Taylor 1984] and measured in two different ways, with the same result, within $\sim 4\%$, whereas errors in the measured $I(E_0)$ would have to be at least $\sim 20\%$ to explain the results obtained. The fact that a number of systems give cross section ratios that are independent of angle, also suggests very strongly that the anomalous cross sections cannot be explained in this way.

The way in which the energy resolution function is incorporated into the data analysis as suggested by Blostein *et al.* [Blostein 2001; Blostein 2003 (a); Blostein 2003 (b)]. Since the results are essentially independent of a wide range of different resolution functions, with or without long wings, it is clear that this suggestion cannot account for the anomalies.

Since the results obtained are independent of sample geometry and scattering power, sample attenuation, multiple scattering and detector dead time effects can all be eliminated as a possible cause of the observed anomalies.

The agreement obtained between neutron and electron-proton scattering [Chatzidimitriou-Dreismann 2003 (a)] measurements also provides strong experimental evidence that the three effects above cannot explain the observed anomalies. The remaining assumptions of the analysis; that the scattering can be described within the Impulse Approximation and that neutron Compton profiles are Gaussian, are shared by the analysis of electron-proton scattering and neutron scattering data and therefore cannot be eliminated quite so conclusively as a possible cause of the observed anomalies. However, simulations using the measured peak shapes indicate that the effect produced by the assumption of Gaussian peak shapes is too small to account for the observations in H_2O / D_2O systems. Furthermore, the trend to lower cross sections for H with increasing q, observed in NbH and Formvar, is masked, rather than enhanced by the effect of peak shapes. Thus, we conclude that there is strong evidence that the observed anomalies are due to a breakdown of standard neutron scattering theory at eV energy transfers and are not an artefact of the VESUVIO instrument or the data analysis procedures employed.

Notes

1. See Fig. 5 a of Ref. [Fielding 2002]
2. Unpublished data from Ref. [Chatzidimitriou-Dreismann 2003 (a)]

QUANTUM ENTANGLEMENT AND DECOHERENCE DUE TO COUPLING OF PROTONS TO ELECTRONIC ENVIRONMENT

T. Abdul-Redah,[1] M. Krzystyniak[2] and C. A. Chatzidimitriou-Dreismann[2]

[1] *ISIS Facility, Rutherford Appleton Laboratory, Chilton / Didcot, OX11 0QX, UK*
t.abdul-redah@rl.ac.uk

[2] *Institut für Chemie, Stranski Laboratorium, Technische Universität Berlin, Straße des 17. Juni 112, D-10623 Berlin, Germany*
krzystyniak@iwan.chem.tu-berlin.de, dreismann@chem.tu-berlin.de

Abstract In recent years many scattering experiments have been performed on hydrogen containing materials which showed a scattering behavior different from the expectation according to standard theories. These scattering anomalies are attributed to the existence of short lived protonic quantum entanglement (QE) and decoherence. It was suggested that also electronic degrees of freedom might be involved in the anomalous scattering behavior. Here, the influence of the electronic structure surrounding the H atoms in different materials on the neutron scattering cross section of hydrogen is investigated. Experimental neutron Compton scattering results of H_2O / D_2O mixtures and LiH at different temperatures are presented. Also LaH_2 which exhibits metallic properties and LaH_3 which is an insulator - *i.e.*, both materials have different electronic structures - are investigated at room temperature. It is found that different electronic environments lead to different scattering behavior, thus strongly supporting the supposition that the electronic degrees of freedom are engaged in the protonic attosecond QE and the decoherence process. Furthermore, we present very recent results of experimental tests of the data analysis procedure which have been criticized recently. We show that the data analysis procedure is correct and criticisms are irrelevant for the experimental setup used in our NCS experiments.

Keywords: Quantum entanglement, attosecond physics, Neutron Compton scattering

Introduction

The research on quantum entanglement (QE) and decoherence, usually focuses on experiments and theory of quantum optics and optical traps involving single or just a few atoms. It is heavily dominated by the potential applica-

V.M. Akulin et al. (eds.), Decoherence, Entanglement and Information Protection in Complex Quantum Systems, 469–482.

bility of QE for the realization of quantum computers and for quantum cryptography. In experiments the involved particles are carefully separated from their environment to minimize decoherence processes in order to preserve QE for a reasonably long time. In condensed phases containing a large number of particles, interactions of a particle with its neighbours can also lead to quantum entanglement. Such effects are theoretically expected to be extremely short-lived, due to environmental disturbances leading to decoherence. Therefore, it has been widely believed that they cannot be experimentally detected. However, recent theoretical work [Chatzidimitriou-Dreismann 1991; Chatzidimitriou-Dreismann 1997 (b)] has suggested that QE may indeed play a significant role in condensed matter physics and that QE features become visible in, *e.g.*, a scattering experiment if the scattering time is short enough. This assumption has been directly confirmed first on H_2O / D_2O mixtures using neutron Compton scattering (NCS) [Chatzidimitriou-Dreismann 1997 (a)] and later on other materials like metallic hydrides [Karlsson 1999; Abdul-Redah 2000; Karlsson 2003 (b)] and organic compounds [Chatzidimitriou-Dreismann 2000 (b); Chatzidimitriou-Dreismann 2001; Chatzidimitriou-Dreismann 2002 (a)]. The effect manifests itself by the reduction of the scattering cross section (SCS) of protons in the system. The NCS method is particularly suitable for this purpose because its time scale lies in the sub-femtosecond range.

The original experiments on H_2O / D_2O mixtures [Chatzidimitriou-Dreismann 1995; Chatzidimitriou-Dreismann 1997 (a)] using laser light Raman scattering and NCS were driven by the assumption that only H, H or D, D entanglement is possible due to the superselection rules of masses and spins. Indeed, the striking experimental result was that the decrease of the SCS ratio σ_H/σ_D is dependent on the D mole fraction of the mixture. Therefore, it was believed that the origin of this effect was the exchange correlation of standard quantum mechanics. However, the D mole fraction independence of the protonic SCS found in later experiments on organic compounds [Chatzidimitriou-Dreismann 2002 (a)], indicated that the anomalous effect may be of intramolecular origin in certain cases. These experimental results suggested that the electronic environment may play a significant role for the QE effect and for decoherence.

In this contribution, we present our most recent results on the QE and decoherence effect in proton containing materials and the dependence of the anomalous proton-neutron SCS on the electronic environment. Concretely, we investigated the temperature dependence of H_2O / D_2O mixtures and LiH. Already note that the different electronic environments of the hydrogen atoms involved in H_2O / D_2O on the one hand and in LiH on the other seem to lead to different temperature dependence behavior.

Furthermore, we show results of the anomalous neutron SCS of the protons in an insulator (LaH_3) and in a material having metallic (LaH_2) properties, respectively, at room temperature. We also show some very recent experimental tests which underline the fact that the found effects are genuine and can not be explained by trivial instrumental or data analysis failures having been suggested recently.

1. Experiments

Experiments were performed using the electron-volt spectrometer (VESUVIO) of the ISIS spallation source (Rutherford Appleton Laboratory, UK). ISIS is the the most powerful neutron spallation source available today and VESUVIO is a unique instrument which exploits the high intensity of high energy neutrons produced at ISIS. VESUVIO (see Fig. 1) is a so-called inverted geometry spectrometer [Seeger 1985; Mayers 1994] which has been designed to measure directly atomic momentum distributions and single particle mean kinetic energies.

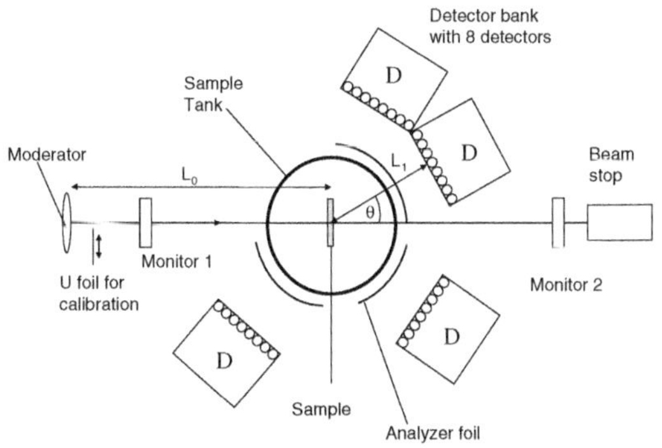

Figure 1 A schematic representation of the VESUVIO spectrometer at ISIS.

In VESUVIO the sample is exposed to a polychromatic neutron beam. In order to analyse the energy of the scattered neutrons, a nuclear resonance difference technique is used, which consists in the following. A gold foil situated between the sample and the detectors strongly absorbs neutrons over a narrow range of energies, centered at a specific Lorentzian shaped nuclear neutron absorption resonance (4908 ± 130 meV). Two measurements are taken: one with the foil between sample and detector and one without foil. The difference between these two spectra gives the final time-of-flight (TOF) spectrum. An example of such a spectrum of LaH_3 recorded in the forward scattering direction is depicted in Fig. 2. The numerical value of TOF refers to the time the neutrons need from the neutron moderator via the sample to the detector. At present four banks of eight detectors each are available plus 44 detectors in the backscattering direction. The backward scattering detectors are used in order to extract the combined scattering signal of, *e.g.*, Li or La from the one of Al can containing sample.

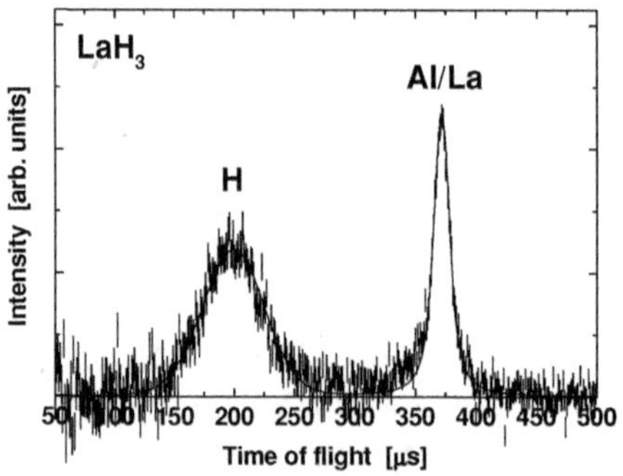

Figure 2 Time of flight difference spectrum of foil in and foil out spectrum of LaH_3.

In contrast to the work on the static structure determination involving X-ray or neutron scattering in which the scattering is elastic and coherent thus giving Bragg peaks, the scattering process in NCS is treated within the *incoherent approximation* [Sears 1984; Watson 1996]. The incoherent approximation - which means that interatomic interference effects are safely neglected and that each atom scatters independently from the other - is valid here because the momentum transfer q (10 - 120 Å^{-1}) or the wavelength of the incoming neutrons $\lambda < 0.1$ Å fulfills the requirement for incoherence, *i.e.*, $q \gg 2\pi/d$ or $\lambda \ll d$, where d is the nearest neighbor distance.

The basic quantity that is measured in a neutron scattering experiment is the partial differential cross section $(d^2\sigma/d\Omega\ dE_1)$. This quantity gives the fraction of neutrons of incident energy E_0 scattered with an energy between E_1 and $E_1 + dE_1$ into a small solid angle $d\Omega$. Applying the *Fermi's Golden Rule* for the transition of the combined system consisting of the scattering target and the neutron from their initial into their final states, the partial differential cross section can be written as [Squires 1996]:

$$\left(\frac{d^2\sigma}{d\Omega\, dE_1}\right)_{\lambda_0\to\lambda_1} = \frac{k_1}{k_0}\frac{m}{2\pi\,\hbar^2}\left|\left\langle \vec{k}_1\,\lambda_1\right| V \left|\vec{k}_0\,\lambda_0\right\rangle\right|^2 \delta(E_{\lambda_0} - E_{\lambda_1} + \omega). \quad (1)$$

in which $|\lambda\rangle$ represents a many body eigenstate of the Hamiltonian of the system. $\vec{k}$ denotes the state of the neutron, m is the neutron mass, $\hbar$ is Planck's constant divided by 2π and the δ-function expresses the energy conservation during the scattering process. Indices "0" and "1" refer to quantities before and after collision, respectively. ω is the energy transfer.

Under the prevalent experimental conditions of NCS, *i.e.*, high energy and momentum transfers, time correlations in the motion of a scatterer can be neglected [Sears 1984; Watson 1996], since the characteristic time of the neutron-scatterer interaction (*i.e.*, the "scattering time") is very short; see below. Thus the dynamic structure factor is accurately described by the *impulse approximation* (IA), *i.e.*, [Sears 1984; Watson 1996]

$$S(q,\, \omega) = \left\langle \delta \left(\omega - \omega_r - \frac{\vec{q}.\vec{p}}{M} \right) \right\rangle = \int n\,(\vec{p})\; \delta \left(\omega - \omega_r - \frac{\vec{q}.\vec{p}}{M} \right)\, d\vec{p}. \quad (2)$$

Here, $\langle \ldots \rangle = \sum p_n \, \langle n|...|n \rangle$ is the appropriate combined quantal and thermodynamic average (over the classical probabilities p_n) related with the condensed matter system. M and $n(\vec{p})$ are the mass and momentum distribution of the scattering nucleus, respectively, and $\omega_r = q^2/2M$ is the recoil energy. For convenience, $\hbar = 1$. Eq. (2) is of central importance in most NCS experiments, since it relates the SCS directly to the momentum distribution. Furthermore, $n(\vec{p})$ is related to the nuclear wave function by Fourier transform and therefore, to the spatial localization of the nucleus. It takes into account the fact that, if the scattering nucleus has a momentum distribution in its ground state, the δ-function centered at ω_r will be Doppler broadened.

Introducing a scaling parameter [Sears 1984] $y = M\ (\omega - \omega_r)/q$ and using the identity $\delta(a\ x) = \delta(x)/a$, the dynamic structure factor (Eq. (2)) can be rewritten as $S(q,\, \omega) = M\ J(y)/q$. For an isotropic harmonically bound nucleus the "neutron Compton profile" $J(y)$ assumes the Gaussian form

$$J(y) = \frac{1}{\sqrt{2\pi\, \sigma_y^2}}\ \exp\left(-\frac{y^2}{2\, \sigma_y^2} \right), \quad (3)$$

which is centered at the recoil energy ω_r; σ_y is the standard deviation of the momentum distribution. When different atomic masses M_i are present in the sample, the scattered intensity consists of different peaks centered at different recoil energies $\omega_{r,\, i}$, which correspond to different TOF values in the spectrum. For each nucleus of mass M the intensity of which is to be analyzed, TOF spectra are fitted with the aid of a Gaussian (Eq. (3)) convoluted with the instrument resolution function. The number of neutrons C scattered by nuclei (of an amorphous, non-magnetic, homogenous and isotropic material) is then proportional to the total SCS σ and to the dynamic structure factor $S(q,\, \omega)$,

$$C_m(t) = \frac{E_0\ I(E_0)}{q} \sum_M B\ b_M^2\ N_M\ A_M\ M\ J_M(y_M) \otimes R_M(t), \quad (4)$$

where B contains instrument parameters and b_M is the scattering length.

In our data analysis only the amplitude A_M is a fitting parameter. Fig. 2 shows an example of a measured TOF spectrum (including error bars) and its fitted spectrum (full line). The area A_i under the Gaussian is determined incorporating the correction for the intrinsic angle (or momentum transfer) dependence of the scattering intensity; for technical details, *cf.* Ref. [Chatzidimitriou-Dreismann 2000 (b)]. The data analysis procedure incorporates also the well known transformation from the "free atom" to the "bound atom" cross section, *i.e.*, $\sigma_{bound} = \sigma_{free}(1+m/M)^2$ (m: neutron mass), which is relevant for light nuclei, like H in order to facilitate the comparison between experimental NCS results and tabulated cross section values [Lovesey 1984]. According to conventional theory, the thus determined peak areas A_i are proportional to the product of the total SCS σ_i and the number density N_i of atoms of mass M_i present in the sample, *i.e.*, $A_i \sim \sigma_i N_i$. Therefore, if the number densities of two atoms with different masses are known, then the following equation holds strictly within conventional theory [Chatzidimitriou-Dreismann 1997 (a)]:

$$\frac{A_H}{A_X} = \frac{N_H \, \sigma_H}{N_X \, \sigma_X}, \tag{5}$$

where N_H/N_X is the ratio of the particle number densities of H and X, precisely known through sample preparation or chemical formulae. σ_H and σ_X are the total neutron SCSs of H and an atom X. Thus, since the conventionally expected values of σ_H and σ_X are given in standard tables [Lovesey 1984], the validity of Eq. (5) is immediately subject to experimental test.

Because of the high transfers of energy and momentum applied on VESUVIO, the recoil peaks of H and other heavier atoms (like D, Li, or La) in the TOF spectra are well separated for a wide range of scattering angles.

Certain hydrogenated materials, *e.g.*, liquids (water) or powders (LiH), necessitate the use of sample cans (aluminum). Since the experiment allows only to determine relative cross sections, σ_H/σ_X, X being the reference signal in the spectrum, it is necessary to nearly resolve the X contribution (Li or La) from any other scattering contribution, for example of the can material. Therefore, it is important to use can material which contains nuclei with significantly different masses than the mass of the nucleus under consideration.

Usually, it is impossible to resolve the heavy atom peak from the peak of the can material in the forward scattering spectra. This is due to the fact that the transfers of energy and momentum in forward scattering are not large enough to achieve a satisfactory peak separation. However, choosing the appropriate can material, it is indeed possible to achieve a satisfactory peak separation if the backscattering detectors are used, *e.g.* putting LiH in an Al can provides a heavy peak overlap in the forward scattering direction, whereas the back scattering spectra show two peaks which can be analyzed independently. This facilitates the determination of σ_{Li}/σ_{Al} which can be used to extract the Li contribution from the joint Li / Al peak of the forward scattering spectra. This in turn makes possible the determination of σ_H/σ_{Li}.

Another crucial feature of the NCS method is the small value of the characteristic time τ_{scatt} of the neutron-nucleus interaction, the so-called "scattering time". It follows from the theory of NCS that each scattering angle θ corresponds to a specific momentum transfer (from the neutron to the struck nucleus, *e.g.* a proton) and to an associated value of τ_{scatt}. According to Sears [Sears 1984] and Watson [Watson 1996], the scattering time may be defined by

$$q(\theta)\ v_0\ \tau_{scatt} \approx 1 \tag{6}$$

where $q(\theta)$ is the momentum transfer depending on the detector angle θ and v_0 is the root-mean-square velocity of the nucleus before collision (*i.e.*, in its initial state). For our purposes, the NCS technique is particularly suitable, because the scattering time - *i.e.*, the interaction time of the (epithermal) neutrons with the scattering nuclei - is sufficiently short, being in the sub-femtosecond time scale. This is a consequence of the large energy (3 - 150 eV) and momentum transfers (10 - 120 Å^{-1}), applied on VESUVIO. The range of scattering times associated with the scattering angles used here is $\tau_{scatt} = 0.05 \ldots 0.6$ fs, for the neutrons scattered on H atoms.

2. Results

2.1 Temperature dependence

The original experiments on H_2O / D_2O mixtures showed a strong decrease of σ_H/σ_D with respect to the value expected to be 10.7. We proceeded by investigating the influence of the environment on this effect. As pointed out already in the introduction, within the framework of modern quantum theory, the environment may be responsible both for the creation and for the destruction of QE. Therefore, if the anomalous effect found in, *e.g.*, H_2O / D_2O arises from QE of protonic degrees of freedom, then a controlled change of the protonic environment should change the scattering behavior. For this purpose we measured the NCS of H_2O / D_2O at different temperatures. The samples were put in flat Al cans. The results of a 50:50 mixture at 15 and 300 K are shown in Fig. 3. Depicted is the experimentally determined quantity $(\sigma_H/\sigma_D)_{exp}$ divided by the tabulated one $(\sigma_H/\sigma_D)_{tab} = 10.7$. As can be seen, $(\sigma_H/\sigma_D)_{exp}$ is anomalously decreased with respect to its tabulated value. Furthermore, it does not exhibit any angle dependence or temperature dependence.

However, the situation changes if the electronic environment of protons is modified. For this purpose we measured the NCS of the ionic hydride LiH at $T = 300$, 20 and 2 K. Note that while the hydrogen is partially positively charged in water, it is almost completely negatively charged in LiH. Furthermore, while protons are subject to rapid exchange between water molecules, they are almost rigidly bonded in a LiH crystal. The experimentally deter-

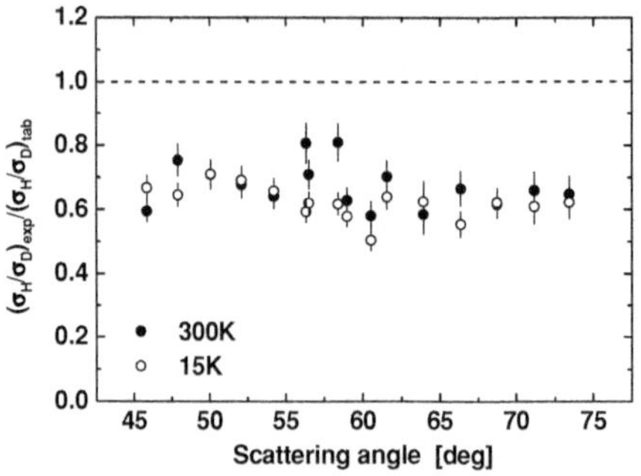

Figure 3. The ratio σ_H/σ_D of a H_2O / D_2O mixture with ($x_D = 0.5$) divided by its tabulated value (10.7) at different temperatures. No dependence on temperature is visible.

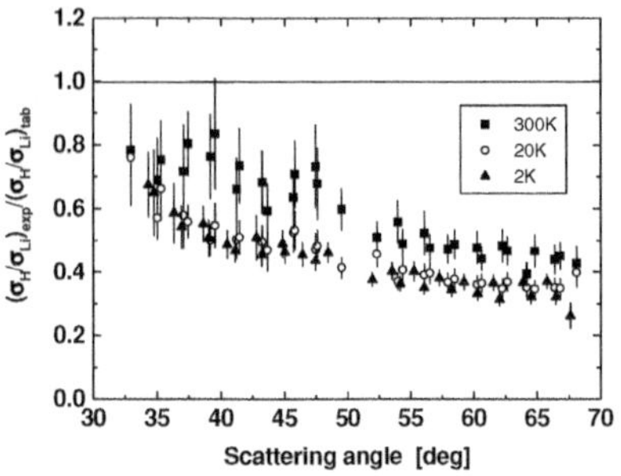

Figure 4. The ratio σ_H/σ_{Li} of LiH divided by its tabulated value (58.3) at different temperatures. The LiH data show the same anomalies at low scattering angles. This anomaly however increases slightly with the scattering angle and a significant difference upon temperature change is observed as well.

mined quantity $(\sigma_H/\sigma_{Li})_{exp}$ divided by the tabulated one $(\sigma_H/\sigma_{Li})_{tab} = 58.3$ is depicted in Fig. 4. All values of $(\sigma_H/\sigma_{Li})_{exp}$ are significantly smaller than the tabulated one and are slightly angle dependent but do not assume $(\sigma_H/\sigma_{Li})_{exp}$ at low angles. The most important result is that in clear contrast to $NbH_{0.78}$ and water data, respectively, LiH shows significant differences between 300 K and 20 K. No changes can be seen by further cooling to 2 K.

This anomaly is much more pronounced than that of the previously measured $NbH_{0.78}$ which showed a strong angle dependence [Karlsson 1999]. The LiH values do not assume $(\sigma_H/\sigma_{Li})_{tab}$ at low angles. The most important difference, however, is that in clear contrast with $NbH_{0.78}$ data, LiH data shows significant differences between 300 K and 20 K. These differences might be caused by different electronic structures of these systems. Whereas H atoms can move between different sites in the niobium lattice, it is rigidly bonded in LiH implying different interactions of H atoms with their environment. If protonic QE exists in these systems and is responsible for the anomalous neutron SCS density of H in LiH, then the reason for the larger anomaly at lower temperature might arise from the fact that at 20 K the QE is not disturbed by the environment to the same extent as at 300 K due to slower movement of the surrounding particles. This interpretation is also supported by the fact that anomalies found in LiH are larger than those found in the interstitial hydrides [Karlsson 1999; Abdul-Redah 2000; Karlsson 2003 (b)].

2.2 Insulating vs. metallic hydride: LaH_2 and LaH_3

To further examine the influence of the electronic environment in the σ_H shortfall we employed LaH_2 and LaH_3. Theses systems have the same nuclei but differ in the H concentration and most important in the electronic properties: whereas LaH_2 has metallic properties, LaH_3 is an insulator.

The results of this experiment are shown in Fig. 5. As can be seen very easily, $(\sigma_H/\sigma_{La})_{exp}$ is strongly reduced with respect to the tabulated value of $(\sigma_H/\sigma_{La})_{tab}$ for both systems. In addition, a slight angle dependence is visible as well: the anomaly is smaller in the low angle region (corresponding to long scattering times) than it is in the high angle one (corresponding to short scattering times). However, most important for the present context is the fact that the anomaly of LaH_3 is different from the LaH_2 one. It is also visible that these differences are more pronounced in the high angle region which - bearing in mind the dependence of the the scattering time on the angle; see Eq. (6) - underlines the dynamic character of the effect.

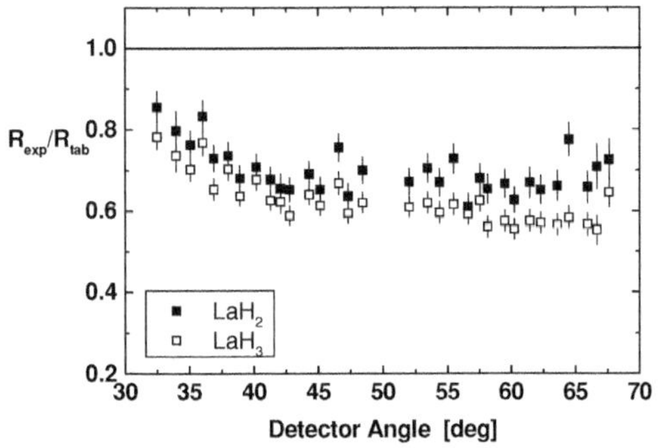

Figure 5. The experimental ratios $R_{exp} = (\sigma_H/\sigma_{La})_{exp}$ of LaH_2 and LaH_3, respectively, divided by the tabulated one $R_{tab} = (\sigma_H/\sigma_{La})_{tab} = 9.1$ as a function of scattering angle. R_{tab}/R_{tab} is strongly reduced and a slight angle dependence is observed for both systems. Most importantly, however, is that LaH_2 and LaH_3 show different anomalies which is attributed to the coupling of the protons to the different electronic environments - LaH_2 being metallic and LaH_3 being an insulator.

2.3 Experimental tests

Recently, the ratios of the neutron total SCS of H and D, $Q = \sigma_H/\sigma_D$, inferred from neutron transmission (NT) experiments on H_2O / D_2O mixtures were found to be in agreement with the tabulated value $Q_{tab} = 81.67/7.63 = 10.7$ [Blostein 2003 (a)]. It was concluded that the strong Q reductions previously observed on VESUVIO [Chatzidimitriou-Dreismann 1997 (a)] are an artifact of the data analysis. Here, we confirm the validity of the DINS data analysis by presenting new experimental results and show that this conclusion is unjustified. The original objection [Blostein 2001] which led to perform the NT experiment [Blostein 2003 (a)] is that, the incorporation of the energy resolution function R_E as a convolution in time of flight is a crude approximation and leads to serious errors due to the long tails of the absorption resonance of gold (Au) used as energy analyzer [Blostein 2001] (full line in Fig. 6). Concretely, the H peak area would be underestimated due to its overlap with, *e.g.*, D thus causing the anomalies [Chatzidimitriou-Dreismann 1997 (a)].

Here, our *working hypothesis* for the following is that *if the data analysis does not account properly for R_E, then a significant improvement of R_E should give different Q when using the same data analysis*. Such an improvement is provided by the so called double difference (DD) technique. In DD one takes three measurements: 1) with no filter, 2) with a filter of thickness τ_1 and absorption A_1 and 3) with a filter of thickness τ_2 and absorption A_2. The results of Ref. [Chatzidimitriou-Dreismann 1997 (a)] were obtained using single difference (SD), *i.e.*, combining only steps 1) and 2). The DD of the three measurements is $R_E^{DD} = A_1(E) - A_2(E)\ \tau_1/\tau_2$. The DD technique relies on the fact that for significant $E - E_0$ (E_0 = foil resonance energy), where the filter absorption cross section $\sigma_a(E)$ is small, it holds:

$$A_{1(2)}(E) = 1 - \exp\left[-N\ \tau_{1(2)}\ \sigma_a(E)\right] \sim N\ \tau_{1(2)}\ \sigma_a(E). \qquad (7)$$

Thus, $R_E^{DD} = 0$ and the wings of A_1 of SD are removed, whatever their functional form (dashed line in Fig. 6). For more details, see Refs. [Seeger 1985; Andreani 2003]. Consequently, overlapping effects are considerably reduced using DD. To test the working hypothesis, we measured liquid 50:50 H_2O / D_2O mixture using both SD and DD. In direct contrast to the working hypothesis, the obtained Q's are: Q_{SD} =6.9(2) and Q_{DD} =6.9(6), *i.e.*, independent of technique, Q is ca. 35% smaller than tabulated (10.7) and in line with Ref. [Chatzidimitriou-Dreismann 1997 (a)] (the error of Q_{DD} is larger because the DD technique reduces the counting statistics [Seeger 1985]).

Summarizing, although the DD technique introduces a significant improvement to the energy resolution function, the ratio Q is here the same as that obtained from the SD technique using the same data analysis. This result clearly demonstrates that the data analysis procedure always accounts properly for the resolution function. Otherwise different results should be obtained for SD and DD. Therefore, the working hypothesis must be rejected. The presented results might be regarded as an additional test to the already existing proofs [Chatzidimitriou-Dreismann 2002 (b); Abdul-Redah 2003] for the validity and correctness of the convolution approximation. Consequently, the conclusion of

Blostein *et al.* [Blostein 2003 (a)] that the validity of the DINS data analysis procedure performed on VESUVIO is limited is incorrect.

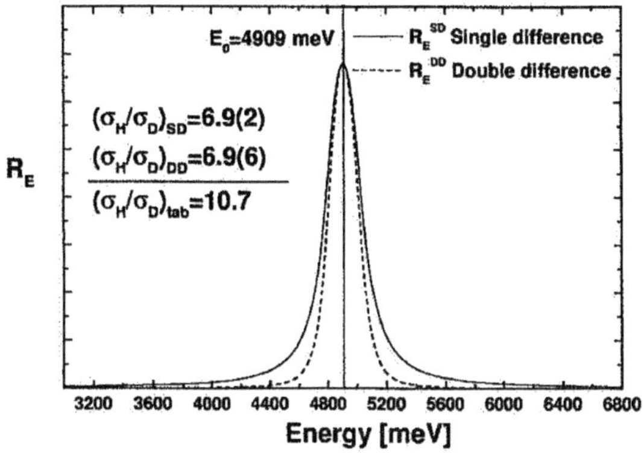

Figure 6 Au foil at resonance energy $E_0 = 4.9$ eV using SD (full line) and DD (dashed line). The wings of R_E^{SD} are completely removed when using DD.

Cowley [Cowley 2003] also criticized the data analysis on VESUVIO and argued that the angular dependent anomaly of σ_H/σ_{Nb} of $NbH_{0.78}$ might be caused by the lack of proper corrections due to the incident neutron flux $I(E_0)$ (see Eq. (4)) or due to a large Jacobian factor

$$J = 1 - \frac{m}{M}\left(1 - \frac{k_1}{k_0}\cos\theta\right) \tag{8}$$

involved in the conversion of a time of flight spectrum into q space. Obviously, this factor differs significantly for largely differing masses, as is the case for H ($M_H = 1$ a.u.) and Nb ($M_{Nb} = 93$ a.u.) thus leading to largely differing correction factors. Our recent experimental results show that the argument of Cowley is not valid and is thus irrelevant for the quantum entanglement effect under consideration. Now, if the intensity distribution $I(E_0)$ of the incident neutrons and the involved Jacobians are not correctly incorporated into the data analysis on VESUVIO then the same angle dependence should be observed also for the data of a different system containing H and another heavy nucleus. Such a system is represented for example by ZrH_2.

It is very important to note that Nb and Zr have almost identical masses, *i.e.*, $M_{Nb} = 93$ and $M_{Zr} = 91$ a.u. This leads to the fact that the same energy transfer is involved for scattering on Zr and Nb, respectively. For this reason and because the final energy E_1 is fixed, the same incident neutron energies E_0 are involved for scattering on Zr and Nb (see Fig. 7 A). Consequently, the same corrections of incident neutron spectrum are involved in the data analysis. In addition, again due to the similarity of the masses of Zr and Nb the same Jacobians are involved for the scattering on those nuclei (see Fig. 7 B).

We performed new NCS measurements on ZrH_2 at room temperature on VESUVIO. The sample was put in a standard flat Al can. Using the back

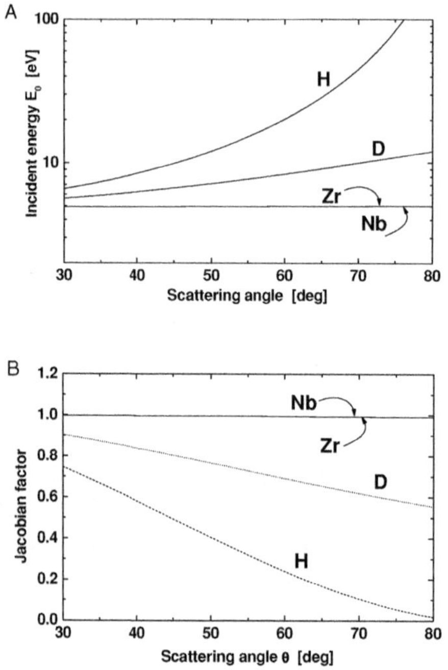

Figure 7 A - Dependence of the involved incident energy on the scattering angle for different masses: H $(M = 1)$, D $(M = 2)$, Zr $(M = 91)$, and Nb $(M = 93)$. The same incident energy is involved for Zr and Nb. B - Dependence of the involved Jacobians (see Eq. (8)) for H, D, Zr, and Nb. The same Jacobian factor is involved for both, Zr and Nb.

scattering detectors - where even higher energy and momentum transfers are involved in the scattering - it is possible to resolve the Zr $(M = 91)$ peak from the Al $(M = 27)$ one. This facilitates the determination of the ratio σ_H/σ_{Zr} from the forward scattering spectra. These ratios have been determined for all available detectors in the forward scattering regime. In contrast to the criticism of Cowley [Cowley 2003], the experimental results (see Fig. 8) show that $NbH_{0.78}$ (open circles) and ZrH_2 (full squares) exhibit completely different results:

(1) while the largest anomaly found in $NbH_{0.78}$ is about 30%, an intensity deficit of ca. 45% is observed for ZrH_2;

(2) more importantly, in contrast to the $NbH_{0.78}$ results, ZrH_2 does not show a significant angle dependence.

Summarizing, neither the angle dependence nor the magnitude of the anomaly of $NbH_{0.78}$ is reproduced by the results of ZrH_2 although the same data analysis procedure is applied. The experimental results reported above clearly refute the criticism of Cowley [Cowley 2003] that the angle dependence of the SCS anomalies of protons in $NbH_{0.78}$ are due to an incorrect account for the spectrum of incident neutrons or for the Jacobians involved when converting the spectra from the TOF into momentum transfer space.

Therefore, as concerns the determined areas from the VESUVIO data, the argument of Cowley does not play any role and the associated sub-femtosecond quantum entanglement effect found in the metallic hydride is not affected at all. The above results demonstrate that the short-lived quantum entanglement effect revealed in the niobium hydride system [Karlsson 1999; Abdul-Redah 2000; Karlsson 2003 (b)] as well as in the other condensed matter systems studied thus far [Chatzidimitriou-Dreismann 1997 (a); Chatzidimitriou-Dreismann 2000 (b); Chatzidimitriou-Dreismann 2001; Chatzidimitriou-Dreismann 2002 (a); Chatzidimitriou-Dreismann 2003 (a)] are real and are not caused by incorrect data reduction.

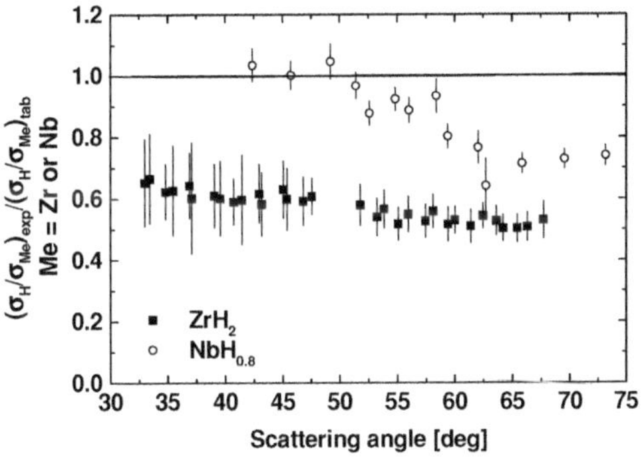

Figure 8. The experimental ratios $(\sigma_H/\sigma_{Nb})_{exp}$ (open circles) and $(\sigma_H/\sigma_{Zr})_{exp}$ (full squares) of the SCS of H to that of Nb and Zr of $NbH_{0.78}$ and ZrH_2, respectively, each one normalized to its tabulated value (*i.e.*, $(\sigma_H/\sigma_{Nb})_{tab}$ and $(\sigma_H/\sigma_{Zr})_{tab}$) as a function of scattering angle. Whereas, the $NbH_{0.78}$ show strong angle dependence the ZrH_2 data are very flat although the same Jacobian (see Fig. 7) and the same incident neutron intensities are involved for both $NbH_{0.78}$ and ZrH_2. $NbH_{0.78}$ data are adapted from Ref. [Karlsson 1999].

Further experimental tests and simulations are presented by J. Mayers and T. Abdul-Redah in the first paper of part VI of these Proceedings.

Very recently, we also succeeded to confirm the NCS results by applying an independent method, namely electron-proton Compton scattering (ECS) [Chatzidimitriou-Dreismann 2003 (a)]. This method is completely different from NCS due to the fact that the Coulomb interaction is involved in contrast to the strong interaction involved in NCS. In addition, the apparatus with which the ECS experiments have been performed is completely different in that the incoming electron beam is monochromatic and the spectra are recorded with respect to energy loss. Therefore, every discussion about the intensity distribution of incident energies or Jacobians is irrelevant.

Conclusion

We presented our most recent NCS results of short lived protonic quantum entanglement in different materials. Quantum entanglement manifests itself by a strong shortfall of the protonic SCS density. The emphasis of the present work is on the different electronic environments the protons are coupled to. As pointed out in the introduction, within the framework of modern quantum theory, the environment may be responsible both for the creation and destruction of QE. The present experimental results on H_2O / D_2O, LiH and LaH_2 and LaH_3 indicate that the electronic environment surrounding the hydrogen atom in the material is responsible for the different features of the cross section anomalies of the proton and thus for different decoherence mechanisms.

We have also clearly shown by employing experimental tests that recently published criticisms of the data analysis are unjustified. In addition to previous work [Chatzidimitriou-Dreismann 2002 (b); Abdul-Redah 2003] we have shown here that the convolution approximation criticized earlier [Blostein 2001] is correct and justified. We also demonstrated here that the neutron incident energy spectrum as well as all involved Jacobians - which have been considered as possible explanations for the anomalies [Cowley 2003] - are accounted properly in the data analysis procedure on VESUVIO.

Thus far, these effects have not found a common explanation based on existing condensed matter theories. Rather, these anomalies are attributed to the existence of short-lived QE of particles involving mainly protons in these materials [Chatzidimitriou-Dreismann 2000 (b); Karlsson 2000; Karlsson 2002 (c); Chatzidimitriou-Dreismann 2003 (b); Karlsson 2003 (a)].

The fact that scattering anomalies of protons are revealed in a time scale similar to that of the electronic rearrangement during the breaking or forming of a chemical bond underlines the importance of the QE and decoherence effect for chemical processes. This conclusion is strongly supported by the fact that similar anomalies have also been found in a chemical reaction, namely during the hydrogen evolution at the mercury drop electrode [Sperling 1999].

Also important is the analogy that every chemical reaction can be regarded as a scattering process with a reaction channel. Bearing this in mind one might appreciate the importance of the above results for the application of the kinetic gas theory to problems in condensed matter physics and chemistry. Also, the experimental and theoretical work of others (see for example Refs. [Prezhdo 1998; Opatrný 2001]) demonstrate that quantum entanglement plays a fundamental role in condensed matter physics and chemistry.

Summarizing, QE and decoherence are not only important for technological applications, like quantum computers and quantum cryptography, but are far more important for condensed matter physics as well as chemistry than it has been realized thus far.

ATTOSECOND EFFECTS IN SCATTERING OF NEUTRONS AND ELECTRONS FROM PROTONS

Entanglement and decoherence in C - H and O - H bond breaking

C. A. Chatzidimitriou-Dreismann,[1] T. Abdul-Redah,[2] M. Krzystyniak,[1] and M. Vos[3]

[1] *Institut für Chemie, Stranski Laboratorium, Technische Universität Berlin Straße des 17. Juni 112, D-10623 Berlin, Germany*
dreismann@chem.tu-berlin.de, krzystyniak@iwan.chem.tu-berlin.de

[2] *ISIS Facility, Rutherford Appleton Laboratory, Chilton / Didcot, OX11 0QX, UK*
t.abdul-redah@rl.ac.uk

[3] *Atomic and Molecular Physics laboratories, Research School of Physical Sciences and Engineering, The Australian National University, Canberra ACT 0200, Australia*
maarten.vos@anu.edu.au

Abstract Application of the neutron Compton scattering (NCS) technique - which operates in the sub-femtosecond timescale - to various materials yields results which indicate the presence of short-lived quantum entanglement between protons and electrons, and reveals novel aspects of attosecond dynamics of chemical bonds, *e.g.*, breaking of covalent C - H bonds. The following striking phenomenon has been revealed: the scattering intensity from H is anomalously decreased - about 20 - 30% of the protons (H-atoms) seem to "disappear". Here we present experimental results from water and H_2O - D_2O mixtures, liquid benzene, amphiphiles (2-isobutoxyethanol dissolved in D_2O), liquid hydrogen and H_2 - D_2 mixtures, and a solid polymer (formvar). Very recently this phenomenon has been confirmed with an independent method: electron Compton scattering from nuclei (ECS). Comparative NCS and ECS results from formvar are presented.

Keywords: Attosecond dynamics, neutron Compton scattering, electron-proton Compton scattering, quantum entanglement, decoherence

Introduction

As a matter of fact, important scientific discoveries have been often made by combining concepts from widely different areas, and fertile ideas are of-

V.M. Akulin et al. (eds.), Decoherence, Entanglement and Information Protection in Complex Quantum Systems, 483–498.

ten formed of elements drawn from domains which are far apart; *cf.* [J. Am. Chem. Soc., Editorial, 1997]. In this spirit, the present paper addresses a topic belonging to the physics of scattering processes, as well as to fundamental chemistry, *i.e.*, the breaking of a covalent C - H or O - H bond in the attosecond time scale. Most of the presented results were obtained with the neutron Compton scattering (NCS) technique [Sears 1984; Watson 1996]. The primary motivation [Chatzidimitriou-Dreismann 1995; Chatzidimitriou-Dreismann 1997 (b)] of our NCS experiments was the following question: Does the fundamental phenomenon of quantum entanglement (QE) [Einstein 1935; Bell 1964]) play a role in condensed molecular matter at ambient experimental conditions? Various experimental results (*e.g.*, see Refs. [Chatzidimitriou-Dreismann 1997 (a); Chatzidimitriou-Dreismann 2003 (a)] and the results presented below) provide strong evidence that the answer is "yes" and also reveal a thus far unknown property of dissociation dynamics of covalent bonds.

Interactions between adjacent particles of condensed phases can lead to quantum correlations, quantum interference, entanglement and decoherence, delocalization and "Schrödinger's cat" states. Such effects are theoretically expected to be extremely short-lived, due to environmental disturbances. Therefore, it has been widely believed that they cannot be experimentally detected. However, based on previous theoretical work (*cf.* [Chatzidimitriou-Dreismann 1995; Chatzidimitriou-Dreismann 1997 (b)]), we proposed to detect QE in condensed systems by means of sufficiently "fast" scattering techniques. Particularly suitable for this purpose is the NCS method. Our NCS investigations (on liquid H_2O - D_2O mixtures [Chatzidimitriou-Dreismann 1997 (a)]) started 1995 and have provided, for the first time, direct experimental evidence of attosecond QE between a proton and its adjacent particles.

To demonstrate this effect, we present in the following NCS and electron-proton scattering (ECS) results from various materials, which reveal the following striking phenomenon [Chatzidimitriou-Dreismann 1997 (a); Chatzidimitriou-Dreismann 2003 (a); Physics Today 2003; Scientific American 2003]:

About 20 - 30% *of the protons (*H *atoms) seem to "disappear" - protons become "invisible" to the probe particles.*

This surprising effect has no explanation within conventional theories of elementary chemical reactions and / or neutron (and electron) scattering theory. It is attributed to very short, attosecond entanglement of a scattering proton with adjacent particles, which can be studied by NCS (and ECS) due to the very short scattering time (*i.e.*, the characteristic time window) of these experimental methods. To date there exist various theoretical models describing this new effect [Karlsson 2000; Karlsson 2002 (c); Karlsson 2003 (a); Chatzidimitriou-Dreismann 2003 (b); Chatzidimitriou-Dreismann 2004 (a)].

Recently [Vos 2002 (a); Vos 2002 (b); Chatzidimitriou-Dreismann 2003 (a); Physics Today 2003; Scientific American 2003] this effect has been independently confirmed at the Australian National University. ECS investigations from a solid polymer showed the same shortfall in scattered electrons (with energy about 15 - 30 keV) from hydrogen nuclei, comparable to the shortfall of scattered neutrons in accompanying NCS experiments on the same polymer. Thus this effect appears to be independent of the two fundamental interactions involved, *i.e.*, the electromagnetic and strong interactions [Chatzidimitriou-Dreismann 2003 (a); Physics Today 2003; Scientific American 2003].

1. Outline of the neutron Compton scattering method

The crucial importance of various neutron scattering techniques for scientific and technological investigations cannot be overemphasized. A specific scattering method is based on the Compton effect. In his original studies, Compton investigated the energy loss of an X-photon colliding with an electron. Recently, the invention of neutron spallation sources opened the way of applying the same effect to neutron-nucleus collision, *i.e.*, the NCS. Usually NCS is utilized in order to measure the momentum distribution of a nucleus in the ground state before collision; *cf.* Ref. [Sears 1984; Watson 1996].

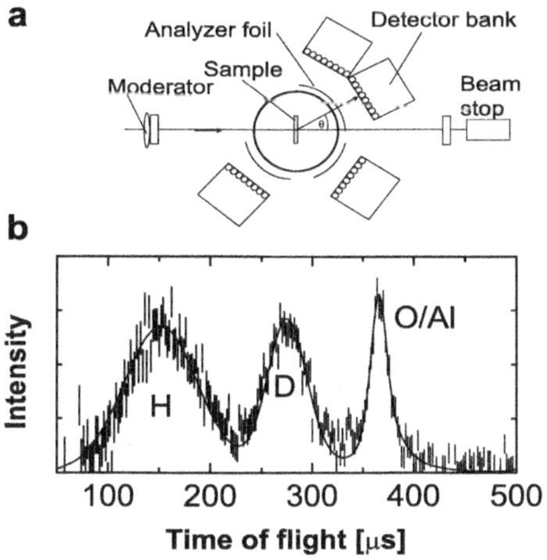

Figure 1 a - Experimental setup: the neutrons leave the moderator, are scattered by the sample under an angle and give a signal in the detector. The analyzer foils are periodically cycled in between sample and detector. b - A typical TOF spectrum of a 20:80 H_2O - D_2O mixture in an Al can (error bars) together with the corresponding fit (full line). The H and D recoil peaks are well separated from each other and from the joint O / Al peak.

For our NCS investigations, the fact that the duration (the scattering time, τ_{sc}) of an H atom-neutron collision amounts to $\tau_{sc} \approx 50$ - 500 attoseconds (1 attosecond $= 10^{-18}$ s) [Karlsson 1999; Abdul-Redah 2000; Chatzidimitriou-Dreismann 2001; Chatzidimitriou-Dreismann 2000 (b); Chatzidimitriou-Dreismann 2002 (a); Karlsson 2003 (b)] is of crucial importance. This estimate follows from the theoretical result of Refs. [Sears 1984; Watson 1996],

$$\tau_{sc} \approx \left(|\vec{q}|\ v_0\right)^{-1}, \tag{1}$$

where v_0 is the root-mean-square velocity of the struck proton in the ground state before collision, and $\hbar\,\vec{q}$ is the momentum transfer from the neutron to the proton. These short scattering times are a consequence of the large energy and momentum transfers attained with the eVS (for the neutron-proton collision): $\Delta E \approx 1 - 100$ eV, $|\vec{q}| \approx 30 - 200\text{Å}^{-1}$. Due to the large energy transfer - depending on the scattering angle - the NCS process causes breaking of covalent C - H (and other) bonds [Chatzidimitriou-Dreismann 2003 (a); Chatzidimitriou-Dreismann 2001; Chatzidimitriou-Dreismann 2002 (a)].

The short scattering time of the NCS process implies that, in this physical context, there is no well-defined time scale separation between the characteristic times of "slow" nuclear and "fast" electronic dynamics. Therefore, the well known Born-Oppenheimer approximation does not apply here (see also below). As a consequence, QE between nuclear and electronic degrees of freedom has been expected to strongly affect the dynamics of H atoms (or protons) [Chatzidimitriou-Dreismann 2003 (a); Chatzidimitriou-Dreismann 2001; Chatzidimitriou-Dreismann 2000 (b); Chatzidimitriou-Dreismann 2002 (a)].

Our NCS studies were carried out with the electron-volt spectrometer Vesuvio (formerly eVS) of the ISIS facility, Rutherford Appleton Laboratory, UK, which is the world's most powerful pulsed spallation neutron source up to date. The instrumental setup [Karlsson 2003 (b); Mayers 1994; Mayers 2004] provides highly energetic incident neutrons with a typical de Broglie wavelength $\lambda_{dB} < 0.1$ Å. Since λ_{dB} is much shorter than the interatomic distances and the scattering angles are large (resulting to very large momentum transfers) one often says that the scattering is incoherent [Sears 1984; Watson 1996], roughly meaning that each detected neutron was scattered from one nucleus only. Consequently, Bragg peaks are completely absent from the recorded time-of-flight (TOF) spectra, and the measured recoil peak of H is well separated from those of heavier nuclei; *cf.* Fig. 1.

It may be appropriate to mention here various criticisms concerning the validity of the analysis of the experimental data [Blostein 2001; Blostein 2003 (b); Cowley 2003]. However, the results of a considerable number of instrumental and experimental tests, as well as related Monte Carlo simulations, have demonstrated the excellent working conditions of Vesuvio and the validity of the data analysis procedure, thus refuting the aforementioned criticisms; for an account in detail, see Ref. [Mayers 2004] and the additional experimental tests presented in the next section.

2. Neutron Compton scattering from water and H_2O - D_2O mixtures

Our first NCS experiments were proposed in 1994 and done on liquid mixtures of H_2O - D_2O at room temperature in 1995 - 1997 [Chatzidimitriou-

Dreismann 1997 (a)]. The importance of this system for condensed matter physics cannot be overemphasized. (Incidentally, one should notice that there is fast H / D exchange, so that the mixtures contain also HDO molecules. This however has no significant effect on the following experimental results.) A sample spectrum has been shown already in Fig. 1 b. The experimental results are shown in Fig. 2.

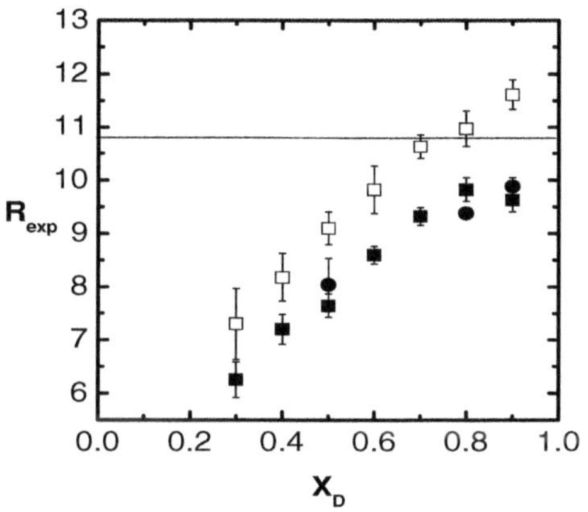

Figure 2 Experimental ratios $R_{exp} = (\sigma_H/\sigma_D)_{exp}$ for various H_2O - D_2O mixtures with D mole fraction x_D. Full (open) symbols: measurements with Au (U) analyzer (full circles: March 1995; full squares: July 1995; open squares: September 1996). Depicted are the means of 16 detectors. Full horizontal line: conventionally expected value $81.67/7.63 = 10.7$ [Chatzidimitriou-Dreismann 1997 (a)].

According to basic NCS theory, the equation $A_H/A_D = (N_H\,\sigma_H)(N_C\,\sigma_D)$ must be strictly valid, *cf.* [Chatzidimitriou-Dreismann 1997 (a); Mayers 2004]. A_i represents the (measured!) integrated area under the recoil peak of nuclei i and σ_i the (tabulated!) total cross-section of those nuclei. N_i is the particle density of atoms i, which is known from sample preparation. It can be seen from Fig. 2 that the experimental ratio $\sigma_H/\sigma_D = (A_H/N_H)/(A_D/N_D)$ of cross sections of H and D is "anomalously" reduced with respect to the tabulated one ($= 81.67/7.63 = 10.7$). The ratio σ_H/σ_D is nearly constant over the detector angles and thus the data represent the average over the angular range $30° < q < 70°$. The reduction of this ratio is strongly dependent on the D mole fraction x_D of the H_2O - D_2O mixtures. This striking result is by many believed to show that the anomalies which we ascribe to QE of particles are changed by the presence of fermions in the vicinity of bosons and vice versa, thus indicating the importance of quantum exchange correlations; a result that becomes more remarkable when compared with the rather "opposite" results obtained from benzene and polystyrene [Chatzidimitriou-Dreismann 2000 (b); Chatzidimitriou-Dreismann 2002 (a)], and also liquid H_2 and HD, see below. A theoretical model of describing the σ_H/σ_D dependence on the D-content has recently been proposed by Karlsson [Karlsson 2003 (a)].

Due to the novelty and implications of our results, it is not surprising that various (experimental as well as theoretical) objections have been raised, *e.g.*, Blostein *et al.* [Blostein 2001; Blostein 2003 (b)] recently criticized the widely used description of the measured spectral intensity as a convolution (see, *e.g.*, [Karlsson 2003 (b); Mayers 2004; Andreani 2003]) and proposed a procedure based on the complete absorption cross section of the used absorption foil (Au in the case of present experiments). They claimed that the "convolution procedure" leads to incorrect values of peak areas in the case of two (or more) overlapping peaks, like, *e.g.*, those of H and D in H_2O - D_2O mixtures. These claims were put to an experimental test [Chatzidimitriou-Dreismann 2002 (b)] and were found to be irrelevant. In particular it was showen [Chatzidimitriou-Dreismann 2002 (b)] that in pure D_2O and in the equimolar H_2O - D_2O mixture (where there is a strong H / D peak overlap), the measured D / O cross section ratios σ_D/σ_O appear to be equal.

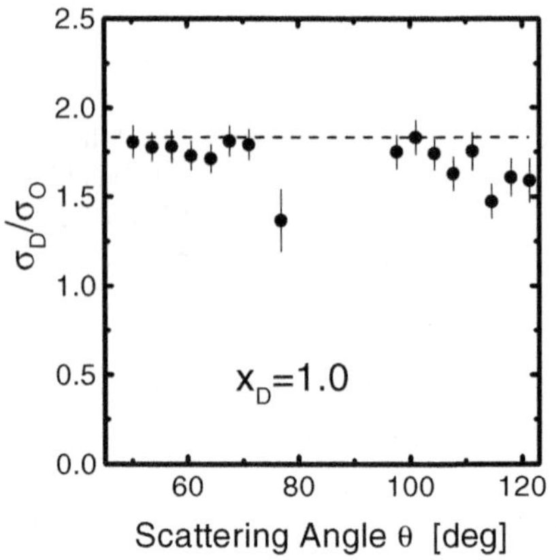

Figure 3 Experimental values of σ_D/σ_O from pure D_2O, for several scattering angles in the "forward" and "backward" directions. Dashed line: conventionally expected value ($=$ 1.80). Note the constancy of the results over θ, within experimental error [Chatzidimitriou-Dreismann 2002 (b)].

To illustrate these crucial results, in Figs. 3 and 4 are depicted the experimentally determined ratios σ_D/σ_O in these two samples. It is clearly seen that, in both samples, both scattering directions yield the same values of this ratio, within experimental error. Additionally, direct comparison of the areas of the H- and O-peaks in the forward direction gave the experimental result $\sigma_H/\sigma_O = 12.5 \pm 1.5$ which is roughly 30% smaller than the conventionally expected value ($= 19.4$). Making the physical assumption that the "heavier" O atoms do not exhibit such "anomalies", all these results imply that the measured considerable anomalies [Chatzidimitriou-Dreismann 1997 (a)] are mainly - if not exclusively - connected with the quantum dynamics of protons.

Moreover, a thorough comparison of the experimental results obtained from foils of metallic hydrides and polymers - both measured under identical experimental conditions - showed that the observed magnitudes of the anomalies are clearly dependent on the physical conditions the hydrogen atoms are involved in [Abdul-Redah 2003]. This result contradicts completely the newest theoretical criticism of Ref. [Blostein 2001; Blostein 2003 (b)]. All the results of these tests prove unequivocally that the aforementioned criticism [Blostein 2001; Blostein 2003 (b)] is unjustified in the context of our NCS experiments.

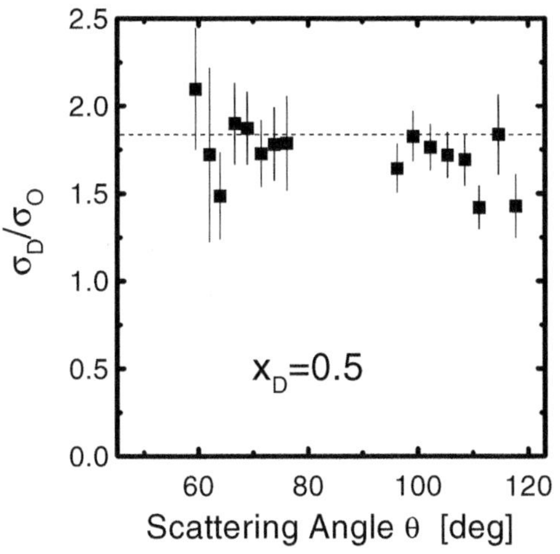

Figure 4 Experimental values of σ_D/σ_O from the equimolar mixture H_2O / D_2O, for several scattering angles in the "forward" and "backward" directions. Dashed line: conventionally expected value (= 1.80). Note the constancy of the results over θ, within experimental error. The values of σ_D/σ_O obtained here appear to be equal, within experimental error; with those of pure D_2O, see Fig. 3 [Chatzidimitriou-Dreismann 2002 (b)].

Furthermore, these results were recently confirmed applying the *double difference method* [Mayers 2004; Andreani 2003], which greatly improves the energy resolution of the Au filter (see Fig. 1 a) by removing the "long tails" of the absorption resonance spectrum. The observed value of σ_H/σ_D in the equimolar H_2O - D_2O mixture was found to be in very good agreement [Mayers 2004] with the original finding [Chatzidimitriou-Dreismann 1997 (a)].

The short-time nature of QE effect revealed by our NCS results is clearly demonstrated by the "negative findings" of a neutron interferometry (NI) experiment by Ioffe *et al.* [Ioffe 1999 (b)] on H_2O / D_2O mixtures, which found no anomaly in the measured coherent scattering length densities of the mixtures. This result was originally misinterpreted as being in clear contradiction to our NCS results [Chatzidimitriou-Dreismann 1997 (a)]. However, such a contradiction does not exist, simply because the NI method (being elastic and coherent) has a characteristic time which is many orders of magnitude larger than that of the NCS method (which is inelastic and incoherent), and thus NI cannot detect the attosecond effect under consideration [Chatzidimitriou-Dreismann 2000 (a); Ioffe 2000].

Additionally, a neutron transmission (NT) experiment on H_2O / D_2O mixtures was recently presented, reporting results in agreement with conventional theory and claiming to present a proof of the absence of the QE effect under consideration [Blostein 2003 (a)]. However, also this experiment does not provide the appropriate, and well defined, sub-femtosecond scattering time [Karlsson 2004]. Moreover, it was discussed by Karlsson and Mayers that the neutron coherence length in this NT experiment is much shorter than that in NCS (being about 2.5 Å, *cf.* [Karlsson 2003 (b)]). Thus there are no reasons whatsoever to expect deviations from the conventional, individual particle cross sections [Karlsson 2004].

3. Neutron Compton scattering from liquid benzene

Recently, we have investigated liquid benzene and various C_6H_6 - C_6D_6 mixtures at $T = 20°C$ [Chatzidimitriou-Dreismann 2002 (a)]. In order to achieve a well defined separation between the carbon and metallic recoil peaks (Fig. 5), the liquids were put in a sample container made of Nb (heavy metal).

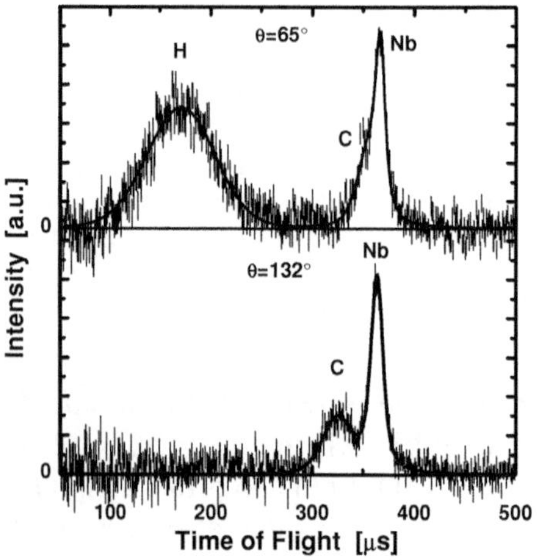

Figure 5 Examples of measured TOF spectra from liquid C_6H_6 in a Nb can, at $T \approx 295$ K. The error bars are due to counting statistics only. The full lines denote the fitted TOF spectra. For the scattering angle $\theta = 65°$, the C- and Nb-recoil peaks overlap. For $\theta = 132°$, however, these two peaks are well resolved, thus facilitating a reliable determination of the C- and Nb-peak intensities.

Here we present new results from pure C_6H_6. According to basic NCS theory, the equation

$$\frac{A_H}{A_C} = \frac{N_H\, \sigma_H}{N_C\, \sigma_C} \tag{2}$$

must be strictly valid [Chatzidimitriou-Dreismann 1997 (a); Mayers 2004]. For benzene one simply has $N_H/N_C = 1$, thus the experimentally determined ratio A_H/A_C can be immediately confronted with standard neutron scattering theory. This is a crucial advantage of the NCS method.

The NCS-results from benzene, see Fig. 6, reveal the following effect: the basic Eq. (2) is strongly violated; the measured ratio A_H/A_C is ca. 25% lower than its conventionally expected value $(\sigma_H/\sigma_C)_{conv} \approx 14.7$ (for the numerical values of cross-sections, *cf.* related textbooks). Also this observation contradicts every conventional expectation. The same anomalous effect has been also found in solid polystyrene and various partially deuterated polystyrene samples [Chatzidimitriou-Dreismann 2002 (a)].

4. Neutron Compton scattering from amphiphilic molecules

Another experiment [Chatzidimitriou-Dreismann 2001] presented here concerns the detection of QE effect in a solution of 2-isobutoxyethanol (in short:

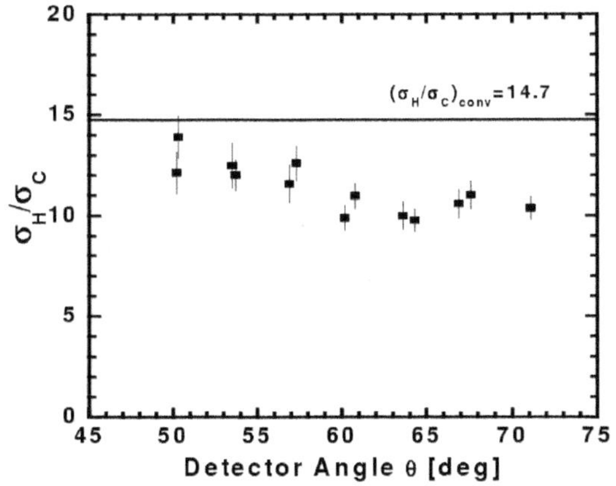

Figure 6 NCS results from liquid benzene. Contrary to conventional theoretical expectations, the measured ratio of peak areas $A_H/A_C \equiv \sigma_H/\sigma_C$ is about 25% lower than the conventionally expected ratio $(\sigma_H/\sigma_C)_{conv} = 14.7$ (horizontal line). The associated scattering times are about $\tau_{sc} \approx 50 - 500 \times 10^{-18}$ s. See Ref. [Chatzidimitriou-Dreismann 2002 (a)] for more details.

iso-C_4E_1) in D_2O, with molar composition iso-C_4E_1:$D_2O = 0.0223 : 1$. Examples of measured TOF spectra are given in Fig. 7 b. For neutron scattering angles in the "forward" direction ($\theta < 90°$), the recoil peaks of C and O of the liquid mixture, and that of the metallic can Nb, do overlap; see Fig. 7 b. Due to this overlapping, a considerable experimental effort has been made to determine the joint intensity of the C- and O-peaks, in relation to the intensity of the Nb-peak. For this, 8 neutron detectors (of 32 available ones) have been positioned in the "backward" scattering regime ($\theta > 90°$), where the momentum transfers are larger than in the "forward" direction ($\theta < 90°$), thus causing a well visible separation of the maxima of the C- and O-peaks from the Nb-peak; see Fig. 7 a. (A separation of the C and O peaks is impossible with the energies of neutrons available at ISIS). Then the ratio of the peak areas $(A_C + A_O)/A_{Nb}$ is determined. Note also that no H-recoil peak exists in the "backward" direction. This follows from trivial kinematics of collision and the fact that neutron and proton have about the same mass.

Having thus determined the joint intensity of the C and O recoil peaks relatively to that of the Nb-peak, the validity of the conventional theoretical expectation can be tested as follows. First, from the NCS spectra in the "forward" scattering direction we determine the ratio $A_H/(A_C + A_O + A_{Nb})$. Second, having determined the ratio $(A_C + A_O)/A_{Nb}$ from the "backward" direction, we can straightforwardly determine the ratio

$$R_{exp}(H) = \frac{A_H}{A_C + A_O}. \tag{3}$$

R_{exp} denotes the experimentally determined ratio to be distinguished from the conventionally expected one, R_{conv}. Note that A_O in the denominator of this ratio refers to iso-C_4E_1 and D_2O as well, since both molecules contain oxygen. Moreover, it should be stressed that the calculations of all these ratios use experimentally determined quantities only, *i.e.*, they contain no additional

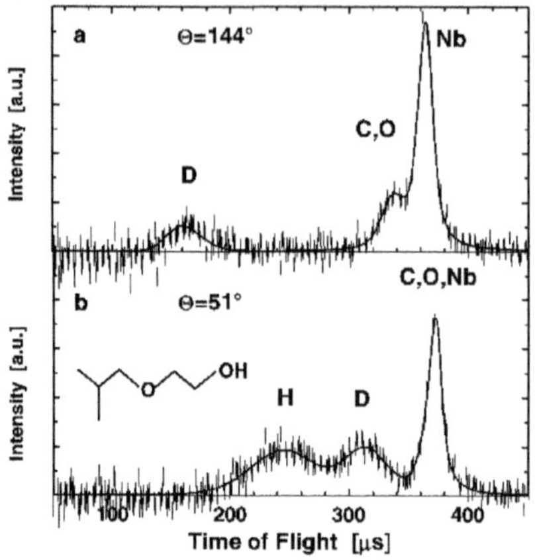

Figure 7 Structure and TOF spectra of iso-C_4E_1 in D_2O measured for scattering angles a - $\theta = 144°$ and b - $\theta = 51°$. Small vertical bars: one standard deviation error due to counting statistics. Full lines: fitted theoretical TOF-spectra to the measured data. a - Note the separation of the joint C, O-peak from the Nb can-peak in the spectrum, which is necessary for data analysis (see text).

fitting parameter. Third, we can immediately calculate the conventionally expected value of this ratio, $R_{conv}(H) = (N_H\ \sigma_H)/(N_C\ \sigma_C + N_O\ \sigma_O)$, (which is equal to 4.95 in our case) since the atom densities N_H, N_C and N_O, are precisely known through sample preparation and chemical formulae.

The result of this experiment is, again, a strong decrease (by about 20%) of the experimentally determined quantity $R_{exp}(H)$ from the conventionally expected $R_{conv}(H)$, *i.e.*, $R_{exp}(H) \approx 0.8\ R_{conv}(H)$; see Fig. 8.

5. Neutron Compton scattering from molecular hydrogen

The hydrogen molecule is the simplest chemical bond and hence it should be of particular interest for the NCS effect under consideration. The microscopic structure of liquid hydrogen has been studied in detail. For the liquid samples (at $T \approx 20$ K; working pressure ≈ 0.58 bar; densities $\approx 21 - 25$ nm^{-3}) we used a flat cell made of Al, providing a sample thickness of about 1 mm (wall thickness: 1 mm; sample volume: 2.6 cm^3). The peak areas A_X of nuclei X in the TOF spectra were determined by the standard data analysis applied at Vesuvio [Mayers 2004]. To test reproducibility, the experiments (always using the Au analyzer) have been repeated three times, also applying certain appropriate variations of the experimental setup; see below. The following interesting results have been obtained.

Fig. 9 shows experimental NCS results of an experiment on the equimolar H_2 - D_2 mixture. Note that there is no H / D exchange in this system, so that the mixtures contain H_2 and D_2 molecules only. The measured values of the ratio σ_H/σ_D of cross-sections show no systematic dependence on scattering angle θ. This finding is similar to that observed in H_2O - D_2O mixtures, but in clear contrast to the corresponding observations in the metallic hydrides Nb - H - D and Pd - H - D [Karlsson 2003 (b)]. In all experiments, the ratio

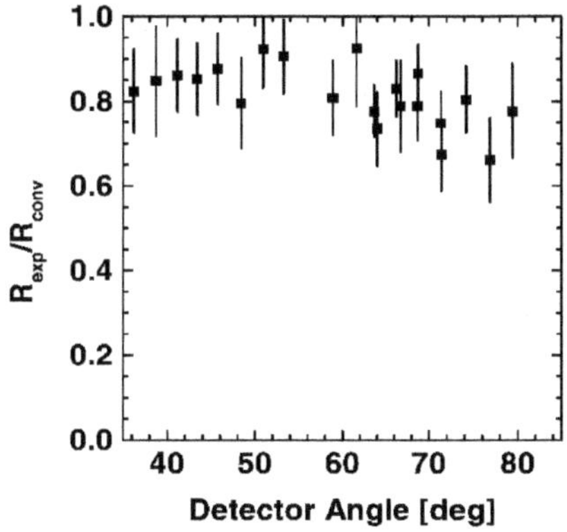

Figure 8 R_{exp}/R_{conv}, *cf.* Eqs. (2) and (3), associated with the neutron scattering intensities from H, C, and O in iso-C_4E_1 and D_2O, as a function of scattering angle. The average of the experimentally determined values of R_{exp} is "anomalously" smaller than the conventionally expected value R_{conv} by ca. 20%.

σ_H/σ_D appeared to be anomalously reduced by ca. 25% (±10%), with respect to the conventionally expected value of about 10.7.

In our NCS investigations on liquid H_2 and various H_2 / D_2 mixtures, we found that the anomaly of the scattering intensity of H is completely independent of the H-particle density N_H and / or the scattering power of the sample, *e.g.*, the anomalies observed in pure H_2 and in the four mixtures with D_2 percentages ranging from 33% to 75% appear to be essentially identical. In Fig. 10 are shown the ratios A_H/A_{Al} of the H peak area A_H to the Al peak area A_{Al}. However, the cell dimensions are not known with sufficient precision. Therefore, to facilitate comparison, the data A_H/A_{Al} for each sample were normalized so that they become equal to unity at the smallest available scattering angles $\theta \approx 35° - 40°$. A strong θ-dependence was found (reminiscent of that in Nb - H [Karlsson 2003 (b)]), which is independent of the atomic H:D composition of the samples. Obviously, these results indicate that the considered effect seems to be of *intramolecular* origin. For the study of the much "weaker" D peak, considerably more experiments would be necessary.

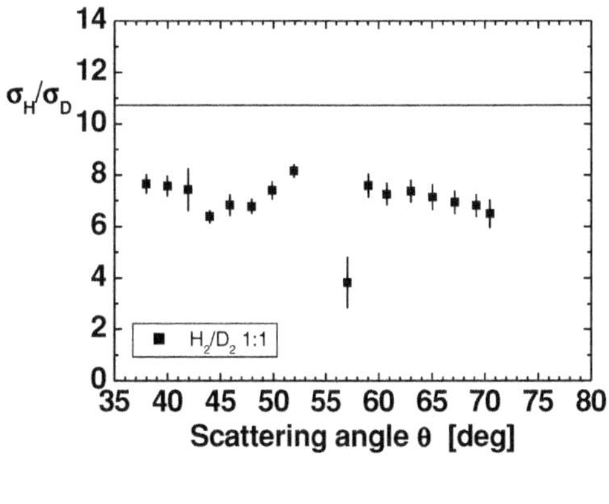

Figure 9 Measured ratios σ_H/σ_D for an equimolar H_2 / D_2 mixture. The observed anomalous decrease of this quantity (with respect to the expected value of 10.7, thin vertical line) amount to about 25%.

However, one may object that the strong θ-dependence of these observations might be due to artifacts (*e.g.*, multiple scattering, "shadowing" *etc.*) of the scattered neutrons at large scattering angles. This is conceivable, since the flat *Al* cell was perpendicular to the incoming neutron beam, so that the count rates at higher scattering angles might be more affected than those at lower angles. To investigate this point, we repeated the measurements on pure H_2 after *rotating* the flat *Al* can by ca. 20° and found that the above findings remained unchanged, see inset of Fig. 10.

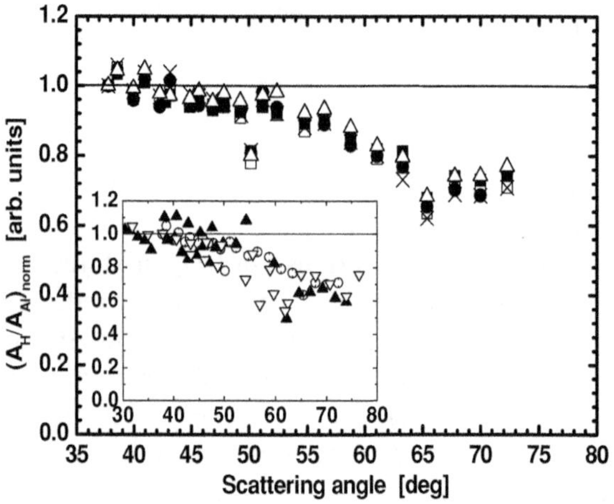

Figure 10 Ratios A_H/A_{Al} of *H* and *Al* peak areas, for H_2 (□) and various H_2:D_2 mixtures (×: 33% D_2; ■: 50% D_2; ●: 66 % D_2; △: 75% D_2) normalized to unity at scattering angles $\theta \approx 35° - 40°$. Inset, with same notations of axes: ratios A_H/A_{Al} for pure H_2 in various setups; d_{SD}: distance sample–detector. With cell perpendicular to beam, ○: $d_{SD} = 60$ cm, and ▲: $d_{SD} = 100$ cm. With cell tilted by 20°, ▽: $d_{SD} = 100$ cm.

Furthermore, we also repeated the measurements on pure H_2 using a new experimental setup, in which the distance between detectors and sample was increased from ca. 60 cm to 100 cm. With this procedure we also tested directly a conceivable influence of possible "dead-time" effects of the detectors for short times on the intensity of the H-signal. Also these additional measurements have clearly confirmed the results presented in Fig. 10; see inset.

All these measurements and / or tests have always reproduced, within experimental error, the scattering angle dependence of A_H/A_{Al} shown in Fig. 10. These results also underline the significance of the anomalously reduced value of σ_H/σ_D (see Fig. 9). Thus we conclude that the effect under consideration represents a real phenomenon, which is also of particular importance for further theoretical investigations due to the simplicity of H_2.

In a very recent experiment we have measured - at $T = 20$ K and with the same experimental setup - (a) the equimolar H_2 - D_2 mixture, and (b) the mixed-isotope system H - D. Both these systems have the same atomic composition, $H : D = 1 : 1$. According to basic NCS theory it should hold $A_H/A_D = \sigma_H/\sigma_D = 10.7$ for both systems. Our current NCS results reveal the following features:

(1) within experimental error, both systems exhibit *identical* anomalies.

(2) the basic equation $A_H/A_D = \sigma_H/\sigma_D$ is strongly violated; in both experiments, the measured ratio A_H/A_D is ca. 25% lower than $\sigma_H/\sigma_D = 10.7$.
(3) the value of A_H/A_D appears to be nearly independent of the scattering angle: $35° < \theta < 75°$.

Result (1) demonstrates that spin entanglement (and / or exchange correlations) play no role in this system. Thus the preceding results cannot be explained with the theoretical model of Refs. [Karlsson 2000; Karlsson 2002 (c); Karlsson 2003 (a)], in which exchange correlations play a crucial role.

6. ECS and NCS from a solid polymer

Vesuvio [Andreani 2003] is a unique instrument, consequently it has been criticized that our NCS effect has not been confirmed by other instruments or methods [Cowley 2003]. Hence it may be appreciated that, recently, the considered effect has been indeed confirmed applying an independent experimental method: *ECS* from protons in a solid polymer [Chatzidimitriou-Dreismann 2003 (a); Physics Today 2003; Scientific American 2003].

Using an electron spectrometer with an improved energy analyzer, Vos observed ECS from protons of formvar [Vos 2002 (a); Vos 2002 (b)]. This is an amorphous polymer widely used in electron microscopy since it makes extremely thin and flexible films and contains no double bonds (producing additional peaks in the energy loss spectra thus obscuring the ECS-peak).

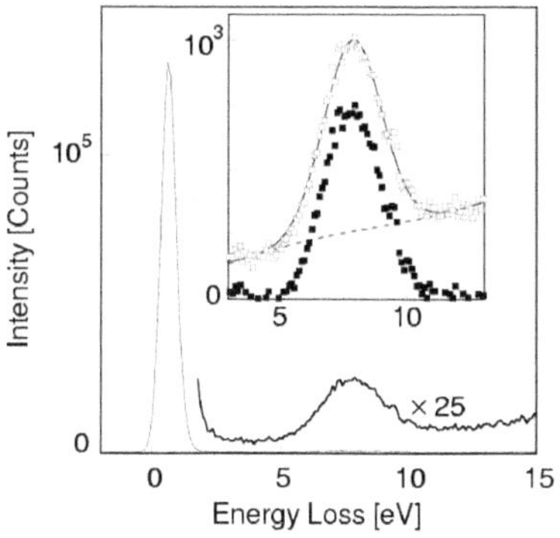

Figure 11. ECS energy-loss spectrum of a formvar film taken using 25 keV electrons, resulting to a mean momentum transfer $\hbar q$ to protons with $q = 61.8$ Å^{-1}. Inset: the fit of the *H*-recoil peak (about 8 eV) with a Gaussian, before and after subtraction of the background; taken from [Chatzidimitriou-Dreismann 2003 (a)].

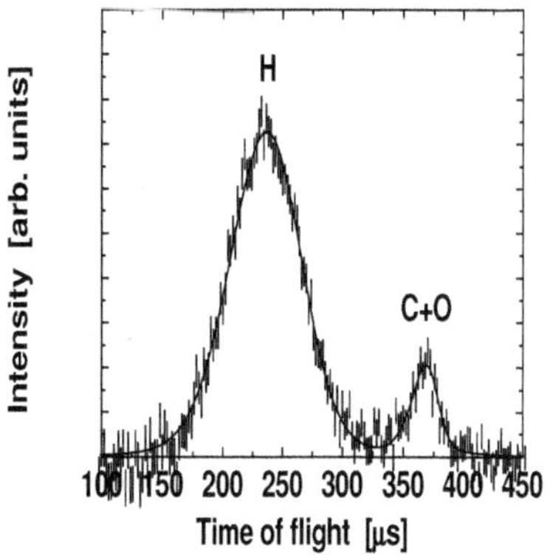

Figure 12. NCS TOF-spectrum of formvar (self-supporting foil, 0.1 mm thick) at scattering angle $\theta = 51.27°$, corresponding to a mean momentum transfer (for the neutron-proton collision) with $q = 60.7$ Å^{-1}; taken from [Chatzidimitriou-Dreismann 2003 (a)].

Using electrons of energy 15 -30 keV and a scattering angle of $\theta = 44.3°$, this instrument achieves electron-proton energy transfers in the range of 2 - 12 eV, and an energy resolution better than 0.4 eV [Vos 2002 (a); Chatzidimitriou-Dreismann 2003 (a)]. The energy loss spectra show that the recoil peak of H is well resolved from the combined peak of the heavier atoms C and O; see Fig. 11. In this context, the ECS method is the electron analog to NCS. We emphasize that, throughout this paper, the abbreviation ECS always refers to *electron-nucleus scattering* only, and not to electron-electron scattering.

Various thin films of formvar (about $50 - 100$ Å thick) were prepared by standard procedures. An ECS spectrum (which is a function of energy transfer ΔE from the electron to the impinging nucleus) is shown in Fig. 11. As is done in NCS investigations [Sears 1984; Watson 1996], the electron spectra were fitted with Gaussians. The "background" of the peaks is mainly due to interactions of the incident electrons with electrons in the sample.

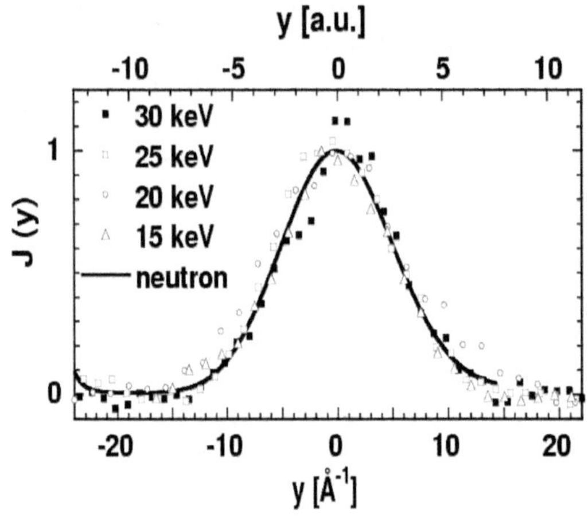

Figure 13 Momentum distributions $J(y)$ derived from the ECS and NCS measurements, with $\hbar y$ being the H-momentum component (before collision) along the direction of momentum transfer $\hbar\vec{q}$. The abscissa is given in $Å^{-1}$ and in atomic units (a.u., on the top). Full line: $J(y)$ derived with NCS at scattering angle $\theta = 66°$. Note the agreement between all distributions determined by ECS and NCS; taken from [Chatzidimitriou-Dreismann 2003 (a)].

According to conventional theory, the equation $R_{exp} = R_{conv}$ should also be valid for ECS. However, the cross section for electron scattering from hydrogen, carbon and oxygen is simply the Rutherford cross section: $\sigma_X \propto Z_X^2$ (Z_X: atomic number of atom X). Calculations of the cross section based on the electronic structure show that screening effects are not important under these conditions [Vos 2002 (a); Chatzidimitriou-Dreismann 2003 (a)]. The results reveal that, as in the case of NCS, the measured ratio R_{exp} of the hydrogen peak and the joint oxygen / carbon peak is considerably decreased: $R_{exp} < R_{conv}$.

Let compare the preceding ECS and NCS results from formvar [Chatzidimitriou-Dreismann 2003 (a)]. A typical NCS TOF-spectrum is presented in Fig. 12. These results, obtained at $T \approx 295$ K, demonstrate two novel features: (i) the ECS technique provides proton momentum distributions which are in quantitative agreement with those obtained with NCS (see Fig. 13);

(ii) the shortfall of NCS-intensity from protons is observed also with ECS and has roughly the same magnitude in both experiments; see below.

The momentum distribution of H is derived from the measured NCS TOF-spectra by standard procedures [Mayers 1994; Mayers 2004]. The distribution $J(y)$ (often called "Compton profile" [Sears 1984; Watson 1996]) is proportional to the density of protons with momentum component $\hbar y$ along the direction of the neutron-proton momentum transfer $\hbar\vec{q}$. $J(y)$ at scattering angle $\theta = 66°$ is shown in Fig. 13 (full line). Here $\hbar y$ is the H-momentum component (before collision) along the direction of momentum transfer $\hbar\vec{q}$.

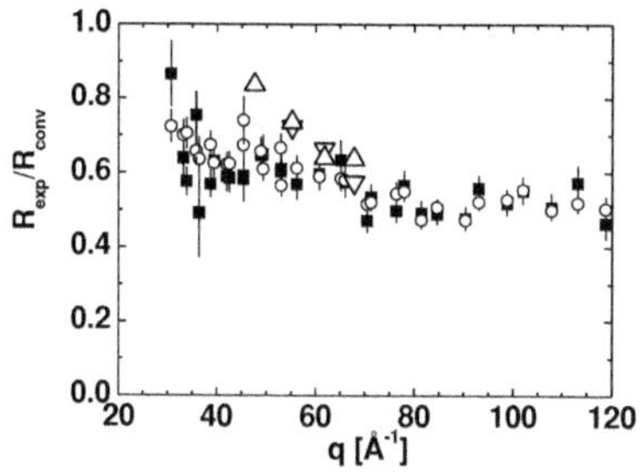

Figure 14. Anomalous reduction of NCS and ECS intensity from H of formvar, as a function of momentum transfer $\hbar q$. The q-range corresponds to scattering times $\tau_{sc} \approx 200 - 1000 \times 10^{-18}$ s. Small squares and circles: values of ratios R_{exp} of NCS peak-areas measured in the detector angular range $32° - 68°$, relative to the conventionally expected value R_{conv}. Full squares (open circles): results for foils of thickness 0.1 mm (0.2 mm). Large open triangles: ratios R_{exp}/R_{conv} measured by ECS from films of 50 - 100 Å thickness, using electrons with kinetic energies 15 - 30 keV [Chatzidimitriou-Dreismann 2003 (a)].

Momentum distributions of protons are derived from ECS spectra. Fig. 13 shows the four measured distributions $J(y)$ obtained from ECS with electrons of 15, 20, 25 and 30 keV, together with NCS data. These energies correspond to momentum transfers of 47.6, 55.1, 61.8 and 67.8 Å^{-1}, respectively. The agreement, within experimental error, confirms that both experiments reveal the same physical quantity: the (projection on the scattering vector $\vec{q}$ of the) momentum density distribution of protons. The NCS and ECS results of Fig. 13 can be compared directly as the experimental energy resolution contributes negligibly to the width of the spectra obtained by other technique.

Fig. 14 shows R_{exp}/R_{conv} as functions of $q = |\vec{q}|$ for ECS and NCS. The effect revealed by NCS is between 25% and 50%, and increases with increasing momentum transfer, (*i.e.* decreasing scattering time); see Eq. (1). The ECS data reveal a corresponding "anomalous" decrease of R_{exp} of 15 - 45%. For the ECS experiment it was checked that no radiation induced modifications of the film occurred for doses required to obtain good quality spectra [Vos

2002 (a)]. The constancy of NCS-results by doubling the thickness of the sample, as shown in Fig. 14, demonstrates that multiple scattering effects play no role here [Chatzidimitriou-Dreismann 2003 (a)]. These combined results demonstrate that the decreasing NCS-intensity from protons is also observable with a different method (ECS). This is important, as our effect appears to be independent of the fundamental interaction involved (electromagnetic versus strong), and also implies that it is a genuine property of matter [Chatzidimitriou-Dreismann 2003 (a); Physics Today 2003; Scientific American 2003].

7. Additional remarks

The effect under consideration, firstly observed in H_2O - D_2O mixtures [Chatzidimitriou-Dreismann 1997 (a)], has been attributed to sub-femtosecond QE of a proton with (i) adjacent particles (electrons and nuclei) caused by mutual Coulombic interactions [Chatzidimitriou-Dreismann 2001; Chatzidimitriou-Dreismann 2003 (b); Chatzidimitriou-Dreismann 2004 (a)], or (ii) an adjacent second proton due to quantum exchange correlations [Karlsson 2000; Karlsson 2002 (c); Karlsson 2003 (a)]. However, the recent NCS results from liquid HD (Sec. 5) clearly contradict the possibility (ii). Results from samples with various isotopic $H : D$ compositions indicate that this effect is mainly of intramolecular origin [Chatzidimitriou-Dreismann 2002 (a)]. We attribute the striking effect to light H atoms, rather than to heavier atoms C or O.

It is important to notice that the neutron-proton scattering time τ_{sc} is roughly of the order of the characteristic time of electronic transitions (or dynamics) in molecules, τ_{el}, *i.e.*, $\tau_{sc} \sim \tau_{el}$. Furthermore, the stuck protons become roughly as fast as the electrons, due to the large energy transfers of these experiments. Thus an indispensable condition for the validity of the usual Born-Oppenheimer approximation - which is the required well-defined time scale separation between nuclear and electronic dynamical time scales - is violated here. This violation, and the fact that electrons and nuclei are strongly coupled due to the Coulombic interactions, imply the existence of quantum correlations and entanglement between electronic and protonic motions [Chatzidimitriou-Dreismann 2001]. Furthermore, for scattering angles $\theta > 50°$, the energy transfer from a neutron to a proton is large enough ($5 - 100$ eV) to break the covalent molecular bond.

Our results presented above raise fascinating questions about the possible role of QE and decoherence in various molecular systems and, more generally, in condensed matter.

Acknowledgments

This work was partially supported by the EU RTN QUACS, and by a grant from the Royal Swedish Academy of Sciences.

MACROSCOPIC QUANTUM ENTANGLEMENT IN THE $KHCO_3$ CRYSTAL

F. Fillaux
LADIR-CNRS and Université Pierre et Marie Curie, UMR 7075, 2 rue H. Dunant, 94320 Thiais, France
fillaux@glvt-cnrs.fr, http//:www.glvt-cnrs.fr/ladir

Abstract After a brief introduction to neutron scattering techniques, illustrated with the scattering function for harmonic oscillators, some new aspects of proton dynamics in the $KHCO_3$ crystal are presented. The full scattering function for the proton modes measured on single crystals provides a graphic view of proton dynamics. Vibrational states are fully characterized with three quantum numbers. The effective oscillator mass of 1 amu confirms the decoupling of protons from the lattice. Combining infrared, Raman and inelastic neutron scattering techniques, the double minimum potential for the transfer of a single proton along hydrogen bonds is totally determined. An outstanding advantage of elastic neutron scattering techniques is to probe vibrational dynamics in the fully-degenerate ground state, which cannot be studied with optical spectroscopy. Decoherence-free quantum entanglement arising from symmetry-related normal coordinates gives rise to quantum interference effectively observed by spectroscopic measurements of elastic scattering. They are due to lines of entangled protons analogous to double slits. With the transversal coherence length of the neutron beam quantum coherence of protons scales to the crystal size, namely $\sim$ 1 cm. With diffraction techniques, the dynamical structure arising from large-scale quantum coherence gives ridges of intensity, in addition to Bragg's peaks. These ridges are fingerprint for macroscopic quantum coherence in two dimensions. The vibrational wave function in the ground state is a superposition of non-factorable macroscopic wave functions. Time independent quantum entanglement, up to macroscopic scale, appears as a natural consequence of the crystal symmetry.

Keywords: Proton dynamics, neutron scattering, hydrogen bond, proton transfer, decoherence-free states, macroscopic quantum entanglement

Introduction

"Quantum mechanics is very much more than just a theory; it is a completely new way of looking at the world, involving a change of paradigm perhaps more radical than any other in the history of human thought" [Leggett 2002]. However, the linear formalism of quantum mechanics extrapolated from the level

V.M. Akulin et al. (eds.), Decoherence, Entanglement and Information Protection in Complex Quantum Systems, 499–528.

of electrons and atoms, for which it was initially conceived, to that of everyday life may lead to conclusions in conflict with our commonsense intuition, such as Schrödinger's cat in a superposition of "alive-dead" states. The nonlocal nature of quantum entanglement is also a source of paradoxes at the heart of the profound difference between quantum mechanics and classical physics [Einstein 1935; Kocher 1967; Freedman 1972; Bell 1964; Aspect 1981]. Such conflicts may lead to a dichotomy of interpretation that while at the microscopic level a quantum superposition indicates a lack of definiteness of outcome, at the macroscopic level a similar superposition can be interpreted as simply a measure of the probability of one outcome or the other, one of which is definitely realized for each measurement of the ensemble (macrorealism). Therefore, whether the linear formalism of quantum mechanics could possibly not apply in unmodified form to macroscopic systems of everyday life in the same way as it does to the microscopic world of electrons and atoms is still a matter of speculation [Laloë 2001; Leggett 2002].

The dichotomy of interpretation is legitimated by the phenomenon of decoherence. In complex systems, an initially entangled subsystem loses its ability to exhibit quantum interference by getting entangled with the many degrees of freedom *via* interaction with the environment. With the words of quantum mechanics, once a superposition of states has progressed to the macroscopic level, so that formally the description is of the form $c_1 \; \Psi_1 + c_2 \; \Psi_2$, where Ψ_1 and Ψ_2 represent macroscopically distinct states, then, because of decoherence, no possible measurement can show the effect of interference between the two states. The quantum mechanics formalism will give for all possible measurements precisely the same predictions as would follow from the assumption that each individual system of the ensemble has realized either state Ψ_1 or Ψ_2 with probability $|c_1|^2$ or $|c_2|^2$, respectively. Therefore, quantum superpositions of macroscopically distinct states are unlikely in the condensed matter, except on a very short timescale, before decoherence moves into the lead [Leggett 1987]. Neutron Compton scattering experiments support the view that quantum entanglement exists in the condensed matter on very short time and spacial scales [Chatzidimitriou-Dreismann 1997 (c); Chatzidimitriou-Dreismann 1997 (a); Karlsson 1999; Karlsson 2000].

However, upon the assumption that decoherence is the only cause preventing the persistence of long-live entanglement, a macroscopic system can possibly behave quantum mechanically in the case it is decoupled from its environment [Caldeira 1983]. Superconductivity, superfluidity of ^4He, and the laser light, are well-known manifestations at a macroscopic level of long-live quantum effects, but the existence of macroscopically distinct states for these systems is debatable [Leggett 2002]. Superconducting quantum interference device (SQUID) loops are by now reference examples of macroscopic quantum behavior. Several groups have established that these quantum objects show macroscopic tunneling of a substantial number of Cooper-pairs ($\sim 10^{10}$) at a very low temperature ($\sim 10^{-3}$ K) [Rouse 1995; van der Wal 2000; Friedman 2000; Makhlin 2001; Grajcar 2004]. However, these experiments are still quite far from the scale of our everyday world as given, for example, by Avogadro's number ($\sim 10^{24}$) and the standard temperature of ≈ 300 K. There is still plenty of room for a possible boundary between quantum and classical worlds.

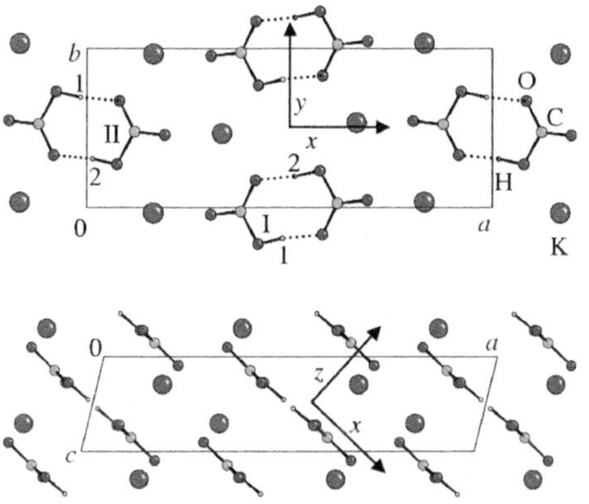

Figure 1 Schematic view of the crystalline structure of $KHCO_3$ at 14 K.

Apart from SQUID devices based on advanced technology, recent observation of macroscopic quantum entanglement of protons in the potassium hydrogen carbonate crystal ($KHCO_3$) has revealed a class of quantum crystals for which quantum entanglement arises naturally from vibrational dynamics of protons in centrosymmetric dimer entities $(HCO_3^-)_2$ linked by moderately strong hydrogen bonds (Fig. 1). Decoherence is cancelled by the perfect separation of dynamics for protons from the rest of the lattice, an inviolable consequence of the Pauli principle [Fillaux 1998] (see below Sec. 4). Decoherence-free quantum entanglement with virtually infinite lifetime was observed at a macroscopic level (namely the crystal size) [Ikeda 1999; Fillaux 2003 (a)]. Instead of a simple product of the vibrational wave functions for each atom site, the macroscopic single-particle ground state corresponds to a superposition of degenerate g and u symmetry species for proton vibrations.

From the standpoint of experiments, both advanced neutron scattering techniques and the particular structure of the $KHCO_3$ crystal were necessary ingredients to uncover the macroscopic quantum coherence of protons. The advantages of neutron scattering can be summarized as follows. First, neutrons can probe vibrational dynamics, namely the distribution of kinetic momentum, in the ground state where degeneracy and quantum statistics are prevailing. According to the measurement theory of quantum mechanics, elastic scattering is ideal to observing quantum coherence, whereas inelastic scattering is largely incoherent in nature, especially for protons. Second, as neutrons probe nuclear spins, interference arising from spin correlation can be observed. Third, quantum coherence increases dramatically the coherent scattering cross-section of protons from ≈ 1.8 to ≈ 81.7 barns. Fourth, with the high flux of epithermal neutrons available at advanced spallation sources measurements over a large range of momentum transfer provide the most detailed information ever obtained on proton dynamics.

Regarding the remarkable structure of the $KHCO_3$ crystal, all protons are crystallographically equivalent and indistinguishable. Planar centrosymmetric dimer entities $(HCO_3)_2^-$ are parallel to the same direction, practically $(30\bar{1})$ planes, throughout the crystal. Thanks to the particular arrangement of hydrogen bonds, all virtually parallel to the same direction, one can probe the anisotropy of proton dynamics and determine directly the orientation of the normal modes [Kashida 1994; Ikeda 2002]. The center of symmetry gives rise to full quantum entanglement for pairs of protons. Vibrational dynamics are represented with symmetric and antisymmetric coordinates totally decoupled from the rest of the crystal lattice [Fillaux 1998].

Quantum entanglement of proton pairs has been probed in different experiments. First, elastic neutron scattering experiments performed with a rather good resolution in energy and a limited coherence length of ≈ 20 Å revealed interference fringes analogous to those observed in double-slit (Young's) experiments [Zeilinger 1988], though significantly different [Ikeda 1999]. It appeared that entangled pairs of protons along a certain normal mode direction (say x) are fully correlated along lines parallel to y (see below Sec. 4 and dotted lines in Fig. 17). Second, diffraction experiments, with a much longer coherence length but without discrimination between the final energies of the scattered neutrons, have revealed rods of intensity specific to the grating-like structure arising from quantum correlation along x of the double lines of protons parallel to y [Fillaux 2003 (a)]. These rods are clearly separate from Bragg's peaks arising from coherent scattering by the lattice in three dimensions. Moreover, these rods have no visible counterpart for the $KDCO_3$ analogue, though the structures of the two crystals are identical. This is a consequence of the different quantum statistics for fermions and bosons.

Macroscopic quantum coherence in $KHCO_3$ was first observed at the rather low temperature of ≈ 15 K at which protons are ordered [Fillaux 2003 (a)]. This temperature is significantly higher than temperatures at which quantum effects are most commonly observed at a macroscopic scale in the condensed-matter and it can be suspected that the quantum entanglement in question is extremely robust with respect to thermal excitation. Upon increasing the temperature one can contemplate two independent mechanisms for decoherence in $KHCO_3$: a breakdown of the symmetry-related quantum entanglement decoupling protons and lattice dynamics, on the one hand, and proton disorder, on the other. In principle, a breakdown of the symmetry is not anticipated at temperatures below the only known phase transition at $T_c = 318$ K [Haussühl

1986; Kashida 1990] and proton disorder could be an alternative decoherence mechanism. Crystal structures determined at various temperatures show that above $\approx$ 150 K protons are distributed over two sites located at $\approx \pm$ 0.3 Å off-center of the hydrogen bond (see below Sec. 3). The occupation ratio for the two sites increases from $\approx 4 : 96$ at 200 K to $\approx 18 : 82$ at 300 K.

Tunneling across a double minimum potential along the hydrogen bond is the most likely mechanism for proton transfer but different models have been proposed [Graf 1981; Meier 1982; Fillaux 1983 (a); Fillaux 1983 (b); Benz 1986; Meyer 1987; Fillaux 1988; Skinner 1988; Stöckli 1990; Eckold 1992; Fillaux 1992; Horsewill 1998; Neumann 1998 (a); Fillaux 2002]. According to models based on the concept of "phonon-assisted incoherent tunneling" [Graf 1981; Meier 1982; Benz 1986; Skinner 1988; Horsewill 1998; Neumann 1998 (a)], dimers are uncorrelated and proton disorder should cancel macroscopic quantum coherence. The rods of intensity should disappear progressively upon increasing the temperature. Alternatively, "coherent tunneling", as proposed for the benzoic acid crystal [Fillaux 2002], is fully compatible with quantum coherence at any temperature. Preliminary measurements of the rods of intensity between 150 and 300 K seem to confirm the existence of macroscopically coherent tunneling states.

This paper is intended to overview experimental and theoretical works yielding evidences for macroscopic quantum entanglement in $KHCO_3$. Vibrational spectroscopy is presented in Sec. 1 with particular emphasis on inelastic neutron scattering techniques. Proton dynamics derived from vibrational spectra of $KHCO_3$ are discussed in Sec. 2. In Sec. 3 it is shown how potential functions for proton transfer *via* tunnelling across the hydrogen bond can be determined entirely from experiments. Finally, evidences for decoherence-free macroscopic quantum entanglement of protons are presented in Sec. 4.

1. Vibrational spectroscopy

A major goal of fundamental research aiming to rationalize the interplay of structure, dynamics, and chemical reactivity, is to determine multidimensional potentials for nuclei in various environments. On the one hand, potential surfaces can be calculated with quantum chemistry methods at various levels of approximation. On the other hand, from the experimentalist viewpoint, vibrational spectroscopy techniques can probe dynamics of atoms, molecules and ions, in various states of the matter. However, there are fundamental and technical limitations to the determination of potential hypersurfaces from vibrational spectra of complex systems, and the confrontation of experiments with theory is far from being free of ambiguities. Consequently, the interpretation of vibrational spectra remains largely based on experiments. Recent progress in neutron scattering techniques have revealed new dynamics, specially for protons in solids, which, apparently, challenge the most advanced techniques in quantum chemistry. These experiments provoke new thoughts regarding foundations of vibrational spectroscopy: normal coordinates, tunnelling and quantum entanglement.

Vibrational dynamics are dominantly represented with normal modes that are coherent harmonic oscillations of all degrees of freedom at the same frequency [Wilson 1964; Califano 1981; Long 2002; Ferraro 2003]. In the classical regime, normal coordinates - the eigenvectors of the dynamical matrix - are determined only to an arbitrary proportionality factor. In other words, the effective mass associated to a normal mode is arbitrary. This is of no consequence for optical spectroscopy techniques (infrared and Raman) that cannot probe masses, because of tiny momentum transfer values. Only recently, effective masses have been determined thanks to vibrational spectroscopy with neutrons [Ikeda 2002]. The existence of well-defined (of course) effective masses should be included in further theoretical developments.

With advanced neutron sources, we are now able to perform vibrational spectroscopy by measuring neutron scattering. Because the neutron-matter interaction is rationalized more easily, neutron scattering spectra can be interpreted more confidently and they provide information complementary to those obtained with optical techniques. A new view of proton dynamics in solids is emerging progressively. It appears that quantum effects have been largely overlooked, specially in the context of hydrogen bonding and proton transfer [Fillaux 1988; Fillaux 2000].

An overview of basic physics necessary to understand neutron scattering experiments is presented below. This section is a brief introduction to vibrational spectroscopy with neutrons. More details can be found in Ref. [Lovesey 1984]. Experienced readers should ignore this rather superficial introduction.

1.1 Momentum transfer

An incident monochromatic beam scattered by a sample is analyzed with a detector at a general position in space (see Fig. 2). Incident and scattered neutrons are regarded as plane waves whose wavevectors are $\vec{k}_i$ and $\vec{k}_f$, respectively. ($|\vec{k}_0| = 2\pi/\lambda_0$ and $|\vec{k}_f| = 2\pi/\lambda_f$, where λ_0 and λ_f are the incident and scattered wavelengths, respectively.) The momentum transfer vector is $\vec{Q} = \vec{k}_i - \vec{k}_f$.

1.2 Nuclear cross-sections

The extremely weak interaction of neutrons and matter is dominated by *spin-spin* interaction with nuclei, whilst interactions with electron-spins are negligible. Nuclear cross-sections for neutron scattering are strictly independent of the electronic structure (ionic or neutral, chemical bonding, *etc*). There-

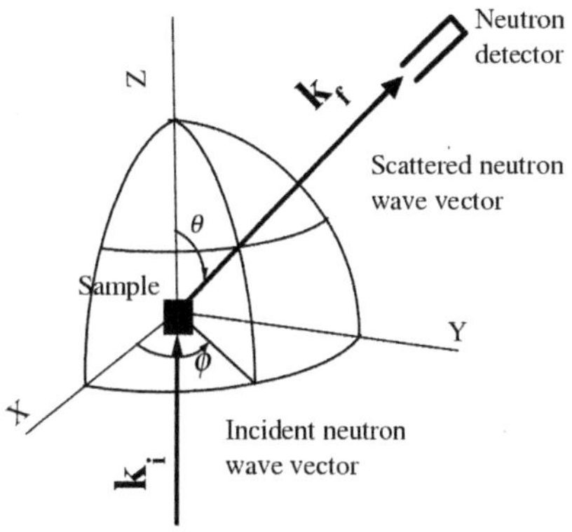

Figure 2 Schematic view of a neutron scattering experiment.

fore, the scattering cross-section of any sample can be calculated exactly, from the known cross-section of each constituent. Compared to optical techniques, neutron scattering intensities can be fully exploited.

1.3 Coherent versus incoherent scattering

For each nucleus, one can distinguish coherent and incoherent scattering cross-sections. Elastic coherent scattering by a crystal gives rise to Bragg diffraction, analogous to what is more routinely measured with X-rays. A great advantage of neutron diffraction is the possibility to locate protons that are normally barely seen with X-rays. Inelastic coherent scattering experiments performed on single crystals probe the frequency dispersion of phonons in momentum space. Inelastic incoherent scattering experiments, namely INS, probe phonon density-of-states over Brillouin-zones (reciprocal space).

1.4 Contrast

The incoherent scattering cross-section of the hydrogen nucleus (proton) is more than one order of magnitude greater than for any other atom. Therefore, neutrons scattered by protons largely dominate INS intensities. In many systems the non-hydrogenous environment can be regarded as virtually transparent. This great sensitivity to proton motions can be further exploited because the incoherent cross-section for deuterons (^2H) is about 40 times weaker. Therefore, isotope substitution at specific sites greatly simplifies the spectra and bands can be assigned with great confidence.

1.5 Penetration depth

Neutrons can penetrate many samples over rather long distances and, therefore, can probe the bulk. As opposed to that, photons in the infrared or visible range can hardly penetrate samples with high refractive indices. Optical techniques probe only a very thin layer at the surface of such samples. The negative consequence of the penetration depth is that neutron scattering measurements require much greater amounts of samples in the beam and / or longer measuring times.

1.6 Wavelength

With optical techniques, vibrational dynamics are probed on spatial scales much greater than molecular sizes or, in crystals, unit cell dimensions commonly encountered. Only a very thin slice of reciprocal space about the center of the Brillouin-zone ($|\vec{Q}| \lesssim 10^{-3}$ Å^{-1}) can be probed. This corresponds to in-phase vibrations of a virtually infinite number of unit cells. Consequently, band intensities largely depend on symmetry related selection rules.

For energy relevant to vibrational spectroscopy, neutron-wavelengths are similar to chemical bonds or unit cell sizes. For example, $|\vec{Q}|$ can be as large as ≈ 30 Å^{-1} for an incident energy of 500 meV (≈ 4000 cm^{-1}). Therefore, the phase correlation of collective oscillations can be probed *via* momentum transfer. Symmetry related selection rules are no longer relevant.

1.7 The scattering function

Using a monochromatic incident beam, the intensity for neutron scattering measured at each frequency (energy transfer $\hbar\omega = h\nu$) depends on the orientation and magnitude of the final wave vector. In one dimension, the scattering function at momentum transfer Q_x and energy transfer $\hbar\omega_{if}$ for a transition between states $|\Psi_i(x)\rangle$ and $|\Psi_f(x)\rangle$ can be written as:

$$S\left(Q_x,\,\omega\right) = \left|\langle\Psi_f\left(x\right)|\exp\left(i\,Q_x\,x\right)|\Psi_i\left(x\right)\rangle\right|^2\,\delta\left(\omega-\omega_{if}\right). \tag{1}$$

The textbook case of the harmonic oscillator in one dimension with mass m presented in this section is meant to feature neutron scattering experiments, compared to infrared and Raman spectroscopy. The Hamiltonian

$$H_x = -\frac{\hbar^2}{2m}\frac{d^2}{dx^2} + \frac{1}{2}\,m\,\omega_{0x}^2\,x^2 \tag{2}$$

has energy levels and wave functions such as [Landau 1966; Cohen-Tannoudji 1977]

$$\begin{aligned} E &= \left(n+\frac{1}{2}\right)\hbar\,\omega_{0x}, \\ \Psi_n\left(x\right) &= \frac{1}{\sqrt{2^n\,n!}}\,H_n\left(\sqrt{\frac{m\,\omega_{0x}}{\hbar}}\,x\right)\exp\left(-\frac{m\,\omega_{0x}}{2\hbar}\,x^2\right). \end{aligned} \tag{3}$$

H_n is the Hermite polynomial of order n. The frequency is $\omega_{0x}/2\pi$. Eigenstates are equidistant in energy (see Fig. 3).

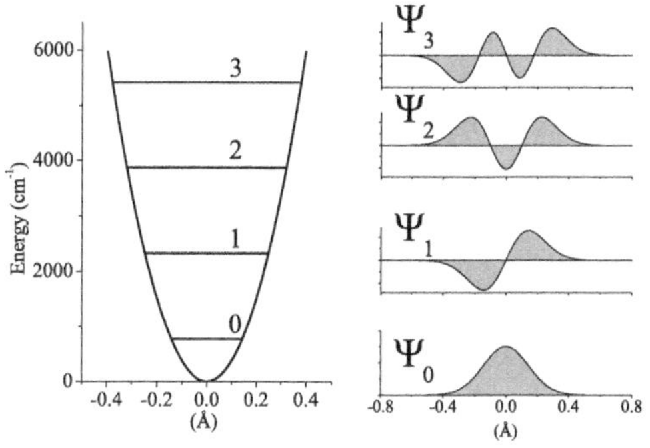

Figure 3 The harmonic oscillator: energy levels and wave functions. $\hbar\omega_0 = 1600\ \mathrm{cm}^{-1}$, $m = 1$ amu.

The scattering function for transitions arising from the ground state is:

$$S_{0\to n}\left(Q_x,\,\omega\right) = \frac{\left(Q_x\,u_x\right)^{2n}}{n!}\;\exp\left(-\,Q_x^2\,u_x^2\right)\;\delta\left(\omega - n\,\omega_{0x}\right), \tag{4}$$

where

$$u_x^2 = \langle\Psi_0(x)\,|x^2|\,\Psi_0(x)\rangle = \frac{\hbar}{2m\,\omega_{0x}} \tag{5}$$

is the mean-square amplitude (MSA) of the oscillator in the ground state.

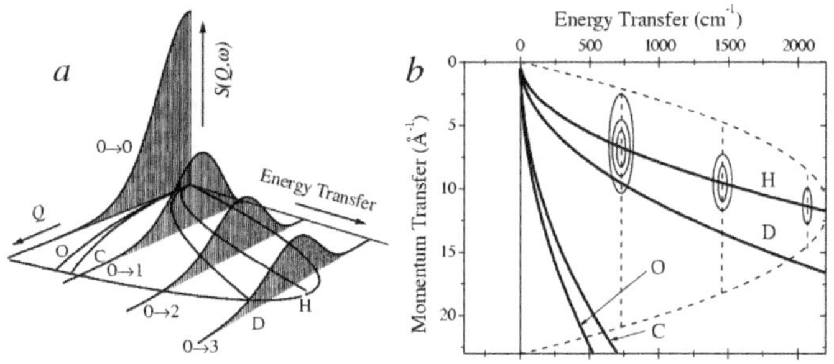

Figure 4. a - Landscape and b - isocontour map representations of the incoherent neutron scattering function for the proton harmonic oscillator in one dimension. The intensity is a maximum along the recoil line labelled H. Recoil lines for oscillators with masses corresponding to D, C and O atoms are shown. For a fixed incident energy, only momentum transfer values inside the parabolic area (dashed lines) can be measured.

All transitions $|0\rangle \to |n\rangle$ arising from the ground state (see Fig. 4) can be probed with NS.[1] In the $S(Q_x,\,\omega)$ map of intensity, each transition appears as an island, thanks to some broadening in energy. Profiles along Q_x are directly related to the MSA, u_x^2. Maxima of intensity occur at $Q_x^2\, u_x^2 = n$, along the *recoil line* for the oscillator mass: $E \approx \hbar^2\, Q^2/2m$.

A distinctive feature of neutron scattering, as compared to optical techniques, is that dynamics can be probed in the ground state with no energy transfer: $\omega_{if} = 0$ in (1) and (4). Such recent experiments are presented below (Sec. 4). They shed light on quantum effects in the ground state.

Eq. (4) and all consequences can be generalized to a set of harmonic oscillators. Then, combination bands can be observed (see below Sec. 3). Both MSAs and effective oscillator masses highlight the very nature of normal modes.

2. Vibrational spectroscopy of protons in the $KHCO_3$ crystal

Potassium hydrogen carbonate ($KHCO_3$) is an ideal system to study proton dynamics with neutrons because spectra are largely dominated by a small number of vibrational modes involving proton displacements with large amplitudes [Fillaux 1988]. The crystal contains centrosymmetric dimer entities $(HCO_3^-)_2$ (Fig. 1) and this structure remains unchanged from 14 K to 298 K [Fillaux 2003 (a); Thomas 1974]. The hydrogen bond with length $O\ \ldots\ O = 2.559$ Å at 14 K (2.607 Å for $KDCO_3$) is moderately strong [Novak 1974].

Protons are perfectly ordered at low temperature. At 298 K they are distributed over two sites located at $\approx \pm\, 0.3$ Å off-center of the hydrogen bond, with population ratio of $\approx 18 : 82$ [Fillaux 2003 (a)]. Needless to say, diffraction studies cannot distinguish statistical and dynamical disordering.

2.1 Normal modes versus localized vibrations

INS experiments evidence decoupling of the proton bending modes from carbonate entities [Fillaux 1988; Kashida 1994]. Simulations of the spectral profile with valence-bond force-field models based on infrared and Raman spectra [Nakamoto 1965], yield spectacular differences between observation and calculations. Discrepancies arise from the force-field representation itself and cannot be eliminated by straightforward adjustment of the force constants. The model protons, bound to oxygen atoms by strong forces, ride displacements at low frequency of carbonate entities, mainly below 200 cm^{-1}. Calculated intensities for these lattice modes are overestimated by at least one order of magnitude.

As opposed to this riding effect, the rather modest intensity observed for lattice modes is clear evidence that protons are almost totally decoupled from carbonate entities. Consequently, conventional force fields must be abandoned. Proton dynamics are better represented with localized modes defined with respect to a "fixed" (laboratory) referential frame. From both viewpoints of spectroscopy and quantum chemistry, this is a dramatic change.

2.2 The proton-crystal model

From the standpoint of INS, $KHCO_3$ should be regarded as a crystal of protons so weakly coupled to the surrounding atoms that the framework of carbonate and potassium ions can be virtually ignored [Fillaux 1988]. Proton modes are parallel to principal axes of dimers, as shown in Fig. 5 [Fillaux 2003 (a)]: x for stretching ($\nu\ OH$), y for in-plane bending ($\delta\ OH$) and z for out-of-plane bending ($\gamma\ OH$).

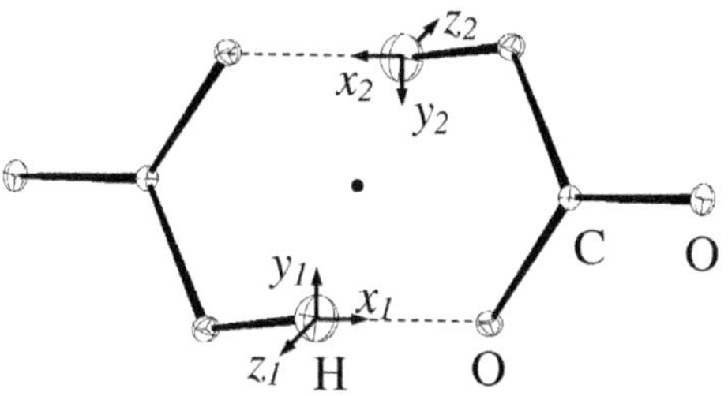

Figure 5 Schematic view of the shape of the refined thermal ellipsoids around the atom positions for the centrosymmetric dimer $(HCO_3^-)_2$ at 14 K. The arrows correspond to vector displacements for proton modes.

Within the harmonic approximation, the proton wave function can be factored as $\Psi_n(\vec{r}) = \Psi_{nx}(x)\ \Psi_{ny}(y)\ \Psi_{nz}(z)$ and the scattering function for $|000\rangle \rightarrow |n_x\ n_y\ n_z\rangle$ transitions can be written as:

$$S(Q_x,\ Q_y,\ Q_z,\ \omega) = \bar{S}_{n_x}(Q_x) \times \bar{S}_{n_y}(Q_y) \times \bar{S}_{n_z}(Q_z) \times \delta\left[\omega - (n_x\,\omega_{0x} + n_y\,\omega_{0y} + n_z\,\omega_{0z})\right], \tag{6}$$

with the scattering amplitude

$$\bar{S}_{n_\alpha}(Q_\alpha) = \frac{\left(Q_\alpha^2\,u_{0\alpha}^2\right)^{n_\alpha}}{n_\alpha!}\ \exp\left(-\,Q_\alpha^2\,u_{0\alpha}^2\right),\ \alpha = x,\ y \text{ or } z. \tag{7}$$

The full scattering function (6) should be represented in five dimensions. Partial landscape views in three dimensions are shown in Fig. 6 [Ikeda 2002]. Energy transfer values correspond to proton modes and the momentum transfer vector spans various planes in reciprocal space, $(Q_i,\ Q_j)$, such as $Q_k = 0$, with i, j or $k = x, y$ or z. The maps of intensity are graphic views of the orientation of the displacement vector for each mode. Visual examination confirms that vibrational coordinates remain unchanged in the various excited states and n_x, n_y, n_z are relevant quantum numbers.

One can compare distributions of intensity for the $|000\rangle \rightarrow |001\rangle$ transition ($\gamma\ OH$ mode) at ≈ 960 cm^{-1} for momentum transfer in the ($\delta\ OH$, $\gamma\ OH$) or ($\nu\ OH$, $\gamma\ OH$) planes. As anticipated, the gaussian profile for the stretching mode ($u_{0x}^2 \approx 10^{-2}$ Å^2) is broader than that of the bending mode ($u_{0y}^2 \approx 1.4 \times 10^{-2}$ Å^2). Similar distributions of intensity are observed for the $\delta\ OH$ mode, $|000\rangle \rightarrow |010\rangle$, at ≈ 1360 cm^{-1} for momentum transfer in the (δ

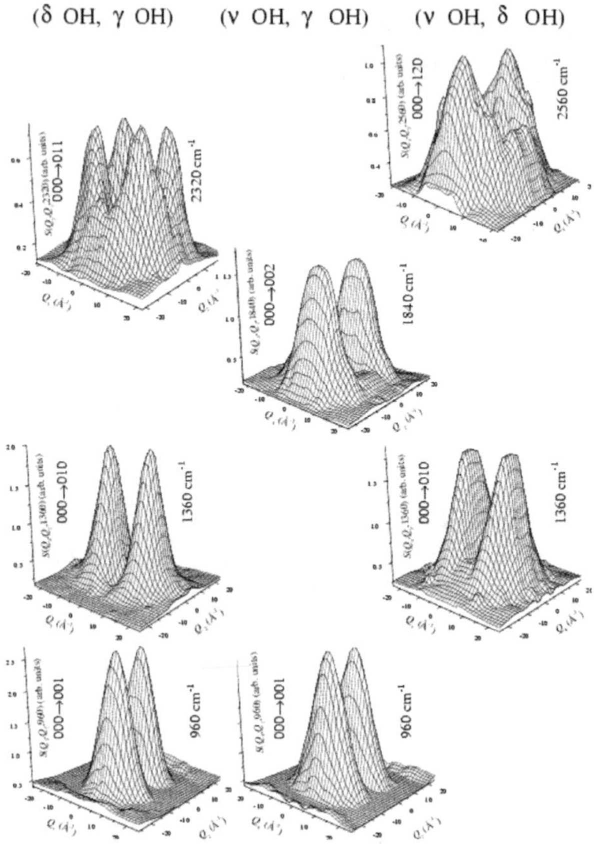

Figure 6. Landscape views of the spatial distribution of the scattering function for proton modes in $KHCO_3$ at 30 K.

$OH, \gamma\, OH)$ or $(\nu\, OH, \delta\, OH)$ planes. The $\gamma\, OH$ overtone, $|000\rangle \rightarrow |002\rangle$, is observed at ≈ 1840 cm^{-1} for momentum transfer in the $(\nu\, OH, \gamma\, OH)$ plane.

The transition at 2320 cm^{-1} in the $(\delta\, OH, \gamma\, OH)$ plane shows four maxima of intensity specific to the combination $|000\rangle \rightarrow |011\rangle$ (this transition is practically invisible in the infrared or Raman). Finally, a superposition of the OH stretching, $|000\rangle \rightarrow |100\rangle$, and $\delta\, OH$ overtone, $|000\rangle \rightarrow |020\rangle$, is observed at $\approx$ 2560 cm^{-1}. The observed intensity ratio for fundamental and overtone is close to that given by Eq. (6). There is no evidence for Fermi resonance mixing the wave functions [Novak 1963; Lucazeau 1973].

Fitting procedures give information on wave functions *via* mean-square displacements (u_α^2) for each vibration and effective oscillator masses. It transpires that proton dynamics for bending modes correspond very closely to isolated harmonic oscillators with a mass of 1 amu [Ikeda 2002]. They are largely decoupled from the lattice. Apparently, the same conclusion holds for the stretching mode but data analysis is hampered by weaker intensity.

In $KHCO_3$, coupling of the two protons of a dimer gives rise to symmetric and antisymmetric modes (see below) that can be distinguished with INS [Fillaux 1988]. Owing to the limited resolution in energy, splittings are not visible in Fig. 6 but further experiments confirm that the effective mass of 1 amu holds for each normal mode, either symmetric or antisymmetric.

As a conclusion for this section, INS studies of $KHCO_3$ single-crystals provide the most detailed, and hopefully the most tutorial, view of proton dynamics ever obtained. The limitation of optical techniques to establishing an unambiguous representation of proton dynamics is emphasized. Effective oscillator masses of 1 amu are determined for each normal mode. Then arises a new fundamental question: *which mechanisms can account for the decoupling of proton dynamics from the lattice?*

3. Proton transfer and dimer interconversion

Vibrations with large amplitudes show great anharmonicity, or even nonlinearity in the extreme cases where perturbation theory no longer holds. This occurs primarily for protons, the lightest nucleus, whose displacements are sufficiently decoupled from the surrounding heavier atoms. In solids, this is observed mainly for hydrogen bonds and for light rotors experiencing weak potential barriers (these rotational dynamics [Fillaux 2003 (b)] are not discussed in this paper).[2] The breakdown of the harmonic approximation gives rise to a manifold of transitions conveying a wealth of information. These systems are unique to understanding nonlinear quantum dynamics in complex environments but the interpretation of infrared and Raman spectra is hampered by the complexity of light-matter interaction.

Another outstanding result of INS studies is the determination of potential functions for proton transfer along hydrogen bonds [Fillaux 1988; Fillaux 2002]. This is of fundamental consequences to many physical, chemical and biochemical processes [Pauling 1960; Pimentel 1960; Vinogradov 1971; Joesten 1974; Schuster 1976; Schuster 1984; Jeffrey 1991; Perrin 1997; Scheiner 1997; Tuckerman 1997]. There is a general agreement that proton transfer dynamics have properties characteristic of a light particle in a heavy framework that can be represented with an effective potential along a local reaction coordinate coupled to the motions of heavy atoms. However, vibrational spectroscopy provides a more focused, and quite different, view.

3.1 The potential function for single-proton transfer

The shape of the quasi-symmetric double minimum potential $V(x)$ for proton transfer along the OH stretching coordinate x (Fig. 7) is dictated by exper-

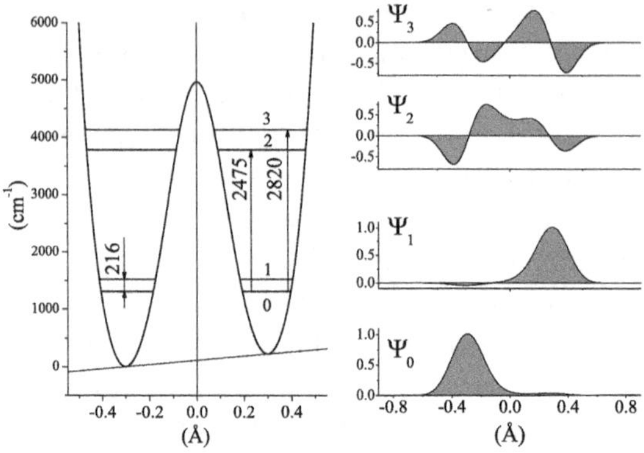

Figure 7 Potential function and wave functions for the OH stretching mode along the hydrogen bond in the $KHCO_3$ crystal at 10 K.

imental data: the distance $2\,x'_0 \approx 0.6$ Å between the minima is known from the crystal structure. $h\nu_{02}$ and $h\nu_{03}$ were determined from the band-shapes of the OH stretching mode in the infrared and Raman [Fillaux 1983 (a); Fillaux 1983 (b)]. The ground state splitting (ν_{01}) associated with quantum transfer of a single proton is not observable with optical techniques, since the transition intensity is too weak. With INS the transition was observed at 216 cm^{-1} (Fig. 8) [Fillaux 1988; Kashida 1994]. The dynamical nature of proton disorder was thus established, and confirmed with NMR for $KDCO_3$ [Benz 1986] and, recently, for $KHCO_3$ [Odin 2004]. If the potential were symmetrical, the splitting of the ground state (the tunnelling frequency) would be $\nu_{0t} \approx 18$ cm^{-1}. The observed splitting of 216 cm^{-1} is largely due to potential asymmetry.

The effective oscillator mass for proton transfer is 1 amu and this conclusion is unquestionable. (For a given set of energy levels, to increase the effective mass would decrease the distance between the minima, and *vice versa*.) The (effective) potential function in Fig. 7 accounts for quantum-transfer dynamics of a single proton virtually uncoupled to the surrounding, since there is no significant mass renormalization, and moving along a linear coordinate. The shape is significantly different for the deuterated derivative [Fillaux 1983 (b)].

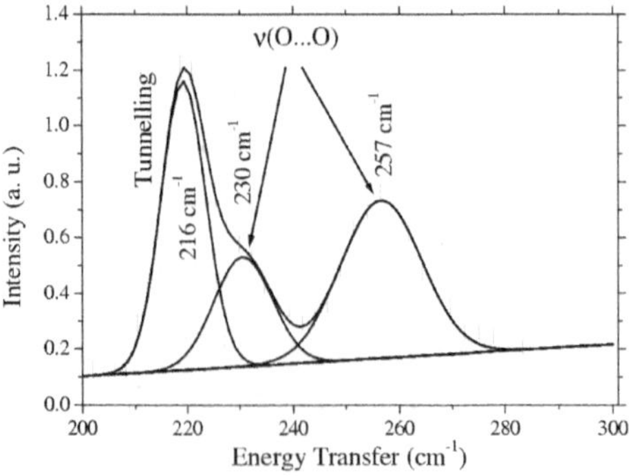

Figure 8 Inelastic neutron scattering spectrum of the $KHCO_3$ crystal at 4 K in the "tunnelling" region and band decomposition into Gaussian profiles.

The $|0\rangle \rightarrow |1\rangle$ tunnelling band in Fig. 8 is rather sharp. The measured full width at half maximum (FWHM) of $\approx$ 10 cm^{-1} is an upper bound for the real bandwidth. This is clear evidence that proton transfer is also totally decoupled from heavy atom dynamics, in line with the remainder of the spectrum. On the time scale of proton tunnelling, dimers should be regarded as rigid entities. Therefore, the prevailing idea that proton transfer occurs along a complex multidimensional reaction path involving rearrangement of heavy atoms [Graf 1981; Meier 1982; Skinner 1988; Stöckli 1990; Tuckerman 1997; Horsewill 1998; Neumann 1998 (a); Neumann 1998 (b); Benderskii 2000; Tuckerman 2001] is not appropriate.

It is worth emphasizing that the $|0\rangle \rightarrow |2\rangle$ and $|0\rangle \rightarrow |3\rangle$ transitions in the 2000 - 3000 cm^{-1} frequency-range impose a rather high potential barrier of $\approx$ 5000 cm^{-1}, owing to the zero-point energy of 1000 - 1500 cm^{-1}. This *unavoidable* conclusion applies to all hydrogen bonded systems with similar O ... O length, like carboxylic acid dimers. For example, a very similar double minimum potential holds for benzoic acid [Fillaux 2002] and previous estimates of the potential barrier (as low as $\approx$ 500 cm^{-1}) [Meier 1982; Stöckli 1990; Horsewill 1998; Neumann 1998 (a); Neumann 1998 (b)] are in dramatic conflict with experimental facts.

3.2 Uncorrelated proton transfer in dimers

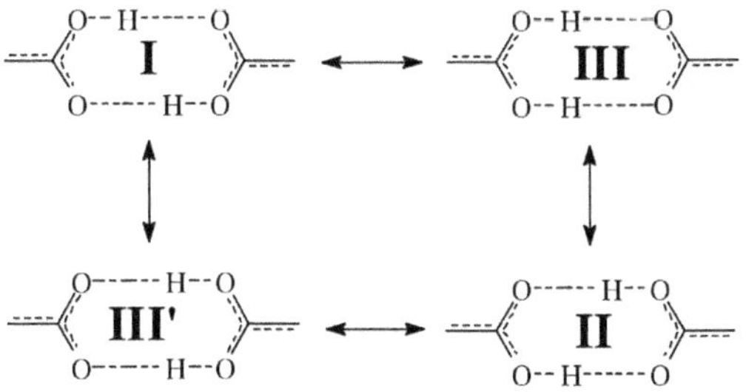

Figure 9 Scheme of the interconversion process in dimers for uncorrelated proton transfers.

As any coupling term between the two protons of a dimer is negligible compared to the barrier height, concerted double proton transfer can be ignored. Interconversion between tautomers I and II (see Fig. 9) occurs *via* III and III' corresponding to the transfer of a single proton along coordinates x_1 or x_2.

The potential surface for these dynamics is $\mathcal{V}(x_1, x_2) = V(x_1) + V(x_2)$ (see Fig. 10), where $V(x_1)$ or $V(x_2)$ is merely the potential function for the transfer of a single proton. In the (x_1, x_2) plane, domains corresponding to tautomers I, centered at $(-x_0', -x_0')$, and II at (x_0', x_0'), or intermediate isomers III and III' at $(\pm x_0', \mp x_0')$, are separated by high potential barriers and can be regarded as well defined entities.

The eigenstates of the corresponding Hamiltonian are tensor products of the eigenstates in one dimension. In addition to the ground state $|0\rangle \otimes |0\rangle$, there are now three tunneling states: $|1\rangle \otimes |0\rangle$ and $|0\rangle \otimes |1\rangle$, degenerate at $h\nu_{01}$, and $|1\rangle \otimes |1\rangle$, at $2h\ \nu_{01}$. The wave functions for these states (see Fig. 11) are simple products of the wave functions along x_1 and x_2, as represented in Fig. 7. In the ground state, the probabilities for tautomers I, III (III') and II are: $\sim 1 : 10^{-3} : 10^{-6}$, respectively. At thermal equilibrium the wave function corresponding to the superposition of tunneling states is

$$\Psi(T) = c_{00}(T)\ \Psi_{00} + \frac{c_{01}(T)}{\sqrt{2}}\ [\Psi_{10} + \Psi_{01}] + c_{11}(T)\ \Psi_{11}, \tag{8}$$

where $\Psi_{ij} = \Psi_i(x_1)\ \Psi_j(x_2)$ and $c_{00}(T)$, $c_{01}(T)$, $c_{02}(T)$, are temperature dependent coefficients. As only one proton occupies the secondary site in states $|1\rangle \otimes |0\rangle$ and $|0\rangle \otimes |1\rangle$, while two protons are transferred in the $|1\rangle \otimes |1\rangle$ state, the relative population of the less occupied site is

$$\mathcal{P}(T) = \frac{p_{01}(T) + 2\,p_{01}^2(T)}{P(T)}, \tag{9}$$

where $p_{01}(T) = \exp(-h\ \nu_{01}/kT)$ and $P(T) = 1 + p_{01}(T) + p_{01}^2(T)$.

In conclusion to this section, band-shape analysis of vibrational spectra and ground state splitting observed with INS demonstrate that proton transfer dynamics are quantal in nature, even at room temperature. Semiclassical models are not relevant. *The dramatic failure of quantum chemistry to account for the observed dynamics should be regarded as one of the major unsolved theoretical problems at the present time.*

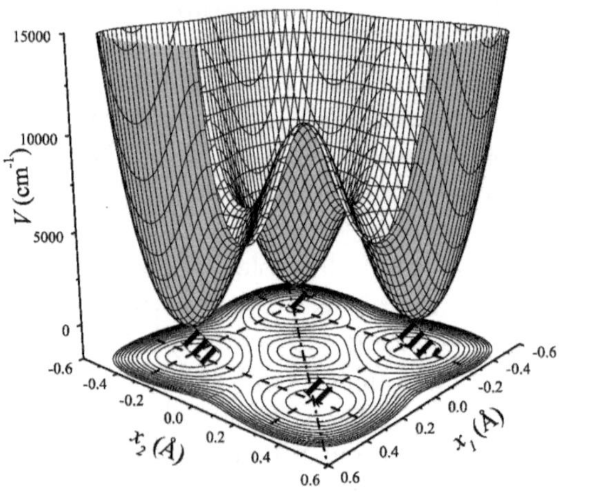

Figure 10 Potential surface for interconversion of uncorrelated protons in a centrosymmetric hydrogen bonded dimer. Coordinates x_1 and x_2 correspond to the stretching modes. The long dash lines are schematic representations of the reaction paths for uncorrelated proton transfers. The potential function is $\mathcal{V}(x_1, x_2) = V_0(x_1) + V_0(x_2) + a_0(x_1 + x_2)$, where $V_0(x)$ is the symmetric part of the double minimum potential represented in Fig. 7.

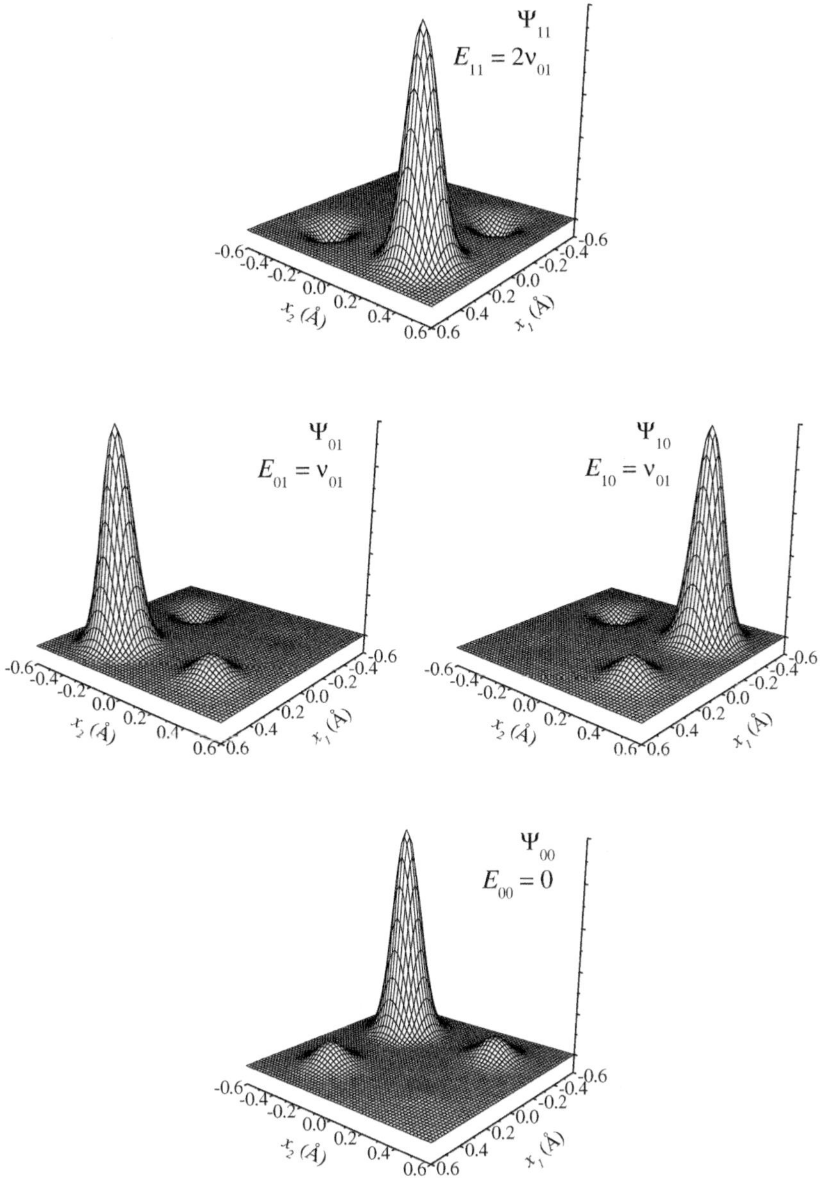

Figure 11. Schematic view in two dimensions of the wave functions for the three lower states of the potential surface represented in Fig. 10. They are simple product in two dimensions of the wave function presented in Fig. 7. The amplitude for the minority entity in each state should be $\approx 1\%$ of the total. For the sake of clarity it was magnified by a factor of 10.

4. Proton dynamics in the ground state: quantum entanglement

As already emphasized, elastic neutron scattering is unique to probing dynamics in the ground state. Then, quantum effects give rise to interferences reminiscent of those observed in optic with double slits or gratings. Moreover, interference patterns impossible to realize with photons are readily observed with neutrons. They reveal periodic structures, superimposed to the crystal lattice, due to quantum-entangled dynamics. The conventional representation of crystals with density probabilities in a periodic lattice of nuclei must be complemented with non factorable vibrational wave functions extending themselves over large domains. Observation of these macroscopic states is a dramatic burst of quantum mechanics in our macroscopic world [Leggett 2002].

4.1 Pairs of coupled oscillators: dynamics

In $KHCO_3$, intra and inter-dimer coupling terms can be distinguished experimentally. Intra-dimer terms give B_u and A_g species observed in the infrared and with Raman, respectively [Novak 1963; Nakamoto 1965; Lucazeau 1973]. INS techniques, on the other hand, probe the density-of-states due to inter-dimer coupling terms. As the observed bandwidths for proton modes are similar with the three techniques, inter-dimer coupling terms can be regarded as weak perturbation giving rise to phonons with negligible dispersion.

The Hamiltonian for a coupled pair of anisotropic harmonic oscillators can be written as $H = \sum_{\alpha} H_\alpha$, $\alpha = x, y$ or z, with [Fillaux 1998]

$$H_\alpha = \frac{P_{1\alpha}^2 + P_{2\alpha}^2}{2m} + \frac{m\,\omega_{0\alpha}^2 \left[(\alpha_1 - \alpha_0)^2 + (\alpha_2 + \alpha_0)^2 + 2\,\lambda_\alpha (\alpha_1 - \alpha_2)^2\right]}{2}. \tag{10}$$

$P_{1\alpha}$ and $P_{2\alpha}$ are kinetic momenta. Coordinates α_1 and α_2 are projections onto the α direction of proton positions with respect to the projection of the dimer center of symmetry (see Fig. 12). The harmonic frequency of the uncoupled oscillators at equilibrium positions $\pm\,\alpha_0$ is $\hbar\,\omega_{0\alpha}$. The coupling potential, proportional to λ_α, depends only on the distance between particles. The equilibrium positions of the coupled oscillators are at $\pm\,\alpha_0' = \pm\,\alpha_0/(1 + 4\,\lambda_\alpha)$.

With normal coordinates corresponding to symmetric and antisymmetric displacements of the particles (see Fig. 13)

$$\alpha_s = \frac{\alpha_1 - \alpha_2}{\sqrt{2}},\ P_{\alpha_s} = \frac{P_{\alpha 1} - P_{\alpha 2}}{\sqrt{2}},\ \alpha_a = \frac{\alpha_1 + \alpha_2}{\sqrt{2}},\ P_{\alpha_a} = \frac{P_{\alpha 1} + P_{\alpha 2}}{\sqrt{2}}, \tag{11}$$

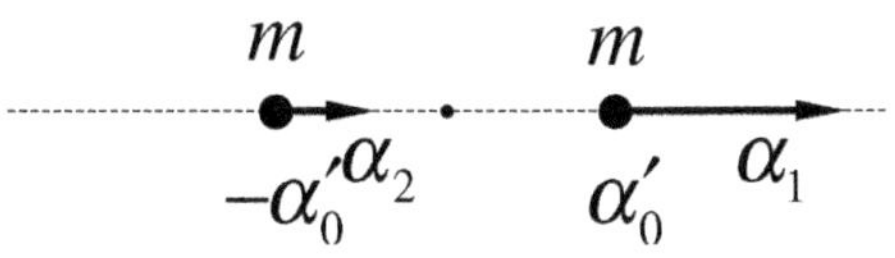

Figure 12. Schematic representation of two coupled linear harmonic oscillators. Equilibrium positions are at $\pm\alpha_0'$. α_1 and α_2 are relative displacements.

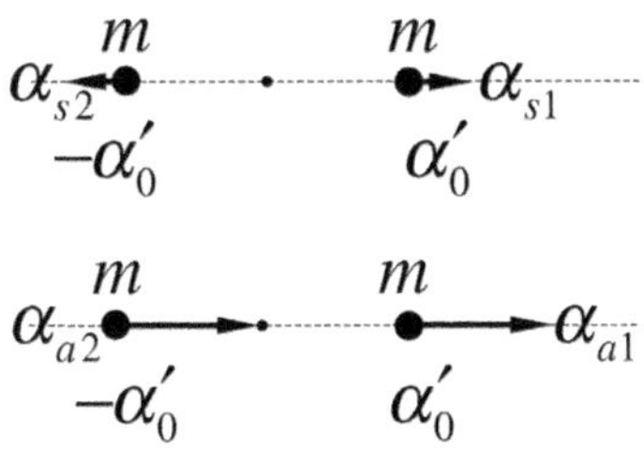

Figure 13. Schematic representation of the symmetric (top) and antisymmetric (bottom) normal coordinates for two coupled linear harmonic oscillators with a center of symmetry.

the Hamiltonian splits into two harmonic oscillators at frequencies $\hbar\,\omega_{s\alpha} = \hbar\,\omega_{0\alpha}\sqrt{1 + 4\,\lambda_\alpha}$ and $\hbar\,\omega_{a\alpha} = \hbar\,\omega_{0\alpha}$, respectively. Quantization gives

$$\begin{aligned} \Psi_{n_a n_s} &= \Psi_{n_a}(\alpha_a)\ \Psi_{n_s}\left(\alpha_s - \sqrt{2}\,\alpha_0'\right), \\ E_{n_a n_s} &= \left(n_a + \frac{1}{2}\right)\hbar\,\omega_{a\alpha} + \left(n_s + \frac{1}{2}\right)\hbar\,\omega_{s\alpha}. \end{aligned} \quad (12)$$

The choice of normal coordinates is arbitrary [Wilson 1964] and definitions at variance from (11) can be found in text books, see for example [Cohen-Tannoudji 1977]. However, the under determination holds only in the classical regime. The effective oscillator mass for coupled proton oscillators is clearly 1 amu according to the scattering function (see above Sec. 3). Only normal coordinates defined in (11) have a physical reality in the quantum regime. At the present time, there is no obvious justification, apart from experiments, for this choice.

4.2 Symmetry-related quantum entanglement in the ground state

For the degenerate ground state, the wave function (10) is relevant only for bosons (deuterons), for the two particles in the same states are indistinguishable. For fermions (protons), spins are correlated in such a way that the total wave function is antisymmetrical with respect to particle permutations, according to the Pauli principle. The spatial wave function can be rewritten as

$$\Theta_{0\pm}(\alpha_1,\,\alpha_2) = \frac{\Psi_0^a(\alpha_a)}{\sqrt{2}}\left[\Psi_0^s\left(\alpha_s - \sqrt{2}\,\alpha_0'\right) \pm \Psi_0^s\left(\alpha_s + \sqrt{2}\,\alpha_0'\right)\right]. \quad (13)$$

The spatially symmetrical wave function (Θ_{0+}) corresponds to the singlet state ($S = 0$) and Θ_{0-} corresponds to the triplet state ($S = 1$). Particle positions and spins are fully entangled. As the ground state is a superposition of Θ_{0+} and Θ_{0-}, there is no preferential spin orientation throughout the crystal.

Wave functions (13) apply to any pair of coupled centrosymmetric protons, irrespective of the distance and of the magnitude of the coupling term. Quantum correlation arises exclusively from normal coordinates representing dynamics of indistinguishable particles. It has been proposed [Keen 2003] that quantum entanglement could be due to the overlap integral of the one-particle wave functions $\psi_0\left(\alpha \pm \alpha_0'\right)$ for protons:

$$\int d\alpha\, \psi_0\left(\alpha + \alpha_0'\right)\, \psi_0\left(\alpha - \alpha_0'\right). \tag{14}$$

However, cursory examination of the crystal structure shows that this integral is zero for protons separated by ≈ 2.2 Å, as in $KHCO_3$ [Fillaux 2004]. This view is totally inappropriate.

Quantum entanglement is observed primarily for simple quantum objects (photons, electrons, atoms, small molecules, *etc*) in environments specially designed to minimize quantum decoherence and dissipation [Bouwmeester 1997; Hagley 1997; Duan 2000 (c); Moore 2000; Pu 2000]. In complex systems, any initially entangled subsystem loses its ability to exhibit quantum interference by getting entangled with the ambient degrees of freedom, *via* interaction with the surrounding environment [Habib 1998; Viola 1999]. Therefore, it has been conjectured that a macroscopic system with many microscopic degrees of freedom can behave quantum mechanically only if it is suitably decoupled from its environment [Caldeira 1983]. In the next section, it is shown that the separation of proton dynamics from the remainder of the crystal lattice occurs naturally, as an inviolable consequence of the Pauli principle.

4.3 Decoupling of the proton dynamics: decoherence-free states

In the harmonic approximation, dynamics of dimers are represented with symmetric and antisymmetric normal coordinates that are linear combinations of atomic displacements (say $\{A_{is},\ A_{ia}\}$, $i = 1,\ 2,\ \ldots,\ N$, where $2N$ is the number of degrees of freedom). For a system exclusively composed of bosons, for example $KDCO_3$, the wave function analogous to (12) is

$$\Xi_0\left(\ldots,\ A_{ia},\ A_{is},\ \ldots\right) = \prod_i^N \Psi_{0i}^a\left(A_{ia}\right)\, \Psi_{0a}^s\left(A_{is} - \sqrt{2}\, A_{i0}'\right) \tag{15}$$

Here, the equilibrium positions for dimer coordinates are at $\pm A_{i0}'$.

For $KHCO_3$, the wave function in the ground state should be antisymmetrical with respect to proton permutation and invariant for permutation of carbon and oxygen atoms (bosons). Therefore, the set of atomic coordinates split into strictly independent subsets, composed of indistinguishable fermions, $\{\alpha_{is}, \alpha_{ia}\}$, on the one hand, and bosons, $\{A_{js}, A_{ja}\}$, on the other. Then, the antisymmetrized wave function $\Xi_{0\pm}$ analogous to (13) can be factored into wave functions Φ_0^F and Φ_0^B, for fermions and bosons, respectively,

$$\Xi_{0\pm}\left(\ldots, \alpha_{is}, \alpha_{ia}, \ldots, A_{js}, A_{ja}, \ldots\right) =$$

$$\prod_i \frac{\Phi_{i0}^{Fa}\left(\alpha_{ia}\right)}{\sqrt{2}} \left[\Phi_{i0}^{Fs}\left(\alpha_{is} - \sqrt{2}\,\alpha'_{i0}\right) \pm \Phi_{i0}^{Fs}\left(\alpha_{is} + \sqrt{2}\,\alpha'_{i0}\right)\right] \tag{16}$$

$$\times \prod_j \Phi_{j0}^{Ba}\left(A_{ja}\right) \Phi_{j0}^{Bs}\left(A_{js} - \sqrt{2}\,A'_{j0}\right).$$

The decoupling of protons from lattice dynamics cancels the main decoherence mechanism and, if there is no further decoherence process, quantum entanglement can last for long time, at least long enough to be observed with elastic neutron scattering techniques. The symmetry-related decoupling of protons arising from the Pauli principle is more fundamental, and certainly more efficient, than any accidental or even specially designed (if this is ever possible) cancellation of off-diagonal coupling terms in the dynamical matrix. This separation is a prerequisite for observing quantum interferences [Fillaux 1998; Ikeda 1999]. Among alternative decoherence mechanisms, spin-spin coupling is on the order of 10^4 Hz. Therefore, quantum decoherence might occur on a time scale as long as $\approx 10^{-4}$ s.

4.4 The elastic scattering function for coupled pairs of protons: quantum interferences

The incoherent elastic neutron scattering function for a pair of coupled bosons obtained from Eqs. (1) and (12)

$$S\left(Q_\alpha, \omega\right) = \left|\left\langle \Psi_{0a}\left(\alpha_a\right) \Psi_{0s}\left(\alpha_s - \sqrt{2}\,\alpha'_0\right)\right| \exp\left[i\, Q_{\alpha_1}\left(\alpha_1 - \alpha'_0\right)\right]\right.$$

$$\left. + \exp\left[i\, Q_{\alpha_2}\left(\alpha_1 + \alpha'_0\right)\right] \left|\Psi_{0a}\left(\alpha_a\right) \Psi_{0s}\left(\alpha_s - \sqrt{2}\,\alpha'_0\right)\right\rangle\right|^2 \delta\left(\omega - \omega_{if}\right), \tag{17}$$

can be written as [Fillaux 1998]

$$S\left(Q_\alpha, \omega\right) = 2\exp\left[-\, Q_\alpha^2\left(\frac{u_{0\alpha}^2}{2\sqrt{1 + 4\,\lambda_\alpha}} + \frac{u_{0\alpha}^2}{2}\right)\right]\delta\left(\omega\right). \tag{18}$$

This profile is very similar to that of a single harmonic oscillator, Eq. (4). Practically, it is impossible to distinguish experimentally gaussian profiles for a single oscillator or for a coupled pair.

In order to calculate the cross-section, the scattering length operator of each atom is expressed in terms of the neutron spin $\vec{s}$ and nuclear spin $\vec{I}$ operators as $b = A + B\,\vec{s}.\vec{I}$. A and B are related to the coherent and total cross-sections as: $\sigma_c = 4\pi\ A^2$ and $\sigma = 4\pi[A^2 + B^2\ I(I+1)/4]$. If there is no spin correlation the intensity is proportional to the "incoherent" scattering cross-section $\sigma_{inc} = \overline{|b|^2} - |\overline{b^2}| = \pi B^2 I(I+1)$, with $I = 1/2$ or 1 for H or D, respectively [Lovesey 1984].

For coupled fermions the singlet $|0+\rangle$ and triplet $|0-\rangle$ states resemble para and ortho species of the hydrogen molecule. For the neutron scattering technique, this is a reference example of quantum interference arising from exchange interaction [Lovesey 1984]. However, for the free H_2 molecule, most of the translational degrees of freedom are unbound. Vibrational dynamics in the ground state are totally represented with the symmetric stretching coordinate invariant upon particle permutation. There is no superposition of the singlet and triplet state and the system is a mixture of the two species with well defined concentrations. The particle positions are distinguishable and quantum interferences arise because the nuclear spins are entangled.

In the present case of two coupled oscillators, the singlet and triplet states do not correspond to different entities and quantum interference arises because both spins and positions of the two particles are entangled. In ordinary words, though counterintuitive, a neutron is scattered coherently by the "same" nucleus with the "same" spin at both sites, either in-phase for the singlet state or anti-phase for the triplet state. In any case, spin-flips at the two sites have opposite signs and compensate exactly one another. The initial and final spin-states are identical. Spin statistics for the singlet and triplet states do not matter and the scattering cross-section is merely proportional to the total cross section $\sigma = \overline{|b|^2}$. For elastic scattering, there is no energy transfer and quantum interferences may arise. The scattering function can be written as [Fillaux 1998]

$$S\left(Q_\alpha,\,\omega\right) = \delta\left(\omega\right)\{|\langle\Theta_{\alpha_0\tau_i}|\exp\left[i\,Q_\alpha(\alpha_1-\alpha_0')\right] + \left(\tau_i\,\tau_f\right)\,\exp\left[i\,Q_\alpha(\alpha_1+\alpha_0')\right]|\,\Theta_{\alpha_0\tau_f}\rangle|^2 + |\langle\Theta_{\alpha_0\tau_i}\,|\exp\left[i\,Q_\alpha(\alpha_2-\alpha_0')\right] + \left(\tau_i\,\tau_f\right)\,\exp\left[i\,Q_\alpha(\alpha_2+\alpha_0')\right]|\,\Theta_{\alpha_0\tau_f}\rangle|^2\Big\}. \tag{19}$$

Indices i and f refer to the initial and final states. τ = "$+$" or "$-$" for the singlet and triplet states, respectively. $(\tau_i\,\tau_f)$ = "$+$" if $\tau_i = \tau_f$ or "$-$" if $\tau_i \neq \tau_f$. The first term of the sum corresponds to scattering by particle 1 at site 1 or site 2. The second term describes the same process for particle 2. Straightforward calculation gives [Fillaux 1998]

$$\begin{aligned} S\left(Q_\alpha,\,\omega\right)_{0+\,0+} &= S_{0\alpha}\left(Q_\alpha,\,\omega\right)\,\cos^2\left(Q_\alpha\,\alpha'_0\right) \\ &\quad\times\left[\cos\left(Q_\alpha\,\alpha'_0\right)+\exp\left(-\frac{\alpha'^2_0}{2\,u^2_{0\alpha}}\right)\right]^2, \\ S\left(Q_\alpha,\,\omega\right)_{0\pm\,0\mp} &= S_{0\alpha}\left(Q_\alpha,\,\omega\right)\,\sin^4\left(Q_\alpha\,\alpha'_0\right), \\ S\left(Q_\alpha,\,\omega\right)_{0-\,0-} &= S_{0\alpha}\left(Q_\alpha,\,\omega\right)\,\cos^2\left(Q_\alpha\,\alpha'_0\right) \\ &\quad\times\left[\cos\left(Q_\alpha\alpha'_0\right)-\exp\left(-\frac{\alpha'^2_0}{2\,u^2_{0\alpha}}\right)\right]^2. \end{aligned} \tag{20}$$

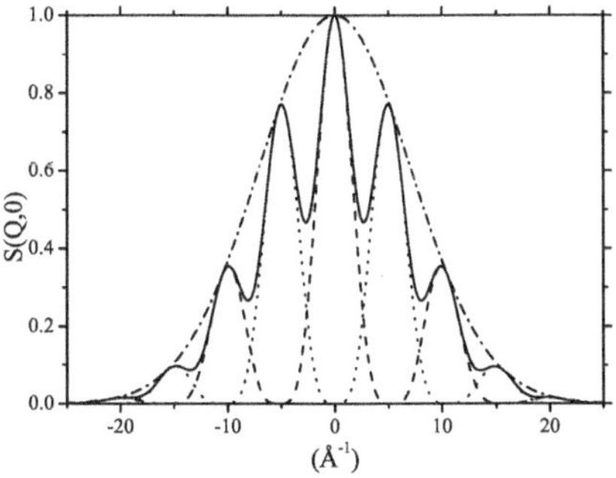

Figure 14. Comparison of the theoretical profiles for a coupled pair of bosons, according to Eq. (18) (dot-dashed line), and a coupled pair of fermions, according to Eq. (20) (solid line). The interference fringes arise from coherent scattering $|0\pm\rangle \rightarrow |0\pm\rangle$ (dashed line) and $|0\pm\rangle \rightarrow |0\mp\rangle$ (dotted line), respectively.

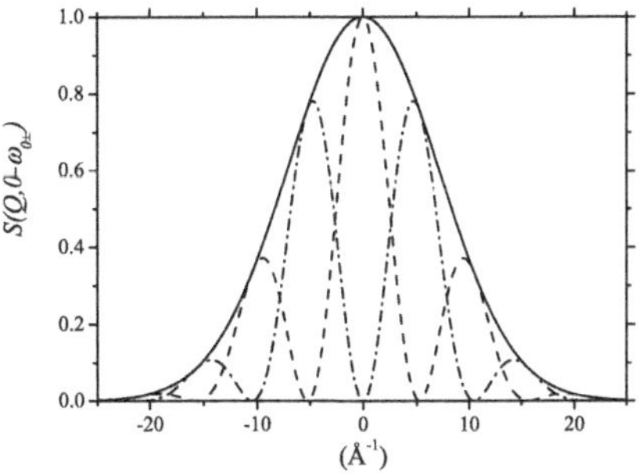

Figure 15. Theoretical profiles for neutron scattering by the tunnelling states of a symmetric double minimum potential, according to Eq. (22). Elastic scattering ($|0\pm\rangle \longleftrightarrow |0\pm\rangle$): dashed line. Inelastic scattering ($|0\pm\rangle \longleftrightarrow |0\mp\rangle$): dot-dashed line. Total scattering function: solid line.

For the dimer of $KHCO_3$, $\alpha'^2_0 \gg u^2_{0\alpha}$. The term $\exp[-\,\alpha'^2_0/(2\,u^2_{0\alpha})]$ is negligibly small and the Gaussian-like profile, analogous to that in Eq. (18), is modulated by $\cos^4(Q_\alpha\,\alpha'_0)$ for $|0\pm\rangle \rightarrow |0\pm\rangle$ transitions, or by $\sin^4(Q_\alpha\,\alpha'_0)$ for $|0\pm\rangle \rightarrow |0\mp\rangle$ transitions (see Fig. 14).

4.5 The scattering function for proton tunnelling in a symmetric double minimum potential

Tunnelling of a single particle in a symmetric double minimum potential is also a reference example of quantum interference for neutron scattering. Even if there is no such symmetrical double minimum potential for $KHCO_3$ [Fillaux 1983 (b)], this case deserves particular attention for further comparison with Eq. (20).

For a single particle in a symmetrical double minimum potential the ground state splits into two sublevels, symmetric $|0+\rangle$ and antisymmetric $|0-\rangle$, respectively. The tunnel splitting is $E_{0-} - E_{0+} = \hbar\omega_{0\pm}$. If the potential barrier is sufficiently high, the wave functions can be written within the two-state approximation as symmetrical and antisymmetrical combinations of harmonic wave functions centered at the potential minima $\pm\,\alpha_0$

$$\Psi_{0\pm}(\alpha) = \frac{[\Psi_0(\alpha - \alpha_0) \pm \Psi_0(\alpha + \alpha_0)]}{\sqrt{2\left[1 \pm \exp\left(-\dfrac{\alpha_0^2}{2\,u_{0\alpha}^2}\right)\right]}}. \tag{21}$$

The scattering function is then [Fillaux 1998]

$$S(Q_\alpha,\omega)_{0+\,0+} = \frac{\left[\cos(Q_\alpha\,\alpha_0) + \exp\left(-\dfrac{\alpha_0^2}{2\,u_{0\alpha}^2}\right)\right]^2}{1 + \exp\left(-\dfrac{\alpha_0^2}{2\,u_{0\alpha}^2}\right)}\,\bar{S}(Q_\alpha)\,\delta(\omega),$$

$$S(Q_\alpha,\omega)_{0\pm\,0\mp} = \frac{\sin^2(Q_\alpha\,\alpha_0)\;\bar{S}(Q_\alpha)\;\delta(\omega_{0\pm} \pm \omega)}{\sqrt{1 - \exp\left(-\dfrac{\alpha_0^2}{u_{0\alpha}^2}\right)}},$$

$$S(Q_\alpha,\omega)_{0-\,0-} = \frac{\left[\cos(Q_\alpha\,\alpha_0) - \exp\left(-\dfrac{\alpha_0^2}{2\,u_{0\alpha}^2}\right)\right]^2}{1 - \exp\left(-\dfrac{\alpha_0^2}{2\,u_{0\alpha}^2}\right)}\,\bar{S}(Q_\alpha)\,\delta(\omega), \tag{22}$$

with $\bar{S}(Q_\alpha) = \exp\left(-\,Q_\alpha^2\,u_{0\alpha}^2\right)$. If $\alpha_0^2 \gg u_{0\alpha}^2$, this Gaussian profile is modulated by either $\cos^2(Q_\alpha\,\alpha_0)$ for $|0\pm\rangle \rightarrow |0\pm\rangle$ or $\sin^2(Q_\alpha\,\alpha_0)$ for $|0\pm\rangle \rightarrow |0\mp\rangle$ (see Fig. 15). These terms arise because neutrons are scattered by the particle delocalized over both sites, either in-phase or out-of-phase.

The scattering function for elastic scattering, $|0\pm\rangle \rightarrow |0\pm\rangle$, is analogous to that for double-slit experiments with neutrons [Zeilinger 1988] or with electromagnetic waves. By contrast, interference fringes for inelastic scattering, $|0\pm\rangle \rightarrow |0\mp\rangle$, cannot be realized with double-slits and neutrons. In principle, this can be observed with electromagnetic waves. However, whereas in-phase and out-of-phase scattering are observed separately with electromagnetic waves, interference fringes arising from tunnelling are observed only if the splitting is greater than the energy-resolution of the neutron scattering experiments. Otherwise, summation of the $\cos^2(Q_\alpha\ \alpha_0)$ and $\sin^2(Q_\alpha\ \alpha_0)$ terms is a constant and only the Gaussian profile, free of quantum interference, corresponding to a simple harmonic oscillator is observed (see Fig. 15).

4.6 Double-slit experiments with neutrons

Interference fringes due to quantum entanglement can be thus clearly distinguished from those arising from other quantum effects, such as spin entanglement or symmetric double minimum potential. They were effectively observed for $KHCO_3$ [Ikeda 1999].

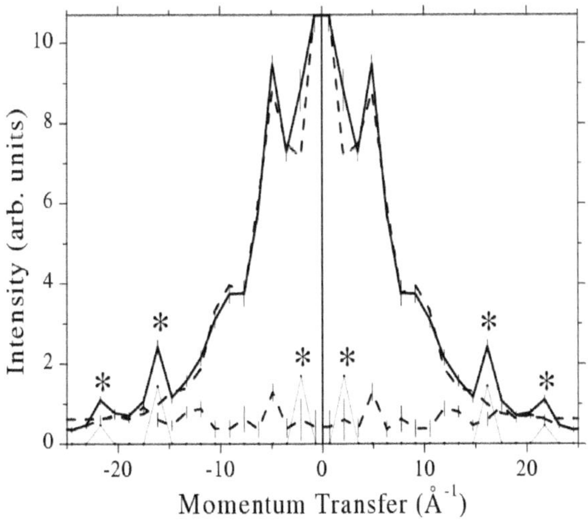

Figure 16 Cut of $S(Q_x,\ Q_y,\ 0)$ along Q_y for a $KHCO_3$ single-crystal at 20 K. Comparison of the measured profile with the MARI spectrometer (solid line with error bars) to the best fit (dash line) obtained with Eq. (16) convoluted with a triangular resolution function. Triangular functions (*) were attributed to other scattering processes. $u_{0y}^2 = 1.26 \times 10^{-2}$ Å^2 and $y_0' = 0.31$ Å. The dash line with error bars is the difference spectrum.

Elastic neutron scattering experiments performed on single crystals with the MARI spectrometer at the ISIS pulsed neutron source (Rutherford Appleton Laboratory, Chilton UK)[3] have evidenced quantum interference in accordance with Eq. (20) [Ikeda 1999]. For example, the best fit to the cut of $S(Q_x,\ Q_y,\ 0)$ along Q_y is presented in Fig. 16.

The profile is quite different from the Gaussian curve Eq. (18). Quantum interferences account for maxima of intensity at $\approx \pm 5$ Å^{-1}, maxima of $S(Q,\ \omega)_{0\pm\ 0\mp}$, and for weaker shoulders at $\approx \pm 10$ Å^{-1}, secondary maxima of $S(Q,\ \omega)_{0\pm\ 0\pm}$. These features cannot be confused with Bragg reflections.

They confirm that any decoherence mechanism is largely cancelled. There is no evidence for significant incoherent scattering, as anticipated for fully entangled protons.

Distances between scatterers of $\approx (0.60 \pm 0.06)$ Å estimated from the interference fringes do not correspond to any visible distance between protons in the crystal structure. However, they correspond closely to distances between projections of proton positions onto directions corresponding to normal coordinates (namely, x, y or z as shown in Fig. 1). It appears that neutrons are scattered coherently by lines of protons quite visible in the crystal structure (see Fig. 17). The elastic scattering function presented in Fig. 16 corresponds to a double-quantum-slits experiment. The interference fringes are quite different from those observed for neutrons scattered by double-classical-slits [Zeilinger 1988]. It can be concluded that quantum entanglement is not limited to single pairs.

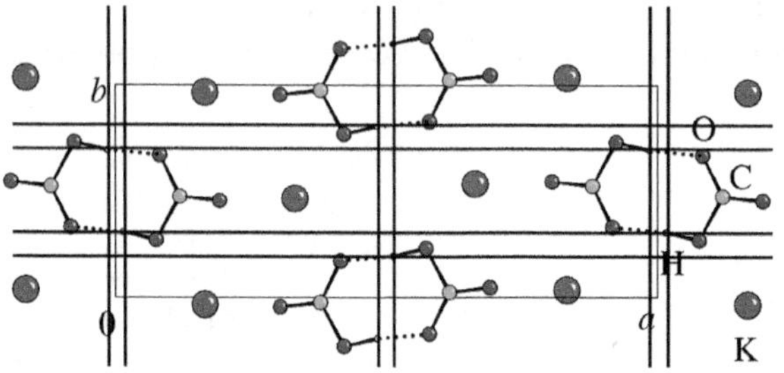

Figure 17 Schematic view of the lines of protons giving rise to quantum interference superimposed to the projection onto the $(a,\ b)$ plane of the $KHCO_3$ crystal at 14 K.

This first evidence for macroscopic quantum coherence of proton pairs has benefited from the transversal coherence length of the neutron beam that is virtually equal to the beam section (namely 3.6×2.4 cm^2). Therefore, elastic neutron scattering experiments carried out with the MARI spectrometer probe long-range quantum coherence perpendicular to the incident beam direction. In these experiments single crystals with cylindrical shape (≈ 1 cm diameter and ≈ 3 cm length) were oriented with lines of protons parallel to the cylinder axis and perpendicular to the beam axis. Therefore, quantum coherence was effectively probed on the cm scale. By contrast, the longitudinal coherence of the neutron beam is limited by the monochromaticity of the beam to ≈ 20 Å, that is similar to the unit cell dimensions [Ikeda 1999]. Consequently, quantum coherence across unit cells along the beam propagation could not be probed. Neutron diffraction experiments utilizing a much longer longitudinal coherence length were performed to probe spatially extended quantum correlation in multi-dimensions.

4.7 The crystal of dimers

In the crystal, quantum entanglement of proton pairs within dimers "hides" the fermionic nature of protons. The singlet or triplet states obey the Bose-

Einstein statistics and collective dynamics are represented with phonons totally decoupled from the lattice. The ground states can be represented as $2A_g + B_g + 2A_u + B_u$ symmetry species [Lucazeau 1973]. The g and u species correspond to collective oscillations of the singlet and triplet states, respectively. The ground state must be regarded as a superposition of coherent states of protons fully entangled with respect to spins and positions. This long-range quantum coherence can be probed with neutron diffraction.

Then, by analogy with the fully entangled pairs, the diffraction intensity should be proportional to the total cross-section for protons ($\approx$ 82 barns). (In the absence of quantum correlation the contribution of H atoms to Bragg-peak intensities is proportional to the coherent scattering cross section - $\sigma_c \approx 1.76$ barns - while the intensity for incoherent scattering is proportional to $\sigma_i \approx 80$ barns.) The dramatic enhancement of intensity offers a better chance for observing diffraction by entangled protons among the Bragg-peaks.

Along the x axis we distinguish two subsets of dimeric entities (labelled I or II in Fig. 1.) They are symmetrical with respect to the binary axis parallel to (a). Phonons for singlet or triplet states correspond to B_g and B_u symmetry species, respectively. For elastic scattering with momentum transfer Q_x, neutrons are scattered coherently by lines of protons parallel to the y direction (see Fig. 17). These lines are arranged in a grating-like structure composed of two subsets of parallel double-slits, with a spacing of 2 x_0' for each double-slits. The distance between equivalent pairs is $D_x \approx a/\cos 42° \approx 20.27$ Å. One subset of double slits is shifted with respect to the other by $D_x/2$ and neutrons are scattered anti-phase by the two subsets. The scattering function for coherent elastic scattering by this grating-like structure can be written as

$$\begin{aligned} S(Q_x,\, \omega)_{0\mathcal{T}\,0\mathcal{T}} &= \delta(\omega) \\ &\times \Bigg| \sum_j \langle \Theta_{x_0\mathcal{T}_j} | \{ \exp[i\, Q_x(x_{1j} - x'_{0j})] + \exp[i\, Q_x(x_{1j} + x'_{0j})] \} \\ &\quad \exp(ij\, Q_x\, D_x) | \Theta_{x_0\mathcal{T}_j} \rangle + \langle \Theta_{x_0\mathcal{T}_j} | \{ \exp[i\, Q_x(x_{2j} - x'_{0j})] \\ &\quad + \exp[i\, Q_x(x_{2j} + x'_{0j})] \}\ \exp(ij\, D_x) | \Theta_{x_0\mathcal{T}_j} \rangle \\ &\quad - \langle \Theta_{x_0\mathcal{T}_{j'}} | \{ \exp[i\, Q_x(x_{1j'} - x'_{0j'})] + \exp[i\, Q_x(x_{1j'} + x'_{0j'})] \} \\ &\quad \exp(ij'\, D_x) | \Theta_{x_0\mathcal{T}_{j'}} \rangle - \langle \Theta_{x_0\mathcal{T}_{j'}} | \{ \exp[i\, Q_x(x_{2j'} - x'_{0j'})] \\ &\quad + \exp[i\, Q_x(x_{2j'} + x'_{0j'})] \}\ \exp(ij'\, D_x) | \Theta_{x_0\mathcal{T}_{j'}} \rangle \Bigg|^2, \end{aligned} \tag{23}$$

where j indexes the lattice sites and $j' = (j + 1/2)$. For waves scattered by equivalent pairs, intensity is a maximum if $D_x/(2\,x'_0) = N_x$ is an integer number. Furthermore, constructive interference for waves scattered by the two subsets of double slits occurs when $D_x/(4\,x'_0) = N_x/2$ is half integer. All together, the intensity is a maximum if N_x is an odd integer number and $Q_x = \pm\, n_x\, \pi/x'_0$, according to (16).

Similar lines of protons parallel to x can be seen in Fig. 17. They form a system of double slits separated by $2\,y'_0$. Phonons along y correspond to A_g and A_u symmetry species for the singlet and triplet states, respectively. The distance between double slits is $D_y = b/2 \approx 2.81$ Å. Phase matching occurs if $N_y = b/(4\,y'_0)$ is an integer number, either odd or even. The intensity is a maximum at $Q_y = \pm\, n_y\, \pi/y'_0$.

With numerical values derived from previous experiments [Ikeda 1999] and from the crystal structure [Fillaux 2003 (a)] we estimate $N_x = 33 \pm 3$ and $N_y = 4.5 \pm 0.5$. These values are quite compatible with the phase matching conditions but, regarding error bars, it is not possible to conclude whether the grating-like structures can be observed. We need direct measurement of the diffraction pattern (see next subsection) to confirm long-range quantum entanglement of protons.

4.8 Diffraction and quantum entanglement

For $KHCO_3$ and $KDCO_3$ at 15 K, rather large volumes of the reciprocal space were measured parallel to $(a^\star,\ c^\star)$ planes with the SXD diffractometer at the ISIS pulsed neutron source.[4] With the time-of-flight technique, the accessible range of reciprocal space is measured all at once for each neutron pulse. This is convenient for seeking signals in addition to the Bragg's peak intensities. Moreover, the high flux of epithermal neutrons delivered by the spallation source is required to probing large $|\vec{Q}|$ values. However, with the limited number of detectors available on SXD at the time of these measurements, only a limited sector of reciprocal space was measured for each crystal orientation. Fig. 18 and 19 are concatenations of data collected for several crystal orientations corresponding to rotations by steps of $\approx 30°$ around a fixed axis perpendicular to the equatorial plane of the detectors.

In Fig. 18, the $(\overline{7}0\overline{1})$ direction corresponds to momentum transfer parallel to the dimer planes ($Q_y = Q_z = 0$). Elastic and inelastic incoherent scattering give the broad signal centered at $\vec{Q} = \vec{0}$, hiding quantum interferences observed in Fig. 16. (With the SXD instrument, neutrons scattered with different final energies are not distinguished.)

The ridges at $\pm(10.25 \pm 0.25)$ Å^{-1} from the center, perpendicular to the $(\overline{7}0\overline{1})$ direction (see arrows in Fig. 18), have all characteristics anticipated for coherent scattering by entangled protons in planes containing the dimer enti-

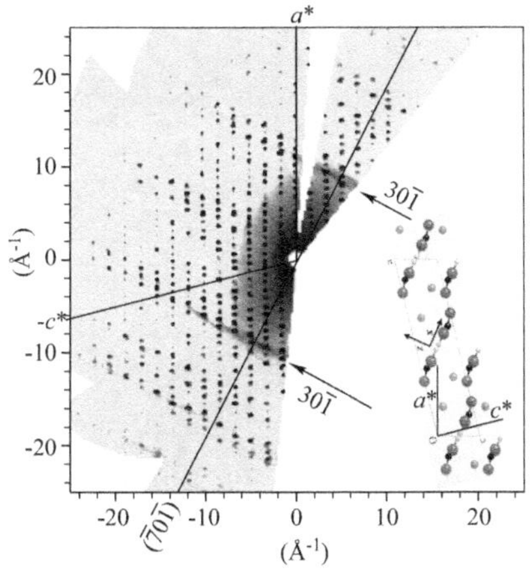

Figure 18 Diffraction pattern of $KHCO_3$ at 15 K in the $(a^\star, c^\star)$ plane. The arrows point to the ridges of intensity. The insert visualizes the correspondence between the direct and reciprocal lattices. The rods of diffuse scattering lie along the $(30\bar{1})$ direction and as such are perpendicular to the plane of the dimers, which contain the x and y directions defined in Fig. 1.

ties:
(i) the ridges are not due to coherent scattering by the whole lattice;
(ii) they are not observed for the deuterated analogue (see Fig. 19);
(iii) they have rod-like shapes;
(iv) orientations and positions are in accord with the structure of the proton sublattice.
Additional ripples of diffuse scattering observed at $\approx \pm 17$ and $\approx \pm 22$ Å^{-1} for $KHCO_3$ are rather weak and broad. As they survive in the deuterated analogue they are not related to quantum statistics. They were tentatively attributed to thermal diffuse scattering [Wilson 2002]. These properties are further documented in Ref. [Fillaux 2003 (a)].

The width of the ridge is related to the spatial extension of quantum coherence. Compared to the half width at half maximum (HWHM) of ≈ 6 Å^{-1} of the Gaussian-like profile for scattering by a single slit, the observed HWHM of ≈ 0.20 Å^{-1} reveals quantum coherence extending itself over more than 30 double slits (more than 300 Å^{-1} along the x direction). This is certainly an underestimate since the instrumental resolution $(\Delta|\vec{Q}|/|\vec{Q}| \approx 1\%)$ is not negligible.

Conclusion

Neutron scattering experiments shed a new light on proton dynamics in the extended arrays of hydrogen bonded centrosymmetric dimer entities of the $KHCO_3$ crystal. Proton dynamics are decoupled from the lattice. Measurements of effective oscillator masses (namely, 1 amu) contribute to full determination of normal coordinates.

The physical meaning of normal coordinates in the quantum regime is further emphasized by interference arising from full quantum entanglement of

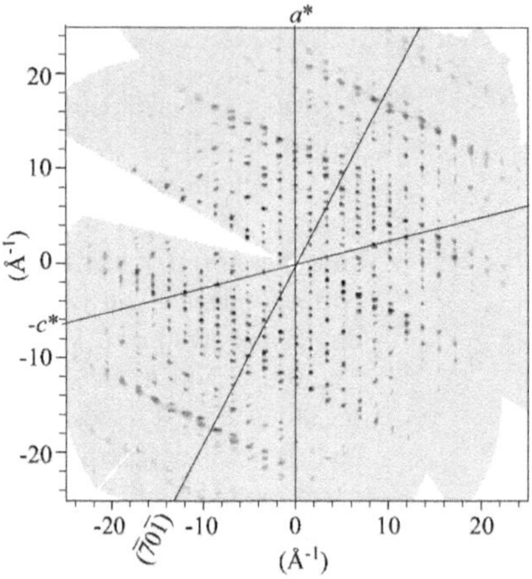

Figure 19 Diffraction pattern of $KDCO_3$ at 15 K in the $(a^\star, c^\star)$ plane.

positions and spins in two dimensions. Regarding the superposition of fully entangled macroscopic vibrational states with long lifetimes, $KHCO_3$ can be termed a "quantal-crystal" (or Q'stal) whose macroscopic quantum effects should be compared to the exotic behavior of superconductivity or superfluidity. Although we do not expect marked transport properties in $KHCO_3$, macroscopic wave functions free of decoherence effects could be of importance for quantum information storage and processing. The quantum character is clearly observed at 15 K, which is already a high temperature compared to normal operation of superconducting devices. This new state of the matter deserves further investigations in $KHCO_3$ and possibly in similar systems.

The double minimum potential for the transfer of a single proton along the hydrogen bond offers a plausible mechanism for proton disordering at high temperature. However, the nature of this disordering is not yet clearly established. It is not known whether this disorder should be regarded as a coherent superposition of states, as in benzoic acid [Fillaux 2002], or as a statistical distribution of probability densities. For coherent disorder quantum entanglement should survive. Otherwise, decoherence could be monitored by the asymmetrical double minimum potential and a transition from Q'stal to C'stal (classical crystal) should occur.

PROBING SHORT-LIVED ENTANGLEMENT WITH INELASTIC X-RAY SCATTERING

Molecular vibrations - First experimental results

H. Naumann,[1] T. Abdul-Redah[2] and C. A. Chatzidimitriou-Dreismann[3]

[1] *Institut für Chemie, Vollmer Laboratorium, PC 14, Technische Universität Berlin, Straße des 17. Juni 135, D-10623 Berlin, Germany*
naumann@chem.tu-berlin.de

[2] *ISIS Facility, Rutherford Appleton Laboratory, Chilton / Didcot, OX11 0QX, UK*
t.abdul-redah@rl.ac.uk

[3] *Institut für Chemie, Stranski Laboratorium, Technische Universität Berlin, Straße des 17. Juni 112, D-10623 Berlin, Germany*
dreismann@chem.tu-berlin.de

Abstract We present preliminary experimental results of inelastic X-ray scattering (IXS) on molecular vibrations of liquid H_2O, D_2O and the equimolar H_2O - D_2O mixture. The data analysis indicates the presence of an anomalous shortfall of scattering intensity from the OH-stretching vibrational modes. This effect has no explanation within the frame of conventional X-ray scattering theory. The possible connection of these observations with recent results of neutron and electron Compton scattering from protons in condensed matter is mentioned, as well as their interpretation in terms of attosecond entanglement.

Keywords: Attosecond dynamics, entanglement, decoherence, inelastic X-ray scattering, molecular vibrations, hydrogen bonds, water

Introduction

In a Raman light scattering (RS) experiment from liquid H_2O, D_2O and their mixtures an anomalous shortfall of scattering intensity from the OH-stretching vibrational modes has been observed [Chatzidimitriou-Dreismann 1995]. Various theoretical analyses showed that this effect could not find a proper interpretation in the frame of standard theory of light scattering. This experiment was motivated by qualitative theoretical and experimental investigations indicating the presence of very short-lived quantum entangled states [Einstein 1935] of H-atoms (or protons) in molecules and condensed mat-

V.M. Akulin et al. (eds.), Decoherence, Entanglement and Information Protection in Complex Quantum Systems, 529–534.

ter; *cf.* Ref. [Chatzidimitriou-Dreismann 1997 (b)]. A related effect was observed in recent neutron-proton (NCS) and electron-proton (ECS) Compton scattering experiments in condensed matter [Chatzidimitriou-Dreismann 1997 (a); Chatzidimitriou-Dreismann 2003 (a); Physics News 2003; Physics Today 2003; Scientific American 2003], in which the experimental scattering time lies in the sub-femtosecond time scale. It was noticed that this experimental time-window coincides with the characteristic time of electronic rearrangements accompanying the breaking (or formation) of a chemical bond. Note also that in these NCS and ECS experiments the energy transferred to a proton is large enough to break the bonds (C - H and O - H).

Very recently, we extended these studies [Chatzidimitriou-Dreismann 1995; Chatzidimitriou-Dreismann 1997 (a); Chatzidimitriou-Dreismann 2003 (a)] into the field of inelastic X-ray scattering (IXS) [Halcoussis 2000; Naumann 2002]. Our experiments were carried out at the beamline ID16 of ESRF (European Synchrotron Radiation Facility) in Grenoble, France. In these IXS investigations, scattering from the molecular stretching vibrations of liquid H_2O, D_2O and the equimolar H_2O - D_2O mixture were studied. In other words, the IXS process does not break the chemical bonds. The analogy of these investigations with those of RS [Chatzidimitriou-Dreismann 1995] is obvious.

1. Theory of inelastic X-ray scattering

The principle of a scattering experiment is illustrated in Fig. 1. A photon with the energy $\hbar\omega_1$, the wave vector $\vec{k}_1$ and the polarization $\vec{\epsilon}_1$ is scattered by the target into a photon with energy $\hbar\omega_2$, the wave vector $\vec{k}_2$ and the polarization $\vec{\epsilon}_2$. The measured quantity is the double differential cross section $d^2\sigma/(d\Omega\, d\omega)$ which is a function of the transferred energy $\hbar\omega = \hbar(\omega_1 - \omega_2)$ and of the transferred momentum $\hbar\vec{q} = \hbar(\vec{k}_1 - \vec{k}_2)$.

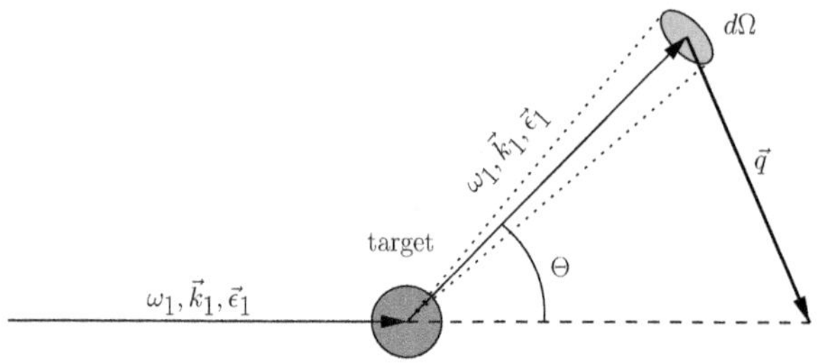

Figure 1 Principle of a scattering experiment

This quantity is given by the following expression [Schülke 1991]

$$\frac{d^2\sigma}{d\Omega\, d\omega} = r_0^2\, (\vec{\epsilon}_1 . \vec{\epsilon}_2)^2 \left(\frac{\omega_2}{\omega_1}\right) S(\vec{q},\, \omega). \tag{1}$$

Here $r_0 = e^2/4\pi\ \epsilon_1\ mc^2$ is the classical electron radius and $S(\vec{q},\ \omega)$ is the scattering function which reflects the properties of the unperturbed scattering system. $S(\vec{q},\ \omega)$, which is also called the dynamic structure factor, can be expressed [van Hove 1954] by

$$S(\vec{q},\ \omega) = \frac{1}{2\pi} \int e^{-i\ \omega t} \sum_I W_I \left\langle I \left| \sum_j e^{-i\ \vec{q}.\vec{r}_j(t)} \sum_k e^{i\ \vec{q}.\vec{r}_k(0)} \right| I \right\rangle\ dt. \quad (2)$$

In the classical limit the probability W_I of initial state I represents essentially the time Fourier transform of the density correlation function. It contains information about the particle density fluctuations in the scattering system.

In the case of IXS from liquid water, the scattering function becomes dominated by the fluctuations of the electronic density surrounding the oxygen atom when $\vec{q}$ increases, (see below). These fluctuations are for example due to collective motions [Sette 1995] or, for energy transfer large enough, vibrational excitations [Halcoussis 2000] in the molecules. From this analysis arises the advantage of IXS over other techniques for studying quantum correlations of vibrational degrees of freedom. In contrast to IXS, inelastic neutron scattering (INS) takes place mainly on protons, whereas RS (with visible light) is caused by the interaction of light with the entire electron distribution of a molecule (or molecular cluster). Additionally, the wavelength of the probing electromagnetic radiation in IXS (with energy 13.8 keV, wavelength 0.9 Å), is of the same length scale as the spatial extension of a water molecule, which opens the possibility to study the spatial dependence of such effects. While in the case of INS the spectra consist of coherent and incoherent scattering as well as multiple scattering contributions, IXS of molecular vibrations is mainly incoherent which make the data analysis much easier.

2. Experimental method and results

To resolve IXS of molecular vibrations one needs an energy resolution of at least 20 meV. Because the energy of the incident beam is in our experiment 13.8 keV, the spectrometer has to provide a relative energy resolution $\Delta E/E_0$ of 10^{-6}. This can be achieved by using high order Bragg diffraction peaks of silicon crystals [Verbeni 1996]. The energy of the diffracted beam is given by $E = hc/\lambda = hc/2d_k\ \sin\Theta_B$. The experimental setup which exploits this relationship is shown in Fig. 2. The energy of the probing beam is sweeped by changing the lattice constant d_k via accurate temperature control of the crystal (monochromator). The resulting monochromatic beam is focused onto the sample and the scattered radiation is then diffracted at another silicon crystal (analyser) into the detector.

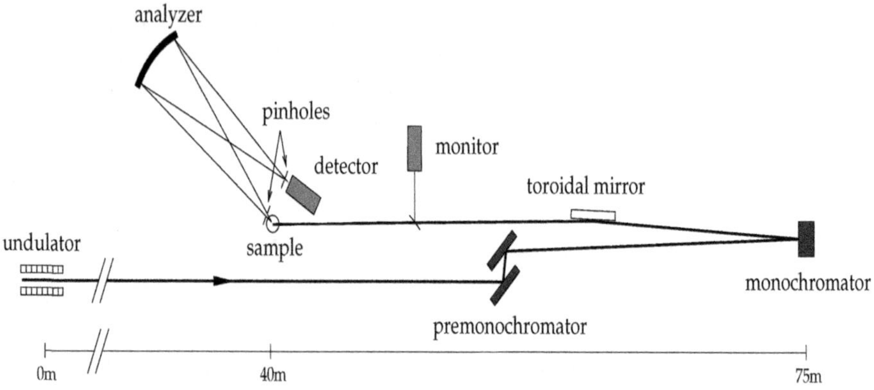

Figure 2. Scheme of the ID16 beamline at ESRF as described in Ref. [Verbeni 1996].

Fig. 3 shows a typical IXS spectrum of H_2O. Experimental details are described in Ref. [Halcoussis 2000; Naumann 2002]. The spectrum shows two distinct features: a peak centered around zero energy transfer due to rototranslational dynamics and another weak but clearly visible band around 430 meV corresponding to the intramolecular OH stretching vibrations of water.

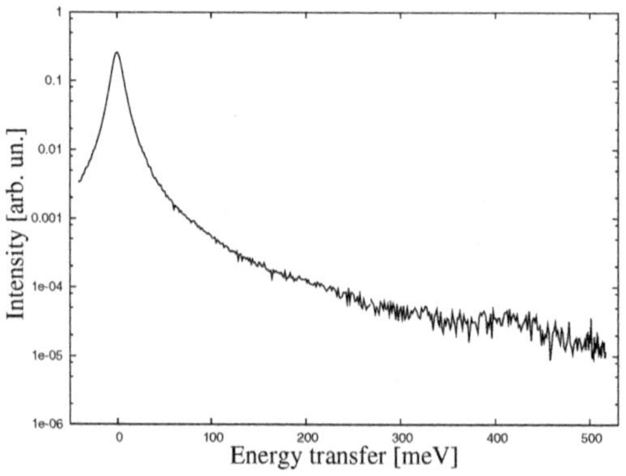

Figure 3 Typical IXS spectrum of H_2O. The stretching vibrational modes are located around 430 meV.

The procedure for probing the quantum entanglement effect is similar to the RS experiment with visible light [Chatzidimitriou-Dreismann 1995]. By comparing the IXS spectra of mixtures of water (H_2O) and heavy water (D_2O) with the pure substances one can obtain an anomalous decrease or increase of the intensities of the OH and OD stretching vibration bands, respectively. The quantity expressing this is defined as follows

$$\Delta\sigma^{OH} = \frac{\sigma^{OH}_{x_H} - \sigma^{OH}}{\sigma^{OH}} \cdot 100\% \tag{3}$$

where σ^{OH} and $\sigma^{OH}_{x_H}$ are the cross sections of the OH stretching band of pure water and the mixture with the molar fraction x_H of H_2O, respectively.

The basis of this analysis is the experimental finding that the microscopic structure of water does not change considerably with the H / D ratio [Turner 1986]. This fact has been recently supported with X-ray absorption spectroscopy (XAS) and non-resonant X-ray Raman scattering (XRS) on H_2O and D_2O [Wernet 2004]. Note that XAS provides very detailed information about the local bonding configurations in the first coordination shell of liquid water by using the near-edge fine structure in X-ray absorption, where a core electron is excited into empty electronic states [Wernet 2004]. Consequently, $\sigma^{OH}_{0.5}$ should be exactly 50% of σ_{OH} of pure H_2O. Similar to Eq. (3) one can define an anomalous scattering cross section for the OD stretching vibration $\Delta\sigma_{OD}$. The analysis of the individual cross sections is possible because the OD and OH band are very well separated as shown in Fig. 4.

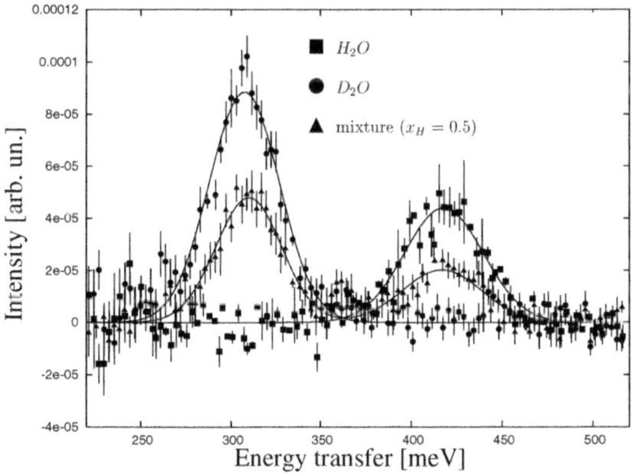

Figure 4 IXS spectra of H_2O, D_2O and a mixture with molar composition of 50% H_2O after subtracting the contributions of the central peak. Gaussians are fitted to the data as a guide to the eye.

3. Additional remarks

The detailed data analysis of our two IXS experiments results revealed that

(A) The IXS-intensity of the OH-peak around 420 meV in the equimolar H_2O - D_2O mixture appears to be ca. 15 - 20% lower than the associated IXS-intensity of the OH-peak in pure H_2O.

(B) The IXS-intensity of the OD-peak around 300 meV in the equimolar H_2O - D_2O mixture seems to be very similar to the associated IXS-intensity of the OD-peak in pure D_2O.

Note that, contrary to the conventional RS using visible light, both the NCS and the IXS techniques apply considerably "larger" momentum transfers (NCS: $q \approx 30 - 200$ Å^{-1}, IXS: $q = 2.9$ Å^{-1} in our experiment), which also implies that the scattering is effectuated by a "smaller" quantum system of linear dimension of the order of $1/q$. Indeed, the IXS-intensity of the OH- and OD-vibrational peaks are known to be caused mainly by scattering on the core electrons of the oxygen atoms. But, from the viewpoint of quantum mechanics, it is also obvious that the O-dynamics is entangled with that of H in H_2O or HDO - and, analogously, with that of D in HDO or D_2O - due to the strong Coulombic interactions. One can also say that the degrees of freedom of O are "dressed" with those of H (and / or D). In other words, the X-rays may scatter from "the core electrons of O", but the O-vibrational degrees of freedom are inextricably intertwined with those of H (and / or D). Interestingly, our results presented above support this viewpoint, since they show a qualitatively different behaviour of the OH- and the OD-vibrational intensities.

Another relevant point concerns the fast H / D exchange in mixtures. As a result, the presence of HDO in the equimolar H_2O - D_2O mixture at room temperature is about 50% (25% is H_2O and 25% is D_2O). Additionally, in IXS (as well as in INS) the intensity is known to be proportional to the square of the amplitudes of the "vibrating units", which here correspond to the vibrational normal modes of the whole molecule. Thus it is conceivable that our experimental findings could be due to a corresponding (unexpected) behaviour of the atomic displacements in the vibrational modes. To investigate this point, we calculated with "GAUSSIAN 03" - applying the two ab initio procedures rb3lyp/6-311g** and ccsd/6-311g** - the atomic displacements of oxygen in H_2O, D_2O and HDO. The results of these calculations indicate that the IXS-intensities of *both* OH- and OD-peaks in the equimolar H_2O - D_2O mixture may be about 5 - 8% lower than the corresponding ones in pure H_2O and D_2O. Additionally, taking into account the associated Debye-Waller factors did not change anything significantly in these calculations. As a consequence, these numerical results cannot explain the experimental observations, which exhibit an opposite behaviour of the OH-peak intensity to that of the OD-peak.

Summarizing, the hitherto existing experimental results seem to be in contrast to well-established conventional expectations. In forthcoming work, we plan to investigate the possible connection and / or common origin of these IXS results with those obtained with NCS and ECS from protons in various systems [Chatzidimitriou-Dreismann 1997 (a); Chatzidimitriou-Dreismann 2003 (a); Physics News 2003; Physics Today 2003; Scientific American 2003]. Novel aspects of attosecond quantum dynamics of hydrogen (and D), as well as hydrogen bonds, in molecules and condensed matter are expected to be revealed.

Acknowledgments

This work was partially supported by the EU RTN QUACS (Quantum Complex Systems: Entanglement and Decoherence from Nano- to Macro-Scales).

PROTON-PROTON CORRELATIONS IN CONDENSED MATTER

E. B. Karlsson
Department of Physics, Uppsala University P. O. Box 530, SE-75121 Uppsala, Sweden
erk@fysik.uu.se

Abstract Certain neutron scattering experiments indicate that protons pairs, or larger clusters, may stay quantum entangled in condensed matter for measurable times. This was first observed in Compton scattering of neutrons, a method which has a time-window of 10^{-16} - 10^{-15} s, but recent experiments with slow neutrons have given supporting evidence that isolated proton dimers in a crystalline material may stay quantum coherent for considerably longer times. Mechanisms leading to local proton entanglement are discussed briefly and a model for neutron scattering on correlated proton pairs is presented, which explains the observed decrease of cross-section as a result of destructive interferences for the particular case when a neutron scatters on two indistinguishable particles. The loss of interference, and the return of the cross-section to its normal value at longer times, is described by a quantitative decoherence model for the protons in water when perturbed by the fluctuating hydrogen bonds.

Introduction

Quantum coherence is extremely sensitive to environmental interactions. This is a main stumbling block in the attempts to build quantum computers, and in spite of the fact that such devices are planned to be based on very weakly interacting systems (entanglement of photons or atoms well isolated in cavities) it is extremely difficult to preserve coherence over a sufficiently large number of basic operations steps. Coherent states in molecules are still more perturbed, as displayed for instance by the difference between the spectra of NH_3 and AsH_3 gases [Omnes 1994]. Here, the H-atom in NH_3 is delocalized in a quantum superposition, being on both sides of the H_3-plane, while the spatial coherence of the heavier As-atom disappears during the time of observation which results in quite different optical properties.

In condensed matter, there are only few examples of coherent states (with the exception of superconductors and superfluids), but certain tunnelling states of protons [Wipf 1987] or positive muons [Karlsson 1998] in metals have been

V.M. Akulin et al. (eds.), Decoherence, Entanglement and Information Protection in Complex Quantum Systems, 535–548.

shown to exhibit local quantum coherence. As illustrated by Fig. 1, coherence is destroyed by the coupling of the tunnelling particle to the phononic and electronic degrees of freedom and survives normally only for times of the order of pico-seconds even when kept at liquid helium temperatures.

It is true that coupling to the phonon bath (which has an approximate T^3-dependence) is strongly reduced at these low temperatures, but the conduction electron coupling is linear in T and continues to give rise to a strong decoherence even at 0.1 K. As was shown by Karlsson *et al.* [Karlsson 1995], such a coupling can only be reduced by going to the superconducting state (which is possible for metal like Al), where the energy gap cuts off the coupling to the electron bath. The coherence of the tunnelling state for a positive muon could then be preserved for as long times as 10^{-6} s.

Coherent states involving two or more atoms (or atomic nuclei) must be expected to be much more sensitive to environmental perturbations than single particle states and it is not surprising that they seem to be totally uncorrelated and can be considered as a collection of independent entities in most scattering experiments. But in the following we provide evidence that quantum entanglement of protons in condensed matter *might* actually be observed, provided the time window for observation is reduced by several orders of magnitude compared to what is used in most common spectroscopies. The reasons for the appearance of local, entangled states of protons in different measurement situations will also be discussed, as well as the mechanisms for their decoherence.

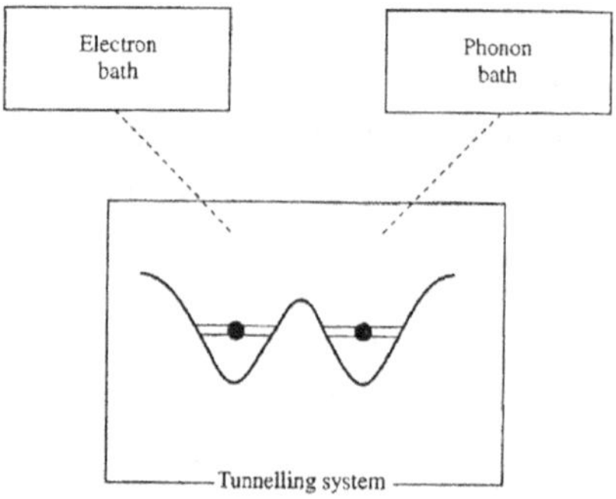

Figure 1. A tunnelling particle (proton or positive muon) in a superposition state in a metal hydride, perturbed by phonons and conduction electrons (from Ref. [Karlsson 1998]).

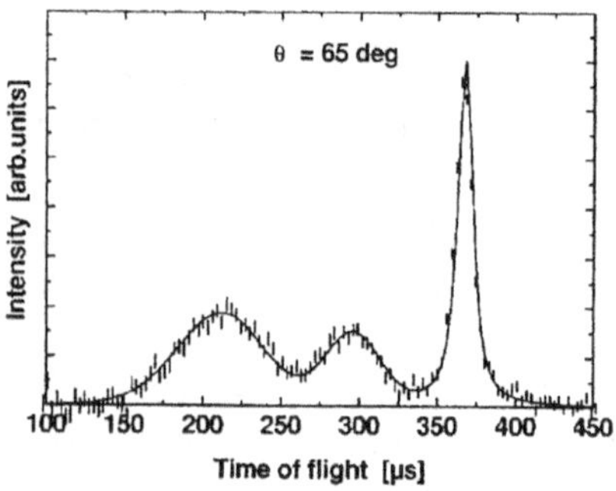

Figure 2. Examples of time-of flight spectra for mixed H- and D-hydrated niobium [Karlsson 1999; Karlsson 2003 (b)].

1. A sub-femtosecond time window

Most common spectroscopies for the investigation of condensed matter work on a time-scale of pico-seconds or slightly below (10^{-13} - 10^{-12} s). When liquid water, for instance, is investigated by NMR or thermal neutron scattering, it is possible to extract details of the dynamics of water molecules, changes in hydrogen bonding, *etc.*, on this time scale. But at least as far as inelastic neutron scattering is concerned, all nuclei in the samples can still be considered as independent scatterers (although, of course, interference effects are observed when the same neutron scatters on two or several neighbouring protons). As will be shown further below, it is plausible that all possible quantum entanglements of neighbouring protons die out very fast in such a fluctuating environment. This situation can be described by stating that decoherence time τ_{coh} is short compared to the observation time τ_{NS} in liquid water. Similarly, very short decoherence times are expected in most solid materials. First estimates for τ_{coh} for an entangled system of extension $|x - x'|$ can be obtained from a formula by Joos and Zeh [Joos 1985], which states that the decoherence rate

$$\tau_{coh}^{-1} \approx \Lambda \, |x - x'|^2, \tag{1}$$

where $\Lambda = N \, \nu \, q^2$, with N being the number of environmental quanta of frequency ν and momentum q that hit the system per second and cm^2. Even for a local system of only a few Angströms size, such estimates end up with decoherence times of the order of 10^{-15} s or less in environments described by typical phonon or conduction electron baths [Karlsson 1999].

Fortunately, one version of neutron scattering has a considerably shorter time window and still maintain the possibility to observe interference from the scattering of neighbouring protons in condensed matter. This is the Compton scattering (also called DINS, Deep INelastic Scattering) method. It is carried out with neutrons of energies in the 10 - 100 eV range, which is about 1000 times higher than in conventional thermal scattering. At these energies the neutron transfers a considerable fraction of its energy to the scattering nucleus (of mass M), which recoils out of its position in molecules or crystal lattices with an energy $E = \hbar^2 q^2 / 2M$, where $\vec{q}$ is the transferred momentum. The velocity of the neutron is measured. The velocity charge obeys the relation

$$\frac{v_{n_1}}{v_{n_0}} = \left(\cos\theta + \sqrt{\left(\frac{M}{m}\right)^2 - \sin^2\theta} \right) \left(\frac{M}{m} + 1 \right)^{-1}, \tag{2}$$

which shows that for each scattering angle θ there will be a peak in the time-of-flight spectrum of the neutrons for each scattering mass M in the sample. An example of such a spectrum is shown in Fig. 2.

The normalized peak areas in this spectrum are proportional to the neutron cross sections (CS) for H, D and Nb in a niobium hydride, with composition NbH_xD_y. The peaks are broadened by the fact that the nuclei have an intrinsic, characteristic spread of momentum Δp when they are hit by the neutron (the original purpose of this method was to deduce the width and shape of the distribution $n(\vec{p})$ for H or D in molecules or crystals [Brugger 1984]).

Another consequence of the momentum spread is that it compresses the time for the formation of the outgoing neutron wave. As first shown by Reiter and Silver [Reiter 1985] and Sears [Sears 1984] and later described in detail by Watson [Watson 1996], this leads to a characteristic neutron Compton scattering time $\tau_{NCS} = M\hbar/[q(\theta)\ \Delta p]$ where $q(\theta)$ is the θ dependent momentum transfer. With typical values for the momentum spread Δp, this leads to the characteristic times τ_{NCS} for scattering on H and D shown in Fig. 3.

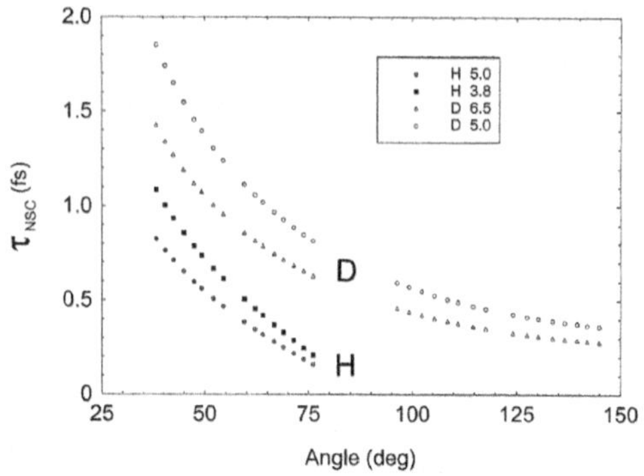

Figure 3 The characteristic time τ_{NCS} for Compton scattering on H and D, as function of scattering angle θ, with energy selection by Au-resonance foils. The widths 3.8 and 5.0 Å^{-1} correspond to proton vibrations in a typical metal hydride and water, respectively. The corresponding widths for deuterons are 5.0 and 6.5 Å^{-1}, respectively.

Compton scattering provides, therefore a sub-femtosecond time window in the range 10^{-16} - 10^{-15} s for sampling the characteristics of H - H systems. For D-scattering, the time window is shifted up into the femto-second range and for heavier atoms to still longer observation times.

In a series of experiments on H-containing materials, starting with an experiment on partially deuterated water [Chatzidimitriou-Dreismann 1997 (a)] and later continued with polymers [Chatzidimitriou-Dreismann 2000 (b)], metal hydrides [Chatzidimitriou-Dreismann 2001] *etc.*, it was found that the normalized area ratios A_H/A_D, A_H/A_O, and A_H/A_{Metal}, did not agree with the expected ratios $A_H/A_D = \sigma_H/\sigma_D$, *etc.*, where the σ_H's are the tabulated neutron total scattering CS, but were reduced by up to 30%. This loss of intensity was interpreted as a short-time effect, mainly involving the H-nuclei (since heavier atoms are observed on a longer time scale and are supposed to be less sensitive to quantum effects).

In experiments on Nb-hydrides [Chatzidimitriou-Dreismann 2001], it was first noted that reduction of the A_H/A_{Nb} ratio was most pronounced for the higher scattering angles θ ($\approx 40\%$ lower), but reached essentially normal ratios for low angles (for water and polymers this dependence had been found to be much weaker). By help of relation (2) the ratio A_H/A_{Metal} could instead be plotted as function of the scattering time τ_{NCS} (see an example from experiments on Pd and Nb-hydrides [Karlsson 2003 (b)] in Fig. 4).

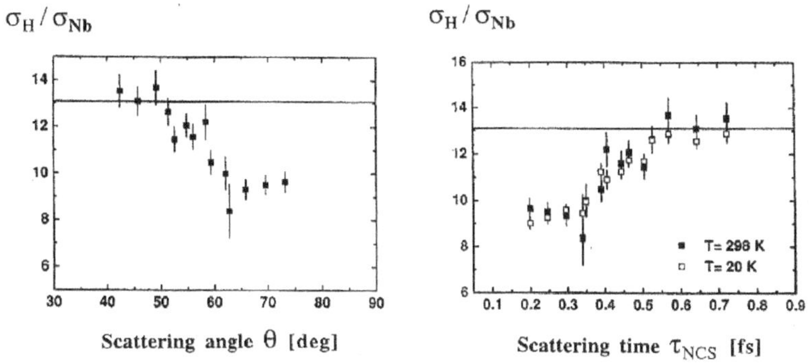

Figure 4. Left panel: scattering angle dependence of H / Nb intensity; right panel: the same quantity as function of scattering time τ_{NCS}. No temperature dependence is expected since the observations times are "subthermal" [Karlsson 1999; Karlsson 2003 (b)].

This kind of plot shows clearly that the anomalous effect is a short-time phenomenon and disappears (in the metal hydrides) in about 5.10^{-16} s.

2. H - H interferences in neutron scattering

The cross-section reductions have been interpreted [Karlsson 2000; Karlsson 2002 (c)] as interference effects, due to short-lived H - H (and also D - D) entanglement during the scattering process, an entanglement that disappears because of decoherence on the femto-second time-scale. With H - H entanglement, the protons can no longer be considered as independent scatterers and new types of interferences appear. But before entering a description of possible quantum correlations effects, it should first be noticed that the Compton scattering method, at least as applied in the experiments cited above, really has a capacity to register interferences between outgoing waves which start from scattering centers with a separation of a few Angströms. The longitudinal correlation length of the neutron wave is determined by the wavelength selection, $l_{coh} \approx \lambda^2/2\ \Delta\lambda$, which with the relatively high energy resolution provided by resonance foils, 4.91 ± 0.14 eV on the outgoing neutrons corresponds to $l_{coh} \approx 2.5$ Å. Since the transverse correlation length is longer, each neutron samples a correlation volume [Mandel 1995] which comprises a few neighbouring H-atoms in the samples investigated.

In a model [Karlsson 2000; Karlsson 2002 (c)], valid for Compton scattering on two identical atoms, it was shown that H cross-sections can be reduced considerably by interference if the proton wavefunctions are not separable. In this model, the non-separability is caused by the indistinguishablity of the particles, just as it is for the thermal neutron scattering on liquid hydrogen, where quantum exchange effects also explain the very low cross-section for para-H_2 (for liquid hydrogen below $T = 10$ K, decoherence seems to be weak enough to preserve H - H entanglement over $\tau_{NS} \approx 10^{-13}$ s, relevant for this method). For indistinguishable particles the initial state must be described by a non-separable, antisymmetrized product wavefunction

$$|i\rangle = \frac{\phi_1(\vec{R}_\alpha)\,\phi_2(\vec{R}_\beta) + \zeta\,\phi_1(\vec{R}_\beta)\,\phi_2(\vec{R}_\alpha)}{\sqrt{2}}\,\chi_M^J(\alpha,\,\beta) \tag{3}$$

where χ_M^J is the coupled spin function and the spatial wave-function is maximally entangled with the phase factor $\zeta = (-1)^J = \exp(\pm\, i\pi/2)$. For Compton scattering, the final state was taken to have the form

$$|f\rangle = \frac{\exp(i\,\vec{p'}.\,\vec{R}_\alpha)\,\psi(\vec{R}_\beta) + \zeta'\,\exp(i\,\vec{p'}.\,\vec{R}_\beta)\,\psi(\vec{R}_\alpha)}{\sqrt{2}}\,\chi_{M'}^{J'}(\alpha,\,\beta) \tag{4}$$

where the recoiling nucleus develops into a plane wave state with total momentum $\vec{p'} = \vec{q} + \vec{p}$, where $\vec{p}$ is the initial momentum of the H-nuclei and the expression $\psi(\vec{R}_i) = [T_i\,\phi_1(\vec{R}_i) + T_2\,\phi_2(\vec{R}_i)]/\sqrt{2}$ stands for the state of the non-recoiling member of the pair immediately after the neutron interaction. Quantum correlations are not yet broken and α and β not yet distinguishable; this occurs only later through a collapse process in the interaction with the condensed matter environment. This is analogous to the case of oxygen dissociation after electron excitation, discussed [Björneholm 2000; Sears 1984] in terms of an intermediate entangled state $O^\star O + OO^\star$ in which the electron hole is not associated with one or the other of the atoms. With this ansatz, and the usual scattering operator $V = b_\alpha\,\exp(i\,\vec{k}.\,\vec{R}_\alpha) + b_\beta\,\exp(i\,\vec{k}.\,\vec{R}_\beta)$ the matrix elements $\langle f|V|i\rangle$ are evaluated as

$$\langle f|V|i\rangle = \frac{\chi_{M'}^{J'}(\alpha,\,\beta)[b_\alpha + \zeta\,\zeta'\,b_\beta]\,\chi_M^J(\alpha,\,\beta)}{\sqrt{2}}\,K(\vec{p})\left[T_2 + \zeta\,e^{-i\,\vec{p}.\vec{d}}T_1\right]. \tag{5}$$

Here, $|K(\vec{p})|^2$ describes the "Compton profile" (shape of the momentum distribution in the outgoing neutron spectrum) and T_i are overlap integrals $\int d\vec{R}\,\psi^\star(\vec{R})\,\phi_i(\vec{R})$, which to a good approximation have the values $1/\sqrt{2}$. The CS are dependent on how the momentum of the local vibration $\vec{p}$ is related to the vector $\vec{d} = \vec{R}_1 - \vec{R}_2$ connecting the two nuclear sites. For half-integer values of particle spins ($I = 1/2$ for protons), this form of the scattering matrix elements leads to the reduction factor [Karlsson 2002 (c)]

$$f_{HH} = \frac{\sigma_{eff}}{\sigma_{sp}} = \frac{1}{2(2I+1)} \left[I \left| T_1 + \exp\left(i\, \vec{p}.\vec{d} \right) T_2 \right|^2 + (I+1) \left| T_1 - \exp\left(i\, \vec{p}.\vec{d} \right) T_2 \right|^2 \right] \tag{6}$$

where σ_{sp} represents the tabulated, independent-particle value for the cross section. The relative signs of the parameters T_1 and T_2 may be discussed. For $T_2 = T_1$ which corresponds to the ground state $\psi(\vec{R}) = [\phi_1(\vec{R}_\beta) + \phi_2(\vec{R}_\beta)]/\sqrt{2}$ of the delocalized non-recoiling proton β, the maximum cross section reduction occurs if $\vec{p}$ is parallel to $\vec{d}$ ($1/4$ of the standard value σ_{sp}), whereas for isotropic vibrations a reduction to $1/3$ is expected. For deuteron pairs, a formula analogous to Eq. 6 shows considerably smaller deviations (f_{DD} minimum $2/3$) from standard values.

We close this section by noting that the destructive interference has its origin in the spatial wavefunction, which gives cancellations in the scattering matrix element if the amplitudes for particle α (or β) has opposite signs when the neutron scatters from sites 1 and 2. Fig. 5 is an attempt to illustrate the scattering from a particle that is delocalized over two different sites.

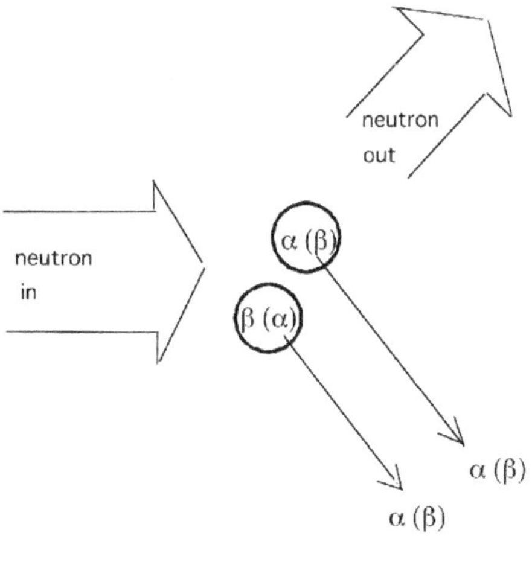

Figure 5 Schematic representation of neutron Compton scattering on an entangled pair of identical particles. The state is a superposition corresponding to particles α and β receiving the recoil during the scattering process.

Similar interferences, not accounted for in independent-particle descriptions, are expected for any kind of non-separable states, for instance entangled states caused by the coupling of two H-nuclei to a common external field at the time of the neutron encounter. For a quantitative description of such situations, particle-field interactions must be specified and amplitudes calculated to insert in the matrix elements (5). Maximal entanglement as in the two-particle, exchange-based model above is not expected (as it is not in the multi-particle exchange-correlated situations perhaps valid for H-configurations in metallic hydrides). Such situations would give reduction factors closer to one.

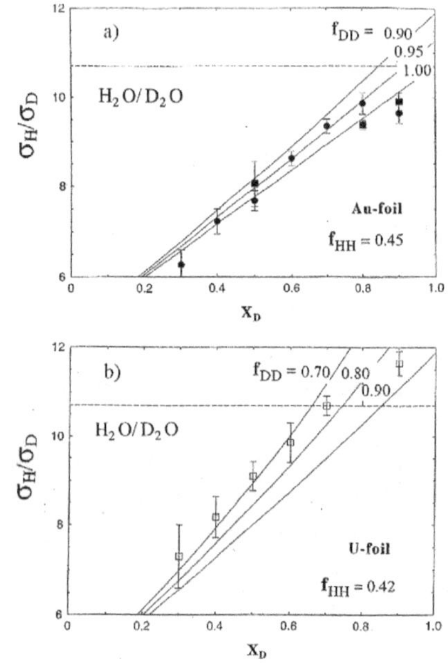

Figure 6 H / D intensity ratios in water observed with neutron energy selection a) by Au-foils, and b) by U-foils, as observed by Chatzidimitriou-Dreismann *et al.* [Chatzidimitriou-Dreismann 1997 (a)]. Lines result from fits to the data obtained in the work of Karlsson [Karlsson 2003 (a)], where the reduction factor f_{DD} was chosen as a free parameter.

A quantitative explanation of data based on the two-particle, exchange-coupled model can be attempted only when protons appear in the form of relatively well-isolated pairs, such as in water. The data recorded for partially deuterated water by Chatzidimitriou-Dreismann *et al.* [Chatzidimitriou-Dreismann 1997 (a)] (see Fig. 6) were interpreted by Karlsson [Karlsson 2003 (a)] in terms of reduced pair CS as given in Eq. (7). With a relative deuterium concentration $X_D = c_D/(c_H + c_D)$, the probabilities for forming H_2O, D_2O and HDO molecules is $p_{HH} = c_H^2$, $p_{DD} = c_D^2$ and $p_{HD} = 2\, c_H\, c_D$, respectively (for $X_D = 0.5$ one has $p_H^2 = 0.25$, $p_D^2 = 0.25$ and $2\, p_H\, p_D = 0.50$, *etc.*). With reduction factors f_{HH} for H - H pairs, f_{DD} for D - D pairs, and $f_{HD} = 1$ (no exchange correlation) for H - D pairs, the model predicts the cross section ratio

$$\left[\frac{\sigma_H}{\sigma_D}\right]_{eff} = \frac{c_D\,[\sigma_H]_{sp}\left(f_{HH}\,p_{HH} + \frac{1}{2}\,p_{HD}\right)}{c_H\,[\sigma_H]_{sp}\left(f_{DD}\,p_{DD} + \frac{1}{2}\,p_{HD}\right)} \tag{7}$$

where $[\sigma_H/\sigma_D]_{sp} = 10.7$. Theoretical values of $f_{HH} = 0.45$ (for Au-foil detection) and $f_{HH} = 0.42$ (for U-foil) were calculated with regard taken to the effect of angle between vibrational momentum $\vec{p}$ and the H - H distance vector $\vec{d}$ (see Eq. (6)), as well as the limited coherence lengths associated with both modes of detection. The quantity f_{DD} was left as a free parameter with values 1.00, 0.90, and 0.80 in the fits to the data in Fig. 6. It is evident that the cross section reductions stem main from H - H interferences, while D - D effects are considerably smaller. Introduction of a factor $f_{DD} \neq 1$ does not improve the fits which indicates that H - D correlations can be neglected in accordance with the assumptions in the exchange correlation model.

3. Other consequences of H - H exchange in scattering

It has recently been pointed out that the indistinguishability of identical particles has also another consequence for neutron scattering. The scattering processes are spin-dependent, which is expressed by writing the scattering lengths b_i (see the scattering operators above) on the form $b_i = A_i + B_i\,\vec{I}\,.\,\vec{s}_n$ where $\vec{I}$ is the angular momentum of the nucleus and $\vec{s}_n$ that of the neutron. For protons, spin-flip scattering dominates, $A_H = -\,0.41 \times 10^{-12}$ cm, $B_H = 5.84 \times 10^{-12}$ cm, while for deuterons non-spin-flip scattering dominates. Common wisdom says that only the A-terms contribute the intensity in diffraction and reflectivity experiments because spin-flips destroy the coherence in the outgoing neutron wave. The coherent cross section for protons is therefore only as low as $\sigma_{coh} = 4\pi\,A_H^2 = 1.8$ barns while the incoherent one is $\sigma_{inc} = 79.8$ barns (and H is often substituted by D to get sufficient intensity in neutron diffraction on hydrogenous systems).

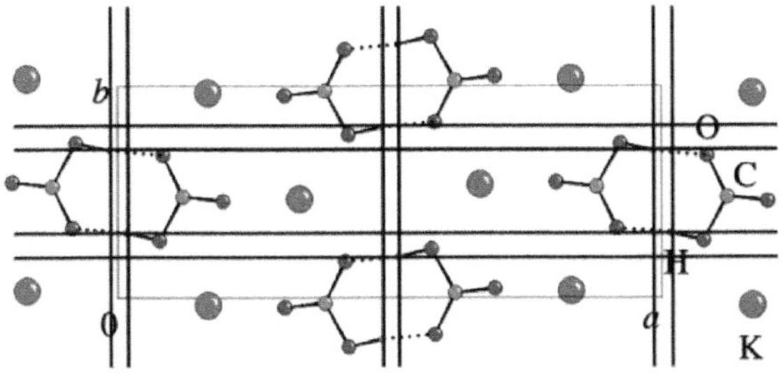

Figure 7 H-dimers in $KHCO_3$ (from Ref. [Fillaux 2003 (a)]. Protons are found along the dotted lines.

However, as first pointed out by Fillaux [Fillaux 1998], this is valid only if the scatterers are quantum mechanically separable. If H - H pairs (or larger systems) are entangled, additional forms of interference will appear (as shown experimentally [Ikeda 1999]) and for protons there will be a contribution from the large B-terms to the diffraction intensity. In a Bragg scattering experiment, Fillaux *et al.* [Fillaux 2003 (a)] found this type of strongly enhanced scattering from close H - H dimers, which form a regular sublattice in the compound $KHCO_3$, as illustrated in Fig. 7. In a later work by Keen and Lovesey [Keen 2003] this was expressed by an additional term in the average $\langle b_\alpha\, b_\beta\rangle$ (which for distinguishable particles would be equal to A_H^2) of the form $\langle b_\alpha\, b_\beta\rangle = B_H^2\, I(I+1)[1 - J(J+1)]/[8I(I+1)]$.

Diffraction experiments are carried out by thermal neutrons and with observation times of 10^{-13} s or more. These experiments indicate that H - H entanglement survives over an unusually long time in this particular compound. The reason for the long decoherence time was discussed in Ref. [Fillaux 1998] as a result of restricted coupling of the H - H dimers to the $KHCO_3$ environment caused by specific fermion / boson superselection rules.

The specific features of neutron scattering on a pair of indistinguishable particles can be described heuristically as kind of "double scattering"; each of the two protons scatters from each of the two sites (compare also Fig. 5). For separable particles, a spin-flip in one of the nuclei identifies the path of the neutron, which makes interference between the two paths impossible, but scattering on entangled particles does not identify the spin-flip site; the spins are delocalized and normally incoherent processes can become coherent. It is interesting to note that the specific feature of the proton, with its exceptionally high ratio $\sigma_{inc}/\sigma_{coh} = 44$, in this way makes it a particularly sensitive tool for investigation of quantum coherence and decoherence.

One example should be mentioned: in a neutron reflectivity experiment on a water / Si interface, Streffer *et al.* [Streffer 1999] found that the H-intensity was enhanced by about 10% with respect to what was expected from standard theory. In the light of the large intensity amplification for quantum correlated protons, such a result is not unreasonable even if H - H pairs keep correlated only over a fraction of the observation time ($\tau_{coh} \approx 2.10^{-14}$ s $< \tau_{NS}$).

4. Decoherence mechanisms

Neutron scattering has provided several examples of situations where it is evident that, on a short time scale, protons in solids or liquids cannot be considered as independent quantum objects. Without going into the detailed reason for the inseparability (except for the obvious correlations imposed when the neutron scatters on a set of indistinguishable particles), we will now consider the mechanisms by which coherence is lost over times much larger than τ_{coh}.

Quantum correlations of the different scattering objects is a necessary condition for reduction of the CS through interference. It is true that coupling to some external degree of freedom at the moment of the neutron encounter may lead to a temporary entanglement of the scattering objects, but environmental fluctuations tend to smear out all quantum phase relations (interferences) over the time of observation and restore the individual particle cross-sections. According to Eq. (1) the correlations decay with a rate proportional to $|x - x'|^2$ which means that only the smallest H-clusters, or single H - H pairs, survive as entangled units over time intervals approaching τ_{coh}. The plot in Fig. 4 can be interpreted as a restoration of the full cross section through such a decay of the interference terms (if on the other hand, the CS were reduced directly by

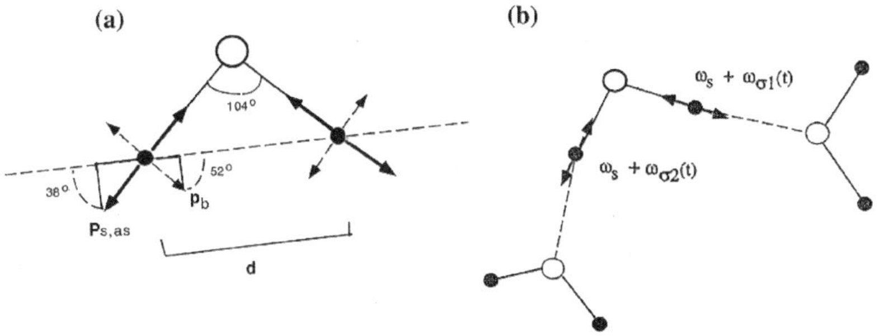

Figure 8. (a) Vibrational modes in a free water molecule and (b) stretching vibrations modified through hydrogen bonds.

$\exp[-\Lambda\,|x-x'|^2 t]$ terms, the most distant H-nuclei would provide the leading terms, which is counterintuitive to a picture of locally entangled protons).

4.1 Natural decoherence mechanisms

We present a realistic decoherence mechanism for an entangled H - H system. The example concerns protons of liquid water. It is inspired by the analogy with T2 relaxation in NMR, where a phase memory is smeared out by random interactions of nuclear spins with surrounding fluctuating fields. The dynamics of H_2O-molecules in water at ambient temperatures has been studied extensively by vibrational spectroscopy [Ware 1968] and neutron scattering [Soper 1986]. The low-energy dynamics of the H-atoms in this situation can be described as OH stretching and bending oscillations, similar to those in free H_2O-molecules, but perturbed by H-bridges to the nearby molecules (Fig. 8). In a simple molecular dimer model, one of the H-atoms is free to vibrate as in the gaseous state, the other is involved in a hydrogen bond. The stretching vibrations ν_s (or ν_{as}) of the latter are coupled [Joesten 1974] to the vibration in the H bond itself (characteristic frequency ν_σ) to bands with frequency $\nu_s \pm\ \nu_\sigma$. These couplings are changing with time during a full vibrational period and give rise to a band of width about 0.03 eV [Bratos 1991].

Decoherence theory states that the entanglement in a system is broken only when the environment vectors, to which it is coupled, form an orthogonal set. This can be expressed as a condition on the reduced density operator $\rho_{S'}$ corresponding to the product wavefunction $|\Psi\rangle = \sum_k |\alpha_k\rangle \sum_l |\phi_{E,\,l}\rangle\langle S_k,\ \phi_{E,\,l}|$ of an entangled two-particle system (with orthonormal basis vectors $\{|\alpha_k\rangle\}$) coupled to an environment E (basis vectors $\{|\phi_{E,\,l}\rangle\}$),

$$\langle l|\rho_{S'}|k\rangle = \sum_l \langle\phi_{E,\,l}|\phi_{E,\,k}\rangle\langle S_l,\ \phi_{E,\,l}|\Psi\rangle^\star\langle S_k,\ \phi_{E,\,l}|\Psi\rangle. \qquad (8)$$

The condition for diagonality, *i.e.*, the disappearance of interference terms, is that the eigenvectors $|\phi_{E,\, l}\rangle$ of the environment form an orthogonal set. This means that, as long as only a few vibrational modes present, as in the OH stretching and bending vibrations in free H_2O-molecule, the entanglement does not vanish but disappears and reappears under the time evolution depending on how orthogonal the environmental vectors are at different times. For complete decoherence it is necessary that the two protons couple to modes whose frequencies change randomly, as will now be discussed for liquid water.

The spatial part of the final state in Eq. (4) can be written as a superposition of states $|f\rangle = \{|f'\rangle + |f''\rangle\}/\sqrt{2}$, where in $|f'\rangle$ the particle α starts from $\vec{R}_\alpha = \vec{R}_1$ (with β staying at site $\vec{R}_\beta = \vec{R}_2$) and α starts from $\vec{R}_\alpha = \vec{R}_2$ (with β remaining at site $\vec{R}_\beta = \vec{R}_1$), and where indices in $|f''\rangle$ are reversed, with

$$|f'\rangle = \frac{1}{\sqrt{2}}\Big\{\exp\Big[i\,\vec{p'}.\Big(\vec{R}_\alpha - \vec{R}_1\Big)\Big]\; T_2\,\phi_2\Big(\vec{R}_\beta\Big) + \zeta'\,\exp\Big[i\,\vec{p'}.\Big(\vec{R}_\alpha - \vec{R}_2\Big)\Big]\; T_1\,\phi_1\Big(\vec{R}_\beta\Big)\Big\}. \tag{9}$$

After extracting a common phase factors $\exp[i\;\omega_1\;t]\;\exp[-i\;\vec{p'}.\vec{R}_i]$, $(i = 1,\;2)$ the following form is obtained,

$$|f'\rangle = \frac{1}{\sqrt{2}}\Big\{\exp\Big[i\,\vec{p'}.\vec{R}_\alpha\Big]\; T_2\,\phi_2\Big(\vec{R}_\beta\Big) + \zeta'\,\exp\left[i\left(\omega_{\sigma_1}+\omega_{\sigma_2}\right)t\right] \times \exp\Big[i\,\vec{p'}.\vec{d}\Big]\;\exp\Big[i\,\vec{p'}.\vec{R}_\alpha\Big]\; T_1\,\phi_1\Big(\vec{R}_\beta\Big)\Big\}. \tag{10}$$

If $e^{i\phi}$ is introduced for $\zeta' = (-1)^{J'}$, where $\phi = \pi$ for J' odd and $\phi = 0$ for J' even, Eq. (10) displays how the phase factor in the spatial part of the two-proton wavefunction disappears and reappears with the factor $e^{i[\phi+(\omega_{\sigma 2}-\omega_{\sigma 1})t]}$.

But in the fluctuating network of hydrogen bonds in liquid water, the factor $(\omega_{\sigma 2} - \omega_{\sigma_1})t$ changes randomly, with an r.m.s.-value of σ_ω on a time-scale which is of the same the order of magnitude as the vibrational period itself. The energy spread of 0.03 eV mentioned above corresponds to $\sigma_\omega \approx 0.5\times10^{14}\ \mathrm{s}^{-1}$. If the distribution in $\omega \approx \omega_s$ can be represented by a Gaussian function one obtains therefore for the mean value of the phase factor,

$$\frac{1}{\sigma_\omega\sqrt{\pi}}\int d\omega'\;\exp\left[\frac{(\omega'-\omega)^2}{4\,\sigma_\omega^2}\right]\;\exp(i\,\omega' t) = \exp\left[-\,\sigma_\omega^2\,t^2\right]\;\exp(i\,\omega t) \tag{11}$$

where the decay factor $\exp[-\,\sigma_\omega^2\,t^2]$ measures the decoherence. From the known value of σ_ω the characteristic decoherence time for the proton-proton correlation water can be estimated to be $\tau_{coh} = 2.10^{-14}$ s. After a complete loss of phase memory the result is a simple product state, for which the tabulated individual particle cross-section is expected. Note that this result is independent of the mechanism by which the entanglement is originally created.

This fast decoherence explains the observation of individual proton CS in thermal neutron scattering on H_2O, for which the time window $\tau_{NS} \gg \tau_{coh}$ (in the absence of decoherence it would be necessary to treat the scattering on para- and ortho-water in analogy with what is done for para- and ortho-H_2 [Schwinger 1937]). It can also be added that Raman scattering data [Ratcliffe 1982] show widths for water superheated to 300°C which are considerably lower, ≈ 5 cm^{-1}, than those measured at room temperature, 12.5 cm^{-1}. Such a narrowing would correspond to a decoherence time of $\tau_{coh} \approx 50$ fs, which indicates that some effects of entanglement might still be observable in water with thermal neutrons under specific circumstances.

This type of estimate of expected decoherence rates could, in principle, be carried out for similar simple systems where the randomness in the interaction with an environment is known from spectroscopic measurements of linewidths. It is valid as long as the measuring probes do not by themselves introduce additional perturbations. The diffraction data mentioned above for H - H dimers in $KHCO_3$ [Fillaux 2003 (a)] are results of elastic neutron interactions, which is a particularly soft way of probing delicate quantum states in condensed matter.

4.2 Measurement induced decoherence

This treatment discussed above is valid if the measurement process itself does not contribute essentially to the decoherence, a condition that would be fulfilled for weak processes like thermal neutron scattering. For Compton scattering, the situation is different because of the strong recoil of the scattering proton. With a typical recoil energy of 10 eV it can be estimated that the proton has moved a distance of 0.3 Å (which means outside its normal vibrational amplitude in its interstitial site) in about 0.5 fs. At this point it is likely that it causes strong local excitations in the crystalline lattice, leading to immidiate loss of quantum coherence. This is also what is observed in the metal hydrides (*cf.* Fig. 4). Weaker decoherence effects of the recoil are expected in liquids with their more flexible bonds, and the time for disruption of the coherent states tends to be longer for larger vibrational amplitudes (larger Δp). Scattering on heavier nuclei (D, He, *etc.*) shifts the observation range upwards (see Fig. 2) such that than $\tau_{NCS} > \tau_{coh}$ where no anomalous effects are seen. The short interaction times in the Compton process also explain the temperature independence of the cross section anomalies, as mentioned for instance in Ref. [Chatzidimitriou-Dreismann 2001]; over intervals of femto-seconds there is no time for the establishment of thermal equilibrium.

Conclusions

Tab. 1 is a collection of scattering results which can be taken as evidence for quantum non-separability of the scattering agents. With neutron scattering, hydrogen-containing systems can be probed over different time-scales and the sub-femtosecond methods provide, most naturally, the majority of the evidence, but there is also clear indication of the survival of H - H entanglement over considerably longer times in particular systems. Information from a couple of other methods have been introduced for comparison in Tab. 1.

Table 1. Experimental data that indicates H - H quantum entanglement in condensed matter. Neutron experiments are described in the text. Raman and electron scattering experiments are published in Refs. [Chatzidimitriou-Dreismann 1995; Chatzidimitriou-Dreismann 2003 (a)] respectively. Raman scattering has also a time window in the fs-range.

	H_2 *liquid*	H_2 $KHCO_3$	H_2O	H_x *polymers*	H_x *hydrides*
Total cross section σ					
Neutrons, thermal					
≈ 0.003 eV; $\tau = 10^{-13}$ s	x				
Neutrons, Compton					
10 - 100 eV; $\tau = 10^{-15}$ s	x	x	x	x	x
Electrons 10 - 30 keV; $\tau = 10^{-15}$ s				x	
Raman			?		
Coherent cross section σ_{coh}					
Neutrons, thermal					
in diffraction		x			
in reflectivity			?		

In the present paper, it has been made plausible that these different effects have a common origin in terms of specific interference phenomena that appear when the scattering systems are quantum correlated. Reasons for the non-separability have been discussed and estimates of the time such states may persist before decoherence occurs have been made for a couple of systems. Although the theoretical discussion centers on entanglement due to indistinguishability of identical particles it does not exclude other entangling mechanisms, but the scattering on exchange-correlated pairs of particles provides a starting point for further discussions. In particular, it makes possible to put up models for decoherence in simple environments, which should be useful also for less clean entanglement situations. A more detailed account of the present problem can be found in Ref. [Karlsson 2004].

IS FERMI'S GOLDEN RULE ALWAYS TRUE FOR COMPTON SCATTERING?

I. E. Mazets, [1, 2] C. A. Chatzidimitriou-Dreismann, [3] and G. Kurizki [1]

[1] *Chemical Physics Department, Weizmann Institute of Science, Rehovot 76100, Israel*

[2] *Ioffe Physico-Technical Institute, St.Petersburg 194021, Russia*

[3] *Institute of Chemistry, Stranski Laboratory, TU Berlin, D-10623 Berlin, Germany*

mazets@astro.ioffe.rssi.ru

Abstract We present an example of a situation where Fermi's golden rule does not apply, even if an unstable quantum system decays exponentially. The work is inspired by the recent experiments by Chatzidimitriou-Dreismann and co-workers on reduction of the neutron Compton scattering cross-section in hydrogenated compounds. Although full quantitative explanation of these experimental results is still to be accomplished, our preliminary results show a possibility of strong modification of the scattering cross-section.

Keywords: Compton scattering, netrons, protons, entanglement, relaxation, electrons, condensed matter, Fermi's golden rule

In numerous experiments on neutron Compton scattering (NCS) a theorctically unexpected shortfall (reduction) of the neutron-proton scattering cross-section (compared to that of free neutron-proton scattering) has been observed in various hydrogenated compounds, such as water [Chatzidimitriou-Dreismann 1997 (a)], niobium and palladium hydride [Karlsson 1999; Karlsson 2003 (b)], and several organic compounds [Chatzidimitriou-Dreismann 2000 (b); Chatzidimitriou-Dreismann 2001; Chatzidimitriou-Dreismann 2002 (a); Chatzidimitriou-Dreismann 2003 (a)]. A similar shortfall has also been observed in electron-proton Compton scattering (ECS) [Chatzidimitriou-Dreismann 2003 (a)]. When protons are substituted by deuterons, the shortfall is considerably less conspicuous [Chatzidimitriou-Dreismann 2000 (b); Chatzidimitriou-Dreismann 2001; Chatzidimitriou-Dreismann 2002 (a)]. In most experiments, the shortfall is nearly independent of, or weakly depending on, inter-proton distance or temperature. Remarkably, and according to present-day experimental accuracy, this shortfall seems to be not accompanied by marked additional spectral broadening (beyond the broadening caused by the momentum width of the localized protons) of the scattering peaks. These facts

V.M. Akulin et al. (eds.), Decoherence, Entanglement and Information Protection in Complex Quantum Systems, 549–554.

call for an explanation of the shortfall that is based neither on specifics of nuclear (strong) interactions, such as spin dependence of nuclear forces [Landau 1997], nor on scattering by correlated proton pairs [Karlsson 2000; Karlsson 2002 (c)], since correlated-pair scattering would be sensitive to inter-proton distances. Here we analyze the possibility that coupling to the environment may change the scattering cross-section (compared to that of free projectile and target particles) without causing an appreciable broadening of the scattering line shape. The analysis is quite general and follows from first principles, even though we specifically refer to neutron projectiles and target nuclei.

Let us consider the scattering of a neutron projectile on a target nucleus of mass m_t. The initial state of the composite system is the direct product of the neutron projectile state in the form of a plane wave with the wave vector $\vec{k}_n$ and the bound state ψ_0 of the target nucleus (*i.e.*, a state localized around a specific lattice site in a crystal or an equillibrium position in a molecule). We consider here Compton-type scattering, so that any influence of the interatomic forces on the wave function of the fast struck target can be neglected, and the final target-nucleus wave function is adequately approximated by a plane wave with the wave vector $\vec{k}'_t$. The final momentum of the projectile neutron is $\hbar\vec{k}'_n$, and $\hbar\vec{q} = \hbar\left(\vec{k}_n - \vec{k}'_n\right)$ is the momentum transferred to the target nucleus. The binding energy of the target nucleus in its localized state ψ_0 is negligible compared to $\hbar^2 q^2/(2m_t)$. Due to the spread of the nucleus momentum in the initial (bound) state, in general $\vec{k}'_t \neq \vec{q}$.

Since we consider the scattering of an unpolarized neutron projectile on an unpolarized target, we omit the spin dependence in the interaction Hamiltonian, which takes the form of the pseudopotential $2\pi\hbar^2\, a\, \mu^{-1}\delta(\vec{r})$. Here the reduced mass is $\mu = m_n\, m_t/(m_n + m_t)$, m_n being the neutron mass, and the spin-averaged scattering length a determines $4\pi\, a^2$, the total cross-section of an unpolarized neutron on an unpolarized, *free* target nucleus. The probability amplitudes for the initial and final states will be denoted by b_i and $c_{\vec{q}}$, respectively. The Schrödinger equation then reads

$$i\,\dot{b}_i = \omega_i\, b_i + \frac{1}{V}\int \frac{d^3k'_n}{(2\pi)^3}\int \frac{d^3k'_t}{(2\pi)^3}\,\frac{2\pi\hbar\, a}{\mu}\,\tilde{\psi}_0^\star\left(\vec{q} - \vec{k}'_t\right)\, c_{\vec{q}}, \tag{1a}$$

$$i\,\dot{c}_{\vec{q}} = \left[(\omega_f + \omega_t) - i\,\Gamma(\omega_t)\right] c_{\vec{q}} + \frac{2\pi\hbar\, a}{\mu}\,\tilde{\psi}_0\left(\vec{q} - \vec{k}'_t\right)\, b_i \tag{1b}$$

where V is the quantization volume,

$$\omega_i = \frac{\hbar\, k_n^2}{2m_n}, \quad \omega_f = \frac{\hbar\, {k'_n}^2}{2m_n}, \quad \omega_t = \frac{\hbar\, {k'_t}^2}{2m_t}, \tag{2}$$

and

$$\tilde{\psi}_0(\vec{Q}) = \int d^3r \, \exp\left(i \, \vec{Q}.\vec{r}\right) \, \psi_0(\vec{r}) \tag{3}$$

is the Fourier transform of the target-nucleus initial wave function. In Eq. (1b) we introduced the relaxation rate $\Gamma(\omega_t)$ of the final state due to interaction of the fast target nucleus with the environment. This working hypothesis is physically related to the introduction of decoherence of a relevant scattering system (*i.e.*, target), which is composed of a proton being entangled with few adjacent particles (electrons and other nuclei), as proposed in Refs. [Chatzidimitriou-Dreismann 2000 (b); Chatzidimitriou-Dreismann 2001; Chatzidimitriou-Dreismann 2002 (a); Chatzidimitriou-Dreismann 2003 (a); Chatzidimitriou-Dreismann 2003 (b)]. This rate depends on ω_t only, and not on ω_f, since, by assumption, the multiple- scattering probability of the neutron projectile is negligible, in contrast to that of the target nucleus.

Using the standard Wigner-Weisskopf procedure [Cohen-Tannoudji 1992; Landau 1997], we find that the decay rate of the initial state, obeying the exponential decay law $|b_i(t)|^2 = \exp(-Wt)$, is equal to

$$W = \frac{2\pi}{V} \int \frac{d^3k'_n}{(2\pi)^3} \int d^3k'_t \left(\frac{2\pi\hbar \, a}{\mu}\right)^2 \tilde{n}_0\left(\vec{q} - \vec{k}'_t\right) \, L\left[\omega_f + \omega_t - \omega_i, \, \Gamma(\omega_t)\right],$$
$$\tilde{n}_0\left(\vec{Q}\right) = \frac{\left|\tilde{\psi}_0\left(\vec{Q}\right)\right|^2}{(2\pi)^3}. \tag{4}$$

Here

$$L\left[\omega_f + \omega_t - \omega_i, \, \Gamma(\omega_t)\right] = \frac{1}{\pi} \, \frac{\Gamma(\omega_t)}{\left(\omega_f + \omega_t - \omega_i\right)^2 + \Gamma^2(\omega_t)} \tag{5}$$

is the generalized (skewed) Lorentzian line shape of the final (target + projectile) state induced by the environment. The total cross-section can be derived from Eq. (4) upon division by the flux density of the incident neutron projectiles: $\sigma = m_n \, V \, W/(\hbar \, k_n)$. Correspondingly, the doubly-differential cross-section is

$$\frac{d^2\sigma}{d\Omega \, d\omega_f} = a^2 \left(\frac{m_n}{\mu}\right)^2 \frac{k'_n}{k_n} \int d^3k'_t \, \tilde{n}_0\left(\vec{q} - \vec{k}'_t\right) \, L\left[\omega_f + \omega_t - \omega_i, \, \Gamma(\omega_t)\right]. \tag{6}$$

Here $d\Omega = 2\pi \, \sin\theta \, d\theta$, θ being the neutron scattering angle. Eqs. (4) and (6) differ from the standard expressions [Watson 1996] only in that the delta-function describing energy conservation is replaced by the Lorentzian Eq. (5).

Eq. (6) is the *most general expression* for the doubly-differential cross-section in terms of the convolution between the initial momentum distribution $\tilde{n}_0$ and the environment-induced broadening line shape L.

Let us assume that the spread of the function $\tilde{n}_0(\vec{q} - \vec{k}_t')$ is much larger than that of $L\left[\omega_f + \omega_t - \omega_i,\ \Gamma(\omega_t)\right]$ (in $\vec{k}_t'$-space), or, equivalently,

$$\Gamma(\omega_t)\ \tau_c \ll 1, \tag{7}$$

where the collision time (also termed scattering time [Watson 1996]) $\tau_c = m_t/(q\ p_t)$, lies in the sub-femtosecond time range [Chatzidimitriou-Dreismann 1997 (a); Karlsson 1999; Karlsson 2003 (b); Chatzidimitriou-Dreismann 2000 (b); Chatzidimitriou-Dreismann 2001; Chatzidimitriou-Dreismann 2002 (a); Chatzidimitriou-Dreismann 2003 (a)], and the initial momentum width p_t is defined by

$$p_t^2 = \frac{1}{3}\int d^3Q\ \tilde{n}_0\left(\vec{Q}\right)\ \left(\hbar\vec{Q}\right)^2. \tag{8}$$

We stress that condition (7) is essential for avoiding extra-broadening of the scattering line shape, beyond that of p_t. Under this condition, the doubly-differential cross-section is related to its standard value [Watson 1996]

$$\frac{d^2\sigma_{std}}{d\Omega\ d\omega_f} = a^2\left(\frac{m_n}{\mu}\right)^2\frac{k_n'}{k_n}\int d^3Q\ \tilde{n}_0\left(\vec{Q}\right)\ \delta\left(\omega_i - \omega_f - \frac{\hbar\vec{Q}.\vec{q}}{m_t - \omega_r}\right), \tag{9}$$

where $\omega_r = \hbar q^2/(2m_t)$ is the mean recoil frequency, via the following expression:

$$\frac{d^2\sigma}{d\Omega\ d\omega_f} = \frac{d^2\sigma_{std}}{d\Omega\ d\omega_f}\ M, \tag{10}$$

where the modification factor is given by

$$M = \frac{1}{\pi}\int_{-\infty}^{\infty} d\omega_t\ \frac{\Gamma(\omega_t)}{(\omega_t - \omega_r)^2 + \Gamma^2(\omega_t)}. \tag{11}$$

The factor M differs appreciably from 1, only if $\Gamma(\omega_t)$ varies significantly as ω_t changes by $\sim \Gamma(\omega_t)$, *i.e.*, if

$$\left|\frac{d\Gamma}{d\omega_t}\right| \gtrsim 1. \tag{12}$$

Note that the derivative sign in Eq. (12) plays no role. If we attempt to satisfy the condition (12) over a wide range of ω_t, then monotonous dependence of Γ on ω_t implies that at a certain energy $\hbar\omega_t$ the relaxation rate $\Gamma(\omega_t)$ becomes so

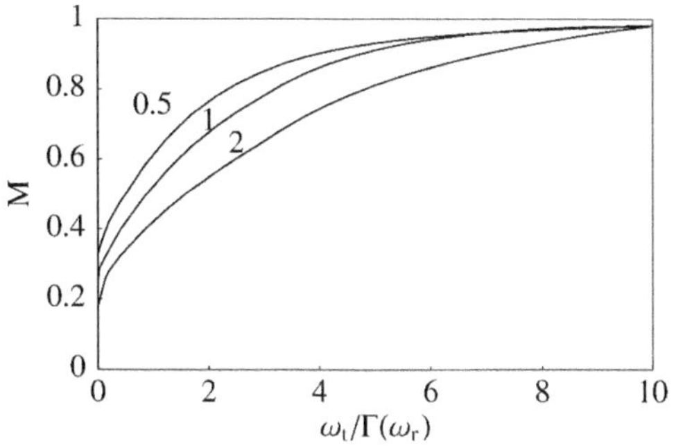

Figure 1 The numerical values of the modification factor for the model of Eq. (13). The values of the parameter κ are indicated above the corresponding curves.

large that the condition (7) is violated, and extra broadening is expected. The numerical results for the power-law model

$$\Gamma(\omega_t) = A_0\, \omega_t^{-\kappa} \tag{13}$$

shown in Fig. 1 confirm this conclusion.

Alternatively, we may introduce a model of *intermittent continuum*, in which $\Gamma(\omega_t)$ is a smooth function *on average* but experiences rapid variation on a smaller scale of ω_t. To this end, we may adopt the following simple model characterized by the two dimensionless parameters η_1, η_2 and the mean relaxation rate $\Gamma_0 \ll \tau_c^{-1}$:

$$\Gamma(\omega_t) = \Gamma_0 \left[1 + \eta_1 \cos\left(\frac{\eta_2\, \omega_t}{\Gamma_0} \right) \right]. \tag{14}$$

Results of our numerical calculations based on the model (14) are excellently fitted by the formula, which is, quite probably, the *exact* solution:

$$M = \frac{1 + \xi \cos\varphi}{1 + 2\xi \cos\varphi + \xi^2}, \quad \xi = \eta_1\, |\eta_2| \exp(-|\eta_2|), \quad \varphi = \frac{\eta_2\, \omega_r}{\Gamma_0}. \tag{15}$$

The maximum possible value of $|\xi|$ is about 0.35. One can see that, according to Eq. (15), M rapidly oscillates between $1/(1+\xi)$ and $1/(1-\xi)$, thus displaying the cross-section reduction as well as enhancement, as ω_t changes, and the averaged value of M over the period is equal to 1.

We note that our analysis bears a certain analogy to that of photon absorption by a narrow-band photodetector [Koshino 2004].

To conclude, our general analysis, from first principles, indicates the remarkable possibility of scattering cross-section modification caused by environmental relaxation of the target states, Eq. (7), without appreciable extra broadening of the scattering line shape. This relaxation is an irreversible process and related to, or accompanied with, fast decoherence of the initial entanglement of the struck proton with its adjacent particles (electrons, and perhaps also other nuclei [Chatzidimitriou-Dreismann 2003 (b)]). However, the condition (7) is very stringent; it limits the energy range $\hbar\omega_r$, in which this behavior exists, to be of order of $\hbar\Gamma$, Γ being the environment-induced relaxation rate. The applicability to Refs. [Chatzidimitriou-Dreismann 1997 (a); Karlsson 1999; Karlsson 2003 (b); Chatzidimitriou-Dreismann 2000 (b); Chatzidimitriou-Dreismann 2001; Chatzidimitriou-Dreismann 2002 (a); Chatzidimitriou-Dreismann 2003 (a)] remains to be examined.

Acknowledgments

The support of the EC (QUACS RTN), ISF and Minerva is acknowledged. I. E. M. also thanks the RFBR (projects 02 - 02 - 17686, 03 - 02 - 17522) and the program Leading Russian Scientific Schools (grant 1115.2003.2).

ON CORRELATION APPROACH TO SCATTERING IN THE DECOHERENCE TIMESCALE

Towards the theoretical interpretation of neutron and electron Compton scattering experimental findings

C. A. Chatzidimitriou-Dreismann[1] and S. Stenholm[2]

[1] *Institut für Chemie, Stranski Laboratorium, Technische Universität Berlin, Straße des 17. Juni 112, D-10623 Berlin, Germany*
dreismann@chem.tu-berlin.de

[2] *Department of Physics, Royal Institute of Technology, SE-10691 Stockholm, Sweden*
stenholm@atom.kth.se

Abstract We provide a "first principles" description of scattering from open quantum systems subject to a Lindblad-type dynamics. In particular we consider the case that the duration of the scattering process is of similar order as the decoherence time of the scatterer. Under rather general conditions, the derivations lead to the the following new result: The irreversible time-evolution may cause a reduction of the system's transition rate being effectuated by scattering. This is tantamount to a shortfall of scattering intensity. The possible connection with striking experimental results of neutron and electron Compton scattering from protons in condensed matter is mentioned.

Keywords: Irreversible dynamics, entanglement, decoherence, neutron Compton scattering, electron-proton Compton scattering

Introduction

The counter-intuitive phenomenon of entanglement [Einstein 1935] between two or more quantum systems has emerged as the most emblematic feature of quantum mechanics. Experiments investigating entanglement, however, are mainly focused on collections of few simple (two- or three-level) quantum systems thoroughly isolated from their environment (*e.g.*, atoms in high-Q cavities and optical lattices). These experimental conditions are necessary due to the decoherence of entangled states. In short, decoherence refers to the suppression of quantum superpositions caused by the environment. By contrast, entanglement in condensed and / or molecular matter at ambient conditions is usu-

V.M. Akulin et al. (eds.), Decoherence, Entanglement and Information Protection in Complex Quantum Systems, 555–562.

ally assumed to be experimentally inaccessible. However, two new scattering techniques operating in the sub-femtosecond time scale provided results indicating that short-lived entangled states may be measurable in condensed matter even at room temperature [Chatzidimitriou-Dreismann 1997 (a); Chatzidimitriou-Dreismann 2003 (a); Physics News 2003; Physics Today 2003; Scientific American 2003].

In this paper we provide a first-principles treatment of scattering from "small" open quantum systems in condense-matter environments, in the "time window" of decoherence of the scattering system. That is, the focus is in "fast" scattering processes with a duration (usually denoted scattering time, τ_{sc}) of the order to the scatterer's decoherence time, τ_{dec}. This may be considered to represent an "extension" of standard scattering theory - as applied, *e.g.*, to neutron physics [van Hove 1954; Squires 1996] or electron scattering [Weigold 1999] - in which the concepts of entanglement and decoherence play essentially no role. The first part of the derivations are analogous to the standard (often denoted) "van Hove formalism" [van Hove 1954]; see also the textbook [Squires 1996]. Then a reduced open quantum system, *i.e.*, a micro- or mesoscopic system characterized by a set of preferred coordinates, is introduced. This corresponds to the "small" physical system that scatters a neutron (electron, *etc.*) with a sufficiently large momentum transfer. Its dynamics is described by a simple Lindblad-type master equation [Lindblad 1976 (b); Barnett 2001] (which, for the sake of simplicity, contains only one Lindblad operator, X), thus including explicitly the effect of decoherence into the formalism.

The striking result of the derivation may be summarized as follows: the irreversible time-evolution (owing to the Lindblad operator X) may cause a reduction of the transition rate of the system (from its initial to its final state). In "experimental" terms, this is tantamount to an effective reduction of the system's cross-section density and thus a shortfall of scattering intensity.

1. Scattering in brief

We assume an N-body Hamiltonian $H_{total} = H_0 + V$ with V of the form

$$V(\vec{r}) = \lambda\, n(\vec{r}) \tag{1}$$

where $n(\vec{r})$ is the particle density operator and λ is the rest of the interaction (contact potential). For example, in the case of neutron scattering from a system consisting of N particles with the same scattering length b one may put

$$n(\vec{r}) = \frac{1}{V} \sum_j \delta(\vec{r} - \vec{R}_j) \tag{2}$$

where V is the volume, $\vec{R}_j$ is the spatial position of the jth particle, and

$$\lambda = \frac{2\pi\,\hbar^2}{m}\,b, \tag{3}$$

m being the neutron mass. (For further details about scattering from "bound" and "free" particles, see the textbook [Squires 1996].)

In the *interaction* picture, the Schrödinger equation is now (putting for simplicity $\hbar = 1$)

$$i\,\partial_t \Psi = \lambda\, n(\vec{r},\, t)\, \Psi \tag{4}$$

with the perturbative solution

$$\Psi(t) = \Psi(0) - i\,\lambda \int_0^t n(\vec{r},\, t')\, dt'\, \Psi(0). \tag{5}$$

We write the transition probability $W(t)$ between initial states ψ_i (with probability P_i that the scattering system is in the state ψ_i) and final states ψ_f of the scattering system to be given by

$$W(t) = \sum_{i,\,f} \left| \langle\psi_f|\,\lambda \int_0^t n(\vec{r},\, t')\, dt'\, |\psi_i\rangle \right|^2 P_i. \tag{6}$$

It should be noted that ψ_i and ψ_f arc eigenstates of the unperturbed N-body Hamiltonian H_0 [Squires 1996; van Hove 1954]. This allows us to write the transition probability in the form

$$W(t) = \lambda^2 \int_0^t dt' \int_0^t dt'' \sum_f \left\langle \psi_f \left| n(\vec{r},\, t')\, \rho\, n(\vec{r},\, t'') \right| \psi_f \right\rangle, \tag{7}$$

where

$$\rho = \sum_i |\psi_i\rangle\, P_i\, \langle\psi_i|, \tag{8}$$

by noting that $n^\dagger(\vec{r},\, t) = n(\vec{r},\, t)$, since $\vec{R}_j$ and $\vec{r}$ are Hermitian operators.

In an actual scattering experiment from condensed matter, we do not measure the cross-section for a process in which the scattering system goes from a specific initial state ψ_i to another state ψ_f, both being unobserved states of the many-body system. Therefore, one takes an appropriate average over all these states [Squires 1996; van Hove 1954], as done in Eq. (6).

Furthermore, the initial ($\vec{k}_0$) and final ($\vec{k}_1$) momenta of an impinging probe particle (neutron) may be assumed to be well defined [Squires 1996; van Hove 1954]. Introducing the momentum transfer $\vec{q} = \vec{k}_0 - \vec{k}_1$ from the probe particle to the scattering system, the Fourier transform of the particle density reads

$$n(\vec{r},\, t) = \frac{1}{(2\pi)^3} \int d\vec{q}\, n(\vec{q},\, t)\, \exp\left(i\, \vec{q}.\vec{r}\right) \tag{9}$$

where, in the case of neutron scattering, *cf.* Eq. (2),

$$n(\vec{q},\, t) = \sum_j \exp\left[-i\, \vec{q}.\vec{R}_j(t)\right]. \tag{10}$$

Since $n(\vec{r},\, t)$ is Hermitian it holds $n^\dagger(\vec{q},\, t) = n(-\vec{q},\, t)$ and one obtains from Eq. (6)

$$W(t) = \lambda^2 \int_0^t dt' \int_0^t dt'' \sum_f \langle \psi_f \left| n(\vec{q},\, t')\, \rho\, n(-\vec{q},\, t'') \right| \psi_f \rangle\,. \tag{11}$$

At this stage one traditionally assumes that the sum over ψ_f runs over all possible eigenstates of H_0 which constitute a complete set, *i.e.*, $\Sigma_f |\psi_f\rangle\langle\psi_f| = \mathbf{1}$; see Refs. [Squires 1996; van Hove 1954]. Hence

$$\sum_f \langle \psi_f \left| n(\vec{q},\, t')\, \rho\, n(-\vec{q},\, t'') \right| \psi_f \rangle = \mathrm{Tr}\left[n(\vec{q},\, t')\, \rho\, n(-\vec{q},\, t'')\right] \tag{12}$$

where Tr[...] denotes the trace operation. As done in standard theory [Squires 1996; van Hove 1954], in Eq. (12) one first sums over all final states, keeping the initial state ψ_i fixed, and then averages over all ψ_i (see, *e.g.*, Ref. [Squires 1996], p. 19). The right-hand-side of Eq. (12) contains the density operator ρ of the system before collision, Eq. (8), which is a well known result.

By introducing a measurement time (the so-called scattering time) τ_{sc}, that is the duration of the scattering process, we find

$$\begin{aligned} W(\tau_{sc}) &= \lambda^2 \int_0^{\tau_{sc}} dt' \int_0^{\tau_{sc}} dt''\, \mathrm{Tr}\left[n(\vec{q},\, t')\, \rho\, n(-\vec{q},\, t'')\right] \\ &= \lambda^2\, \tau_{sc} \int_0^{\tau_{sc}} d\tau\, \mathrm{Tr}\left[n(\vec{q},\, t')\, \rho\, n(-\vec{q},\, t'+\tau)\right], \end{aligned} \tag{13}$$

where the stationary property of the correlation function has been used [Squires 1996]. Now one can introduce the transition rate, say $\dot{W}$, which is defined as

$$\begin{aligned} \dot{W} \equiv \frac{W(\tau_{sc})}{\tau_{sc}} &= \lambda^2 \int_0^{\tau_{sc}} d\tau\, \mathrm{Tr}\left[n\left(\vec{k},\, t'\right)\, \rho\, n\left(-\vec{k},\, t'+\tau\right)\right] \\ &\equiv \lambda^2 \int_0^{\tau_{sc}} d\tau\, C\left(\vec{q},\, \tau\right). \end{aligned} \tag{14}$$

Here the correlation function

$$C(\vec{q},\, t) = \mathrm{Tr}\left[n(\vec{q},\, 0)\;\rho\, n(-\vec{q},\, t)\right] \tag{15}$$

is introduced, which is analogous to the so-called intermediate function of neutron scattering theory [Squires 1996].

2. Irreversible dynamics

We now introduce a set of preferred coordinates $\{|\xi\rangle\}$, *cf.* Refs. [Kübler 1973; Zeh 1973; Zurek 1981; Zurek 1982; Zurek 2003]. These are the relevant degrees of freedom coupled to the neutron probe. The density matrix needed in (13) is then the *reduced* one in the space spanned by these states, and it is obtained by tracing out the (huge number of the) remaining degrees of freedom belonging to the "environment" of the microscopic scattering system (*e.g.*, a proton and its adjacent particles). To simplify notations, we denote this reduced density matrix by ρ too.

In the *subspace* spanned by the preferred coordinates (also denoted "pointer basis"), we assume the relevant density matrix to obey a Lindblad-type equation of the form [Lindblad 1976 (b); Barnett 2001]

$$\partial_t\, \rho = -i\,[H,\; \rho] + \mathcal{R}\,\rho \equiv \mathcal{L}\,\rho \tag{16}$$

with the formal solution

$$\rho(t) = e^{\mathcal{L}t}\;\rho(0). \tag{17}$$

Let us look at a time-dependent expectation value

$$\langle A(t)\rangle \equiv \mathrm{Tr}\left[\rho(t)\; A\right] = \mathrm{Tr}\left[e^{\mathcal{L}t}\;\rho(0)\; A\right] = \mathrm{Tr}\left[\rho(0)\; e^{\mathcal{L}^{\dagger}t}\; A\right], \tag{18}$$

where we define $\mathcal{L}^{\dagger}$ by setting

$$\mathrm{Tr}\left[(\mathcal{L}X)\, Y\right] = \mathrm{Tr}\left[X\left(\mathcal{L}^{\dagger}Y\right)\right]. \tag{19}$$

Thus we obtain a Lindblad time evolution for the operators too by writing

$$\partial_t\; A(t) = \mathcal{L}^{\dagger}\; A(t). \tag{20}$$

This form was actually the original Lindblad result. Note that this works as long as $\mathcal{L}$ does not depend on time. For time-dependent generators of the evolution, a somewhat more elaborate scheme is needed.

Now we find that we may use this formalism to calculate correlation functions like the one in (14). We write

$$\langle A(t)\, B\rangle = \mathrm{Tr}\left[\rho(0)\left(e^{\mathcal{L}^{\dagger}t}A\right)B\right] = \mathrm{Tr}\left[A\, e^{\mathcal{L}t}\left(B\,\rho(0)\right)\right] \equiv \mathrm{Tr}\left[A\,\rho_B(t)\right], \tag{21}$$

where $\rho_B(t)$, as defined in Eqs. (21), obeys the equation

$$\partial_t\, \rho_B(t) = \mathcal{L}\, \rho_B(t) \tag{22}$$

and the initial condition

$$\rho_B(0) = B\,\rho(0). \tag{23}$$

Thus, except for the initial condition, we have to solve the same equation of motion as for the density matrix (16).

3. Application to scattering

We here assume a simple Lindblad-type ansatz for the master equation in the relevant subspace. We set

$$\partial_t\, \rho = -i\,[H,\, \rho] - K\,[X,\, [X,\, \rho]] = \mathcal{L}\,\rho, \tag{24}$$

where $K > 0$ and H is the reduced (or relevant Hamiltonian) of a microscopic or mesoscopic scattering system and the double commutator term describes decoherence. For simplicity of the further calculations, we here assume that

$$\begin{aligned} H\,|\xi\rangle &= \mathcal{E}_\xi\,|\xi\rangle \\ X\,|\xi\rangle &= \xi\,|\xi\rangle. \end{aligned} \tag{25}$$

With Eq. (15) we have

$$\begin{aligned} C\,(\vec{q},\,\tau) &= \mathrm{Tr}\left[n\,(\vec{q},\,0)\;\rho\, n\,(-\vec{q},\,\tau)\right] = \mathrm{Tr}\left[n\,(\vec{q},\,0)\;\rho\, e^{\mathcal{L}^{\dagger}t}\, n\,(-\vec{q},\,0)\right] \\ &= \mathrm{Tr}\left[n\,(-\vec{q},\,0)\;e^{\mathcal{L}t}\;\left(n\,(\vec{q},\,0)\;\rho\right)\right]. \end{aligned} \tag{26}$$

This is equivalent with the expression

$$C\,(\vec{q},\,\tau) = \sum_{\xi,\,\xi'} \langle\xi\,|n\,(-\vec{q},\,0)|\,\xi'\rangle\,\langle\xi'\,|\rho_n(t)|\,\xi\rangle\,. \tag{27}$$

With Eq. (24), one easily finds the well known solution

$$\begin{aligned} \langle\xi'\,|\rho_n(t)|\,\xi\rangle = \;& \exp\left[-i\left(\mathcal{E}_{\xi'} - \mathcal{E}_\xi\right)t\right]\,\exp\left[-K\,(\xi'-\xi)^2\,t\right] \\ & \times\;\langle\xi'\,|n\,(\vec{q},\,0)\;\rho(0)|\,\xi\rangle\,. \end{aligned} \tag{28}$$

Inserting this into the expression (14) for the transition rate we find

$$\dot{W} = \lambda^2 \int_0^{\tau_{sc}} d\tau \sum_{\xi,\, \xi'} \exp\left[-i\left(\mathcal{E}_{\xi'} - \mathcal{E}_{\xi}\right) t\right] \exp\left[-K\left(\xi' - \xi\right)^2 t\right] \times \left\langle \xi \left| n\left(-\vec{q},\, 0\right) \right| \xi' \right\rangle \left\langle \xi' \left| n\left(\vec{q},\, 0\right)\; \rho(0) \right| \xi \right\rangle . \tag{29}$$

4. Decoherence and decrease of transition rate

Obviously, the decoherence-free limit of this result, *i.e.*, with $K = 0$, corresponds to the conventional result of scattering theory.

The oscillating factors $\exp\left[-i\left(\mathcal{E}_{\xi'} - \mathcal{E}_{\xi}\right) t\right]$ are characteristic for the "unitary-type" dynamics caused by the commutator part $-i\left[H,\, \rho\right]$ of the master Eq. (24) for the reduced (or: relevant) density matrix ρ. These factors have the absolute value 1 and do not affect the numerical value of the transition rate.

On the other hand, the restrictive factors $\exp(-K\left(\xi' - \xi\right)^2 t) \leq 1$, which are due to the decoherence, can be seen to cause a decrease of the transition rate and thus of the associated cross-section. This can be illustrated in physical terms as follows.

Let us first assume that the reduced density operator $\rho(0)$ can be chosen to be *diagonal* in the preferred ξ-representation (which corresponds to the usual random phase approximation at $t = 0$). Then each term of Eq. (29) is of the form

$$\begin{aligned} &\left\langle \xi \left| n\left(-\vec{q},\, 0\right) \right| \xi' \right\rangle \left\langle \xi' \left| n\left(\vec{q},\, 0\right)\; \rho(0) \right| \xi \right\rangle \\ &= \left\langle \xi \left| n\left(-\vec{q},\, 0\right) \right| \xi' \right\rangle \left\langle \xi' \left| n\left(\vec{q},\, 0\right) \right| \xi \right\rangle \left\langle \xi \left| \rho(0) \right| \xi \right\rangle \\ &= \left| \left\langle \xi \right| n\left(-\vec{q},\, 0\right) \left| \xi' \right\rangle \right|^2 \left\langle \xi \left| \rho(0) \right| \xi \right\rangle \geq 0. \end{aligned} \tag{30}$$

The last inequality is valid because it holds $\langle \xi | \rho(0) | \xi \rangle \geq 0$.

If the assumed diagonal form of $\rho(0)$ would be considered as being "too strong", one may note the following. The exponentials $\exp(-K\left(\xi' - \xi\right)^2 t)$ due to decoherence imply that only terms with $\xi \approx \xi'$ contribute significantly to the transition rate. Thus we may conclude that, by continuity, all associated terms with $\xi \approx \xi'$ in Eq. (29) should be positive, too. The further terms with ξ being much different from ξ' can be positive or negative. But they may be approximately neglected, since they decay very fast and thus contribute less significantly to $\dot{W}$.

The main conclusion from the preceding considerations is that the time average in Eq. (29) always decreases the numerical value of $\dot{W} \equiv W(\tau_{sc})/\tau_{sc}$, due to the presence of the exponential factors $\exp(-K\left(\xi' - \xi\right)^2 t) \leq 1$. In other words, the effect of decoherence during the experimental time window

τ_{sc} plays a crucial role in the scattering process and leads to an "anomalous" decrease of the transition rate and the associated scattering intensity. This result is in line with that of Ref. [Chatzidimitriou-Dreismann 2003 (b)], which investigated the standard expression of the double differential cross-section of neutron scattering theory by ad hoc assuming decoherence of final and initial states of the scatterer.

Very interesting is also the conclusion that, in the limit of very slow decoherence ($K \to 0$), this 'anomaly' disappears, *i.e.*, the scattering results are expected to agree with conventional theoretical expectations. This is contrary to the associated prediction of the theoretical model of Refs. [Karlsson 2000; Karlsson 2002 (c)].

5. Additional remarks

A related effect (*i.e.*, a shortfall of scattering intensity) was observed in recent neutron-proton Compton scattering (NCS) and electron-proton Compton scattering (ECS) experiments in condensed matter [Chatzidimitriou-Dreismann 1997 (a); Chatzidimitriou-Dreismann 2003 (a); Physics News 2003; Physics Today 2003; Scientific American 2003], in which the experimental scattering time lies in the sub-femtosecond time scale. This coincides with the characteristic time of electronic re-arrangements accompanying the breaking (or formation) of a chemical bond. Note that in these experiments the energy transferred to a proton is large enough to break the bond (C - H and O - H).

Some remarks about the possible selection, definition and / or physical meaning of the preferred coordinates may be appropriate. In the case of conventional NCS theory, for example, one uses momentum eigenstates of the scattering particle (*e.g.*, proton) - as well as for the neutron - as the appropriate basis [Watson 1996]. In the light of the preceding derivations, however, one may observe the following. Due to the strong (Coulomb) interactions of the scattering proton with its adjacent particles (electrons, and probably also other nuclei), $\{|\xi\rangle\}$ can not be one-body states but they should rather be considered to represent momentum states being strongly "dressed" (and entangled) with degrees of freedom of adjacent particles.

Further work will deal with the more general - and experimentally relevant - case, in which the preferred states $\{|\xi\rangle\}$ are not eigenstates of the "reduced" energy and Lindbald operators, Eqs. (25). In that case, the result of Eq. (29) will become less simple.

Acknowledgments

This work was partially supported by the EU RTN QUACS (Quantum Complex Systems: Entanglement and Decoherence from Nano- to Macro-Scales).

VII

COHERENCE AND ENTANGLEMENT IN MESOSCOPIC SYSTEMS

COHERENCE AND ENTANGLEMENT IN MESOSCOPIC SYSTEMS

M. Blaauboer,[1] N. Davidson,[2] M. Heiblum,[3] G. Kurizki,[4] D. O'Dell,[5] A. M. Dykhne[6] and E. Sarnelli,[7]

[1]*Kavli Institute of Nanoscience, Delft University of Technology, Lorentzweg 1, 2628 CJ Delft, The Netherlands*

[2]*Department of Physics of Complex Systems, Weizmann Institute of Science, Rehovot 76100, Israel*

[3]*Department of Condensed Matter Physics, Weizmann Institute of Science, Rehovot 76100, Israel*

[4]*Chemical Physics Department, Weizmann Institute of Science, Rehovot 76100, Israel*

[5]*Department of Physics & Astronomy, University of Sussex, Falmer, Brighton BN1 9QH, England*

[6]*TRINITI, 142092 Troitsk, Russia and Moscow Institute of Physics and Technology, 141700 Dolgoprudny, Russia*

[7]*Istituto di Cibernetica "E.Caianiello" del CNR, Via Campi Flegrei 34, I-80078 Pozzuoli (Naples), Italy*

Introduction

This part gives an overview of recent progress in the area of coherent dynamics and quantum correlations in systems of mesoscopic and nanoscopic dimensions. On a broad level, the systems considered can be divided in two classes: atomic ensembles and Bose-Einstein condensates (BECs), and superconducting and molecular junctions. Even though these are two different systems that involve different particles (atoms *vs.* electrons), their basic Hamiltonians are similar, and analogous effects have been observed in both. A striking example hereof is for example the observation of Josephson-like current-phase effects in a trapped atomic condensate [Andrews 1997]. By exploring this multiatom-condensed matter analogy new and interesting insight into the physical properties of both kinds of systems may be gained. Here we are mainly concerned with deviations from the mean field approximation in both types of systems which are related to either quantum correlations and statistics (the papers by Blaauboer *et al.*, O'Dell *et al.*, Katz *et al.*, Ji *et al.*, and Blaauboer) or

V.M. Akulin et al. (eds.), Decoherence, Entanglement and Information Protection in Complex Quantum Systems, 565–572.

to decoherence and its prevention (the papers by Barone *et al.*, Esteve, Sarnelli *et al.*, and Dykhne *et al.*). The motivation for these studies is both fundamental and practical: on the fundamental side, mesoscopic systems are the bridge between quantum and classical domains; on the applied side, these systems are candidates for quantum computing and quantum information processing. Below a short introduction to each of the specific topics is given.

1. Atomic ensembles and Bose-Einstein condensates

The subject of the first paper (Blaauboer *et al.*) is the scattering of a probe by fluctuating multiatom ensembles in an optical lattice. So far, the transition between "insulating" (localized) and "metallic" (superfluid) phases have been studied in the framework of the Bose-Hubbard model [Jaksch 1998] and recently been observed experimentally [Greiner 2002]. As confirmed by Ref. [Jaksch 1998], the mean-field approximation is inadequate for small numbers of interacting atoms in lattices due to the presence of large quantum fluctuations. Still more difficult is the analysis of fluctuations in systems of cold atoms coupled by long-range ($1/r$ or $1/r^3$) field-induced interactions [O'Dell 2000]. This leads to the intruiging question: Is there a way to circumvent the formidable task of treating the full dynamics of such systems and still infer their important characteristics, *e.g.*, their dependence on temperature, quantum statistics (Bose or Fermi), number of atoms and lattice parameters? And: Are there universal measurable features which can be a "signature" of the statistical ensemble (distribution function) of such systems?

Here the authors consider the possibility of inferring such statistical characteristics from the spectral features of probe photons or particles that are scattered by the density fluctuations of trapped atoms, notably in optical lattices, in two hitherto unexplored scenarios. (a) The probe is weakly (perturbatively) scattered by the local atomic density corresponding to the random occupancy of different lattice sites. (b) The probe is multiply scattered by an arbitrary (possibly unknown a priori) multi-atom distribution in the lattice. The highlight of the analysis, which is based on this random matrix approach, is the prediction of a semicircular spectral lineshape of the probe scattering in the large-fluctuation limit of trapped atomic ensembles. Thus far, the only known case of quasi-semicircular lineshapes in optical scattering has been predicted [Akulin 1993] and experimentally verified [Ngo 1994] in dielectric microspheres with randomly distributed internal scatterers.

In the second paper, O'Dell *et al.* discuss Bose-Einstein condensates with laser-induced dipole-dipole interactions. A Bose-Einstein condensate formed in a trapped ultracold atomic vapour constitutes an example of macroscopic

quantum coherence akin to that found in superfluid helium, superconductors, or lasers. Following the first experimental realization of atomic BECs in 1995 in rubidium, cesium and lithium [Anderson 1995], there have been many fundamental experiments involving these systems (see, *e.g.*, Ref. [Ketterle 1999]). For the most part these experiments have demonstrated the dramatic *first order* coherence of BECs, namely that they behave essentially like classical waves, described by a macroscopic wavefunction $\Psi(\vec{r}) = A(\vec{r}) \exp i\phi(\vec{r})$. The amplitude A depends on the total atom density $\sqrt{N/V}$, and so the interference fringes formed when two BECs overlap [Andrews 1997] have magnitudes dependent upon the N atom amplitude (*i.e.*, not just that of a single atom). The macroscopic wavefunction $\Psi(\vec{r})$ also gives rise to quantized vortices and irrotational flow familiar from superfluids.
The subject of this contribution (which is largely based on Ref. [O'Dell 2003]) is however, BECs with enhanced *second order* coherence. This requires interparticle correlations, but these are generally rather weak in atomic vapours because they are very dilute with respect to the range of the interactions. This situation should be contrasted to that found in liquids such as superfluid helium II say, where strong correlations play an important role. In some senses a BEC is in fact defined by vanishing or weak correlations amongst the atoms - Einstein's original proposal was for an ideal gas with no interactions at all - and first and second order coherence might even be considered to be mutually opposing. The relatively strong interactions found in liquid helium deplete the condensate and form a cloud of "above-condensate" atoms which are highly correlated, leaving only about 10% of the atoms in the condensate fraction even at temperatures approaching absolute zero. This means that descriptions of helium II in terms of a mean-field uncorrelated wavefunction $\Psi(\vec{r})$ become problematic. Authors interest in the second contribution is to probe the intermediate regime between a weakly interacting BEC and more strongly correlated system by introducing controllable long-range interactions. These interactions are induced by an off-resonant laser which slightly polarizes all the atoms and the ensuing dipole-dipole interactions amongst the atoms are of a very different nature to the short-range van der Waals interactions usually found in atomic gases. In particular the authors predict that one might be able to induce a "roton" minimum in the excitation spectrum of a gaseous BEC. This "roton" minimum, which is well-known in the physics of helium II, is the hallmark of significant second order correlation.

The paper by Katz *et al.* also focuses on BECs and presents new experimental and theoretical results on excitation evolution and decay in a BEC. As the field of Bose-Einstein condensation is rapidly maturing, new regimes of interest are now coming under scrutiny. These include trapping in various optical lattice potentials - leading to squeezed atomic states and the Mott insulator state, ultracold fermionic systems, Fermi-Bose mixtures and molecular condensates. In general the trend is to move away from mean-field effects towards "quantum atom optics". In such systems the mature ideas from conventional quantum optics are immediately applicable and produce many new effects.

The starting point of most discussions about BEC is the Gross-Pitaevskii equation (GPE) [Ginzburg 1958; Gross 1963]:

$$i\hbar\,\frac{\partial\Psi}{\partial t} = \left(-\frac{\hbar^2}{2m}\nabla^2 + V_{ext}(\vec{r},\,t) + g\,|\Psi|^2\right)\Psi \tag{1}$$

which contains a kinetic energy term $-\hbar^2\,\nabla^2/2m$, the external potential $V_{ext}(\vec{r},\,t)$ and the self interaction term $g\,|\Psi(\vec{r},\,t)|^2$. The interplay between these three components of the GPE gives rise to subtle and rich behavior, notably the well-known Bogoliubov excitation spectrum [Bogoliubov 1947], which was measured by authors group [Steinhauer 2002]. This spectrum results in a linear excitation slope at low momenta and a shifted quadratic excitation spectrum at higher excitation momentum. The transition occurs at the inverse healing length,

$$\xi = \frac{\hbar}{\sqrt{2m\,g\,n}}, \tag{2}$$

where $n = |\Psi|^2$ is the condensate density. The healing length is the minimal distance over which the condensate wave function can vary its value significantly. The GPE also explains much of modern nonlinear physics, namely solitons, wave mixing and matter-wave amplifications which have all been observed in the lab. However, despite all this, it remains a classical nonlinear wave equation. Higher order terms, truncated in the derivation of the GPE from the many-body Hamiltonian, lead to an even richer picture of the BEC as a many-body quantized field. These higher order terms in the many-body theory of interacting Bose gases are the motivation for most of recent research results discussed in the third paper of this part. These terms lead to damping and mixing of excitations, which cannot be completely described by mean-field pictures. The authors develop a theoretical many-body dressed state model that deals with the mixing of excitations, both as a coherent three-wave-mixing process (3WM) and as an incoherent decay mechanism. They also present four different approaches to reduce the various dephasing mechanisms of excitations. Namely, the existence of discrete radial Bogoliubov excitation modes in elongated traps, the use of a momentum echo, non-linear effects of strong excitations and more uniform density distributions obtained for flat-bottom traps are all proposed and shown to offer significant suppression of the various dephasing mechanisms. These methods open access to measuring the homogeneous broadening of condensate excitations.

In an optical lattice the state of the system is expanded using the single particle Bloch states. In this picture the collisional decay of strongly driven condensates is then calculated, leading to a remarkable splitting / shift of the collisional products, away from the usual spherical s-wave.
Beyond mean-field effects are then explored in a theoretical calculation of the spatial correlations in a boson gas where the interactions are no longer weak. The authors find an enhancement in the response of the system, related to a possible roton in such system, as observed in liquid ^4He.
In the last section results of a matter wave interference method are presented which has been developed to probe very weak excitations at ultra-low momentum. The interference between the excitations and the condensate leads to strong density modulations after the release of the cloud, whose contrast is a heterodyne probe of the excitation strength.

2. Two-dimensional electron gases, superconducting and molecular junctions

The next five papers deal with coherence and entanglement in condensed matter mesoscopic systems, in particular two-dimensional electron gases (2DEGs) and superconducting nanosystems. Ji *et al.* discuss the first Mach-Zehnder interferometer for electrons, realized in a 2DEG. Double-slit electron interferometers, fabricated in high mobility 2DEG, proved to be very powerful tools in studying coherent wave-like phenomena in mesoscopic systems [Yacobi 1995; Schuster 1997; Buks 1998; Ji 2000]. However, such interferometers have their disadvantages. They support multiple channels in each slit and consequently suffer from a small fringe visibility [Schuster 1997]. Their open geometry, required to eliminate multiple paths interference, allows only a small fraction of the injected current to be collected [Büttiker 1986; Schuster 1997; Buks 1998; Ji 2000]. Moreover, they do not function in a high magnetic field, which imposes a strong Lorentz force on the electrons and destroys the symmetry between the left and right slits. Hence, they are limited in their applications and cannot be employed, for example, in the quantum Hall effect (QHE) regime. The authors have fabricated and measured a novel, single channel, two-path electron interferometer, that functions in a high magnetic field. It is the first electronic analog of the well-known optical Mach-Zehnder (MZ) interferometer [Born 1999]. Based on a single edge state transport in the QHE regime, the interferometer collects all the injected electrons, hence having extremely high visibility (up to 62%) and high sensitivity to a small number of injected electrons. It is found, unexpectedly, that the interference pattern is extremely sensitive to the electron temperature or energy. By performing shot noise measurements of the interfering electrons the authors show that the observed loss of interference results from phase averaging among electrons and is not due to incoherent scattering processes.

In the paper by Blaauboer coherent transport by adiabatic pumping is studied. Interest in this phenomenon is based on the well-known fact that the quantum transport properties of a mesoscopic system are modified in the presence of a superconducting interface, due to interference between normal and Andreev reflections. Andreev reflection (AR) [Andreev 1964] is the electron-to-hole reflection process which occurs when an electron with energy slightly above the Fermi energy is incident on the boundary between a normal metal and a superconductor: the electron enters the superconductor after forming a Cooper pair, and leaves a hole in the normal metal with energy slightly below the Fermi level which travels back along (nearly) the same path where the electron came from. Because of the phase-coherent character of AR, it is interesting to study its effect on transport in mesoscopic systems, where phase coherence plays an important role. In the last decade, this has led to the discovery of a wealth of quantum interference effects in mesoscopic normal-metal-superconductor (NS) structures [Beenakker 1997], such as the observation of a large narrow peak in the differential conductance of a disordered NS junction ("reflectionless tunneling") [Kastalski 1991], and the discovery of novel Kondo phenomena in quantum dots attached to a normal and a superconducting lead [Kang 1998]. In addition, investigations of the conductance in superconductor-carbon-nanotube devices have recently appeared [Morpurgo 1999; Wei 2001 (a)], which indicate that also in these devices resonant behavior due to AR occurs.
In this paper the effects of AR on adiabatic quantum pumping are considered, see also Ref. [Blaauboer 2002], and the limiting time for fully coherent transport of electrons is discussed. Quantum pumping involves the generation of a d.c. current in the absence of a bias voltage by periodic modulations of two or more system parameters, such as, *e.g.*, the shape of the system or a magnetic field. The idea was pioneered by Thouless for electrons moving in an infinite one-dimensional periodic potential [Thouless 1983]. In recent years, adiabatic quantum pumping in quantum dots has attracted a lot of attention [Aleiner 1998; Brouwer 1998; Switkes 1999]. Quantum dots are small metallic or semiconducting islands, confined by gates and connected to electron reservoirs (leads) through quantum point contacts (QPCs) [Kouwenhoven 1997]. In addition to investigations of pumping in quantum dots, theoretical ideas have been put forward for charge pumping in carbon nanotubes [Wei 2001 (b)], and for pumping of Cooper pairs [Zhou 1999]. Here the author considers a mesoscopic system consisting of an arbitrary normal-metal region, *e.g.*, a quantum dot or QPC, coupled to a superconductor. To start with, a general formula for the pumped current through this NS system in terms of its scattering matrix is

derived. This is the N-mode generalization of the result of Wang *et al.* for a NS system with single-mode leads [Wang 2001]. This result is then used to calculate the current pumped through a quantum dot in the Coulomb blockade regime coupled to a superconducting lead. Comparison with the pumped current in the corresponding system attached to normal leads only, shows that AR enhances quantum pumping by up to a factor of 4 at low temperatures and for (nearly) symmetric coupling to the leads, while it reduces quantum pumping in the opposite situation of strongly asymmetric coupling.

The subjects of the paper by Barone *et al.* are the Zeno and anti-Zeno effects in driven Josephson junctions and their relation to the control of macroscopic quantum tunneling. It is known that frequent perturbations (or measurements) of quantum states decaying into an energy continuum can either cause slow-down of the decay (the quantum Zeno effect - QZE) [Misra 1977; Kofman 2000; Kofman 2001 (a)] or, conversely, its speedup (the anti-Zeno effect - AZE) [Lane 1983; Facchi 2000 (a); Kofman 2000; Fischer 2001; Kofman 2001 (a)]. Here the authors show that the QZE and AZE are realizable for *macroscopic* quantum states that decay via macroscopic quantum tunneling (MQT) in a superconducting current-biased Josephson junction (JJ) [Barone 1982; Clarke 1988; Fisher 1988; Leggett 1987] and its analogs in ultracold atomic condensates [Smerzi 1997; Anderson 1998 (a)], upon varying the rate of the bias-current modulation. The present theory extends previous treatments of tunneling through time-dependent barriers [Fisher 1988; Ivlev 2002], revealing unknown aspects of MQT dynamics, particularly, the short-time *reversibility* of decay to the continuum. It may also be useful for optimizing JJ-based schemes of quantum computing [Averin 2000].

The last two papers are also related to quantum computing issues, in particular the characterization of various kinds of qubits based on Josephson junctions.

In the paper by Sarnelli *et al.*, the employment of submicron $YBa_2Cu_3O_{7-x}$ grain boundary junctions for the fabrication of "quiet" superconducting flux-qubits is considered. The implementation of superconducting flux-qubits represents a challenge for the future. Superconductors offer both the scalability, typical of solid state technology, and the properties of a macroscopic quantum system. However, the real employment of superconducting devices passes through the reduction of decoherence time of a single qubit as well as a number of entangled qubits. Decoherence can be reduced by using appropriate technological choices, like the ones minimizing the coupling of the quantum system to the environment. The discovery of unconventional superconductors offered the possibility to study new Josephson junction based circuitries that, in principle, can reduce the coupling to external room temperature electronics. The authors re-examine the five junction flux qubit proposed in 2001 by Blatter *al.*, composed of a superconducting loop with four conventional junctions and one π-junction, characterized by an energy minimum shifted by π with respect to conventional Josephson junctions. In this scheme, the π-junction acts as a simple phase shifter, frustrating the superconducting loop, and allowing the formation of a double degenerate ground state. In the proposal presented in the fourth article of this part, the π-junction is a particular device based on the properties of Andreev reflections to form bound states at the Fermi energy.

Finally, the paper by Dykhne *et al.* considers broken symmetry and coherence of vibrations in molecular tunnel junctions. In particular, the authors examine the Breit-Wigner resonances that ensue from field effects in molecular single electron transistors (SETs). Adiabatic dynamics of a quantum dot that is elastically attached to electrodes is treated in the Born-Oppenheimer approach. The relation between thermal and shot noises induced by source-drain voltage V_{bias} is found when the SET operates tending to thermodynamic equilibrium far from resonance. The equilibration of electron-phonon subsystems produces broadening and doublet splitting of transparency resonances helping to explain a negative differential resistance (NDR) of current versus voltage (I - V) curves. Mismatch between the electron and phonon temperatures brings out the bouncing-ball mode in crossover regime close to internal vibrations mode. The shuttle mechanism occurs at a threshold V_{bias} of the order of Coulomb energy U_c. The charge accumulated is followed by the Coulomb blockade and broken symmetry of a single or double well potential. The Landau bifurcation cures the shuttling instability and resonance levels of the quantum dot get splitting because of molecular tunneling. The authors calculate the tunnel gaps of conductivity and propose a tunneling optical trap (TOT) for quantum dot isolation permitting coherent molecular tunneling in virtue of the Josephson oscillations in charge Bose gas. They also discuss experimental conditions for which the above theory can be tested.

PROBE SCATTERING BY FLUCTUATING MULTI-ATOM ENSEMBLES IN OPTICAL LATTICES

M. Blaauboer,[1] G. Kurizki[2] and V. M. Akulin[3]

[1] *Kavli Institute of Nanoscience, Delft University of Technology, Lorentzweg 1, 2628 CJ Delft, The Netherlands*
m.blaauboer@tnw.tudelft.nl

[2] *Chemical Physics Department, Weizmann Institute of Science, Rehovot 76100, Israel*

[3] *Laboratoire Aimé Cotton, Bâtiment 505, Campus d'Orsay, 91405 Orsay Cedex, France*

Abstract We investigate probe scattering by fluctuating ensembles of atoms trapped in optical lattices: weak scattering by effectively random atomic density distributions and multiple scattering by arbitrary atomic distributions. Both regimes are predicted to exhibit a universal semicircular scattering lineshape for large density fluctuations, which depend on temperature and quantum statistics.

Keywords: Quantum scattering, optical lattices.

Introduction

The transition between "insulating" (localized) and "metallic" (superfluid) phases have been studied in the framework of the Bose-Hubbard model [Jaksch 1998] and recently been observed experimentally [Greiner 2002]. As confirmed by Ref. [Jaksch 1998], the mean-field approximation is inadequate for small numbers of interacting atoms in lattices due to the presence of large quantum fluctuations. Still more difficult is the analysis of fluctuations in systems of cold atoms coupled by long-range ($1/r$ or $1/r^3$) field-induced interactions [O'Dell 2000]. This leads to the intruiging question: Is there a way to circumvent the formidable task of treating the full dynamics of such systems and still infer their important characteristics, *e.g.*, their dependence on temperature, quantum statistics (Bose or Fermi), number of atoms and lattice parameters? And: Are there universal measurable features which can be a "signature" of the statistical ensemble (distribution function) of such systems?

We consider the possibility of inferring such statistical characteristics from the spectral features of probe particles scattered by the density fluctuations (DF) of trapped atoms, notably in optical lattices (OL), in two hitherto unex-

V.M. Akulin et al. (eds.), Decoherence, Entanglement and Information Protection in Complex Quantum Systems, 573–580.

plored scenarios:
(a) The probe is weakly scattered by the local atomic density corresponding to the random occupancy of different lattice sites (Fig. 1 - inset (a)).
(b) The probe is multiply scattered by an arbitrary (possibly unknown a priori) multi-atom distribution in the lattice (Fig. 1 - inset (b)).

The key idea of our analysis is that the Green function (GF)of the scattered photon or particle, which embodies the relevant spectral information, can be qualitatively estimated without resorting to cumbersome perturbative calculations of the probe-multiatom interaction by replacing this interaction Hamiltonian by an equivalent random matrix. The random matrix approach, successfully applied to various disordered systems [Mehta 1991], allows the evaluation of probe spectra to *all orders of scattering*, expressing them by means of *only the first two moments* (the mean and variance of the random interaction, averaged over the statistical ensemble of the multiatom system (MS). The highlight of our analysis, based on this random matrix approach, is the prediction of a semicircular spectral lineshape of the probe scattering (PS) in the large-fluctuation limit of trapped atomic ensembles. Thus far, the only known case of quasi-semicircular lineshapes in optical scattering has been predicted [Akulin 1993] and experimentally verified [Ngo 1994] in dielectric microspheres with randomly distributed internal scatterers.

1. Analysis

The GF of the probe (P) at energy $\epsilon = \hbar\omega$ is given by

$$G_P(\epsilon) = \mathrm{Tr}_S \left[\frac{1}{\epsilon - \hat{H}_P - \hat{V}} \hat{\rho}_S \right], \tag{1}$$

where $\hat{H}_P$, $\hat{V}$ and $\hat{\rho}_S$ are, respectively, the unperturbed probe Hamiltonian, the probe-system interaction Hamiltonian and the density operator for the ensemble of the MS (S). We assume that the following conditions hold.
(i) There is no appreciable back-effect of the probe on the MS (otherwise it is no longer a probe).
(ii) The state of the MS does not change during the interaction time with the probe, *i.e.*, the MS remains "frozen", as is applicable for optical or atomic probing. This situation then cannot be described as Markovian relaxation (exponential decay) of the probe state into the multiatom reservoir, since the correlation time of this reservoir is now much longer than that of the probe, in contrast with the basic assumption of relaxation.
(iii) The probe spectrum is broadband, *i.e.*, it encompasses many of its eigenstates.

For an ensemble "frozen" during the interaction time, the tracing in (1) implies statistical averaging over repeated realizations of the MS, each time the

PS is recorded, or taking the expectation value with respect to the quantum state of the system. For simplicity, let us explicitly consider elastic scattering (the extension to inelastic scattering is straightforward), for which

$$\hat{V} = \sum_{\vec{k}} f_{\vec{k}}\, \hat{\rho}^{\dagger}_{P\vec{k}}\, \hat{\rho}_{S\vec{k}} + \text{h.c.} \tag{2}$$

or

$$\hat{V} = \sum_{\vec{k}} f_{\vec{k}}\, a^{\dagger}_{\vec{k}}\, \hat{\rho}_{S\vec{k}} + \text{h.c.}\,. \tag{3}$$

Here $f_{\vec{k}}$ is the scattering amplitude for momentum exchange $\hbar\vec{k}$ between the probe and the system and the $\vec{k}$-mode Fourier components of the probe (system) density operators $\hat{\rho}_{P\vec{k}}$ ($\hat{\rho}_{S\vec{k}}$) are defined in terms of their respective creation and annihilation operators $\hat{\rho}_{P\vec{k}} = \sum_{\vec{q}} a^{\dagger}_{\vec{q}}\, a_{\vec{q}+\vec{k}}$, $\hat{\rho}_{S\vec{k}} = \sum_{\vec{q}} c^{\dagger}_{\vec{q}}\, c_{\vec{q}+\vec{k}}$. Eqs. (2) and (3) stand, respectively, for bilinear and linear probe-system coupling. For optical probes (2) and (3) correspond to Raman and single-photon scattering, respectively. For atom or neutron probes (2) is appropriate.

The GF (1) is obtainable, to all orders in $\hat{V}$ [Anderson 1961], by solving the set of equations for its diagonal elements

$$G_{\vec{k}\vec{k}}(\epsilon) = \left[\epsilon - \epsilon_{\vec{k}} - \sum_{\vec{k}'} \left\langle \hat{V}^2_{\vec{k}\vec{k}'} \right\rangle G_{\vec{k}'\vec{k}'}(\epsilon)\right]^{-1}, \tag{4}$$

where $\epsilon_{\vec{k}}$ are the probe energy eigenvalues in the absence of potential fluctuations and pointed brackets denote the expectation value. The spectral information contained in these $G_{\vec{k}\vec{k}}$ is given by the density of states (DOS) of the probe $g(\epsilon) = -\text{Im} \sum_{\vec{k}} G_{\vec{k}\vec{k}}(\epsilon)/\pi$.

In order to extract information on the system we shall make two simplifying assumptions regarding the probe and the coupling potential (3):
(i) $f_{\vec{k}}$ is flat in $\vec{k}$ (the coupling is strongly localized in space) within a band exceeding the relevant band of the system, so that $f_{\vec{k}} \approx f$;
(ii) the probe statistical distribution is also flat in $\vec{k}$ and its second moment in $\langle \hat{V}^2 \rangle$ is replacable by the square of its mean flux (or density) $\bar{n}^2_P$ for the bilinear coupling (2) or by its mean flux (density) $\bar{n}_P$ for the linear coupling (3).
Under these assumptions we can write the squared coupling potential in (4) as

$$\left\langle \hat{V}^2_{\vec{k}\vec{k}'} \right\rangle = \left\langle \hat{V}_{\vec{k}\vec{k}'} \right\rangle^2 + \mathcal{F}_P\, S_{\vec{k}\vec{k}'}. \tag{5}$$

Here $\langle \hat{V} \rangle$ is the mean coupling potential and $\mathcal{F}_P \sim |f|^2\, \bar{n}^2_P$ or $\mathcal{F}_P \sim |f|^2\, \bar{n}_P$ in the case of (2) and (3), respectively. The quantity of interest for the system is the Fourier-transformed density-density correlation of the atomic system

$$\mathcal{S}_{\vec{k}\vec{k'}} = \left\langle \hat{\rho}^{\dagger}_{S\vec{k}}\, \hat{\rho}_{S\vec{k'}} \right\rangle + \text{c.c.}\,. \tag{6}$$

Its diagonal element $\mathcal{S}_{\vec{k}\vec{k}}$ is the static structure factor $\mathcal{S}_{\vec{k}}$, which is the Fourier transform of the van-Hove correlation function $\langle \hat{\rho}^{\dagger}_{S}(\vec{r},\, t=0)\; \hat{\rho}_{S}(\vec{r'},\, t'=0)\rangle$ for the spatial DF of the "frozen" atomic ensemble.

The difficulty of having to evaluate or measure the matrix elements $\mathcal{S}_{\vec{k}\vec{k'}}$ is avoided for a spatially random density distribution of the atomic system, due to random site occupancy (Fig. 1, inset (a)) and short-range interaction with the probe (*e.g.*, a neutron or thermal atom). The elements $\mathcal{S}_{\vec{k}\vec{k'}}$ in (5) and (6) can then be replaced by the average of the structure factor over all relevant $\vec{k}$:

$$\mathcal{S}_{\vec{k}\vec{k'}} \rightarrow \bar{\mathcal{S}} = \int d\vec{k}\; \mathcal{S}_{\vec{k}} \sim \langle \hat{n}^2_S\rangle - \langle \hat{n}_S\rangle^2, \tag{7}$$

where the right-hand side of $\bar{\mathcal{S}}$ denotes the local atomic density or number variance averaged over the ensemble. The implications of evaluating the probe DOS $g(\epsilon)$ using Eq. (4) to (7) will be examined for random fluctuations about a mean scattering potential $\langle V_S(x)\rangle$ (corresponding to the mean atomic density distribution) that is 1D-periodic. The "unperturbed" probe dispersion associated with $\langle V_S(x)\rangle$ is $\epsilon_{\vec{k}} = -2J\,\cos(k_x d) + A$, J being the hopping frequency, d the lattice period and A the band energy offset. This gives rise to the following expression for the GF (4)

$$G(\epsilon) = \left\{ \epsilon - \epsilon_{\vec{k}} - \langle W^2\rangle \sum_{\vec{k'}} \left[\epsilon - \epsilon_{\vec{k'}} - \Lambda(\epsilon) + i\Delta(\epsilon)\right]^{-1} \right\}^{-1}. \tag{8}$$

Here

$$\begin{aligned}
\langle W^2\rangle &\equiv \mathcal{F}_P\, \bar{\mathcal{S}}, \\
\Lambda(\epsilon) &= \frac{\langle W^2\rangle}{\sqrt{(\epsilon - A)^2 - 4J^2}} \quad \text{for } |\epsilon - A| > 2J, \\
&= 0 \quad \text{otherwise}, \\
\Delta(\epsilon) &= \frac{\langle W^2\rangle}{\sqrt{4J^2 - (\epsilon - A)^2}} \quad \text{for } |\epsilon - A| < 2J, \\
&= 0 \quad \text{otherwise}.
\end{aligned} \tag{9}$$

Fig. 1 shows how the probe DOS $g(\epsilon)$ changes from that of a periodic band structure corresponding to the mean potential $\langle V_S(x)\rangle$ to a semicircular shape as the amount of fluctuations measured by $\langle W^2\rangle$ increases.

In the multiple-scattering scenario, which pertains to resonantly scattered atomic probes or to intracavity optical probes (Fig. 1, inset (b)), semicircular lineshapes are obtained even when the $\mathcal{S}_{\vec{k}\vec{k}'}$ cannot be claimed to belong to a random distribution (Fig. 2, inset). In the case of strongly-interacting atoms within a lattice site or longe-range intersite density correlations [O'Dell 2000] the distribution may be quite intricate, corresponding to sharp peaks of $\mathcal{S}_{\vec{k}\vec{k}'}$. Nevertheless, the universal spectral trends of Fig. 1 can be shown to hold in this scenario, provided $\langle\hat{V}^2\rangle^{1/2} g_0(\epsilon) \gg 1$, $g_0(\epsilon)$ denoting the "unperturbed" probe DOS. This condition allows us to estimate $G_{\vec{k}\vec{k}}$ in (4) to all orders in $\hat{V}$, upon replacing the state of the atomic system by a gaussian random ensemble [Mehta 1991; Akulin 1993]. The result is the following universal formula [Akulin 1993] for the renormalized probe energy $\tilde{\epsilon}$ at a given input energy ϵ

$$\epsilon = \tilde{\epsilon} + \langle W^2 \rangle \ \mathrm{Tr}_P \left(\frac{1}{\tilde{\epsilon} - \hat{H}_P - i0} \right). \tag{10}$$

The use of (10) leads to a semicircular lineshape similar to the one in Fig. 1, as if the potential were random.

2. Results for limiting cases

In order to illustrate the role of temperature, quantum statistics and the mean lattice potential in producing the semicircular lineshape, we proceed to evaluate $\langle W^2\rangle$ for several simple models.

2.1 The isolated-site limit

The tightly-bound Bose or Fermi distributions in a lattice can be estimated by taking the potential of each site to be that of a harmonic well of depth V_0. The isolated-site approximation holds for atoms in the lowest vibrational band, when the coupling energy is much smaller than the excitation energy to the next band [Jaksch 1998], $\sqrt{\langle W^2\rangle} \ll \hbar\omega_\nu = 2\pi\ \hbar/\lambda\ \sqrt{2V_0/m}$, λ being the wavelength of the laser light.

In the absence of additional external perturbations, the coupling $\langle W^2\rangle$ arises because of temperature-dependent fluctuations in the site-occupancy of the optical lattice, which has an approximately gaussian distribution [DePue 1999]. The resulting random coupling energies averaged over all states yield

$$\begin{aligned} \langle W^2\rangle \ &\sim \mathcal{F}_P \left\langle \langle n_{\vec{k}},\, n_{\vec{k}'}\rangle - \langle n_{\vec{k}}\rangle\langle n_{\vec{k}'}\rangle \right\rangle_{\vec{k}\vec{k}'} \\ &= \mathcal{F}_P \left\langle \langle n_i,\, n_j\rangle - \langle n_i\rangle\langle n_j\rangle \right\rangle_{ij} \approx \mathcal{F}_P \left(\langle n_S^2\rangle - \langle n_S\rangle^2 \right). \end{aligned} \tag{11}$$

Here $n_{\vec{k}} \equiv c^\dagger_{\vec{k}}\, c_{\vec{k}}$, i and j label atomic sites, and $\langle n_S\rangle$ is the average number of atoms per site. The last step applies whenever $n_i \approx n_j\ \forall i,\ j$ and the DF are approximately site-independent. We have verified this by numerical simulation, considering 2 to 4 identical atoms on a 1D lattice with 6 sites and calculating the DF if the probability of an occupied site is 1/10 of the probability of an empty site. In all cases the maximum relative difference between $\mathcal{F}_P\, S_{\vec{k}\vec{k}'}$ and $\langle W^2\rangle$ was less than 10 %. The kinetic contribution to $\langle W^2\rangle$ due to evaporation of atoms from the lattice is the dominant one at high temperatures, regardless of the statistics. If all the atoms are in the lowest energy band, we may adopt the rate equation used to describe the formation of electron-hole clusters in a plasma [Klingshirn 1981] and find

$$\langle W^2\rangle_{\text{evap}} = a\,\langle n_S\rangle\, T^2\, e^{-\beta V_0}. \tag{12}$$

Here $a = k_B\, m\, c_p$, with k_B the Boltzmann constant, m the mass of the atoms and c_p their specific heat, T denotes the temperature, $\beta^{-1} \equiv k_B T$ and V_0 is the optical lattice potential. The influence of evaporation becomes the dominant effect for $T \sim 25\ \mu$K. Around $T \sim 300\ \mu$K these fluctuations become comparable in size to the square of the optical lattice potential ($\langle W^2\rangle \sim V_0^2 \sim 100(\text{neV})^2$) and atoms then largely escape from the lattice.

At low temperatures (well below 100 μK) the density-density fluctuations (DDF) depend on whether the atoms in the lattice are bosons or fermions. For bosonic atoms in the lowest vibrational state we obtain [Reichl 1998]

$$\langle W^2\rangle_{\text{stat, Bose}} = \frac{z}{1-z} + \left(\frac{z}{1-z}\right)^2 + \left(\frac{d}{\lambda_T}\right)^3 \sum_{\alpha=1}^{\infty} \frac{z^\alpha}{\alpha^{1/2}}. \tag{13}$$

Here we have approximated the motion of the atoms in the potential wells by a harmonic oscillation with frequency ω_v [Kastberg 1995], $z \equiv e^{-k_B T/\hbar\omega_v}$, d denotes the average lattice spacing and $\lambda_T = (2\pi\, \hbar^2/m\, k_B T)^{1/2}$, the thermal wavelength, is the length scale separating quantum statistical behavior (for $\lambda_T \sim d$) from classical Maxwell-Boltzmann behavior (for $\lambda_T \ll d$). For fermionic atoms in an optical lattice [Andrews 1999] one starts with the analog of the couplings (3) and (2) for particles with spin, using creation and annihilation operators $c^\dagger_{\vec{k}\sigma}$ and $c_{\vec{k}\sigma}$ and performing an additional sum over the spin index σ, and follows the same analysis as above. One then finds

$$\langle W^2\rangle_{\text{stat, Fermi}} = \frac{z}{1+z} + \left(\frac{z}{1+z}\right)^2 + 2\,\frac{d^3}{\lambda_T^2\,\lambda_F} \sum_{\alpha=1}^{\infty} \frac{z^\alpha}{\alpha^{1/2}}, \tag{14}$$

with λ_F the Fermi wavelength. At high temperatures $z \to 0$ and both (13) and (14) reduce to the classical Maxwell-Boltzmann result $\langle W^2\rangle_{\text{stat, clas}} = z$. At low temperatures, fermionic fluctuations approach a constant value, whereas bosonic fluctuations become very large as T decreases below $\sim 1\mu$K, marking the Bose-Einstein condensation.

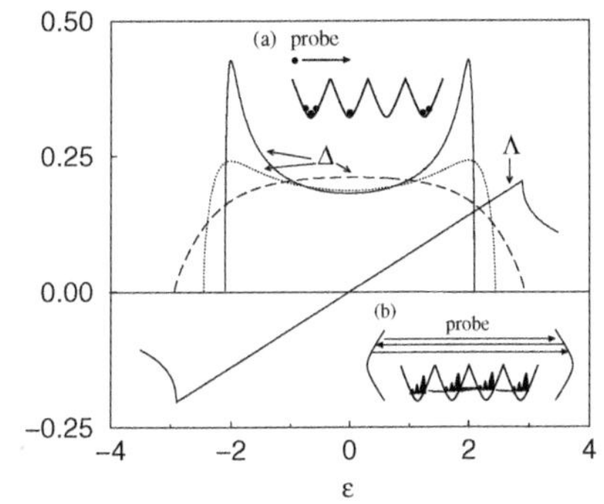

Figure 1. Density of states $g(\epsilon)$ of a probe scattered by bosonic atoms in a 1D optical lattice. Solid, dotted and dashed curves stand for $\Delta(\epsilon)$ and the thick curve stands for $\Lambda(\epsilon)$ (dispersion), see text. All curves are numerically computed from $G(\epsilon)$ and correspond to average random couplings $\langle W^2 \rangle$ = 0.4, 2 and 10 respectively. The hopping frequency $J = 1$, and for all curves $\int d\epsilon\, g(\epsilon) = 1$. Inset (a): a probe weakly scattered by a randomly occupied lattice. Inset (b): a probe multiply scattered by a regular atomic distribution.

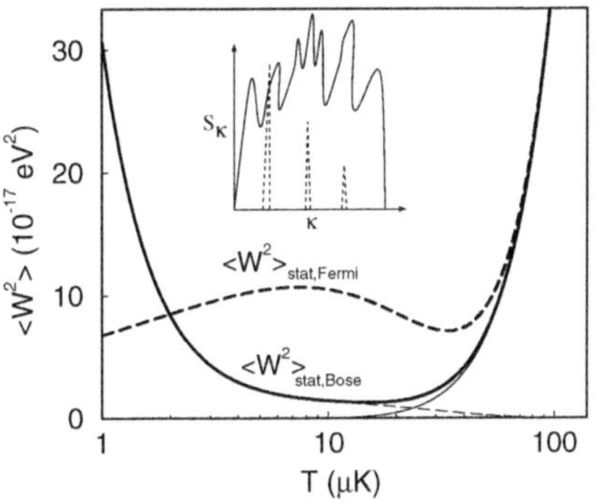

Figure 2. DDF vs. temperature for bosonic and fermionic Li atoms in an optical lattice. Thin solid line: fluctuations due to evaporation (12) (scaling factor: 7.8), thin dashed line: statistical fluctuations (13). Thick solid (dashed) line: total fluctuations $\langle W^2 \rangle$ for bosonic (fermionic) Li atoms. Parameters: $V_0 = 5$ neV, $\langle n_S \rangle = 0.1$, d = 0.1 μm, $c_p(Li) \sim 3.6 \times 10^6$ J.kg^{-1}.K^{-1}, $\lambda_F(Li) = 6.\,10^{-10}$ m and $\omega_v(Li) \sim 2.\,10^6$ s^{-1} [Kastberg 1995]. Inset: solid (dashed) line: static structure factor vs. κ for phonons (nearly-free fermions) in a lattice at finite T.

In Fig. 2 we have taken typical parameters for available OL [Birkl 1995; Kastberg 1995; DePue 1999] to show how $\langle W^2 \rangle$ evolves as a function of temperature for both bosonic and fermionic Li atoms. The total DDF consist in the sum of (12) and either (13) or (14), depending on the statistics. The isolated-site condition is satisfed for the entire temperature range displayed in Fig. 2. Since the hopping frequency $J \sim V_0$, the random coupling for bosonic Li atoms changes from $\langle W^2 \rangle / J^2 \sim$ 10 to $\langle W^2 \rangle / J^2 \sim$ 100, when going from $T \sim 8$ μK to $T \sim 100$ μK. Simultaneously the DOS then evolves from the periodic to the semicircular shape as in Fig. 1. The behavior for Na or Rb atoms is found to be similar to that of bosonic Li atoms, apart from their different mass values in (12) and (13). Employing currently achievable Bragg scattering techniques [Birkl 1995] with a far off-resonant laser ($1\ \mathrm{mW/cm^2}$ intensity, 5.2 MHz detuning) and a lattice with a sufficiently high atomic filling factor [DePue 1999], the scattering spectrum is expected to evolve with T, in the microkelvin range, from the discrete to the semicircular regime.

2.2 The nearly-free limit

A Bose or Fermi gas weakly modulated by the lattice potential yields $\mathcal{S}_{\vec{k}\vec{k}'} = \mathcal{S}_{\vec{\kappa}=\vec{k}-\vec{k}'} = |\phi_{\vec{\kappa}}|^2 \mathcal{S}_{\vec{\kappa}}^{(\text{free})}$. Here $\vec{\kappa}$ is a reciprocal lattice vector, $\phi_{\vec{\kappa}}$ is the corresponding Fourier harmonic of the lattice potential (normalized to 1) and $\mathcal{S}_{\vec{\kappa}}^{(\text{free})}$ is the structure factor for momentum transfer $\hbar\kappa$ in a free Bose or Fermi gas. For a Fermi gas $\mathcal{S}_{\vec{\kappa}}^{(\text{free})} = \pm\Theta(k_f - \kappa)$, the Fourier transform of pair correlations with parallel or anti-parallel spins (which determines the sign): it is the well-known step function which vanishes for κ larger than the Fermi wavevector k_f. At finite temperatures this distribution broadens. The replacement of the nearly-free fermionic $\mathcal{S}_{\vec{k}\vec{k}'}$ by the average value (7) is then justifiable only in the multiple-scattering scenario, while in the weak-scattering one the lattice potential harmonics $\phi_{\vec{k}}$ pick out well-defined $\mathcal{S}_{\vec{k}-\vec{k}'=\vec{\kappa}}$ (ig. 2, inset).

2.3 The phonon regime

Excitations at frequencies below the chemical potential of a Bose condensate trapped in a lattice can produce collective phonon modes [Stamper-Kurn 1999] whose "frozen" spectrum is characterized by

$$\mathcal{S}_{\vec{\kappa}} = \sum_{\vec{q}} \left[\left(\langle n_{\vec{q}} \rangle + 1\right) \sum_{\vec{G}} \delta\left(\vec{\kappa} - \vec{q} - \vec{G}\right) + \langle n_{\vec{q}} \rangle \sum_{\vec{G}} \delta\left(\vec{\kappa} + \vec{q} + \vec{G}\right) \right], \tag{15}$$

where $\langle n_{\vec{q}} \rangle$ is the mean number of phonons at temperature T with wavevector $\vec{q}$, and $\vec{G}$ denotes the reciprocal lattice vector. The phonon mode spectrum includes quasi-local modes in the case of fluctuating atomic distributions. This leads to the limit (7) and an effectively random coupling (Fig. 2, inset).

Conclusions

We have identified new regimes of PS by atoms trapped in OL in the random-density and multiple-scattering regimes. These regimes cannot be treated by the mean-field approximation, but are characterized by a universal feature of large DF, namely, semicircular scattering lineshapes. This is the atom-optical analog of the semicircular broadening of the DOS in disordered electronic systems, which has not yet been observed unambiguously. This atom-optical counterpart presents a nontrivial but feasible challenge for experimentalists.

Acknowledgments

This work was supported by US-Israel BSF, the Israel Council for Higher Education, Minerva, Arc-en-Ciel and the EU RTN No. HPRN-CT-2002-00309.

BOSE-EINSTEIN CONDENSATES WITH LASER-INDUCED DIPOLE-DIPOLE INTERACTIONS

D. O'Dell,[1] S. Giovanazzi[2] and G. Kurizki[3]

[1] *Department of Physics & Astronomy, University of Sussex, Falmer, Brighton BN1 9QH, England*

[2] *School of Physics & Astronomy, University of St Andrews, North Haugh, St Andrews KY16 9SS, Scotland*

[3] *Chemical Physics Department, Weizmann Institute of Science, Rehovot 76100, Israel*

Abstract We consider the interparticle correlations in a gaseous Bose-Einstein condensate which has laser-induced dipole-dipole interactions. These correlations, which are tunable and occur at the length scale of the laser wavelength, can lead to a 'roton' minimum in the excitation spectrum.

Introduction

A Bose-Einstein condensate (BEC) formed in a trapped ultracold atomic vapour constitutes an example of macroscopic quantum coherence akin to that found in superfluid helium, superconductors, or lasers. Following the first experimental realization of atomic BECs in 1995 in rubidium, cesium and lithium [Anderson 1995; Bradley 1995; Davis 1995], there have been many fundamental experiments involving these systems (see, *e.g.*, [Ketterle 1999]). For the most part these experiments have demonstrated the dramatic *first order* coherence of BECs, namely that they behave essentially like classical waves, described by a macroscopic wavefunction $\Psi(\vec{r}) = A(\vec{r}) \exp \mathrm{i}\phi(\vec{r})$. The amplitude A depends on the total atom density $\sqrt{N/V}$, and so the interference fringes formed when two BECs overlap [Andrews 1997] have magnitudes dependent upon the N atom amplitude (*i.e.*, not just that of a single atom). The macroscopic wavefunction $\Psi(\vec{r})$ also gives rise to quantized vortices and irrotational flow familiar from superfluids.

The subject of this contribution (which is largely based on reference [O'Dell 2003]) is however, BECs with enhanced *second order* coherence. This requires interparticle correlations, but these are generally rather weak in atomic vapours

V.M. Akulin et al. (eds.), Decoherence, Entanglement and Information Protection in Complex Quantum Systems, 581–588.

because they are very dilute with respect to the range of the interactions. This situation should be contrasted to that found in liquids such as superfluid helium II say, where strong correlations play an important role. In some senses a BEC is in fact defined by vanishing or weak correlations amongst the atoms - Einstein's original proposal was for an ideal gas with no interactions at all - and first and second order coherence might even be considered to be mutually opposing. The relatively strong interactions found in liquid helium deplete the condensate and form a cloud of 'above-condensate' atoms which are highly correlated, leaving only about 10% of the atoms in the condensate fraction even at temperatures approaching absolute zero. This means that descriptions of helium II in terms of a mean-field uncorrelated wavefunction $\Psi(\vec{r})$ become problematic. Our interest here will be to probe the intermediate regime between a weakly interacting BEC and more strongly correlated system by introducing controllable long-range interactions. These interactions are induced by an off-resonant laser which slightly polarizes all the atoms and the ensuing dipole-dipole interactions amongst the atoms are of a very different nature to the short-range van der Waals interactions usually found in atomic gases. In particular we predict one might be able to induce a "roton" minimum in the excitation spectrum of a gaseous BEC. This 'roton' minimum, which is well-known in the physics of helium II, is the hallmark of significant second order correlation.

1. Ground state of a weakly interacting Bose gas at zero temperature: the Bogoliubov transformation and two-mode squeezing

For a gas of non-interacting identical bosons at zero temperature the ground state is a product state of the form $|\Psi\rangle = |\psi_{\rm s}(1)\rangle\, |\psi_{\rm s}(2)\rangle \cdots |\psi_{\rm s}(N)\rangle$, where each atom is in the ground state $|\psi_{\rm s}\rangle$ of the single particle Hamiltonian. There is no correlation, or equivalently, second order coherence present. The same is true of the Hartree self-consistent mean-field ground state which includes the effects of interactions. Again one assumes a product state in which each atom has the same wave function, but now these wave functions are solutions of the Gross-Pitaevskii equation [Dalfovo 1999]. For an inhomogeneous system the solution of the Gross-Pitaevskii equation already presents a non-trivial problem. For the homogenous case, however, it is straightforward to go the next level of sophistication which is the Bogoliubov solution [Dalfovo 1999] for the ground state, which does include correlations. In principle one can also combine the Gross-Pitaevskii approach with that of Bogoliubov to obtain the Bogoliubov-de Gennes equations describing an inhomogeneous system with correlations, but we shall limit ourselves here to the homogeneous case where the simple Bogoliubov theory applies.

Assuming a macroscopic population, N_0, of the zero momentum state (condensate), then crudely speaking the Bogoliubov prescription replaces the boson operators $\hat{c}_0^\dagger$, $\hat{c}_0$ with the c-number $\sqrt{N_0}$, and then drops all terms in the interaction part of the Hamiltonian that don't contain N_0 to at least the first power. The resulting quadratic Hamiltonian may then be diagonalized via the Bogoliubov transformation $\hat{c}_k = U_k\,\hat{b}_k + V_k\,\hat{b}^\dagger_{-k}$ from atomic ($\hat{c}$) to quasi-particle ($\hat{b}$) operators. The Bogoliubov ground state turns out to be a superposition of paired 'back-to-back' atomic momentum states $|n_{k_j}\rangle|n_{-k_j}\rangle$, with $n_{k_j} = n_{-k_j}$, summed over all momenta, and occupation numbers [Huang 1987]

$$|0_{\mathrm{B}}\rangle = \mathcal{N} \sum_{n_{k_1}=0}^{\infty} \sum_{n_{k_2}=0}^{\infty} \ldots [(\beta_{k_1})^{n_{k_1}} (\beta_{k_2})^{n_{k_2}} \ldots]\,|n_{k_1}, n_{k_2}, \ldots\rangle \tag{1}$$

where for economy each occupation number n_{k_j}, denotes an identical number of k_j and $-k_j$, atoms. The superposition amplitudes are given by the ratio of coefficients that appear in the Bogoliubov transform

$$\beta_k = \frac{V_k}{U_k} = \frac{E_{\mathrm{B}}(k) - mv^2 - \dfrac{\hbar^2 k^2}{2m}}{mv^2} \tag{2a}$$

$$\text{in which } E_{\mathrm{B}}(k) = \sqrt{(v\,\hbar k)^2 + \left(\frac{\hbar^2 k^2}{2m}\right)^2} \tag{2b}$$

where E_{B} is the Bogoliubov dispersion relation, which depends on the number density of atoms $n = N/V$, the atomic mass m, and $v \equiv \sqrt{\widetilde{U}(k)n/m}$. $\widetilde{U}(k)$ is the Fourier transform of the effective interatomic potential. Note v defined above corresponds to the velocity of sound in the case where $\widetilde{U}(k)$ is a constant. However, for the long-range dipole-dipole interactions we consider here, $\widetilde{U}(k)$ is not a constant and the sound velocity must be obtained as the derivative of the dispersion relation with respect to k.

The physical interpretation of the ground state (1) is that it arises from the mutual scattering of pairs of atoms out of the condensate ($k = 0$ state), momentum being conserved during these virtual scatterings. The degree of correlation present in the Bogoliubov state can be evaluated via the *static structure factor* $S(k)$, which is a density-density correlation $S(k) \equiv \langle 0|\hat{\rho}_k\,\hat{\rho}_k^\dagger|0\rangle N$ where $|0\rangle$ is the ground state of the system. $\hat{\rho}_k$ is the density fluctuation operator, which is the Fourier transform of the particle density operator $\hat{\rho}(x) = \hat{\Psi}^\dagger(x)\,\hat{\Psi}(x)$, where the field operators $\hat{\Psi}(x)$ are defined by

$$\hat{\Psi}(x) \equiv \sum_k \exp(\mathrm{i}\, kx)\, \hat{c}_k. \tag{3}$$

Thus

$$\hat{\rho}_k \equiv \int d^3x \, \exp(-\mathrm{i}\, kx)\, \hat{\rho}(x) = \sum_q \hat{c}_q^\dagger\, \hat{c}_{q+k}. \tag{4}$$

Being of fourth order in the particle operators, the static structure factor provides a measure of second order correlation. Defining the normalized second order correlation function as

$$g(x_1,\, x_2) \equiv \langle 0|\hat{\Psi}^\dagger(x_1)\, \hat{\Psi}^\dagger(x_2)\, \hat{\Psi}(x_2)\, \hat{\Psi}(x_1)|0\rangle n^2, \tag{5}$$

then $S(k)$ and $g(x_1,\, x_2)$ are related via a Fourier transformation [Pitaevskii 2003]

$$g(x) = 1 + \frac{1}{n(2\pi)^3} \int dk\, [S(k) - 1] \exp(-\mathrm{i}\, kx) \tag{6}$$

where we have assumed a uniform system so that $x = x_1 - x_2$. The static structure factor is a very important expression of second order correlation since it can be measured directly in scattering experiments. For example, when light is scattered off the system, then the total intensity (*i.e.*, integrated over all wavelengths) scattered into a particular direction is proportional to the static structure factor evaluated at the momentum corresponding to the deflection angle [Pines 1999].

$S(k)$ is easily calculated within the Bogoliubov approximation. Retaining only terms that involve the condensate one can approximate the density operator as [Cohen-Tannoudji 2001]

$$\begin{aligned} \hat{\rho}_k \approx \hat{c}_0^\dagger\, \hat{c}_k + \hat{c}_{-k}^\dagger\, \hat{c}_0 \quad &\approx \sqrt{N}\left(\hat{c}_k + \hat{c}_{-k}^\dagger\right) \\ &= \sqrt{N}\left(U_k\, \hat{b}_k + V_k\, \hat{b}_{-k}^\dagger + U_k\, \hat{b}_{-k}^\dagger + V_k\, \hat{b}_k\right) \end{aligned} \tag{7}$$

where the Bogoliubov transformation $\hat{c}_k = U_k\, \hat{b}_k + V_k\, \hat{b}_{-k}^\dagger$ has been made in the last step. Taking the system ground state $|0\rangle$ to be the Bogoliubov ground state $|0_\mathrm{B}\rangle$, which is destroyed by $\hat{b}_k$, so that $\hat{\rho}_k^\dagger|0_\mathrm{B}\rangle = \sqrt{N}(U_k + V_k)|n_k = 1\rangle$, where $|n_k = 1\rangle$ is the state containing a single quasi-particle, one finds

$$S(k) = (U_k + V_k)^2 = \frac{\hbar^2 k^2}{2m\, E_\mathrm{B}(k)}. \tag{8}$$

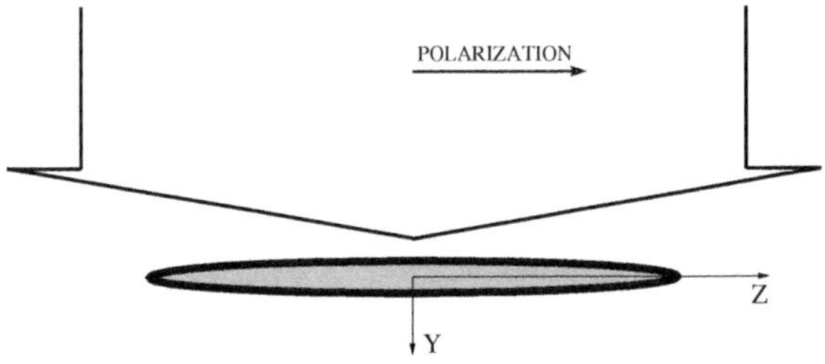

Figure 1 The laser beam and condensate geometry.

We note that the approximations made in Eq. (7) of this derivation are consistent with the Bogoliubov approximation outlined above. Here we have Feynman's [Feynman 1954] famous formula for the excitation spectrum of liquid helium II in terms of the static structure factor. Rearranged as

$$E_{\rm B}(k) = \frac{\hbar^2 k^2}{2m\, S(k)} \tag{9}$$

we see immediately that the enigmatic roton minimum in the excitation spectrum (see later) arises from a peak in $S(k)$. When combined with the normalization condition, $U_k^2 - V_k^2 = 1$, for the Bogoliubov amplitudes [Lifshitz 1998], one obtains the Bogoliubov amplitudes solely in terms of the directly experimentally measurable quantity $S(k)$

$$U_k = \frac{S(k)+1}{2\sqrt{S(k)}}, \quad V_k = \frac{S(k)-1}{2\sqrt{S(k)}} \tag{10}$$

(a related method of obtaining the Bogoliubov amplitudes from $S(k)$ is proposed in Ref. [Brunello 2000]). Note that the Feynman formula, which usually only provides an upper bound to the energy for strongly correlated liquids like helium II, is actually exact within the Bogoliubov approximation that applies to weakly interacting gases.

There is another measure of the degree of correlation between pairs of $\pm k$ modes in the Bogoliubov state (1) which is frequently used in the field of quantum optics, which is called "squeezing". A natural setting for squeezing arises in the down-conversion of a photon of frequency ν by a non-linear crystal into two photons of frequencies ν_1 and ν_2, respectively, where $\nu_1 + \nu_2 = \nu$ (the special case $\nu_1 = \nu_2$ like we have in the Bogoliubov state is known as degenerate down conversion). In quantum optics the perfect correlation between the pairs of modes in the Bogoliubov state is referred to as perfect two-mode squeezing [Gasenzer 2002; Rogel-Salazar 2002; Barnett 1997]: the dispersion in the difference of occupation between two paired modes is zero. The magnitude $r(k)$ of the complex squeezing parameter $\zeta(k) = r(k)\exp(\mathrm{i}\theta)$, appearing in the two-mode squeezing operator [Barnett 1997]

$$\hat{S}_{\pm k}(\zeta) \equiv \exp\left(\zeta^\star(k)\,\hat{c}_k\,\hat{c}_{-k} - \zeta(k)\,\hat{c}_k^\dagger\,\hat{c}_{-k}^\dagger\right) \tag{11}$$

provides a measure of the degree of second order (atom-atom) correlation. In terms of $\hat{S}_{\pm k}(\zeta)$, the Bogoliubov transformation becomes

$$\hat{b}_k = \hat{S}_{\pm k}(\zeta)\,\hat{c}_k\,\hat{S}_{\pm k}^\dagger(\zeta) = \cosh\left[r(k)\right]\,\hat{c}_k + \exp(\mathrm{i}\theta)\,\sinh\left[r(k)\right]\,\hat{c}_k^\dagger \tag{12}$$

and one must set $\theta = \pi$ for consistency with the usual definition of the Bogoliubov transformation. The Bogoliubov ground state is then given by the action of the squeezing operator for each paired mode $\pm k$ upon the fully condensed unperturbed ground state

$$|0_{\mathrm{B}}\rangle = \prod_{k\neq 0} \hat{S}_{\pm k}(\zeta)\,|n_0 = \infty,\; n_{k\neq 0} = 0\rangle. \tag{13}$$

For the Bogoliubov ground state, the magnitude of the squeezing parameter is related to static structure factor via the formula

$$\tanh\left[r(k)\right] = \beta_k = \frac{S(k)-1}{S(k)+1}. \tag{14}$$

One also sees how the expansion coefficients β_k, or if one prefers, the squeezing parameter $\zeta(k)$, of the Bogoliubov ground state are experimentally accessible since as noted above $S(k)$ can in principle be determined directly from a scattering experiment.

2. Roton minimum in a gaseous BEC via laser-induced dipole-dipole interactions

We are interested in fully retarded *dynamic* dipole-dipole interactions, such as those induced by an electromagnetic wave (*e.g.*, laser beam). Let take the case of a ^{87}Rb BEC irradiated by a laser detuned by 6.5 GHz (1134 linewidths) below the D1 line. We then calculate the polarizability, α, to be $\alpha \approx 5.0 \times 10^{-35}\mathrm{cm}^2/\mathrm{V}$ (*cf.* the static value $5.3 \times 10^{-39}\mathrm{cm}^2/\mathrm{V}$). Consider a cigar-shaped BEC tightly confined in the radial x-y plane so that radial excitations are frozen out, irradiated by a plane-wave laser as in Fig. 1. The laser polarization is along the long z-axis of the condensate to suppress collective ("superradiant") Rayleigh scattering [Inouye 1999; Moore 1999] or coherent atomic recoil lasing [Piovella 2001] that are forbidden along the polarization direction. A convenient dimensionless measure of the laser intensity I is given by parameter

$$\mathcal{I} = \frac{I\,\alpha^2(\omega)\,m}{8\pi\,\epsilon_0^2\,c\,\hbar^2\,a}. \tag{15}$$

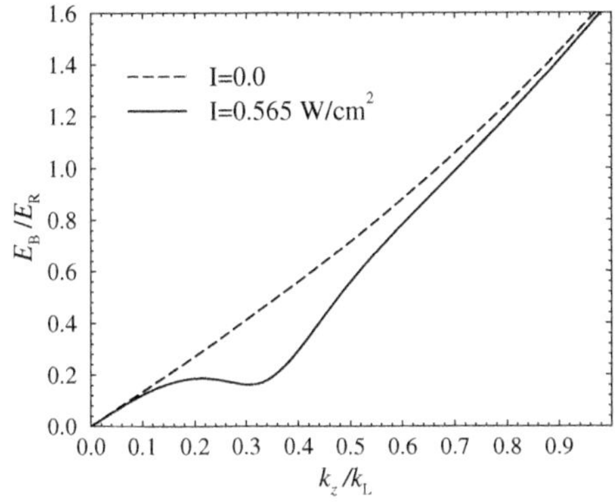

Figure 2. The Bogoliubov dispersion relation (scaled by the recoil energy of an emitting atom) for short-range s-wave scattering (dashed line) can be modified by the dipole-dipole interaction to the extent that is displays a roton minimum (solid line). Parameters: laser wavelength and scaled intensity $\lambda_L = 795$ nm, $\mathcal{I} = 0.057$, BEC radius and density $w_r = 3.5\,\lambda_L$, $8.\,10^{20}$ atoms.m^{-3}, s-wave scattering length due to van der Waals interaction 5.5 nm.

The dipole-dipole potential between atoms induced by far-off resonance electromagnetic radiation of intensity I, wave-vector $\vec{k}_{\mathrm{L}} = k_{\mathrm{L}}\,\vec{\hat{y}}$, and polarization $\vec{\hat{e}} = \vec{\hat{z}}$ (along the z-axis) is [Thirunamachandran 1980; Craig 1984]

$$U_{\mathrm{dd}}(\vec{r}) = \frac{\mathrm{I}\,\alpha^2(\omega)}{4\pi\,c\,\varepsilon_0^2}\,V_{zz}\,(k_{\mathrm{L}},\,\vec{r})\cos\,(k_{\mathrm{L}}\,y)\,. \tag{16}$$

Here $\vec{r}$ is the interatomic axis, and V_{zz} is the component of the retarded dipole-dipole interaction tensor generated by the linearly $\hat{z}$-polarized laser light

$$V_{zz} = \frac{1}{r^3}\Big[(1 - 3\cos^2\theta)(\cos k_{\mathrm{L}}\,r + k_{\mathrm{L}}\,r\,\sin k_{\mathrm{L}}\,r) - \sin^2(\theta)\,(k_{\mathrm{L}}\,r)^2\cos k_{\mathrm{L}}\,r\Big] \tag{17}$$

where θ is the angle between the interatomic axis and the z-axis. The far-zone ($k_{\mathrm{L}}\ r \gg 1$) behavior of (16) along the z-axis is proportional to $-\sin(k_{\mathrm{L}}\ r)/(k_{\mathrm{L}}\ r)^2$. Taking as $\widetilde{U}(k)$ the sum of the Fourier transforms of $U_{\mathrm{dd}}(\vec{r})$ and $U_{\mathrm{s}}(r) = (4\pi\,a\,\hbar^2/m)\,\delta(r)$ (which represents the short-range van der Waals interaction), we can calculate the Bogoliubov dispersion relation $E_{\mathrm{B}}(k)$ defined in (2b), is shown in Fig. 2, for the cigar shaped BEC. Note that by naively inserting the Fourier transform of the dipole-dipole interaction directly into the Bogoliubov approximation we are assuming the Born approx-

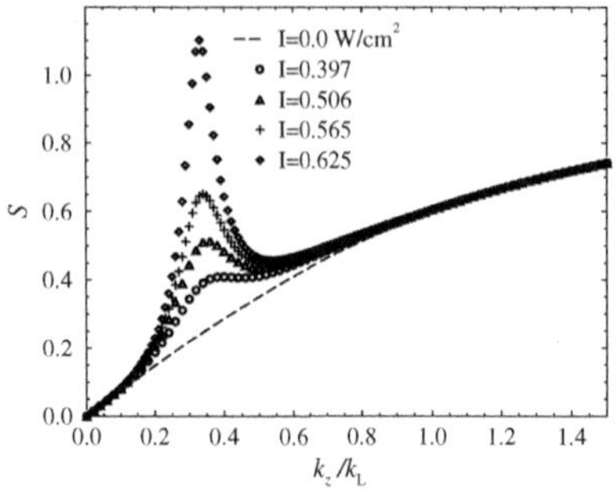

Figure 3. The static structure factor, $S(k)$, for various laser intensities. The peak in $S(k)$ corresponds to the roton minimum in the energy spectrum. The paramaters are the same as those used in Fig. 2.

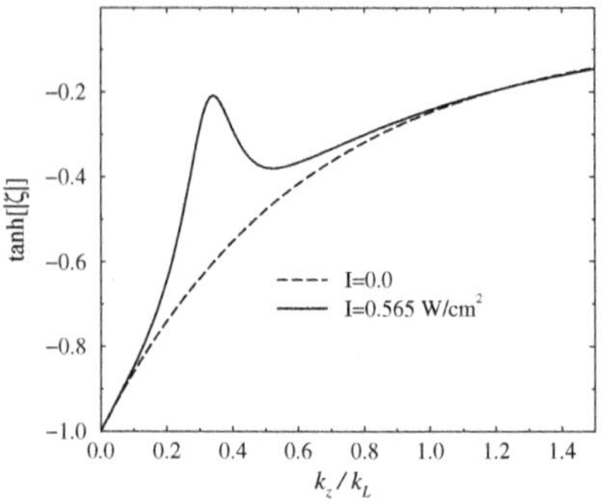

Figure 4. The superposition coefficient $\beta_{k_z} = \tanh|\zeta(k_z)|$, where $\zeta(k_z)$ is the squeezing parameter. The paramaters are the same as those used in Fig. 2.

imation for atom-atom scattering by this interaction. This is in contrast to the short-range interaction coupling $(4\pi\ a\ \hbar^2/m)$ which is the result of solving the short-range scattering problem to higher order (T-matrix in the ladder approximation) [Pitaevskii 2003]. We are also assuming that the dipoles are unscreened, *i.e.*, there are no intermediate interactions of the photons with other atoms located in between two atoms located at $\vec{r}$ and $\vec{r'}$ for which we are evaluating the dipole-dipole interaction. This is a reasonable first approximation for far detuned laser photons traversing a thin sample of dilute gas.

According to formula (9) the roton minimum in the dispersion of the gaseous BEC arises because of laser-induced correlations giving a peak in $S(k)$, see Fig. 3. The magnitude of the $\beta(k)$ coefficient appearing in the Bogoliubov ground state (1) is shown in Fig. 4. $\beta(k)$ controls the amplitude of the correlated states that are mixed in with the uncorrelated BEC state in the superposition, and is also directly related to the squeezing parameter, see Eq. (14).

Conclusion

The roton minimum in the excitation spectrum of helium II is one of the most famous paradigms of quantum liquids, and can be understood as having its origin in strong second-order correlations. The roton minimum is absent in conventinal gaseous BECs because of the short range of the interactions, but can be introduced using long-range interactions such as those induced by an off-resonant laser. If this could be implemented experimentally it would give us a method to microscopically and directly tune the correlation properties of quantum gases.

ATOM OPTICS WITH BOSE-EINSTEIN CONDENSATION USING OPTICAL POTENTIALS

N. Katz, E. Rowen, R. Ozeri*, J. Steinhauer†, E. Gershnabel and N. Davidson
Department of Physics of Complex Systems, Weizmann Institute of Science, Rehovot 76100, Israel
nadav.katz@weizmann.ac.il

Abstract Using our Bose-Einstein condensation (BEC) machine and the Bragg spectroscopy technique we study excitation evolution and decay in BEC. New results have been achieved with this system, and are reported here. We also develop various theoretical models for simulating atomic optical behavior in dynamically changing trapping schemes.

Keywords: Atom optics, Bose-Einstein condensation, optical potentials

Introduction

Atomic vapor Bose-Einstein condensation (BEC) has proven to be a fascinating meeting point of several diverse disciplines, such as condensed matter theory and experiment, atomic physics, nonlinear optics and quantum statistical mechanics. The field is rapidly maturing, and new regimes of interest are now coming under scrutiny. These include trapping in various optical lattice potentials - leading to squeezed atomic states and the Mott insulator state, ultracold Fermionic systems, Fermi-Bose mixtures and molecular condensates. In general the trend is to move away from mean-field effects towards "quantum atom optics". In such systems the mature ideas from conventional quantum optics are immediately applicable and produce many new effects.

The starting point of most discussions about BEC is the Gross-Pitaevskii equation (GPE) [Gross 1963; Ginzburg 1958]:

$$i\hbar \frac{\partial \Psi}{\partial t} = (-\frac{\hbar^2}{2m}\nabla^2 + V_{ext}(\vec{r},\ t) + g\,|\Psi|^2)\,\Psi \tag{1}$$

*Current address: Time and Frequency Division NIST 325 Broadway Boulder, Colorado 80305, USA
†Current address: Department of Physics, Technion, Haifa, Israel

V.M. Akulin et al. (eds.), Decoherence, Entanglement and Information Protection in Complex Quantum Systems, 589–600.

which contains a kinetic energy term $-\hbar^2\nabla^2/2m$, the external potential $V_{ext}(\vec{r},\ t)$ and the self interaction term $g\,|\Psi(\vec{r},\ t)|^2$. The interplay between these three components of the GPE gives rise to subtle and rich behavior, notably the well-known Bogoliubov excitation spectrum [Bogoliubov 1947], which was measured by our group [Steinhauer 2002]. This spectrum results in a linear excitation slope at low momenta and a shifted quadratic excitation spectrum at higher excitation momentum. The transition occurs at the inverse healing length, $\xi = \hbar/\sqrt{2m\,g\,n}$ where $n = |\Psi|^2$ is the condensate density. The healing length is the minimal distance over which the condensate wave function can vary significantly. The GPE also explains much of modern non-linear physics, namely solitons, wave mixing and matter-wave amplifications which have all been observed. However, despite all this, it remains a classical nonlinear wave equation. Higher order terms, truncated in the derivation of the GPE from the many-body Hamiltonian, lead to an even richer picture of the BEC as a many-body quantized field. These higher order terms in the many-body theory of interacting Bose gases are the motivation for most of our recent research results discussed below. These terms lead to damping and mixing of excitations, which cannot be completely described by mean-field pictures.

We develop a theoretical many-body dressed state model that deals with the mixing of excitations, both as a coherent three-wave-mixing process (3WM) and as an incoherent decay mechanism. We also present here four different approaches to reduce the various dephasing mechanisms of excitations. Namely, the existence of discrete radial Bogoliubov excitation modes in elongated traps, the use of a momentum echo, non-linear effects of strong excitations and more uniform density distributions obtained for flat-bottom traps are all proposed and shown to offer significant suppression of the various dephasing mechanisms. These methods open access to measuring the homogeneous broadening of condensate excitations. In an optical lattice we expand the state of the system using the single particle Bloch states. In this picture we can calculate the collisional decay of strongly driven condensates, leading to a remarkable splitting / shift of the collisional products, away from the usual spherical s-wave (described below). Beyond mean-field effects are also explored in a theoretical calculation of the spatial correlations in a Boson gas where the interactions are no longer weak. We find an enhancement in the response of the system, related to a possible roton in such system, as observed in liquid ^4He. In the last section we present results of a matter wave interference method we develop to probe very weak excitations at ultra-low momentum. The interference between the excitations and the condensate leads to strong density modulations after the release of the cloud, whose contrast is a heterodyne probe of the excitation strength.

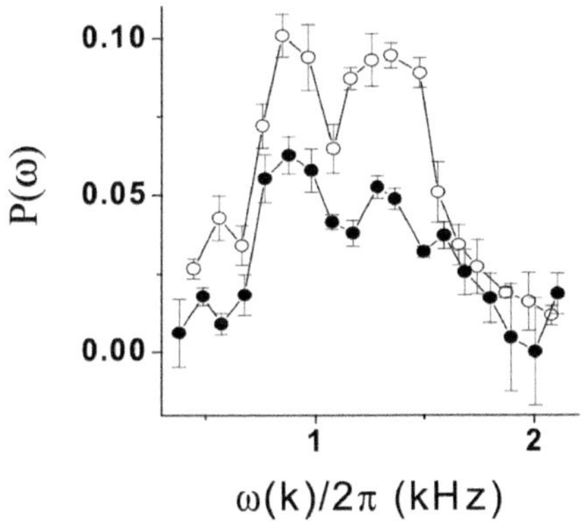

Figure 1. High resolution Bragg spectra at $k = 3.1\ \mu\text{m}^{-1}$. Excitation fraction $P(\omega)$ as a function of the excitation frequency $\omega/2\pi$. Open and closed circles: same measurement at different Bragg laser intensities. A splitting in the spectrum is clearly observed at both intensities. Expansion of the excitation state in a Bogoliubov mode basis of an infinite cylindrical condensate, gives good agreement with these measured resonances, and characterizes them as radial Bogoliubov modes.

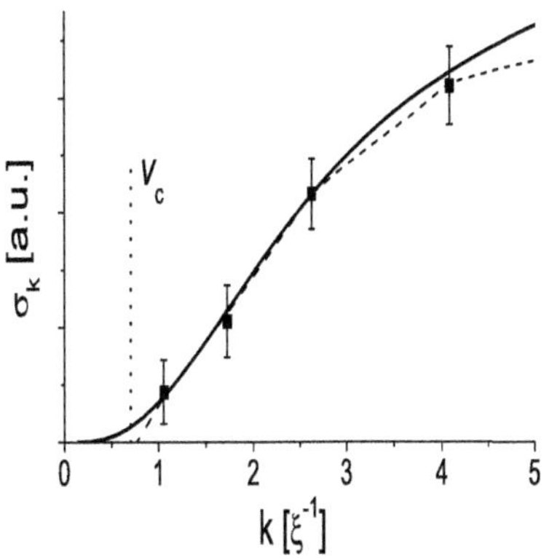

Figure 2. Suppression of identical particle collisions. Full squares: measured scattering cross-section for Beliaev damping as a function of the excitation wavenumber in units of the inverse healing length. The assumptions of our analysis are tested using hydrodynamic simulations (dashed line), and found to agree with Beliaev damping theory (solid line) and the experimental data. Corrections observed in the hydrodynamic simulation take into account the full inhomogeneity and finite size of the experimental system, and validate the approximations of our analysis.

Experimental set-up

In brief, our set-up is based on the double magneto-optical trap (MOT) scheme [Raab 1987]. We apply an experimental scheme including Doppler cooling, compression, polarization gradient cooling, magnetic trapping and RF forced evaporation. Almost pure ^{87}Rb (more than 95% of the atoms are in the $|F,\ m_f\rangle = |2,\ 2\rangle$ ground state of the system) condensates of at least $200,000$ atoms are routinely achieved. The condensate is formed in an elongated cylindrically-symmetric harmonic trap with axial frequency $\omega_z = 2\pi \times$ 25 Hz and radial frequency $\omega_\rho = 2\pi \times 220$ Hz. The Thomas-Fermi radii of the condensate are typically of the order of $3\ \mu$m and $30\ \mu$m in the radial and axial dimensions. This leads to a chemical potential $\mu/h \sim 2$ kHz, and an average healing length of $\xi = 0.24\ \mu$m. For details see Ref. [Ozeri 2003].

Absorption images of the condensate are a basic tool for analyzing the distribution of atoms in our experiments. Typically, after a few tens of msec of ballistic expansion, an on-resonance beam with the $5S_{1/2},\ F = 2 \longrightarrow$

$5P_{3/2}$, $F' = 3$ transition of the ^{87}Rb atoms in the $|F, m_F\rangle = |2, 2\rangle$ ground state, passes through the atomic cloud. We image the absorbed beam on a high resolution CCD digital camera. Analysis of this image compared with an unabsorbed beam allows us to calculate the atomic column density.

We excite the condensate using Bragg spectroscopy [Stenger 1999; Kozuma 1999]. This is a two photon process in which an atom absorbs and emits a photon in a stimulated process, returning to the same internal quantum state. The change is only in the external degrees of freedom, *i.e.*, in the momentum of the atom. This can be viewed as a form of Raman spectroscopy. This coupling can be strong (*i.e.*, Rabi regime) or perturbative.

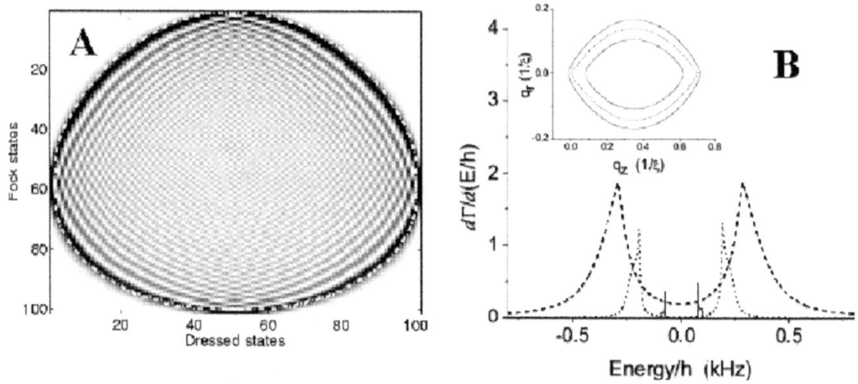

Figure 3. Dressed state basis for atomic collisions. A - The square of the transfer matrix between the excitation Fock state and the dressed state bases for $N = M = 100$. Darker areas correspond to larger probability. B - Damping spectrum between the $N = M = 5000$ manifold and the $N = 4999$, $M = 5000$ manifold. Dashed line: $k\,\xi = 3.2$, dotted line: $k\,\xi = 1.6$ and solid line: $k\,\xi = 0.7$, $q = k/\sqrt{2}$. Inset: energy-conserving surfaces for the two center frequencies of the solid line and for elastic damping from mode k (dashed line). The splitting in the spectrum is due to the nonlinear population oscillations due to three-wave mixing of the modes in the time domain. This behavior is analogous to that of a strongly driven two level atom (Mollow splitting).

1. Multi-branch Bogoliubov spectrum due to radial modes

We measure the response of an elongated BEC to a two-photon Bragg pulse. If the duration of the pulse is long, the total momentum transferred to the condensate exhibits a nontrivial behavior which reflects the structure of the underlying Bogoliubov spectrum (see Fig. 1). It is thus possible to perform a spectroscopic analysis in which axial phonons with a different number of radial nodes are resolved [Steinhauer 2003]. The local density approximation is shown to fail in this regime, while the observed data agree well with the numerical solution of the GPE, and also with a theoretical expansion in terms of Bogoliubov radial modes [Tozzo 2003]. The observation and theoretical explanation of the radial Bogoliubov modes leads us to explore the possible decay mechanisms of these modes. This includes a suppression of the inhomogeneous dephasing, and possibly, due to level repulsion, a modification of the superfluid properties of the lowest branch excitation.

2. Beliaev and Landau damping of excitations

In order to study the decoherence of quasi-particles within BEC, we use Bragg spectroscopy and Monte Carlo hydrodynamic simulations of the system [Castin 1996], and confirm recent theoretical predictions of the identical particle collision cross-section within a Bose-Einstein condensate. We use computerized tomography [Ozeri 2002] of the experimental images in determining the exact distributions. We then conduct both quantum mechanical and hydrodynamic simulation of the expansion dynamics, to model the distribution of the atoms, and compare theory and experiment [Katz 2002] (see Fig. 2).

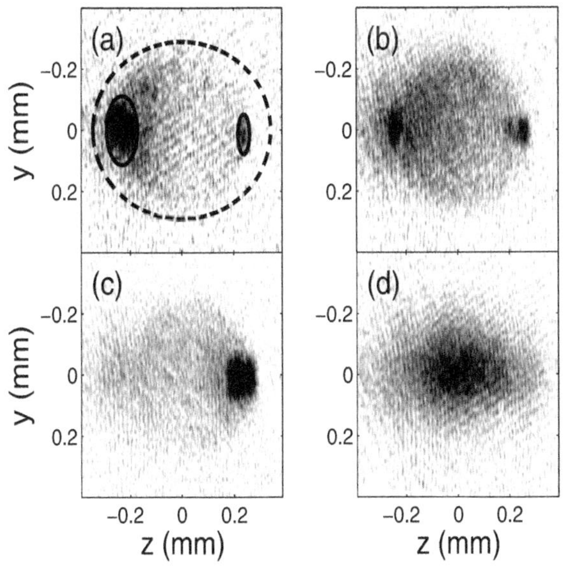

Figure 4 Time of flight images of oscillating and colliding BECs. (a) A weak perturbation of the BEC (left cloud) to the excited population (right cloud). Dashed line: region of interest to measure the average momentum P_{tot}. (b) A $\pi/2$ pulse, note the strong collisional sphere. (c) Almost a π pulse, note the weak collisional sphere, indicating collisions between the excitation and the zero momentum BEC. (d) After further oscillation ($> 10\pi$ at 8.6 kHz), the BECs are completely decohered, and the Bragg coupling no longer affects the system.

3. Dressed state basis for coherent scattering processes

We introduce a basis of states which are dressed by the interaction between three, largely populated, modes of Bogoliubov quasi-particles in a Bose-Einstein condensate at zero temperature [Ozeri 2003]. The dressed basis is obtained by diagonalizing the N×M matrix, which represents the coupling between mode $\vec{k}$ with N excitations and mode $\vec{q}$ with M excitations and an initially unoccupied mode $\vec{k} - \vec{q}$ (see Fig. 3 A). Using this basis of quantum states we calculate the time evolution of the system. The excited population exhibits non-linear oscillations between the different momentum modes.

We also calculate the energy spectrum of Beliaev damping for this system. A transition from elastic to inelastic damping is found, and we predict a splitting of the spectrum into a doublet of resonance energies (see Fig. 3 B), in close analogy with the Mollow splitting in the florescence spectrum of a strongly driven two-level atom.

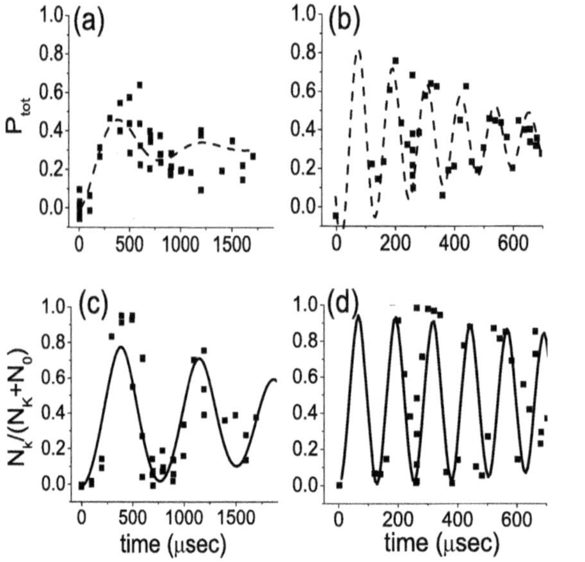

Figure 5 Solid boxes: experimental Rabi oscillations between two coupled BECs. (a)-(b) Average momentum P_{tot} at (a) 1.2 and (b) 8.6 kHz oscillations. The oscillations are damped due to collisions. Dashed lines: exponentially decaying oscillation fits to the points. (c)-(d) Post-selection of the uncollided fraction $N_k/(N_0 + N_k)$ for the (c) 1.2 and (d) 8.6 kHz oscillations. Solid lines: GPE simulations with no fitting parameters. The decay of the GPE simulation oscillations due to dephasing effects is thus distinguished from the decoherence due to collisions.

4. Dephasing and decoherence of strongly driven colliding condensates

We observe the decoherence and wavepacket dephasing between two colliding, strongly coupled, identical BECs (see Fig. 4) [Katz 2003]. We measure, in the strong excitation regime, a suppression of the mean-field shift, compared to the shift which is observed for a weak Bragg excitation. This suppression is explained by applying the Gross-Pitaevskii energy functional. Next, by selectively counting only the non-decohered fraction in a time of flight image we observe Rabi-like oscillations between the two BECs for which both inhomogeneous and Doppler broadening are suppressed, in quantitative agreement with a full GPE simulation (see Fig. 5). If no post selection is used, the decoherence rate due to collisions can be extracted, and is found to be in agreement with the local density average calculated rate. In the future we hope to measure a shift in the energy of the excitations due to off-resonance Bragg pulses, in analogy with the ac Stark shift. This shift may even modify the collision rate by shifting the resonance energy sufficiently to influence the interaction with the finite width collisional quasi-continuum.

5. Collisional and spectral splitting in the excitations of a strongly driven BEC

In Sec. 4 the time domain evolution and decay of strongly driven condensates was discussed. The strong oscillations in time should lead to an observable spectral splitting in the excitation spectrum. We observe [Rowen 2004] this splitting by adding an additional Bragg probe to the strongly oscillating system, coupling to a third, initially unpopulated state and measuring the excitation spectrum of the dressed (oscillating) condensate state (see Fig. 6).

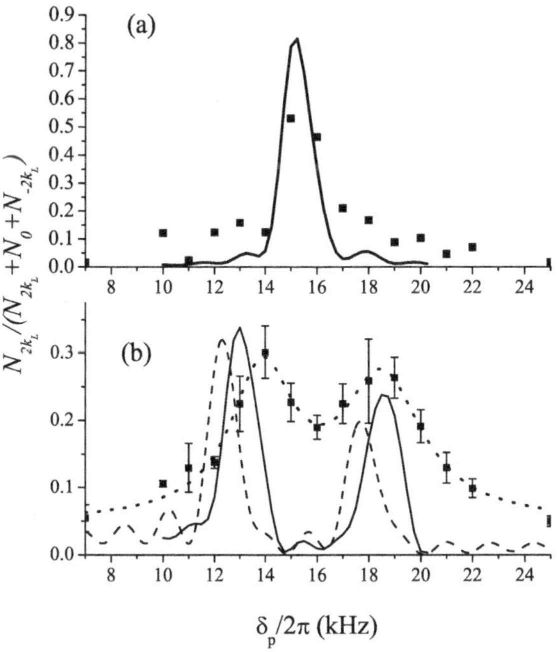

Figure 6 Bulk excitation spectrum of BEC with and without strong Rabi driving. The plotted quantity is the fraction of atoms in $2k_L$ state. (a) No lattice ($\Omega = 0$), solid line: GPE simulation. (b) $\Omega = 5.5$ kHz, solid line: GPE simulation, dotted line: Lorentzian fit to the data giving a splitting of $\Delta E = h \times 4.7(\pm 0.2)$ kHz, dashed line: noninteracting model. The time domain Rabi oscillations transform into a splitting in the spectral domain. The mean-field energy of the atoms contributes a shift of the non-interacting model spectra, but otherwise does not influence the dynamics significantly.

In order to treat this system theoretically, the Bloch states of the strong optical lattice must be considered. In this picture we develop a model to explain the collisional decay of the oscillations as a two-particle collision of Bloch-states (and no longer free atoms). There are various quantum paths for this collision (since every lattice momentum has several relevant branches), leading to a destructive interference of the central s-wave collisional sphere, and a splitting in the resulting collisional shell, related to the observed spectral splitting. The matrix elements for this process lead to a suppression of the outward driven shell and enhance the centrally driven collisional shell, which is no longer spherical. Fig. 7) shows a clear inward driving of our collisional decay sphere, in agreement with this simplified two-particle collision model.

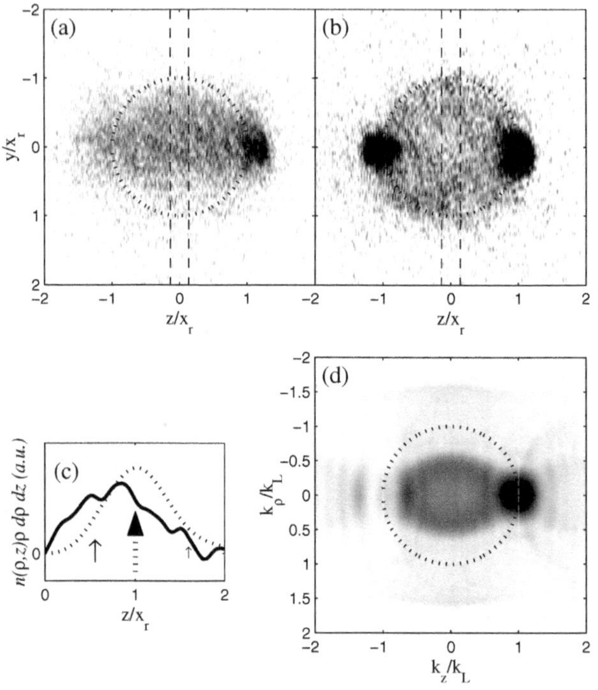

Figure 7. (a)-(b) Absorption images after $t_{TOF} = 38$ ms time of flight following a resonant Bragg dressing pulse. $x_r = t_{TOF}\,\hbar\,k_L/m$ is the ballistic expansion distance of an atom with wavenumber k_L in the lattice frame of reference. (a) Strong dressing pulse (8.6 kHz potential depth), pulse width $t = 660\mu$s. (b) Weak pulse (< 2 kHz potential depth), pulse width $t = 370\ \mu$s. (c) Radial density distribution obtained by computerized tomography of the data in (a)-(b) averaged over a slice marked by vertical dashed lines. Solid line: strongly driven BEC (a), dashed line: weakly excited BEC (b). Dashed arrow: predicted location of s-wave shell, solid arrows: locations of the split shells. A clear shift of the collisional products is observed. (a), (b) and (d): dotted circles: predicted s-wave scattering shell. (d) Simulation of predicted column momentum distribution for the same parameters as (a). The inward shift of the collisional products from the s-wave shell is observed in both. The momentum distribution is only roughly equivalent to the spatial distributions of (a) due to interactions during expansion.

6. Echo spectroscopy of excitations

We propose and demonstrate [Gershnabel 2003] an echo method of reducing the inhomogeneous broadening of Bogoliubov excitations in a harmonically-trapped BEC. Our proposal includes transfer of $+q$ to $-q$ momentum excitations, using a double two-photon Bragg process, in which a substantial reduction of the inhomogeneous broadening is calculated. Furthermore, we predict an enhancement in the method efficiency for low momentum due to many-body effects. The echo can also be implemented by using a four photon process, as is demonstrated in a proof-of-principle experiment (see Fig. 8).

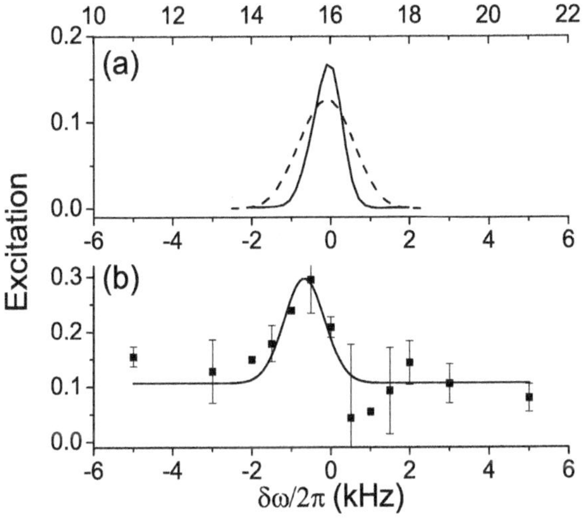

Figure 8 (a) GPE predictions. Solid curve and lower axis: four-photon echo spectrum for the smoothly varying 1 ms pulse used in the experiment (linewidth of 0.9 kHz). Dashed curve and upper axis: Bragg spectrum for the same pulse (linewidth of 1.63 kHz). (b) Experimental echo spectrum (dots) and a Gaussian fit (solid line), yielding a linewidth of 1.21 kHz, providing a linewidth substantially narrower than the Bragg linewidth, and approaching the predicted narrowing in (a).

7. Anharmonic trap geometries

In order to resolve beyond mean-field processes in BEC, we must overcome inhomogeneous broadening, which masks homogeneous lineshapes we want to measure. One possibility is to resolve the radial modes (Sec. 1), or for appropriate parameters, the use of our echo spectroscopy scheme (Sec. 6).

In this section we propose to trap the BEC in a non-harmonic trap along the radial dimension [Gershnabel 2004]. Such traps are possible either by dark trap (blue detuned) holographic methods, or by rotating beam time averaged billiards, both pioneered by our group [Friedman 2002]. These traps approach a flat trap radially, and will lead to increasingly uniform density distributions. However, on the boundary of the BEC, a layer of healing length ξ thickness is expected, which have an important influence on the dynamics of the excitations. In order to study these effects quantitatively, we modify and apply the Bogoliubov quasi-particle projection method [Tozzo 2003]. First we find the ground state of the BEC in such a trap by imaginary time evolution, and next we solve the linearized inhomogeneous Bogoliubov eigenvalue problem, to find the excitations of the system (see Fig. 9). We find that if the boundary of the BEC becomes sufficiently narrow, then the effective potential experienced by the excitations does not contain any bound states on the boundary. This results in a highly uniform excitation which overlaps with all the condensate.

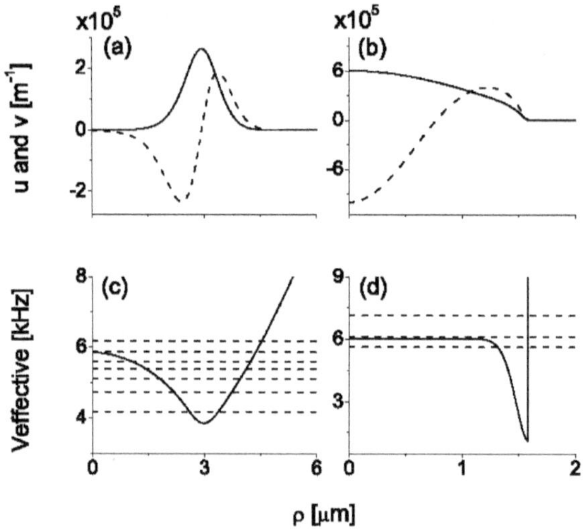

Figure 9 Quasi-particles amplitudes $u_{n,k}$ and effective potential (small hole amplitudes $v_{n,k}$ are not shown). (a)-(b) solid line: $u_{0,k}$, dashed line: $u_{1,k}$. (a) Harmonic trap. (b) Flat-bottom trap, the large overlap of the lowest branch with the entire ground-state explains the suppression of the coupling of the Bragg excitation to all higher modes. (c)-(d) Effective potential and Bogoliubov frequencies with an harmonic trap (c) or a flat-bottom trap(d). (c) Note the large number of trapped states, leading to a multi-branch spectrum.

The excitation spectrum for excitations travelling along the axial direction will contain only one peak, and all other modes will be suppressed (see Fig. 10). When the effective potential contains a few trapped states, the excitation spectrum will contain several radial modes, and we return to the multi-mode state measured in the harmonic trap (Sec. 1).

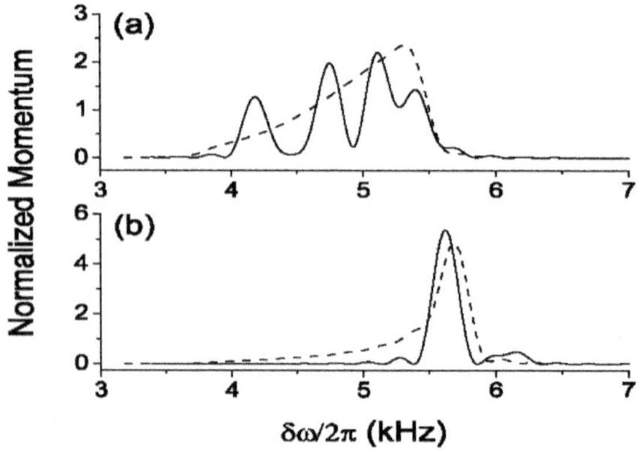

Figure 10 Bragg spectra for Bogoliubov (extremely weak Bragg pulses) excitations. (a) Harmonic trap. (b) Flat-bottom trap. Solid line: linearized GPE solution, dashed line: LDA. y axis: $P_z/(N\ V_B^2\ \hbar q)$ in units of $(\hbar\omega_\rho)^{-2}$. The effective potential picture of Fig. 9 is closely linked with the number of observed modes in the spectrum.

8. The roton in atomic BEC

We calculate the roton [Steinhauer 2004] in atomic vapor BEC using a Jastrow wavefunction, which allows us to calculate beyond mean-field correlations in the quantum state, giving rise to a roton peak in the structure factor at a momentum $k \propto a^{-1}$ (where a is the s-wave scattering length of the atoms). The relevant parameter $\epsilon = \sqrt[3]{na^3}$ (n is the density of the atoms in the BEC), is calculated over a range of values from $\epsilon \ll 1$ up to $\epsilon \sim 1$.

We apply a low density approximation to the Jastrow ansatz and find analytical results for the roton height and position. These low density results compare favorably to a Monte-Carlo quantum (MCQ) calculation of the wavefunction up to $\epsilon = 0.1$. Using the MCQ simulations we find a many-body enhancement of the roton at higher ϵ, leading to a roton peak of almost 8% at $\epsilon = 0.22$.

We note that this calculation elucidates the physics behind the roton from a complementary point of view to that of liquid ^4He. In this strongly interacting superfluid all the relevant length scales (healing length, interaction potential size and interparticle spacing) are roughly equal, leading to some confusion as to what length scale determines the location of the roton peak.

9. Detection of excitations using matter-wave interference

We study the low momentum excitations of a Bose-Einstein condensate using a novel matter-wave interference technique [Katz 2004]. We observe in time-of-flight expansion images a strong matter-wave fringe pattern for very low momentum excitations (see Fig. 11 (c)). The contrast of these fringes is a sensitive spectroscopic probe of the excitation strength, and is explained by use of the Bogoliubov excitation projection method applied to the rescaled expanding condensate. Rigorous GPE simulations confirm the validity of this new theoretical method (see Fig. 11 (d)). We show that the high sensitivity of this detection scheme gives access to the quantized quasiparticle excitation regime.

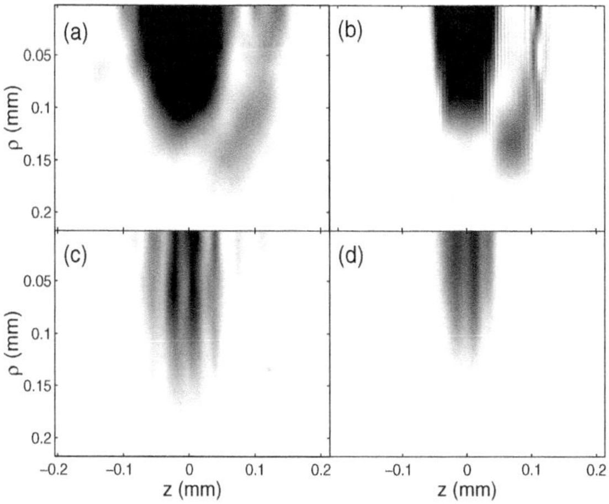

Figure 11 Cylindrical density $n(z, \rho)$, calculated by computerized tomography of experimental TOF column density absorption images. (a) $k\,\xi = 0.6$, note the clearly separated excitation shell. (c) $k\,\xi = 0.1$, note the strong density modulation, with no significant outcoupled fraction. The modulation is due to a matter-wave interference between the travelling excitation and the groundstate wavefunction. (b)-(d) Simulation of (a)-(b), confirming the experimental observation.

As a side-note we recall that in the phonon regime, using computerized tomography [Ozeri 2002] (see Fig. 11 (a)), the freely expanding excitation cloud was measured to be carrying almost all the momentum and excess energy of the system due to excitation. By solving the time dependent GPE for the expansion dynamics, we now confirm this observation quantitatively. The excess energy is observable in the larger radial size of the excitation cloud to the right, both in the experiment (Fig. 11 (a)) and in simulation (Fig. 11 (b)).

We analyze the fringes by taking the spatial Fourier transform of the absorption image and measuring the area of the sidelobes as a function of the excitation frequency. We observe clear resonances at the predicted excitation energy, with an unprecedented signal to noise (Fig. 12).

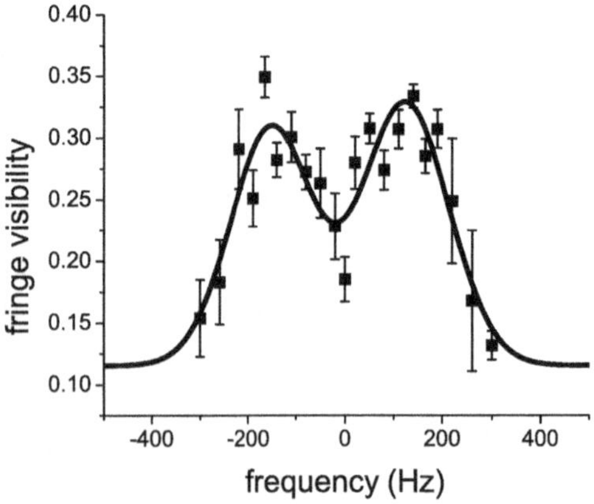

Figure 12 Boxes: fringe visibility as a function of Bragg excitation frequency. Error bars: uncertainty due to four measurements. We observe a clear double-peaked spectrum, which is finite-time broadened. Solid line: double peaked Gaussian fit. The peaks are found at $\pm 139 \pm 10$ Hz, near the expected Bogoliubov local density approximation average excitation energy (138 ± 5 Hz).

The high sensitivity of this method is related to the fact that we observe a matter-wave interference between the excitation and the condensate, *i.e.*, a heterodyne measurement. Expansion in the inhomogeneous Bogoliubov projection basis confirm this picture [Tozzo 2004] We estimate that this improved sensitivity should give us access to the singly quantized excitation regime.

Conclusion

In conclusion we present various experimental schemes for overcoming dephasing, in order to study the intrinsic decoherence of BEC. Using some of these methods, we quantify nontrivial decoherence mechanisms both for weak excitations in BEC and in strongly excited condensates. We calculate the effects of a stronger inter-atomic interaction, namely, the appearance of a peak in the static structure factor of the system. Matter wave interference spectra are shown to be a highly sensitive probe of condensate response, with the sensitivity of this method approaching the few excitation limit.

Acknowledgments

This work was supported in part by the Israel Ministry of Science, the Israel Science Foundation and Minerva. We thank C. Tozzo and F. Dalfovo for fruitful discussions and collaborations.

A MESOSCOPIC MACH-ZEHNDER INTERFEROMETER

Y. Ji, Y. Chung, D. Sprinzak, F. Portier and M. Heiblum
Braun Center for Submicron Research, Department of Condensed Matter Physics, Weizmann Institute of Science, Rehovot 76100, Israel

Keywords: Mach-Zehnder, electron interferometer

Double-slit electron interferometers, fabricated in high mobility two-dimensional electron gas (2GES), proved to be very powerful tools in studying coherent wave-like phenomena in mesoscopic systems [Yacobi 1994; Yacobi 1995; Schuster 1997; van der Viel 1997; Buks 1998; Ji 2000]. However, such interferometers have their disadvantages. They support multiple channels in each slit and consequently suffer from a small fringe visibility [Schuster 1997]. Their *open geometry*, required to eliminate multiple paths interference, allows only a small fraction of the injected current to be collected [Büttiker 1986; Schuster 1997; Buks 1998; Ji 2000]. Moreover, they do not function in a high magnetic field, which imposes a strong Lorentz force on the electrons and destroys the symmetry between the left and right slits. Hence, they are limited in their applications and cannot be employed, for example, in the quantum Hall effect (QHE) regime. We have fabricated and measured a novel, single channel, two-path electron interferometer, that functions in a high magnetic field. It is the first *electronic analog* of the well-known optical Mach-Zehnder (MZ) interferometer [Born 1999]. Based on a single *edge state* transport in the QHE regime, the interferometer collects all the injected electrons, hence having extremely high visibility (up to 62%) and high sensitivity to a small number of injected electrons. We find, unexpectedly that the interference pattern is extemely sensitive to the electron temperature or energy. By performing shot noise measurements of the interfering electrons we show that the observed loss of interference results from phase averaging among electrons and not due to incoherent scattering processes.

An optical MZ interferometer is described schematically in Fig. 1 a. In the electronic counterpart, depicted in Fig. 1 b, a *quantum point contact* (QPC) functions as a beam splitter and an *Ohmic contact* serves as a detector. The QPC is formed in the 2DEG by depositing split metallic gates, separated by a small gap, on the surface of the semiconduuctor and biasing them negatively

V.M. Akulin et al. (eds.), Decoherence, Entanglement and Information Protection in Complex Quantum Systems, 601–604.

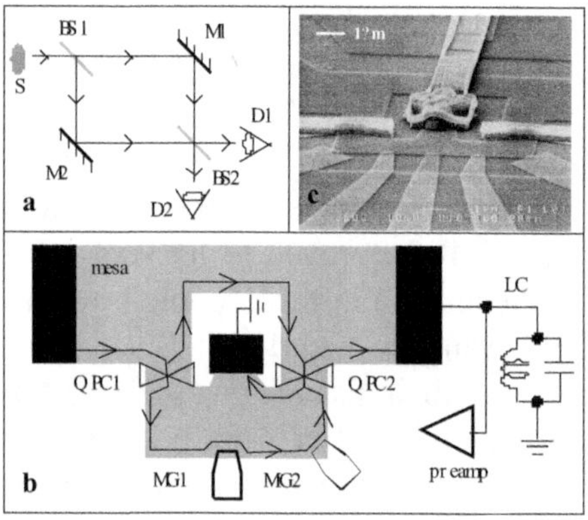

Figure 1. a - Schematics of an optical Mach-Zehnder interferometer. D1 and D2 are detectors, BS1 and BS2 are beam splitters, and M1 and M2 are mirrors. With 0 (π) phase difference between the two paths, D1 measures maximum (zero) signal and D2 zero (maximum) signal. The sum of the signals in both detectors is constant and equals to the input signal. b - Schematics of the electronic Mach-Zehnder interferometer and the measurement system. Edge states are formed in a high perpendicular magnetic field. The incoming single edge state from S is split by QPC1 (quantum point contact) to two paths, of which one moves along the inner edge and the other along the outer edge of the device. The two paths meet again at QPC2, interfere, and result in two complementary currents in D1 and D2. Via changing the contours of the outer edge state and thus the enclosed area between the two paths, the modulation gates (MG) tune the phase difference beween the two paths via the Aharonov-Bohm effect. A high signal-to-noise ratio measurement of the current in D1 is performed at 1.4 MHz with a cold LC resonant circuit as a band pass filter and a cold, low noise, preamplifier. c - SEM picture of the device. A centrally located small Ohmic contact ($3 \times 3\ \mu m^2$), serves as D2, is connected to the outside via a long metallic air-bridge. Two smaller metallic air-bridges bring the voltage to the inner gates of QPC1 (2), both serve as beam splitters for edge states. The five metallic gates (at the lower part of the figure) are modulation gates (MG).

with respect to the 2DEG. As shown schematically in Fig. 1 b, QPC1 splits the incoming current from S into an *inner path* abd *outer path* (traveling under the gates MG) currents, both meeting at QPC2 and interfering there to result in two edge currents that are being drained by D1 and D2. The actual device, seen in Fig. 1 c, is composed of a ring-shape mesa, 3 μm in width, defined by plasma etching with Ohmic contacts (for S, D1 and D2) connected to the inner and outer moving paths in the ring. The inner contac, D2, and the two QPCs are connected to the outside circuit via air bridges that float above the mesa. A phase difference ϕ beween the two paths is introduced via the Aharonov-Bohm (AB) effect [Aharonov 1959; Aronov 1987], $\phi = 2\pi\ B\ A/\varphi_0$, with B the magnetic field, A the area enclosed by the two paths ($\sim$ 45 μm^2), $\varphi_0 = h/e = 4.14 \times 10^{-15}$ T.m^2 the flux quantum, with h the Planck constant and e the electron charge. A few *modulation gates* (MG) are added to tune the phase difference between the two paths by changing the enclosed area A of the magnetic flux.

The conductance, from source to drain, is determined by the corresponding transmission probability T_{SD}. Neglecting decoherence processes, with transmission (reflection) amplitude t_i (r_i) of the ith QPC fulfilling $|r_i|^2 + |t_i|^2 = 1$, the collected currents at D1 and D2 are:

$$\begin{aligned} I_1 \propto T_{SD1} &= |t_1\, t_2 + r_1\, r_2\, e^{i\,\phi}|^2 \\ &= |t_1\, t_2|^2 + |r_1\, r_2|^2 + 2\,|t_1\, t_2\, r_1\, r_2|\, \cos\phi \end{aligned} \tag{1}$$

and

$$\begin{aligned} I_2 \propto T_{SD2} &= |t_1\, r_2 + r_1\, t_2\, e^{i\,\phi}|^2 \\ &= |t_1\, r_2|^2 + |r_1\, t_2|^2 - 2\,|t_1\, t_2\, r_1\, r_2|\, \cos\phi \end{aligned} \tag{2}$$

respectively. Note that the two currents oscillate out of phase as function of ϕ and $T_{SD1} + T_{SD2} = 1$, as expected. The visibility of the oscillation is defined as:

$$v = \frac{I_{max} - I_{min}}{I_{max} + I_{min}}, \tag{3}$$

and for example, when QPC2 is tuned so that $T_2 = 0.5$, the visibility is $v = 2\sqrt{T_1(1 - T_1)}$ with $T_i = |t_i|^2$.

Measurements were done at filling factor 1 (magnetic field $\sim$ 5.5 T) and also at filling factor 2 with similar results at electron temperature $\sim$ 20 mK. High sensitivity measurements of the interference pattern and the current noise were conducted at a frequency of 1.4 MHz[1]. The interfering signal was measured by two methods. The first as function of time when the superconducting

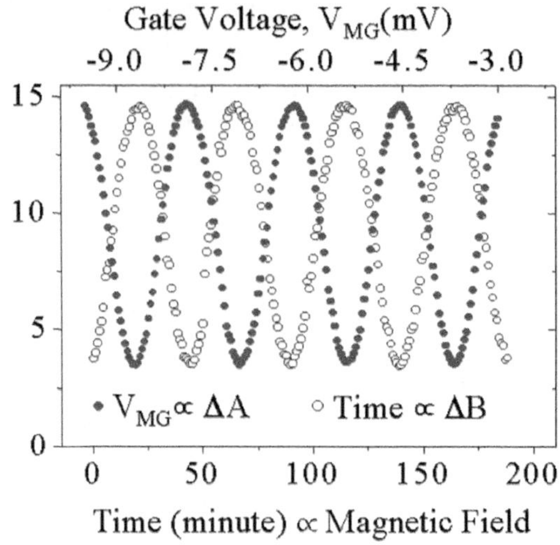

Figure 2 The current collected by D1 plotted as function of the voltage on a modulation gate (red dots) and as function of the magnetic field (blue dots). The visibility of the interference is 0.62.

magnet was set in its *persistent mode*; with the magnetic field decaying naturally at a rate of ~ 0.12 mT/hour. And the second, via changing the voltage on one of the modulation gates (MG). Pronounced interference patterns in D1 or in D2 were observed as a function of magnetic field or MG voltage (see Fig. 2. For both QPCS set to $T_1 = T_2 \sim 1/2$ a visibility as high as 0.62 was measured.

However, the visibility was found to drop precipitately with increasing the energy spread of the injected electrons, whether by increasing the temperature of the applied voltage in the source S. Measurements are conducted now to understand this effect.

Acknowledgments

The work was partly supported by the MINERVA foundation, the German Israeli Project Cooperation (DIP), and the EU QUACS project.

Notes

1. Because the capacitance of the wire connecting the sample to the outside world could not be measured precisely, we can only estimate the AC voltage at the source to be $\sim 0.5\ \mu V$. We have also used the standard Lock-in technique with a low-frequency signal (7 MHz, 10 μV RMS) and observed similar interference patterns; however, the measurement lasted much longer and was prone to sample's instability.

COHERENT TRANSPORT BY ADIABATIC PUMPING

An application to electrons in a quantum dot coupled to a superconducting lead

M. Blaauboer
Kavli Institute of Nanoscience, Delft University of Technology, Lorentzweg 1, 2628 CJ Delft, The Netherlands
m.blaauboer@tnw.tudelft.nl

Abstract We derive a general scattering-matrix formula for the pumped current through a coherent mesoscopic region attached to a normal and a superconducting lead. As an application of this result we calculate the current pumped through a quantum dot coupled to a superconducting lead. Andreev reflection is shown to enhance the pumped current by up to a factor of 4 in case of equal coupling to the leads. We find that this enhancement can still be further increased for slightly asymmetric coupling and show that the pumping cycle is completed well before decoherence sets in. Adiabatic pumping of electrons can thus in principle be used to coherently transfer quantum information across a quantum dot structure.

Keywords: Coherent transport, quantum dots, Andreev reflection

Introduction

It is well-known that the quantum transport properties of a mesoscopic system are modified in the presence of a superconducting interface, due to interference between normal and Andreev reflections. Andreev reflection (AR) [Andreev 1964] is the electron-to-hole reflection process which occurs when an electron with energy slightly above the Fermi energy is incident on the boundary between a normal metal and a superconductor: the electron enters the superconductor after forming a Cooper pair, and leaves a hole in the normal metal with energy slightly below the Fermi level which travels back along (nearly) the same path where the electron came from. Because of the phase-coherent character of AR, it is interesting to study its effect on transport in mesoscopic systems, where phase coherence plays an important role. In the last decade,

V.M. Akulin et al. (eds.), Decoherence, Entanglement and Information Protection in Complex Quantum Systems, 605–614.

this has led to the discovery of a wealth of quantum interference effects in mesoscopic normal-metal - superconductor (NS) structures [Beenakker 1997], such as the observation of a large narrow peak in the differential conductance of a disordered NS junction ("reflectionless tunneling") [Kastalski 1991; Volkov 1993; Nazarov 1994; Beenakker 1994 (b)], and the discovery of novel Kondo phenomena in quantum dots attached to a normal and a superconducting lead [Kang 1998; Clerk 2000; Cuevas 2001; Sun 2001]. In addition, investigations of the conductance in superconductor - carbon-nanotube devices have recently appeared [Morpurgo 1999; Wei 2001 (a)], which indicate that also in these devices resonant behavior due to AR occurs.

In this contribution we study the effects of AR on adiabatic quantum pumping, see also Ref. [Blaauboer 2002], and the limiting time for fully coherent transport of electrons. Quantum pumping involves the generation of a d.c. current in the absence of a bias voltage by periodic modulations of two or more system parameters, such as, *e.g.*, the shape of the system or a magnetic field. The idea was pioneered by Thouless for electrons moving in an infinite one-dimensional periodic potential [Thouless 1983]. In recent years, adiabatic quantum pumping in quantum dots has attracted a lot of attention [Aleiner 1998; Brouwer 1998; Switkes 1999; Zhou 1999; Avron 2000; Vavilov 2001; Cremers 2002]. Quantum dots are small metallic or semiconducting islands, confined by gates and connected to electron reservoirs (leads) through quantum point contacts (QPCs) [Kouwenhoven 1997]. In addition to investigations of pumping in quantum dots, theoretical ideas have been put forward for charge pumping in carbon nanotubes [Wei 2001 (b)], and for pumping of Cooper pairs [Aunola 2001; Zhou 2001]. Here we consider a mesoscopic system consisting of an arbitrary normal-metal region, *e.g.*, a quantum dot or QPC, coupled to a superconductor, as schematically depicted in Fig. 1 (a). We start by deriving a general formula for the pumped current through this NS system in terms of its scattering matrix. This is the N-mode generalization of the result of Wang *et al.* for a NS system with single-mode leads [Wang 2001]. We then use this result to calculate the current pumped through a quantum dot in the Coulomb blockade regime coupled to a superconducting lead. Comparing with the pumped current in the corresponding system attached to normal leads only, shows that AR enhances quantum pumping by up to a factor of more than 4 at low temperatures and for (nearly) symmetric coupling to the leads, while it reduces quantum pumping in the opposite situation of strongly asymmetric coupling. The distribution of pumped current peak heights at higher temperatures is qualitatively similar to the one in case of two normal leads, involving only quantitative changes such as a larger mean. We conclude with a discussion of the decoherence time in quantum dots and show that many pumping cycles can be completed before decoherence sets in. Adiabatic quantum pumping is thus a suitable candidate for coherently transferring quantum information in a semiconducting quantum dot microstructure.

1. Derivation of NS pumping formula

Consider the system in Fig. 1 (a). The normal region (which may, *e.g.*, be disordered, or contain a constriction) is coupled to ideal normal leads 1 and 2 containing N modes each. No bias voltage is applied to the system, so all reservoirs are held at the same potential. We assume a constant pair potential $\Delta(\vec{r}) = \Delta_0\, e^{i\phi}$ in the superconductor, which is applicable for wide junctions [Likharev 1979] and has previously been used to derive the conductance through a NS junction [Beenakker 1992]. We also assume that the NS interface is ideal, *i.e.*, no specular reflection occurs for energies $0 < \epsilon < \Delta_0$, with ϵ the energy measured from the Fermi energy ϵ_F. The scattering matrix $S_{\rm NS}$ of the entire system is given by [Beenakker 1994 (a)]

$$S_{\rm NS}(\epsilon) = \begin{pmatrix} S^{ee}(\epsilon) & S^{eh}(\epsilon) \\ S^{he}(\epsilon) & S^{hh}(\epsilon) \end{pmatrix}, \tag{1}$$

where S^{ee}-S^{hh} are $N \times N$ scattering matrices given by

$$S^{ee}(\epsilon) = r_{11}(\epsilon) + \alpha^2\, t_{12}(\epsilon)\, r^{\star}_{22}(-\epsilon)\, M_e\, t_{21}(\epsilon), \tag{2a}$$

$$S^{eh}(\epsilon) = \alpha\, e^{i\phi}\, t_{12}(\epsilon)\, M_h\, t^{\star}_{21}(-\epsilon), \tag{2b}$$

$$S^{he}(\epsilon) = \alpha\, e^{-i\phi}\, t^{\star}_{12}(-\epsilon)\, M_e\, t_{21}(\epsilon), \tag{2c}$$

$$S^{hh}(\epsilon) = r^{\star}_{11}(-\epsilon) + \alpha^2\, t^{\star}_{12}(-\epsilon)\, r_{22}(\epsilon)\, M_h\, t^{\star}_{21}(-\epsilon). \tag{2d}$$

Here $r_{ii}(\epsilon)$ $[t_{ij}(\epsilon)]$, $i,\ j = 1,\ 2$, denotes the reflection [transmission] amplitude for electrons at energy ϵ, with $0 < \epsilon < \Delta_0$, traveling from lead i $[j]$ to lead i, $\alpha \equiv \exp[-i \arccos(\epsilon/\Delta_0)]$, $M_e \equiv [1 - \alpha^2\, r_{22}(\epsilon)\, r^{*}_{22}(-\epsilon)]^{-1}$, and $M_h \equiv [1 - \alpha^2\, r^{*}_{22}(-\epsilon)\, r_{22}(\epsilon)]^{-1}$. The scattering matrix (1) is unitary and satisfies the symmetry relation $S_{\rm NS}(\epsilon,\ B,\ \phi)_{ij} = S_{\rm NS}(\epsilon,\ -B,\ -\phi)_{ji}$ for time reversal invariance. Adiabatic quantum pumping in this NS junction is obtained by slow and periodic variations of two external parameters X_1 and X_2 as $X_1(t) = \bar{X}_1 + \delta X_1 \sin(\omega t)$ and $X_2(t) = \bar{X}_2 + \delta X_2 \sin(\omega t + \phi)$. The frequency ω has to be such that $\omega \ll \tau_{\rm dwell}^{-1}$, with $\tau_{\rm dwell}$ the time particles spend in the system, in order for equilibrium to be maintained throughout the entire pumping cycle. The net charge $\delta Q(t)$ emitted into lead 1 due to the modulations δX_1 and δX_2 consists of the amount of negative charge carriers (electrons) minus the amount of positive charge carriers (holes) emitted into lead 1. For infinitesimal variations δX_i, $i = 1, 2$, this charge is given by

$$\delta Q(t) = \frac{e}{2\pi} \sum_{\alpha,\,\beta \in 1} \left[\mathrm{Im} \left(\frac{\partial S^{ee}_{\alpha\beta}}{\partial X_1} S^{ee\,\star}_{\alpha\beta} - \frac{\partial S^{he}_{\alpha\beta}}{\partial X_1} S^{he\,\star}_{\alpha\beta} \right) \delta X_1(t) \right. \\ \left. + \mathrm{Im} \left(\frac{\partial S^{ee}_{\alpha\beta}}{\partial X_2} S^{ee\,\star}_{\alpha\beta} - \frac{\partial S^{he}_{\alpha\beta}}{\partial X_2} S^{he\,\star}_{\alpha\beta} \right) \delta X_2(t) \right], \tag{3}$$

where the indices α and β are summed over all N modes in lead 1. This expression is obtained along the same lines as the pumped current in a quantum dot coupled to two normal leads [Brouwer 1998] and based on a formula derived by Büttiker *et al.* [Büttiker 1994; Gramespacher 2000], see also Ref. [Wang 2001]. The total charge emitted into lead 1 during one period $\tau \equiv 2\pi/\omega$ is found by integrating Eq. (3) over time,

$$Q(\tau) = \frac{e}{2\pi} \int_0^{\tau} dt \sum_{\alpha,\,\beta \in 1} \left[\mathrm{Im} \left(\frac{\partial S^{ee}_{\alpha\beta}}{\partial X_1} S^{ee\,\star}_{\alpha\beta} - \frac{\partial S^{he}_{\alpha\beta}}{\partial X_1} S^{he\,\star}_{\alpha\beta} \right) \frac{dX_1}{dt} \right. \\ \left. + \mathrm{Im} \left(\frac{\partial S^{ee}_{\alpha\beta}}{\partial X_2} S^{ee\,\star}_{\alpha\beta} - \frac{\partial S^{he}_{\alpha\beta}}{\partial X_2} S^{he\,\star}_{\alpha\beta} \right) \frac{dX_2}{dt} \right], \tag{4}$$

and rewriting this as an integral over the area A that is enclosed in parameter space $(X_1,\, X_2)$ during one period. We then find that the total current $I_{\mathrm{NS}} \equiv \omega\, Q(\tau)/2\pi$ pumped into lead 1 is given by

$$I_{\mathrm{NS}} = \frac{\omega\, e}{2\pi^2} \int_A dX_1\, dX_2 \sum_{\alpha,\,\beta \in 1} \Pi_{\alpha\beta}(X_1,\, X_2) \tag{5a}$$

$$\approx \frac{\omega\, e}{2\pi}\, \delta X_1\, \delta X_2\, \sin\phi \sum_{\alpha,\,\beta \in 1} \Pi_{\alpha\beta}(X_1,\, X_2), \tag{5b}$$

with

$$\Pi_{\alpha\beta}(X_1,\, X_2) \equiv \mathrm{Im} \left[\frac{\partial S^{ee\,\star}_{\alpha\beta}}{\partial X_1} \frac{\partial S^{ee}_{\alpha\beta}}{\partial X_2} - \frac{\partial S^{he\,\star}_{\alpha\beta}}{\partial X_1} \frac{\partial S^{he}_{\alpha\beta}}{\partial X_2} \right]. \tag{6}$$

Eqs. (5a) and (5b) are valid at zero temperature and to first order in the frequency ω. For $N = 1$ it reduces to the single-mode result of Ref. [Wang 2001]

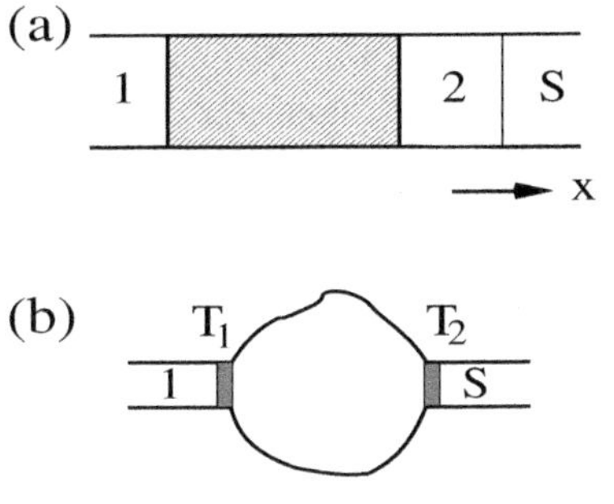

Figure 1 (a) Normal-metal region (hatched) adjacent to a superconducting lead (S). The normal leads 1 and 2 contain N modes each. (b) Quantum dot coupled via tunneling barriers T_1 and T_2 to a normal (1) and a superconducting (S) lead. Adapted from Ref. [Blaauboer 2002].

(apart from a factor of 2 for spin degeneracy in the latter). Eq. (5b) applies for bilinear response in the parameters X_1 and X_2, in which case the integral in (5a) becomes independent of the pumping contour. $I_{\rm NS}$ is of similar generality as the expression for the pumped current in the presence of two normal-metal leads [Brouwer 1998]

$$I_{\rm N} = \frac{\omega\, e}{2\pi^2} \int_A dX_1\, dX_2 \sum_{\alpha\in 1} \sum_{\beta\in\{1,\,2\}} {\rm Im}\left(\frac{\partial S^{\star}_{\alpha\beta}}{\partial X_1} \frac{\partial S_{\alpha\beta}}{\partial X_2} \right). \tag{7}$$

Here $S_{\alpha\beta}$ denotes the $2N \times 2N$ scattering matrix of the system. Note that the index β in this case is summed over the modes in both lead 1 and lead 2, since electrons can be incident from either lead. In our NS junction, the charge pumped into the right lead is converted into a supercurrent.

In order to illustrate the result (5a), we now proceed to apply it to several NS configurations. Unless otherwise noted, we restrict ourselves to the linear response regime corresponding to weak pumping. In that regime, only the scattering matrix at the Fermi level $\epsilon = 0$ is needed.

2. Application: nearly-isolated quantum dot

As an example [Blaauboer 2002], we study quantum pumping in a quantum dot which is coupled via two tunneling barriers with transmission probabilities T_1 and T_2 to a normal and a superconducting single-mode lead, see Fig. 1 (b). Pumping is achieved by periodic variations of the strength of the two tunneling barriers V_1 and V_2 (with $V_{1,2} \equiv \sqrt{(1-T_{1,2})/T_{1,2}}$ for delta-function barriers) as $V_1(t) = \bar{V}_1 + \delta V_1 \sin(\omega t)$ and $V_2(t) = \bar{V}_2 + \delta V_2 \sin(\omega t + \phi)$. We are interested in the regime of high barriers (where transmission is low, $T_1,\ T_2 \ll 1$), and weak pumping ($\delta V_m \ll \bar{V}_m$, $m = 1,\ 2$). At low temperatures such that $k_B T \ll \Delta < E_C$ [with Δ the single-particle level spacing and $E_C = e^2/C$ the charging energy of the dot, C being the total capacitance] the quantum dot then remains in the Coulomb blockade regime during the whole pumping cycle and transport through the dot is mediated by resonant transmission through a single level [Beenakker 1991]. Substituting the appropriate scattering matrix [Blaauboer 2001; Levinson 2001] into Eq. (5b) yields, up to lowest order in T_1 and T_2 and for thermal energies less than the total decay width Γ into the leads $k_B T < \Gamma \ll \Delta$ [with $\Gamma = \Gamma_1 + \Gamma_2 \equiv \hbar\nu(T_1 + T_2)$ and ν the attempt frequency, the inverse of the round-trip travel time between the two barriers],

$$I_{\rm NS} = \frac{\omega\, e}{4\pi}\, \delta V_1\, \delta V_2\, \sin\phi\, \frac{T_1^{7/2}\, T_2^{5/2}\, (\sqrt{T_1} + \sqrt{T_2})}{\left[\frac{1}{4}\,(T_1^2 + T_2^2) + \left(\frac{\epsilon - \epsilon_{\rm res}}{\hbar\nu}\right)^2\right]^3}. \tag{8}$$

Here $\epsilon_{\rm res}$ denotes the resonance energy for a completely isolated dot ($T_1 = T_2 = 0$). Note that Eq. (8) is not symmetric with respect to T_1 and T_2, in contrast with the conductance [Beenakker 1992]

$$G_{\rm NS} = \frac{e^2}{h}\, \frac{T_1^2\, T_2^2}{\left[\frac{1}{4}\,(T_1^2 + T_2^2) + \left(\frac{\epsilon - \epsilon_{\rm res}}{\hbar\nu}\right)^2\right]^2} \tag{9}$$

through this system. This is due to the fact that $I_{\rm NS}$ depends on $\partial S/\partial X$, whereas $G_{\rm NS}$ depends on the transmission eigenvalues, the eigenvalues of the matrix $t_{12}^\dagger\, t_{12}$. Compared to the pumped current in a dot coupled to two normal leads [Blaauboer 2001]

$$I_{\rm N} = \frac{\omega\, e}{4\pi}\, \delta V_1\, \delta V_2\, \sin\phi\, \frac{(T_1 T_2)^{3/2}(\sqrt{T_1} + \sqrt{T_2})(T_1 + T_2)}{\left[\frac{1}{4}(T_1 + T_2)^2 + \left(\frac{\epsilon - \epsilon_{\rm res}}{\hbar\nu}\right)^2\right]^2}, \tag{10}$$

we find that

$$\frac{I_{\rm NS}}{I_{\rm N}} = \frac{T_1^2\, T_2 \left[\frac{1}{4}(T_1 + T_2)^2 + \left(\frac{\epsilon - \epsilon_{\rm res}}{\hbar\nu}\right)^2\right]^2}{(T_1 + T_2)\left[\frac{1}{4}(T_1^2 + T_2^2) + \left(\frac{\epsilon - \epsilon_{\rm res}}{\hbar\nu}\right)^2\right]^3}. \tag{11}$$

For symmetric coupling ($T_1 = T_2$) close to resonance, $(\epsilon - \epsilon_{\rm res}) \ll \hbar\nu T_{1,2} \equiv \Gamma_{1,2}$, AR enhances the pumped current by a factor of 4.[1] In case of strongly asymmetric coupling close to resonance, on the other hand, AR reduces the pumped amplitude: $I_{\rm NS}/I_{\rm N} \sim (T_1/T_2)^2 \ll 1$ for $T_1 \ll T_2$, and $I_{\rm NS}/I_{\rm N} \sim T_2/T_1 \ll 1$ for $T_2 \ll T_1$. The enhancement by a factor of 4 for symmetric barriers consists of 2 contributions of a factor of 2: one factor of 2 is due to the contribution of both electrons and holes to the current, which is also responsible for the doubling of conductance $G_{\rm NS}/G_{\rm N} = 2$ in this NS structure [Beenakker 1992]. A second factor of 2 comes from the asymmetry of the NS dot with respect to injection of charge carriers into the leads, since electrons can only leave the system through the left, normal lead. This leads to an extra doubling of the pumped current compared to the normal case where electrons are injected into both the left and the right leads. This extra factor of 2 does not occur in the presence of an applied bias, as for conductance, since the bias causes charge carriers to flow from one side to the other in both the normal and the NS system. Note that due to the asymmetry of Eq. (11) with respect to T_1 and T_2 the maximum attainable enhancement is even larger than 4: for a slightly asymmetric junction (with $T_1/T_2 \sim 1.26$) one obtains $I_{\rm NS}/I_{\rm N} \sim 4.23$. In this case quantum interference between electrons and holes in the NS system is maximal. If the barrier asymmetry is further increased, $I_{\rm NS}/I_{\rm N}$ decreases and eventually becomes less than 1 for strongly asymmetric coupling, when pumping is dominated by one barrier only.

At temperatures higher than the decay width, $\Gamma \ll k_B T \ll \Delta$, the pumped current exhibits Coulomb oscillations as a function of an applied gate voltage [Blaauboer 2001]. The peak heights of these oscillations can be obtained by thermally averaging Eq. (8) as

$$I_{\rm NS,\ peak} \equiv -\int d\epsilon\ I_{\rm NS}\ f'(\epsilon,\ T) \approx \frac{1}{4k_B T}\int_{\epsilon_{\rm res}-\frac{1}{2}\hbar\nu\sqrt{T_1^2+T_2^2}}^{\epsilon_{\rm res}} d\epsilon\ I_{\rm NS}, \qquad (12)$$

where $f(\epsilon,\ T) \equiv [1+\exp(\epsilon/k_B T)]^{-1}$ denotes the Fermi function. We obtain

$$I_{\rm NS,\ peak} = \frac{\omega\ e\ (8+3\pi)\hbar\nu}{16\pi\ k_B T}\ \delta V_1\ \delta V_2\ \sin\phi\ \frac{T_1^{\frac{7}{2}}\ T_2^{\frac{5}{2}}\ \left(T_1^{\frac{1}{2}}+T_2^{\frac{1}{2}}\right)}{(T_1^2+T_2^2)^{\frac{5}{2}}}. \qquad (13)$$

This thermal average does not explicitly include the effect of the charging energy E_C on the pumped current. A full linear response theory for Coulomb blockade conductance oscillations including charging energy was developed in Ref. [Beenakker 1991]. There it was shown that for temperatures $k_B T \ll \Delta$ only one level $N_{\rm min}$ participates in the transport [$N_{\rm min}$ is defined as the level which minimizes the energy $E_{\rm N} + {\rm U(N)} - {\rm U(N-1)} - \epsilon_{\rm F}$, with $E_{\rm N}$ the energy of the Nth level of the dot, and U(N) the electrostatic energy of a dot containing N electrons], and the oscillation peaks are well described by the thermal average. One can show that for the same reason the pumped current peaks in this temperature range are well described by the thermal average (13), with the understanding that $T_{1,2}$ in Eq. (13) refer to the level $N_{\rm min}$. From (13) and the analogous normal-state result [Blaauboer 2001]

$$I_{\rm N,\,peak} = \frac{\omega\, e\,(2+\pi)\hbar\nu}{16\pi\, k_B T}\,\delta V_1\,\delta V_2\,\sin\phi\,\frac{(T_1T_2)^{\frac{3}{2}}\left(T_1^{\frac{1}{2}}+T_2^{\frac{1}{2}}\right)}{(T_1+T_2)^2}, \tag{14}$$

we obtain

$$\frac{I_{\rm NS,\,peak}}{I_{\rm N,\,peak}} = \left(\frac{8+3\pi}{2+\pi}\right)\frac{T_1^2\,T_2(T_1+T_2)^2}{\left(T_1^2+T_2^2\right)^{5/2}}. \tag{15}$$

Also here, AR enhances the pumped current in case of symmetric tunnel barriers, while a reduction occurs for asymmetric barriers. Maximum enhancement of $I_{\rm NS,peak}/I_{\rm N,peak} \sim 2.55$ is reached for $T_1/T_2 \sim 1.292$. This factor is less than 4, because the average over energy ϵ from which the peak heights (13) are obtained also involves contributions of $I_{\rm NS}$ [Eq. (8)] further away from resonance, for which $I_{\rm NS}/I_{\rm N}$ is much less than 4 [consider, *e.g.*, Eq. (11) for $\epsilon - \epsilon_{\rm res} = \hbar\nu\sqrt{T_1^2+T_2^2}/2$]. This results in lower maximal enhancement of the pumped current peaks (15) at higher temperatures $k_BT \gg \Gamma$.

From the distributions of the decay widths Γ_1 and Γ_2, which for a chaotic dot are given by the Porter-Thomas distribution [Jalabert 1992; Prigodin 1993]

$$P_\beta(\Gamma_m) = \left(\frac{\beta}{2\bar\Gamma}\right)^{\beta/2}\frac{1}{G(\beta/2)}\,\Gamma_m^{\frac{\beta}{2}-1}\,e^{-\beta\,\Gamma_m/2\bar\Gamma}, \tag{16}$$

$[m = 1,\ 2]$, with G the Gamma function and the symmetry index $\beta = 1(2)$ in the presence (absence) of time-reversal symmetry, one can obtain the distributions of the pumped current peak heights (13) and (14) [Blaauboer 2001]. For symmetric coupling to the leads $\Gamma_1 = \Gamma_2 \equiv \Gamma$ and in zero magnetic field these are given by

$$P_{\rm N}(\alpha) = \frac{4}{3\sqrt{2\pi}\,(2\alpha)^{2/3}}\,e^{-\frac{1}{2}(2\alpha)^{2/3}}, \tag{17}$$

and

$$P_{\rm NS}(\alpha) = \frac{4}{3\sqrt{2\pi}\,\mu^{1/3}\,(2\alpha)^{2/3}}\,e^{-\frac{1}{2}(2\alpha/\mu)^{2/3}}, \tag{18}$$

with $\alpha \equiv \Gamma\sqrt{\Gamma}/\bar\Gamma^{3/2}$, $\mu \equiv (8+3\pi)/(\sqrt{2}(2+\pi))$ and $\bar\Gamma$ the mean resonance width. The distributions (17) and (18) are plotted in Fig. 2.

They have the same qualitative form, but Andreev reflection enhances the average peak heights by a factor of $\mu \sim 2.4$: $\langle\alpha\rangle = \sqrt{2/\pi}$ for NN and $\langle\alpha\rangle = \mu\sqrt{2/\pi}$ for NS.

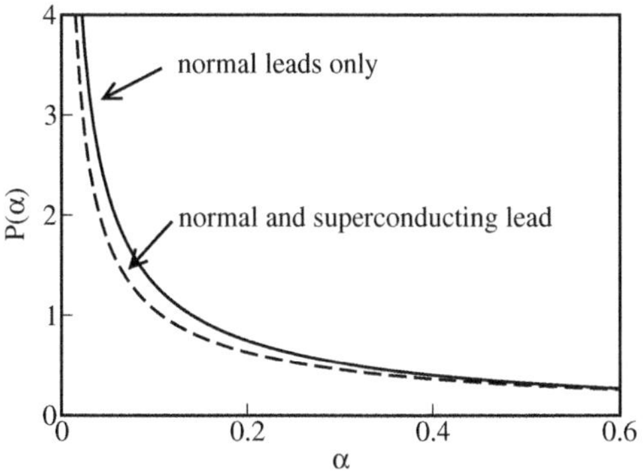

Figure 2 Distribution of the pumped current peak heights (17) and (18) for two normal leads (solid line), and one normal and one superconducting lead (dashed line).

Another interesting result is obtained in this system by relaxing the assumption of weak pumping and considering quantum pumping by varying the two tunneling barriers in such a way that the loop which describes the pumping cycle in parameter space encircles the entire resonance line. For normal-metal contacts this problem has recently been studied [Levinson 2001] and led to the prediction that at zero temperature the charge transferred during one pumping cycle is quantized, $Q = e$ (for spinless electrons). The transferred charge in our NS system is obtained by substituting the scattering matrix of a 1D double-barrier junction given in Ref. [Blaauboer 2001] into Eq. (5a) and integrating over the resonance line $V_1^{-1} + V_2^{-1} = |(\epsilon - \epsilon_{\rm res})/\hbar\nu| \ll 1$, with $V_i^{-1} \equiv \sqrt{T_i/(1-T_i)}$ for $i = 1, 2$. We then obtain

$$Q = \frac{3\sqrt{2}\,e}{2} \int_{-1}^{1} dz\, \frac{(1+z)^2(1-z^2)^3}{[1+6z^2+z^4]^{5/2}} = e, \tag{19}$$

so AR neither enhances nor reduces quantum pumping. This occurs because charge is effectively transferred by a shuttle mechanism (first through one barrier and then through the next), which is unaffected by Andreev interference effects and fixed by the pumping loop. Since only pairs of electrons can enter the superconductor, but the strong Coulomb interaction (charging energy) forbids simultaneous pumping of 2 electrons with opposite spin, this pumping process is *not allowed* in a nearly-closed NS quantum dot. As pointed out in Ref. [Wang 2002 (b)] it can, however, occur in a double-barrier junction in which electron-electron interactions may be neglected. Both in case of two normal and in case of one normal and one superconducting single-mode lead a quantized amount of charge of 2e is then transferred during each pumping cycle [Wang 2002 (b)].

Conclusion

We have studied adiabatic quantum pumping in mesoscopic NS systems. Compared to the conductance in these systems we predict two striking differences:
(1) For a nearly-isolated quantum dot with symmetric ($T_1 = T_2$) tunneling barriers AR enhances the pumped current by a factor of 4, which is twice the maximum enhancement of the conductance in these systems.
(2) In case of quantum pumping this enhancement is not an absolute maximum, whereas in case of conductance it is.
These differences are due to, resp., the absence of an external bias and the asymmetric dependence on the tunnel barriers in case of quantum pumping. Finally, let us address the question whether adiabatic quantum pumping can be used to transfer quantum information in a semiconducting microstructure consisting of quantum dots. This question is relevant for the current effort to develop qubits consisting of single electrons confined in a quantum dot [Vandersypen 2003]. In order to transport these entangled states coherently and reliably, one needs a controllable way of coherent transfer, for which adiabatic quantum pumping may be a suitable candidate. The decoherence time T_2 for a single spin in a quantum dot has not been measured yet, but is expected to be more than 0.1 μs based on measurements of ensembles of spins [Kikkawa 1999]. A recent theoretical study [Golovach 2004] suggests that $T_2 > T_1$, with T_1 the relaxation time of a single spin in a quantum dot. The latter has recently been measured and was found to be $T_1 \geq 50$ μs [Hanson 2003]. These findings suggest that T_2 is at least of order 10 μs and most likely longer. The time τ it takes to complete one pumping cycle depends on the pumping frequency ω, which is $\sim$ 10 MHz [Switkes 1999]. This yields $\tau \sim 0.6$ μs and hence at least 15 cycles can be coherently completed before decoherence sets in. Even this careful estimate of T_2 thus already shows that adiabatic quantum pumping can be used to coherently transfer quantum information. We hope that these fascinating effects of Andreev reflection on quantum pumping will find experimental confirmation, *e.g.*, in nearly-closed quantum dots [Folk 2001].

Acknowledgments

Stimulating discussions with C. M. Marcus are gratefully acknowledged. This work was supported by the Stichting voor Fundamenteel Onderzoek der Materie (FOM), and by the EU's Human Potential Research Network under contract No. HPRN-CT-2002-00309 ("QUACS").

Notes

1. Enhancement by a factor of 4 has also been predicted for the acconductance in the presence of interactions in a NS wire, see Ref. [Pilgram 2002].

ZENO AND ANTI-ZENO EFFECTS IN DRIVEN JOSEPHSON JUNCTIONS: CONTROL OF MACROSCOPIC QUANTUM TUNNELING

A. Barone,[1] A. G. Kofman[2] and G. Kurizki[2]

[1] *INFM Coherentia, Dept. of Physical Sciences, University of Naples Federico II, Italy 80125*

[2] *Department of Chemical Physics, Weizmann Institute of Science, Rehovot 76100, Israel*

gershon.kurizki@weizmann.ac.il, abraham.kofman@weizmann.ac.il

Abstract We show that the slowing down or speeding up of the decay of a quasibound state to the continuum are realizable by current-bias modulation in Josephson junctions (and their analogs in atomic condensates).

Keywords: Zeno, Anti-Zeno, Tunneling, Josephson Junctions

Frequent perturbations (or measurements) of quantum states decaying into an energy continuum can either cause slowdown of the decay (the quantum Zeno effect - QZE) [Misra 1977; Kofman 2000; Kofman 2001 (a)] or, conversely, its speedup (the anti-Zeno effect - AZE) [Lane 1983; Facchi 2000 (a); Kofman 2000; Fischer 2001; Kofman 2001 (a)]. Here we show that the QZE and AZE are realizable for *macroscopic* quantum states that decay via macroscopic quantum tunneling (MQT) in a superconducting current-biased Josephson junction (JJ) [Barone 1982; Clarke 1988; Fisher 1988; Leggett 1987] and its analogs in ultracold atomic condensates [Smerzi 1997; Anderson 1998 (a)], upon varying the rate of the bias-current modulation. The present theory extends previous treatments of tunneling through time-dependent barriers [Fisher 1988; Ivlev 2002], revealing unknown aspects of MQT dynamics, particularly, the short-time *reversibility* of decay to the continuum. It may also be useful for optimizing JJ-based schemes of quantum computing [Averin 2000].

Consider a system ruled by hamiltonian $H = H_0 + \epsilon(t)V$. We take V to be a perturbation that couples the initial state $|n\rangle$ only to eigenstates $|f\rangle$ of H_0 with energies in the *continuous spectrum* and $\epsilon(t)$ to express the time-dependent modulation of the perturbation. Under these assumptions, the probability amplitude $\alpha_n(t)$ (in the interaction representation) of the initial state $|n\rangle$, which

V.M. Akulin et al. (eds.), Decoherence, Entanglement and Information Protection in Complex Quantum Systems, 615–622.

has the energy eigenvalue $\hbar\omega_n$, obeys the following *exact* integro-differential equation [Kofman 2000; Kofman 2001 (a)]:

$$\dot{\alpha}_n = -\int_0^t dt'\ \epsilon^\star(t)\ \epsilon(t')\ \mathcal{G}_n(t-t')\ e^{i\omega_n(t-t')}\ \alpha(t'). \tag{1}$$

Here $\mathcal{G}_n(t) = \hbar^{-2}\langle n|V\ e^{-iH_0t/\hbar}\ V|n\rangle = \hbar^{-2}\int_0^\infty d\omega_f\ \rho(\omega_f)|V_{nf}|^2\ e^{-i\omega_f t}$ is the memory (correlation) function of the coupling to the energy continuum, which is expressed by the coupling strength $|V_{nf}|^2 = |\langle n|V|f\rangle|^2$ and the density of continuum states $\rho(\omega_f)$. For sufficiently short times one can set $\alpha_n(t') \approx 1$ in the integral in (1), resulting in the *time-dependent decay rate of* $|\alpha(t)|^2$:

$$\frac{1-|\alpha_n(t)|^2}{Q(t)} \simeq R_n(t) = \frac{2}{Q(t)}\ \mathrm{Re}\left[\int_0^t dt'\ \epsilon^\star(t')\int_0^{t'} dt''\ \epsilon(t'')\right.$$
$$\left.\mathcal{G}_n(t'-t'')\ e^{i\omega_n(t'-t'')}\right] = 2\pi\int_{-\infty}^{\infty} d\omega\ F_t(\omega-\omega_n)\ G_n(\omega), \tag{2}$$

where $Q(t) = \int_0^t d\tau\ |\epsilon(\tau)|^2$ is the effective time. In the frequency domain, $R_n(t)$ is seen from Eq. (2) to be the convolution of two spectral functions: (i) $G_n(\omega) = \rho(\omega)\ |V_{ef}(\omega)|^2$, the Fourier transform (FT) of the continuum memory function $\mathcal{G}_n(t)$, and (ii) $F_t(\omega) \sim |\int_0^t \epsilon(t')\ e^{i\omega t'} dt'|^2$, the spectral density (SD) of the modulation function (normalized to 1).

The *universal* Eq. (2) [Kofman 2000; Kofman 2001 (a)] has much broader applicability than its Golden-Rule counterpart, $R_{\mathrm{GR}} = 2\pi\ G_n(\omega_n)$, obtained by standard perturbation theory [Landau 1977]: $R_n(t)$ may decrease with the modulation period τ, thus obeying the QZE, but only if τ is shorter than the memory (correlation) time τ_c (the time over which $\mathcal{G}(t)$ changes). Equivalently, the rate $1/\tau$ and the corresponding spectral width of $F_t(\omega)$ must exceed the width of the coupling spectrum $G_n(\omega)$. The opposite, AZE-like, behavior, *i.e.*, an increase of $R_n(t)$ with $1/\tau$, obtains for $F_t(\omega)$ with a narrow (but *non-negligible*) spectral width compared to $1/\tau_c$.

The foregoing results purport to be *universal*, but their applicability to time-modulated tunneling to the continuum is far from obvious, as detailed below. We shall consider a *low-temperature, non-dissipative* JJ (with negligible thermal effects) driven by time-dependent bias current $I_b(t)$. This system is adequately described by the following Hamiltonian, in terms of the magnetic-flux variable Φ [Leggett 1987]:

$$H(t) = -\frac{\hbar^2}{2C}\frac{\partial^2}{\partial\Phi^2} - I_b(t)\ \Phi - E_J\ \cos\frac{2\pi\ \Phi}{\Phi_0}, \tag{3}$$

where C is the junction capacitance, $E_J = \Phi_0\, I_c/2\pi$ is the Josephson energy, $\Phi_0 = \hbar/2e$ is the flux quantum, and I_c is the critical current. This Hamiltonian is equivalent to that of a fictitious particle of "mass" $m = C$ and "momentum" $p = -i\hbar\, \partial/\partial\Phi$ moving along the coordinate $x = \Phi$ in a tilted (washboard) potential, consisting of a time-dependent tilt term and a sinusoidal "lattice" potential:

$$H(t) = \frac{p^2}{2m} - ma(t)\, x + U_0 \cos 2k_L\, x; \;\; a(t) = \bar{a} + b\, \epsilon(t). \qquad (4)$$

Here the time-dependent "acceleration" is $a(t) = I_b(t)/C$, $U_0 = E_J$, $m = C$ and $2k_L = 2\pi/\Phi_0$. We shall assume *abrupt* (step-like) changes of the tilt, causing the acceleration to *periodically* alternate between $\bar{a} + b$ ($\epsilon = 1$), within time intervals of length τ_1, and $\bar{a}$ ($\epsilon = 0$), within time intervals of length $\tau_0 - \tau_1$. This time dependence is realizable in a JJ by rapid ($\lesssim 0.1$ ns) up-down ramping of the bias current. For atomic Bose condensates trapped in optical lattices [Smerzi 1997; Anderson 1998 (a)] we can turn the coupling and tilt between adjacent wells up and down by fast ($\lesssim 10$ ps) modulation of the laser intensity.

There are several queries regarding the suitability of Eq. (2) for the system at hand:

(a) How to separate $H(t)$ in Eq. (4) into H_0, whose eigenstates are either bound (with discrete energies) or unbound (with energies in the continuum), and a time-dependent perturbation $\epsilon(t)V$ that couples them? The difficulty therein is that, even for a constant tilt ($\epsilon = 1$), Eq. (4) has only quasibound eigenfunctions with tails in the region of unbound motion.

(b) Are abrupt changes of the tilt compatible with the assumption that $|\alpha_n(t)|^2$ evolves slowly enough to warrant the use of Eq. (2) or with the *impulse* (shock) approximation [Landau 1977; Ivlev 2002], whereby the pre-shock wavefunction is "projected" onto post-shock wavefunctions, resulting in sudden population loss (mainly to the continuum)?

(c) How to compare Eq. (2) with previous treatments [Ivlev 2002] of periodically modulated barriers (with sinusoidal $\epsilon(t)$), which have predicted tunneling-rate *enhancement*, whereas Eq. (2) can yield either enhanced or suppressed tunneling rates?

Before addressing these queries in detail, we may be reassured that Eq. (2) is indeed valid in a similar (albeit not identical) situation. The hamiltonian in Eq. (4) is analogous to that describing cold atoms in a repeatedly accelerated (tilted) optical washboard potential, resulting in periodically interrupted tunneling from a quasibound state to the continuum [Fischer 2001]. The experiment in Ref. [Fischer 2001] has demonstrated good agreement with Eq. (2) and has provided the only convincing proof to date of both the QZE and AZE in decay to a continuum.

We shall first address query (a), noting the important differences between the system in Ref. [Fischer 2001], characterized by a slightly tilted, shallow sinusoidal potential supporting a *single* quasibound band, and a biased JJ, describable by a *strongly tilted* potential, characterized by a near-critical acceleration ($a \lesssim a_c = 2k_L\, U_0/m$) and supporting *many* bound levels [Barone 1982; Leggett 1987; Clarke 1988; Fisher 1988]. The sinusoidal potential in Eq. (4) can be effectively replaced in a biased JJ by the cubic form (Fig. 1-upper inset)

$$U(q) = \frac{U_0}{6}\, q^2(q_0 - q), \tag{5}$$

whose maximum $U_m = (2/81)U_0\, q_m^3$ is at $q_m = 2q_0/3 = \sqrt{8(1 - a/a_c)}$ and $q = 2k_L\, x - \pi/2 + q_m/2$. We shall consider a quasibound level n localized in the well on the left (around the minimum $U = 0$ at $q = 0$). The effective hamiltonian for such a quasibound level [Gurvitz 1987] can be written as

$$H = \tilde{H}_0 + \epsilon(t)V,$$

$$\tilde{H}_0 = \frac{p^2}{2M} + U(q)\,\theta(q_m - q) + U_m\theta(q - q_m), \tag{6}$$

$$V = [U(q) - U_m]\,\theta(q - q_m).$$

Here θ is the Heaviside step function and the effective mass is $M = m/4k_L^2$. In (6) we have divided $U(q)$ between $\tilde{H}_0$, whose eigenstates are strictly bound if their energy $\hbar\omega_n < U_m$, and a potential $V < 0$ in the time-modulated perturbation $\epsilon(t)V$. This perturbation allows the particle to tunnel periodically from the bound state to the the unbound (continuum-energy) eigenstates of H whenever $\epsilon(t) = 1$ in Eq. (4). This form conforms to Eq. (1) and answers query (a) (Fig. 1-insets).

Let us next consider query (b). Under the constraint $R_n\tau_c \ll 1$, which implies that the (modified) decay rate R_n of level n is slow on the memory-(correlation-) time scale of the continuum, the approximation Eq. (2) should hold, at least whenever the tilt is constant (either $\bar{a}$ or $\bar{a} + b$). We shall denote the corresponding survival probability by [Kofman 2000; Kofman 2001 (a)] $P_n(t)|_{\text{modul}} \simeq e^{-R_nT}$, where T is the total time in the interval $(0,\ t)$, during which the tunneling is switched on ($\epsilon = 1$). By contrast, the impulse (shock) approximation should apply during the much shorter ramping times $\tau_r \ll \tau_1$, if the sudden-change condition holds [Landau 1977]: $\tau_r\, \omega_{nn'} \ll 1$, $\omega_{nn'}$ being the transition frequencies from level n to all possible (discrete or continuous) levels n'. According to the impulse (shock) approximation [Landau 1977], the abrupt ramping of the tilt down and up causes the higher-tilt ("pre-shock") wavefunction $|\psi_n^{(+)}(t)\rangle$ to be projected onto its lower-tilt ("post-shock") coun-

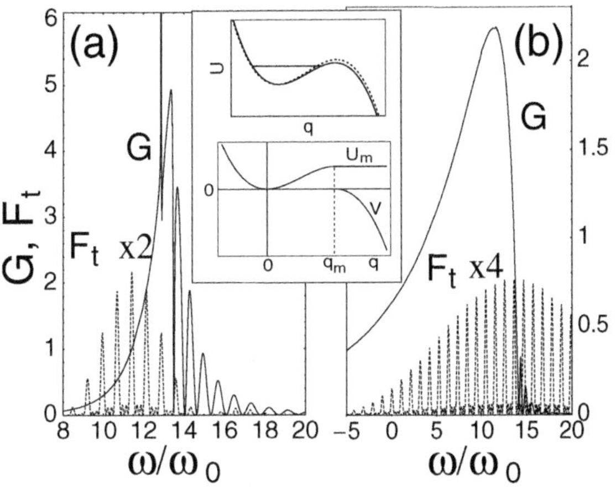

Figure 1. Upper inset: tilted potential $U(q)$ [Eq. (5)]; solid: maximal tilt, $\epsilon = 1$; dashed: minimal tilt, $\epsilon = 0$ [Eq. (4)]. Lower inset: U $(\epsilon = 0)$ is approximated by a binding potential for energies below U_m, since tunneling is negligible; U $(\epsilon = 1)$ is divided into the same binding potential and a perturbation $V < 0$, allowing tunneling. Main figure: (a) the coupling spectrum $G_{n=12}(\omega)$ and the modulation function $F_{t=4\tau_0}(\omega)$ (multiplied by 2) with $\tau_1 = 1/\omega_0$ and $\tau_0 = 5\tau_1$, where ω_0 is the fundamental (harmonic) oscillation frequency in the well; (b) idem, with $G_{n=15}(\omega)$, $\tau_1 = 0.3/\omega_0$, and $F_{t=4\tau_0}(\omega)$ (times 4).

terpart $|\psi_n^{(-)}(t)\rangle$ at time τ_1, then back again after time $\tau_0 - \tau_1$. This yields the following estimate for the survival probability of the nth level after time τ_0:

$$P_n|_{\text{impulse}} = |\langle\psi_n^{(-)}|\psi_n^{(+)}\rangle|^2 \, |\langle\psi_n^{(+)}|\psi_n^{(-)}\rangle|^2 = |\langle\psi_n^{(-)}|\psi_n^{(+)}\rangle|^4. \tag{7}$$

The answer to query (b) is therefore that Eqs. (7) and (2) apply at different time intervals.

We to calculate $G_n(\omega)$, $F_t(\omega)$ and their convolution, to obtain $R_n(t)$:

1) *The nth-level coupling to the continuum* corresponding to Eq. (6) with $\epsilon(t) = 1$ must be evaluated for the entire energy spectrum $-\infty < \hbar\omega < \infty$. It has the form $G_n(\omega) = |\int_{-\infty}^{\infty} dq\; \psi_\omega^\star(q)\; V(q)\; \psi_n(q)|^2$, where $\psi_n(q)$ and $\psi_\omega(q)$ are the initial (bound) and final (continuum)-state wavefunctions, respectively. The initial wave function $\psi_n(q)$ was calculated semiclassically [Landau 1977], within an error $< 1\%$ for $n \geq 4$. The final wave functions $\psi_\omega(q)$ were approximated in three overlapping regions of $\hbar\omega$. For energies below and not too close to the top ($\hbar\omega \lesssim U_m$), as well as for energies well above the barrier, $\psi_\omega(q)$ were semiclassically approximated. By contrast, for energies sufficiently close to the top ($\hbar\omega \simeq U_m$), where the semiclassical approximation fails, the barrier was considered as parabolic and the wavefunctions were obtained in terms of confluent hypergeometric functions ${}_1F_1$,

$$\psi_\omega(q) = e^{-i\xi^2/2}[A\,\xi\,{}_1F_1(3/4 + i\eta,\ 3/2,\ i\xi^2) \\ + B\,{}_1F_1(1/4 + i\eta,\ 1/2,\ i\xi^2)], \tag{8}$$

where $\xi = \sqrt{M\omega_0/\hbar}\,(q - q_m)$, $\eta = (\hbar\omega - U_m)/2\hbar\,\omega_0$, $\omega_0 = \sqrt{U_0\,q_m/2M}$ is the oscillation frequency near the bottom of the well, and the coefficients A and B are obtained from the boundary conditions. Fig. 1 shows $G_n(\omega)$, calculated for the states $n = 12$ and $n = 15$ (for the parameters given below). A crucial parameter is Γ_R, the total width of G, which defines the *shortest correlation time* $\tau_c \sim 1/\Gamma_R$, and is related to the energy distance from the top $\hbar\Gamma_R \sim \hbar\omega_m = U_m - E_n$. Several peaks can be discerned in $G_n(\omega)$ of Fig. 1 (a) ($n = 12$). The narrow peaks below the barrier height represent enhanced coupling to the continuum (*tunneling resonances*) via quasidiscrete levels. The broader, progressively diminishing peaks at $\hbar\omega > U_m$, whose separation scales as $\omega^{1/2}$ at high ω, are above-the-barrier resonances, indicating constructive interference of waves transmitted to the right and those reflected from the "wall" on the left (Fig. 1-inset), with phase differences of 2π, 4π, *etc.* By contrast, $G_n(\omega)$ of Fig. 1 (b) ($n = 15$) is smooth, without resonances.

2) *The step-like periodic modulation function* in Eq. (6): $\epsilon(t) = 1$ for $j\,\tau_0 < t < j\,\tau_0 + \tau_1$, $\epsilon(t) = 0$ for $j\,\tau_0 + \tau_1 < t < (j+1)\,\tau_0$ $(j = 0,\ 1,\ \ldots)$, has the spectral density

$$F_{t=N\tau_0}(\omega) = \frac{2\sin^2(\omega\,\frac{\tau_1}{2})\,\sin^2(N\,\omega\,\frac{\tau_0}{2})}{\pi\,N\,\tau_1\,\omega^2\sin^2(\omega\,\frac{\tau_0}{2})}. \tag{9}$$

The function (9) consists of a "comb" of bell-shaped (sinc^2) spectral peaks with the width $1/t$ separated by $2\pi/\tau_0$, whose weights diminish with $k\pi\,\tau_1/\tau_0$ [Kofman 2000; Kofman 2001 (a)]. When the alternating-tilt intervals satisfy $\tau_1 \lesssim \tau_c$ and $\tau_1 \sim \tau_0 - \tau_1$, the modulation $\epsilon(t)$ causes successive tunneling events to be strongly correlated (within the memory time τ_c). The spectral peaks of $F_t(\omega)$ are then sparse and may coincide with the peaks of $G_n(\omega)$ (tunneling resonances), as discussed below. By contrast, when the low-tilt interval $\tau_0 - \tau_1 \gg \tau_c$, the system effectively loses its "memory" between consecutive tunneling events, which then *resemble irreversible measurements* that interrupt the evolution (although the evolution remains unitary, in reality). We may then approximate $F_t(\omega)$ in (9) by *smoothing* out the "comb" peaks separated by $2\pi/\tau_0$ to become $F_t(\omega) \approx (\tau_1/2\pi)\ \mathrm{sinc}^2[(\omega - \omega_n)\,\tau_1/2]$, where $\mathrm{sinc}(x) = \sin x/x$. This form of $F_t(\omega)$ effectively amounts to *spectral spread (broadening)* of $G_n(\omega)$ over a frequency range $\sim 1/\tau_1$ in the convolution integral (2). In this case R_n is *insensitive* to narrow resonant peaks of $G_n(\omega)$.

3) *The convolution of* $F_t(\omega - \omega_n)$ and $G_n(\omega)$ yields, at t large enough to replace each peak in (9) by $\delta(\eta_k)$:

$$R_n \approx \frac{2\pi\,\tau_1}{\tau_0} \sum_{k=-\infty}^{\infty} \operatorname{sinc}^2\left(\frac{k\pi\,\tau_1}{\tau_0}\right) G\left(\omega_n + \frac{2k\,\pi}{\tau_0}\right). \tag{10}$$

We may discern three limiting regimes in the dependence of the decay rate R_n on the control parameters τ_0 and τ_1 in $F_t(\omega)$, which determine the kth peak weight and peaks' spacing in Eq. (10):

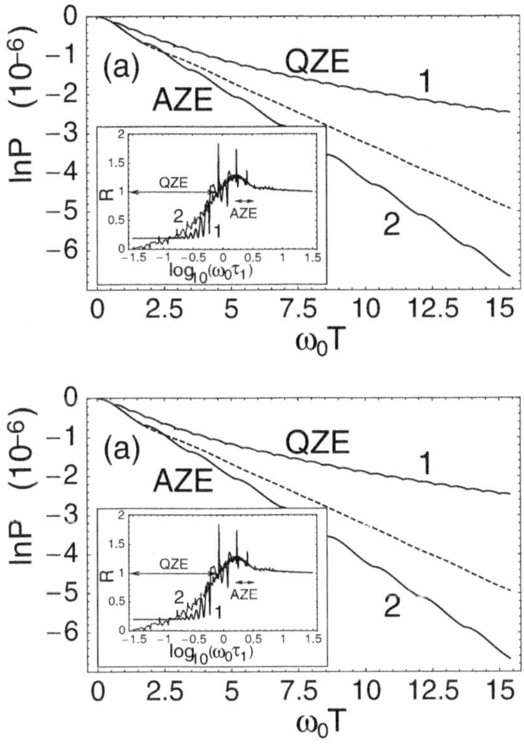

Figure 2 (a) Inset: the decay rate R (in units of the unperturbed, Golden-Rule rate R_{GR}), as obtained from the convolution of $F_t(\omega)$ and $G_n(\omega)$ in Fig. 1 a, for $n = 12$. The rate is plotted as a function of the interruption time τ_1 (in units of $1/\omega_0$) on a log scale for $\tau_0 = 5\tau_1$ (curve 1) and $\tau_0 = 50\tau_1$ (curve 2). The domains of QZE, QZE scaling and AZE are marked. Main figure: the evolution of $\ln P(t)$ for level $n = 12$: solid 1 - QZE-like decay; solid 2 - AZE-like decay; dashed - unperturbed decay. (b) Idem, for level $n = 15$, corresponding to Fig. 1 b. Lower inset: decay rate R shows only QZE behavior. Upper inset: P vs. total time t (including the lower-tilt period), showing impulsive jumps.

i) The QZE (*i.e., reduction of the decay rate R_n with the modulation rate*) is obtained when $1/\tau_1 \gtrsim \omega_m, \Gamma_R$. We may graphically infer from Fig. 1 that the large width of the modulation spectrum $F_t(\omega)$ (compared to the spectral width of $G_n(\omega)$) in the convolution (2) is then the origin of the QZE. The physical sense is that when the system is perturbed (interrupted) frequently enough, the QZE arises since the energy uncertainty incurred by the perturbations causes the effective decay rate to be *averaged over the entire spectrum of continuum states*, most of which do not contribute to the decay ($G_n(\omega) \simeq 0$). The corresponding time-domain behavior (Fig. 2) is one of *repeatedly interrupted evolution* of the initial-state population $P_n(t)$, attesting to the oscillatory character (*short-time reversibility*) of the tunneling at $\tau_1 \ll \tau_c$. This evolution is in sharp contrast to the effectively irreversible step-like population loss associated with $P_n|_{\mathrm{impulse}}$ (Eq. (7), Fig. 2 (b)-upper inset). The loss due to $P_n|_{\mathrm{impulse}}$ should be used to calibrate the oscillatory $P_n(t)|_{\rm modul} \simeq e^{-R_n T}$ (solid curves), thereby allowing an experimentally distinct signature of the QZE.

ii) The AZE (*i.e., decay speedup as the modulation rate increases*) is seen from Fig. 2 (a) to arise when the unperturbed energy is strongly detuned from the maximum of the coupling spectrum $G_n(\omega)$, *i.e.*, $G_{12}(\omega_{12}) \ll G_{12}(\omega_m)$, and the modulation rate satisfies $1/\tau_1 < \omega_m$. It implies that the decay rate R_n *grows* with $1/\tau_1$, since the modulation function $F_t(\omega)$ is then probing more of the *rising* part of $G_n(\omega)$ in the convolution (2). Physically, it means that, as the energy uncertainty grows with the modulation rate $1/\tau_1$, the state decays into *more and more channels, whose weight* $G_n(\omega)$ *is progressively larger.*

iii) For low lying nearly harmonic levels ($0 \leq n \leq 12$) $G_n(\omega)$ has distinct tunneling resonances if $2\pi\ k/\tau_0 \simeq n\ \omega_0$. A periodic modulation corresponding to narrow spectral peaks $F_t(\omega) \sim \delta(\omega - 2k\ \pi/\tau_0)$ would excite such a resonance and give rise to *AZE-like resonantly enhanced* tunneling (see spikes in Fig. 2 (a)-inset). In this regime, our theory qualitatively reproduces the resonant enhancement predicted by Ref. [Fisher 1988], thus settling query (c).

For numerical examples we chose $C = 58$ pF and $I_c = 15\ \mu$A (Figs. 1 (a) and 2 (a)) and $I_c = 16\ \mu$A (Figs. 1 (b) and 2 (b)), yielding $\omega_0 = 3.85 \times 10^9$ s^{-1} and $\omega_0 = 3.97 \times 10^9$ s^{-1}, respectively, whereas $\omega_m = 7.8 \times 10^9$ s^{-1} for $n = 12$ and $\omega_m = 1.1 \times 10^9$ s^{-1} for $n = 15$. The QZE for $n = 12$ and 15 then requires $\tau_1 \lesssim 0.1$ ns and $\tau_1 \lesssim 1$ ns, respectively, while the AZE requires $\tau_1 > 0.1$ ns for $n = 12$ and is almost absent for $n = 15$ when I_b is modulated between $0.99265\ I_c$ and $0.993\ I_c$.

It should be mentioned that significant variation of the decay rates to the continuum of a JJ as a function of the bias-current rate of modulation has been observed by Silvestrini *et al.* [Silvestrini 1997]. This observation was made under rather high-temperature conditions, where many (bound or quasibound) levels are populated in the JJ potential and the present picture does not hold.

To sum up, our treatment has elucidated the *short-time* dynamics of low-temperature MQT through time-modulated barriers. Current-bias modulation has been shown to *imitate either frequent measurements or correlated perturbations* of a decaying state, between successive impulses (shocks) [Landau 1977; Ivlev 2002]. Such modulation has been demonstrated to either enhance or suppress the MQT rate (causing the AZE or QZE, respectively). Remarkably, quantum gates based on JJ qubits [Averin 2000] or their atomic-condensate counterparts [Smerzi 1997; Anderson 1998 (a)] may benefit from the ability to suppress the decoherence due to MQT to the continuum.

Acknowledgments

This work was supported by the EC (QUACS RTN), ISF, Minerva and the Israel Ministry of Absorption.

EMPLOYMENT OF SUBMICRON $YBA_2CU_3O_{7-X}$ GRAIN BOUNDARY JUNCTIONS FOR THE FABRICATION OF "QUIET" SUPERCONDUCTING FLUX-QUBITS

E. Sarnelli,[1] G. Testa,[1] A. Monaco,[1] M. Adamo[2] and D. Perez de Lara[3]

[1] *Instituto di Cibernetica "E.Caianiello" del CNR, Via Campi Flegrei 34, I-80078 Pozzuoli, Italy*

[2] *Promete srl, Via Buongiovanni 49, S. Girogio a Cremano (Naples), Italy*

[3] *INFM - Monte S. Angelo, I-80126 - Naples, Italy*

Abstract

The implementation of superconducting flux-qu-bits represents a challenge for the future. Superconductors offer both the scalability, typical of solid state technology, and the properties of a macroscopic quantum system. However, the real employment of superconducting devices passes through the reduction of decoherence time of a single qu-bit as well as a number of entangled qu-bits. Decoherence can be reduced by using appropriate technological choices, like the ones minimizing the coupling of the quantum system to the environment. The discovery of unconventional superconductors offered the possibility to study new Josephson junction based circuitries that, in principle, can reduce the coupling to external room temperature electronics. We re-examined the five junction flux qu-bit proposed in 2001 by Blatter and coworkers, composed of a superconducting loop with four conventional junctions and one π-junction, characterized by an energy minimum shifted by π with respect to conventional Josephson junctions. In this scheme, the π-junction acts as a simple phase shifter, frustrating the superconducting loop, and allowing the formation of a double degenerate ground state. In our proposal, the π-junction is a particular device based on the properties of Andreev reflections to form bound states at the Fermi energy.

Keywords: Quantum bit, Josephson junction, d-wave superconductors.

Introduction

The simplest superconducting flux qubit is a superconducting loop interrupted by one Josephson junction (radiofrequency rf-SQUID - SQUID is an acronym for superconducting quantum interference device). The potential energy of such a devices is described by the equation:

V.M. Akulin et al. (eds.), Decoherence, Entanglement and Information Protection in Complex Quantum Systems, 623–633.

$$\frac{U(\phi)}{U_0} = \frac{\pi}{\beta}\,(\phi - \phi_e xt)^2 - \cos(2\pi\;\phi) \tag{1}$$

where $U_0 = I_c\,\phi_0/2\pi$ is the Josephson coupling energy, $\beta = L\,I_c/\phi_0$ the normalized loop inductance L, with I_c the Josephson critical current, $\phi_0 = 2.07 \times 10^{-15}$ Weber the magnetic flux quantum, and ϕ and ϕ_{ext} the total flux and the external applied magnetic flux normalized to ϕ_0 respectively. U(ϕ) shows periodic minima when $\beta > 1$ (see Fig. 1), that implies that the device is characterized by large inductance loop. A qubit made by such concept is achieved when an external magnetic flux ϕ_{ext} = $\phi_0/2\pi$ is applied to the loop. In this case, the potential energy shows a twofold degenerate ground state, (solid line in Fig. 1).

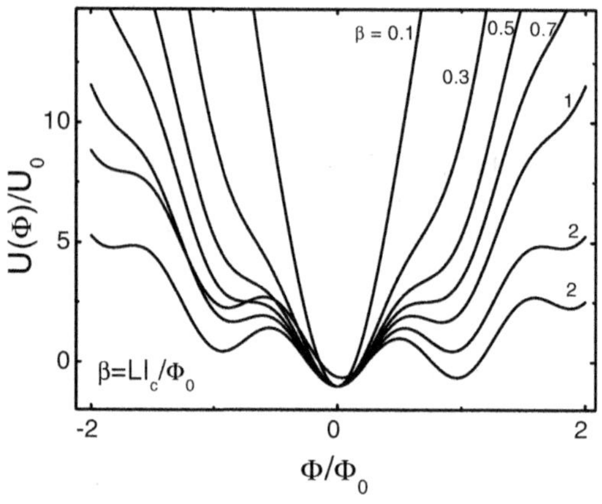

Figure 1 Potential energy of a conventional rf-SQUID, *i.e.*, a superconducting loop interrupted by one s-wave Josephson junction. Full line indicates the potential energy when an external magnetic flux $\phi = \phi_0/2$ is applied. A double degenerate ground state is obtained.

The two states are distinguished by a clockwise or counterclockwise circulating current, both characterized by relatively large values. These currents, in turn, can interact with other qubits, reducing the coherence time of the entire system. Similar argumentations can be produced also in the case of a qubit composed of a superconducting loop interrupted by two Josephson junctions (dc SQUID). Large currents describing the twofold degenerate ground state can be avoided by increasing the number n of the Josephson junctions inside the superconducting loop ($n > 2$) [Mooij 1999]. In this case, the loop inductance can be kept small ($\beta \ll 1$) and it is more appropriate to talk about "phase qubit", rather than "flux qubit". In a phase qu-bit, the magnetic coupling with the environment is weak, as a consequence of the decoupling between flux and phase, with the ground state distinguished by minima in the phase space, as shown in Fig. 2.

We point out that the two states defining the ground state differentiate by current circulating clockwise or counter clockwise, same as in the flux qu-bit, the main difference consisting of the amplitude of such circulating currents,

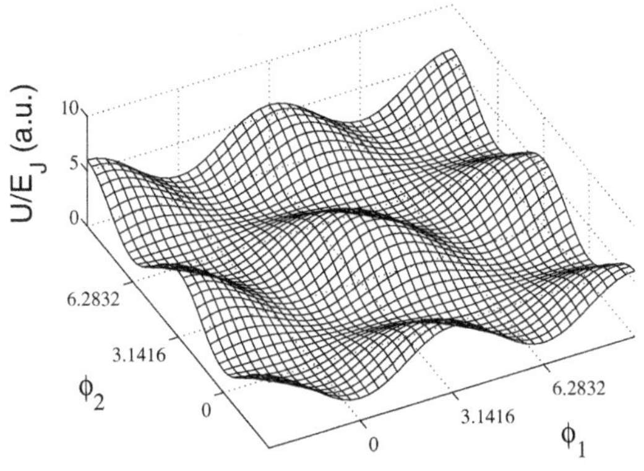

Figure 2 Phase space dependence of the potential energy for a three-junction superconducting loop. U is normalized to the Josephson coupling energy E_J.

arbitrarily low in the case of a phase qu-bit. A phase qubit needs to be stable with respect both to static and dynamic charge and flux fluctuations [Blatter 2001]. While the three junction loop described by Mooij *et al.* [Mooij 1999] can be made stable against charge fluctuations by breaking the symmetry between the junctions [Blatter 2001], it is worth discussing the flux stability. Indeed, the phase qu-bit is very sensitive to magnetic flux fluctuations, because the magnetic field is the external variable used for both manipulating the qubit and setting the degenerate ground state. By simple calculations [Blatter 2001], it can be shown that a magnetic flux precision $\delta\phi/\phi_0 \approx 10^{-6} - 10^{-7}$ is required. This condition implies a stringent magnetic flux noise condition $S_\phi \sim \phi_0/(U_0\,\hbar\,\Delta/N)^{1/2} \sim 10^{\ 8} - 10^{-9}(\phi_0/Hz^{1/2})$, which is very difficult to achieve, either with conventional or superconducting circuits. An extremely energetically stable qu-bit can be obtained by introducing a strong π-junction in the loop [Blatter 2001]. Such a junction is characterized by an energy minimum at $\phi = \pi$ instead of $\phi = 0$, as shown in Fig. 3.

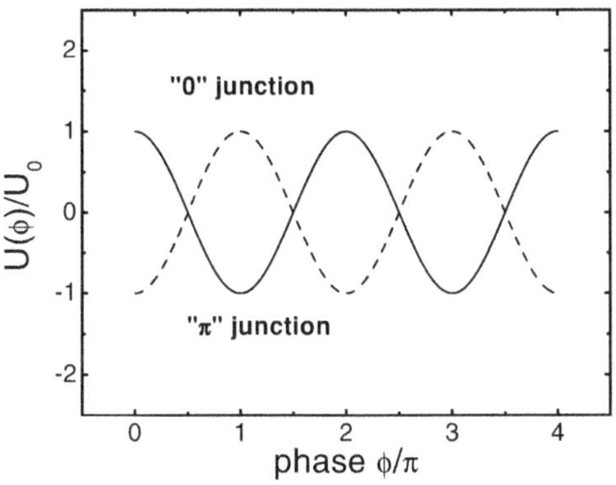

Figure 3 Potential energy of a single conventional "0" junction (dashed line) and a single "π" junction (solid line).

It means that the ground state of a loop containing a π-junction is equivalent to a loop containing a single conventional junction with an external applied magnetic field, corresponding to half flux quantum. In this way, a π-junction frustrates the loop by the right amount, allowing a high phase stability. In principle, a π-junction can be fabricated by different approaches, employing non conventional Josephson junctions. For instance, junctions made with ferromagnetic barriers [Ryazanov 2001], or based on non conventional high-temperature superconductors (HTS) [Smilde 2002], under particular conditions can show the energy minimum at π.
In this paper, we describe advantages and disadvantages of possible HTS devices for the fabrication of π-junction. In the second part, a particular junction category, showing a "0" to "π" transition at low temperatures, is described in detail. In particular, the inversion of the current direction, characterizing the transition of the junction from "0" to "π", is the effect of the formation of Andreev bound states at the Fermi level, the so-called midgap states (MGS) [Lofwander 2001]. It is worth noting that such MGS are the effect of a sign change of the order parameter symmetry, and they cannot be formed in conventional superconductors.

1. The "quiet" qu-bit

In the proposals of "quiet" qu-bits by Ioffe *et al.* [Ioffe 1999 (a)] and Blatter *et al.* [Blatter 2001], different schemes employing non-conventional junctions have been discussed. In the following, a brief list of such schemes, indicating the main problematic aspects and possible solutions, is presented.

1.1 Two SD junction loop

If a conventional superconductor (S) described by a s-wave order parameter symmetry (OPS) is put together with a non conventional superconductor (D), described by a pure d-wave OPS, to form two junctions in a superconducting loop, as indicated in Fig. 4, a self $\pi -$"$frustrated$" loop is achieved [van Harlingen 1995]. Indeed, one of the two SD junctions behaves as a conventional "0" junction, since the Josephson coupling is between the positive lobe of the d-wave superconductor (white color in Fig. 4) and the S electrode; on the contrary, the other junction is a "π" junction, because the coupling is now between the S electrode and the negative lobe. As a consequence, a shift of π along the loop is achieved and the device is self-frustrated by a half flux quantum.

The potential energy is characterized by a twofold degenerate ground state, as shown in Fig. 5, without an external applied magnetic field. Although such a solution seems very convenient, it brings to some difficulties. Indeed, a SD junction is the product of the coupling of two electrodes with different order parameter symmetries.

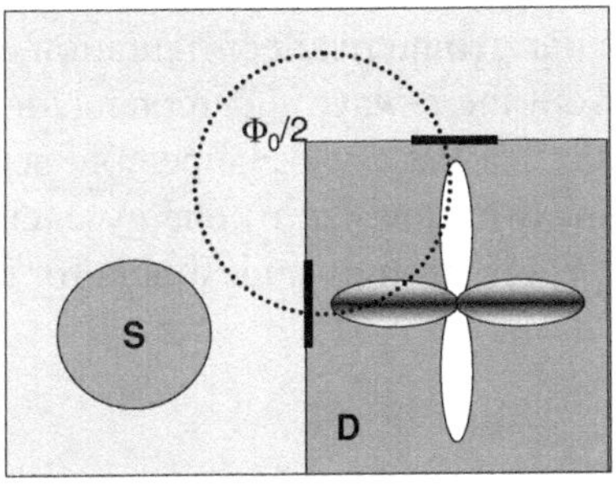

Figure 4. SD π-loop obtained by a s-wave and a d-wave superconductor. The two junctions composing the loop are indicated by black rectangles. The positive lobe of the d-wave is white.

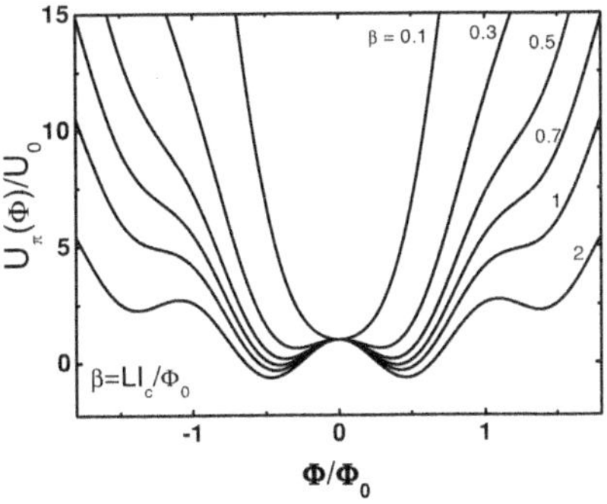

Figure 5. Potential energy as a function of the total flux in the case of a rf-SQUID frustrated by a single π-junction.

It has been shown [Huck 1997] that at the surface of a d-wave superconductor, the S and D states form a hybrid d + i s state, characterized by an imaginary order parameter. In such cases, the junction can undergo a time reversal symmetry breaking, with typical spontaneous currents associated to it. Such currents are definitely non negligible, being the associated magnetic field of the order of the critical field H_{c1}. These currents, even though partially compensated by the Meissner screening currents, still remain large and can disturb the coherence of the qubit or interact with other qubits.

1.2 Asymmetric SD junction

Another possible "quiet" qubit can be made by using again a SD junction, this time with the interface normal directed toward the node of the d-wave electrode, see Fig. 6 a. In this particular configuration, for symmetry reasons, the first harmonic of the Josephson current, proportional to $\sin\phi$, where ϕ is the phase difference across the junction, cancels out, and the second harmonic, proportional to $\sin 2\phi$, becomes the dominant term. In principle, this junction represent a qubit by itself, without the need of a superconducting loop. Indeed, the ground state is twofold degenerate, presenting minima at $\pm\pi/2$ [Ioffe 1999 (a); Blatter 2001]. This is a consequence of the fact that the ground state is obtained at the phase value for which the tunneling current is zero. However, also in the present case, the ground state may undergo a braking of the time reversal symmetry state. Such occurrence is highly probable, because only the first harmonic perpendicular to the junction barrier is negligible. On the other hand, the component parallel to the junction interface is dominated by a $\sin\phi$ term, which is clearly non zero in the ground state [Huck 1997]. As a consequence, spontaneous currents parallel to the interface are generated, also in the absence of tunneling currents. Moreover, the particular geometrical configuration described so far involves a direct tunneling from the s-wave superconductor to the node of a d-wave electrode. In this direction, the d-wave superconductor shows gapless superconductivity, accompanied by low (zero) energy excitations [Fominov 2003; Amin 2004]. This aspect is strongly undesirable for the fabrication of superconducting qubits.

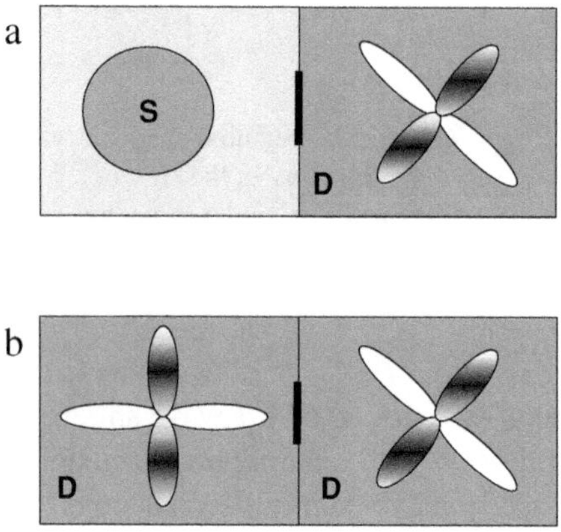

Figure 6 Schematic of a - SD asymmetric junction; b - DD asymmetric junction.

1.3 Asymmetric d-wave / d-wave junction

In analogy to what described in the previous section, similar results may be obtained with DD junctions. In particular, an asymmetric configuration (see Fig. 6 b) has to be used in order to get a potential energy with minima at $\pm\pi/2$ [Il'ichev 2001], as described in the previous subsection. Also in this case, low-energy excitations are present, and a time reversal symmetry broken state is expected as well [Lofwander 2000].

2. Symmetric d-wave / d-wave junction

2.1 General properties related to the order parameter symmetry: "0" to "π" transition in symmetric DD junction

Before going into details of the possible use of symmetric DD junctions ($\alpha_1 = -\alpha_2$, see Fig. 7) for the implementation of superconducting quiet qubit, in this section we describe general aspects of symmetric DD grain boundary junctions. In the case of a symmetric junction, the direction perpendicular to the interface does not point toward a node. In this sense, no low-energy exci-

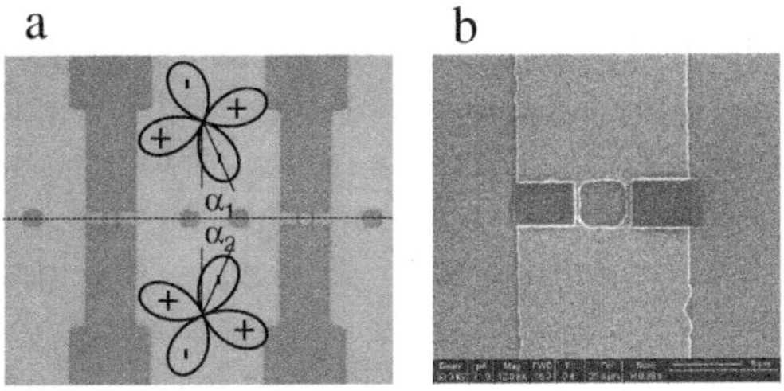

Figure 7 a - Photograph at an optical microscope of two sub-micron YBCO dc-SQUIDs, fabricated on symmetric 45° [001] tilt bicrystal substrates ($\alpha_1 = -\alpha_2$); b - FIB image of one dc-SQUID.

tations are expected, especially in the case of pure tunnel regime (narrow cone around the interface normal for the transport current). It is worth noting that a pure tunnel regime is well described by a junction characterized by a low-transparency barrier. Moreover, for symmetry reasons, no breaking of the time reversed state is expected. For these two main reasons, a symmetric DD junction can be considered a very quiet system, with no low-energy phenomena, usable for its employment in quantum computing.
Although symmetric d-wave / d-wave junctions might be useful for the implementation of superconducting qubits, as discussed before, in principle, they behave as a conventional one, showing properties typical of "0" junctions. In this picture, a symmetric DD junction does not add anything new with respect to the more controllable $low - T_c$ devices. However, under particular conditions, a symmetric DD junction can undergo a transition from the "0" state to the "π" state. In the next paragraphs we present an experiment showing the "0" to "π" transition in a symmetric 45° [100] grain boundary junction [Testa 2004 (b)].

2.2 Midgap states in symmetric DD junctions

At the surface of a superconductor, electrons are reflected as holes and holes are reflected as electrons-Andreev reflection [Andreev 1964]. d-wave superconductors can form superconducting bound surface states at the Fermi energy, the so-called mid-gap states (MGS) [Lofwander 2001]. In conventional s-wave superconductors the formation of MGS is forbidden. On the other hand, in d-wave superconductors, the order parameter symmetry changes its sign after 90° in-plane rotation. As a consequence, under certain conditions, Andreev bound states are formed at the Fermy energy (MGS) [Lofwander 2001]. Considering a symmetric DD junction (Fig. 7), MGS on each electrode overlap, forming states of the entire junction, contributing to the electrical transport. In order to verify the occurrence of a "0" to "π" transition, the presence of prevalent electrical conduction through MGS needs to be detected. In order to clearly observe this effect, some experimental precaution has to be taken.

Indeed, the interface in GB junctions is not a straight line, but it is affect by a meandering-like structure (facets) [Hilgenkamp 1996; Mannhart 1996]. Considering the non-conventional OPS in HTS materials, the presence of faceting along the junction interface disturbs the formation of MGS, and it becomes very difficult to observe their effect. Moreover, the presence of defects at the barrier, typical of HTS GBJs [Hilgenkamp 2002], represents another source of disturb for the formation of MGS. For these two reasons, we decided to drastically reduce the dimensions down to one order of magnitude with respect typical junction dimensions we used for previous experiments [Sarnelli 2002]. Indeed, the reduction of the junction width minimizes both the number of defects and the influence of faceting.

3. Fabrication process

To fabricate our samples, thin films were deposited by pulsed laser deposition (PLD) from polycrystalline $YBa_2Cu_3O_{7-x}$ targets onto 45° symmetric [001] tilt bicrystalline $SrTiO_3$ substrates. The films were grown at a temperature of 765 °C, an oxygen pressure of 0.2 mbar and a laser fluence of 2 $J.cm^{-2}$. The films had a thickness of 120-140 nm and exhibited a superconducting transition temperature of 89-91 K. An amorphous 50 nm thick $SrTiO_3$ layer was then deposited, by pulsed laser deposition over the bicrystal line to passivate the GBJs and protect the $YBa_2Cu_3O_{7-x}$ surface from gallium contamination during the focused ion beam (FIB) etching. Finally, a gold film, 20-30 nm thick, was deposited by magnetron sputtering as a conductive layer for the FIB process. It also prevented degradation of the $YBa_2Cu_3O_{7-x}$ surface in the contact area during the photolithographic processes.
The device geometry was initially achieved by standard photolithography and a water-cooled argon ion milling etch. Several GBJs, with widths ranging from 2 to 20 μm, were defined along the bicrystal line. Some of the junction widths were then reduced by using the technique described above. Fig. 7 b shows the FIB picture of one submicron dc-SQUID. We have fabricated submicron GBJs with different junction dimensions, down to 0.3 μm. Very high quality junctions have been achieved, demonstrating the feasibility of our technology for the fabrication of submicron HTS GBJs using FIB processes [Testa 2004 (a)].

4. Phase sensitive measurements

In the following, we describe the two phase sensitive tests we have done in order to check the "0" to "π" transition of our sample. In particular, we choose to fabricate a dc-SQUID composed by two submicron 300 nm wide junctions on a 45° [001] symmetric tilt bicrystal substrate, see Fig. 7. Although we are interested in the single junction behavior, we decided to investigate the effect in a dc-SQUID because in this particular structure it is possible to investigate the "0" to "π" transition of the SQUID as a whole. In the following, the two experiments are described in details. In a symmetric $45°$ [001] tilt GBJ, in the case of low-transparency barrier, a "0" to "π" transition is expected at about $T^\star = T_c/2$ [Tanaka 1997]. At this threshold temperature, an inversion of the critical current sign should be measurable, as a consequence of a phase shift in the current-phase relation [Tanaka 1997].

4.1 Temperature dependence of the critical current

In the inset of Fig. 8 the I-V characteristics of a superconducting loop with two junctions (dc-SQUID) in the temperature range 15 K - 44 K are shown. I-V curves have been shifted along the x-axis to allow a better comparison. In the figure, the Josephson current becomes almost zero at a temperature $T^\star$ between 25 K and 28 K, and increases for either higher or lower temperatures. This non-monotonic temperature dependence of the critical current is consistent to what predicted in the litterature [Tanaka 1997; Barash 2000], as the result of a "0" to "π" transition. In Fig. 8, the dependence of the normalized Josephson current on the temperature, $I_c(T)/I_c(0.3K)$ is reported. Both data sets show a minimum at the same temperature $T^\star \approx 0.5\ T_c$, in perfect agreement with theory [Tanaka 1997]. For completeness, the critical temperature of such a device was about 56 K, much lower than the film T_c, as expected because of the thermally activated phase slippage caused by the reduced Josephson coupling energy at the grain boundary interface [Gross 1990].

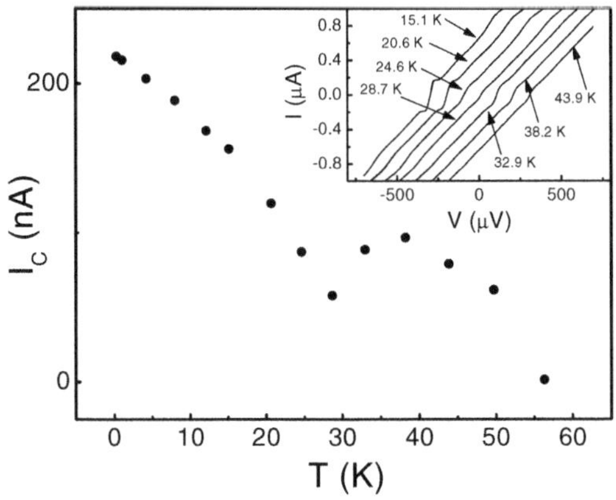

Figure 8 Temperature dependence of the critical current of a submicron YBCO dc-SQUIDs, fabricated on symmetric $45°$ tilt bicrystal substrate. The occurrence of a "0" to "π" transition is evident. In the inset, the current-voltage characteristics of the same sample as a function of the temperature are reported.

4.2 Magnetic field dependence of the critical current

In order to detect the presence of midgap state preferential conduction and, as a consequence, the occurrence of a "0" to "π" transition, we have investigated the SQUID magnetic properties in a narrow temperature range close to $T^{\star}$. In Fig. 9, we show the dependences of the critical current on the external magnetic field from 21 K to 33 K. Curves have been shifted along the y-axis for sake of clarity. Diffraction patterns with a period of 0.2 G, corresponding to an effective area of about 100 μ^2 and a flux focusing $f = A_{eff}/A_{geom}$ of about 6, can be observed.

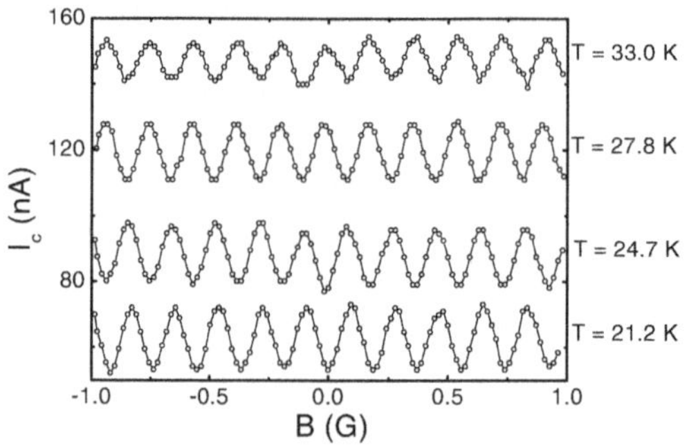

Figure 9 Dependence of the critical current on the external magnetic field from 21 K to 33 K, for the sample of Fig. 8. A clear phase shift of π is observed at the same threshold temperature for which a minimum of the Josephson current is achieved.

The $\beta = 2L\ I_c/\phi_0$ value is of the order of 10^{-3} and so flux trapping, self field effects and asymmetries are completely negligible. It is worth noting that the residual field in our cryostat is lower than 1 mG, corresponding to less than $0.01\ \phi_0$. The most interesting feature observed is the crossover from a minimum to a maximum zero-field critical current, at a temperature $T_1^{\star}$ between $T = 24.7$ K and $T = 27.8$ K. This transition temperature is perfectly consistent with the $T^{\star}$ value extrapolated by the I-V characteristics. The half flux quantum shift is a strong evidence that one of the two single junctions undergoes a "0" to "π" phase transition close to $T_1^{\star}$, frustrating the superconducting loop. It is worth noting that at $T = 4.2$ K the SQUID shows again a maximum of the critical current at zero external magnetic field, indicating that, at that temperature, also the second junction is a π-junction, making the SQUID a "0" SQUID. The difference in the transition temperature between the two junctions can be mainly related to the non homogeneity of the barrier transparency along the bicrystal line, typical of HTS grain boundary junctions. On the other hand, it looks like that a clear "0" to "π" transition is achieved as far as the junction dimension is in the submicron regime, and the temperature is lowered enough. It is possible to assure that these GBJs are reasonable all in the π state at the typical operating temperatures of superconducting qubit, of the order of a few mK.

Conclusions

In conclusions, the possibility to implement qubit devices by using properties of non-conventional superconductors has been discussed. Different technological alternatives employing high-temperature superconductors have been presented. We tried to point out advantages and disadvantages of different technical solutions. In particular, we proposed the possible use of a symmetric DD junctions, for the implementation of "quiet" qubits composed of five junction loops. We have presented experimental results showing the "0" to "π" transition taking place in symmetric d-wave / d-wave grain boundary junctions. The effect is related to the properties of d-wave superconductors to form Andreev bound states at the Fermi energy (midgap states). Under particular conditions, represented essentially by low enough temperatures, clean interfaces, low barrier transparency, and appropriate relative electrode orientations, the current carried by such MGS overcomes the conventional Josephson current, and a current sign change occurs. In these conditions, the junction shows an energy minimum at "π", instead of "0", and a π-junction is achieved. Such π-junctions may be of interest for the fabrication of so-called "quiet" qubit, because their symmetric configuration should prevent the occurrence of low-energy excitations, that in turn are one of the main source of decoherence.

Acknowledgments

Samples have been fabricated at Dept. of Materials Science of the Cambridge University, in collaboration with D. J. Kang, E. J. Tarte, S. H. Mennema, C. Bell, and M. G. Blamire. This work has been partially supported by the ESF Network "Pi-shift", the project DG236RIC "NDA", the TRN "DeQUACS", and "Proprietà di trasporto e di interfaccia in giunzioni Josephson HTc su scala submicrometrica".

BROKEN SYMMETRY AND COHERENCE OF MOLECULAR VIBRATIONS IN TUNNEL TRANSITIONS

A. M. Dykhne[1, 2] and A. G. Rudavets[2]

[1]*TRINITI, 142092 Troitsk, Russia*

[2]*Moscow Institute of Physics and Technology, 141700 Dolgoprudny, Russia*

Arudavets@mics.msu.su

This paper is dedicated to the blessed memory of A. P. Kazantsev.

Abstract We examine the Breit-Wigner resonances that ensue from field effects in molecular single electron transistors (SETs). The adiabatic dynamics of a quantum dot elastically attached to electrodes are treated in the Born-Oppenheimer approach. The relation between thermal and shot noise induced by the source-drain voltage V_{bias} is found when the SET operates in a regime tending to thermodynamic equilibrium far from resonance. The equilibration of electron-phonon subsystems produces broadening and doublet splitting of transparency resonances helping to explain a negative differential resistance (NDR)of current versus voltage (I - V) curves. Mismatch between the electron and phonon temperatures brings out the bouncing-ball mode in the crossover regime close to the internal vibrations mode. The shuttle mechanism occurs at a threshold V_{bias} of the order of the Coulomb energy U_c. An accumulation of charge is followed by the Coulomb blockade and broken symmetry of a single or double well potential. The Landau bifurcation cures the shuttling instability and the resonance levels of the quantum dot become split because of molecular tunneling. We calculate the tunnel gaps of conductivity and propose a tunneling optical trap (TOT) for quantum dot isolation permitting coherent molecular tunneling by virtue of Josephson oscillations in a charged Bose gas. We discuss experimental conditions when the above theory can be tested.

Keywords: Nanoelectromechanics, Coulomb blockade, broken symmetry bonding, electro-optical traps

Introduction

In this new millennium, we are witnessing the birth of molecular electronics, which can now operate with a single atom or a molecular nano-scale cluster

V.M. Akulin et al. (eds.), Decoherence, Entanglement and Information Protection in Complex Quantum Systems, 635–676.

called a quantum dot. Each quantum state of the dot can be characterized by electron tunneling, since the position of its discrete level is not averaged due to thermal spreading k_BT, especially at cryogenic temperatures T such that $\Delta E \gg k_BT$, where the ΔE is the energy spacing between levels and k_B denotes the Boltzmann constant. The quantum dot is confined between electrodes in a composite nano-scale system. Typically, the molecular cluster is trapped on the surface of a lead by a Lennard-Jones or van der Waals like potential. The electronic reservoirs are set out of equilibrium by applying a bias voltage V_{bias}, which modifies the electrochemical potentials (Fermi levels) of each electrode μ_σ driving a current through the quantum dot. The question arises: What kind of the coherent properties one would expect by measuring the electron transport in such system?

The experiments with nanoclusters, *e.g.* including fullerene molecules C_{60} [Park 2000] and C_{140} [Park 2003; Pasupathy 2003], which were addressed in a single electron transistor (SET), reveal a plethora of new behavior, which cannot be explained within the framework of solid-state nanostructures [Datta 1995; Datta 2000] theory. For instance, there are rich structures of conductivity resonance in a magnetic field [Park 2002] and anomalously high Kondo temperatures, above 50 K, found in [Yu 2003]. The conductivity gap is well correlated with the resonant frequency $\Omega = 5$ mev of the "bouncing-ball" mode of C_{60} and manifests itself as an enhancement of the conductance. There are two schools of thought, which attribute this enhancement either to Frank-Condon transitions or to the shuttling instability [Gorelik 1998; Altshuler 1999; Lubkin 1999; Switkes 1999; Tuominen 1999; Weiss 1999; Boese 2001; Dykhne 2001; Krotkov 2001; Fedorets 2002; Braig 2003; Flensberg 2003; Braig 2004] discussed below. For both schools the issue of quantum mechanical coherence is the matter of a considerable concern. We reconcile the above points of view in the adiabatic approximation, *i.e.* by assuming that electronic degrees of freedom are much faster than that of the dot. This is the case of local equilibrium in phase space between the electronic flows and the dot dynamics. Can the electron flow dephase the quantum dot by scattering off it?

From first sight, there is a physical constraint for dephasing, since an electron passing through the dot carries information in one direction, which is fixed by the bias voltage. This produces an asymmetry (chirality mapping) that accompanies the charging, by virtue of the electron affinity to the quantum dot. The charging modifies the Coulomb energy that competes with the elastic deformation potential of the dot. Thence, the Coulomb potential breaks the original symmetry of the molecular bonds in such a way that the growth of V_{bias} produces a bifurcation of the bonding potential. The broken symmetry signifies the existence of a border for the non-demolition quantum measurements. The quantum-classical transition can thus be detected by the broken phase of the oscillating current.

We have seen that the phase coherence of the oscillating current is affected by the tunnel coupling to thermal reservoirs due to the itinerant electrons. The phase-breaking process also changes the interface states as a result of interaction with other electrons via global Coulomb blockade. The Coulomb energy of electrons and holes forces their coherent reorganization. The electron coherence length L_e increases with decreasing temperature $k_B T$. As a result of freezing, $k_B T \leq 1$ mK, a solid-state system becomes mesoscopic on scales $L_e \leq 1$ μm.[1] Under this condition, the description of transport in terms of the local conductivity breaks down. The transport in a ballistic regime is dominated by electron scattering from the interface but not from the quantum dots. Most investigations of nanoscale systems are carried out for ballistic dots, whose theory and experiment have together formed a new realm of mesoscopic physics [Datta 1995; Datta 2000]. For instance, observation of the Kondo effect in single-atom and single-molecular junctions has led to a promising field called spintronics [Wolf 2001; Wolf 2002].

The SET device reported in Ref. [Park 2000] is an example of a double tunnel junction system in which the quantum dot self-oscillates between the leads. The mode softness significantly influences the electronic transport due to the effect of mechanical deformation on the electrical properties. Nanoelectromechanics (NEM) ensures that the electron's phase can be preserved over distances larger than the dot size, thus giving rise to a quantum interference which cannot be observed in macroscopic conductors. The conductance of dots could inherit this quantum coherence, which can manifest itself in superconducting current echoes [Vion 2002; Collin 2004 (b)], for instance. It means that a single molecular NEM-SET could, in principle, display a high charge sensitivity, enabling non-demolition measurements at the quantum threshold.

The experiments going under this title encompass a wide range of the electromechanical devices from the macro- to nano-scale. Coupling of a mechanical oscillator to non-equilibrium baths is accompanied by stiff dynamics of elastic self excitations and brings the charge transfer into the shuttle regime as a result. The evident advantage of shuttling lays in avoiding the tunnel coupling bottleneck. The phenomenon of the shuttle instability for quantum dots in double junctions resembles a more general class of the adiabatic quantum pump for electroacoustic and/or photovoltaic effects [Altshuler 1999; Lubkin 1999; Switkes 1999].

The current control at the one-by-one electron accuracy level is feasible in mesoscopic devices due to quantum interference. Though the electric charge is quantized in units of e, the current is not quantized, but behaves as a continuous fluid according to the jellium electron model of metals. The prediction of the current quantization dates back to 1983 when D. Thouless [Thouless 1983] found a direct current induced by slowly-traveling periodic potential in a 1D gas model of non-interacting electrons. The adiabatic current is the charge pumped per period, I, and has to be multiple of the electron charge, *i.e.* $I = e\,\nu\,N$, where the frequency $\nu = V/A$ is related to the traveling wave velocity V and the wavelength A. Then the charge transmitted in the adiabatic pump is period-independent.

The frequency dependence of the current I holds true for a dot shuttling between the leads and thereby modulating in phase the conductivity of the sequential tunnel junctions. The electron interference is manifested in an instability of the dot charge, which is subject to either stochastic or periodic oscillations. These shuttle dynamics allow one to find a new compromise between tunnel charging and Coulomb forces, rigidity and elasticity of dot bonds. Yet, a simple scattering theory that would help to estimate a measurements of current induced by the shuttling is absent, to our knowledge. Even for the usual I-V curve, characterized by regions of negative differential resistance and observed experimentally in Ref. [Park 2000], a mutual consensus between the Franck - Condon picture of electron transport and the shuttling mechanism has not been found.

In the quest for a new functionality of NEM SET, the shuttling mechanism has attracted a considerable interest as an effective method of control of electron transport, whose current depends on the frequency ν. A description of the shuttling instability can be based on a general master equation [Gorelik 1998] and Green's function [Fedorets 2002] methods of coarse grained dynamics over a scattering spectrum, without paying special attention to resonance field effects. Almost all of the obtained results are strongly model dependent and do not shed light on the underlying physics. For example, remaining unexplained are the higher Fano factors in the shuttling regime as compared to that obtained in the tunneling regime [Isacsson 2004]. This is one reason why it is worthwhile to develop a conceptually clear picture of the phenomenon, in the spirit of the Breit-Wigner theory of resonance cross sections. Another reason for this is that many results obtained both in the framework of the scattering approach and by classical methods in mesoscopic physics are applicable on an equal footing to the shuttling process[Blanter 2000].

The universality of both the Breit-Wigner method [Kadigrobov 1998; Blanter 2000] and the shuttling is manifested by studies of the NEM Josephson junctions [Gorelik 1998; Gorelik 2001; Isacsson 2002]. The latter belong to the mesoscopic system wherein the Cooper-pair box is shuttled between remote electrodes in the superconducting SET (SSET). The NEMs favor coherent coupling [Gorelik 2001; Isacsson 2002] and allow the suppression of quantum fluctuations of dissipationless persistent current in the ground state of the system. The shuttle mechanism reduces the Fano factor at low temperatures of about 1 mK [Gorelik 1998]. This is in accordance with the general rule [Blanter 2000] that the voltage, the current, and the charge oscillations due to

Josephson plasmons are less noisy and more entangled for strongly correlated system.

The shuttle mechanism based on the tunnel Hamiltonian [Gorelik 1998; Altshuler 1999; Lubkin 1999; Switkes 1999; Tuominen 1999; Weiss 1999; Boese 2001; Dykhne 2001; Krotkov 2001; Fedorets 2002; Braig 2003; Flensberg 2003; Braig 2004] is equivalent to the simplest possible Holstein-type polaron models.[2] These models ignore all complexity of the real molecular SETs: A detailed understanding of the charge screening, geometry of electrodes, hybridization with continuous and bound surface states, scattering off impurities is either absent or presented in a fragmented manner. The lack of knowledge about frequency shifts of the scattering resonances is filled by a phenomenological approach. Taking for granted their argumentation, we tackle the tunnel resonances using a common theory of resonance scattering in the Breit-Wigner approximation with a more pragmatic goal. By developing the Born-Oppenheimer adiabatic strategy in forbidden (for electrons) inter-electrode zone, where quantum dots are allowed to move classically, we present here the current and the quantum-dot charge in a self-consistent, model independent, and tractable form.

In Sec. Section2, we discuss the Landauer formula for the current, deriving it in parallel with inferring the mean charge from detailed balance conditions. This is followed by deducing the resonant tunneling in the Breit-Wigner approximation, which provides a convenient framework for description of quantum transport phenomena. In Sec. 2, the dissipative tunneling is taken into account with the help of a phenomenological model. In Sec. 3 we make use of adiabatic dot dynamics in order to describe field splitting and broadening of the resonance levels. The main aim is to show how do the dot oscillations between the leads influence markedly the current-voltage curves. With an increasing voltage the charging regimes change. Sec. 4 is dedicated to the Coulomb blockade modified by the adiabatic motion in phase space. We calculate the self-consistent charge accumulated in resonance windows of conductivity in order to illustrate the tunnel term transformation from a single well symmetry to a double well symmetry. We analyze how the shuttling instability depends on bias voltage. In Sec. 5 we estimate the shot noise of the shuttling adopting the resonance scattering approach. In Sec. 6 we justify the assumption of broken symmetry of the tunnel terms by employing a quantum treatment of the SET setup in the Born-Oppenheimer adiabatic approach. In Sec. 7 we propose the use of TOT protection of a quantum dot from parasite hybridization with the substrate surface in order to avoid dissipative tunneling roadblocks. Coherence of electron transport via double wells is sketched briefly in Sec. 8.

1. The Landauer formula.

The Landauer's seminal suggestion, that the current is transmission, dominates in mesoscopic physics and has applications to a variety of systems, including the electron transport in solids, liquids, quantum wires and dots. First and foremost, this theory describes conductance by purely dissipationless electrons scattering. Pursuing the NEM phenomena, we shall follow a similar reasoning, omitting the non-elastic effects on the microscopic scales.

Consider a ballistic quantum dot between two metallic terminals. Their electronic reservoirs are held at the thermal equilibrium described by the Fermi-Dirac distribution

$$f_\sigma(E) = \left[1 + \exp\left(\frac{E - \mu_\sigma}{k_B T}\right)\right]^{-1}, \tag{1}$$

where the chemical potentials $\mu_{l(r)} = E_F \pm eV/2$ correspond to a shifted Fermi energy, E_F, while eV is the biased electron potential across the source and drain leads. Electrons flow from the high potential μ_l to the low potential μ_r passing the Fermi level E_F. The electronic scattering state ψ with the Fermi energy E is normalized to unit flux in the lead $\sigma \in (l,\, r)$ at far asymptotic distances $\chi = \pm\infty$ from the quantum dot placed at a fixed coordinate x. The partial current of electrons, averaged over the momenta $p(E) = \sqrt{2m\, E}$, where m is electron mass, is defined as

$$I_\sigma(\chi,\, t) = \frac{e}{m}\,\mathrm{Im}\int_0^\infty \frac{dp(E)}{2\pi}\,\psi_\sigma^\star(\chi,\, t)\,\frac{d\psi_\sigma(\chi,\, t)}{d\chi}\,f_\sigma(E). \tag{2}$$

The measured current from the source to the drain electrode is proportional to the energy integral of squared modulus of the scattering matrix $|S(E)|^2$, whose integrand is called the quantum dot transparency $\Upsilon(E) = |S(E)|^2$, overlapped with the difference of Fermi functions of the electrons in the right and the left leads

$$\begin{aligned} I &= I_r(\chi = \infty,\, t) + I_l(\chi = -\infty,\, t) \\ &= \frac{\pi\, e}{\hbar}\int_0^\infty dE\,\Upsilon(E)\,[f_r(E) - f_l(E)], \end{aligned} \tag{3}$$

where $\hbar$ is Plank's constant. The only quantum property playing a role in the average current I is the Pauli principle, which dictates that each quantum state in the Fermi sea has to be occupied by a single electron. This means that only a fixed number of electrons can be accumulated in the scattering sector at a fixed energy E, thus setting a limit on the average current flow I. Factor 2 in Eq. 3, representing the electron spin degeneracy, has to be carefully taken into account, especially for the detailed balance conditions, as demonstrated in Sec. 2 and used in Sec. 4. At small bias, that is at $V \ll k_B T$, one writes $I = GV$, by introducing the linear conductivity G defined as

$$G = \frac{2\,e^2}{h} \int_0^\infty dE\ |S(E)|^2\ \frac{df(E)}{dE}. \tag{4}$$

The Fermi-Dirac distribution, $f(E)$, is the step-like function of energy E, while its derivatives are delta functions in energy. One therefore obtains $G = g_0|S(E_F)|^2$. The conductance quantum $g_0 = 2\,e^2/h$ is a universal factor of maximum conductivity for a single scattering channel at unit transmission. The minimum quantum resistance is the inverse quantity $g_0^{-1} = 12.9$ KΩ, which implies that the dissipation of an ideally transparent quantum dot occurs due to the scattering at the interface with electronic thermal reservoirs. This means that even for the system in a perfectly ballistic condition, the coupling of the electron with the reservoir induces decoherence. Irreversibility of an open system arises from uncorrelated "itinerant" electrons broadening the resonant state. "Itineracy" destroys the unitarity of the quantum mechanical evolution, and the scattering matrix S in the Landauer theory is reduced to phenomenological determination. Below, we present the Breit-Wigner resonance approximation for the scattering matrix as an example of such an approach.

1.1 Resonant tunneling in Breit-Wigner approximation

Let us consider the tunnel resistance of a $\chi = 1$ nm-broad vacuum gap between two gold electrodes. For the Fermi energy $E_F \sim 8$ eV and the work function $W \sim 5$ eV, the dominating electron wave function exponentially decreases in vacuum with the rate $\gamma = \sqrt{2m\,W}/\hbar \sim 1.3\text{Å}^{-1}$, where m is the electron mass. The enormous resistance $R_{1nm} = e^{-2\,\chi\gamma}/g_0 \sim 2.5 \times 10^{15}\ \Omega$ prohibits electron transport in vacuum in the absence of a quantum dot.

For a dot fixed between two electrodes, a weak electron coupling results from the tunnel current passing through the dot and broadens the resonant dot levels. The broadening width exponentially decreases as function of the distance χ between the dot and the electrode surface. The typical tunnel probability for an electron to jump from the electrode to the quantum dot formed by a C_{60} molecule [Park 2000] placed at the distance $\chi \approx 6$Å from the lead, can be estimated as $\Gamma_0 \sim E_F\, e^{-2\,\gamma\chi} \sim 0.1 - 1\ \mu$eV. In the situation when the quantum dot can move, its hybridization with the lead depends on the dot coordinate $\chi = \pm\, x$: $\Gamma(\chi) = \Gamma_0\, e^{-2\,\gamma x}$. The dot is characterized by the two parameters $\Gamma^r = \Gamma_0\, e^{2\,\gamma x}$ and $\Gamma^l = \Gamma_0\, e^{-2\,\gamma x}$ describing the exponentially decreasing couplings $\Gamma^{l(r)}$ with the reservoirs l or r. The total broadening $\Gamma = (\Gamma^r + \Gamma^l)/2$ denotes the width of the energy level E_{dot}. The broadening Γ may vary with the level energy and depends on the electron-electron correlation, although not strongly. In the weak-coupling limit, the resonance width Γ

is much smaller than the average spacing between the energy levels Δ, $\Delta \gg \Gamma$. In this limit only the scattering energies E which are nearest neighbours to the dot energy level E_{dot} contribute to conductance. In this case, the Breit-Wigner resonance of the scattering matrix $|\mathrm{S}(E)|^2$ or the transparency $\Upsilon(E)$ reads

$$\Upsilon(E) = |S(E)|^2 = \frac{\Gamma^r\,\Gamma^l}{(E - E_{dot})^2 + \Gamma^2}. \tag{5}$$

The scattering matrix elements implicitly depend on x via the parametric coordinate dependence of E_{dot} and Γ. The tunnel coupling Γ homogeneously broadens the resonance energy levels E_{dot}. Therefore, the Lorentzian distribution of the scattering matrix element implies an exponential decay via the coupling of electrons to the environment across only a resonance window of density of states $\rho(E)$

$$\rho(E) = \frac{1}{\pi}\,\frac{\Gamma}{(E - E_{dot})^2 + \Gamma^2}. \tag{6}$$

By substituting the matrix $\mathrm{S}(E)$ from 5 into Landauer's formula 3 at zero-temperature limit, the tunneling conductance is exactly represented as

$$G_L(E_F) = g_0\,\frac{\Gamma^r\,\Gamma^l}{(E_F - E_{dot})^2 + \Gamma^2}. \tag{7}$$

For the most of the experiments of interest, the temperature k_BT typically exceeds the resonance width $\Gamma \ll k_BT \ll \Delta$, and hence the interference effect of adjacent levels separated by the Δ is inessential. The conductance G from 4 is given by the integral of the Breit-Wigner transparency 5 yielding

$$G(E_{dot},\,T) \approx \frac{G_0}{\cosh^2(\Delta_T)},\quad \Delta_T = \frac{E_{dot} - E_F}{2\,k_BT},\quad G_0 = \frac{\pi\,g_0}{4\,k_BT}\,\frac{\Gamma^r\,\Gamma^l}{\Gamma}, \tag{8}$$

where Δ_T is offset from the resonance energy from the Fermi level in units of k_BT. The relationships 8 between the conductance G and the tunneling broadening Γ solve the electron transport problem for the static dot in the ohmic regime.

The concept of the static dot is justified by the fact that the dot's translational and rotational degrees of freedom vary slowly as compared to the fast motion of the electrons. The parameters E_{dot} and Γ of the scattering matrix $\mathrm{S}(E)$ represent adiabatic variables, validated by the Born-Oppenheimer approach. An instant conductivity of a movable dot is to be computed in a phase space of the NEM oscillator's coordinates x and momenta p. The x, p point plays the role of a partial scattering channel in which the Wigner delay time of electron tunneling $\tau = \partial S(E)/\partial E$ is the shortest time scale. The marginal delay τ

justifies the picture of an instant scattering, where the current is obtained by averaging over the phase space. This reasoning will thread throughout this chapter after a short discussion of the proximity effects and electron balancing flows in Sec. 2.

1.2 The proximity effects and dissipative tunneling

If the dot sticks to a metal lead and forms an "adatom", its energy levels E_{dot} get broadened by hybridization with the surface continuum states. Since the electron affinity is different for the dot and the host surface, the charge transfer causes a change in the electrostatic potential inside the dot and shifts the energy levels by a contact potential. The result is that the adatom could be directly charged classically, by the electrons flowing in and out of the conduction bands. For example, Rubidium (Rb) is electropositive on a gold (Au) surface, since its ionization potential $I_i \sim 4.2$ eV is smaller than the Au work function $W \sim 5$ eV. But it remains neutral on an isolator, such as glass, because the bonding originates from the van der Waals covalent coupling. The dot's charging depends on conditions of "quasi-equilibrium" in the open driven system "dot + leads".

For molecular dots placed far from the surfaces, the tunnel coupling is rather weak, as compared to the mechanical and electrostatic energies. Under these circumstances, the transport is usually subject to global Coulomb blockade. When the charge attempts to tunnel, it strongly perturbs the surface states, inducing a coherent reorganization of electrons and holes generating phonons or plasmons on the substrates and leads [Braig 2003; Flensberg 2003; Braig 2004]. Fig. 13 shows how the Fermi-edge is disturbed due to the tunneling.

A priori, when the phonon relaxation is faster than the tunneling rates, thermodynamic equilibrium should hold at the temperature of the host reservoir. However, for the nano-junctions the local surface temperature may differ from the bulk equilibrium temperature. This is due to the Anderson orthogonality catastrophe (AOC)[3] associated with interplay between the van der Waals and the electrostatic forces. The electron tunneling affects the overlap between differently shifted phonon ground states of the surface. The faster the tunneling rate, the closer is the phononic overlap to zero, and that hinders relaxation of the surface temperature. AOC presents the mechanism also affecting the thermal state of the electronic reservoir due to electron-phonon coupling. In Sec. 3, from comparison of our theoretical I-V curves at different electron-phonon temperatures and the experimental data [Park 2000] we infer that AOC exists.

In Sec. 2, we make use of the detailed balance conditions for derivation of a generic expression for the mean charge of the quantum dot. We formulate the field effect on the splitting and broadening of the tunnel resonance for adiabatic evolution. Here we want to emphasize that the irreversible decay is especially important for establishing steady states in the non-equilibrium system of the dot at contact with an external infinite bath. Besides the aforementioned AOC, damping of the shuttle motion of charged particles between metallic electrodes can be related to radiative decay mechanisms, discussed long ago in the seminal paper [Kazantsev 1963] by Kazantsev and Surdutovich.

2. Current at detailed balance

The system made-up of a quantum dot and two leads of different Fermi energies experiences electron flows tending to bring the system to thermodynamic equilibrium. In the steady-state regime, the net current summed over the electrodes is zero, that is the incoming and the out-going flows of electrons through the quantum dot compensate each other. The balancing process is provided by the tunnel rates Γ^σ weighted by the Heisenberg time needed for an electron of energy E to escape into the electrode $\sigma =\in (r,\ l)$, $\Gamma^\sigma = \hbar/\tau^\sigma$. If we denote the electron distribution function inside the quantum dot by $f_{dot}(E)$, the out-flow current at energy E to the right electrode is

$$I_{out}(E) = \frac{f_{dot}(E)}{\tau^r} = \frac{\Gamma^r\, f_{dot}(E)}{\hbar}. \tag{9}$$

The electrons populate the level E of the quantum dot by backward transitions from the lead with an incoming flow

$$I_{in}(E) = \frac{2}{\tau^r}\, \rho(E)\, f_r(E) \tag{10}$$

occurring through the resonance window $\rho(E)$ over the time τ^r , while the factor 2 in the Eq. 10 accounts for the spin degeneracy in the Fermi sea. The difference of the in- and out-going electron flows at energy E produce the net current

$$I_r(E) = e\,\frac{\Gamma^r}{\hbar}\,\left[2\,\rho(E)\,f_r(E) - f_{dot}(E)\right], \tag{11}$$

Analogously, the expression for the current on the left lead reads

$$I_l(E) = e\,\frac{\Gamma^l}{\hbar}\left[2\,\rho(E)\,f_l(E) - f_{dot}(E)\right]. \tag{12}$$

Since the detailed balance principle implies that in equilibrium the electron population of the dot has to be at a steady-state, $I_l + I_r = 0$, we immediately arrive at the relation

$$\Gamma^r\,\left[2\,\rho(E)\,f_r(E) - f_{dot}(E)\right] = \Gamma^l\,\left[f_{dot}(E) - 2\,\rho(E)\,f_l(E)\right] \tag{13}$$

which yields the distribution $f_{dot}(E)$

$$f_{dot}(E) = 2\,\rho(E)\,\frac{\Gamma^r\, f_r(E) + \Gamma^l\, f_l(E)}{\Gamma^r + \Gamma^l}. \tag{14}$$

The net current in the scattering channel E is represented as

$$\begin{aligned} I(E) &= \frac{e\,\Gamma^l}{\hbar}\,[2\,\rho(E)\,f_l(E) - f_{dot}(E)] \\ &= \frac{e\,\Gamma^r\,\Gamma^l}{\hbar\Gamma}\,\rho(E)\,[f_l(E) - f_r(E)] \\ &= \frac{2e}{h}\,\Upsilon(E)\,[f_l(E) - f_r(E)]\,. \end{aligned} \tag{15}$$

We see that the detailed balance is equivalent to Landauer's formula Eq. 3. The generic method of the steady state regime is utilized here in order to emphasize a common meaning of Landauer's ansatz of Eq. 3 corresponding to Eq. 15. The current flows and electronic distribution $f_{dot}(E)$ in the quantum dot are self-consistently related provided

$$Q = 2e\int_0^{\infty} dE\;\rho(E)\,\frac{\Gamma^r\, f_r(E) + \Gamma^l\, f_l(E)}{\Gamma^r + \Gamma^l}. \tag{16}$$

The total charge accumulated in the resonance window is a trade-off between the Fermi seas of the electrodes, allowing for the establishment of equilibrium between the continuous spectra of their conduction bands and the intermediate resonance state of the quantum dot. The steady-state kinetics refers to diverse transport problems, and it can be derived by more subtle techniques such as the density matrix approach in the framework of the Keldysh representation [Keldysh 1964-65] on a closed time path [Gogolin 2002], or the nonlinear Green function methods [Wingreen 1994; Datta 1995; Datta 2000]. But our aim is more pragmatic: we apply Eq. 16 in the Breit-Wigner approximation for adiabatic motion of the quantum dot and calculate the average current.

2.1 Field splitting and broadening of the resonance level

Due to the huge difference of the molecular and the electron mass, slow molecular vibrations can be considered as quasistatic when compared to the fast electron motion. The slowness of the molecular vibrations justifies the Born-Oppenheimer adiabatic approach. This strategy provides us with a paradigm useful for consideration of electron tunneling through a movable quantum dot. By the analogy to the Born-Oppenheimer molecular terms, we use the concept of tunnel curves, representing a total electronic energy as a function of the dot coordinate in the inter-electrode regions, forbidden for the electrons. However, the massive dot could move classically in this region, thus transporting an attached electron and having a potential energy, called the tunnel curve.

Note that this tunnel term, being very similar to molecular one, includes both the potential and kinetic electron energies as a function of the dot center-of-mass coordinate x.

Having this in mind, the tunnel term can simply be regarded on phenomenological grounds. From the other hand, the quantum-mechanical method of Sec. 6 exemplifies an "ab-initio" approach employed here for the calculating the tunneling current. The tunnel term's topology defines the adiabatic dot dynamics and the average position of the measured electron. Thus, the averaging of the instant scattering S-matrix or the resonance transparency $\Upsilon(E(t))$ over the adiabatic paths becomes a basic ingredient of the electron transport problem.

Without loss of generality, we set the resonance energy level ε_0 equal to the Fermi energy, $E_F = \varepsilon_0$. In equilibrium, the dot assists the tunneling in remaining neutral, since the electro-chemical potentials of the leads and the dot are identical. The self-consistent "electro-chemical" potential for charging is $U_{mean} = U_c(Q - Q_0) = 0$, where U_c is the Coulomb energy cost per one electron, and $Q - Q_0$ is an extra charge with respect to unbiased electronic reservoirs. To estimate the field effect we write down the mechanical and electrostatic energy E_{dot} as follows

$$E_{dot} = \varepsilon_0 + U_{mean} + E_{vib} + U_{ext} \tag{17a}$$

$$E_{vib} = \frac{p^2}{2M} + \frac{M}{2}\,\Omega^2\,x^2, \quad U_{ext}(x) = -Fx \tag{17b}$$

where E_{vib} is the vibrational energy of the dot of mass M. The "bouncing-ball" mode consists of the kinetic part $p^2/2M$, and the potential part $M\,\Omega^2\,x^2/2$, with Ω being the vibration frequency. The external field force is $F = e\,V/D$, where D is the tunnel (inter-electrodes) gap, and V is the bias voltage.

If discharging and relaxation are faster than charging, the quantum dot remains neutral. The main role of the intermediate quantum dot is to assist a virtual tunneling through its resonance state. The quantum dot in a simple harmonic model just oscillates over a closed trajectory having the coordinate $x_n(t) = x_{n0}\,\cos\Omega t$ and momentum $p_n = p_{n0}\,\sin\Omega t$ where the $x_{n0} = x_0\,\sqrt{2n}$ and $p_{n0} = p_0\,\sqrt{2n}$ are integrals of motion of nth vibrational energy eigenstate. The zero-point amplitude $x_0 = \sqrt{\hbar/M\,\Omega}$ and momentum $p_0 = \hbar\,x_0$ correspond to the vibrational ground state. By denoting the frequency shift due to external potential field as $\nu_n(t) = F\,x_n(t)/\hbar$, we count off the dot energy levels $E_{dot}(t) = \hbar(n\,\Omega - \nu_n(t))$, from the resonance reference point $\varepsilon_0 = E_F$. The detuning frequencies $n\,\Omega - \nu_n$ define the energy positions of the transparency resonances, required for calculation of the instant electron current by the Landauer's formula Eq. 3. A thermodynamic average of the electron transport over the dot vibrations has to be taken in order to obtain the mean current and to compare the calculated I-V curves with experimental current-voltage characteristics.

At increasing voltage, the coupling between the dot and the leads tears the phonon temperature away from the equilibrium temperature T. Thermalization of this driven open system involves complicated dynamics of the electron-phonon interactions. As an estimate, we use the Callen-Welton theorem in order to represent the fluctuation-dissipation relation between the noise power and the effective temperature of the system "leads + dot". On the one hand, the Johnson noise[4] reads $S_{John} = 4\,k_B T\,G$, where G is the conductivity [Blanter 2000]. The parameter T has the meaning of an effective temperature T_{eff} for the non-equilibrium system. On the other hand, the voltage V produces the Schottky's noise that gives the main contribution to the power spectrum at a low temperature, except of the transparency resonances [Blanter 2000]

$$S_{Schottky} = 2e\,I, \tag{18}$$

where Ohm's law reads $I = VG$. By equating the power spectra of the thermal noise and the non-equilibrium shot noise $S_{John} = S_{Schottky}$, we immediately relate the effective temperature T_{eff} with the voltage V, which drives the current and heats the SET to the temperature

$$k_B T_{eff} = \frac{e\,V}{2}. \tag{19}$$

This fundamental relationship has been obtained from the density matrix of the quantum point contact coupled with mechanical oscillator in [Mozyrsky 2002]. Therein quantum heating and damping of the oscillator manifest themselves in an induced quantum-classical transition. In fact, our simple FDT arguments also indicate that heating by the electron-phonon interplay can be much larger than heating by the thermal reservoirs. The source-drain voltage V adversely affects the current and may result in undesirable artifacts, such as electro-migration, SET disruption, etc. Therefore, a non-demolished experiment requires a small voltage and low temperature, where the adiabatic approximation holds. In this limit, we can ignore the dependence of tunneling probability Γ both on the energy E and the coordinate x, setting $\Gamma \sim \Gamma_0$ near $x = 0$, since $\gamma x_0 \ll 1$. For the coherent model, the instant value of $\Upsilon(E,\,t)$ is given by

$$\Upsilon(E,\,t) = \frac{1}{Z}\sum_{n=1}^{\infty} \frac{\Gamma_0^2\,\exp\left(-\dfrac{n\,\hbar\Omega}{k_B T}\right)}{[E - E_d(n,\,t)]^2 + \Gamma_0^2} \tag{20}$$

where $Z = \sum_{n=1}^{\infty} \exp[-n\,\hbar\Omega/(k_B T)]$ is the vibrational partition function. The path averaging yields the mean transparency

$$\overline{\Upsilon(E)} = \frac{\Omega}{2\pi} \oint dt\, \Upsilon(E,\, t)$$

$$= \frac{\Gamma_0}{Z}\, \mathrm{Im} \sum_{n=1}^{\infty} \frac{\mathrm{csgn}(c)\, \exp\left(-\frac{n\,\hbar\Omega}{k_B T}\right)}{\sqrt{(E - \hbar\, n\Omega - i\Gamma_0)^2 - 2n\,(\hbar\nu_0)^2}} \qquad (21)$$

where $\mathrm{csgn}(c)$ [Abramowitz 1972] denotes a signum function of complex argument $c = e^{-i\phi}$, and $\phi = \arg(E - n\,\hbar\Omega + \sqrt{2n}\,\hbar\nu_0 - i\,\Gamma_0)$ is used to keep continuity of the square root of the complex-valued function. The root function originates from the uniform distribution of the particles along the adiabatic path in the phase space x, p. The main contribution to the resonances gives the phase space sector $p = 0$, since the electron has a higher probability to scatter off the "static" dot, than at the dot moving in-between the turning points. The shifted frequencies ν_n are maximum at the turning points, that provides the doublet splitting of the vibrational resonances $n\Omega$. Since the frequency shift depends on the x_n, the spectral broadening increases with the vibrational energy $n\Omega$. The inset in Fig. 1 shows the resonance transparency 21 averaged over the dot path.

In order to make comparison with the experiment [Park 2000], we chose the frequency $\Omega = 5$ meV, which corresponds to a C_{60} quantum dot interacting with gold electrodes via the Lennard-Jones potentials. For the electron temperatures $k_B T = 0.4$ meV, the vibrational amplitude is $x_0 \sim 0.03$Å, and hence only the zero-point mode is active, provided $\Omega \gg k_B T$. The equilibrium distance $D/2 \sim 6.2$Åseparates the center-of-mass of the quantum dot from both leads. When $e\,V \leq \Omega$, the frequency shifts ν_n are negligible, since the system "leads + dot" is cold $k_B T \ll \Omega$, and therefore $\nu_n = F\,x_n = \Omega\,x_n/D \ll n\Omega$.

However, on average, the current through the vibrating dot differs significantly from the "static" dot transmission. At growing effective temperature 19, the bouncing-ball mode produces *an inhomogeneous* broadening. This field effect results from averaging over the frequency shifts $\nu_n(t)$ that depend on the location of the quantum dot bouncing between the electrodes via the source-drain voltage characteristic, while the tunnel coupling broadens the resonance line *homogeneously*.

The Breit-Wigner approximation allows one to represent the resonance tunneling spectrum as superposition of Lorentzian lines. In the limit of narrow electronic levels $\Gamma_0 \to 0$, the resonance transparency obeys the equation

$$\begin{aligned}\overline{\Upsilon(E)} &= \frac{\Omega\,\Gamma_0}{2Z}\sum_{n=1}^{\infty}\oint dt\,\delta\left[E - E_{dot}(n,\,t)\right]\,\exp\left(-\frac{n\,\hbar\Omega}{k_B T}\right) \\ &= \frac{\Omega\,\Gamma_0}{2Z}\sum_{n=1}^{\infty}\exp\left(-\frac{n\,\hbar\Omega}{k_B T}\right)\,\frac{dt}{dE}\bigg|_{E=E_{dot}(n,\,t)}.\end{aligned} \tag{22}$$

This equation states that the resonance transparency of trapped dots, atoms or molecules, is formed in the neighborhood of the turning points, where $\dot{E} = 0$. At these points, $p = 0$, and the resonance transparency $\Upsilon(E)$ becomes infinite in the limit $\Gamma_0 \to 0$. However, the coupling with the environment produces a finite broadening Γ that negates the divergence of the transparency $\Upsilon(E)$.

3. Current voltage curves

The adiabatic treatment of the dot's motion can be used to clarify as yet unexplained features in the experimental I-V curves [Park 2000]. To this end one not only needs to find the adiabatic paths but also to perform averaging over their thermal perturbations. In our first example, the electron reservoirs are strongly coupled with the dot vibrations that are characterized by the effective temperature $k_B T_{eff} = e\,V/2$. The quantum dot can locally heat the surface of the leads up to the same temperature $T = T_{eff}$. The reason behind the anomalous electron heating is the electron tunneling perturbing the surface. The local quasi-particle perturbations, *e.g.* plasmons, cannot relax fast because of the AOC. The surface plasmons create no overlap between the shifted phonon states. As a result of phonon-plasmon interplay [Braig 2003; Flensberg 2003; Braig 2004] the electron temperature T equilibrates with the phonon temperature T_{eff}.

The electronic current versus the source-drain voltage shown in Fig. 1 displays the typical features of C_{60} SET [Park 2000] at the open gate $V_{gate} \ll \Omega$. At a small gate voltage, surface charge [Braig 2003; Flensberg 2003; Braig 2004] is allowed, and the AOC mechanism may take place. For the C_{60} SET, the phonons manifest themselves in the step-plateau-like current, with the conductivity gap corresponding to the phonon frequency $\Omega = 5$ meV, and with a negative differential resistance (NDR) regions.

The NDR can be explained by the field effect, which splits the resonance scattering lines because of the frequency shifts at the turning points, and therefore the electron flow only partly contributes to the current. In accordance with the Pauli principle, the Fermi energy spectrum window given by the difference of Fermi functions results in a step-like increase of the current, when being overlapped with the transparency resonance at the bias voltage equals a multiple of the phonon quantum $n\Omega$, as shown in Fig. 1. Since the transparency doublets are split faster than the Fermi windows are broadened out, after a

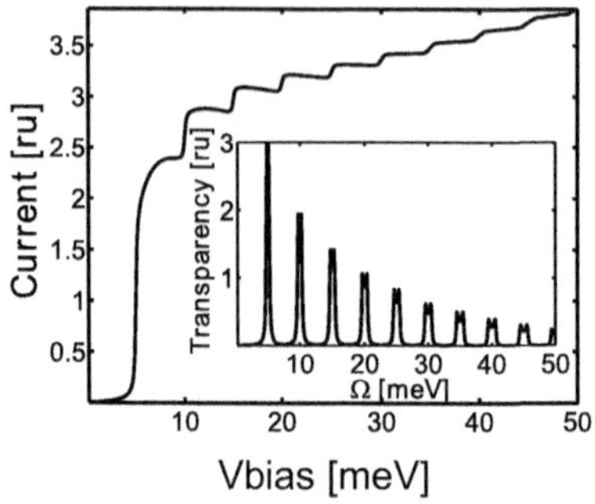

Figure 1. Intensity-voltage (I - V) curve at the same temperatures of electron and phonon reservoirs due to AOC effect. Inset: the doublet spectra caused by vibrating dot with $\Omega = 5$ meV and T_{eff} at the end point of the I - V curve.

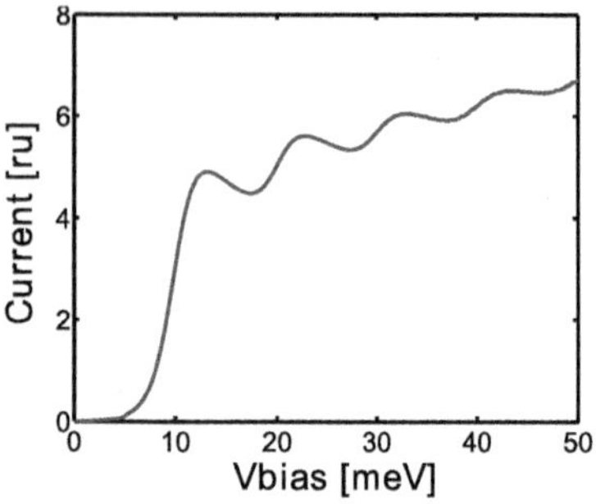

Figure 2. Intensity-voltage (I - V) curve of "dot + leads" system at the equilibrium electron temperature $k_BT_e = 0.4$ meV, but at a different effective phonon temperature $k_BT_{ph} = e\,V/2$

step-like increase the transmitted current slightly decreases with the increasing voltage V, until the next vibrational quantum, $n + 1$, provides an additional channel for tunneling.

As the second example, we consider a model of the C_{60} SET at a large gate voltage V_{gate} that stops the tunneling current. The V_{gate} eliminates the AOC and eliminates the charge disturbances at the lead surfaces. Under these conditions, the local electronic reservoirs remain at the equilibrium temperature $k_BT \sim 0.5$ meV. If the Fermi distributions of electrons (see Eq. 3) have sharp edges at $E = e\,V/2$, the step of the current occurs when crossing the tunneling resonance $E = \Omega$ at $e\,V = 2\Omega$, (see Fig. 2). The conductivity gap at double frequency 2Ω has been observed in Berkeley [Park 2000] by measuring the differential conductance $\partial I/\partial V$ at the frequency domain far from the bouncing-ball and the breath modes of the C_{60} transistor. The crossover regime to the 2Ω frequency has been found at large V_{gate} that suppresses the current through the C_{60} transistor, as shown in Fig. 2 on the right hand side of the $\partial I/\partial V$ plot[5]. On the contrary, a small V_{gate} opens the transistor. Then, the current produces a heating of the electronic reservoirs that smears the 2Ω line.

The present scattering theory seems to work reasonably well, considering its simplicity. The adiabatic picture explains the major features of the I-V curves and especially their NDR behaviour, displayed in the Figs. 1 and 2. The bias voltage therein is below of the Coulomb energy, and the charging of the quantum dot can be disregarded. But the broken Coulomb blockade is the basic tenet of the electron transport, whose current growth can be related to the electron affinity of quantum dot at $e\,V_{bias} \sim U_c$. In the next section we demonstrate that the sequential tunneling can affect the quantum dot paths by *breaking* symmetry of the adiabatic potential.

4. Charging and discharging in adiabatic theory of Coulomb blockade

From the detailed balance we have learned that the charging is a trade-off between the broadening Γ and the equilibration of the electron flows. The source-drain voltage can result in a sudden change of this balance and the corresponding current. Indeed, turning on the shuttle can maximize the transparency and hence the conductivity. It can also affect the noise characteristics. Qualitatively new effects such as low-frequency coherent oscillations, electric echoes, and quantum entanglement may arise due to coherence of the charge states.

In order to pick up a charge from the electronic terminal, several mechanisms have to be involved in parallel. One- and many-body effects (tunneling, screening, AOC etc.) are among them. Earlier, considering the proximity surface effects, we have noticed that in the equilibrium conditions the dot charge depends on the contact potential. This is ensured by the electrochemical potentials μ_σ, $\sigma \in (l,\ r)$ and μ_{dot} in the leads and the dot correspondingly, when these subsystems do not interact and can be considered separately. The difference of the electrochemical potentials creates a driving force for the electron flow, which obeys the detailed balance conditions. The dot charges positively or negatively, depending on whether μ_{dot} greater or less than μ_σ, and it remains neutral when $\mu_{dot} = \mu_\sigma$.

Accordingly, the formulas for current and self-consistent charge employ the electro-chemical potentials for the control of the electron distributions in the leads and dot. The electro-chemical potentials of the electrodes are defined as $\mu_\sigma = E_F \pm e\, V/2$, where E_F is the Fermi energy of electrons (including the self-energy of pair interaction) and V the external potential. The electrochemical potential $\mu_{dot} = U_{mean}(x,\ p)$ is characterized, in turn, by the electron affinity given by the charging potential $U_{mean}(x,\ p)$ so that the electronic energy of the quantum dot is

$$E_{dot} = E(x,\ p) + U_{mean}(x,\ p), \tag{23}$$

where $E(x,\ p)$ is the energy of the resonance electron level. For the quantum dot between leads, the role of electron affinity and ionization potential are played by the LUMO and HOMO state energies, respectively. For the self-consistent electron interaction, the mean-field potential $U_{mean} = U_c(Q - Q_0)$ denotes the charging (discharging) energy. The coefficient U_c is the cost of Coulomb energy when the dot gains (loses) one electron, Q is the self-consistent charge and Q_0 is the background charge, respectively. In the microscopic description, calculation of the mean Coulomb energy relies on the Hartree-Fock terms. The Coulomb energy integral depends on the wave-function overlap accounting for the electron-electron interactions between the electronic reservoirs and the dot.

This overlap measures the rate of tunnel transitions that entails the detailed balance condition for the SETs. The transition dynamics fall into the generic class of two-level-band systems [Akulin 2004]. According to the common quantum-mechanical rules [Landau 1977] we identify the following cases:

- Coherent tunneling via the virtual state of the dot: The resonance level assists instantaneous transitions between electrodes, while the resonance state remains empty.

- Sequential tunneling via the real state of the dot: The dot charge is not exhausted on the time scale of adiabatic dynamics.[6] The charge transport could be driven by actual shuttling.

Virtual and real tunneling transitions between the leads are similar, in a sense, to Raman scattering and resonance fluorescence for optical transitions. The critical dependence of the shuttling on the bias voltage, that counteracts the relaxation due to itinerant electrons and other dissipation mechanisms, implies the importance of the latter for the onset of the steady-state regime. This is similar to Vavilov's rule in optics, stating that transitions through a real state are characterized by their quenching ability. This analogy holds true for resonance tunneling, in which the electron states and transitions are treated quantum mechanically, while the adiabatic dynamics may remain classical. The notion of the shuttle instability is related to the dynamical symmetry of the bond potential (attractor) called as the tunnel term, in which the dot resides. The critical dependence of the tunnel term on bias voltage is demonstrated below in the framework of a phenomenological approach. The quantum description of the voltage-driven bifurcation is briefly sketched in Sec. 6.

In the Breit-Wigner approximation, the charging - discharging transitions are accounted for by the offset of the energy (or the frequency shifts) dependence on the dot location relative to the leads. According to Eq. 17a it is

$$E(x,\, p) = \varepsilon_0 + E_0(x,\, p) = \varepsilon_0 + K_{in}(p) + U_{pot}(x) + U_{ext}(x). \qquad (24)$$

The first part of the energy $E_0(x,\, p)$ is the kinetic energy $K_{in}(p) = p^2/2M$. For the matter-field interaction, this energy produces the well-known Doppler effect of inhomogeneous broadening of optical transitions. The second part is the interaction with electrodes, *i.e.* the Lennard-Jones potential near the equilibrium point $x_0 = 0$ taken in the harmonic approximation as $U_{pot}(x) = M\,\Omega^2\,x^2/2$. The third part is the electrostatic interaction of the dot electron in the external field between the electrodes $U_{ext}(x) = -Fx$. In order to formally compute the mean charge during the resonance tunneling, we substitute the dot energy Eq. 23, 24 into the steady-state condition Eq. 16. This yields

$$Q(x,\, p) = \frac{1}{\pi} \int_0^\infty dE\, \frac{\Gamma^r\, f_r(E) + \Gamma^l\, f_l(E)}{[E - E_{dot}(x,\, p,\, Q)]^2 + \Gamma^2}. \tag{25}$$

The Franck-Condon principle implies locality in phase space x, p. It states that the coordinate and momentum of a massive dot stay unchanged during the electron transitions, *i.e.* electron tunneling in our case. For detailed balance, the tunnel coupling Γ_σ has to be invoked as the main prerequisite for the Coulomb blockade. The additional relaxation due to molecular collisions and electromagnetic radiation may have to be favored for the steady state. This is similar to the dissipation necessitated in the orthodox theory of the Coulomb blockade [Kulik 1975; Averin 1991].

The shuttling [Gorelik 1998] also requires Coulomb blockade, which is a combined effect of the single-electron charging and the quantized spectrum of a bound system. The charging is produced by the bias voltage V_{bias}, and hence it can be accompanied by a shuttle instability with threshold behaviour. The physical meaning of the criticality is shown to be the Landau bifurcation of the bonding potential. The vibrational energy, the temperatures, and the charge dissipation prevent the shuttle mechanism taking effect before the threshold V_{bias} is reached.

4.1 The self-consistent charge

Mathematically speaking, we have arrived at a nonlinear integro-functional equation, for which an exact analytical solution is impossible, although its analysis within physically meaningful approximations remains instructive. The integral over the spectrum corresponds to the electron equilibrant flows ensued by contact with two electron reservoirs. However, when the inter-electrode distance is much longer than the Fermi length $1/\gamma$ and the dot is trapped in the middle, $\gamma D \gg 1$, the resonance window of the dot state density is extremely narrow: $\Gamma(D/2) \approx E_F\, e^{-\gamma D} \to 0$. Then, the integral over the continuous spectrum of conducting electrons is collapsed by the delta functions of the dot state density:

$$Q(x,\, p) = \frac{1}{\Gamma(x)} [\Gamma^r(x)\, f_r(E_{dot}) + \Gamma^l(x)\, f_l(E_{dot})]. \tag{26}$$

In Sec. 2 we have already used this limit of narrow levels. The resonance spectra of the electron tunneling without charging are explained by virtual transitions. For real transitions through a charge state, the convoluted integral 25 can be reduced to the nonlinear equation 26 for the distribution $Q(x,\, p)$. The nonlinearity emerges from the condition of detailed balance between the dot and the Fermi reservoirs f_σ, which interact via irreversible coupling Γ_σ.

The mean charge Q can be computed with the help of library routines, *e.g.* by using either the Newton-Raphson algorithm or a globally convergent one. Figs. 3 and 4 display distributions of the charge of the molecular C_{60} transistor, for which the thermal spread of the Fermi functions is much larger than the tunnel broadening $k_B T \gg \Gamma$ at bias voltages exceeding the Coulomb energy.

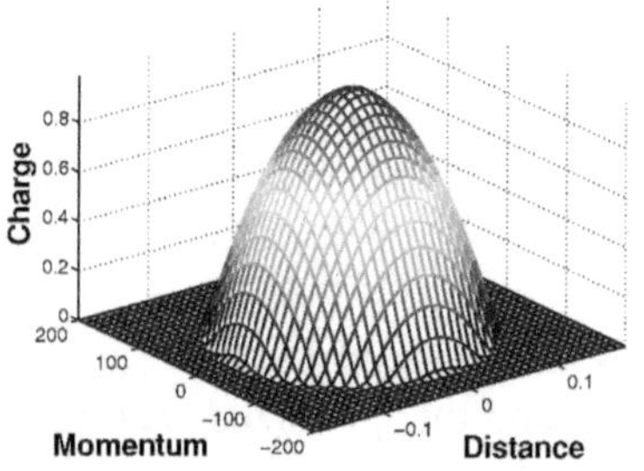

Figure 3. Charge distribution in the phase space of x and p, of the vibrator with zero point oscillation amplitude $x_0 = 0.03$ and momentum $p_0 = 1/x_0$, acquires almost the Gaussian shape for the bias voltage up to $V = 100$ meV.

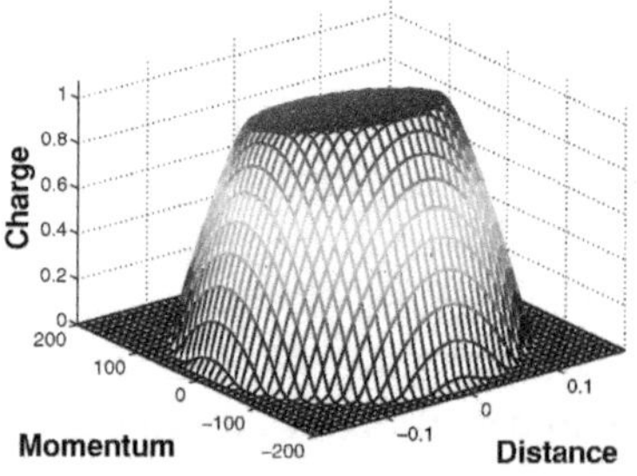

Figure 4. On increasing voltage beyond $V = 100$ meV, the excess charge grows from its background value to the saturated value allowed by Pauli principle. The saturation characterized by the voltage bias $V = 150$ meV which is accompanied by the triple well potential.

The charge yield (shown on the plots 3 and 4) is related to the mean field that detunes the dot level from the resonance. The detuning also includes the frequency shifts due to the bonding interaction in the harmonic approximation near the potential minimum. The use of the Lennard-Jones potential could be made at longer distances, where the attraction to the surface is created by the van der Waals or Casimir-Polder potentials. This "spontaneous" interaction is relevant for the shuttle with a large amplitude of vibrations close to electrodes' surfaces.

At moderate deviations from the equilibrium, we can chose the adiabatic potential of the movable quantum dot as detailed above. The first part is the potential energy U_{pot} of the dispersion interaction in the harmonic approximation. The second part accounts for the electrostatic energy of charging. By definition, it is the minimum energy required for adding charge Q to the quantum dot:

$$U_{adia}(Q) = U_{pot} + U_Q + U_{ext} \tag{27}$$

where the Coulomb and the external potentials are $U_Q = U_c\, Q^2$ and $U_{ext} = -\, Fx\, Q$, respectively. The electro-chemical potential of the dot is the difference between the adiabatic potentials of the adjacent charge states, different by one electron

$$\mu_{dot} = U_{adia}(Q+1) - U_{adia}(Q) = 2\, U_c\, Q - F\, x + U_c. \qquad (28)$$

The factor 2 in the electro-chemical potential of the quantum dot is to be absorbed in the mean charge Q , which accounts for the spin degeneracy in the Eq. 26. The energy U_c shifts the resonance level. Without loss of generality, we assume that the dot energy level ε_0 aligns with the Fermi energy of the reservoirs. Then from Eq. 28, we obtain $U_{mean} = U_c\, Q - F\, x$. For the self-consistent potential of the mean field. The adiabatic potentials modify the conductivity (see Eq. 4) in the x, p channel such that Landauer's formula relates the average charge with the average current through the quantum dot. This relationship is the central aim of any transport measurement theory. Eq. 26 achieves this aim in the mean field approximation, in the spirit of the famous Weiss theory of magnetic susceptibility.

Eq. 26 has two mathematical properties important for the adiabatic motion. Firstly, the dot charge in units e has, for sure, to be between zero and one. This circumstance helps one find an approximate solution in the form of a converging power series over Q, provided $u = U_c/k_B T \leq 1$. On the other hand, the decomposition over exponent power e^{uQ} can be employed for $u \geq 1$. Secondly, if the thermal smearing is marginal, the dot population at the point x, p of the phase space switches quickly between the equilibrium states 0 and 1. The solution is therefore expected to be expressed in terms of the step-like Fermi functions or their derivatives.

One totally disregards the Coulomb blockade by neglecting the mean field of charging in Eq. 27. In the zeroth approximation, the $Q(x,\, p)$ is given by the sum of two Fermi distributions in the reservoirs

$$Q^{(0)}(x,\, p) = \frac{1}{\Gamma}\left\{\Gamma^r(x)\, f_r\left[E_0(x,\, p)\right] + \Gamma(x)^l\, f_l\left[E_0(x,\, p)\right]\right\}. \qquad (29)$$

each of which contributes proportionally to its coupling Γ^σ with the dot. Hence the charging becomes considerable for $V \geq \Omega$, provided $V,\ \Omega \gg kT_B$. This means that the magnitude of the bias voltage $e\, V$ has to be greater than the zero-point energy of the dot vibrations (with amplitude $x = x_0$ and $p = p_0$). Therefore

$$Q^{(0)}(x,\, p) \approx \exp\left(\frac{\dfrac{e\, V}{2} - E_0(x,\, p)}{k_B T}\right). \qquad (30)$$

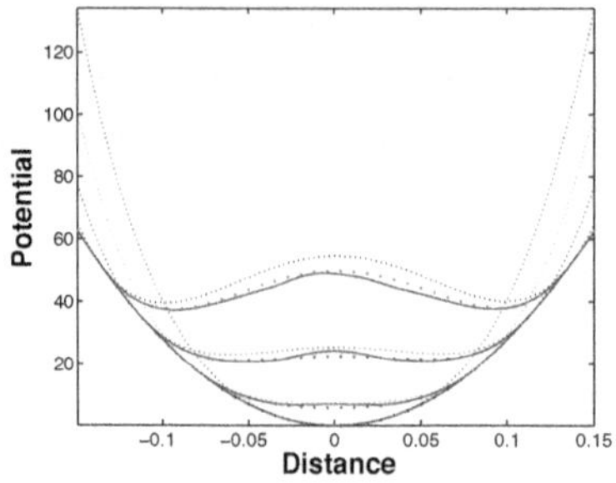

Figure 5. The tunnel terms at $\Omega = 5$ meV, $k_BT = 0.5$ meV, $U_c = 50$ meV. The four curves correspond to $V = 0$, 25, 50, 75, 100 meV. Thin lines refer to the exact numerical solution, dotted lines to Eq. (32) without logaritmic correction.

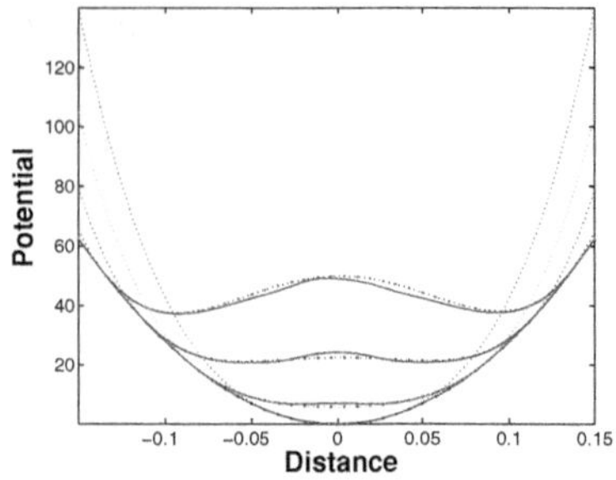

Figure 6. The same graphs but including the first logarithmic corrections for Coulomb blockade. The points on both the plots correspond to the numerical solution in the "Boltzmann limit" where the saturation of the resonance level is ignored.

This necessary condition corresponds to the onset of shuttling and has been found in Ref. [Fedorets 2002]. For $u \ll 1$, however, Eq. 30 can be utilized only at the beginning of the instability, because in this approximation the charge is independent of U_c, which does not account for the Coulomb blockade effect.

The molecular SETs are characterized by quite the opposite condition of strong blockade, $u \gg 1$, where the Coulomb energy $U_c \sim e^2/(4\pi\ r)$ is dominant. For instance, for C_{60} molecular radius $r \sim 3.5$ nm, we obtain that the Coulomb energy $U_c \sim 100$ meV is larger than the vibrational energy $E_0 \sim 5$ meV, and the thermal spread $k_BT \sim 0.5$ meV. We have therefore to keep the charge Q in the exponent, instead of casting it into a power series. This approximation is equivalent to the Boltzmann limit for the Fermi reservoirs, where the saturation of population $f \sim 1$ is disregarded and the charge distribution obeys the nonlinear equation

$$Q(x,\,p)\,e^{\frac{U_c\,Q(x,\,p)}{k_BT}} = 2\,\frac{\cosh\left(\dfrac{e\,V}{2\,k_BT} - 2\,\gamma x\right)}{\cosh 2\,\gamma x}\,e^{-\frac{E_0(x,\,p)}{k_BT}}. \tag{31}$$

An iterative solution can be obtained with the help of a converging series of logarithmic corrections, the first two of which read

$$Q(x,\,p) \approx \frac{e\,V - 2\,E_0(x,\,p)}{2U_c} - \frac{k_BT}{U_c}\ln\frac{e\,V - 2\,E_0(x,\,p)}{2U_c}, \tag{32}$$

being derived from Eq. 31 under the typical conditions of molecular SET $e\,V \gg k_BT$, $\gamma\,x \ll 1$. Comparison of the approximative solutions with the exact numerical calculations of Eqs. 31 and 26, shown in Figs. 5 and 6, demonstrates the good accuracy of the Boltzmann limit. This concerns the logarithmic corrections Eq. 32 as well, until the bias voltage reaches a value of the order of the Coulomb energy, and $E_0 \sim e\,V/2$.

4.2 The adiabatic Hamiltonian

Dynamics of mechanical motion of the dot in x, p phase space is governed by the Hamiltonian

$$H(x,\,p) = K_{in}(p) + U_{adia}(Q) = \frac{p^2}{2M} + \frac{M}{2}\,\Omega^2 x^2 + U_c\,Q^2 - F\,x\,Q. \quad (33)$$

By substituting the Eq. 32 into Eq. 33 and omitting the small logarithmic correction one obtains the adiabatic potential

$$U_{adia}(x) = \frac{M\Omega^2}{2}\left(1 - \frac{e\,V}{U_c}\right)x^2 - \frac{F\,x}{2U_c}\left(e\,V - M\,\Omega^2\,x^2\right) + \frac{M^2\Omega^4}{4U_c}\,x^4, \quad (34)$$

for $p = 0$, which has either a single- or double-well shape, depending on the bias voltage. The threshold voltage, at which the frequency of vibrating dot becomes zero, coincides with the Coulomb energy U_c. At this threshold, the original solution of the classical equations of motion near the potential energy minimum at the point $x = 0$ becomes instable. The bond symmetry is therefore broken by the bias voltage due to the dot charging and the Coulomb blockade. The single-well potential is hence transformed into a double-well potential with new stable points locating far from the original point of equilibrium. This broken symmetry is shown in Figs. 7 and 8, obtained from the exact nonlinear Eq. 26. Figs. 5 and 6 also illustrate this effect in the Boltzmann

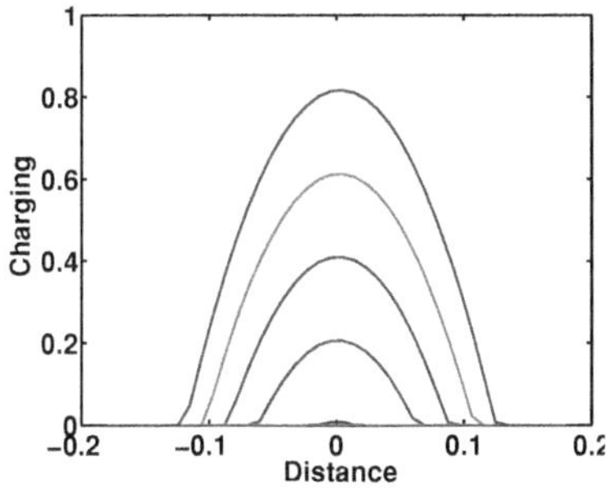

Figure 7. The charge distribution (at the zero p slice) at the bias voltages at $V = 0$, $25, 50, 75, 100$ meV

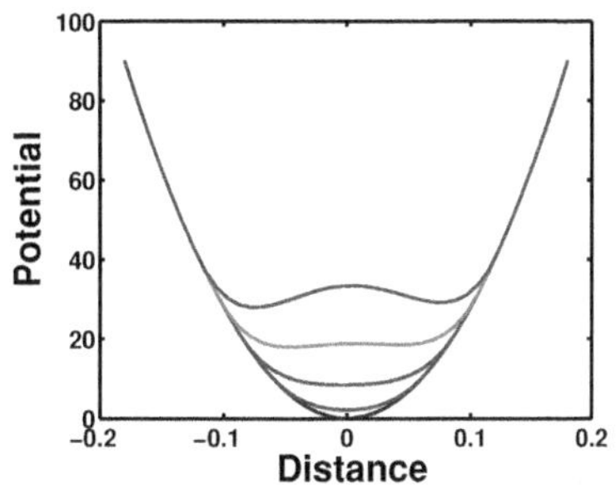

Figure 8. The adiabatic potential $U_c(Q)$ at the same voltages $\Omega = 5$ meV, $k_B T = 0.4$ meV and $U_c = 50$ meV.

limit of Eq. 31, where the analytical solution of Eq. 32 is compared with the logarithmic approximation.

4.3 Tunnel terms symmetry and phase space structure of the dot motions

Now we are in a position to discuss the classical trajectories of the quantum dot driven by the bias voltage. To this end, the charge distributions $Q(x,\ p)$, plotted in Figs. 3 and 4 are substituted into Eq. 33, and the resulting adiabatic Hamiltonians in phase space are shown in Figs. 9 and 10. In the thermodynamic limit, the dot tends to occupy the minimum-energy valley due to irreversible relaxation. The valley topology strictly depends on the bias voltage. The dot trajectories are running towards the center of the circle, $x = 0$ and $p = 0$, until the first threshold is reached.

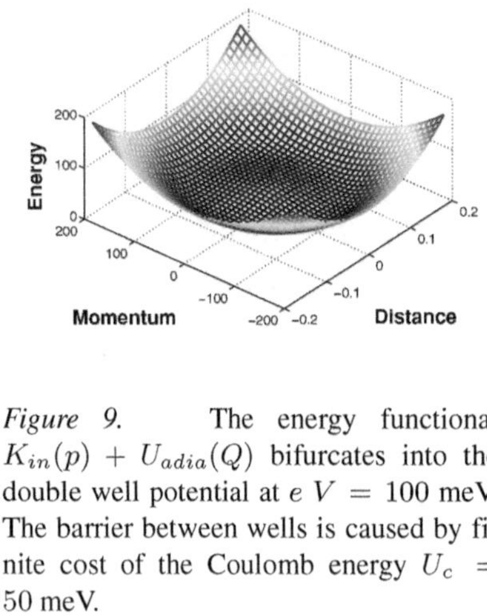

Figure 9. The energy functional $K_{in}(p) + U_{adia}(Q)$ bifurcates into the double well potential at $e\ V = 100$ meV. The barrier between wells is caused by finite cost of the Coulomb energy $U_c = 50$ meV.

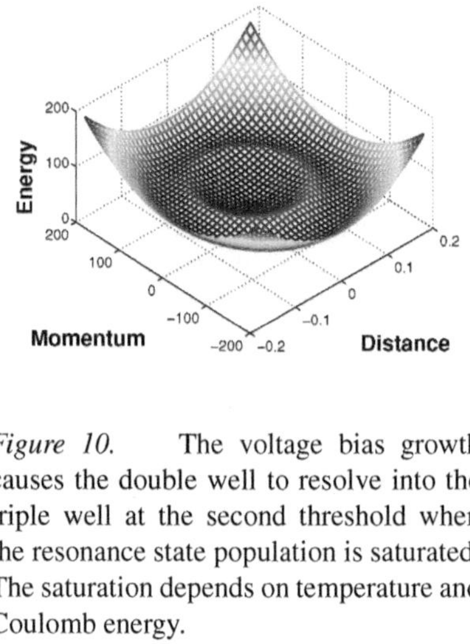

Figure 10. The voltage bias growth causes the double well to resolve into the triple well at the second threshold when the resonance state population is saturated. The saturation depends on temperature and Coulomb energy.

The source-drain voltage specified below breaks the symmetry of the single well potential (for the case of the ground state frequency $\Omega = 5$ meV it is shown in Fig. 9). The center of attraction at the point $x = 0$, $p = 0$ becomes unstable: the fixed points settle outside the circle, where the dot trajectories are winding. The initially compact dot distribution in the x, p phase space plane spreads into a crescent-shaped "moon", and, then into a ring distribution of a radius found from Eq. 34 by setting $\partial U_{adia}(Q)/\partial x = 0$. If the bias voltage is larger than the Coulomb energy U_c, this condition yields new stationary points of equilibrium $x^2 = (V - U_c)/(2M\ \Omega^2)$, corresponding to the minimum potential energy $U_{adiab}(Q)$. Furthermore, with the increase of the bias voltage the bond symmetry is broken again. The persistent Coulomb blockade produces the triple-well potential by combining the single and double wells into unified potentials as shown in Figs. 9 and 10. The multiple-well structures can not be inessential for the quantum dynamics. The coherent tunneling of electrons entails coherence between the wells, as will be discussed in Sec. 8. However, before plunging into the quantum theory we wish to turn our attention to the classical aspects of the system noise.

5. The shot noise of the shuttling instability

At the voltage threshold $V \sim \Omega$ the dot begins to oscillate with a larger amplitude between electrodes thus augmenting the current and its fluctuations. For the molecular SET, an explanation of the current-voltage steps has been attempted in terms of the Frank - Condon transitions [Boese 2001]. The shuttling mechanism in the coherent regime [Fedorets 2002] has been advocated in the works of Shekhter's group in Chalmer's [Gorelik 1998]. However, it seems that the kinetic models of incoherent tunneling offer the best fitting [Braig 2003; Flensberg 2003; Braig 2004] of the current-voltage curves for the C_{60} SET [Park 2000].

To resolve the dilemma, one needs an additional analysis of the conduction mechanisms. For instance, the amount of electrons transported per period has been estimated on the basis of the known source-drain current $I \sim 100$ pA [Park 2000]. However, when taking into account the frequency $\Omega \sim 1.2$ THz of the "bouncing-ball mode", it turned out that the C_{60} transistor transmits only a small fraction $q \sim 10^{-3}\ e$ of the elementary charge e per vibrational period $\tau = 2\pi/\Omega \sim 10^{-12}$ s. This excess charge is too small to compete with and to affect the shuttling dynamics of the C_{60} SET.

In principle, an objection based on the broken bond symmetry model can be raised, because the frequency of dot vibrations is softened to zero at the onset of the shuttling instability. However, as shown above, this could occur at a bias voltage of the order of the Coulomb energy $V \sim U_c \gg \Omega$, which is considerably larger than the observed conductivity gap Ω [Park 2000]. The current by itself cannot instruct us unambiguously about its transport mechanisms. This suggests one should try more advanced tools dealing with fluctuations rather then just with the mean current. For instance, the noise spectrum or a more general full-counting statistics could reveal more subtleties. In particular, this also concerns the Fano factor, which measures the ratio of the actual noise of fluctuation spectrum and the Poisson shot noise produced due to the tunneling of isolated independent electrons $F_{Fano} = S_{noise}/S_{Schottky}$. This Poissonian limit, also termed as Schottky noise, has been obtained for a very low transparency $\Upsilon \ll 1$ and is proportional to the conductivity Eq. 8. We consider the resonance case, for which the dot oscillates on an adiabatic trajectory $x = A\ \cos \Omega t$ with the transparency

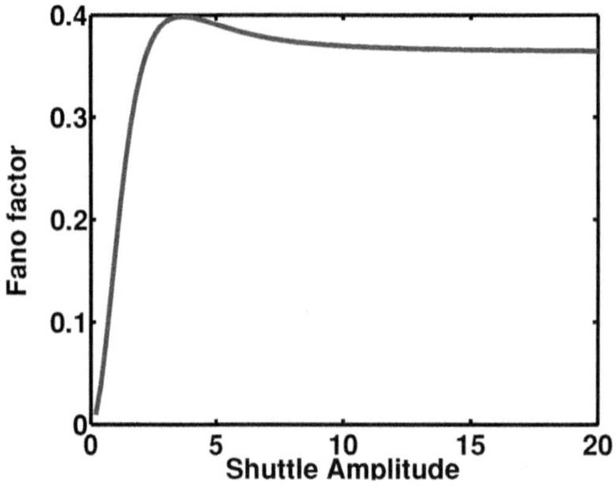

Figure 11 The Fano factor as function of amplitude A of the dot oscillation. The static dot ($A = 0$) has the resonance transparency $\Upsilon = 1$ and is characterized by zero noise. The Fano-factor grows fast until the A reaches the threshold value of $A_{thr} \sim \gamma^{-1}$ the Fermi length. After, that the shuttle mechanism dominates enhancing the transparency and correspondingly decreasing the noise slightly to keep it below of the Poissonian limit.

$$\Upsilon(\varphi) \sim \frac{\Gamma^r(\varphi)\,\Gamma^l(\varphi)}{\Gamma(\varphi)} = \frac{\Gamma_0}{\cosh(2\,A\,\gamma\,\cos\varphi)} \tag{35}$$

where $\phi = \Omega t$. The Schottky noise is proportional to the mean current (see Eq. 18) averaged over the dot trajectory:

$$\langle I \rangle = \frac{e^2 V}{h} \int_0^{2\pi} \frac{d\varphi}{2\pi}\,\Upsilon(\varphi). \tag{36}$$

The main contribution to this integral comes from the neighborhood of $\phi = \pi/2$ where the transparency is maximum, and the noise is minimum according to

$$S = \frac{2\,e^3 V}{h} \int_0^{2\pi} \frac{d\varphi}{2\pi}\,\Upsilon(\varphi)\,[1 - \Upsilon(\varphi)]\,. \tag{37}$$

The Fano factor F_{Fano}, (see Fig. 11) as a function of the shuttle amplitude is not a monotonic curve. At rest, $A = 0$, the noise is minimum for resonance tunneling since the dot energy level is not broadened and the transparency is ideal $\Upsilon = 1$. With increasing A the F-factor rapidly grows towards the threshold value $A \sim 1/\gamma$, and thereafter one is in the shuttling regime. The electron transport favors a decrease in the F-factor and augments the noise amplitude A by a value larger than the Fermi length $\lambda \sim 1/\gamma \sim 1$Å. Fig. 11 shows the Fano factor as a function of shuttling amplitude and illustrates that the better the transparency the lower the noise, as it should be for a quiet electronic sea.

This analysis shows that shuttling is hardly feasible in the reported experiments [Park 2000]. The first reason is the AOC that is caused by the tunneling at a small gate voltage. Growth of the local environment temperature destroys the Coulomb blockade, which could lead to charge accumulation and shuttled transport. The second reason is a large gating of the C_{60} transistor that removes the AOC and thermalizes the SET at low temperature $T \sim 4$ K. Nevertheless, the amplitude of the thermal vibration is still too small to activate the shuttle mechanism at distances comparable with the Fermi length $\lambda \sim 1Å$. The threshold of shuttling $A_{thr} \sim \lambda$ can not be reached even at $e\,V \sim \Omega = 5$ meV for the C_{60} transistor.

If zero-point vibration amplitudes of the dot are comparable with the Fermi length of the electrons, the shuttling takes place at small bias voltage. This is the case for cold dots. The constructive interference of electron waves in the tunnel gap center effectively charges the dot. In the quantum limit, this charging requires a justification of the tunnel-term concept based on the Schrödinger equation. In next section we address a more rigorous quantum mechanical picture based on the "ab-initio" SET model.

6. The Born-Oppenheimer approximation for the tunnel curves

In previous section, by considering the electrostatic energy of the quantum dot charging we have determined the tunnel curves using the phenomenological approach. A strict definition of the tunnel curves as total electronic energy at a fixed dot location between leads is implied by the Born-Oppenheimer adiabatic strategy. For the quantum-mechanical computation of the tunnel curves, the information about (1) the spatial profile of electrostatic potential and (2) the electron and ion distributions of the SET is required as an input.

The electrostatic interaction of the quantum dot with the leads is not weak due to charge images. Metallic leads screen their Coulomb interaction by creating charge holes in the electron reservoirs and by breaking the translation symmetry of the electron-electron and the electron-hole interactions. The quantum dot is subject to the dipole and multipole forces, which derive from the tunnel terms. Besides that, the screening diminishes the threshold of cold emission. The tunneling rate increases by three-orders of magnitude [Flügge 1971] as a result. We therefore have to consider the quantum dot and the leads as a strongly interacting combined system when computing the resonance tunneling as a function of the dot position in the tunnel gap.

The Green's function of the electrostatic field inside the double junction obeys the 3D-Poisson equation:

$$\left(\partial_x^2 + \partial_y^2 + \partial_z^2\right) G(x,\, y,\, z,\, x_1,\, y_1,\, z_1) = \delta(x - x_1)\, \delta(y - y_1)\, \delta(z - z_1). \quad (38)$$

The metallic electrodes are implied to be equipotential surfaces at which the boundary conditions read $G(\pm D/2,\, y,\, z,\, \pm D/2,\, y_1,\, z_1) = 0$, for all points y, z, y_1, z_1. The solution can be represented as a sum of two components. The first one satisfies the homogeneous Poisson equation with the boundary condition provided by the voltage applied. This part gives us an external potential resulting in a Stark shift. The second part is responsible for the charging and for the Coulomb electron-electron interaction. The potential field of the charge obeys the Poisson equation, which has a delta-source on the right hand side of Eq. 38 at zero boundary conditions. The solution of the inhomogeneous equation is given by a two-fold Fourier integral. The transversal isotropy of the tunneling electron flow allows one to reduce it to a one-fold integral

$$\varphi(\chi,\, \chi_1) = 4\, e^2 \int_0^\infty d\lambda \left[\frac{J_1(\lambda\, a)}{\lambda\, a}\right]^2 \frac{\sinh\left[\lambda\left(\frac{D}{2} + \chi_<\right)\right]\, \sinh\left[\lambda\left(\frac{D}{2} - \chi_>\right)\right]}{\sinh(\lambda D)}, \tag{39}$$

which describes the field between the tips with a charge located at the point χ_1, where $\chi_> = \max(\chi,\, \chi_1)$, $\chi_< = \min(\chi,\, \chi_1)$. Here $J_1(a)$ is the Bessel function, and the tunneling radius a equals the molecular van der Waals radius of the dot. The electron tunneling from the electrode tips occurs uniformly over a sphere of this radius.

The charge density of the dipole of the length a is $n_d(\chi,\, x) = F[\delta(\chi - x + a) - \delta(\chi - x - a)]$ where the $F = \alpha V/D$ and α is the dot polarizability. The electronic density $n_e(\chi,\, x)$ of scattering waves can be found from the wave function $\psi_{E,\,\sigma}(x)$ constructed from the transmitted and the reflected plane waves

$$\begin{aligned}
\psi_{k,\,l}\left(x < -\frac{D}{2}\right) &= e^{ik\left(x+\frac{D}{2}\right)} + R_{k,\,l}\, e^{-ik\left(x+\frac{D}{2}\right)}, \\
\psi_{k,\,r}\left(x > \frac{D}{2}\right) &= e^{-i\,k_r\left(x-\frac{D}{2}\right)} + R_{k,\,r}\, e^{i\,k_r\left(x-\frac{D}{2}\right)}, \\
\psi_{k,\,l}\left(x > \frac{D}{2}\right) &= T_{k,\,l}\, e^{i\,k_l\left(x-\frac{D}{2}\right)}, \\
\psi_{k,\,r}\left(x < -\frac{D}{2}\right) &= T_{k,\,r}\, e^{-ik\left(x+\frac{D}{2}\right)},
\end{aligned} \tag{40}$$

where the partial channels $k_l = k_r = \sqrt{2E + e\,V}$ and $k = \sqrt{2E}$ are averaged over the thermal reservoirs of the leads. This yields

$$n_e(\chi,\, x) = \sum_{\sigma=r,\, l} \int_0^\infty dE\, f_\sigma(E)\, \left|\psi_{E,\,\sigma}(\chi,\, x)\right|^2. \tag{41}$$

The net charge distribution $\rho(\chi,\, x) = n_e\,(\chi,\, x) - n_d(\chi,\, x)$ can be used for computing the energy functional of the electron-electron and electron-hole interactions in the tunnel gap. The potential energy of the induced charges as a function of the dot location is proportional to the integral

$$P_H(x) = \frac{1}{2}\int_{-D/2}^{D/2} d\chi \int_{-D/2}^{D/2} d\chi_1\, \rho(\chi,\, x)\, \varphi(\chi,\, \chi_1)\, \rho(\chi_1,\, x). \tag{42}$$

which measures the electron and hole overlap. It is related to the rate of the tunnel transitions by the detailed balance condition between the Fermi reservoirs and the dot. The kinetic energy also depends on the dot location and can be calculated as

$$K_{el}(x) = \sum_{\sigma=r,\, l} \int_0^{\infty} \frac{dE}{2}\, f_\sigma(E) \int_{-D/2}^{D/2} d\chi \left|\frac{\partial\, \psi_{E,\,\sigma}(\chi,\, x)}{\partial\chi}\right|^2. \tag{43}$$

Both the electrostatic and the kinetic parts combine into the total electron energy. As a function of the coordinate x, this energy plays the role of a mechanical potential

$$U_T(x) = [K_{el}(x) + P_H(x)] \left[\int_{-D/2}^{D/2} d\chi\, n_e(\chi,\, x)\right]^{-1}. \tag{44}$$

of the quantum dot in the classically forbidden region of the electron motion. This adiabatic potential is called a tunnel curve, by analogy to molecular curves in the Born-Oppenheimer theory. The tunnel curve depends on the location of the quantum dot in the tunnel gap and on the Fermi distribution of the electronic reservoirs. The electron density in the quantum dot is obtained by solving the Hartree equation for the electron wave function

$$-\frac{1}{2}\frac{\partial^2\, \psi_{E,\,\sigma}(\chi,\, x)}{\partial\chi^2} + [U_{ext}(\chi) + U_H(\chi,\, x)]\, \psi_{E,\,\sigma}(\chi,\, x) = E\, \psi_{E,\,\sigma}(\chi,\, x) \tag{45}$$

with the boundary conditions at the electronic reservoirs, where the potential $U_H(\chi,\, x)$ reads

$$U_H(\chi,\, x) = \int_{-D/2}^{D/2} d\chi_1\, \varphi(\chi,\, \chi_1)\, \rho(\chi_1,\, x). \tag{46}$$

One should also invoke Fermi statistics. A typical tunnel curve is shown in Fig. 12 for SET model with $D = 14$ a.u., $a = 1$ a.u., the work function of electrodes $W = 0.4$ a.u., the Fermi energy $E_F = 0.2$ a.u., and the polarizability $\alpha = 200$ a.u. (of Na atom). The potential drops near the interface of the source-drain electrodes, as it should for the ballistic regime. The tunnel curve has a single shallow well at a small bias voltage. When the latter increases, the well becomes deeper, and the dot is attracted to the inter-electrode gap center

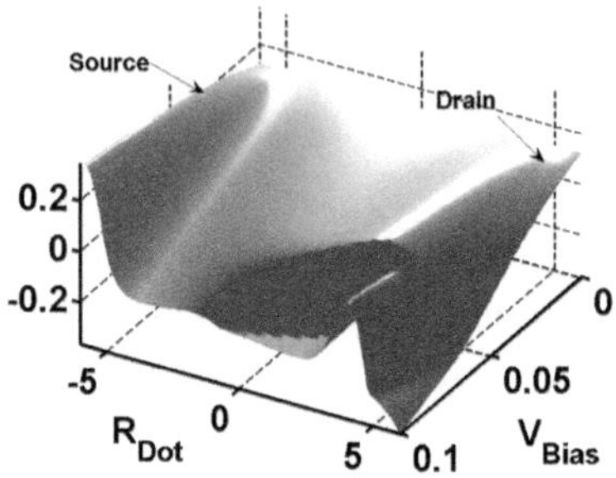

Figure 12. The typical tunnel curve of the dipole dot as function of its coordinate and the source-drain voltage bifurcates at a threshold voltage: a single well is replaced by a double well followed by a wide well.

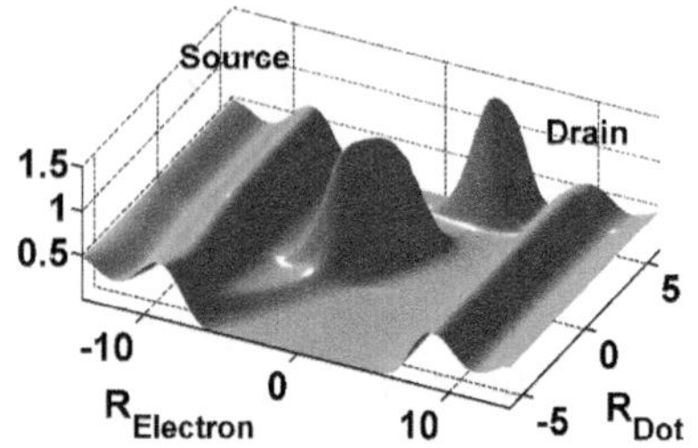

Figure 13. The electron density distribution inside SET displays the Friedel oscillations on the Fermi edges and the peak of resonance tunneling in classically forbidden region in the inter-electrode gap.

due to resonant tunneling. When the bias voltage grows further, the second well appears near the drain electrode. Then, the wells are broadened and combined into a single wide potential thus giving rise to the shuttle mechanism of conductivity, as displayed in Fig. 12. This figure justifies, in principle, the broken-symmetry mechanism discussed in Sec. 4 in the framework of a phenomenological model. The electron density as a function of the electron and the dot coordinates is plotted in Fig. 13.

The edge electron density is subject to Friedel oscillations due to interference between the incident and the reflected electron waves. The electrons in the conducting bands of the leads are treated within the jellium model as shown in Fig. 13. The transmitted electronic waves of the source and drain electrodes can interfere constructively on the quantum dot, which is located near the inter-electrode center of the sequential tunnel junctions. The Fermi electrons create a maximum density on the dot, provided the condition of resonance with the dot quantum level is fulfilled. This effect dominates in the tunnel gap center, as well as in close proximity to the drain electrode, which screens a negative charge more effectively thus serving as a barrier to electron tunneling. When the bias voltage increases, the resonance level of the dot may shift downward the transparency window allowed by the Pauli principle, and that may entail a negative differential conductivity in concert with suppression of the electron interference, as featured in Figs. 14 and 15.

7. Tunneling optical traps

Recent progress in the atomic microchips industry, has stimulated great interest in studies of neutral ultracold gases [Lin 2004]. The ultra cold atomic samples are typically produced in magneto-optical traps, then loaded into ei-

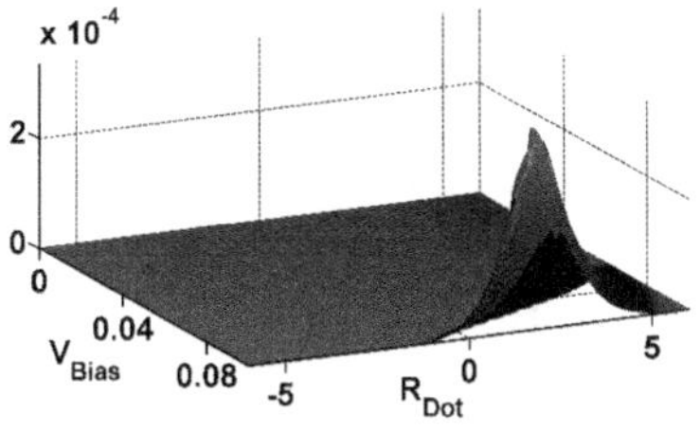

Figure 14. The current through the dipole as function of it position inside the tunnel gap and bias voltage. The maximum tunneling current is mounted at the gap center where the constructive interference of electron flows takes place.

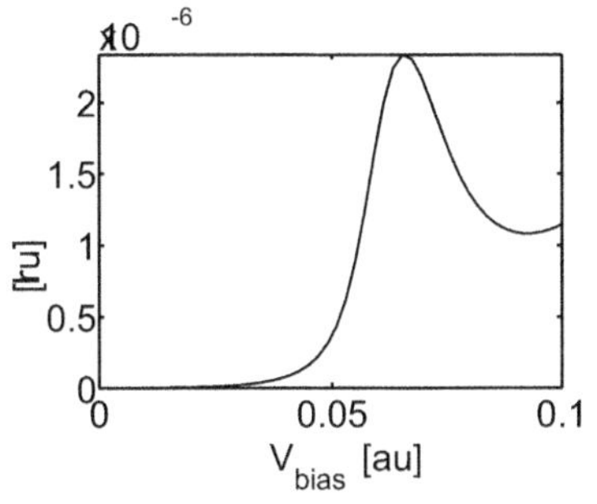

Figure 15. The current voltage curve of the tunnel transition via the dipole. Averaging of it coordinate between electrodes is given classically with the Gibbs distribution. The step increase of current is due to shuttling.

ther stable microtraps or atomic guides, which employ microfabricated structures of current-carrying wires at surface substrates. These chips offer great promise for continuously improving the functioning of a variety of atomic-optic devices, including matter-waves interferometers, double well potentials for atomic Josephson junctions, and Fabry-Perot resonators for coupling cavities with atoms. However, a number of physical mechanisms stipulate fundamental limitations and hinder immediate applications of these devices.

The micro-devices require large gradients of magnetic fields for manipulation of cold atoms. Thus, large currents, up to 1 A, create significant magnetic forces in close proximity ($\sim 1\ \mu$m) to surfaces. Apart from collisional losses in a traped thermal atomic cloud, strong thermal- and Johnson-noises due to current fluctuations inevitably induce additional losses, by heating and pushing atoms to the surfaces. Recently created combined magnetic and electrostatic traps [Kruger 2003] suffer from similar shortcomings: up to 100 V-high bias voltages applied between μm-scale distant electrodes are also accompanied by trap fluctuations that prevent an accurate deposition of the atoms close to the electrode surface.

Theoretically, the electro-optical trapping (TOT), shown in Figs. 16 and 17, is a much more appealing scheme for maintaining and manipulating atoms as compared to the MOT, since the electric coupling dE_s or $e\ \phi$ in such traps does not involve the atomic fine structure and is much stronger than the magnetic coupling. This reduces by many orders of magnitude the required voltage applied between the electrodes in order to create a trap of a same depth. Moreover, the TOT is free from the Majoranna spin-flip losses. These facts could enable one to eliminate the most of noises[7] and to improve the trap control.

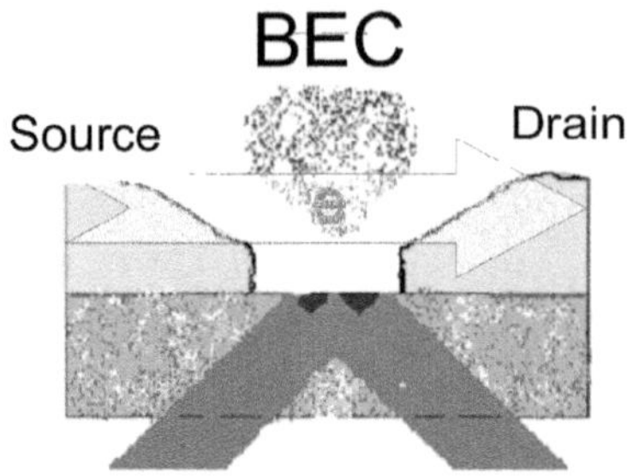

Figure 16. Schematics of tunneling optical trap (TOT): the evanescent field of the blue detuned light repels the atoms from the laser irradiated substrate and the electrical terminals.

Figure 17. The current carried terminals on surface tend vice versa to attract the atoms to the inter-electrodes zone of maximum field. The TOT current is easily controlled by the evanescent light that plays the role of gate electrode.

Contemporary standards for labs and industry require construction of TOT in planar geometry based on the evanescent laser fields. The dipole potential for evanescent waves reads:

$$U = -\,0.5\,\alpha_s\,E_s^2. \tag{47}$$

where the sign of the polarizability α_s depends on the sign of the detuning from resonance. At blue detuning, the field tends to expel the irradiated atoms from the region with high intensity of the light (see appendix). The advantage of blue detuning the evanescent field is that the atoms spend most of the time in the field-free region, and hence they are less affected by spontaneous radiation and heating. The attraction of atoms to the tip's neighbourhood is due to the electrostatic source-drain field. The separation 0.1 - 10 μm between the source-drain electrodes allows the elimination of leakage currents and to facilitate cooling by electron transport through the resonance states of the TOT atoms.

Up-to-date techniques of cooling, including radio-frequency evaporation, optimal control of current, degenerative feedback, and, last but not the least, adiabatic laser cooling [Chen 1992], have been purposely developed, and now they enable one, in principle, to maintain the ultra-cold atoms in a TOT. Sisyphus cooling by the blue-detuned light may provide the required dissipation in the TOT and further reduce a diffusive heating produced by the electron current and atomic collisions. With the repulsive light forces, which push atoms towards the dark regions near the trap center where the radiation losses should be minimal, the cold atoms can be significantly compressed adiabatically, thus yielding a background-free sub-10-nm spot [Khaykovich 2000]. In addition, constructive electron interference in these regions provides a maximum transparency of the resonant tunneling, which is required in order to minimize losses due to current driven heating of the atomic cloud.

Moreover, delivering the already cooled atoms close to the surface does not pose a serious problem. The most routine method of loading is cold-cloud transport directly from MOT to TOT. Loading atoms on the fly by photodesorption from the surfaces of a glass cell has been demonstrated recently [Atutov 2003]. In order to minimize radiation heating and collisional losses, the ultra cold atoms can be isolated in a dark spot near the tunnel gap leads and the substrate. In fact, a thermal cloud exhibits loss at a distance larger than the size of a compact condensate, because the proximity to the surface can provoke cloud evaporation[Harber 2003; McGuirk 2004]. Then, the interface can be used to selectively absorb higher energy atoms. Recent MOT experiments [Lin 2004] demonstrate that by tuning the potential it is possible to bring ultra-cold atoms to a distance 0.5 μm-close to the surface. This is the case where the TOT can manifest its excellence.

In addition, the evanescent field of the TOT is concentrated near the electrode apex, thus providing an additional gain in repulsion of the cold atoms to the dark spot between the electrode tips. Therefore, the TOT protects the quantum dots better against surface losses. Moreover, if the de Broglie wave length of atoms is larger than the correlation length of the surface roughness, reflection of the cold remnants occurs elastically. A movable quantum dot starts to oscillate, being trapped between the voltage biased tips. The adiabatic dynamics, in which the repulsion of the evanescent laser field compensates the electrostatic attraction to the tips, mimics Franklin's Bell oscillations with the shuttling mechanism of conductivity through the charge BEC.

The strength of attraction between the dot and the source-drain electrodes depends on the ground state polarizability, α_s. For example, the α_s is about $80[\mathrm{mHz/V^2/cm^2}]$ for Rb^{87}. For a μm-scale inter-electrode gap, $D \sim 1$ μm, the voltage bias $V \sim 30$ mV produces an external field pulling the atoms into a $U_{trap} = \alpha_s(V/D)^2 \sim 8$ KHz-depth trap. For an atom de Broglie length $\lambda_{dB} \sim \hbar/\sqrt{2M\, U_{trap}} \sim 0.1$ μm and a trap size of the order of D, we expect that approximately $N = (D/\lambda_{dB})^3 \sim 10^3$ atoms can be put into the quantum degenerate regime, provided their protection against sticking and colliding with surfaces is efficient. The repulsion potential of the atoms is due to the blue-detuned evanescent field. The corresponding dipole force compensates for the electrostatic attraction of the charged ultra-cold particle to the leads and to the substrate.

The density of the atoms, which should approach the leads as close as possible without sticking, significantly influences the TOT resistance R. Therefore, the repulsion of atoms has to be controlled by the attenuation of the evanescent

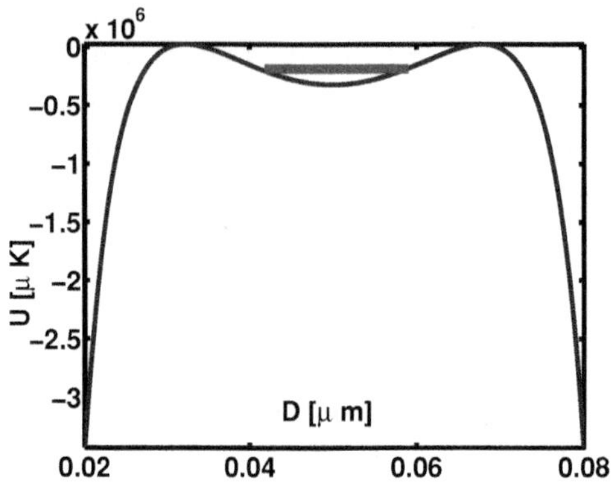

Figure 18 Schematics of electro-optical potential including the Casimir-Polder attraction of Eq. (48) for ^{87}Rb atom between leads. The evanescent field, that decreases exponentially on it half wave-length expels atoms away of the surface to the location where attractive forces balance the light pressure. The line in the well cartoons the trap occupation.

laser field. Typically the laser beam of 1 - 100 mW power can be focused on the interface to allow ultra-cold atoms to levitate above it.

Levitation of Cs atoms in the evanescent field has been demonstrated in Ref. [Hammes 2002], where the the optical dipole potential created by 1 W-laser has been utilized for trapping the atoms far from the surface. The exponential profile of the potential decreases at a half-wave length $\lambda_{blue}/2 \sim 250$ nm. At these distances the optical field compensates for the long-range attraction induced by the electrostatic polarization and the Casimir-Polder potential

$$U_{Casimir-Polder} \simeq \frac{c_4}{R^4}, \quad c_4 \sim 1 \text{ nK.} \, \mu\text{m}^4, \tag{48}$$

which exists owing to spontaneous electromagnetic field fluctuations for neutral atoms and ensures the existence of a stationary point in the net potential field. The Casimir-Polder interaction for a Rb^{87} atom located at a distance ~ 0.5 μm from the surface is equal to the polarization potential Eq. 47 induced by 15 mV voltage of the source-drain terminals. The surface repulsion due to the evanescent field is of the same order of magnitude for other alkali atoms K, Na, Rb, provided the corresponding matter constants and the optical wave lengths are taken into account. For the reader's convenience, in an Appendix we quote the formula for the dipole potentials in the laser field.

Both the optical and the electrical trapping have already been demonstrated separately. The double evanescent wave trap for atoms has been proposed in Ref. [Ovchinnikov 1991] and has been demonstrated in Ref. [Hammes 2002] for Cs atoms. Bezryadin and coworkers [Bezryadin 1997] have reported a method of fabrication for nanoscale controllable break junctions, in which the polarized nano-clusters, such as Pd clusters, have been trapped between the electrodes. The nano-clusters were self-assembled in the region of the maximum field in order to produce a wire connecting the tips [Bezryadin 1997].

Both types of trapping, by optical and by electrical potentials, eventually can be combined into a common protocol of cross-coupling ultra-cold atoms and electrical circuits. Such a cross coupling of the electron current and the ultra-cold atoms in matter wave-guides is proposed here using the example of hollow optical fibre guiding. In this scheme, a laser field, detuned either to the blue or to the red wing of the atomic transition, allows the atoms to be guided through the fibre capillary. The technique has already been routinely used [Dall 2001].

Here we suggest that experimentalists address the ultra-cold metallic gas cloud in the capillary channel *electrically* in order to measure a tunneling I-V curve. To this end, the voltage bias could be led to the cold cloud through a lateral wire, which crosses the bulk of the optical-guide wall and immediately breaks-off inside the capillary. The blue-detuned evanescent field repels the ultra-cold atoms towards the capillary axis. The ponderomotive field potential accelerates the tunneling electrons and facilitates charging of the trapped atoms. This scheme features an effective reduction of the collisional loss and protection from leakage of currents. The optical field protects the electrical reservoirs of the source-drain leads from any disturbances by tunnel electrons. The last but not least advantage is that the ultra-cold atoms are trapped inside the capillary in the dark spot, which is affected neither by heating caused by the spontaneous decay nor by induced radiation processes.

The TOT scheme relying on the wave-guides also has a practical advantage over the planar design, since it requires a smaller number of microfabricated components. The heart of the setup is the quantum dot, which is created by self-assembly of the ultra-cold cloud in a dark spot between the electrodes. The self-assembly due to the optical dipole force and the electrical polarization potential is very useful for fabrication of TOT devices at such a small length scale. This scheme of cross-coupling merits a detailed consideration, since it should allow one to discover the true electrical resistance of atomic BEC.[8] Theoretically, the BEC collective state could be described by a composite order parameter, which is formed by the diatomic molecules and superconducting BCS pairs. It would be advantageous to map the atomic wave coherences to the coherent electric responses (echoes) of the normal or superconducting leads by applying the voltage bias to the quantum degenerate ultra-cold gas in the TOT.

We conclude this section by listing the evident TOT advantages:

- Small density of states, typical of the TOT atoms, reduces the leakage currents, the dissipative tunneling, and noises in measurements.
- High protection of atoms in the TOT from the influence of the environment favors coherence of their electrical response (nutations and echoes) to pulsed bias.

- Controllable double-, triple-, or, in principle, multiple-well potential as well as current oscillations in the atomic Josephson transitions can be achieved.
- Precise control of matter wave interferometry is foreseen.
- Combined addressing of the quantum states of the TOT both optically and electrically is possible.

In brief, we propose the TOT electro-optical setup in which electrical measurements have a high quality factor combined with the coherence of an all optical experiment. The TOT can be easily incorporated in electrical circuits as a nonlinear element ensuring a scalability of the architecture. The quantum dot isolation in the TOT will protect entanglement of quantum states thus permitting field programmable gates arrays.

8. Coherence of electron transport via double wells

Thus far we have made no reference to a wave-like motion of the NEM-SET devices. However, the coherent behaviour of the trapped quantum dots will become increasingly important with the NEM-SET made smaller and colder. The wave function Ψ formalism is useful as a description of quantum dots only when their de Broglie wavelength L_d is comparable with the Fermi length of electrons, that is $L_d\ \gamma\ \sim\ 1$, and we can rely on the tunnel curves. At the temperature when quantum degeneracy occurs, the self consistent charge Eq. 16 transported through the SET junctions via the shuttle mechanism is not smeared over the tunnel terms. As demonstrated in Secs. 6 and 7, a peak of the charge distribution is centered between the electrodes and the Coulomb blockade abruptly breaks this symmetry of the tunnel term when the bias voltage $V \geq U_c$. The single-well bond then bifurcates into a double-well potential, because the barrier between the wells brings a gain in the Coulomb energy. The corresponding self-consistent charge is imposed by the condition of detailed balance between the electron flows and an irreversible coupling Γ_σ.

Tunneling of non-correlated itinerant electrons through the lead interfaces at the rate Γ_σ provides irreversible dissipation, which is required in our approach[9] by the detailed balance condition. The electron flows charge the resonant state of the dot (see Eq. 26) thus forming the tunnel curves U_σ (where $\sigma = l,\ r$), provided the charge is located in the center between the electrodes. The electron scattering and tunneling are the dissipating mechanisms that ensure stability of the adiabatic dots' dynamics. The dots can also relax their kinetic energy by collisional losses, scattering, spontaneous emission, or laser cooling. In quantum gases, binary scattering of atoms usually results in a shift of the energy level, which is proportional to the product of s-scattering length a and the gas density ρ. The interplay of the quantum spreading and scattering of the dots sitting in a single-well potential $U(x)$ is accounted for by the Gross-Pitaevskii equation for the "condensate" wave function Ψ [Dalfovo 1999] as

$$i\hbar\,\dot{\Psi} = \frac{p^2}{2\tilde{M}}\,\Psi + U(x)\,\Psi + \frac{4\pi\,\hbar^2 a}{\tilde{M}}\,|\Psi|^2\,\Psi \tag{49}$$

where the $\hat{p} = i\hbar\,\partial/\partial x$ is the momentum operator, the $\tilde{M}$ is the effective mass of a dot. The scattering length a has to be smaller than the average distance between dots in the trap, this serves as a criterion of validity of the Gross-Pitaevskii equation. The quantum dispersion of the dot $\hat{p}^2/2\tilde{M}$ brings about the major spreading mechanism in the degenerate quantum regime. The electron transport critically depends on the bias voltage V: for $V < U_c$ the adiabatic potential U features a single well, which for $V > U_c$ bifurcates into a double-well potential with degenerate states located in the well minima.

Should one disregard the s-scattering off a small density of states $|\Psi|^2$ in a dilute quantum gas, the resulting linear Schrödiger equation would describe the dynamics of the Ψ-state in the potential U. The Landau bifurcation then emerges from casting the potential $U(x)$ of Eq. 34 into a Taylor series over a small deviation x from the bifurcation point: $U(x) = \tilde{M}(\Omega^2 - \Sigma^2)\,x^2 + \Xi\,x^4$. Parameters Σ and Ξ are defined by the condition of Coulomb blockade. They depend on the temperature, the bias voltage, and the coupling to electron reservoirs, that all influence the criticality of quantum dynamics. The quantum degenerate regime is meaningful when the dot temperature is smaller than the Coulomb energy. The heat supplied to the dots by the bias voltage has to be dissipated in order to reach the degeneracy point. The required relaxation is provided by the radiation channels via the Josephson plasmons of the dot tunneling across the Coulomb barrier. The tunnel transitions split the degenerate energy levels corresponding to the dot motion in the wells' minima of the bifurcated potential. The frequency of splitting is just the Josephson frequency quantum.

Alternatively the quantum dynamics on the tunneling curve can be invoked by the wave functions Ψ_l and Ψ_r of the dots, each of which resides in its own well. The functions Ψ_l and Ψ_r obey the system of equations

$$i\hbar\,\dot{\Psi}_{l(r)} = \hat{H}_{l(r)}\,\Psi_{l(r)} + \Delta_{l(r)}\,\Psi_{r(l)} \tag{50}$$

where $\sigma = l,\ r$, and the quantities $g_\sigma = 4\pi\,\hbar^2\,a_\sigma/\tilde{M}$ are related to the corresponding s-scattering lengths of the matter waves with Hamiltonians $\hat{H}_\sigma = \hat{p}^2/2\tilde{M} + U_\sigma + g_\sigma\,|\Psi_\sigma|^2$. Eq. 50 can be considered as a generalization of the Gross - Pitaevskii equation of the interacting charged Bose gas. Alternatively, it can be obtained from the many-body Hamiltonian of the field theory. The mean field Hartree dynamics is presented in Eq. 50. The tunnel terms $U_{l,\,r}$ are modified by the collisional shifts due to s-scattering of particles in the same well. The $\Delta_{l(r)}$ defined below implies the inter-wells' scattering. For the charged Bose gas the interaction Hamiltonian of the field theory reads

$$\hat{H}_{\text{int}} = \int dx_1 \int dx \; \hat{\Psi}^\dagger(x)\, \hat{\Psi}^\dagger(x_1)\, \varphi(x,\, x_1)\, \hat{\Psi}(x_1)\, \hat{\Psi}(x) \tag{51}$$

where $\varphi(x,\, x_1)$ denotes the Green's function of the Poisson equation for the electrostatic potential in the inter-electrodes zone. We represent the field operator of the Bose field as $\hat{\Psi}(x) = \Psi_l(x) + \Psi_r(x) +$ fluctuating fields, where the average of the fluctuating part tends to zero in the thermodynamic limit. The overlap integral Δ measures the rate of particle tunneling through the barrier. The integral Δ is obtained from the variational derivative of the interaction Hamiltonian Eq. 51

$$\frac{\delta \hat{H}_{\text{int}}}{\delta \hat{\Psi}^\dagger(x)} = \int dx_1\, \varphi(x,\, x_1) \left\langle \hat{\Psi}^\dagger(x_1)\, \hat{\Psi}(x_1) \right\rangle . \tag{52}$$

The dependence of φ on x is slow and we can disregard it as compared to the function of the difference $x - x_1$ modelled by the Dirac δ-function, so we have

$$\Delta_{l(r)} \approx \int dx_1\, \varphi(x_{l(r)},\, x_1)\, \Psi_r^\star(x_1)\, \Psi_l(x_1) \approx \frac{4\pi\, \hbar^2 a}{\tilde{M}}\, \Psi_r^\star(x_{l(r)})\, \Psi_l(x_{l(r)}). \tag{53}$$

The two states Ψ_l and Ψ_r are macroscopically distinct if their separation is of the order of the wavelength. For instance, ultra-cold rubidium Rb atoms oscillating in a TOT with frequency 1 kHz have a wavelength of about 1 μm. The coherence length limits the maximum barrier width that allows atoms to be hybridized between the wells. The model describes screening in the charged Bose gas with Josephson plasmons. The current through the degenerate quantum gas reads

$$J_l = \frac{2e}{h} \left\{ \Psi_l^\star\, \frac{d\Psi_l}{dx} - \Psi_l\, \frac{d\Psi_l^\star}{dx} \right\} . \tag{54}$$

The Josephson current is obtained by multiplying Eq. 50 by the conjugated wave function $\Psi_l^\star$ and integrating by parts. This yields the flow of the charged Bose gas with the current

$$J_l = \frac{2e}{h}\; \Delta\, \text{Im} \left\{ \int dx\; \Psi_l^\star(x,\, t)\, \Psi_r(x,\, t) \right\} . \tag{55}$$

The overlap of the wave functions oscillates at a frequency $\Delta = \Delta_l \approx \Delta_r$. For weak superconductors, the tunneling frequency Δ is controlled by the source drain voltage. The alternating current (ac) is zero below the threshold voltage and it oscillates after the threshold with an amplitude growing with the bias voltage. Existence of ac at the frequency Δ immediately indicates the broken symmetry of the bond potential. The frequency of the current is proportional to the source-drain voltage and, thus, the SET radiates an electromagnetic field. This radiation causes relaxation that ensures a back-action mechanism for establishing an equilibrium in the composite system.

It is worthwhile to note that the charged Bose gas trapped in the double well potential U_σ of Eq. 50 behaves as an inverted Josephson junction (N-S-S-N). The super-current, which accompanies the matter wave coherence, is induced between the degenerate resonance states of the adjacent wells at the frequency of the tunnel splitting $\Delta \ll \Omega$. Note that the shuttling frequency Ω cannot be displayed directly because of a large response time, as is typical of tunnel junctions (whose frequency cutoff is much smaller than the vibrational frequency even for nano junctions). The coherent oscillations of the Josephson current can be observed by virtue of their slow frequency $\Delta \sim V$ which is robustly controlled by the bias voltage.

Summary

The step-wise and negative differential resistance regions of the current-voltage curves observed in the molecular C_{60} transistor [Park 2000] are explained by the field effect, in which the voltage bias of the source-drain leads intensifies the NEM effective temperature. With increasing bias, the field splitting and the inhomogeneous broadening modify the transparency spectra of electron scattering. We have obtained a good qualitative agreement between our simple adiabatic models and the differential conductance observed experimentally. Our formula also displays the crossover between the internal mode ($\sim$ 33 meV) and the bouncing-ball mode ($\Omega \approx 5$ meV) of the differential conductivity in the molecular SET. With further increase of the electrostatic interaction, the shuttling mechanism of conductivity replaces the tunneling regime. Then, the Coulomb energy growth destroys the bonding symmetry of the single well potential. The broken symmetry of the molecular vibrations in back-action cures the shuttling instability. The shuttle regime is characterized by shot noise reduction and by coherence of the oscillating current. The primary quantization picture demonstrates that the current oscillates at the frequency of the ground state splitting Δ and the amplitude of this oscillation grows proportionally to Δ.

We have developed an unified adiabatic approach allowing one to tackle transport problems in traps of different geometry. The magnetic and electrical fields, charge screening, and other factors (a spin-orbit interaction, hyperfine structure, *etc*) can influence the quantum dot paths within an easily tractable Breit-Wigner-resonance approximation for the electron scattering. The utility and universality of the tunnel terms concept are confirmed for the phenomenological and "ab-inition" theories of the shuttling instability.

The shuttling in a quantum gas is relevant to electron transport in the presence of relaxation and the Coulomb blockade. The Coulomb blockade entails broken symmetry of molecular potentials U_σ ($\sigma = l,\ r$) when a single well bifurcates into a double well. Tunneling of the charged Bose gas in a double well between biased electrodes creates a current which is subject to Josephson oscillations. This ac generates an electro-magnetic field and thus providing an additional mechanism of dissipation. Thence, the broken symmetry and coherent oscillation due to molecular vibrations ensure the necessary and sufficient conditions legitimizing the present scenario of bond bifurcation.

The adiabatic theory of electron transport in the Breit-Wigner approximation may be of more than academic interest. It can help one to devise the TOT protection of the NEM - SET systems against decoherence. The TOT technology is better suited for molecular optoelectronics due to a low noise in combination with protection control. The connection between the TOT conductivity and quantum Franklin's Bell paradigm is discussed. The TOT design for avoiding the dissipative "roadblocks", could serve as a road map toward a new generation of optical SET, that should enable electrical non-demolition measurements on the quantum threshold. It would be of great interest to measure the resistance of a TOT comprising of BEC molecules and BCS atomic pairs.

Acknowledgments

With the present paper we pay a tribute to our friend and teacher A. P. Kazantsev. He made the seminal contribution to the now enormous field of activity of mechanical action of light on neutral atoms. He often prophesied the future development for years ahead. Four decades ago he recognized the significance of radiation from accelerated charged particles near metallic surfaces [Kazantsev 1963] that appears to be of importance for dissipative mechanisms in TOT electron transport.

The authors are indebted to colleagues for numerous discussions. In particular we thank A.Bezryadin for his preprint [Bezryadin 1997] which he sent to us, to I. Novobrantsev for keen interest and encouragement, and G. Surdutovich for his valuable remarks and references. We are thankful to RFBR and the program of scientific schools for the financial support. The hospitality of INF in Ferrara (Italy) is greatly acknowledged by one of the authors (A. R.), especially, to S. Atutov and R. Calabrese for a kind invitation. The collaboration with the experimental team at the winters 2000-2001 has led to the ideas developed therein.

GLOSSARY

AOC Anderson orthogonality catastrophe [Anderson 1967]. Zero overlap between ground states of surface vibrations: *e.g.* the original state and charge induced configurations, where phonons are shifted by the tunneling electron.

BEC Bose-Einstein condensate. Degenerate state of an ultra-cold ensemble, *i.e.* a coherent, single-mode, bright atomic source of zero momentum $p = 0$.

MOT Magneto Optical Trap. The well established technique for Doppler cooling and trapping of a thermal cloud of cold atoms.

NDR Negative differential resistance.

NEMs Nanoelectromechanical systems. Composite mesoscopic and nano-scale devices designed for a new functionality.

SET Single-electron transistor. Double tunnel junctions with a central island serving as a gate electrode.

SSET Superconducting Single-electron transistor. The same as SET, but the bulk of the electrodes are superconducting.

TOT Tunneling Optical Trap.

Appendix: Radiation pressure

Demonstration of levitation of micron-sized latex particles by radiation pressure dates back to 1970 in the experiments reported by Ashkin [Ashkin 1970]. The average force accelerating (or slowing down) atoms in a laser field was derived by A.Kazantsev in 1972 [Kazantsev 1974]. Later in 1972-1974 he classified the optical forces as spontaneous, induced and mixed. In particular, it was he who first presented the dipole potentials for velocity broadened lines of resonance atoms in the logarithmic form

$$U = \frac{\delta}{2} \ln \left[1 + \frac{G^2}{\delta^2 + \gamma^2} \right] \tag{A.1}$$

here Ω is the frequency of light, Ω_0 the resonance transition, $\delta = \Omega - \Omega_0$ the detuning from resonance, γ the atomic linewidth, d the dipole moment of the transition (in units $\hbar = c = 1$), E is the laser field amplitude and $G = d\ E$ the Rabi frequency. At large detunning, the dipole potential takes the canonical form (Askaryan, 1962) $U = -\ 0.5\ \alpha\ E^2$, where the optical polarization is $\alpha = -\ d^2/\delta$. This potential repels the atom from antinodes of the blue detuned field $\Omega \geq \Omega_0$. The potential of red detuned light, $\Omega \leq \Omega_0$, vice versa attracts the dots to the field antinodes. Use of both methods allows guiding and trapping of atomic matter waves.

Notes

1. When the electron's phase coherence length L_e, *i.e.* the typical distance for the electron to travel without losing its phase, is of the order of the dot mean free path L_d, such a system is called mesoscopic. The regime ensuring $L_e \gg L_d$ is called ballistic.

2. The coupling of Einstein phonons to finite electron density in conduction band.

3. The notion of the OC [Anderson 1967] was introduced by P. W. Anderson with respect to the Fermi-edge singularity where the overlap of states which differ in the number of holes by one tends to zero in the thermodynamical limit.

4. The fluctuations at thermodynamic equilibrium are related in a universal way to the kinetic response according to the fluctuation-dissipation theorem (FTD).

5. The gap of $2\Omega = 10$ meV on plots $2d$ and 3 in Ref. [Park 2000] still remains to be undocumented in details and unexplained in bulk of theoretical papers.

6. *E.g.* by occasional hybridization with the surface impurities.

7. The smaller the current and voltage applied, the smaller the Schottky shot noise.

8. Prof. V. L. Ginzburg once said that any synthesized atomic or molecular combination is a valid candidate in . . . the quest for a superconductive state.

9. As well as in the orthodox theory of Coulomb blockade [Kulik 1975; Averin 1991]

References

Abdul-Redah T. *et al.*, Physica B **276-278**, 824 (2000).

Abdul-Redah T. and Chatzidimitriou-Dreismann C. A., J. Alloys Comp. **356**, 249 (2003).

Abdul-Redah T. and Chatzidimitriou-Dreismann C. A., Physica B **350**, E1035 (2004).

Abouraddy A. F. *et al.*, Phys. Rev. Lett. **87**, 123602 (2001).

Abramowitz M. and Stegun I., *Handbook of Mathematical Functions*, (National Bureau of Standards, Washington, DC 1964).

Abramowitz M. and Stegun I. A., *Handbook of Mathematical Functions*, (Dover, 1972).

Adamyan H. H. and Kryuchkyan G. Yu., Phys. Rev. A **69**, 053814 (2004).

Adamyan H. H. and Kryuchkyan G. Yu., NAS of Armenia, Physics,**N2**, 39 (2004).

Adamyan H. H. and Kryuchkyan G. Yu., quant-ph/0404110 (2004).

Agarwal G. S., *Quantum Statistical Theories of Spontaneous Emission*, (Springer, Berlin, 1974).

Aharonov Y. and Bohm D., Phys. Rev. **115**, 485 (1959).

Aichele T., Lvovsky A. I. and Schiller S., Eur. Phys. J. D **18**, 237 (2002).

Akopyan M. E., Bibnikov N. K., Koth D. B., Pravilov A. M. and Stevonov M. B., Chem. Phys. **242**, 253 (1999).

Akulin V. M. and Kurizki G., Phys. Lett. A **174**, 267 (1993); Akulin V. M., Phys. Rev. A **48**, 3532 (1993).

Akulin V. M., de Tomasi F., Mourachko I. and Pillet P., Physica D **131**, 125 (1999).

Akulin V. M., Gershkovich V., and Harel G., Phys. Rev. A **64**, 012308 (2001).

Akulin V. M., *Dynamics of Complex Quantum systems*, Chapt. 4, (Springer, 2004).

Aleiner I. L. and Andreev A. V., Phys. Rev. Lett. **81**, 1286 (1998).

Alexandrov E. *et al.*, Phys. Rev. A **66**, 042903 (2002).

Alhassid Y. and Levine R. D., Chem. Phys. Lett. **72**, 401 (1980).

Allen L. and Eberly J. H., *Optical resonances and two-level atoms*, (J. Wiley, New York 1975).

Altshuller B. L., Lee P. A. and Webb R. A., *Mesoscopic Phenomena in Solids*, (Elsevier, Amsterdam, 1991).

Altshuler B. and Glazman L., Science **283**, 1864 (1999).

Ambegaokar V., Ber. Bunsenges Phys. Chem. **95**, 400 (1991).

Amin M. H. S. and Smirnov A. Yu., Phys. Rev. Lett. **92**, 017001 (2004).

Anderson P. W., Phys. Rev. **124**, 41 (1961).

Anderson P. W., Phys. Rev. Lett. **18**, 1049 (1967).

Anderson M. H., Ensher J. R., Matthews M. R., Wieman C. E., and Cornell E. A., Science **269**, 198 (1995).

Anderson B. P. and Kasevich M. A., Science **282**, 1686 (1998).

Anderson W. R., Veale J. R. and Gallagher T. F., Phys.Rev. Lett. **80**, 249 (1998).

Anderson W. R., Robinson M. P., Martin J. D. D. and Gallagher T. F., Phys. Rev. A **65**, 063404 (2002).

Andre A. and Lukin M. D., Phys. Rev. Lett. **89**, 143602 (2002).

Andreani C. *et al.*, Nucl. Instr. Meth. A **497**, 535 (2003).

Andreev A. F., Zh. Eksp. Teor. Fiz. **46**, 1823 (1964) [Sov. Phys. JETP **19**, 1228 (1964)]; *ibid.* **51**, 1510 (1966) [**24**, 1019 (1967)].

Andreev A. F., Zh. Eksp. Teor. Fiz. **51**, 1510 (1966) [**24**, 1019 (1967)].

Andrews M. R. *et al.*, Science **275**, 637 (1997).

Andrews M. R., Nature **398**, 195 (1999).

Argyres P. N., *Lectures in Theoretical Physics*, Summer Institute for Theoretical Physics, vol. VIII, ed. by W. E. Britten, (University of Colorado Press 1966), page 183.

Arndt M. *et al.*, Nature **401**, 680 (1999).

Aronov A. G. and Sharvin Yu. V., Rev. Mod. Phys. **59**, 755 (1987).

Ashkin A., Phys. Rev. Lett. **24**, 156 (1970).

Aspect A., Grangier P. and Roger G., Phys. Rev. Lett. **47**, 460 (1981).

Aspect A., Grangier P. and Roger G., Phys. Rev. Lett. **49**, 91 (1982).

Aspect A., Dalibard J. and Roger G., Phys. Rev. Lett. **49**, 1804 (1982).

Assion A. *et al.*, Science **282**, 919 (1998).

Atutov S. N. *et al.*, Phys. Rev. A **67**, 053401 (2003).

Auffeves A. *et al.*, Phys. Rev. Lett. **91**, 230405 (2003).

Aunola M., Phys. Rev. B **63**, 132508 (2001).

Autler S. H. and Townes C. H., Phys. Rev. **100**, 703 (1955).

Averbukh I. Sh. and Shapiro M., Phys. Rev. A **47**, 5086 (1993).

Averin D. V. and Likharev K. K., *Mesoscopic Phenomena in Solids*, ed. by Altshuler B. L., Lee P. A. and Webb R. A., (Elsevier, Amsterdam, 1991), p. 173.

Averin D. V., Fortschr. Phys. **48**, 1055 (2000).

Avron J.E. *et al.*, Phys. Rev. B **62**, 10618 (2000).

Aytur O. and Kumar P., Phys. Rev. Lett. **65**, 1551 (1990).

Babichev S. A., Brezger B., Lvovsky A. I., Phys. Rev. Lett. **92**, 047903 (2004); Babichev S. A., Appel J., Lvovsky A. I., *ibid.* **92**, 193601 (2004).

Bacher R. F. and Goudsmit S., *Atomic energy states*, (Mc Graw-Hill, New York, 1932).

Baev A. *et al.*, Phys. Rev. A **66**, 022509 (2002).

Banaszek K. and Wódkiewicz K., Phys. Rev. A **58**, 4345 (1998).

Banaszek K. and Wódkiewicz K., Phys. Rev. Lett. **82**, 2009 (1999).

Barash Yu. S., Phys. Rev. B **61**, 678 (2000).

Barenco A. et al., Phys. Rev. A. **52**, 3457 (1995).

Barenco A., Proc. Roy. Soc. London A **449**, 679 (1995).

Barett M. D. *et al.*, Nature **429**, 737 (2004).

Barnett S. M. and Radmore P. M., *Methods in Theoretical Quantum Optics*, (Clarendon press, Oxford, 1997).

Barnett S. M. and Stenholm S., Phys. Rev. A **64**, 033808 (2001).

Barnett S. M., private communication; to be published (2004).

Bartlett S. D., Sanders B. C., Braunstein S. L. and Nemoto K., Phys. Rev. Lett. **88**, 097904 (2002).

Barone A. and Paternò G., *Physics and Applications of the Josephson Effect*, (Wiley, New York, 1982).

Beenakker C. W. J., Phys. Rev. B **44**, 1646 (1991).

Beenakker C. W. J., Phys. Rev. B **46**, 12841 (1992).

Beenakker C. W. J., *Mesoscopic Quantum Physics*, Les Houches, Session LXI, Ed. by E. Akkermans, G. Montambaux, and J. L. Pichard (North-Holland, Amsterdam, 1994).

Beenakker C. W. J. *et al.*, Phys. Rev. Lett. **72**, 2470 (1994).

Beenakker C. W. J., Rev. Mod. Phys. **69**, 731 (1997).

Bell J., Physics **1**, 195 (1964).

Belyaev S. T., Zh. Eksp. Teor. Fiz. **34**, 4334 (1958), [Sov. Phys. JETP **7**, 299 (1958)].

Benderskii V. A., Vetoshkin E. V., Irgibaeva I. S. and Trommsdorff H. P., Chem. Phys. **262**, 393 (2000).

Bennett C. H., Bessette F., Brassard G., Savail L., and Smolin J., J. Cryptology **5**, 3 (1992).

Bennett C. H. *et al.*, Phys. Rev. Lett. **70**, 1895 (1993).

Bennett C. H. *et al.*, Phys. Rev. Lett. **87**, 077902 (2001).

Bennik R. S., Bentley S. J. and Boyd R. W., Phys. Rev. Lett. **89**, 113601 (2002).

Bennik R. S., Bentley S. J., Boyd R. W. and Howell J. C., Phys. Rev. Lett. **92**, 033601 (2004).

Benz S. and Haeberlen U., J. Mag. Res. **66**, 125 (1986).

Bergmann K., Theuer H., and Shore B. W., Rev. Mod. Phys. **70**, 1003 (1998).

Berman P. R., editor, *Atom Interferometry*, (Academic Press, New York, 1997).

Bernstein E. and Vazirani U., SIAM J. Comput. **25**, 1411 (1997).

Bertet P. *et al.*, Phys. Rev. Lett. **89**, 200402 (2002).

Bethe H. A., Phys. Rev. **72**, 339 (1947).

Beveratos A., Brouri R., Gacoin T., Poizat J. P., and Grangier P., Phys. Rev. A **64**, 061802 (2001).

Bezryadin A., Dekker C. and Schmid G., Appl. Phys. Lett. **71**, 1273 (1997).

Birkl G. *et al.*, Phys. Rev. Lett. **75**, 2823 (1995).

Bixon M. and Jortner J., J. Chem. Phys. **48**, 715 (1968).

Björneholm O. *et al.*, Phys. Rev. Lett. **84**, 2826 (2000).

Blaauboer M. and Heller E. J., Phys. Rev. B **64**, 241301 (2001).

Blaauboer M., Phys. Rev. B **65**, 235318 (2002).

Blanchard Ph. *et al.*, *Decoherence*, (Springer, 2000).

Blanco A. *et al.*, Appl. Phys. Lett. **73**, 1781 (1998).

Blanter Ya. M. and Buttiker M., Phys. Rep. **336**, 1 (2000).

Blatter G., Geshkenbein V. B., and Ioffe L., Phys. Rev. B **63**, 174511 (2001).

Blinov B. B., Moehring D. L., Duan L. M., and Monroe C., Nature **428**, 153 (2004).

Blostein J. J., Dawidowski J. and Granada J. R., Physica B **304**, 357 (2001).

Blostein J. J., Dawidowski J., Ibañez S. A. and Granada J. R., Phys. Rev. Lett. **90**, 105302 (2003).

Blostein J. J., Dawidowski J. and Granada J. R., Physica B **334**, 257 (2003).

Blostein J. J., Dawidowski J. and Granada J. R., Nucl. Inst. Meth. B **217**, 333 (2004).

Boese D. and Schoeller H., Europhys. Lett. **54**, 668 (2001).

Bogoliubov N. N., J. Phys. (Moscow) **11**, 23 (1947).

Bohm D., *Quantum Theory*, (Prentice-Hall, Englewood Cliffs, NJ., 1951).

Bohr N., Phys. Rev. **48**, 696 (1935).

Boller K. J., Imamoglu A., and Harris S. E., Phys. Rev. Lett. **66**, 2593 (1991).

Bordé Ch. J., Courtier N., Burck F. Du, Goncharov A. N. and Gorlicki M., Phys. Lett. A **188**, 187 (1994).

Born M. and Wolf E., *Principles of Optics*, 7th (expanded) version, (Cambridge University Press, 1999).

Bose S. *et al.*, Philos. Trans. R. Soc. London A **356**, 1823 (1998).

Bouwmeester D. *et al.*, Nature **390**, 575 (1997).

Bouwmeester D., Ekert A. and Zeilinger A., *The Physics of Quantum Information*, (Springer, Berlin, 2001).

Bowen W. P., Schnabel R., Lam P. K., and Ralph T. C., Phys. Rev. Lett. **90**, 043601 (2003).

Bowen W. P. *et al.*, Phys. Rev. A **67**, 032302(2003).

Bowen W. P. *et al.*, Phys. Rev. A **69**, 012304 (2004).

Boyd R. W., *Nonlinear Optics*, (Academic Press, San Diego, CA, 1992).

Bradley C. C., Sackett C. A., Tollett J. J., and Hulet R. G., Phys. Rev. Lett. **75**, 1687 (1995).

Braig S. and Flensberg K., Phys. Rev. B **68**, 205324 (2003).

Braig S. and Flensberg K., cond-mat/0401347 (2004).

Braje D. A., Balić V., Yin G. Y., and Harris S. E., Phys. Rev. A **68**, 041801 (2003).

Bratos S., Tarjus G., Diraison M., and Leicknam J Cl., Phys. Rev. A **44**, 2745 (1991).

Braunstein S. L., Phys. Rev. Lett. **80**, 4084 (1998); Nature **394**, 47 (1998).

Braunstein S. L. and Kimble H. J., Phys. Rev. Lett. **80**, 869 (1998).

Braunstein S. L. *et al.*, Phys. Rev. Lett. **83**, 1054 (1999).

Braunstein S. L., and Kimble H. J., Phys. Rev. A **61**, 042302 (2000).

Braunstein S. L., Fuchs C. A. and Kimble H. J., J. Mod. Opt. **47**, 267 (2000).

Braunstein S. L. and Pati A. K., *Quantum Information Theory with Continuous Variables*, (Kluwer, Dordrecht, 2003), and references therein.

Bravyi S. and Kitaev A., quant-ph/0003137 (2000).

Brendel J. *et al.*, Phys. Rev. Lett. **82**, 2594 (1999).

Brennen G. K., Caves C. M., Jessen P. S., and Deutsch I. H., Phys. Rev. Lett. **82**, 1060 (1999).

Breuer H. P. and Petruccione F., *Theory of Open Quantum Systems*, (Oxford University Press, Oxford 2002).

Brezger B. *et al.*, Phys. Rev. Lett. **88**, 100404 (2002).

Brezger B., Arndt M. and Zeilinger A., J. Opt. B: Quantum Semiclass. Opt. **5**, S82 (2003).

Briegel H. J., Dur W., van Enk S. J., Cirac J. I., and Zoller P., *The Physics of Quantum Information*, ed. by D. Bouwmeester, A. Ekert, and A. Zeilinger, (Springer, Berlin, 2000).

Brion E., Harel G., Kébaïli N., Akulin V. M. and Dumer I., Europhys. Lett. **66**, 157 (2004).

de Broglie L., Nature **112**, 540 (1923).

Brouwer P. W., Phys. Rev. B **58**, R10135 (1998).

Bruch L. W., Schöllkopf W. and Toennies J. P., J. Chem. Phys. **117**, 1544 (2002).

Brugger R. M., Taylor A. D., Olsen C. E., Goldstone J. A., and Soper A. K., Nucl. Instr. & Meth. in Phys. Research **221**, 393 (1984).

Brühl R. *et al.*, Europhys. Lett. **59**, 357 (2002).

Brukner C. and Zeilinger A., Phil. Trans. R. Soc. Lond. A **360**, 1061 (2002).

Brune M. *et al.*, Phys. Rev. Lett. **77**, 4887 (1996).

Brunello A., Dalfovo F., Pitaevskii L., and Stringari S., Phys. Rev. Lett. **85**, 4422 (2000).

Brunello A., Dalfovo F., Pitaevskii L., Stringari S., and Zambelli F., Phys. Rev. A **64**, 063614 (2001).

Budker D., Kimball D., Rochester S., and Yashchuk V., Phys. Rev. Lett. **83**, 1767 (1999).

Budker D., Kimball D., Rochester S., Yashchuk V., and Zolotarev M., Phys. Rev. A **62,** 043403 (2000).

Budker D. *et al.*, Rev. Mod. Phys. **74,** 1153 (2002).

Budker D., Kimball D., Yashchuk V., and Zolotarev M., Phys. Rev. A **65**, 055403 (2002).

Buks E., Schuster R., Heiblum M., Mahalu D., and Shtrikman H., Nature **391**, 871 (1998).

Büttiker M., Phys. Rev. Lett. **57**, 1761 (1986).

Büttiker M., Thomas H., and Prêtre A., Z. Phys. B **94**, 133 (1994).

Buzek V., Moya-Cessa H., Knight P. L. and Phoenix S. J. D., Phys. Rev. A **45**, 8190 (1992).

Buzek V., Phys. Rev. A **58**, 1723 (1998).

Calarco T., Datta A., Fedichev P., Pazy E., and Zoller P., Phys. Rev. A **68**, 012310 (2003).

Caldeira A. O. and Leggett A. J., Ann. Phys. **149**, 374 (1983).

Califano S., Schettino V. and Neto N., *Lattice Dynamics of Molecular Crystals*, (Springer Verlag, Berlin, 1981).

Campbell E. E. B., Ulmer G. and Hertel I. V., Phys. Rev. Lett. **67**, 1986 (1991).

Carnal O. and Mlynek J., Phys. Rev. Lett. **66**, 2689 (1991).

Casimir H. B. G. and Polder D., Phys. Rev. **73**, 360 (1948).

Castin Y. and Dum R., Phys. Rev. Lett. **27**, 5315 (1996).

Caves C. M., Thorne K. S., Drever R. W. P., Sandberg V. D., and Zimmermann M., Rev. Mod. Phys. **52**, 341 (1980).

Chapman M. S. *et al.*, Phys. Rev. Lett. **75**, 3783 (1995).

Chapman M. S. *et al.*, Phys. Rev. Lett. **74**, 4783 (1995).

Chapman M. S. *et al.*, Phys. Rev. A **51**, R14 (1995).

Chaturvedi S., Dechoum K. and Drummond P. D., Phys. Rev. A **65**, 033805 (2002).

Chatzidimitriou-Dreismann C. A., Adv. Chem. Phys. **80**, 201 (1991).

Chatzidimitriou-Dreismann C. A., Krieger U. K., Möller A., and Stern M., Phys. Rev. Lett. **75**, 3008 (1995).

Chatzidimitriou-Dreismann C. A., Abdul-Redah T., Streffer R. M. F. and Mayers J., Phys. Rev. Lett. **79**, 2839 (1997).

Chatzidimitriou-Dreismann C. A., Adv. Chem. Phys. **99**, 393 (1997).

Chatzidimitriou-Dreismann C. A., *Advances in Chem. Phys. Resonances, Instability and Irreversibility*, ed. by Prigogine I. and Rice S. A., (John Wiley & Sons, New-York, 1997), Vol. XCIX, pp. 393 - 430.

Chatzidimitriou-Dreismann C. A., Abdul-Redah T. and Streffer R. M. F., Ber. Bunsenges. Phys. Chem. **83**, 544 (1998).

Chatzidimitriou-Dreismann C. A., Abdul-Redah T., Streffer R. M. F. and Hessmo B., Phys. Rev. Lett. **84**, 2036 (2000).

Chatzidimitriou-Dreismann C. A., Abdul-Redah T. and Sperling J., J. Chem. Phys. **113**, 2784 (2000).

Chatzidimitriou-Dreismann C. A., Abdul-Redah T. and Kolaric B., JACS **123**, 11945 (2001).

Chatzidimitriou-Dreismann C. A., Abdul-Redah T., Streffer R. M. F. and Mayers J., J. Chem. Phys. **116**, 1511 (2002).

Chatzidimitriou-Dreismann C. A., Abdul-Redah T. and Mayers J., Physica B **315**, 281 (2002).

Chatzidimitriou-Dreismann C. A., Vos M., Kleiner C. and Abdul-Redah T., Phys. Rev. Lett. **91**, 057403 (2003).

Chatzidimitriou-Dreismann C. A., J. Alloys Comp. **356-357**, 244 (2003).

Chatzidimitriou-Dreismann C. A., J. Mod. Opt., to be published (2004).

Chatzidimitriou-Dreismann C. A. and Abdul-Redah T., Physica B **350**, 239 (2004).

Chen T. *et al.*, J. Vac. Sci. Technol. B **9**, 2461 (1991).

Chen J., Story J. G., Tollett J. J. and Hulet R. G., Phys. Rev. Lett. **69**, 1344 (1992).
Chikkatur A. P. *et al.*, Phys. Rev. Lett. **85**, 483 (2000).
Chiorescu I., Nakamura Y., Harmans C. J. P. M. and Mooij J. E., Science **299**, 1869 (2003).
Choi S., Burnett K., Friesch O. M., Kneer B., and Schleich W. P., Phys. Rev. A **63**, 065601 (2001).
Chu S., Rev. Mod. Phys. **70**, 685 (1998).
Chuang I. L. and Yamamoto Y., Phys. Rev. A **52**, 3489 (1995).
Chuang I. L., Vandersypen L. M. K., Zhou X. L., Leung D. W. and Lloyd S., Nature **393**, 143 (1998).
Cirac J. I. and Zoller P., Phys. Rev. Lett. **74**, 4091 (1995).
Cirac J. I., Zoller P., Kimble H. J. and Mabuchi H., Phys. Rev. Lett. **78**, 3221 (1997).
van Cittert P. H., Physica **1**, 201 (1934).
Clarke J., Cleland A. N., Devoret M. H., Esteve D. and Martinis J. M., Science **239**, 992 (1988).
Clauser J. F., Horne M. A., Shimony A. and Holt R. A., Phys. Rev. Lett. **23**, 880 (1969).
Clauser J. F. and Reinsch M. W., Appl. Phys. B **54**, 380 (1992).
Clauser J. F. and Li S., Phys. Rev. A **49**, R2213 (1994).
Clauser J. F., *Experimental Metaphysics*, ed. by Cohen R. S., Horne M. and Stachel J., (Kluwer Academic, 1997).
Clerk A. A. *et al.*, Phys. Rev. B **61**, 3555 (2000).
Cohen-Tannoudji C., Diu B. and Laloë F., *Mécanique Quantique*, (Hermann, Paris, France, 1977).
Cohen-Tannoudji C., Dupont-Roc J., and Grynberg G., *Atom-Photon Interactions*, (Wiley, New York, 1992).
Cohen-Tannoudji C. N., Rev. Mod. Phys. **70**, 707 (1998).
Cohen-Tannoudji C., and Robilliard C., C. R. Acad. Sci. Paris, Série IV, t. 2, (2001), p.445.
Collin E. *et al.*, cond-mat/0404503 (2004).
Collin E. *et al.*, submitted to Phys. Rev. Lett. (2004).
Colognesi D., Physica B, Oct. 2003.
Condon E. U. and Gurney R. W., Nature **122**, 439 (1928).
Cook R. J., Phys. Scr. **T21**, 49 (1988).
Cornwell J. F., *Group Theory in Physics*, (Academic Press, London, 1984), vol. 2, p. 378.
Cowley R. A., J. Phys. Condens. Matter **15**, 4143 (2003).
Craig D. P. and Thirunamachandran T., em Molecular Quantum Electrodynamics, (Academic Press, London, 1984).
Cremers J. N. H. J. and Brouwer P. W., Phys. Rev. B **65**, 115333 (2002).
Csesznegi J. R. and Grobe R., Phys. Rev. Lett. **79**, 3162 (1997).
Cuevas J. C. *et al.*, Phys. Rev. B **63**, 094515 (2001).
Dalfovo F., Giorgini S., Pitaevskii L. P. and Stringari S. Rev. Mod. Phys. **71**, 463 (1999).
Dall R., Hoogerland M., Baldwin K. and Buckman J., C.R. Acad. Sci. Paris, tome 2, (Serie IV), p. 595 (2001).
Datta S., *Electronic Transport in Mesoscopic Systems*, (Cambridge, UK, 1995).
Datta S., Super-lattices and Microstructures **28**, 253 (2000).
Davies H. J. and Adams C. S., Phys. Rev. A **55**, R2527, (1997).
Davis K. B. *et al.*, Phys. Rev. Lett. **75**, 3969 (1995).
Davisson C. J. and Germer L. H., Nature **119**, 558 (1927).
Dechoum K. *et al.*, quant-ph/0310129 (2003).

Dennis E., Kitaev A., Landahl A. and Preskill J., J. Math. Phys. **43**, 4452 (2002).

DePue M. T. et al., Phys. Rev. Lett. **82**, 2262 (1999).

Deutsch I. H. and Brennen G. K., Fortschr. Phys. **48**, 925 (2000).

Dicke R. H., Phys. Rev. **93**, 99 (1954).

Diedrich F., Peik E., Chen J. M., Quint, W. and Walther H., Phys. Rev. Lett. **59**, 2931 (1987).

Dieks D., Phys. Lett. A **92**, 271 (1982).

DiVincenzo D. P., Fortschr. Physik **48**, 771 (2000).

Dorner U., Fedichev P., Jaksch D., Lewenstein M. and Zoller P., Phys. Rev. Lett. **91**, 073601 (2003).

Doucot B., Ioffe L. B. and Feigelman M. V., Phys. Rev. Lett. **90**, 107003 (2003).

Doucot B., Ioffe L. B. and Vidal J., Phys. Rev. B, (in press, 2004); cond-mat//0302104.

Doucot B., Feigelman M. V., Ioffe L. B. and Ioselevich A. S., (unpublished, 2004).

Dresselhaus M. S., Dresselhaus G. and Eklund P. C., *Science of Fullerenes and Carbon Nanotubes*, (Academic Press, San Diego, 1998), 2nd edition.

Drummond P. D., and Gardiner C. W., J. Phys. A **13**, 2353 (1980).

Drummond P. D. and Raymer M. G., Phys. Rev. A‘ **44**, 2072 (1991).

Drummond P. D., Dechoum K. and Chaturvedi S., Phys. Rev. A **65**, 033806 (2002).

Duan L. M. *et al.*, Phys. Rev. Lett. **84**, 2722 (2000).

Duan L. M., Giedke G., Cirac J. I. and Zoller P., Phys. Rev. Lett. **85**, 2722 (2000)

Duan L. M., Sørensen A., Cirac J. I. and Zoller P., Phys. Rev. Lett. **85**, 3991 (2000).

Duan L. M., Cirac J. I., Zoller P., and Polzik E. S., Phys. Rev. Lett. **85**, 5643 (2000).

Duan L. M., Lukin M. D., Cirac J. I., and Zoller P., Nature **414**, 413 (2001).

Duan L. M., Phys. Rev. Lett. **88**, 170402 (2002).

Duan L. M., Cirac J. I. and Zoller P., Phys. Rev. A **66**, 023818 (2002).

Dunningham J. A., Burnett K., and Barnett S. M., Phys. Rev. Lett. **89**, 150401 (2002).

Dupont E. *et al.*, Phys. Rev. Lett. **74**, 3596 (1995).

Dykhne A. M. and Zosimov V. V., JETP lett. **74**, 332 (2001).

Eberly J. H., Narozhny N. B. and Sanchez-Mondragon J. J., Phys. Rev. Lett. **44**, 1323 (1980).

Eckold G., Grimm H. and Stein-Arsic M., Physica B **180 - 181**, 336 (1992).

Einstein A., Podolsky B., and Rosen N., Phys. Rev. **47**, 777 (1935).

Ekert A. and Jozsa R., Rev. Mod. Phys. **68**, 733 (1996).

Elattari B. and Gurvitz S. A., Phys. Rev. Lett. **84**, 2047 (2000).

Esry B. D., Phys. Rev. A **55**, 1147 (1997).

Estermann I. and Stern O., Z. Phys. **61**, 95 (1930).

Evans A. C., Timms D. N., Mayers J. and Bennington S., Phys. Rev. B **53**, 3023 (1996).

Fabre C., Mason E. I., Wong N. C., Opt. Comm. **170**, 299 (1999).

Facchi P. and Pascazio S., Phys. Lett. A **241**, 139 (1998).

Facchi P. and Pascazio S., Phys. Rev. A **62**, 023804 (2000).

Facchi P., Nakazato H., and Pascazio S., arXiv:quant-ph/0006094 (2000).

Facchi P., Nakazato H., and Pascazio S., Phys. Rev. Lett. **86**, 2699 (2001).

Facchi P., Pascazio S., Scardicchio A., and Schulman L. S., Phys. Rev. A **65**, 012108 (2001).

Facchi P. and Pascazio S., Prog. in Opt. **42**, 147 (2001).

Facchi P. and Pascazio S., Phys. Rev. Lett. **89**, 080401 (2002).

Falcone R. W., Green W. R., White J. C., Young J. F. and Harris S. E., Phys. Rev. A **15**, 1333 (1977).

Fano U., Phys. Rev. **96**, 869 (1954).

Fano U., Phys. Rev. **124**, 1866 (1961).

Fano U., Phys. Rev. **131**, 259 (1963).

Fedorets D., Gorelik L.Y., Shekhter R. I. and Jonson M., Europhys. Lett. **58**, 99 (2002).

Feigelman M. V., Ioffe L. B., Geshkenbein V. B., Dayal P and Blatter G., Phys. Rev. Lett. **92**, 098301 (2004).

Feigelman M. V., Ioffe L. B., Geshkenbein V. B., Dayal P. and Blatter G., (submitted to Phys. Rev. B, 2004).

Feld J., Han K., and Harris S., Phys. Rev. Lett. **67**,3062 (1991).

Feng S. and Pfister O., J. Opt. B **5**, 262 (2003).

Feng S. and Pfister O., Phys. Rev. Lett. **92**, 203601 (2004).

Fermi E., *Nuclear Physics*, (Univ. of Chicago Press, Chicago, 1950).

Ferraro J. R., Nakamoto K. and Brown C. W., *Introductory Raman Spectroscopy*, (Academic Press, Amsterdam, 2003), 2nd Edition.

Feynman R. P., Phys. Rev. **94**, 262 (1954).

Feynman R. P. and Hibbs A. R., *Quantum Mechanics and Path Integrals*, (McGraw-Hill, New York, 1965).

Feynman R. P., *Lectures on Computation*, (Addison-Wesley, New York 1996).

Fielding A. L. and Mayers J., Nucl. Inst. Meth. A **480**, 680 (2002).

Fillaux F., Chem. Phys. **74**, 395 (1983).

Fillaux F., Chem. Phys. **74**, 405 (1983).

Fillaux F., Tomkinson J. and Penfold J., Chem. Phys. **124**, 425 (1988).

Fillaux F. and Tomkinson J., J. Mol. Struct. **270**, 339 (1992).

Fillaux F., Physica D **113**, 172 (1998).

Fillaux F., Solid State Ionics **125**, 69 (1999).

Fillaux F., Int. Rev. in Phys. Chem. **19**, 553 (2000).

Fillaux F., Limage M. H. and Romain F., Chem. Phys. **276**, 181 (2002).

Fillaux F., Cousson A. and Keen D. A., Phys. Rev. B **67**, 054301 and 189901 (E) (2003).

Fillaux F., Nicolaï B. and Cousson A., *Proceedings of the* 3rd *conference localization & energy transfer in nonlinear systems*, ed. by Vásquez L., MacKay R. and Zorzano M. P., (World Scientific Publishing Co., Inc, New Jersey, London, Singapore, Hong-Kong, 2003).

Fillaux F. and Cousson A., J. Phys.: Cond. Matter **16**, 1007 (2004).

Fioretti A., Comparat D., Drag C., Gallagher T.F. and Pillet P., Phys.Rev. Lett. **82**, 1839(1999).

Fisher M. P. A., Phys. Rev. B **37**, 75 (1988).

Fischer M. C., Gutierrez-Medina B. and Raizen M. G., Phys. Rev. Lett. **87**, 040402 (2001).

Fisk P. T. H., Bachor H. A. and Sandeman R. J., Phys. Rev. A **33**, 2418 and 2424, **34**, 4762 (1986).

Fitzgerald R., Physics Today **53**, 20 (2000).

Fleischhauer M. and Lukin M. D., Phys. Rev. Lett. **84**, 5094 (2000).

Fleischhauer M. and Lukin M. D., Phys. Rev. A **65**, 022314 (2002).

Flensberg K., Phys. Rev. B **68**, 205323 (2003).

Flügge S., *Practical Quantum Mechanics*, (Springer-Verlag, 1971).

Folk J. A., Marcus C. M., and Harris J. S. (Jr)., Phys. Rev. Lett. **87**, 206802 (2001).

Fominov Ya. V., Golubov A. A., and Kupriyanov M. Yu., JETP Letters **77**, 587 (2003).

Fonda L., Ghirardi G. C., Rimin A., and Weber T., Nuovo Cim. **15A**, 689 (1973).

Fonda L., Ghirardi G. C., and Rimini A., Rep. Prog. Phys. **41**, 587 (1978).

Ford G. W., Kac M., and Mazur P., J. Math. Phys. **6**, 504 (1965).

Franson J. D., Phys. Rev. Lett. **62**, 2205 (1989).
Frasier J. S., Celli V. and Blum T., Phys. Rev. A **59**, 4358 (1999).
Freedman S. J. and Clauser J. F., Phys. Rev. Lett. **28**, 938 (1972).
Frerichs V., and Schenzle A., Phys. Rev. A **44**, 1962 (1991).
Friedman J. R., Chen V. P. W., Tolpygo S. K. and Lukens J. E., Nature (London) **406**, 43 (2000).
Friedman N., Kaplan A., N. Davidson, Adv. At. Mol. Opt. Phys. **48**, 99 (2002).
Friedrichs K., Commun. Pure Appl. Math. **1**, 361 (1948).
Furusawa A. *et al.*, Science **282**, 706 (1998).
Gähler R. and Zeilinger A., Am. J. Phys. **59**, 316 (1991).
Gallager R. G., *Information Theory and Reliable Communication*, (John Wiley & Sons, New York 1968).
Gallagher T. F. *et al.*, Phys. Rev. A **25**, 1905 (1982).
Gallagher T. F., Phys. Rep. **210**, 319 (1992).
Gallagher T. F., *Rydberg Atoms*, (Cambridge University Press, Cambridge, 1994).
Gamow G., Z. Phys. **51**, 204 (1928).
García-Ripoll J. J., Zoller P., and Cirac J. I., Phys. Rev. Lett. **91**, 157901 (2003).
Gardiner C. W., *Quantum Noise*, (Springer-Verlag, Berlin 1991).
Garraway B. M., Phys. Rev. A **55**, 4636 (1997).
Garraway B. M., Phys. Rev. A **55**, 2290 (1997).
Gasenzer T., Roberts D. C., and Burnett K., Phys. Rev. A **65**, 021605 (2002).
Gea-Banacloche J., Phys. Rev. Lett. **65**, 3385 (1990).
Gershnabel E. *et al.*, cond-mat/0309584 (2003).
Gershnabel E., Katz N., Rowen E., and Davidson N., cond-mat/0404080 (2004).
Gerry C. C., Phys. Rev. A **59**, 4095 (1999).
Giedke G., Kraus B., Lewenstein M., and Cirac J. I., Phys. Rev. Lett. **87**, 167904 (2001).
Ginzburg V. L. and Pitaevskii L., Zh. Eksp. Teor. Fiz. **34**, 1240 (1958).
Gisin N., Phys. Lett. A **154**, 201 (1991).
Gisin N. and Peres A., Phys. Lett. A **162**, 15 (1992).
Gisin N., Ribordy G., Tittel W. and Zbinden H., Rev. Mod. Phys. **74**, 145 (2002).
Glyde H. R., Phys. Rev. B **50**, 6726 (1994).
Gnutzmann S. and Haake F., Z. Phys. B **101**, 263 (1996).
Gogolin A. O. and Komnik A., cond-mat/0207513 (2002).
Goldberger M. L. and Watson K. M., *Collision Theory*, (Wiley, New York 1964), chapter 8.
Golovach V. and Loss D., Phys. Rev. Lett. **93**, 016601 (2004).
Goodson B. M., Goswami D., Rabitz H. and Warren W. S., J. Chem. Phys. **112**, 5081 (2000).
Gorelik L. Y. *et al.*, Phys. Rev. Lett. **80**, 4526 (1998).
Gorelik L. Y., Isacsson A., Galperin Y. M., Shekhter R. I. and Jonson M., Nature **411**, 454 (2001).
Gorini V., Kossakowski A., and Sudarshan E. C. G., J. Math. Phys. **17**, 821 (1976).
Gorlitz A. *et al.*, Phys. Rev. Lett. **90**, 090401 (2003).
Gottesman D., Phys. Rev. A **54**, 1862 (1996).
Gottesman D., *Stabilizer Codes and Quantum Error Correction*, (PhD thesis, California Institue of Technology, Pasadena CA, 1997).
Grabert H., Talkner P., and Hänggi P., Z. Phys. B **26**, 389 (1977).
Grabert H., Talkner P., Hänggi P., and Thomas H., Z. Phys. B **29**, 273 (1978).
Graf R., Meyer R., Ha T. K. and Ernst R. R., J. Chem. Phys. **75**, 2914 (1981).

Graham R., Zeit. Phys. **210**, 319 (1968).
Grajcar M. *et al.*, Phys. Rev. B **69**, 060501(R) (2004).
Gramespacher T. and Büttiker M., Phys. Rev. B **61**, 8125 (2000).
Grangier P., Roger G. and Aspect A., Europhys. Lett. **1**, 173 (1986).
Grosshans F. *et al.*, Nature **421**, 238 (2003).
Greenberger D. M., Horne M. A. and Zeilinger A., *Bell's Theorem, Quantum Theory, and Conceptions of the Universe*, (Kluwer, Dordrecht, 1989).
Greenberger D. M., Horne M. A., Shimony A. and Zeilinger A., Am. J. Phys. **58**, 1131 (1990).
Greentree A. *et al.*, Phys. Rev. A **67**, 023818 (2002).
Greiner M. *et al.*, Nature **415**, 39 (2002).
Grisenti R. E., Schöllkopf W., Toennies J. P., Hegerfeldt G. C., and Köhler T., Phys. Rev. Lett. **83**, 1755 (1999).
Grisenti R. E. *et al.*, Phys. Rev. Lett. **85**, 2284 (2000).
Gross E. P., J. Math. Phys. **4**, 195 (1963).
Gross R., Chaudhari P., Dimos D., Gupta A., and Koren G., Phys. Rev. Lett. **64**, 0228 (1990).
Grosshaus F. *et al.*, Nature **421**, 238 (2003).
Grover L., *Proceedings of the Annual ACM Symposium on the theory of Computing*, (New York, ACM Press 1996).
Grover L. K., Phys. Rev. Lett. **79**, 4709 (1997).
Gupta S., Dieckmann K., Hadzibabic Z., and Pritchard D. E., Phys. Rev. Lett. **89**, 140401 (2002).
Gurvitz S. A., Phys. Rev. A **38**, 1747 (1988).
Gurvitz S. A., Phys. Rev. B **54**, 15215 (1997).
Haake F. and Lewenstein M., Phys. Rev. A **28**, 3606 (1983).
Haake F. and Reibold R., Phys. Rev. A **32**, 2462 (1985).
Haake F., private communication (1994).
Habib S., Shizume K. and Zurek W. H., Phys. Rev. Lett. **80**, 4361 (1998).
Hackermüller L., Hornberger K., Brezger B., Zeilinger A. and Arndt M., Appl. Phys. B **77**, 781 (2003).
Hackermüller L., Hornberger K., Brezger B., Zeilinger A. and Arndt M., Nature **427**, 711 (2004).
Hagley E. *et al.*, Phys. Rev. Lett. **79**, 1 (1997).
Hakim V. and Ambegaokar V., Phys. Rev. A **32**, 423 (1985).
Halban Jr H. and Preiswerk P., Comptes Rendus Acad. Sci. Paris **203**, 73 (1936).
Halcoussis C., Abdul Redah T., Naumann H., Monaco G. and Chatzidimitriou-Dreismann C. A., ESRF Newsletter **34**, 17 (2000).
Hald J., Sørensen J. L., Schori C., and Polzik E. S., Phys. Rev. Lett. **83**, 1319 (1999).
Hammes M., Rychtarik D., Engeser B., Nagerl H. C. and Grimm R., physics/0208065 (2002).
Hänsch T. W. and Schawlow A. L., Opt. Commun. **13**, 68 (1975).
Hansen K. and Campbell E. E. B., Phys. Rev. E **58**, 5477 (1998).
Hanson R. *et al.*, Phys. Rev. Lett. **91**, 196802 (2003).
Harber D. M., McGuirk J. M., Obrecht J. M. and Cornell E. A., J. Low Temp. Phys. **133**, 229 (2003).
Harel G. and Akulin V. M., Phys. Rev. Lett. **82**, 1 (1999).
van Harlingen D. J., Rev. Mod. Phys. **67**, 515 (1995).
Harris S. E., Phys. Today **50**, 36 (1997).

Harris S. E., Phys. Today **50**, 36 (1997).
Harris S. E. and Yamamoto Y., Phys. Rev. Lett. **81**, 3611 (1998).
Harris S. E. and Hau L. V., Phys. Rev. Lett. **82**, 4611 (1999).
Hasegawa Y., Loidl R., Badurek G., Baron M. and Rauch H., Nature **425**, 45 (2003).
Hau L. V., Harris S. E., Dutton Z., and Behroozi C. H., Nature **397**, 594 (1999).
Haussühl S., Solid State Comm. **57**, 643 (1986).
Hecht E., *Optik*, (Addison-Wesely, 1994).
Heidmann A., Horowicz R. J., Reynaud S., Giacobino E., and Fabre C., Phys. Rev. Lett. **59**, 2555 (1987).
Heitler W., *The Quantum Theory of Radiation*, (Clarendon Press, Oxford, 1954).
von Helden G. *et al.*, Phys. Rev. Lett. **81**, 1825 (1998).
Henrikson N. E. and Møller K. B. J. Chem. Phys. **119**, 2569 (2003).
Hessel M. M., Smith E. W. and Drullinger R. E., Phys. Rev. Lett. **33**, 1251 (1974).
Heszler P., Carlsson J. O. and Demirev P., Chem. Phys. **107**, 10440 (1997).
Hilgenkamp H., Mannhart J., and Mayer B., Phys. Rev. B **53**, 14586 (1996).
Hilgenkamp H. and Mannhart J., Rev. Mod. Phys. **74**, 485 (2002).
Hilgevoord J. and Uffink J. B. M., Phys. Lett. A **95**, 474 (1983).
Hodby E., Marago O. M., Hechenblaikner G., and Foot C. J., Phys. Rev. Lett. **86**, 2196 (2001).
Hofmann H. F. and Takeuchi S., Phys. Rev. Lett. **88**, 147901 (2002).
Hong C. K. and Mandel L., Phys. Rev. Lett. **56**, 58 (1986).
Hong C. K., Ou Z. Y. and Mandel L., Phys. Rev. Lett. **59**, 2044 (1987).
Hornberger K. *et al.*, Phys. Rev. Lett. **90**, 160401 (2003).
Hornberger K. and Sipe J. E., Phys. Rev. A **68**, 012105 (2003).
Hornberger K., Sipe J. E. and Arndt M., submitted to Phys. Rev. A, (2004).
Horsewill A. J. *et al.*, Ber. Bunsenges. Phys. Chem. **102**, 317 (1998).
van Hove L., Phys. Rev. **95**, 249 (1954).
Howell J. C. *et al.*, Phys. Rev. Lett. **92**, 210403 (2004).
Huang K., *Statistical Mechanics*, (Wiley, New-York 1987).
Huck A., van Otterlo A., Sigrist M., Phys. Rev. B **56**, 14163 (1997).
Hughes R. J., Morgan G. L., and Peterson C. G., J. Mod. Opt. **47**, 533 (2000).
Idziaszek Z., Rzazewski K., and Wilkens M., J. Phys. B **32**, L205 (1999).
Ikeda S. and Fillaux F., Phys. Rev. B **59**, 4134 (1999).
Ikeda S. *et al.*, Phys. Rev. B **66**, 184302 (2002).
Il'ichev E. *et al.*, Phys. Rev. Lett. **86**, 5369 (2001).
Imamoğlu A., Schmidt H., Woods G., and Deutsch M., Phys. Rev. Lett. **79**, 1467 (1997).
Inouye S. *et al.*, Nature **392**, 151 (1998).
Inouye S. *et al.*, Science **285**, 571 (1999).
Ioffe L. B., Geshkenbein V. B., Feigel'man M. V., Fauchere A. L., and Blatter G., Nature **398**, 678 (1999).
Ioffe A., Arif M., Jacobson D. L. and Mezei F., Phys. Rev. Lett. **82**, 2322 (1999).
Ioffe A. *et al.*, Phys. Rev. Lett. **84**, 2037 (2000).
Ioffe L. B. *et al.*, Nature **415**, 503 (2002).
Ioffe L. B. and Feigelman M. V., Phys. Rev. B **66**, 224503 (2002).
Ioffe L. B., Geshkenbein V. B., Helm Ch. and Blatter G., (unpublished, 2004).
Isacsson A., Gorelik L. Y., Shekhter R. I., Galperin Y. M. and Jonson M., cond-mat/0106597 (2002).

Isacsson A. and Nord T., cond-mat/0402228 (2004).

Isihara A., *Statistical Physics*, (Academic Press, New York, 1971).

Itano W. M., Heinzen D. J., Bollinger J. J., and Wineland D. J., Phys. Rev. A **41**, 2295 (1990).

Ivlev B. I., Phys. Rev. A **66**, 012102 (2002).

Jaksch D. *et al.*, Phys. Rev. Lett. **81**, 3108 (1998).

Jaksch D. *et al.*, Phys. Rev Lett. **85**, 2208 (2000).

Jalabert R. A. *et al.*, Phys. Rev. Lett. **68**, 3468 (1992).

Jeffrey G. A. and Saenger W., *Hydrogen Bonding in Biological Structures*, (Springer-Verlag, Berlin, 1991).

Jennewein T. et al., Phys. Rev. Lett. **84**, 4729 (2000).

Ji Y., Heiblum M., Sprinzak D., Mahalu D., and Shtrikman H., Science **290**, 779 (2000).

Jin D. S., Matthews M. R., Ensher J. R., Wieman C. E., and Cornell E. A., Phys. Rev. Lett. **78**, 764 (1997).

Joannopoulos J. D., Meade R. D., and Winn J. N., *Photonic Crystals: Molding the Flow of Light*, (Princeton University Press, 1995).

Joannopoulos J. D., Villeneuve P. R., and Fan S., Nature **386**, 143 (1997).

Joesten M. D. and Schaad L. J., *Hydrogen Bonding*, (M. Dekker, New-York, 1974).

John S. and Wang J., Phys. Rev. Lett. **64**, 2418 (1990).

Johnston C. W. and van der Mullen J. J. A. M., J. Chem. Phys. **119**, 2057 (2003).

Jones J. A., Mosca M. and Hansen R. H., Nature **393**, 344 (1998).

Jönsson C., Brandt D. and Hirschi S., Am. J. Phys. **42**, 4 (1974).

Joos E., Phys. Rev. A **29**, 1626 (1984).

Joos E. and Zeh H. D., Z. Phys. B. **59**, 223 (1985).

Joos E. *et al.*, *Decoherence and the Appearance of a Classical World in Quantum Theory*, (Springer, Berlin, 2003), 2nd edition.

J. Am. Chem. Soc. **119**, 6213 (1997), editorial.

Jozsa R. and Linden N., Proc. Roy. Soc. London A **459**, 2001 (2003).

Judson R. S. and Rabitz H., Phys. Rev. Lett. **68**, 1500 (1992).

Julsgaard B., Kozhekin A., and Polzik E. S., Nature **413**, 400 (2001).

Julsgaard B., Schori C., Sørensen J. L. and Polzik E. S., Quant. Inform. and Comp., spec. issue, **3**, 518 (2003).

Julsgaard B., Sherson J., Sørensen J. L. and Polzik E. S., J. Opt. B: Quantum Semiclass. Opt. **6**, 5 (2004).

Kachru R., Tran N. H. and Gallagher T. F., Phys. Rev. Lett. **49**, 191 (1982).

Kadigrobov A., Gorelik L. Y., Shekhter R. I. and Jonson M., cond-mat/9811212 (1998).

Kane B. E., Nature **393**, 133 (1998).

Kang K., Phys. Rev. B **58**, 9641 (1998).

Kang H. and Zhu Y., Phys. Rev. Lett. **91**, 093601 (2003).

Kant I., *Critique of Pure Reason*, (Hartknoch, Riga, 1781).

Karlsson E. *et al.*, Phys. Rev. B **52**, 6417 (1995).

Karlsson E., Phys. Scr. T **76**, 179 (1998).

Karlsson E. B. *et al.*, Europhys. Lett. **46**, 617 (1999).

Karlsson E. B. and Lovesey S. W., Phys. Rev. A **61**, 062174 (2000).

Karlsson E. B. and Lovesey S. W., Phys. Rev. A **61**, 062714 (2002).

Karlsson E. B., Abdul-Redah T., Udovic T. J., Hjörvarsson B., and Chatzidimitriou-Dreismann C. A Appl. Phys. A **74**, S1203 (2002).

Karlsson E. B. and Lovesey S. W., Phys. Scr. **65**, 112 (2002).

Karlsson E. B., Phys. Rev. Lett. **90**, 095301 (2003).

Karlsson E. B. *et al.*, Phys. Rev. B **67**, 184108 (2003).

Karlsson E. B. and Mayers J., Phys. Rev. Lett. **92**, 249601 (2004), comment to Ref. [Blostein 2003 (a)].

Karlsson E. B., Mod. Phys. Lett. B **18**, 247 (2004).

Kash M. *et al.*, Phys. Rev. Lett. **82**, 5229 (1999).

Kashida S. and Yamamoto K., Solid State Chem. **86**, 180 (1990).

Kashida S., Ikeda S. and Nakai Y., J. Phys. Soc. Jpn **63**, 4643 (1994).

Kastalsky A. *et al.*, Phys. Rev. Lett. **67**, 3026 (1991).

Kastberg A. *et al.*, Phys. Rev. Lett. **74**, 1542 (1995).

Katz N., Steinhauer J., Ozeri R., and Davidson N., Phys. Rev. Lett. **89**, 2120401 (2002).

Katz N., Ozeri R., Rowen E., Gershnabel E., and Davidson N., cond-mat/0308492 (2003).

Katz N. *et al.*, cond-mat/0405222 (2004).

Kaulakys B. and Gontis V., Phys. Rev. A **56**, 1131 (1997).

Kazantsev A. P. and Surdutovich G. I., Sov. Phys. Dokl. **7**, 990 (1963).

Kazantsev A. P., Zh. Eksp. Teor. Fiz. **66**, 1599 (1974); [Sov. Phys. JETP **39**, 784 (1974)].

Keen D. A. and Lovesey S. W., J. Phys.: Condens. Matter **15**, 4937 (2003).

Keldysh L. V., Zh. Eksp. Teor. Fiz. **47**, 1515 (1964); [Sov. Phys. JETP **20**, 1018 (1965)].

Kempe J., Bacon D., Lidar D. A., and Whaley K. B., Phys. Rev. A **63**, 042307 (2001).

Ketterle W., Durfee D. S. and Stamper-Kurn D. M., Proc. Int. Sch. Phys. "Enrico Fermi", Course CXL, Ed. by M. Inguscio, S. Stringari and C. E. Wieman, (IOS Press, Amsterdam, 1999).

Khalfin L. A., Zh. Eksp. Teor. Fiz. **33**, 1371 (1957); [Sov. Phys. JETP **6**, 1053 (1958)].

Khalfin L. A., JETP Lett. **8**, 65 (1968).

Khaykovich L. and Davidson N., Appl. Phys. B **70**, 683 (2000).

Kheruntsyan K. V., Krähmer D. S., Kryuchkyan G. Yu., Petrosyan K. G., Opt. Comm. **157**, 139 (1997).

Kheruntsyan K. V., Petrosyan K. G., Phys. Rev. A **62**, 015801 (2000).

Kikkawa J. M. and Awschalom D. D., Nature **397**, 139 (1999).

Kim M. S., Son W., Buzek V., and Knight P. L., Phys. Rev. A **65**, 032323 (2002).

Kitaev A. Yu., preprint http://xxx.lanl.gov/abs/quant-ph/9707021 (1997).

Kitaev A., cond-mat/0010440 (2000).

Kitaev A. Yu., Ann. Phys. **303**, 2 (2003).

Klingshirn C. and Haug H., Phys. Rept. **70**, 315 (1981).

Klyshko D. N., Phys. Lett. A **128**, 133 (1988).

Klyshko D. N., Phys. Lett. A **132**, 299 (1988).

Klyshko D. N., Sov. Phys. Usp., **31**, 74 (1988).

Knight P. L., Nature (London) **344**, 493, (1990).

Knill E. and Laflamme R., Phys. Rev. A **55**, 900 (1997).

Knill E., Laflame R. and Zurek W. H., Proc. R. Soc. London A **454**, 365 (1998).

Knill E., Laflamme R., and Viola L., Phys. Rev. Lett. **84**, 2525 (2000).

Knill E., Laflamme R., and Milburn G. J., Nature **409**, 46 (2001).

Knize R., Wu Z., and Happer W., Adv. At. Mol. Phys. **24**, 223 (1988).

Kocher C. A. and Commins E. D., Phys. Rev. Lett. **18**, 575 (1967).

Kofman A. G., Kurizki G., and Sherman B., J. Mod. Opt. **41**, 353 (1994).

Kofman A. G. and Kurizki G., Phys. Rev. A **54**, R3750 (1996).

Kofman A. G. and Kurizki G., Nature (London) **405**, 546 (2000).

Kofman A. G. and Kurizki G., Phys. Rev. Lett. **87**, 270405 (2001).

Kofman A. G., Kurizki G., and Opatrný T., Phys. Rev. A **63**, 032108 (2001).

Kokorowski D. A., Cronin A. D., Roberts T. D. and Pritchard D .E., Phys. Rev. Lett. **86**, 2191 (2001).

Koshino K. and Shimizu A., Phys. Rev. Lett. **92**, 030401 (2004).

Kouwenhoven L. P., Schön G., and Sohn L. L., *Proceedings of the Advanced Study Institute on Mesoscopic Electron Transport*, (Kluwer, Dordrecht, 1997).

Kozuma M. *et al.*, Phys. Rev. Lett. **82**, 871 (1999).

Krotkov R. V., Tuominen M. T. and Breuer M. L., Am. J. Phys. **69**, 50 (2001).

Kroto H. W., Heath J. R., Ot'Brian S. C., Curl R. F. and Smalley R. E., Nature **318**, 162 (1985).

Kruger P. *et al.*, Phys. Rev. Lett. **91**, 233201 (2003).

Kryuchkyan G. Yu., Kheruntsyan K. V., Zh. Eksp. Teor. Fiz. **103**, 18 (1993).

Kryuchkyan G. Yu., Kheruntsyan K. V., Zh. Eksp. Teor. Fiz. **104**, 1161 (1994).

Kryuchkyan G. Yu. and Kheruntsyan K. V., Opt. Comm. **127**, 230 (1996).

Kryuchkyan G. Yu., Petrosyan K. G. and Kheruntsyan K. V., JETP Lett. **63**, 526 (1996).

Kryuchkyan G. Yu. and Muradyan N. T., Phys. Lett. A **286**, 113 (2001). Kryuchkyan G. Yu., Manukyan L. A. and MuradyanN. T., Opt. Comm. **190**, 245 (2001).

Kryuchkyan G. Yu. and Manvelyan S. B., Phys. Rev. Lett. **88**, 094101 (2002).

Kryuchkyan G. Yu. and Manvelyan S. B., Phys. Rev. A **68**, 013823 (2003).

Kryuchkyan G. Yu. and Manukyan L. A., Phys. Rev. A **69**, 013813 (2004).

Kübler O. and Zeh H. D., Ann. Phys. (NY) **76**, 405 (1973).

Kubo R., J. Math. Physics **4**, 174 (1963).

Kuhn A., Hennrich M., and Rempe G., Phys. Rev. Lett. **89**, 067901 (2002).

Kulik I. O. and Shekhter R. I., Sov. Phys. JETP **41**, 308 (1975).

Kurizki G., Giovanazzi S., O'Dell D., and Artemiev A. I., *Springer Lecture Notes in Physics*, (Springer-Verlag, Heidelberg, 2002), vol. 602, p. 369.

Kurizki G. and Kofman A. G., *Quantum optics in photonic band gap materials*, Encycl. of Opt. Engin. (Dekker, 2003).

Kurtsiefer C., Mayer S., Zarda P., and Weinfurter H., Phys. Rev. Lett. **85**, 290 (2000).

Kuzmich A., Mølmer K. and Polzik E. S., Phys. Rev. Lett. **79**, 4782 (1997).

Kuzmich A., Bigelow N. P. and Mandel L., Europhys. Lett. **42**, 481 (1998).

Kuzmich A. and Polzik E. S., Phys. Rev. Lett. **85**, 5639 (2000).

Kuzmich A. *et al.*, Nature (London) **423**, 731 (2003).

Kwiat P. G. *et al.*, Phys. Rev. Lett. **75**, 4337 (1995).

Laloë F., Am. J. Phys. **69**, 655 (2001).

Landau L. and Lifchitz L., *Mécanique quantique, Vol. III of Physique Théorique*, (Editions MIR, Moscou, 1966).

Landau L. D. and Lifshitz E. M., *Quantum Mechanics*, (Pergamon, Oxford, 1977).

Landau L. D. and Lifshitz E. M., *Quantum mechanics: non-relativistic theory*, (Pergamon Press, New York, 1977).

Landau L. D. and Lifshitz E. M., *Quantum Mechanics*, (Butterworth-Heinemann, Oxford, 1997).

Lane A. M., Phys. Lett. A **99**, 359 (1983).

Lane A. S. *et al.*, Phys. Rev. A **38**, 788 (1988).

Leggett A. J. et al., Rev. Mod. Phys. **59**, 1 (1987).

Leggett A. J., J. Phys.: Condens. Matter **14**, R415 (2002).

Leibscher M. and Stenholm S., submitted to Phys. Rev. A (2004).

Levinson Y., Entin-Wohlman O., and Wölfle P., Physica A **302**, 335 (2001).

Levitov L. S., Orlando T. P., Majer J. B., and Mooij J. E., cond-mat/0108266 (2001).

Levy M., *Lectures on Field Theory and the Many-Body Problem*, Ed. by E. R. Caianello, (New York, Academic Press, 1961), page 47.

Lewenstein M. and Rzazewski K., Phys. Rev. A **61**, 022105 (2000).

Li Y. and Xiao M., Phys. Rev. A **51**, R2703 (1995).

Li X. *et al.*, Phys. Rev. Lett. **88**, 047904 (2002).

Li X. *et al.*, Phys. Rev. Lett. **88**, 047904 (2002).

Lidar D. A., Chuang I. L., and Whaley K. B., Phys. Rev. Lett. **81**, 2594 (1998).

Lidar D. A., Bacon D., and Whaley K. B., Phys. Rev. Lett. **82**, 4556 (1999).

Lidar D. A. and Whaley K. B., *Irreversible Quantum Dynamics*, ed. by F. Benatti and R. Floreanini, (Springer Lecture Notes in Physics, Berlin, 2003), vol. 622, pp. 83-120.

Lifshitz E. M. and Pitaevskii L. P., *Statistical Physics*, part 2, (Butterworth-Heinemann, Oxford, 1980).

Lifshitz E. M. and Pitaevskii L. P., *Statistical Physics*, Part 2, (Butterworth-Heinemann, Oxford 1998).

Likharev K. K., Rev. Mod. Phys. **51**, 101 (1979).

Lin Y., Teper I., Chin C. and Vuletic V., Phys. Rev. Let. **92**, 050404 (2004).

Lindblad G. S., Commun. Math. Phys. **40**, 147 (1976).

Lindblad G. S., Commun. Math. Phys. **48**, 119 (1976).

Linden N. and Popescu S., Phys. Rev. Lett. **87**, 047901 (2001).

Lindgard A. and Nielsen S. E., At. Data Nucl. Data Tables **19**, 533 (1977).

Lisdat Ch., Frank M., Knöckel H., Almazor M. L. and Tiemann E., Eur. Phys. J. D **12**, 235 (2000).

Liu C., Dutton Z., Behroozi C. H., and Hau L. V., Nature **409**, 490 (2001).

Lloyd S. and Slotine J. J. E., Phys. Rev. Lett. **80**, 4088 (1998).

Lloyd S. and Braunstein S. L., Phys. Rev. Lett. **82**, 1784 (1999).

Lofwander T., Shumeiko V. S., and Wendin G., Phys. Rev. B **62**, R14653 (2000).

Lofwander T., Shumeiko V. S., and Wendin G., Supercond. Sci. Technol. **14**, R53 (2001).

Long D. A., *The Raman effect. A unified treatment of the theory of Raman scattering by molecules.*, (John Wiley & Sons, Ltd., New York, 2002).

Longchambon L., Laurat J., Coudreau T. and Fabre C., quant-ph/0310036 (2003).

Longchambon L., Laurat J., Coudreau T. and Fabre C., quant-ph/0311123 (2003).

van Loock P. and Braunstein S. L., Phys. Rev. Lett. **84**, 3482 (2000).

Loss D. and DiVincenzo D. P., Phys. Rev. A **57**, 120 (1998).

Lovesey S. W., *Theory of Neutron Scattering from Condensed Matter*, Vol. 1, (Oxford University Press, New York, 1984).

Lubkin G. B., Physics Today **52**, 19 (1999).

Lucazeau G. and Novak A., J. Raman Spect. **1**, 573 (1973).

Luis A. and Sánchez-Soto L. L., Prog. Opt. **41**, 421 (2000).

Lukin M.D. *et al.*, Phys. Rev. Lett. **79**, 2959 (1997).

Lukin M., Matsko A., Fleischhauer A., and Scully M., Phys. Rev. Lett. **82**, 1847 (1999).

Lukin M. D. and Imamoğlu A., Phys. Rev. Lett. **84**, 1419 (2000).

Lukin M. D., Yelin S. F., and Fleischhauer M., Phys. Rev. Lett. **84**, 4232 (2000).

Lukin M. D. *et al.*, Phys.Rev. Lett. **87**, 037901 (2001).

Lukin M. D., Rev. Mod. Phys. **75**, 457 (2003).

Lvovsky A. I. *et al.*, Phys Rev. Lett. **87**, 050402 (2001).

Lvovsky A. I., Mlynek J., Phys. Rev. Lett. **88**, 250401 (2002).

Magnier S., Aubert-Frécon M., and Millié Ph., J. Mol. Spectrosc. **200**, 96 (2000).

Mair A. *et al.*, Nature **412**, 313 (2001).

Maitland G. C., Rigby M., Smith E. B. and Wakeham W. A., *Intermolecular Forces - Their Origin and Determination*, (Clarenden Press, Oxford, 1981).

Makhlin Y., Schön G. and Shnirman A., Rev. Mod. Phys. **73**, 357 (2001).

Malakyan Yu., Rochester S., Budker D., Kimball D., and Yashchuk V., Phys. Rev. A **69,** 013817 (2004).

Mandel L. and Wolf E. *Optical Coherence and Quantum Optics*, (Cambridge University Press, Cambridge, 1995).

Mandel O. *et al.*, Phys. Rev. Lett. **91**, 010407 (2003).

Mann A. and Revzen M., *The Dilemma of EPR - 60 Years Later*, (Institute of Physics, 1996).

Mannhart J., Hilgenkamp H., Mayer B., and Gerber Ch., Phys. Rev. Lett. **77**, 2782 (1996).

Manz J. and Ludger W., *Femtosecond Chemistry*, (VCH, Weinheim, 1995).

de Marco B. and Jin D. S., Science **285**, 1703 (1999).

Marks R. W., *The Dymaxion World of Buckminster Fuller*, (Reinhold, New York, 1960)

Martinis J. M., Nam S., Aumentado J. and Urbina C., Phys. Rev. Lett. **89**, 117901/1 (2002).

Martorell J. and Lawandy N. M., Phys. Rev. Lett. **65**, 1877 (1990).

Mason E. I., Wong N. C., Opt. Lett. **23**, 1733 (1998).

Matsko A. B., Rostovtsev Yu. V., Cummins H. Z., and Scully M. O., Phys. Rev. Lett. **84**, 5752 (2000).

Matsko A., Novikova I., Welch G., and Zubairy M., Opt. Lett. **28**, 96 (2003).

Matt S. *et al.*, Chem. Phys. Lett. **303**, 379 (1999).

Mattle K., Weinfurter H., Kwiat P. G. and Zeilinger A., Phys. Rev. Lett. **76**, 4656 (1996).

Mayers J., Andreani C. and Baciocco G., Phys. Rev. B **39**, 2022 (1989).

Mayers J., Phys. Rev. B **41**, 41 (1990).

Mayers J., Burke T. M. and Newport R. J., J. Phys.: Condens. Matter **6**, 641 (1994).

Mayers J., Fielding A. L. and Senesi R., Nucl. Inst. Meth. A **481**, 454 (2002).

Mayers J. and Abdul-Redah T., J. Phys.: Condens. Matter **16**, 4811 (2004).

Mazets I. E. and Shifrin L. B., Phys. Lett. A **229**, 73 (1997).

McGuirk J. M., Harber D. M., Obrecht J. M. and Cornell E. A., cond-mat/0403254 (2004).

McKeever J. *et al.*, Science **303**, 1992 (2004).

McNeil K. J. and Gardiner C. W., Phys. Rev. A **28**, 1560 (1983).

Mehta M. L., *Random Matrices*, (Academic, New York, 1991).

Meier B. H., Graf F. and Ernst R. R., J. Chem. Phys. **76**, 767 (1982).

Mertz J., Debuisschert T., Heidman A., Fabre C., and Giacobino E., Opt. Lett. **16**, 1234 (1991).

Meunier T. *et al.*, submitted to Phys. Rev. Lett. (2004).

Meyer R. and Ernst R. R., J. Chem. Phys. **86**, 784 (1987).

Mezard M., Parisi G. and Virasoro M., *Spin Glass Theory and Beyond*, (Word Scientific, Sigapore 1997).

Michler P. *et al.*, Science **290**, 2282 (2000).

Milburn G. J., Phys. Rev. A **33**, 674 (1986).

Milburn G. J. and Holmes C. A., Phys. Rev. Lett. **56**, 2237 (1986).

Milman P. *et al.*, submitted to Eur. Phys. J. D (2004).

Milonni P. W., Nature (London) **405**, 525 (2000).
Misra B., and Sudarshan E. C. G., J. Math. Phys. **18**, 756 (1977).
Mitzner R. and Campbell E. E. B., J. Chem. Phys. **103**, 2445 (1995).
Mølmer K. and Sørensen A., Phys. Rev. Lett. **82**, 1835 (1999).
Mooij J. E. *et al.*, Science **285**, 1036 (1999).
Moore M. G. and Meystre P., Phys. Rev. Lett. **83**, 5202 (1999).
Moore M. G. and Meystre P., Phys. Rev. Lett. **85**, 5026 (2000).
Morigi G., Solano E., Englert B. G. and Walther H., Phys. Rev. A **65**, 040102 (2002).
Morpurgo A. F.*et al.*, Science **286**, 263 (1999).
Moses H. E., Lett. Nuovo Cim. **4**, 51 (1972).
Mourachko I. *et al.*, Phys. Rev. Lett. **80**, 253 (1998).
Mozyrsky D. and Martin I., Phys. Rev. Lett. **89**, 018301 (2002).
Mughabgab S. F., *Neutron Cross Sections*, (Academic Press, Orlando, Florida, 1984).
Muller A., Zbinden H., and Gisin N., Europhys. Lett. **33**, 335 (1996).
Munro W. J. and Gardiner C. W., Phys. Rev. A **53**, 2633 (1996).
Myatt C. J. *et al.*, Nature **403**, 269 (2000).
Nabors C. D., Yang S. T., Day T., and Byer R. L., J. Opt. Soc. Am. B **7** 815 (1990).
Naik D. S. et al., Phys. Rev. Lett. **84**, 4733 (2000).
Nairz O., Arndt M. and Zeilinger A., J. Mod. Opt. **47**, 2811 (2000).
Nairz O., Arndt M. and Zeilinger A., Am. J. Phys. **71**, 319 (2003).
Nakajima S., Progr. Theoret. Phys. **20**, 948 (1958).
Nakamoto K., Sarma Y. A. and Ogoshi H., J. Chem. Phys. **43**, 1177 (1965).
Nakamura Y., Pashkin Yu. A. and Tsai J. S., Nature **398**, 786 (1999).
Naumann H., Diploma Thesis, Technical University of Berlin (2002).
Nazarov Yu. V., Phys. Rev. Lett. **73**, 134 and 1420 (1994).
von Neumann J., *Mathematical foundations of Quantum Mechanics*, (Princeton, 1955).
Neumann M. *et al.*, J. Chem. Phys. **109**, 7300 (1998).
Neumann M. A. *et al.*, Ber. Bunsenges. Phys. Chem. **102**, 325 (1998).
Ngo D. and Pinnick R. G., J. Opt. Soc. Am. A **11**, 1352 (1994).
Nielsen M. A., Knill E. and Laflamme R., Nature **396**, 52 (1998).
Nielsen M. A. and Chuang I. L., *Quantum Computation and Quantum Information*, (Cambridge University Press, Cambridge, UK, 2000).
Novak A., Saumagne P. and Bock L. D. C., J. Chim. Phys-Chim. Biol. **60**, 1385 (1963).
Novak A., Struct. and Bonding (Berlin) **18**, 177 (1974).
O'Brien J. L. et al., Nature **426**, 264 (2003).
O'Dell D. H. J., Giovanazzi S., Kurizki G. and Akulin V. M., Phys. Rev. Lett. **84**, 5687 (2000).
O'Dell D. H. J., Giovanazzi S. and Kurizki G., Phys. Rev. Lett. **90**, 110402 (2003); J. Mod. Opt. **50**, 2655 (2003).
Odin C., J. Phys. Chem. B **108**, 7402 (2004).
Ogawa T., Ueda M., and Imoto N., Phys. Rev. A **43**, 6458 (1991).
Omnes R., *The Interpretation of Quantum Mechanics*, (Princeton University Press, 1994), p. 301.
Opatrný T. and Kurizki G., Phys. Rev. Lett. **86**, 3180 (2001).
Opatrny T., Deb B., and Kurizki G., Phys. Rev. Lett. **90**, 250404 (2003).
Ottaviani C., Vitali D., Artoni M., Cataliotti F., and Tombesi P., Phys. Rev. Lett. **90**, 197902 (2003).
Ou Z. Y., Wang L. J. and Mandel L., Phys. Rev. A **40**, 1428 (1989).

Ou Z. Y., Pereira S. F., Kimble H. J., and Peng K. C., Phys. Rev. Lett. **68**, 3663 (1992).

Ou Z. Y., Qu. Semiclass. Opt. **9**, 599 (1997).

Ou Z. Y. and Mandel L., Phys. Rev. Lett. **61**, 50 (1998).

Ovchinnikov Y. B., Shul'ga S. V. and Balykin V. I., J. Phys. B: At. Mol. Opt. Phys. **24**, 3173 (1991).

Ozeri R., Steinhauer J., Katz N., and Davidson N., Phys. Rev. Lett. **88**, 220401 (2002).

Ozeri R., PhD. thesis (2003).

Ozeri R., Katz N., Steinhauer J., Rowen E., and Davidson N., Phys. Rev. Lett. **90**, 170401 (2003).

Ozhigov Y., quant-ph/0201132 (2002).

Ozhigov Y. and Fedichkin L., quant-ph/0202030 (2002).

Pan J. W. *et al.*, Phys. Rev. Lett. **80**, 3891 (1998).

Pan J. W. et al., Nature **421**, 721 (2003).

Panov A. D., Ann. Phys. **249**, 1 (1996).

Park H. *et al.*, Nature **407**, 57 (2000).

Park J. *et al.*, Nature **417**, 722 (2002).

Park J. *et al.*, Thin Solid Films **438-439**, 457 (2003).

Parkins A. S. and Kimble H. J., Phys. Rev. A **61**, 052104 (2000).

Paspalakis E., Kylstra N. J., and Knight P. L., Phys. Rev. A **65**, 053808 (2002).

Pasupathy A. N. *et al.*, cond-mat/0311150, (2003).

Pati A. K. and Braunstein S. L., Quant. Inf. Commun. **2**, 399 (2002).

Patorski K., *Progress in Optics XXVII*, ed. by Wolf E., (Elsevier Science Publishers B. V., Amsterdam, 1989), pp. 2 - 108.

Pauling L., *The nature of the chemical bond*, (Cornell University Press, Ithaca, New York, 1960).

Peierls R. E., *Proceedings of the Glasgow Conference*, (Pergamon Press, Oxford, 1955), p. 296.

Peng K. C. *et al.*, Appl. Phys. **66**, 755 (1998).

Pereira S. F., Ou Z. Y. and Kimble H. J., Phys. Rev. A **62**, 042311 (2002).

Perrie W., Duncan A. J., Beyer H. J. and Kleinpoppen H., Phys. Rev. Lett. **54**, 1790 (1985).

Perrin C. L. and Nielson J. B., Annu. Rev. Phys. Chem. **48**, 511 (1997).

Petrosyan D. and Kurizki G., Phys. Rev. Lett. **89**, 207902 (2002).

Petrosyan D. and Malakyan Yu., Phys. Rev. A **70**, to be published (2004); quant-ph/0402070.

Pfau T., Spälter S., Kurtsiefer Ch., Ekstrom C. and Mlynek J., Phys. Rev. Lett. **73**, 1223 (1994).

Phillips W. D., Rev. Mod. Phys. **70**, 721 (1998).

Phillips D. F., Fleischhauer A., Mair A., Walsworth R. L., and Lukin M. D., Phys. Rev. Lett. **86**, 783 (2001).

Physics News (The AIP Bulletin of), Physics New Update No. 648, 31 July 2003.

Physics Today, Sept. 2003, p. 9.

Pilgram S., Schomerus H., Martin A. M., and Büttiker M., Phys. Rev. B **65**, 045321 (2002).

Pillet P., Kachru R., Tran N. H., Smith W. W. and Gallagher T. F., Phys. Rev. Lett. **50**, 1763 (1983).

Pillet P., Kachru R., Tran N. H., Smith W. W. and Gallagher T. F., Phys. Rev. A **36**, 1132 (1987).

Pimentel G., McClellan C. and Lester A., *The hydrogen bond*, (W. H. Freeman, San Francisco, 1960).

Nozières P. and Pines D., *The Theory of Quantum Liquids*, (Perseus, Cambridge, Massachusetts, 1999).

Piovella N., Bonifacio R., McNeil B. W. J., and Robb G. R. M., Opt. Commun. **187**, 165 (2001).

Pistolesi F., Phys. Rev. B, in press; cond-mat/0401361 (2004).

Pitaevskii L. and Stringari S., Phys. Rev. Lett. **83**, 4237 (1999).

Pitaevskii L. and Stringari S., *Bose-Einstein Condensation*, (Oxford, 2003).

Pittman T. B. *et al.*, Phys. Rev. A **52**, 3429 (1995).

Pittman T. B. *et al.*, Phys. Rev. A **53**, 2804 (1996).

Pittman T. B., Jacobs B. C. and Franson J. D., Phys. Rev. Lett. **88**, 257902 (2002).

Pittman T. B., Fitch M. J., Jacobs B. C. and Franson J. D., Phys. Rev. A **68**, 032316 (2003).

Polzik E. S., Phys. Rev. A **59**, 4202 (1999).

Polzik E. S., Julsgaard B., Sherson J. and Sørensen J. L., Phil. Trans. Roy. Soc. A **361**, 1391 (2003).

Preskill J., Proc. Roy. Soc. A **454**, 385 (1998).

Preskill J., Proc. R. Soc. London A **454**, 469 (1998).

Preskill J., *Lecture Notes for Physics 229: Quantum Information and Computation*, (California Institute of Technology, 1998), http://www.theory.caltech.edu/~preskill/ph229.

Prezhdo O. V. and Rossky P. J., Phys. Rev. Lett. **81**, 5294 (1998).

Prezhdo O. V., Phys. Rev. Lett. **85**, 4413 (2000).

Prigodin V. N. *et al.*, Phys. Rev. Lett. **71**, 1230 (1993).

Pu H. and Meystre P., Phys. Rev. Lett. **85**, 3987 (2000).

Qi J. *et al.*, Phys. Rev. Lett. **83**, 288 (2000).

Quang T., Woldeyohannes M., and John S., Phys. Rev. Lett. **79**, 5238 (1997).

Raab E. L., Prentiss M., Cable A., Chu S., and Pritchard D. E., Phys. Rev. Lett. **59**, 2631 (1987).

Raimond J. M., Brune M., and Haroche S., Rev. Mod. Phys. **73**, 565 (2001).

Raithel G., Benson O., and Walther H., Phys. Rev. Lett. **75**, 3446 (1995).

Ralph T. C. and Huntington E. H., Phys. Rev. A **66** 042321 (2002).

Rarity J. G. *et al.*, Appl. Phys. B **55**, 250 (1992).

Ratcliffe C. E. and Irish D. E., J. Phys. Chem. **86**, 4897 (1982).

Raymer M. G., Walmsley I. A., Mostowski J., and Sobolewska B., Phys. Rev. A **32**, 332 (1985).

Raymer M. G. and Walmsley I. A., *Progress in Optics*, ed. by , E. Wolf, (North-Holland, Amsterdam, 1996), vol. 28, p. 181.

Reichl L. E., *A modern Course in Statistical Physics*, (Wiley, New York, 1998).

Reid M. D. and Drummond P. D., Phys. Rev. Lett. **60**, 2731 (1988).

Reid M. D., Phys. Rev. A **40**, 913 (1989).

Reid M. D. and Drummond P. D., Phys. Rev. Lett. **60**, 2731 (1988).

Reiter G. and Silver R., Phys. Rev. Lett. **54**, 1047 (1985).

Reiter G. F., Mayers J. and Noreland J., Phys. Rev. B **65**, 104305 (2002).

Renn M. J. and Gallagher T. F., Phys. Rev. Lett. **67**, 2287 (1991).

Renn M. J., Thomson D. S. and Gallagher T. F., Phys. Rev. A **49**, 409 (1994).

Reynaud S., Fabre C., and Giacobino E., J. Opt. Soc. Am. **B4**, 1520 (1987).

Reynaud S., Lamine B., Lambrecht A., Neto P. M. and Jaekel M. T., Int. J. Mod. Phys. A **17**, 1003 (2002).

Rice S. A. and Zhao M., *Optical Control of Molecular Dynamics*, (John Wiley & Sons, New York, 2000).

Riebe M. *et al.*, Nature **429**, 734 (2004).

Roati G., Riboli F., Modugno G., and Inguscio M. Phys. Rev. Lett. **89**, 150403 (2002).

Robinson M. P., Laburthe Tolra B., Noel M. V., Gallagher T. F. and Pillet P., Phys.Rev. Lett. **85**, 4466 (2000).

Rogel-Salazar J., New G. H. C., Choi S. and Burnett K., Phys. Rev. A **65**, 023601 (2002).

Roos C. F. *et al.*, Science **304**, 1478 (2004).

Rouse R., Huan S. and Lukens J. E., Phys. Rev. Lett. **75**, 1614 (1995).

Rowe M. A. *et al.*, Nature **409**, 791 (2001).

Rowen E., Katz N., Ozeri R., Gershnabel E., and Davidson N., cond-mat/0402225 (2004).

Ryazanov A. A. *et al.*, Phys. Rev. Lett. **86**, 2427 (2001).

Rzazewski K., Lewenstein M., and Eberly J. H. J. Phys. B **15**, L661 (1982).

Sackett C. A. *et al.*, Nature (London) **404**, 256 (2000).

Saffaro L., *Proceedings of the First Italian Workshop on Fullerenes: Status and Perspectives*, ed. by Taliani C., Ruani G. and Zamboni R., (World Scientific, Singapore, 1992), p. 55.

Safinya K. A., Delpech J. F., Gounand F., Sandner W. and Gallagher T. F., Phys. Rev. Lett. **47**, 405 (1981).

Sakurai J. J., *Modern Quantum Mechanics*, (Addison-Wesley, Reading, 1994), pp. 484-486.

Sanaka K., Kawahara K. and Kuga T., Phys. Rev. A **66** 040301, (2002).

Sanaka K. and Resch K., quant-ph/0312226 (2004).

Sanders B. C. and Milburn G. J., Phys. Rev. A **45**, 1919 (1992).

Sanders B. C., Lee K. S. and Kim M. S., Phys. Rev. A **52**, 735 (1995).

Santori C., Fattal D., Vučković J., Solomon G. S., and Yamamoto Y., Nature **419**, 594 (2002).

Sarnelli E. and Testa G., Physica C **371**, 10 (2002).

Savas T. A., Shah S. N., Schattenburg M. L., Carter J. M. and Smith H. I, J. Vac. Sci. Technol. B **13**, 2732 (1995).

Scheiner S., *Hydrogen Bonding: A Theoretical Perspective*, (Oxford University Press, Oxford, 1997).

Schieve W. C., Horwitz L. P., and Levitan J., Phys. Lett. A **136**, 264 (1989).

Schmidt H. and Imamoğlu A., Opt. Lett. **21**, 1936 (1996).

Schmidt H. and Imamoglu A., Opt. Lett. **23**, 1007 (1997).

Schmidt A. U., math-ph/0307044 (2003). To appear in Prog. Math. Phys. Res. (2004).

Schmidt-Kaler F. et al., Nature **422**, 408 (2003) and references therein.

Schöllkopf W. and Toennies J. P., Science **266**, 1345 (1994).

Schöllkopf W. and Toennies J. P., J. Chem. Phys. **104**, 1155 (1996).

Schori C., Julsgaard B., Sørensen J. L. and Polzik E. S., Phys. Rev. Lett. **89**, 057903 (2002).

Schori C., Sørensen J. L. and Polzik E. S., Phys. Rev. A **66**, 033802 (2002).

Schrödinger E., *Naturwissenschaften* **23**, 807, 823, 844 (1935).

Schrödinger E., Naturwissenschaften **48**, 807; **49**, 823; **50**, 844 (1935).

Schülke W., *Handbook on Synchrotron Radiation*, 1st edition, Vol. 3, Chapt. 15, pp. 565 - 637 (1991).

Schulman L. S., J. Phys. A **30**, L293 (1997).

Schulman L. S., Phys. Rev. A **57**, 1509 (1998).

Schumacher B. and Caves C. M., Phys. Rev. A **31**, 3093 (1985).

Schuster P., Zundel G. and Sandorfy C., *The hydrogen bond. Recent developments in theory and experiments*, (North-Holland Pub. Co., Amsterdam, 1976), Vols. I, II and III.

Schuster P., *Hydrogen Bonds*, (Springer-Verlag, Berlin, 1984).

Schuster R. *et al.*, Nature **385**, 417 (1997).

Schwinger J. and Teller E., Phys. Rev. **52**, 296 (1937).

Scientific American, Oct. 2003, p. 20.

Scully M. O. and Zubairy M. S., *Quantum Optics*, (Cambridge University Press, Cambridge, UK, 1997).

Scully M. O., submitted to Phys. Rev. Lett. (2004).

Search C. and Berman P. R. arXiv:physics/003003.

Sears V. F., Phys. Rev. B. **30**, 44 (1984).

Seeger P. A., Taylor A. D. and Brugger R. M., Nucl. Instr. Meth. A **240**, 98 (1985).

Seife C., Science **288**, 1564 (2000).

Sette F. *et al.*, Phys. Rev. Lett. **75**, 850 (1995).

Shapiro M. and Brumer P., *Principles of the Quantum Control of Molecular Processes*, (Wiley, 2003).

Sherson J. and Mølmer K., in preparation, to be submitted to Phys. Rev. A (2004).

Shih Y. H. and Alley C. O., Phys. Rev. Lett. **61**, 2921 (1988).

Shor P. W., *Proceedings of the* 35th *Symposium on the Foundations of Computer Science*, Ed. by S. Goldwasser, (Los Alamitos, CA:IEEE Press, 1994).

Shor P. W., Phys. Rev. A **52**, 2493 (1995).

Silberhorn C. *et al.*, Phys. Rev. Lett. **86**, 4267 (2001).

Silberhorn C., Korolkova N. and Leuchs G., Phys. Rev. Lett. **88**, 167902 (2002).

Silvestrini P. *et al.*, Phys. Rev. Lett. **79**, 3046 (1997).

Simon R., Phys. Rev. Lett. **84**, 2726 (2000).

Skinner J. L. and Trommsdorff H. P., J. Chem. Phys. **89**, 897 (1988).

Slater J. C., Phys. Rev. **87**, 807 (1952).

Sleator T. and Weinfurter H., Phys. Rev. Lett. **74**, 4087 (1995).

Smerzi A. *et al.*, Phys. Rev. Lett. **79**, 4950 (1997).

Smilde H. J. H. *et al.*, Phys. Rev. Lett. **88**, 057004-1 (2002).

Smithey D. T., Beck M., Belsley M., and Raymer M. G., Phys. Rev. Lett. **69**, 2650 (1992).

Soper A. K. and Phillips M. G., Chem. Phys. **107**, 47 (1986).

Special Issue on Experimental Proposals for Quantum Computation, Fort. der Physik **48**, No. 9-11 (2000).

Sperling J., Tributsch H., Streffer R. M. F., Abdul-Redah T. and Chatzidmitriou-Dreismann C. A., J. Electroanal. Chem. **477**, 62 (1999).

Squires G. L., *Introduction to the Theory of Thermal Neutron Scattering*, (Dover Publications, Mineola, 1996).

Stamper-Kurn D. M. *et al.*, Phys. Rev. Lett. **83**, 2876 (1999).

Steane A., Rept. Prog. Phys. **61**, 117 (1998).

Steger H., de Vries J., Kamke B., Kamke W. and Drewello T., Chem. Phys. Lett. **194**, 452 (1992).

Steinhauer J., Ozeri R., Katz N., and Davidson N., Phys. Rev. Lett. **88**, 120407 (2002).

Steinhauer J. *et al.*, Phys. Rev. Lett. **90**, 060404 (2003).

Steinhauer J., Ozeri R., and Davidson N., cond-mat/0303375 (2004).

Stenger J. *et al.*, Phys. Rev. Lett. **82**, 4569 (1999).

Stenholm S., Physica Scripta **T12**, 56 (1986).

Stenholm S., *Quantum Dynamics of Simple Systems*, ed. by G. L. Oppo, S. M. Barnett, E. Riis, and M. Wilkinson, (Institute of Physics Publishing, Bristol, 1994), p. 267.

Stenholm S. and Paloviita A., J. mod. Optics **44**, 2533 (1997).

Stöckli A., Meier B. H., Meyer R. and Ernst R. R., J. Chem. Phys. **93**, 1502 (1990).

Streffer R. M. F. *et al.*, Physica B **266**, 190 (1999).

Strunz W. T., Haake F., and Braun D., Phys. Rev. A **67**, 022101 and 022102 (2003).

Sudarshan E. C. G., Phys. Rev. A **46**, 37 (1992).

Sun Q. *et al.*, Phys. Rev. Lett. **87**, 176601 (2001).

Switkes M., Marcus C. M., Campman K., and Gossard A. D., Science **283**, 1905 (1999).

Takahashi Y. *et al.*, Phys. Rev. A **60**, 4974 (1999).

Talbot H. F., Philos. Mag. **9**, 401 (1836).

Tan S. M., Holland M. J. and Walls D. F., Opt. Comm. **77**, 285 (1990).

Tan S. M., Walls D. F. and Collet M. J., Phys. Rev. Lett. **66**, 252 (1991).

Tanaka Y. and Kashiwaya S., Phys. Rev. B **56**, 892 (1997).

Tannor D. J. and Rice S. A., J. Chem. Phys. **83**, 5013 (1985).

Taylor A. D., RAL-84-120 (1984).

Testa G. *et al.*, Supercond. Sci. Technol. **17**, 287 (2004).

Testa G. *et al.*, to be published in Appl. Phys. Lett. (2004).

Thirunamachandran T., Mol. Phys. **40**, 393 (1980).

Thomas J. O., Tellegren R. and Olovsson I., Acta Cryst. B **30**, 1155 and 2540 (1974).

Thomson D. S., Renn M. J. and Gallagher T. F., Phys. Rev. Lett. **65**, 3273 (1990).

Thomson D. S., Renn M. J. and Gallagher T. F., Phys. Rev. A **45**, 358 (1992).

Thouless D. J., Phys. Rev. B **27**, 6083 (1983).

Timmermans E. and Côté R., Phys. Rev. Lett. **80**, 3419 (1998).

Tittel W., Brendel J., Zbinden H. and Gisin N., Phys. Rev. Lett. **81**, 3563 (1998).

Tozzo C. and Dalfovo F., New J. of Phys. **5**, 54 (2003).

Tozzo C. and Dalfovo, F., cond-mat/0401359, (2004).

Tuckerman M. E., Marx D., Klein M. L. and Parinello M., Science **275**, 817 (1997).

Tuckerman M. E. and Marx D., Phys. Rev. Lett. **86**, 4946 (2001).

Tuominen M. T., Krotkov R. V. and Breuer M. L., Phys. Rev. Lett. **83**, 3025 (1999).

Turchette Q. A., Hood C. J., Lange W., Mabuchi H., and Kimble H. J., Phys. Rev. Lett. **75**, 4710 (1995).

Turner J. and Soper A., Chem. Phys. **107**, 3049 (1986).

Uffink J. B. M. and Hilgevoord J., Phys. Lett. A **105**, 176 (1984).

Uffink J. B. M. and Hilgevoord J., Found. Phys. **15**, 925 (1985).

Ullersma P., Physica **32**, 27, 56, 74, 90 (1966).

Unanyan R. G. and Fleischhauer M., Phys. Rev. Lett. **90**, 133601 (2003).

Unruh W. G. and Zurek W. H., Phys. Rev. D **40**, 1071 (1989).

Vaidman L., Phys. Rev. A **49**, 1473 (1994).

Vandersypen L. M. K. *et al.*, Nature **414**, 883 (2001).

Vandersypen L. M. K. *et al.*, *Quantum computing and quantum bits in mesoscopic systems*, (Kluwer, New York, 2003).

Varshalovich D. A., Moskalev A. N. and Khersonsky V. K., *Quantum Theory of Angular Momentum*, (World Scientific, Singapore, 1988).

Vavilov M. G. *et al.*, Phys. Rev. B **63**, 195313 (2001).

Vedral V., Int. J. Quantum Inf. **1**, 1 (2003), and references therein.

Velicky B. *et al.*, Phys. Rev. **179**, 747 (1968).

Verbeni R. *et al.*, J. of Synchroton Radiation **3**, 62 (1996).

Vewinger F., Heinz M., Garcia Fernandez R., Vitanov N., and Bergmann K., Phys. Rev. Lett. **91**, 213001 (2003).

Vidal G., Phys. Rev. Lett. bf 91, 147902 (2003).

van der Viel W. G. *et al.*, Science **289**, 2105 (1997).

Vinogradov S. N. and Linnell R. H., *Hydrogen Bonding*, (Van Nostrand-Reinhold, New York, 1971).

Viola L. and Lloyd S., Phys. Rev. A **58**, 2733 (1998). Viola L. *et al.*, Phys. Rev. Lett. **82**, 2417 (1998).

Viola L., Knill E., and Lloyd S., Phys. Rev. Lett. **82**, 2417 (1999).

Viola L., Knill E., and Lloyd S., Phys. Rev. Lett. **85**, 3520 (2000).

Vion D. *et al.*, Science **296**, 886 (2002).

Vitali D. and Tombesi P., Phys. Rev. A **59**, 4178 (1999).

Volkov A. F. *et al.*, Physica C **210**, 21 (1993).

Vos M., Phys. Rev. A **65**, 012703 (2002).

Vos M., Ultramicroscopy **92**, 143 (2002).

Waks E., Diamanti E., Sanders B. C., Bartlett S. D., and Yamamoto Y., Phys. Rev. Lett. **92**, 113602 (2004).

van der Wal C. H. *et al.*, Science **290**, 773 (2000).

van der Wal C. *et al.*, Science **301**, 196 (2003).

Walls D. F. and Milburn G. J., *Quantum Optics*, (Springer, Berlin, 1994).

Wang J., Wei Y., Wang B., and Guo H., Appl. Phys. Lett. **79**, 3977 (2001).

Wang X. B., Phys. Rev. A **66**, 024303, 064304 (2002).

Wang J. *et al.*, Phys. Rev. B **65**, 153311 (2002).

Wannier G. H., Phys. Rev. **52**, 191 (1937).

Ware M. J., *Physical Methods in Advanced Inorganic Chemistry*, ed. by Hill H. A. O. and Day P., (Interscience Publishers, London, 1968), p. 214.

Watson G. I., J. Phys.: Condens. Matter **8**, 5955 (1996).

Wei Y. *et al.*, Phys. Rev. B **63**, 195412 (2001).

Wei Y., Wang J., Guo H., and Roland C., Phys. Rev. B **64**, 115321 (2001).

Weigold E. and McCarthy I. E., *Electron Momentum Spectroscopy*, (Kluwer Academic / Plenum, New York, 1999).

Weihs G., Jennewein T., Simon C., Weinfurter H. and Zeilinger A., Phys. Rev. Lett. **81**, 5039 (1998).

Weiss C. and Zwerger W., Europhys. Lett. **47**, 97 (1999).

Wen X. G. and Niu Q., Phys. Rev. B **41**, 9377 (1990).

Wen X. G., Phys. Rev. B **44**, 2664 (1991).

Wenger J. *et al.*, Opt. Lett. **29**, 1267 (2004).

Werner R. F. and Wolf M. M., Phys. Rev. Lett. **86**, 3658 (2001).

Wernet Ph. *et al.*, Science **304**, 995 (2004).

Wilkinson S. R. *et al.*, Nature (London) **387**, 575 (1997).

Wilson E. B., Decius J. C. and Cross P. C., *Molecular Vibrations The Theory of Infrared and Raman Spectra*, (McGraw-Hill Book Company, INC., New-York, 1964).

Wilson C. C., *Single Crystal Neutron Diffraction from Molecular Materials*, (World Scientific Co. Pte. Ltd., Singapore, 2002).

Windsor C., *Pulsed Neutron Scattering*, (Taylor and Francis, London, 1981), Eq. (9.30).

Wineland D. and Dehmelt H., Bull. Am. Phys. Soc., Ser. II, **20**, 637 (1975).

Wineland D. J., Bergquist J. C., Itano W. M., Bollinger J. J., and Manney C. H., Phys. Rev. Lett. **59**, 2935 (1987).

Wingreen N. S. and Meir Y., Phys. Rev. B **49**, 11040 (1994).

Winter R. G., Phys. Rev. **123**, 1503 (1961).
Wipf H. *et al.*, Europhys. Lett. **4**, 1379 (1987).
Woerdman J. P., J. Chem. Phys. **75**, 5577 (1981).
Woldeyohannes M. and John S., J. Opt. B **5**, R43 (2003).
Wolf S. A. *et al*, Science **294**, 1488 (2001).
Wolf S. A., *Semiconductor Spintronics and Quantum Computation*, ed. by Awschalom D. D., Samarth N. and Loss D., (Springer, Berlin, 2002).
Wootters W. K. and Zurek W. H., Nature **299**, 802 (1982).
Wu L. A. and Lidar D. A., Phys. Rev. Lett. **88**, 207902 (2002).
Yacoby A., Heiblum M., Umansky V., Shtrikman H., and Mahalu D., Phys. Rev. Lett. **73**, 3149 (1994).
Yacoby A., Heiblum M., Mahalu D. and Shtrikman H., Phys. Rev. Lett. **74**, 4047 (1995);
Yariv A. and Yeh P., *Optical Waves in Crystals*, (Wiley, 1984).
Yashchuk V. V. *et al.*, Phys. Rev. Lett. **90**, 253001 (2003).
Yoshino K. *et al.*, Appl. Phys. Lett. **73**, 3506 (1998).
Yu L. H. and Natelson D., cond-mat/0310625 (2003).
Yurke B. and Stoler D. Phys. Rev. Lett. **57**, 13 (1986).
Zanardi P. and Rasetti M., Phys. Rev. Lett. **79**, 3306 (1998).
Zanardi P., Phys. Lett. A **258**, 77 (1999).
Zanardi P. and Lloyd S., Phys. Rev. Lett. **90**, 067902 (2003).
Zeh H. D., Found. Phys. **3**, 109 (1973).
Zeilinger A., Gähler R., Shull C. G., Treimer W. and Mampe W., Rev. Mod. Phys. **60**, 1067 (1988).
Zeilinger A., Nature **408** 639 (2000).
Zernike F., Physica **5**, 785 (1938).
Zewail A. H., *Femtochemistry*, (World Scientific, Singapore, 1994).
Zewail A. H., Angew. Chem. Int. Ed. **39**, 2586 (2000).
Zhang Y. *et al.*, Phys. Rev. A **62**, 023813 (2000).
Zhang T. C. *et al.*, Phys. Rev. A **67**, 033802 (2003).
Zhang J., Peng K. and Braunstein S. L., Phys. Rev. A **68**, 013808 (2003).
Zhou F. *et al.*, Phys. Rev. Lett. **82**, 608 (1999).
Zhou F., cond-mat/9905190 (2001).
Zoller P., Cirac J. I., Duan L. M., Garcia-Ripoll J. J., quant-ph/0405025 (2004).
Zondy J. J., Douillet A., Tallet A., Ressayre E. and Berre M. L., Phys. Rev. A **63**, 023814 (2001).
Zubairy M. S., Matsko A. B., and Scully M. O., Phys. Rev. A bf 65, 043804 (2002).
Zukowski M., Zeilinger A. and Weinfurter H., Ann. NY Acad. Sci. **91**, 755 (1995).
Zumino B., *Lectures on Field Theory and the Many-Body Problem*, ed. by E. R. Caianello, (New York, Academic Press, 1961).
Zurek W. H., Phys. Rev. D **24**, 1516 (1981).
Zurek W. H., Phys. Rev. D **26**, 1862 (1982).
Zurek W. H., Phys. Today **44**, 36 (1991).
Zurek W. H., Ann. Phys. (Leipzig) **9**, 855 (2000).
Zurek W. H., Rev. Mod. Phys. **75**, 715 (2003).
Zwanzig R., J. Chem. Phys. **33**, 1338 (1960).
Zwanzig R., J. Chem. Phys. **33**, 1338 (1960).
Zwanzig R., *Nonequilibrium Statistical Mechanics*, (Oxford, New York, 2001).

Index

GPSR Compliance

The European Union's (EU) General Product Safety Regulation (GPSR) is a set of rules that requires consumer products to be safe and our obligations to ensure this.

If you have any concerns about our products, you can contact us on ProductSafety@springernature.com

In case Publisher is established outside the EU, the EU authorized representative is:

Springer Nature Customer Service Center GmbH
Europaplatz 3
69115 Heidelberg, Germany

Batch number: 10873394

Printed by Printforce, the Netherlands